Chordate Structure and Function

CHORDATE STRUCTURE

AND FUNCTION

Second Edition

ARNOLD G. KLUGE, *University of Michigan*

IN COLLABORATION WITH

B. E. Frye, *University of Michigan*

Kjell Johansen, *University of Aarhus, Denmark*

Karel F. Liem, *Harvard University*

Charles R. Noback, *Columbia University*

Ingrith D. Olsen, *University of Washington*

Allyn J. Waterman, *National Institute of Health*

Macmillan Publishing Co., Inc.

NEW YORK

Collier Macmillan Publishers

LONDON

Printed in the United States of America

Macmillan Publishing Co., Inc.
866 Third Avenue, New York, New York 10022

Collier Macmillan Canada, Ltd.

Library of Congress Cataloging in Publication Data

Kluge, Arnold G
Chordate structure and function.

Published in 1971 by A. J. Waterman.
Includes bibliographies and index.
1. Anatomy, Comparative. 2. Vertebrates—Anatomy.
I. Waterman, Allyn Jay, (date) Chordate structure and function. II. Title.
QL805.W37 1977 596'.04 75-45173
ISBN 0-02-364800-7

Printing: 2 3 4 5 6 7 8 Year: 8 9 0 1 2 3

Preface

Chordate Structure and Function is designed for those undergraduate courses that, at least traditionally, have been thought of as comparative anatomy. A cursory examination of our book reveals that it differs in many ways from the textbooks that have been and are now being used in these courses. *Chordate Structure and Function* begins with descriptions of the theories of evolution, speciation, and phylogeny (Chapter 1). We believe that these theories must be considered in some detail at the outset because they are the bases of most of the deductions that are made in our later chapters. In our second chapter the general form, biology, and relevant fossil record of the major groups of chordates are briefly described. Knowledge of these few facts helps the reader to evaluate more critically the specific conclusions that arise in the remaining chapters. In Chapters 3 through 12, we illustrate, by numerous examples, the importance of (1) integrating structural and functional facts and (2) rigorously analyzing these data by the comparative and experimental scientific methods. Limitations of space have forced us to sacrifice many structural and functional details. We have only sought a balance between a knowledge of the more important facts and the logical and methodological bases of their study.

We want to express our gratitude to many biologists for their pointed comments, helpful suggestions, and new insights. Among those who offered their suggestions were Dr. Daniel A. Guthrie of The Claremont Colleges, Dr. Albert E. Feldman of Dutchess Community College, Dr. Reinhard Harkema of North Carolina State University, Dr. Ellis A. Hicks of Iowa State University, Dr. Richard C. Snyder of the University of Washington, Dr. John S. Stephens, Jr., of Occidental College, and Dr. Keith S. Thomson of Yale University. We incorporated many of the changes they suggested. Any errors are our own.

Miss Rosalind DeMatteis also deserves our thanks for cheerfully typing and retyping the more difficult parts of the manuscript.

For Further Reading

The following books are general anatomical and physiological references.

[1] Bolk, L., E. Göppert, E. Kallius, and W. Lubosch, *Handbuch der Vergleichende Anatomie der Wirbeltiere.* Berlin: Urban und Schwarzenberg, 1936. (Reprinted in 1967 by A. Asher and Company, Amsterdam.)

[2] Fawcett, D. W. *An Atlas of Fine Structure.* Philadelphia: W. B. Saunders Company, 1966.

[3] Florey, E. *General and Comparative Animal Physiology.* Philadelphia: W. B. Saunders Company, 1966.

[4] Gordon, M. S., G. A. Bartholomew, A. D. Grinnell, C. Barker Jørgensen, and F. N. White. *Animal Function: Principles and Adaptations.* New York: Macmillan Publishing Co., Inc., 1968.

[5] Grassé, P. P., ed. *Traité de Zoologie, Anatomie Systématique, Biologie.* Paris: Masson et Cie., 1948–present.

[6] Hoar, W. S. *General and Comparative Physiology.* Englewood Cliffs, N.J.: Prentice-Hall, Inc., 1966.

[7] Jollie, M. *Chordate Morphology.* New York: Van Nostrand Rheinhold Company, 1962.

[8] Porter, K. R., and M. A. Bonneville. *Fine Structure of Cells and Tissue.* 3rd ed. Philadelphia: Lea & Febiger, 1968.

[9] Prosser, C. L., and F. A. Brown. *Comparative Animal Physiology.* Philadelphia: W. B. Saunders Company, 1961.

[10] Romer, A. S. *The Vertebrate Body.* 3rd ed. Philadelphia: W. B. Saunders Company, 1962.

[11] Schmidt-Nielsen, K. *Animal Physiology.* 2nd ed. Englewood Cliffs, N.J.: Prentice-Hall, Inc., 1965.

[12] Young, J. Z. *The Life of Vertebrates.* 2nd ed. New York: Oxford University Press, 1962.

Contents

1

Concepts and Principles of Morphologic and Functional Studies

Purpose and Scope

The purpose of this book is to present the history of structural systems, their functions, and the whole organisms of which they are a part. Its scope is confined to selected examples of structures and functions of chordate animals, and organic evolution is the principle biologic theory guiding their analysis.

Reasoning and Methodology

Examples of any subject presented without rhyme or reason lose much of their interest and intrinsic value. Only certain kinds of reasoning and methodology have gone into the original discovery and subsequent elaboration of the examples described in this book, and the reader is introduced to these in the first three chapters.

Basic to any scientific inquiry is the scientific method itself. Although it may sound formal and possibly suggest an elaborate formula, the scientific method can be reduced to a simple procedure and to a set of rules. To practice the method in its most elementary form one need only ask sensible questions and pursue sensible answers. The more sophisticated practice of the method, however, entails (1) formulating the problem, (2) proposing the hypothesis that best explains the specified phenomena of that problem, and (3) testing the premises and conclusions of that hypothesis.

The formulation of the problem requires first an awareness that the problem exists and then its description. The greater the detail in which the problem is described, the more precise are the questions that can be asked as well as the resulting answers. A hypothesis is a provisional explanatory statement that proposes a probable relationship of an unsolved problem to other, already tested and accepted, statements. In order to be scientific, the statement must be tested, either directly or indirectly, by observation. The rejection of authority—that is, the refusal to accept a statement simply because someone says it is so—should be a matter of routine in science. In testing a hypothesis, one seeks exceptions to the premises embodied in the proposition or to the conclusions themselves. The testing is accomplished by the experimental and comparative approaches described subsequently. If exceptions are found, then either the observations or the hypothesis itself is incorrect. If the observations are repeatable, then the hypothesis has been falsified and must be either discarded or modified. The hypothesis that has met extended tests and is the most probable explanation is more appropriately called a theory or principle. A general theory, like that of evolution, is one that continues to draw together previously unrelated facts and in this way exemplifies and clarifies our understanding of all natural phenomena. Nevertheless, observation of a particular object or occurrence is the fact from which all science stems and to which all scientific theories must return for verification.

Deductive reasoning is applied to many of the examples cited in *Chordate Structure and Function.* Deduction, in a logical sense, is the form of reasoning by which a conclusion is derived or inferred from a physical law, biological theory or principle. Organic evolution is the theory on which the deductions in this text rest. Of course, there are many principles attendent to, or derived from, the theory of evolution that also form the basis for certain deductions (e.g., speciation and phylogeny). In deductive reasoning one begins with a general truth and seeks to connect it to some individual case by a logical extension or by means of an intermediate set of cases known to be connected to both. One reasons from the general to the individual case, affirming the latter as well as distinctive qualities of the former [4].

Induction, in a logical sense, is the alternative form of reasoning to deduction. Here a general conclusion is drawn from particular bases. One observes sufficient individual facts in inductive reasoning and extends to others of the same class what is true of them, thereby identifying general principles or laws.

Parsimonious reasoning is a rule of logic and it is fundamental to the scientific method. The *rule of parsimony* was first promulgated by the scholastic and great Franciscan opponent of the Papacy, William of Ockham (1270–1349). The dictum ascribed to him, "Entities are not to be multiplied beyond necessity," was a response to the unnecessary and irrelevant hypotheses given in explanation of phenomena. This principle, usually referred to as Ockham's razor, simply means that of several possible explanations the one that is the simplest—that is, has the fewest assumptions and is the most consistent with the data at hand—is the most acceptable one until evidence to the contrary is presented.

A scientific hypothesis can be tested in two ways, experimentally and/or comparatively. Both approaches have been used in the research of the examples described in the latter chapters of this book. The experimental approach begins with observations that suggest a causal relationship between variables. For example, variable *X* is hypothesized to produce some predictable change in variable *Y* because of their known relationship to variable *Z*. Based on this suspicion, an experiment is formulated that allows the investigator to attempt to falsify the hypothesized affect that variable *X* has on *Y*. There are two general ways the experimentalist can test such a hypothesis. The simple "all-or-none" form of experiment involves testing the hypothesized effect that the presence or absence of variable *X* will have on *Y*. For example, the scientist removes a gland from several individuals—the experimental group—and the effect of the absence of that tissue on some variable, say bone growth, is observed. The experimenter must have a control group of like individuals on whom the identical operation is performed but from whom the glandular tissue is not removed. The conditions of variable *Y* in the experimental and control groups are contrasted and those of the former group are compared to the conclusion of the hypothesis. The assumption is made that only *X* has been varied and that it alone is responsible for variation in *Y*. In the second experimental approach, *X* is varied continuously, or nearly so, and the corresponding conditions of *Y* are observed. Occasionally, the field scientist finds naturally occurring population variation in space and/or time, and the ideal experimental situation is available for study. Unfortunately, these ideal situations are usually discovered first and then a relevant hypothesis is constructed. Much research that passes for experiment is nothing more than mere description. For example, to feed a group of fish a constant source of food and record their individual variations in growth is not an experiment. In summary, the experimental approach usually involves (1) inductive reasoning, (2) living systems, (3) manipulation of variables, and (4) the replication of observations. The correlations that the experimentalist observes must be reproducible. In the jargon of the statistician, the experimentalist's data exist in sample sizes greater than one.

In general, the comparative approach to hypothesis testing involves (1) deductive reasoning, (2) nonliving systems—usually, although not always, (3) variables that cannot be manipulated directly, and (4) observations that are usually impossible to replicate, at least exactly. The observed correlative relationship between variables is almost always inferred from an example of one. The investigative power of the comparative approach is derived from the predictive nature and universality of evolutionary theory and its attendent principles and not from the increasing probability that accompanies increasing sample size.

The hypotheses to which the comparative approach are applied can be summarized in terms of the following two related classes of questions: What was the most likely ancestral condition of a given structure or function? What advantage is conferred on individuals who possess a given derived structure or function? According to theory, evolutionary change is continuous and involves only slight modification from one generation to the next, and, therefore, the first question might appear to be trivial. However, extinction tends to eliminate most of the relevant descendent lineages, and the fossil record itself is too fragmentary and discontinuous to provide the direct observations required. To be able to answer the first class of questions one must find among the living or fossil record the closest ancestral condition and then reconstruct the probable intermediate evolutionary stages between the two known extremes. Evolutionary theory dictates that hypothetical intermediate conditions must form a gradual series of modifications and that we must be able to postulate a probable adaptation and biological context for each stage. Answers to the second class of questions primarily come from observations of the living extremes and secondarily from what we propose to be true of the hypothetical intermediates. The actual hypothesis testing related to both classes of questions comes into play when the dictates of evolutionary theory are violated or equally or more probable ancestral conditions and advantages can be identified.

History of Evolutionary Morphology

The study of the evolution of structure and function, the central theme of this book, is the product of at least a thousand years of endeavor by both scientists and nonscientists, and, like most sciences, it has proceeded through a definite sequence of levels of study. The sequence first began with the description of separate phenomena; next came comparative studies of groups of phenomena; then generalizations as to relationships among phenomena; and finally the explanation of those generalized relationships. It should be noted that this trend closely follows the sequence described for the scientific method of inquiry. As in that method, the earlier levels of investigation are still pursued, and rightfully so, because they will always be a necessary preliminary to a better understanding of the final product.

In 1555, the French anatomist Pierre Belon du Mans [1] (1517–1564) published a detailed comparison of the skeleton of a human and of a bird (Figure 1-1). This research, generally acknowledged as the first demonstration of the importance of comparative studies, marks the departure from purely descriptive anatomy. From Belon's work, which clearly suggests the equivalence between parts of different organisms, stems the most important generalization of anatomy: the *principle of homology*. From its relatively unassuming origin in sixteenth-century anatomy, the principle of homology has become one of the most important concepts of comparative and evolutionary biological studies.

From 1555 to 1859 the main tenet of the study of morphology was that the equivalences between different organisms were but minor variations of an ideal state (Plato's meaning of the word) or a basic plan or pattern (Aristotle's use).

Figure 1-1 The skeletons of a man and a bird were compared in 1555 by Pierre Belon du Mans [1]; a plate from his book is reproduced here. His study is believed to be the first detailed work on the equivalence between numerous parts of different organisms.

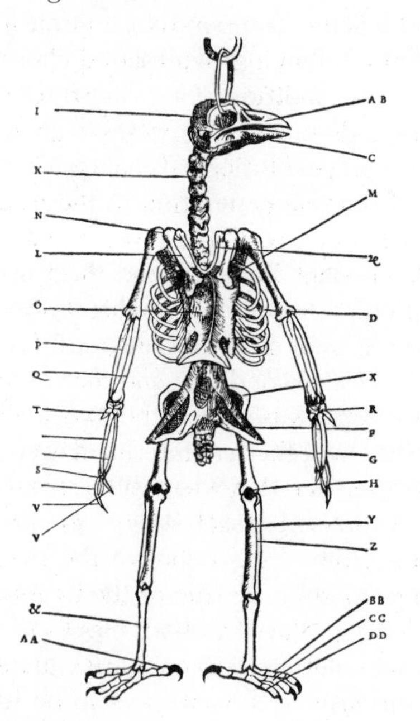

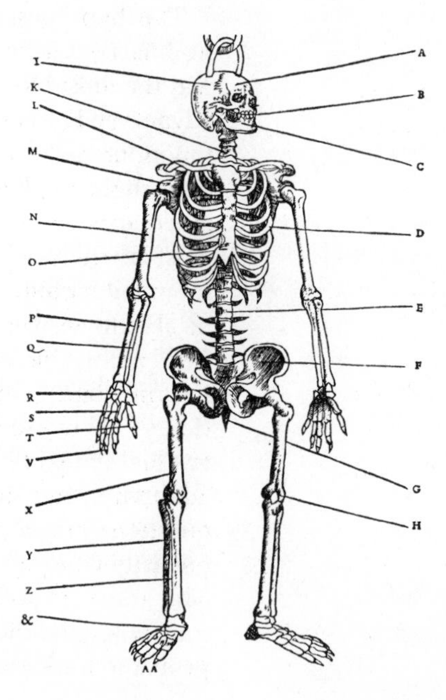

In either case, "ideal state" or "plan" are mental constructs without reality. The Kantian transcendental method of philosophy dominated much of this period of history, and it is clearly responsible for maintaining this explanatory theory. The leaders of this idealistic morphologic, or typologic, school of thought were Johann Wolfgang von Goethe (1749–1832), Lorenz Oken (1779–1851), and Richard Owen (1804–1892).

Owen, who formed generalizations about the relationships of equivalent or similar structures, in his later works recognized three kinds of homology: (1) general homology, which was the correspondence between the structure of an actual organism and an ancestral archetype. (2) special homology, which was the correspondence between the structures of actual organisms, and (3) serial homology, which corresponded to the anatomical relationship among repetitive structures within a single organism, such as vertebrae or ribs. Owen [8] stated that homologous attributes were recognizable from their similar form and their similar connection to other similar parts.

The influence of the typologic school was interrupted in 1859 when Charles Darwin published his first edition of *On the Origin of Species by Means of Natural Selection.* The historian Francis J. Cole [2] succinctly described the meaning of the break with the pre-Darwinian school of thought: "Until an evolutionary *principle* was demonstrated, further random research could but swell the accumulation of data which awaited integration into a science. In this inchoate state anatomy remained until it was quickened by the publication of the *Origin of Species,* which forthwith raised this uneventful record of factual competence to the dignity of a learned discipline." * Darwin devoted an entire chapter of the *Origin of Species* to morphology and embryology wherein he stressed the importance of an evolutionary interpretation. He also, of course, examined these subjects for support for his general theory of evolution. His most important conclusion in that chapter was that the similarity of homologous structures, the equivalences between different organisms, was due to the fact they were inherited from a similar or identical condition in a common ancestor. As a result of Darwin's explanation of the cause of homology and related phenomena, the final level of the historical sequence leading to evolutionary morphology was entered. This phase is the explanation of generalized relationships.

Major Tenets of Evolutionary Theory

Many students begin the study of structure and function with little or no understanding of organic evolution. Obviously, as a result of this deficiency, they cannot hope to evaluate critically the examples described or the comparative approach in general. In the following paragraphs the basic processes of evolution and speciation in sexually reproducing diploid organisms are summarized briefly, with particular emphasis on those aspects that are essential to a better understanding of this course of study. Most chordate animals reproduce sexually and possess a diploid set of chromosomes.

Broadly defined, evolution simply means change, and broadly applied, all matter, animate and inanimate, undergoes this process. The details and the consequences of the evolution of living and nonliving matter are, however,

* *A History of Comparative Anatomy From Aristotle to the Eighteenth Century* (London: Macmillan & Company, Ltd., 1944), p. 471.

totally different. In this study only organic evolution pertains, and it may be defined formally as a change in the genetic composition of a population of organisms from generation to generation. The fossil record provides overwhelming evidence that the *phenotype* of former life was different from that of the present day, and both laboratory and field studies clearly demonstrate the details of changes in gene frequency from one generation to another.

The Raw Material

In order for organic evolution to occur there must be available a permanent source of changeable, hereditable matter, capable of being transferred from one generation to the next. The chromosomes and their genes possessed by all organisms are the changeable matter, and they constitute a pool of hereditable information in the interbreeding population. The relevant initial changes occur in the gonadal germinal tissue of the individual organism as a result of the intrinsic mutability of gene and chromosomal materials. During gametogenesis, the paired chromosomes and genes are sorted independently into haploid sex cells, the sperm and eggs. With fertilization these genotypically highly varied haploid gametes are randomly recombined into a wider variety of diploid products. In summary, the initial changes are *mutation*, which in turn are greatly amplified in the interbreeding population as a whole by the process of genetic recombination. Also increasing the variability of the gene pool of a population is the addition of immigrant individuals with different genotypes. Here again recombination operates in the manner just described.

Recent biochemical-genetic studies indicate that more similar *genotypes* usually produce more similar phenotypes. Given this causal relationship and Darwin's historical theory of the origin of homologues, the student of evolutionary morphology and function might well be concerned with genotypic homology. Extrapolating still further, the students' ultimate concern might be the interaction between the replicative capacity of the genetic system and natural selection—that interaction that determined genotypic homology. This progression of potential study forces us to recognize an important concept. If replication were perfect, evolution almost certainly would not occur. In almost all morphologic and functional studies it is phenotypic similarity that is used as evidence of common descent, albeit an indirect one.

It is very important to recognize in making the concept homology operational that the same genotype may produce different phenotypes under different environmental conditions. An extreme environment may bring out developmental potencies that are not expressed under usual conditions; such an environment permits genetic factors that do not usually reach the threshold of phenotypic expression to manifest themselves. The researcher in evolutionary morphology and function must constantly remain aware of this problem.

The Agent: Natural Selection

Although mutation and recombination are the sources of variability, it is *natural selection* that determines the direction of change. All parts of the environment, biotic and abiotic, are potentially capable of directing the course of change. Natural selection may be defined as the sum of the environmental factors that determine the relative success of different phenotypes and their corresponding genotypes. The environment is like a filter; it preserves individuals that are better adapted to it and eliminates by means of differential reproduction those less well adapted. No thinking or planning is involved. The filtering operation is based solely on natural and physical laws.

The Criterion of Success: Adaptation

Adaptation can be defined as the hereditary adjustment of an organism to its environment by means of natural selection. Although adaptation may be viewed as the total fit of the phenotype to its environment, it is clear that each organism is a complex of more specific adaptations that enable it to perform its numerous life functions. The environment to which an organism is adapted is a varied series of many factors (e.g., physical, social, biotic, and so on). The adaptation produced by selection for one aspect of the total environment need not necessarily be useful and may even be detrimental in relation to other environmental parameters. Moreover, selection is opportunistic in that it produces adaptations to existing environmental conditions; such adaptations may or may not be beneficial to descendents in future environments. The collective forces of natural selection that promote diverse kinds of adaptations do not guarantee evolutionary success in the constantly changing environment.

Evolution is opportunistic and experimental. Natural selection operates only on organisms extant during a period of environmental change. Although there is no evidence of anticipating future need—namely, the operation appears to be totally mechanistic—the sequence of organism succession lends itself readily to the idea of a "master planner" and the existance of a "divine design." Perhaps the most concise statement of this thesis is that of George Gaylord Simpson [11], one of the outstanding students of evolutionary morphology: "Adaptation is real, and it is achieved by a progressive and directed process. This process achieves the aspect of purpose without the intervention of a purposer, and it has produced a vast plan, without the concurrent action of a planner. It may be that the initiation of the process and the physical laws under which it functions had a Purposer and that this mechanistic way of achieving a plan is the instrument of a Planner—of this still deeper problem the scientist, as scientist, cannot speak."*

The Product: Speciation

Strictly speaking, the formation of new species, *speciation*, is not evolution but may be considered a very frequent adjunct to it. Evolution is simply hereditable change through time, from generation to generation, and in space (that is, geographically), at one horizontal plane in time. With the interruption of the continuity of the common gene pool of an interbreeding population and the maintenance of the discontinuity, almost invariably by geographic isolation, the resultant populations often shift to different genotypic modes. The differences in the derived populations are the result of inescapable mutation and recombination and, most important, of the different forces of selection imposed on the populations by their different environments. The genotypic differences are at first subtle. They gradually accumulate, however, and may eventually manifest themselves, singly or collectively, in some mechanism (structural, behavioral, ecological, physiological, and so on) that prevents the divergent populations from interbreeding should they come into contact when the geographic barrier disappears. Such derived populations are *species* (Figure 1-2). In summary, a biological species may be defined as groups of actually or potentially interbreeding natural populations that share a common gene pool and that are reproductively isolated from other such groups.

When divergent evolution occurs, as just described, all of the derived populations tend to evolve, although not necessarily equally so, in response to

* *The Meaning of Evolution* (New Haven, Conn.: Yale University Press, 1949).

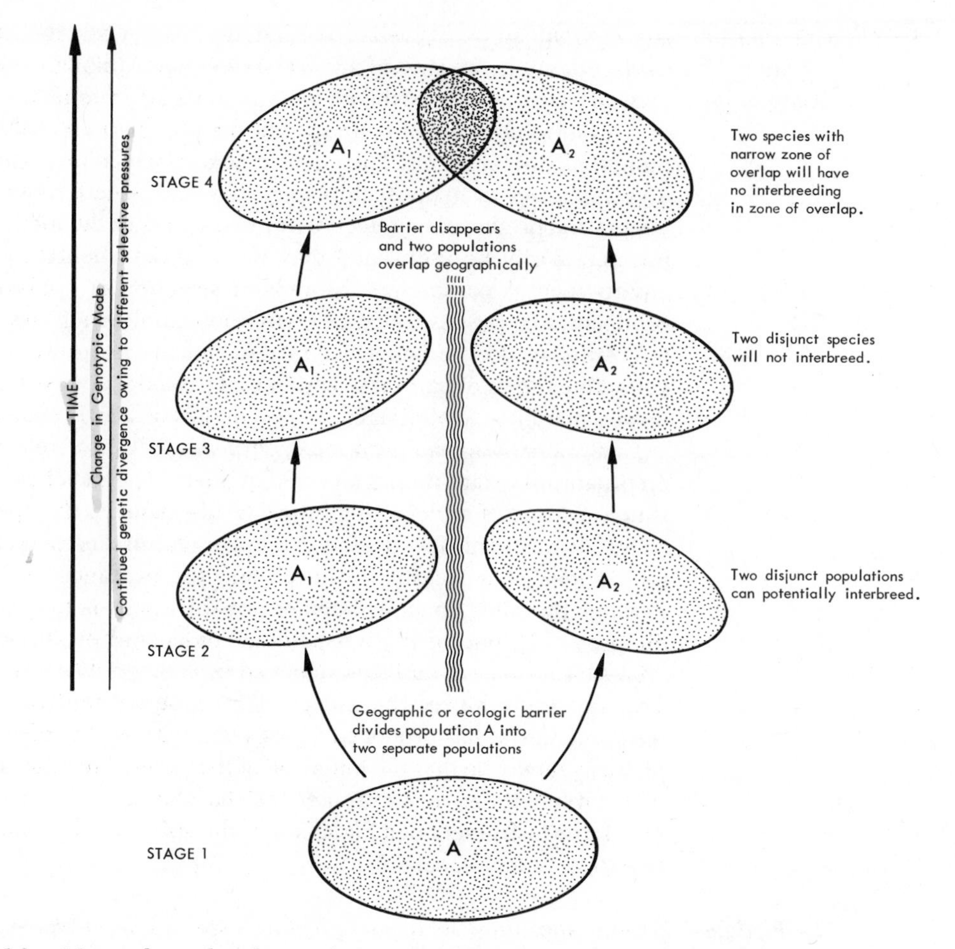

Figure 1-2 Major physical and spatial events in speciation. A = single population.

the constantly changing environment. For example, it is highly probable that reptiles evolved from amphibians, but it is fruitless to look for the specific reptilian ancestor among modern amphibians. Both groups have diverged from their common ancestral population.

With a knowledge of evolutionary theory as now generally accepted, it is easy to visualize how all attributes, morphological or functional, have originated and have been modified through time. Jean Baptiste de Lamarck (1744–1829) was the first biologist to maintain clearly and consistently that all species have arisen by evolution and form a continuum—that is, they exhibit *phylogeny*. The process of evolution tends to produce groups of organisms, however, because of different rates of change and because lineages become extinct.

The very nature of an evolutionary continuum appears to have contributed to at least one major misconception. This is the irreversibility of evolution, which is most often referred to as Dollo's law (after Louis Dollo, a famous Belgian paleontologist, 1857–1931). It is now generally recognized that there is no genetic or evolutionary phenomenon that of itself absolutely prohibits reversal or the independent origin of like genotypes and phenotypes. Certainly, back mutations occur, and even natural selection acting on the usual amount of genetic variation that exists in small populations can cause identical genotypes to arise independently. Another possible source of reversal, although not yet studied adequately in a wide variety or organisms, is the reconstitution of a recombinate that has been lost. It is generally agreed that populations that exhibit greater genetic variability produce more successful descendent lineages than those gene pools with little or no variability. In the usual condition, then, *heterozygosity* is a probable condition in derived phyletic lines, and the greater the length of time the heterozygosity persists, the greater chance there will be for identical genotypes to be produced by segregation and recombination within the descendent gene pools. In short, even identical genotypes can appear independently in species derived from species that did not show the phenotype because the genotype for the characteristic was potentially within their common gene pool [13].

In spite of all the possible ways in which evolutionary reversals can be realized, the actual phenomenon appears to be extremely rare. The statistical probability of a complete or nearly complete reversal to a previous condition, particularly a remote one, is very small for most structures and functions. This probability is related to three factors: (1) the length of time necessary for a change, such as a reverse mutation, to occur; (2) genes usually have the capacity to affect several different aspects of the total phenotype (*pleiotropy*); and (3) virtually all structures and functions appear to have at least a moderately broad genetic foundation (*polygenic*).

Morphologic and Functional Data and Their Interpretation

The relationship between structure and function in organisms, to which most of the references in the later chapters of this book are made, resembles the relationship between a well-designed instrument or machine and its action. This relationship is a product of evolution, but as even a casual reading of any one of the following chapters will indicate, many of the relationships described have no obvious bearing on *mortality* and *natality*, the demographic parameters usually referred to in discussions of natural selection. How, then, is the relationship between structure and function involved in the process of natural selection in these cases? Many of the relationships that will be described can be thought of as affecting in some way the amount of energy that an organism utilizes in its metabolism. The organism that has a relatively smaller energy consumption for a given set of conditions will be more likely to increase the number of offspring reaching sexual maturity and, consequently, will be favored by selection. This point of view explains the evolution of adaptations in terms of small savings of energy.

In the study of the relationship between structure and function one must consider the interaction between different functions of a single structure—that is, all the implications of a single aspect of design. Equally important is that the evolution of a structure very often results in the modification of other structures that serve very different functions. An obvious example is the change in the size of the eye in the course of bony fish evolution, which in turn appears to have resulted in changes in the shape of the skull, jaw muscles, and suspensorium, and even in the brain [6]. A structure can take on a new function, or functions, in the course of its evolution, and may even lose its original function altogether. New designs can evolve only by modification of existing designs. The evolution of structures with new functions depends on the opportunities offered by existing structures.

The data derived from structural and functional studies are very often used in taxonomy. The term *taxonomy* is defined as the study of those theoretical bases, principles, and procedures that are necessary to understand the relationships among organisms. Modern taxonomy is evolutionary. The basic parameter is phylogeny, which cannot be directly observed and very often is inferred only from *neontologic* data, which lacks the important dimension of time.

The data derived from structural and functional research are also used in *classification.* Classification is the formal arrangement of similar kinds of organisms into groups (*taxa*) and the application of *nomenclature* to the groups that are recognized. Thus, classification is an extension of taxonomy, and both classification and taxonomy are parts of the more inclusive discipline of *systematics,* which is, in its broadest context, the study of the evolutionary relationships of organisms.

Morphologic characters long have been used as the data base for estimating phylogeny. It has been only recently that biochemistry, cellular and systemic physiology, developmental dynamics, ecology, behavior, and so on, have been considered as evidence. These newer areas of investigation in no way supersede the morphologic studies; their use in modern systematics reflects the current advances in instrumentation and methodology in these areas and is consistent with the generally accepted thesis that some large part of the biology of organisms should be analyzed within the framework of evolutionary theory if a definitive conclusion is to be approached.

Relative recency of common ancestry, which is crucial in recognizing related individuals or groups of individuals, is judged on similarities among, as well as differences between, the individuals or groups. This judgment is made on attributes of many different kinds. The taxonomist compares organisms in terms of the presence, absence, or degree of expression of specific attributes of their phenotypes. Attributes in different organisms are termed *character states;* a set of attributes is said to be a *character,* which is, thus, only an abstraction. For example, the variable number of teeth on the mandible of adult mammals is a character, and the real numbers of mandibular teeth exhibited by different mammals are the states of that character. Character, as it is used in this context, is defined as any variable structure, substance, or action of an organism that is the product of the underlying genome and the interaction of the environment. Even the habitat of the organism can be considered indirectly to be a character. There are two basic classes of characters: those that are continuously variable, such as *morphometric* characters, and those that are discontinuously variable—that is, *meristic* characters. Characters also are considered either as quantitative or qualitative. Quantitative characters consist of naturally occurring sets of states to which numbers logically apply. The number of mandibular teeth cited here is such a character. Qualitative characters are those whose states are defined arbitrarily and numbered by the investigator. Characters may vary ontogenetically in the individual and within and between taxonomic groups of organisms. In addition, they may vary both geographically and from generation to generation.

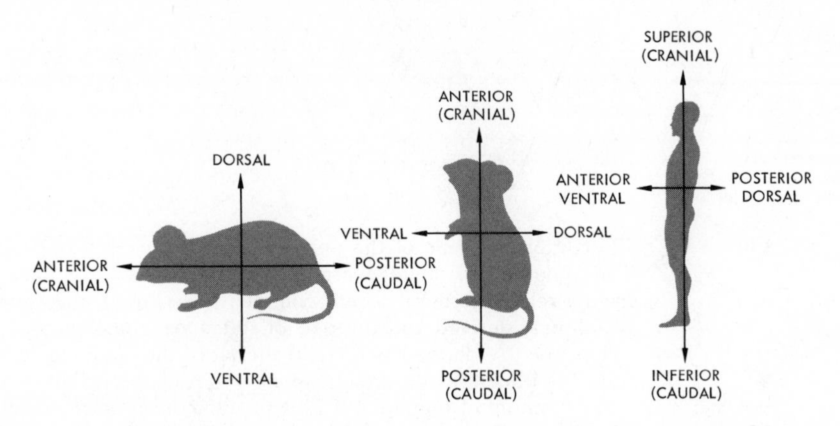

Figure 1-3 An illustration of some alternative positional terms in the mouse and human. (Modified after Romer.)

Character descriptions usually contain terms that indicate relative anatomical position. The more useful ones are defined here; some are illustrated in Figure 1-3.

(1) *Transverse:* A vertical section across the long axis of the organism. *Cross section* is an often used equivalent term.
(2) *Sagittal:* Any vertical section along the long axis of the organism. Mediosagittal is a section along the middle of the organism, whereas parasagittal is any other section.
(3) *Frontal:* Any horizontal section along the long axis of the organism.
(4a) *Anterior:* The direction, part, or surface toward the head end of the organism, or the direction in which it faces. *Cranial* and *superior* are frequently used synonyms.
(4b) *Posterior:* The direction, part, or surface toward the tail end of the organism. *Caudal* and *inferior* are frequently used synonyms.
(5a) *Dorsal:* The direction, part, or surface toward the upper aspect or back of the organism.
(5b) *Ventral:* The direction, part, or surface toward the lower aspect, or belly, of the organism.
(6a) *Medial:* A direction, part, or surface toward the midline of the organism.
(6b) *Lateral:* A direction, part, or surface toward the side of the organism.
(7a) *Proximal:* A direction, part, or surface close to some reference point.
(7b) *Distal:* A direction, part, or surface far from some reference point.
(8a) *Superficial:* A direction, part, or surface located near or on the exterior of the organism.
(8b) *Deep:* A direction, part, or surface located within the organism.

Criteria Derived from Similarity

THE CONCEPT OF HOMOLOGY. Are all similar character states inherited from a similar or identical condition in the most recent common ancestor? Do organisms independently evolve similar or even identical states from a dissimilar state in their most recent common ancestor? The answers to the questions are no and yes, respectively, and, therefore, uniquely and independently evolved kinds of similarities must be recognized (Figure 1-4). Furthermore, it is convenient to distinguish the structural aspects of an organism from its

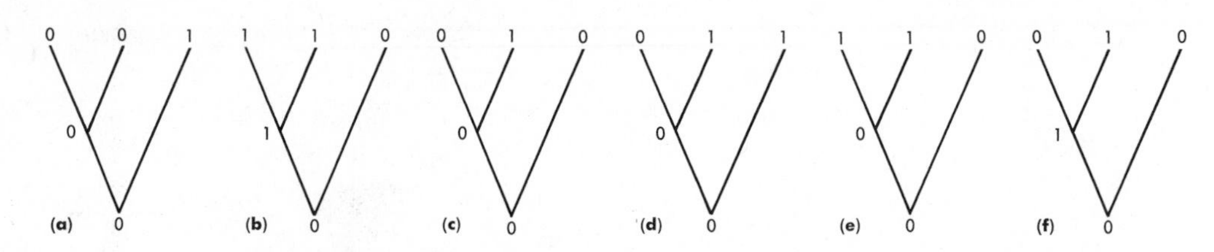

Figure 1-4 Illustrations of some of the possible evolutionary histories of binary characters. The hypothetical phylogenetic model consists of three lineages, two of which have a relatively more recent common ancestor. In illustrations (a) to (c) state 1 is uniquely derived and the pair of states are homologous. In (d) to (e) state 1 is independently derived twice and therefore the character contains a homoplasious state. (f) illustrates the independent origin of the primitive 0 state and it is another example of homoplasy. This peculiar form of homoplasy is called evolutionary reversal.

functional features. When the two kinds of characters, structural and functional, and the two kinds of similarity, uniquely and not uniquely evolved, are integrated, the following classification and modified definitions result.

I. Structural similarity of two or more states.
   A. *Homology* is uniquely evolved structural similarity in different organisms. The commonness of embryonic origin (namely, from the same anlage) of two or more structures is used very often as explicit evidence of homology. The similar states are *homologous,* and the noun for them is *homologues.* The degree of similarity between homologous states is generally considered proportional to the amount of evolutionary change that has occurred since they diverged from their common ancestral state.
   B. *Homoplasy* is independently evolved structural similarity in different organisms. The similarity is not due directly to inheritance from a common ancestor or development from a common anlage. The similar states involved are *homoplastic;* there is no noun for them. There are several kinds of homoplasy recognized, the following two being the most commonly mentioned in research on structure. In both cases, the homoplasious similarities are due to adaptations to environments that have identical or similar selective forces.
      1. *Parallelism* is homoplasy involving the independent evolution of similar states in organisms having lineages of relatively recent common ancestry. The similarity is not a direct result of that ancestry. The similar states are *parallelisms;* there is no noun for them.
      2. *Convergence* is homoplasy involving the independent evolution of similar states in organisms not having lineages of recent common ancestry. The similar states are *convergent;* there is no noun for them.

II. Functional similarity of two or more states.
   A. *Analogy* is uniquely evolved functional similarity in different organisms. The similar functional states are *analogous,* and the noun for them is *analogues.* Analogues have corresponding homologues, however, it does not necessarily follow that homologous states have proportionately

similar functions. The definition of analogy has been restricted because of the phylogenetic context in which it is used here.

B. *Anaplasy* is independently evolved functional similarity in different organisms. The similarity may or may not be due to the homoplasy of the structures involved. The similar characters are *anaplastic;* there is no noun for them.

The term *serial homology,* defined by Owen as the anatomical relationship among repetitive structures within or on a single organism, is mentioned occasionally. It must be emphasized, however, that this definition carries with it a connotation that is different from that of homology between different organisms; therefore the term is not particularly relevant to evolutionary studies. For example, the *homonomous series* as a whole, such as the vertebral column, can be homologous among different organisms, but a specific vertebra of the series need not be homologous with a similarly positioned element in the sequence of the other organism. As a case in point, with few exceptions the nearly constant number of cervical vertebrae of mammals are considered homologues, both individually and as a series. On the other hand, the homology of mammalian cervical vertebrae to those of amphibians and reptiles, which usually have very different numbers of cervicals, applies only to the series as a whole, or to the embryonic field producing it, but not to the individual vertebrae.

THE OPERATIONAL BASIS OF HOMOLOGY. Given that the actual course of ancestor-descendent relationships (phylogeny) cannot be observed directly due to the passage of time, is it possible to distinguish between homologous and homoplasious similarities? The answer is yes, but it must be emphasized at the outset that the distinction can never be made with certainty. The procedure for doing so begins with the choice of that attribute (e.g., whole bone, a process of a bone, or a foramen in a bone), which is to be compared to its "equivalent" in another organism. In order to be able to identify the most likely equivalent, the maximum degree of similarity between whole organisms, or their parts (e.g., skulls), must be established first. Such a maximum can be estimated only when whole organisms, or their parts, are orientated in such a way that they lie in the same plane and direction. Obviously, it would be illogical to compare cranial bones with one of the skulls upside down and backward with respect to the other.

The next step in the procedure is to identify the specific equivalent in the second organism that has the greatest degree of similarity to the originally chosen attribute. Specifically, the identification is made on the basis of the similarity of form—which is measured in architectural detail, size, and shape—and in terms of relative position and attachment to other elements. At the conclusion of this detailed level of comparison, the originally chosen attribute is estimated to be homologous with a particular state in the second organism because they are the most similar among all of the possible attributes that can be discerned. It is usually assumed that the reliability of two attributes being correctly identified as homologues—that is, derived from a similar condition in their most recent common ancestor—is related to how similar they are. However, no matter how similar a particular pair of attributes may be,

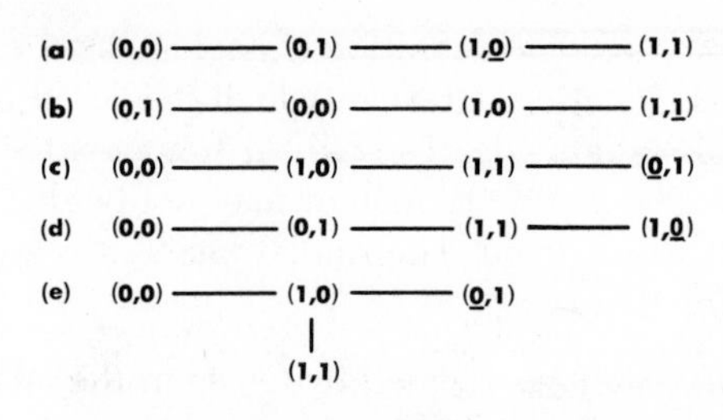

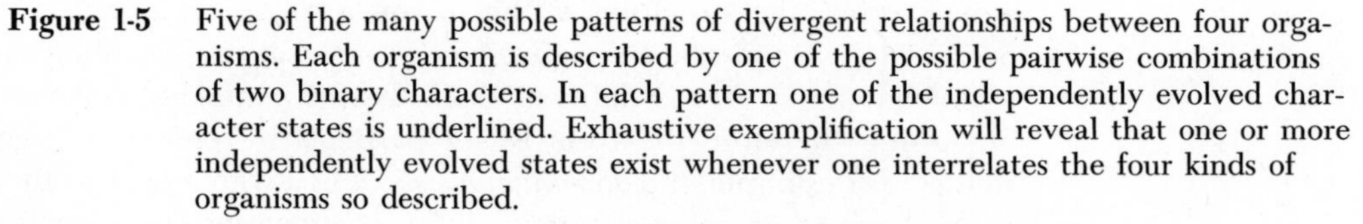

Figure 1-5 Five of the many possible patterns of divergent relationships between four organisms. Each organism is described by one of the possible pairwise combinations of two binary characters. In each pattern one of the independently evolved character states is underlined. Exhaustive exemplification will reveal that one or more independently evolved states exist whenever one interrelates the four kinds of organisms so described.

they are not likely to be homologues when the organisms are otherwise dissimilar.

A TEST FOR HOMOPLASIOUS STATES. Homoplasy can be predicted provided that the states of the characters are coded in a binary form—that is, but two forms of expression (e.g., 0 or 1). Because even a multistate character can be reduced to an equivalent series of binary expressions, the test is generally applicable. "Color of eyes," where only blue-(0) and brown-(1) eyed individuals are observed, is a simple example of a binary character. Binary characters, taken two at a time, yield four possible character state combinations, (0,0), (0,1), (1,0) and (1,1). These four combinations of states can be thought of as partial descriptions of four different organisms. The test rests on the fact that at least one homoplasious event exists when these four partially described organisms are part of a real data set. They cannot be linked together in any divergent direction without the independent origin of at least one of the character states. Exhaustive trial-and-error ordering of the four partially described organisms is sufficient to reveal the truth of this statement (Figure 1-5). Only if fewer than the four kinds of binary combinations exist in a real data set is it possible that the states of each of the compared pair of characters are uniquely derived and, consequently, homologous.

Table 1-1 A Data Matrix

Organisms	*Characters* A	B	C	D	E
I	1	1	1	0	1
II	1	0	1	1	0
III	1	1	0	1	0
IV	0	0	0	0	0
V	0	0	0	0	0

Table 1-2 A Character by Character Matrix of Observed Character State Combinations. Based on Data Matrix Presented in Table 1-1.

Characters				
B	(1,1)(0,0) (1,0)			
C	(1,1)(0,0) (1,0)	(1,1)(1,0) (0,1)(0,0)		
D	(1,0)(0,0) (1,1)	(1,0)(1,1) (0,1)(0,0)	(1,0)(0,1) (1,1)(0,0)	
E	(1,1)(0,0) (1,0)	(1,1)(1,0) (0,0)	(1,1)(0,0) (1,0)	(0,1)(0,0) (1,0)
	A	B	C	D

Characters

It must be emphasized that when the four possible combinations are present the investigator still does not know which character contains the homoplasious state. The test only denotes the incompatibility of the *pair* of characters being compared at that time. In the hypothetical data set exemplified in Table 1-1, there are five organisms (I–V) described by five binary characters (A–E). In Table 1-2 the observed combination of states is listed, and in Table 1-3 the incompatibility that each character has with all others is expressed.

It is assumed that the larger the number of incompatibilities a character exhibits the more likely it is to be homoplasious. This assumption is based on the fact that a homologous character is only incompatible with homoplasious ones, whereas a homoplasious character can be incompatible with other homoplasious characters as well. Although this test is a powerful analytic tool, you can never assume that all of the incompatibilities in a data set have been discovered. The addition of a new character can reveal a previously unnoticed incompatibility in the original set of data. However, such a discovery is unlikely when the original data set consists of a large number of characters because the incompatibility of the added attribute is more likely to denote its own homoplasy rather than an unnoticed one among the original set.

CHRONOCLINES. As stated earlier, much of evolution has the appearance of being an ordered series of steps and sustained in one direction. The linearity

Table 1-3 A Character by Character Matrix of Character Compatibilities. Based on Character State Combination Matrix Presented in Table 1-2.
0 = compatible.
X = incompatible.

Characters				
B	0			
C	0	X		
D	0	X	X	
E	0	0	0	0
	A	B	C	D

Characters

and orientation are a consequence of persistent forces of natural selection. Almost all fossil lineages that are well represented stratigraphically exhibit these evolutionary trends in single or correlated groups of characters, and the term *chronocline* is applied to them.

It is unusual for an evolutionary trend to involve only a single characteristic; more often it consists of a complex of attributes. Although our knowledge is not based entirely on evidence derived from fossils, two specific kinds of trends involve character complexes. First, it has been demonstrated frequently that when a group of numerous and similar morphologic parts tends to be reduced in number, the remaining parts become more differentiated from each other. Secondly, when a series of relatively few dissimilar morphologic parts tends to increase in number, the differences among the parts tend to become less. Both of these trends, in all the many organisms that exhibit them, involve a complex of characters that have a broad functional interrelationship. As a corollary to the foregoing statement, it can be said that a given function may be performed by a series of similar morphologic parts acting as a whole, or the function may be performed by one or more different parts acting as separate units or limited sets of units.

Chronoclines consisting of many distinct states have proven to be very useful in phylogenetic inference. However, their value does not come from the successive manner in which the fossils are laid down, as is so often asserted, but rather from the improved knowledge of structure and variation that they provide. Phylogeny cannot be observed in a fossil sequence, nor does a sequence provide unequivocal evidence of homology. More will be said about this later.

A single character chronocline is shown [7] in Figure 1-6. This particular trend consists of five character states (a, A, A^{*}, A^{**}, and A^{***}), and the hypothetical example depicts fossilized right hindlimbs that were collected in successive geologic strata. The character states are those stages of modification, here the loss of entire toes, that were coincidentally fossilized and collected. The character state A^{***} can be considered that of a living species.

All chronoclines, as exemplified by Figure 1-6, exhibit *polarity;* that is, one character state is evolved sequentially from another. The derived character state is said to be advanced relative to the character state from which it evolved, whereas the character state that gives rise to a derived state is primitive relative to that state. Of the character states expressed in the cline in Figure 1-6, state a is the most primitive and A^{****} the most advanced. The relative terms *specialized* and *generalized* have been reserved for other phenomena, and they should not be considered synonyms of *advanced* or *derived* and *primitive,* respectively. Again, referring to Figure 1-6, it should be noted that, at least theoretically, a much larger number of intermediate states exists between primitive state a and the derived extreme A^{***} than the three shown (A, A^{*}, and A^{**}). These "missing" states might correspond to the different stages in reduction and the ultimate loss of the individual phalangeal elements of each toe. The actual number of intermediate states in a cline depends on the duration of the trend and the number of relevant mutations and recombinations that occurred. The completeness of the collection of material from the continuum and the degree of refinement of the investigator's observation also are involved.

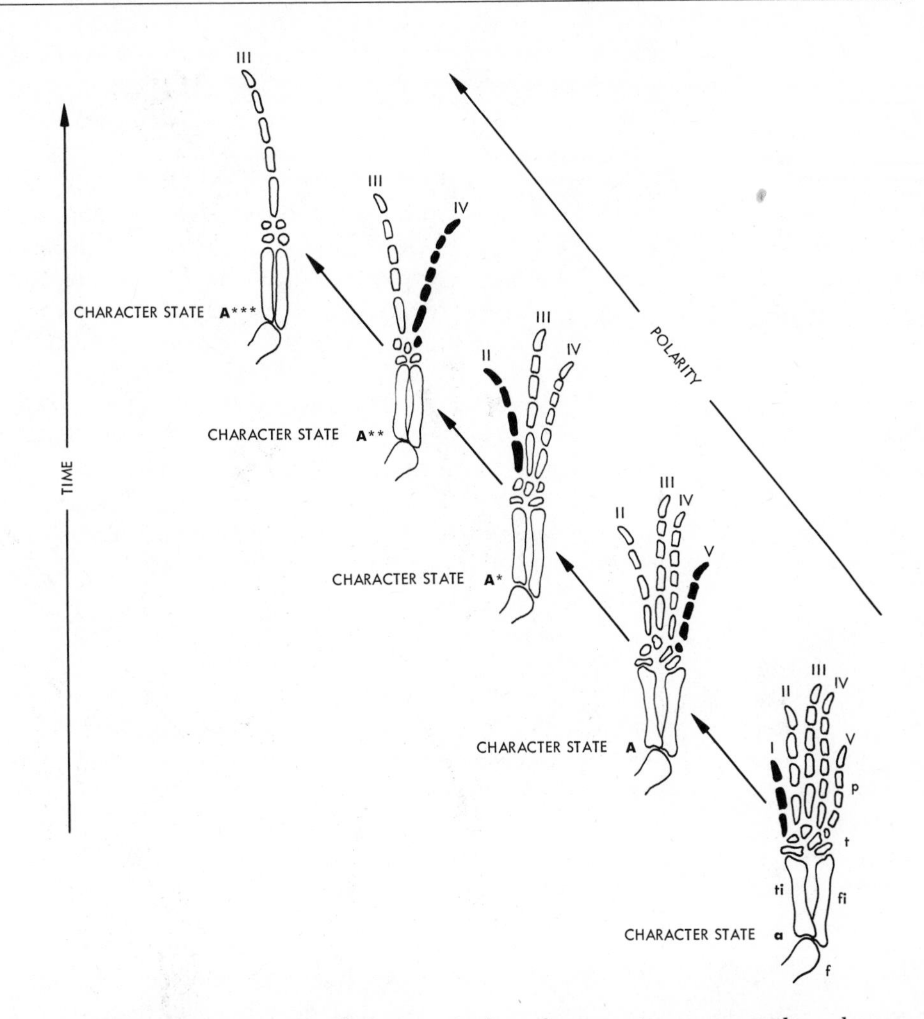

Figure 1-6 A chronocline in a single character, number of toes, as it progresses through time. The actual character states are 5, 4, 3, 2 and 1. The diagrammatic illustrations consist of femur (f), tibia (ti), fibula (fi), tarsals and metatarsals (t), and five phalanges (p) numbered I–V.

When a character state of any chronocline is similar to an extreme of another trend, then the first cline is primitive and that character state forming an extreme of the second trend is its primitive extreme. This criterion is visualized in Figure 1-7. Here the primitive and derived states can be determined because two of the clines are successively dependent on character states of the preceding cline for their origin. The characters in this example are the number of toes, a, A, A*, A**, and A***; the number of terminal T-shaped phalangeal elements, A, A′*, A′**, and A′***; and the number of bifurcated T-shaped bones, A′**, A″***, and A″****.

MORPHOCLINES. In Figure 1-8 a contemporary series of homologues is illustrated that is identical to the character states of its ancestral chronocline. Such

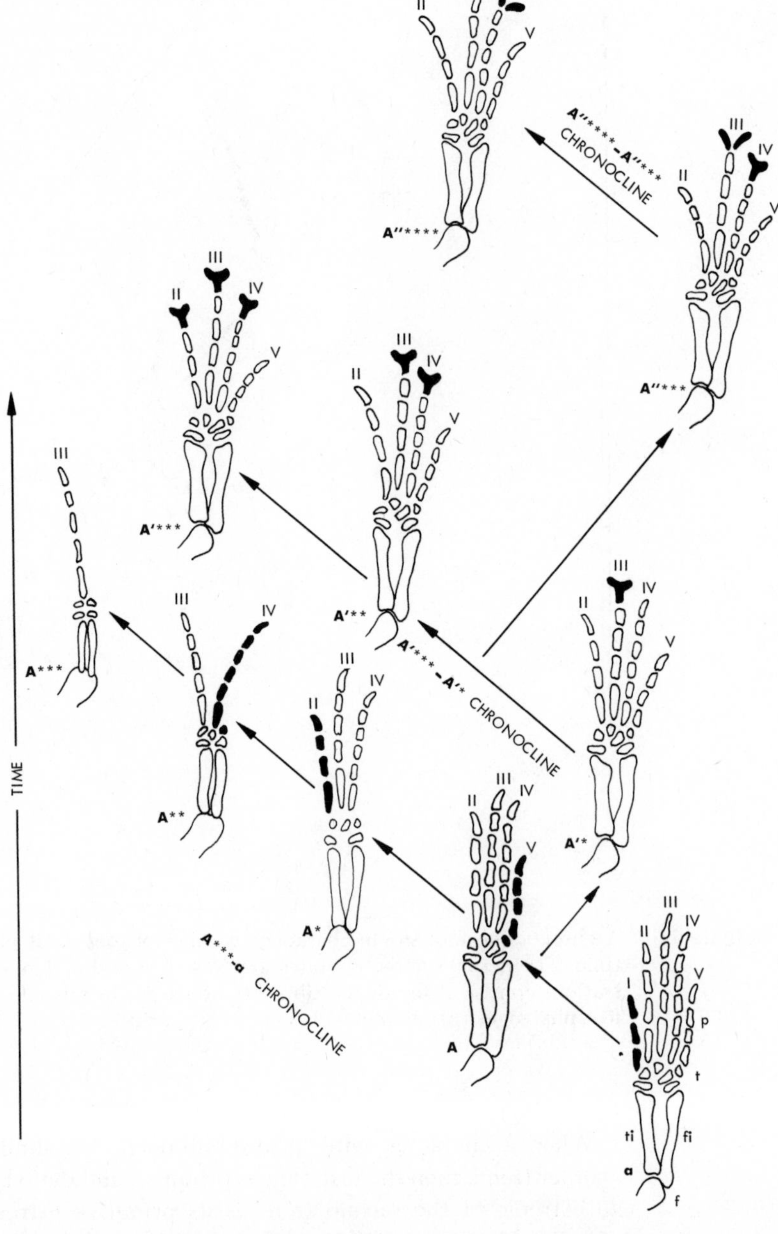

Figure 1-7 An example of derived chronoclines. The characters are number of toes (a-A-A*-A**-A***), number of terminal T-shaped phalangeal elements (A-A′*-A′**-A′***), and number of bifurcated T-shaped elements (A′**-A″***-A″****). See the text and legend of Figure 1-6 for further explanation.

a contemporaneous sequence of stages of modification is called a morphocline and, like the chronocline, exhibits polarity. This model assumes that evolution is divergent and that adaptations persist unchanged for varying lengths of time. The degree of resemblance between a morphocline and its corresponding chronocline depends on the amount and direction of evolution. Also affecting

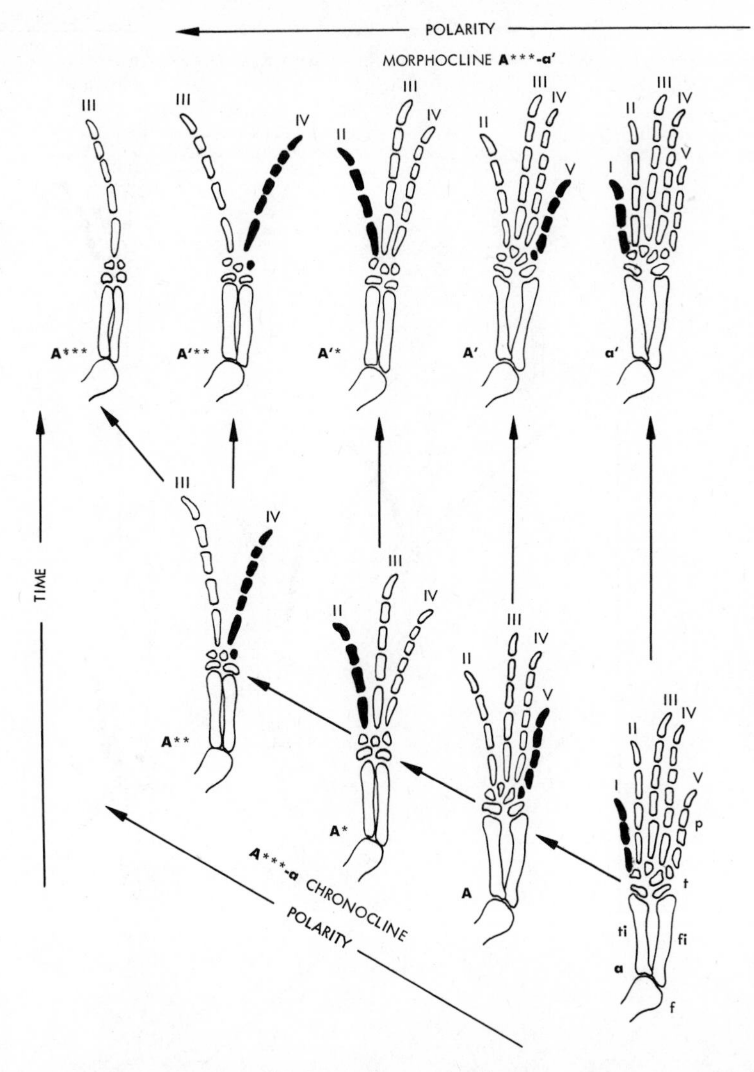

Figure 1-8 An example of a morphocline. The series of contemporaneous examples that form the morphocline exhibit the identical character states of their ancestral chronocline. No change in the character state has occurred in each divergent line. See the text and legend of Figure 1-6 for further explanation.

the correspondence between the two is the fact that some of the divergent lineages become extinct at different levels in time.

It is often observed that one member of a group of related organisms is conspicuously different from others of its group in one or more characters (Figures 1-9 and 1-10). This degree of difference in a character results from the absence of organisms exhibiting intermediate character states. The intermediate organisms were present in the chronocline, but the divergent lineages (other chronoclines) did not materialize, or if they did they became extinct before reaching the level of time in which the morphocline in question occurs. The number of "absentees" determines the degree of continuity of the morphocline. A relatively discontinuous morphocline is difficult to recognize as

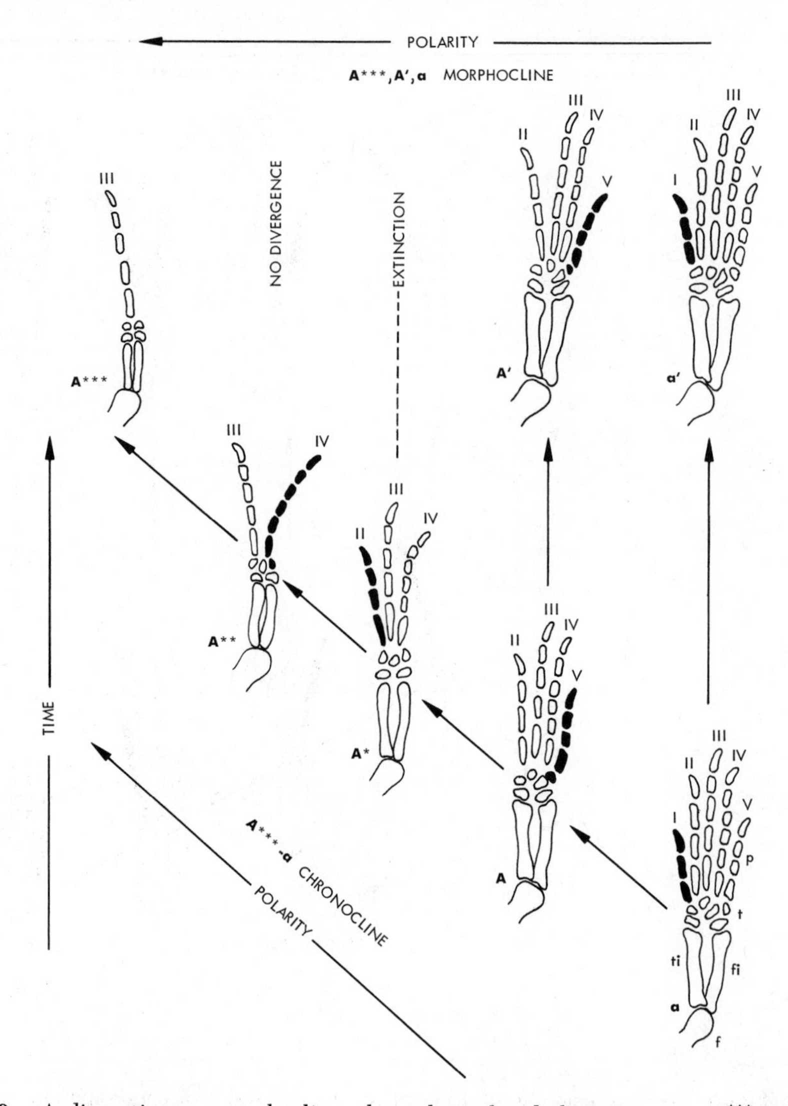

Figure 1-9 A discontinuous morphocline where the isolated character state A*** is the most highly modified or advanced. See the text and legend of Figure 1-6 for further explanation.

the product of a single evolutionary trend because too few of the intermediate character states persist. The same problem of discontinuity also is observed in chronoclines, where it is probably more common. The discontinuities in chronoclines are in large part abrupt and are probably a result of the fortuitous nature of the process of fossilization and the more difficult collection of fossil material. It has been observed in a number of combined paleontological-neontological studies of a related group of organisms that character discontinuities may exist in both the morphocline and the chronocline of a particular character. These same studies also show that considerable reciprocal evidence for the actual presence of absentees in a character trend can be obtained by using both the living organisms and the relevant fossil record.

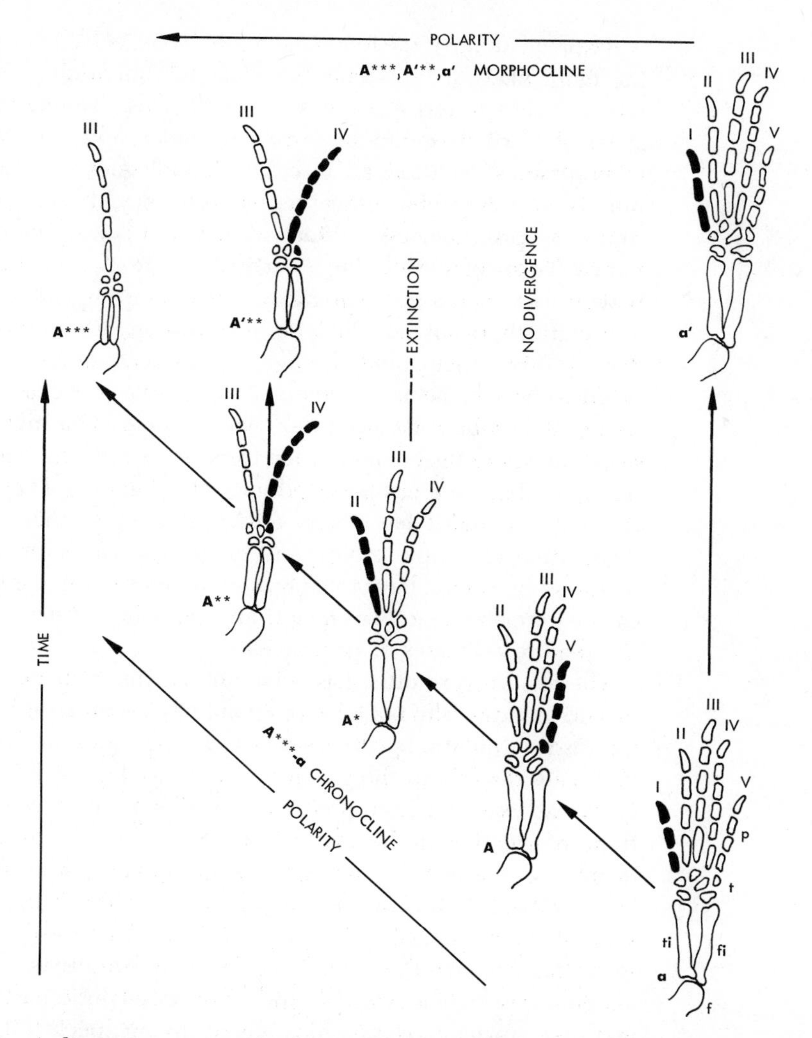

Figure 1-10 A discontinuous morphocline where the isolated character state a′ is the most primitive or least modified. See the text and legend of Figure 1-6 for further explanation.

Postulates for the Estimation of Polarity

A synthesis of the criteria derived from similarity and chronoclines and morphoclines, coupled with parsimonious reasoning, has led to the formulation of the following probabilistically based postulates, which can be used to estimate which of two or more character states of a cline is the most primitive.

A character state is considered primitive when it is the one most widely distributed among the members of (1) other similar groups of organisms, including fossils; (2) dissimilar subgroups within the group being studied; and/or (3) when it represents a structural-functional condition from which all other homologues must have evolved by reason of mechanical necessity.

Postulates 1 and 2 can be rigorously applied, and they are generally regarded as the best estimators of primitiveness. The third postulate is not as practical as the others, and one usually finds its users asserting, rather than demonstrating, mechanical necessity. Postulates 1 and 2 rest on the assumption that the more widespread a state is the less likely it is to have evolved independently—that is, the more likely it represents the identical condition possessed by the most recent common ancestor. These postulates also assume that some character states, the primitive ones in particular, undergo little or no change, and that homoplasious conditions are rarer than homologous ones. Considerable caution must be exercised when using relative frequency of character-state occurrence to predict primitiveness. "Widely distributed" means more than just the raw number of species exhibiting a particular state. A character state is said to be widespread when it occurs in many species or groups, at least several of which are relatively dissimilar. Phyletic lines vary greatly in the number of species derived from them, and it might be misleading to base the decision of primitiveness on the mere numerical superiority of the most successful lineage, particularly when its species are very similar. One must try to avoid the situation where the common ancestor of the successful lineage evolved a new state and that state has persisted unchanged among its many similar species. Postulates 1 and 2 have been criticized because they require an a priori recognition of related groups and subgroups. However, the details of these relationships do not have to be known in order to use the postulates, and they can be estimated exactly by quantitative measures of over-all similarity that do not require estimates of primitiveness.

The fossil record often has been cited as the ultimate basis for identifying primitive states. This is a falacious point of view unless it is known a priori that the fossil in question occupies a phyletic position *exactly* within the lineage that leads *directly* to the group under study (Figure 1-6). But how is one to know the exact phyletic position of fossils? Such knowledge does not come from information on the absolute or relative age of a fossil! If a fossil is not a member of the lineage leading to the group being studied then it, too, might have evolved. In this presumably usual situation it is the unknown character state of the most recent common ancestor of the fossil and the group under study that provides the ultimate evidence for primitiveness. Evidence for such ancestor-descendent relationships is not an intrinsic part of the fossil record and, like phyletic relationships among living species, it must be estimated indirectly. Nevertheless, fossils are extremely important. They very often provide real observations on unknown portions of a morphocline or chronocline. Moreover, the fossil record usually demonstrates greater diversity than is present among living species. Consider how much our knowledge of reptilian diversity has increased with the study of dinosaurs.

Phylogeny: A Summation of Data and Hypotheses

Phylogeny is defined as the continuous historical product of evolution and speciation, and it has the following parameters: (1) relative position of speciational events, (2) amount of evolutionary change, and (3) direction of evolutionary change. Paleontologists are very much concerned with the dimension of real time, which often is referred to as a fourth parameter. Real time is not a property of phylogeny itself, and no biological theory, such as constancy of evolutionary rates, has proven to be a good predictor of it. It follows from evolutionary and speciational theories that each and every species, living and dead, is linked by genealogical descent and common ancestry. We must conclude that there is but one true and estimable phylogeny for any group of species. The *results* of evolution and speciation can be observed, but rarely if ever can the actual causes and actions because of the great amounts of time that pass before perceptible and sustained differences occur.

For many years taxonomists have graphically represented phylogeny as a continuous series of diverging species lineages (Figure 2-1). Each pair of lineages is represented by lines that diverge from its most recent common ancestor, and the relative positions of the common ancestors are sequentially displayed in a vertical plane according to their relative time of appearance. The oldest divergence forms the base, or root, of the phylogenetic tree, and the tips of the branches are the most recently derived evolutionary products. The parameter of direction of evolutionary change was referred to previously as polarity, and it is characterized by the upward diverging nature of the lines of a phylogenetic tree. The study of the parameter of relative position of speciational events is referred to as cladistics, and it is represented in a tree by the relative position of the common ancestors from which the diverging lines arise. Patristics is the study of the third parameter, amount of evolution, and it can be represented by the relative or absolute lengths of the lines. Rarely are all of the products of evolution and speciation available in a given study. At the very best, one has access to the living end products plus a few fragmentary fossils. Nothing even remotely approximating a complete phylogenetic history is available. Horse and higher primate phylogenies, although very incomplete, are probably the best among chordates. Logically, this is not a real deficiency because even the complete record of fossils and living species would not provide the essential description of matings, births, and the origin of reproductive isolating mechanisms. These dynamic phenomena do not fossilize!

Almost all published phylogenetic trees are the visual product of the feeling, or gestalt, of the investigator. Such an expression of the taxonomists understanding of the phylogeny of a group of organisms tends to be a useless hypothesis because the raw data (homologous sets of character states and their polarity) are not sufficiently precisely described and interpreted to permit evaluation and further use. Perhaps a more serious criticism of the intuitive approach is the absence of any accompanying reference to how and why the data were analyzed and interpreted to mean certain historical relationships but not other possibilities. The supportive aspects of evolutionary and speciational theories, the assumptions, and the general methodologies usually are not cited. Only in the last few years, with the concentrated study of old and new theory and the development of quantitative methods, has the reconstruction of phylogeny become a part of modern science. Logically and scientifically pursued, a phylogenetic tree is the final and most encompassing summation of all data gathered in a comparative study. Even prior to this stage the different kinds of research that necessarily precede the reconstruction of a phylogeny provide numerous avenues of feedback, or counterchecks and balances, on the investigator's experimental and comparative approaches and his working premises of evolutionary and speciational theories. Furthermore, an objectively constructed phylogeny provides a framework from which new hypotheses can be realized more readily and against which previously formulated hypotheses can be tested more rigorously. Above all else the predictive powers of phylogenies reconstructed by the newer quantitative methods have been found to be so great that comparative biologists can no longer fail to take advantage of them in their research and teaching, regardless of whether they compare and contrast sets of related phenomena or whole organisms or populations.

Phylogenetic Relationships

A theoretically sound and well-known methodology exists for estimating the cladistic aspect of phylogeny. The patristic parameter, on the other hand, is a very hotly debated area of phylogenetic inference. Patristic methodology is not well delineated, nor are the available procedures easy to understand, and it will not be considered further. Correct execution of the method of cladistic inference requires that the raw data consist of sets of homologous character states whose chronoclines are specified. Species described by these data are clustered into groups (pairs of species and pairs of species pairs) according to the greatest number of shared derived character states. In actual practice two species, or two such groups of species, are shown graphically diverging from a common ancestor when that particular association exhibits more shared derived states than any other possible grouping. The cladistic method is most readily applied to a set of binary characters. The primitive state of each character can be estimated from the polarity postulates, and characters containing homoplasious conditions can be identified, with application of the homoplasy test, and eliminated. The postulates and test are described on pages 21 and 14, respectively.

Classification: A Product of Phylogeny

Why Classify Organisms?

The classification of organisms and the application of scientific nomenclature to the units of classification are extremely important to all biologists. A classification provides the necessary framework for the mental storage of vast amounts of information into a condensed and readily retrievable form. The nomenclature associated with the classification provides a standardized terminology that can be used to communicate objectively with others about the information embodied in the classification. In addition to these practical aspects, a classification may in itself provide the stimulus for further conceptualization. The way in which the information is ordered in a classification may lead to new, and probably otherwise unidentifiable, meaningful concepts.

To be of value a classification system must be both practical and meaningful. In the field of biology, where knowledge continues to increase very rapidly, the system must be able to incorporate changing data and ideas with facility and yet maintain some over-all stability. Because organic evolution pervades all fields of biology and related disciplines, it is agreed generally that the most meaningful classificatory system is one that at least in part is based on this common theme. The Linnaean binomial hierarchical system of classification can be related to phylogeny, and it is used widely by both evolutionary and nonevolutionary biologists.

What Is a Phylogenetic Classification?

A phylogenetic classification is a hierarchical arrangement of the way evolutionary lineages are linked together in phylogeny; therefore, evolutionary and speciational theories are its interpretative basis. The units of a phylogenetic classification are equivalent to lineages that have common ancestry. The more inclusive the classificatory unit, the relatively older are the included common ancestors.

RULE OF MONOPHYLY. The only scientific rule that must be followed in translating a phylogenetic tree into a classification is that the units be *monophyletic*. A monophyletic unit consists of one or more lineages and the most recent common ancestor from which the lines diverge. The rule is violated when a classificatory unit contains a lineage that does not diverge from the included most recent common ancestor.

The cladistic parameter of phylogeny can be translated into classification by strict adherence to the rule of monophyly. The aspect of polarity is specified by the hierarchical nature of the classificatory system. No generally satisfactory way has yet been found to translate the patristic parameter of phylogeny into classification.

The classificatory category of species, as defined earlier on the basis of other criteria, is on occasion exceptional to the thesis of monophyly. For example, a species may be a hybrid—that is, a product resulting from interbreeding between two species when their usual isolating mechanisms have temporarily and locally broken down. The phenomenon, which appears to be extremely rare in animals, is known to be very common in plants. The strict application of the rule of monophyly to the category of species that are not hybrid products and to all other taxonomic units ensures that the relative recency of common ancestry estimated by phylogeny is translated into the classification.

CLASSIFICATORY UNITS AND NOMENCLATURE. In spite of the many advances that have been made in taxonomic theory and procedure in the last century, the general form of classification has changed very little from the pre-Darwinian hierarchical binominal system of nomenclature used by Carl von Linné in 1758 in his tenth edition of *Systema Naturae Regnum Animale.* In this edition Linné used only five units—species, genus, order, class, and kingdom—to classify all the animals known to him. Correlated with the many species that have been recognized since 1758, with the attempt to relate classification to phylogeny, and with ever-increasing fineness of discrimination, the number of classificatory units has been expanded to approximately twenty-one levels. Only those categories most frequently used today are listed in Table 1-4 in order of their taxonomic inclusiveness (from bottom to top). In this example of the system, the nomenclatures of the domestic cat, *Felis catus,* and man, *Homo sapiens,* are compared relative to these categories. The scientific names applied to the different classificatory units in a particular study are formed and further recognized and synonymyzed according to the International Rules of Zoological Nomenclature. These purely legalistic conventions help to insure nomenclatural consistency among scientists. They do not provide ways of evaluating the validity of a given phylogenetic tree or the monophyletic units that are recognized.

Table 1-4 Classificatory Categories.

Categories	*Domestic Cat*	*Man*
Kingdom	Animalia	Animalia
Phylum	Chordata	Chordata
Subphylum	Vertebrata	Vertebrata
Class	Mammalia	Mammalia
Order	Carnivora	Primates
Family	Felidae	Hominidae
Genus	*Felis*	*Homo*
Species	*catus*	*sapiens*

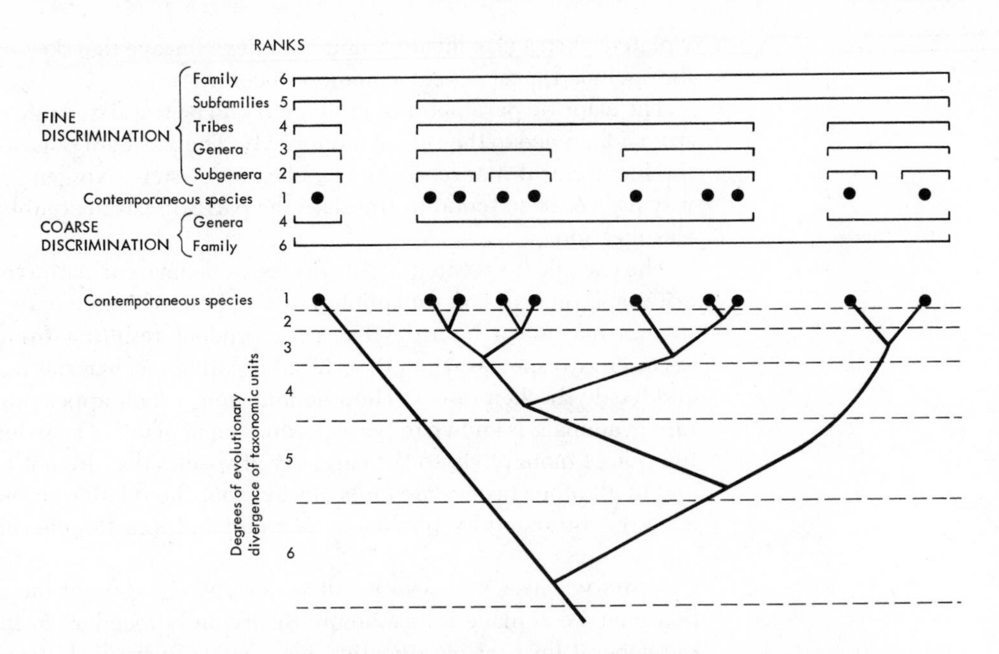

Figure 1-11 Correspondence between degrees of evolutionary divergence of taxonomic units and classification. The degrees of evolutionary divergence below are translated into Linnean classificatory categories above. The brackets at each taxonomic level indicate the taxa that are included. (Modified after G. G. Simpson, *Principles of Animal Taxonomy*. New York: Columbia University Press, 1962).

In spite of the rigorous rules of nomenclature and monophyly, there is a certain amount of flexibility in classifying organisms. This flexibility is demonstrated in Figure 1-11 above the hypothetical phylogeny where either a course or a fine discrimination of the same tree can be used. Both classificatory products are consistent with the rule of monophyly. Usually, the more objective the assessment of phylogeny and the more divergent the phylogenetic history, the finer is the classificatory discrimination employed. New discoveries, and often only restudy, will fill in phylogenetic gaps. It must be remembered that phylogeny is a continuum, and the gaps that we predict are only convenient results of the processes of speciation, extinction, and/or fossilization. As a corollary, it can be stated that the continuum of phylogeny causes difficulty in classification, and that continued difficulty is logical evidence of evolution.

References for Chapter 1

[1] Belon du Mans, Pierre. *L'Histoire de la Nature des Oyseaux.* Paris: 1555.

[2] Cole, F. J. *A History of Comparative Anatomy from Aristotle to the Eighteenth Century.* London: Macmillan & Company, Ltd., 1944. This book offers a detailed treatment of the origin and evolution of concepts pertaining to comparative anatomy.

[3] Darwin, Charles. *On the Origin of Species by Means of Natural Selection, or the Preservation of Favoured Races in the Struggle for Life.* London: John Murray, 1859. (Cited from a reprint of the sixth edition; this has all additions and corrections. New York: A. L. Burt Company)

[4] HULL, DAVID L. "Certainty and Circularity in Evolutionary Taxonomy." *Evolution,* **20**(1967), 174–189.

[5] LEQUESNE, W. J. "A method of Selection of Characters in Numerical Taxonomy." *Systematic Zoology,* **18**(1969), 201–205.

[6] MCNEILL, A. R. *Functional Design in Fishes.* Hutchinson University Library, 1967. McNeill's short essay on aspects of fish design is presented and interpreted in the context of natural selection. For this reason, the book will very likely serve as a model for further research into form and function.

[7] MASLIN, T. PAUL. "Morphological Criteria of Phyletic Relationships." *Systematic Zoology,* **1**(1952), 49–70. Maslin's paper provides an introduction to the concepts and methodologies of delimiting chronoclines and morphoclines and their polarity.

[8] OWEN, R. *Lectures on the Comparative Anatomy and Physiology of Invertebrate Animals.* London: Longman, Brown, Green and Longmans, 1843.

[9] SIMPSON, G. G. *Principles of Animal Taxonomy.* New York: Columbia University Press, 1962.

[10] ———. "Anatomy and Morphology: Classification and Evolution: 1859 and 1959." *Proceedings of the American Philosophical Society,* **103**(1959), 286–306. Simpson examines the relationship between anatomical observations and evolutionary theory.

[11] ———. *The Meaning of Evolution.* New Haven, Conn.: Yale University Press, 1949.

[12] STEBBINS, G. L. *Processes of Organic Evolution.* Englewood Cliffs, N.J.: Prentice-Hall, Inc., 1966. This is a short paperback book that will serve adequately as a concise overview of organic evolution.

[13] THROCKMORTON, L. H. "Similarity Versus Relationship in Drosophila." *Systematic Zoology,* **14**(1965), 221–236.

2

A Framework for the Further Study of Chordate Animals

Introduction It will be valuable at this point to review briefly some observations and hypotheses. This will give readers a preliminary perspective that will increase their understanding of the more detailed discussions of individual morphologic parts and even of the whole systems and related functions that are to follow. The perspective depends on a general mental image of the taxa considered. For example, when a particular function of a teleostean fish is described and compared with that of a chondrostean fish, the reader must be able to relate automatically the information to (1) the organism's general body form and its major kinds of variability; (2) the characters that diagnose it taxonomically as an adapted kind of animal; (3) its phylogenetic position—that is, its usually accepted intertaxonomic and intrataxonomic relationships; and (4) aspects of its general biology, such as environments occupied, feeding and reproductive habits, and so on. The following descriptions of the major kinds of chordate taxa attempt to provide this necessary perspective.

In an effort to further the student's understanding of the more isolated, and as yet unrelated, subject matters, the generally accepted and largely subjectively constructed phylogeny of chordates is given in Figure 2-1. This hypothesis of chordate relationships can be tested by comparing it to the results obtained from the application of objective phyletic methods to the data on structure and function presented in later chapters. In addition, the usually accepted classification of chordates is presented as another kind of framework which can be tested objectively (Table 2-1). Similarly, the tabularized summary of the major features of biotic and abiotic evolution through geologic history (Table 2-2) gives a rigid framework within which conclusions can be displayed relative to absolute time. The table also draws attention to the correspondence between major changes in animal and plant life, long-term climatic conditions, and major geologic events. In many ways these correspondences permit the

Table 2-1 A Classification of Chordates Transcribed in Part from the Intuitively Constructed Phylogeny Shown in Figure 2-1. *The asterisks indicate extinct taxa; all other taxa have living representatives.

- Phylum Hemichordata
  - Class Pterobranchia
  - Class Enteropneusta
  - Class Planctosphaeroidea
- Phylum Chordata
  - Subphylum Tunicata
    - Class Ascidiacea
    - Class Thaliacea
    - Class Larvacea
  - Subphylum Cephalochordata
  - Subphylum Vertebrata
    - Class Agnatha
      - Subclass Monorhina
        - Order Osteostraci*
        - Order Anaspida*
        - Order Cyclostomata
          - Suborder Petromyzontoidei
          - Suborder Myxinoidei
      - Subclass Diplorhina*
        - Order Heterostraci
        - Order Coelolepida(?)
    - Class Placodermi*
      - Order Petalichthyida
      - Order Rhenanida
      - Order Arthrodira
      - Order Phyllolepida
      - Order Ptyctodontida
      - Order Antiarchi
    - Class Chondrichthyes
      - Subclass Elasmobranchii
        - Order Cladoselachii*
        - Order Pleuracanthodii*
        - Order Selachii
        - Order Batoidea
      - Subclass Holocephali
        - Order Chimaeriformes
    - Class Osteichthyes
      - Subclass Acanthodii*
      - Subclass Actinopterygii
        - Infraclass Chondrostei
          - Order Palaeonisciformes*
          - Order Polypteriformes
          - Order Acipenseriformes
        - Infraclass Neopterygii
          - Division Ginglymodi
            - Order Semionotiformes
            - Order Pycnondontiformes*
          - Division Halecostomi
            - Subdivision Halecomorphi
              - Order Amiiformes
              - Order Aspidorhynchiformes*
              - Order Pholidophoriformes*
            - Subdivision Teleostei
      - Subclass Sarcopterygii
        - Order Crossopterygii
          - Suborder Rhipidistia*
          - Suborder Coelacanthini
        - Order Dipnoi
    - Class Amphibia
      - Subclass Labyrinthodontia*
      - Subclass Lepospondyli*
      - Subclass Lissamphibia
        - Superorder Salientia
          - Order Anura
        - Superorder Caudata
          - Order Urodela
          - Order Apoda
    - Class Reptilia
      - Subclass Anapsida
        - Order Cotylosauria*
        - Order Mesosauria*
        - Order Chelonia
      - Subclass Lepidosauria
        - Order Eosuchia*
        - Order Squamata
          - Suborder Lacertilia
          - Suborder Ophidia
        - Order Rhynchocephalia
      - Subclass Archosauria
        - Order Thecodontia*
        - Order Crocodilia
        - Order Pterosauria*
        - Order Saurischia*
        - Order Ornithischia*
      - Subclass Euryapsida*
      - Subclass Ichthyopterygia*
      - Subclass Synapsida*
    - Class Aves
      - Subclass Archaeornithes*
      - Subclass Neornithes
    - Class Mammalia
      - Subclass Prototheria
        - Order Monotremata
      - Subclass Allotheria*
      - Subclass Theria
        - Infraclass Trituberculata*
        - Infraclass Metatheria
          - Order Marsupialia
        - Infraclass Eutheria
          - Order Insectivora
          - Order Tillodontia*
          - Order Taeniodontia*
          - Order Chiroptera
          - Order Primates
          - Order Creodonta*
          - Order Carnivora
          - Order Condylarthra*
          - Order Amblypoda*
          - Order Proboscidea
          - Order Sirenia
          - Order Desmostylia*
          - Order Hyracoidea
          - Order Embrithopoda*
          - Order Notoungulata*
          - Order Astrapotheria*
          - Order Litopterna*
          - Order Perissodactyla
          - Order Artiodactyla
          - Order Edentata
          - Order Pholidota
          - Order Tubulidentata
          - Order Cetacea
          - Order Rodentia
          - Order Lagomorpha

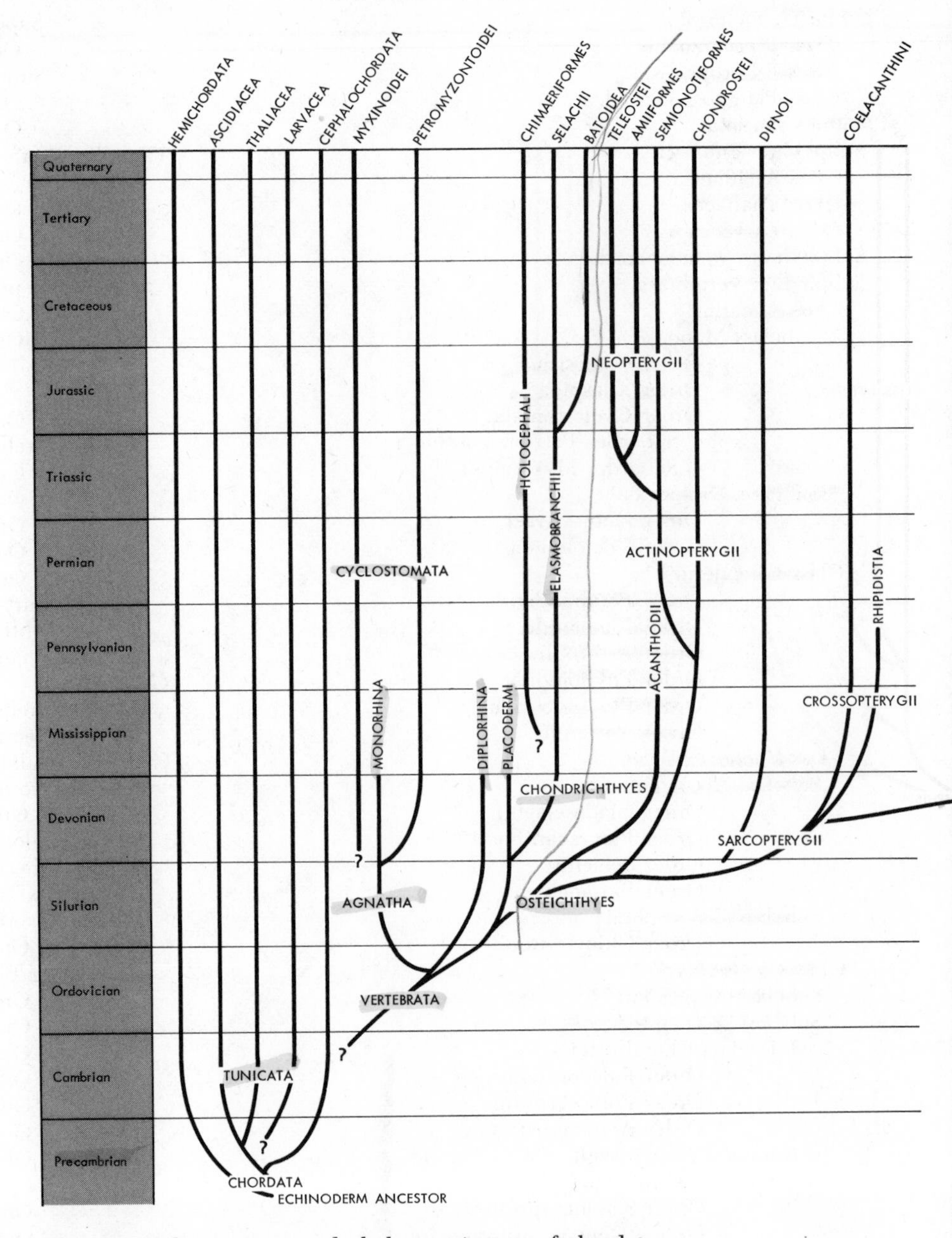

Figure 2-1 An intuitively reconstructed phylogenetic tree of chordates.

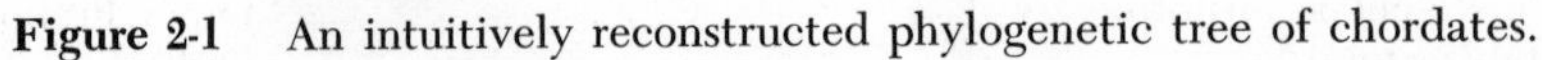

reader to determine more accurately cause and effect and to consider the likelihood of postulated relationships and adaptations.

What Are Chordates?

Chordates are those animals that at some time during their life history possess the following conditions: (1) a pharyngeal wall containing bilaterally symmetrical openings, gill pouches, and associated slits; (2) a mid-dorsal hollow cylinder of nervous tissue, the central nerve cord; (3) a single mid-dorsal axial supportive structure, the notochord, located below the nerve cord; (4) a

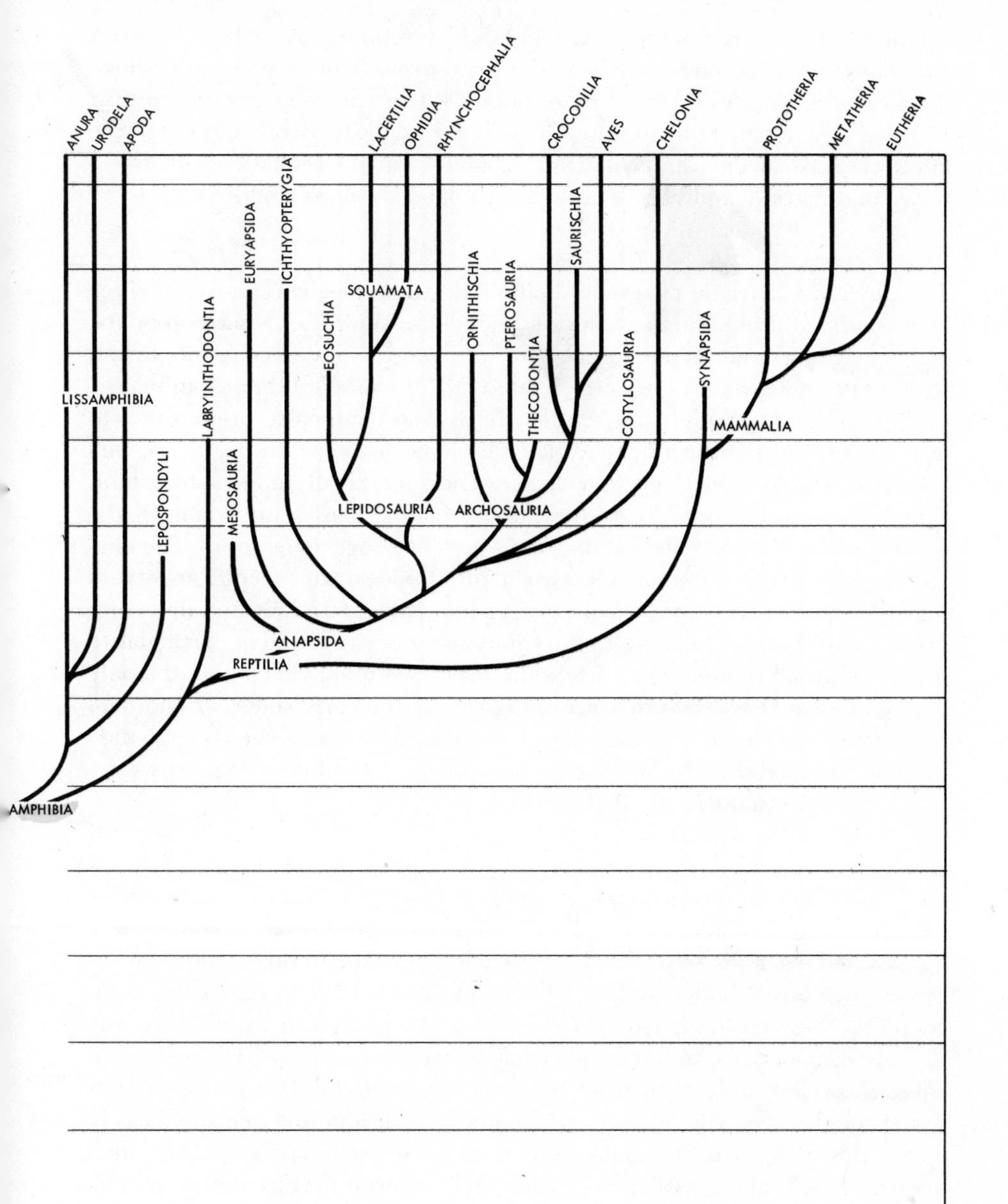

discrete subpharyngeal gland that can bind iodine and synthesize thyroxine and related substances; (5) a body segmented throughout or in part; (6) an early stage of embryogenesis in which the organism consists of three well-defined germ layers: ectoderm, mesoderm, and endoderm; (7) an embryonic blastopore that may become the anus of the adult; (8) coelomic (body) cavities lined with tissue of mesodermal origin; (9) a postanally located muscularized tail at some stage in the life history of the organism; (10) structural bilateral symmetry; (11) cephalization; and (12) an internal skeleton derived from mesoderm. Characters 1 through 4 are restricted to the chordates, whereas the remaining conditions are shared with a variety of invertebrates. As whole animals, chordates vary in size from microscopic larval forms to quadruped and fishlike mammals that reach lengths of more than 100 feet and weights of more than 150 tons. The diversity of size and shape corresponds relatively closely to the kinds of physical environments they inhabit—namely acquatic, both freshwater and marine, and aerial, both strictly land dwelling and flying.

Vertebrate Ancestors

The phylum Chordata, as usually delimited taxonomically, consists of three major units at the classificatory rank of subphylum: Tunicata, Cephalochordata, and Vertebrata. The Hemichordata are now rarely considered a subphylum within the Chordata, but are related to them. Vertebrate animals—cyclostomes, fish, amphibians, reptiles, birds, and mammals—are familiar to nearly everyone, but the other chordate subphyla, on the contrary, are not well known. Both have relatively few species and superficially appear to exhibit little diversity of general body form and function. Closer inspection of the Tunicata, however, reveals that they are a very heterogeneous group. The kind and degree of the differences between the Tunicata and Cephalochordata suggest that they are survivors of lineages that have had a long and divergent evolutionary history. Relative to the Chordata as a whole, and in particular to the major group therein, the Vertebrata, these two subphyla contain the only available evidence of invertebrate ancestry and the early stages of chordate phylogeny. The Hemichordata will be treated in some detail, even though they are not considered a true chordate, because they exhibit so many invertebrate-chordate transitional characteristics.

Hemichordata

The Hemichordata are solitary or colonial vermiform animals with bodies conspicuously divided into three parts; proboscis (protosome), collar (mesosome), and trunk (metasome).* Directly corresponding to this subdivision are the major internal body cavities, which also exhibit three divisions. In hemichordates these coelomic cavities develop by the process of enterocoely, i.e., they are evaginations from the embryonic gastrocoel (see page 31). The body is covered by cilia, and a cuticlelike covering appears to be absent. The nervous system of the group is epidermal in embryonic origin and remains largely within the integument in adults. Both large open spaces and relatively small discrete vessels, along with a contractile sac, comprise the circulatory system. The excretory system includes a number of nephridiumlike structures. A projection at the anterior end of the digestive tract, the supposed notochord, is usually referred to as the buccal diverticulum. The sexes are separate. Fertilization and development are external, and the bilaterally symmetrical larva is encircled by an irregularly placed band of cilia. Hemichordates are exclusively marine, occurring throughout a wide range of depths and in numerous different habitats. With few exceptions they are rarely encountered in large numbers.

The Hemichordata are divided into three taxonomic groups; the class Enteropneusta, class Pterobranchia, and class Planctosphaeroidea. The enteropneusts, more commonly referred to as acorn worms, are those wormlike hemichordates with numerous gill slits, a straight digestive tract, and a termi-

* Unknown for the Planctosphaeroidea.

nal anus. They lack tentaculated arms (Figure 2-2), are solitary, reproduce sexually, and the adults grow from moderate to considerable lengths (2 centimeters to 2.5 meters). They are found buried in the mud, under rocks, or in masses of vegetation. Locomotion is sluggish and much like that of earthworms. Food is obtained by filtering nutriment out of the water and mud as it is ingested and passed out through the gill slits or on through the digestive tract.

The pterobranchs are small to minute mosslike animals (Figure 2-3). They have a U-shaped digestive tract, may or may not possess gill slits, and exhibit one to nine pairs of ciliated tentaculated arms or lophophores. Pterobranchs are colonial, living within a common or individually secreted encasement, the coenecium. Some colonies are made up of a closely compacted mass of separate individuals and encasements; others include within a common encasement numerous individuals joined by a continuous organic tube, the stolon. Asexual budding appears to be the rule among pterobranchs. The colonies are usually found on the surfaces of dead parts of corals, rocks, shells, and the like; unlike many hemichordates they do not occur below the substrate. They feed by trapping small organisms in the mucus of their tentacles and passing the food-ladened mucus in deep ciliated grooves along the ventral surfaces of the arms to the mouth where ingestion takes place.

The Planctosphaeroidea is known only from a few transparent, spherical larvae. In some respects these resemble the larvae of the other two classes, but their numerous differences suggest a very different adult form; accordingly they are placed in a different taxon.

Together the Enteropneusta and Pterobranchia exhibit three of the unique characters of chordates: (1) pharyngeal gill slits and pouches, (2) a notochord, and (3) a hollow dorsal nerve cord. However, the homology of the latter two character states with those of chordates may be questioned. For example, in the hemichordates there are both large dorsal and ventral solid nerve cords that are connected to each other by a ring of nervous tissue at the junction of the collar and trunk (Figure 2-4). It is only within the collar that the dorsal cord continues anteriorly and may have a central cavity, or cavities, that approximates the usual chordate state of "hollow dorsal nerve cord." Moreover, this area of the nervous system does not appear to integrate nervous stimuli, as does the supposedly equivalent central nervous system of vertebrates. It seems difficult if not impossible to justify the often-described equivalence of the spinal nerves of typical chordates and that condition found in the collar of hemichordates. Secondly, it seems highly questionable whether the supportive structure in the collar of the hemichordates (buccal diverticulum or stomochord) is homologous with the notochord of vertebrates. They differ in gross and microscopic appearance, pattern of development, and apparently in embryologic origin. A similarity of function is also difficult to accept.

Even from this brief discussion it seems reasonable to assume that the relationships of hemichordates to chordates is very distant in time; the only character that relates them conclusively is the detailed similarity of the gill slits and pouches. If confirmed by additional study, the homology of the proboscis pore of hemichordates and the adenohypophysis of vertebrates may add further support to that thesis. Other evidence that hemichordates are only distantly related to chordates obtains from the following characters: (1) a postanal tail is absent in most larvae; (2) the direction of the flow of blood is annelidlike—namely, anteriorly in the dorsal vessel and posteriorly in the ventral vessel; and

Table 2-2 The major features of biotic and abiotic evolution through geologic time.

Eras (major vertebrate radiations and durations)	**Precambrian**	Great Revolution (considerable destruction of fossils)	**Paleozoic** ⟶ (Age of Fish; 340,000,000 years)	
Periods			*Cambrian*	*Ordovician*
Epochs				
Duration (years)			*70,000,000*	*50,000,000*
Beginning (years ago)	*3,000,000,000*		*570,000,000*	*500,000,000*
Animal life	First known fossils; primitive marine invertebrates, nonsegmental and segmental worms, and algae and fungi		Establishment of most modern phyla; abundance of marine invertebrates, particularly trilobites and brachiopods	Evolution of jawless fish, probably in fresh water
Plant life			Diversification of marine algae, especially blue-green algae and probably green algae	Marine algae abundant; early stages of terrestrial plant evolution
Climate and climatic conditions			Rapid rise in temperatures from conditions much like today	Warming trend increases and is uniform over most of earth.
Geologic and physical events			Lands low; formation of major geosynclines and adjacent archipelagos	Extensive submergence of land; continents low with shallow seaways; one of greatest marine inundations of all time

(3) segmentation of the major functional systems, such as muscular, nervous, circulatory, and excretory, is absent.

The relatively close relationship of hemichordates to echinoderms is easily demonstrable. It involves the considerable similarity of their tornarian and bipennarian larvae, the details of development of the coelom, the madreporic vesicle and heart vesicle, muscle chemistry and blood proteins, and probably

Paleozoic ⟶			
Silurian	*Devonian*	*Mississippian (Carboniferous)*	*Pennsylvanian (Carboniferous)*
50,000,000 *450,000,000*	*60,000,000* *400,000,000*	*30,000,000* *340,000,000*	*30,000,000* *310,000,000*
Evolution of jawed acanthodian fish; eurypterid arthropods at their peak of radiation; arthropods invade land	Origin of amphibians; lungfish, sharks abundant; beginning of radiation of teleosts; appearance of first winged insects	Radiation of primitive amphibians	Considerable specialization in amphibians; origin of reptiles; gigantic terrestrial arthropods abundant
Radiation of terrestrial plants, psilophytes, club and spike "mosses," horsetails, sphenopsids, and ferns	Considerable diversity in terrestrial plants; appearance of forests; evolution of primitive seed plants, gymnosperms and liverworts	Club and spike mosses horsetails, and sphenopsids dominant; gymnosperms increasingly widespread; extensive lowland forests	Extensive coal swamp forests; primitive tropical fern and seed fern forests; evolution of true mosses
Peak of warming trend reached	Cooler at beginning of period, then warmer; trend toward increasing aridity	Warm and humid at first of period, cooler later with local dry conditions	Warm and very humid
Continents relatively flat; inland seas still widespread; land slowly uplifted	Smaller inland seas; land higher; glacial mountain building; filling in some of geosynclines; intermountain fresh-water basins; extensive inland seas	Widespread inland seas; mountain building; beginning of great coal measures	Filling in more of great geosynclines; formation of coal in swamps; shallow inland seas; some glaciation in Southern Hemisphere

the axial gland and glomerulus in the respective groups. It is generally agreed that the hemichordates are more closely related to echinoderms than they are to chordates. And, it also seems likely that the hemichordates have retained more primitive states than have the echinoderms. It is for these reasons that the hemichordates seem best treated as a separate phylum, phyletically near both chordates and echinoderms, but closer to the latter.

Table 2-2 (continued).

Paleozoic		**Mesozoic** ⟶ (*Age of Reptiles; about 167,000,000 years*)		
Permian		*Triassic*	*Jurassic*	*Cretaceous*
Epochs				
50,000,000 *280,000,000*		*50,000,000* *230,000,000*	*50,000,000* *180,000,000*	*67,000,000* *130,000,000*
Reptiles radiate: displacement of amphibians as dominant group; establishment of modern insect orders; appearance of first mammal-like reptiles	Appalachian Revolution (some destruction of fossils)	Origin of mammals; dominance of mammal-like reptiles; appearance and radiation of dinosaurs, turtles, ichthyosaurs, and plesiosaurs; extinction of primitive amphibians	Origin of birds; peaks of reptile evolution on land, in sea and air	Extinction of large land and marine reptiles at end of period; appearance of a few modern orders and families of mammals; archaic mammals common; presence of modern groups of insects; extinction of primitive birds
Decline of club and spike mosses, horsetails, and sphenopsids; extensive development of glossopteris forests		Gymnosperms, cycads, ginkgos, and conifers dominant; evolution of angiosperms; extinction of seed ferns.	Increase in dicotyledon angiosperms; cycads and ginkgos common; appearance of modern genera of conifers	Dominance of angiosperm plants commences; flowering plants; monocotyledons appear; first oak and maple forests; gymnosperms decline
Extreme drop in temperature at beginning of period, followed by high temperatures; trend of increasing aridity		Rising temperatures; with arid and semiarid environments	Uniformly high temperatures with aridity	Warm at beginning of period, cool at its close
Widespread glaciation; elevation of continents; elimination of geosynclines		Continents exposed; widespread deserts	Continents with only moderate elevations; shallow seas restricted over some parts of Northern Hemisphere	Mountain building toward close of period; last extensive spread of oceanic waters over the continents; swamps common

Pterobranchs are said to be more primitive than the enteropneusts, and the argument for this supposition appears to be based largely on the degree of continuity of the collar portion of the nervous system with the epidermis. Given this phyletic relationship, it follows that enteropneusts have lost the tentaculated arms that are characteristic of pterobranchs and primitive echinoderms, and the postanal tail seen in some juvenile enteropneusts is probably

Rocky Mountain Revolution (little destruction of fossils)

Cenozoic ⟶
(Age of Mammals; about 63,000,000 years)

Tertiary ⟶

Paleocene	*Eocene*	*Oligocene*	*Miocene*
7,000,000 *63,000,000*	*22,000,000* *56,000,000*	*9,000,000* *34,000,000*	*13,000,000* *25,000,000*
Primitive mammals dominant; evolution of modern groups of birds and many new groups of marine invertebrates	Rise of many modern orders and suborders of mammals; continued presence of some primitive mammals	Origin of more modern families of mammals; extinction of primitive mammals	Radiation of anthropoid apes; appearance of modern subfamilies of mammals; rise and rapid evolution of grazing mammals
Trending toward subtropical plants	Subtropical forests dominant	Evolution of plants toward temperate kinds; maximum spread of forests; rise of monocotyledons, flowering plants	Evolution of grasslands; temperate kinds of plants dominant; retreat of polar flora
Cool at beginning of period, becoming warmer; relatively dry	Warmer—heavy rainfall, little or no frost	Warm and humid to cooler and drier	Cooler; moderate seasonal climates
Culmination of late Cretaceous mountain building; trend toward transgression of marine waters into epicontinental embayments	Mountains eroded; no continental seas	Holarctic land connection; some mountain building; restriction of inland embayments; lands lower	Development of plains; invasion of marine water into epicontinental embayments; continental erosion

homologous with the stalk of pterobranches and primitive echinoderms. The general similarity of the arms of pterobranchs and the tentaculated projections of lophophorate invertebrates—namely, phoronid worms, the mosslike ectoprocts, and the lampshells or brachiopods—may indicate a close phyletic relationship.

Table 2-2 (continued).

Eras (major vertebrate radiations and durations)	**Cenozoic**		
Periods		**Quaternary** ⟶	
Epochs	*Pliocene*	*Pleistocene*	*Recent*
Duration (years)	*10,500,000*	*1,500,000*	*10,000*
Beginning (years ago)	*12,000,000*	*1,500,000*	*10,000*
Animal life	Evolution of man; appearance of many modern genera of mammals	Evolution of most modern species; decline of large mammals; evolution of human social life	Rise of man through stone, bronze, and iron ages
Plant life	Decline of forests, spread of grasslands	Extensive extinction of species	Decline of woody plants; rise of herbaceous ones
Climate and climatic conditions	Semiarid—dryer and cooler	Fluctuating, cold and mild	Warmer
Geologic and physical events	Continued rise of continental elevations; retreat of inland marine embayments; volcanic activity	Repeated widespread glaciation; four major glacials; high relief of continents	End of last Ice Age; extensive continents

Tunicata The Tunicata are soft-bodied solitary or colonial, relatively inactive animals that as a group of filter feeders exhibit a wide variety of sizes, shapes, colonial structure, and life histories (Figure 2-5). They range from solitary sessile or free-floating individuals that are microscopic or large to forms connected by stolons or encased within a common outer covering or tunic. Some species are larviform, and they can be either pelagic or nectonic when reproductively

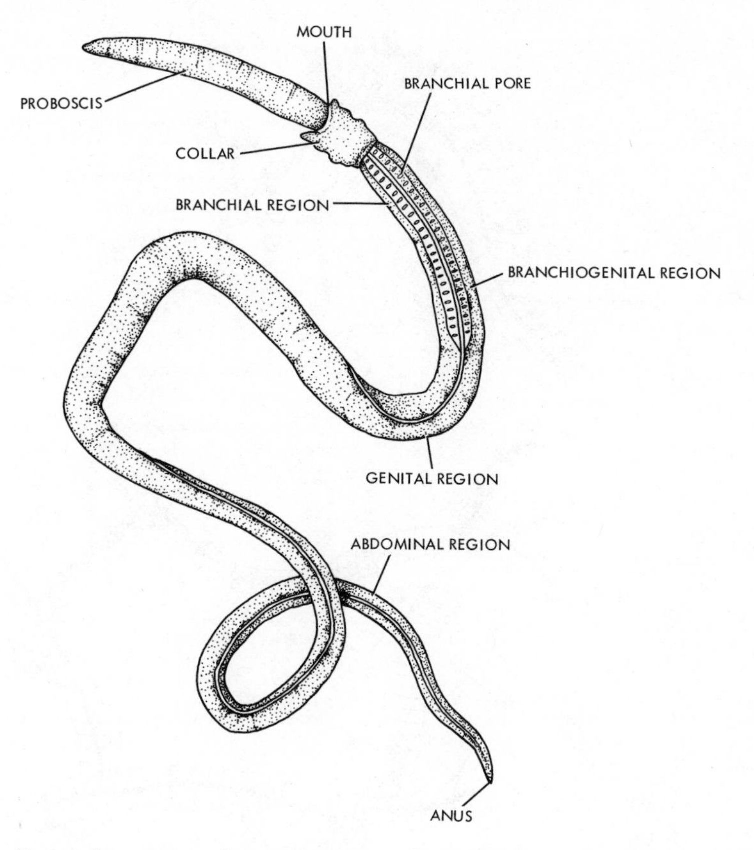

Figure 2-2 External anatomy of an enteropneust hemichordate (genus *Balanoglossus*). (Modified after Crouch.)

mature. As sessile adults other species are saclike in appearance. Despite a general diversity of form, they have a large number of character states in common that suggest that they are a monophyletic group. As either free-swimming larvae or sessile adults, tunicates almost uniformly exhibit the following states: (1) absence of coelomic cavities, and unlike other chordates the tail muscles and body mesenchyme develop directly from bands of mesoderm without passing through the usual embryonic stage of coelom formation; (2) bilateral symmetry; (3) a thin, secreted mantle forming an outer acellular fibrous covering or tunic containing a celluloselike substance called tunicin; (4) incurrent (branchial pore or mouth) and excurrent (atrial pore) siphons that together with the cilia control the circulation of water into the much enlarged pharynx and through its ciliated apertures or stigmata into the atrial cavity; (5) large atrium surrounding the pharynx that channels water to the exterior via the excurrent siphon; (6) thyroid gland homologue, the endostyle, located in the wall of the pharynx; (7) in the adult a large solid neural ganglion connected to nerve fibers that appear to control the muscular activity associated with the siphons as well as the lateral body musculature; and (8) a conspicuous neural gland closely associated with the neural ganglion, the function of which is now in dispute. Formerly, the gland was widely thought to be a primitive state of the pituitary gland (hypophysis) of vertebrates and of Hatschek's pit, which is its homologue in cephalochordates. Tunicates are marine exclusively, ranging

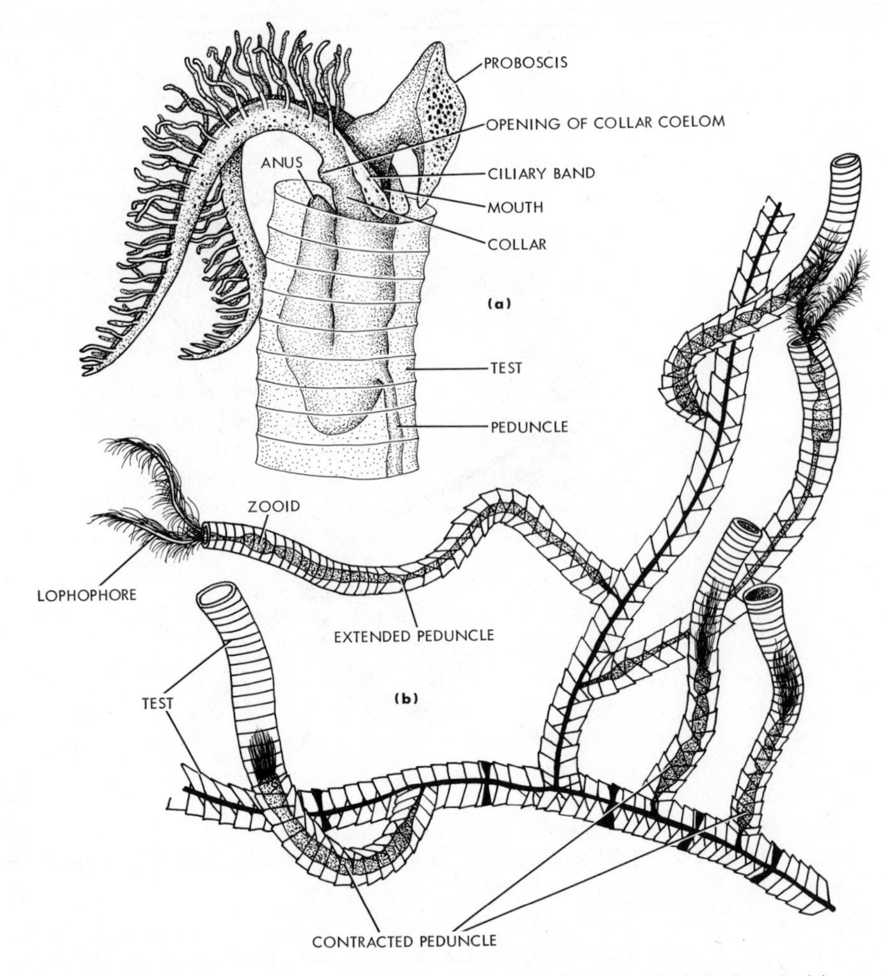

Figure 2-3 External anatomy of a pterobranch hemichordate (genus *Rhabdopleura*): (a) an individual extending from the aperture of the test; (b) part of a colony. (Modified after Dawydoff.)

in distribution from the Arctic to Antarctic seas and from tidepools to great depths in most oceans.

The commonly encountered sessile adult stage of life history of tunicates exhibits little or no resemblance to vertebrates, e.g., *Molgula manhattensis* found along the Atlantic Coast. On the other hand, the tadpolelike stage shares numerous similarities with vertebrates. This stage (Figure 2-6) exhibits (1) a muscularized tail, but with unsegmented muscles (this view may be in error); (2) a hollow dorsal nerve cord in the tail, which is widened into a brainlike vesicle in the body portion of the organism; and (3) a notochord located completely within the tail and composed of large vacuolated cells encased within a thin fibrous sheath. In addition to these vertebrate characteristics, it possesses a relatively large pharynx perforated by ciliated stigmata and also rather well-developed equilibrium and photosensory structures. When the tail is free from the tunic encasement it may be used as a propulsive organ and thereby aid in larval dispersal. If there is an adult sessile stage in the life history, the larva swims for a few hours before settling and attaching to a suitable object by means of small cephalic adhesive projections. Following attachment, the larva rapidly undergoes a very dramatic metamorphosis, during which the tail, which houses the vertebrate characteristics of nerve cord and notochord, decreases and is absorbed into the body. In addition, the brain vesicle, located between the larval incurrent and excurrent siphons, becomes reduced to the neural ganglion of the adult, and the associated sensory structures of a larva appear to be lost entirely. In those species with a sessile adult stage the larva does not feed. It is in the larva-sessile adult type of life history that the pharynx becomes greatly enlarged at metamorphosis. In the sessile stage, feeding involves this complex pharyngeal structure associated with a ciliary-mucoid feeding action. The sessile adults may be hermaphroditic or they may produce offspring by simple budding. Only those individuals that result from the union of sperm and egg go through a larval stage.

Within the Tunicata three distinct lineages are usually recognized: the classes Ascidiacea, Thaliacea, and Larvacea (Figure 2-5). The ascidiaceans, or sea squirts as they are commonly referred to, exhibit the greatest degree of diversity; about two thousand living species are now recognized. The ascidiacean life history includes a motile larva and a sessile adult. About one hundred living species are referred to the Thaliacea. In this group the adults are usually barrel-shaped and nectonic. Larvae may or may not be present (Figure 2-6). The remaining class, called the Larvacea or Appendicularia, consists of about seventy-five living species. Larvaceans are readily distinguished from the other tunicates by the presence of a free-swimming adult that differs little from its larval stage.

The body of a solitary sea squirt is ovoid and covered by an opaque leathery-textured tunic. In some species the animal is fastened to the substrate directly, in others it is fixed at the end of a long stalk. The shape of the individuals of a colonial form (Figure 2-7), whether joined by stolons or massed within a common tunic, is largely influenced by the kind and degree of contact

Figure 2-4 The anterior end of an enteropneust hemichordate (sagittal section). The proboscis is highly contracted in this view. (Modified after Dawydoff.)

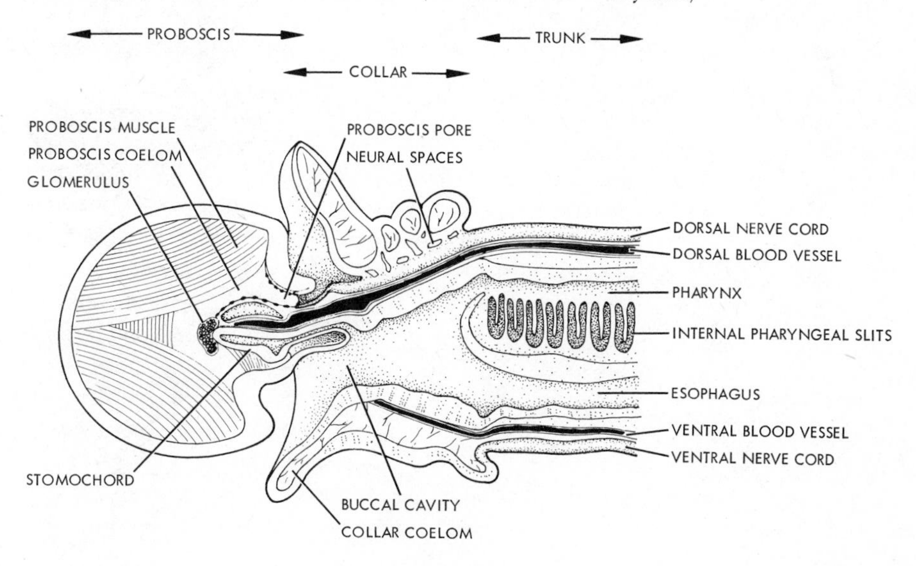

between the individuals. In any case, the siphons are directed upward, opposite the point of attachment, and the pharynx is extremely large and perforated by numerous ciliated stigmata. As would be expected, food is filtered out of the incurrent flow of water at the level of the stigmata. Coelomic cavities are absent. A few large vessels or haemocoels conduct the blood, containing variously structured and pigmented cells, from the heart, which alternately

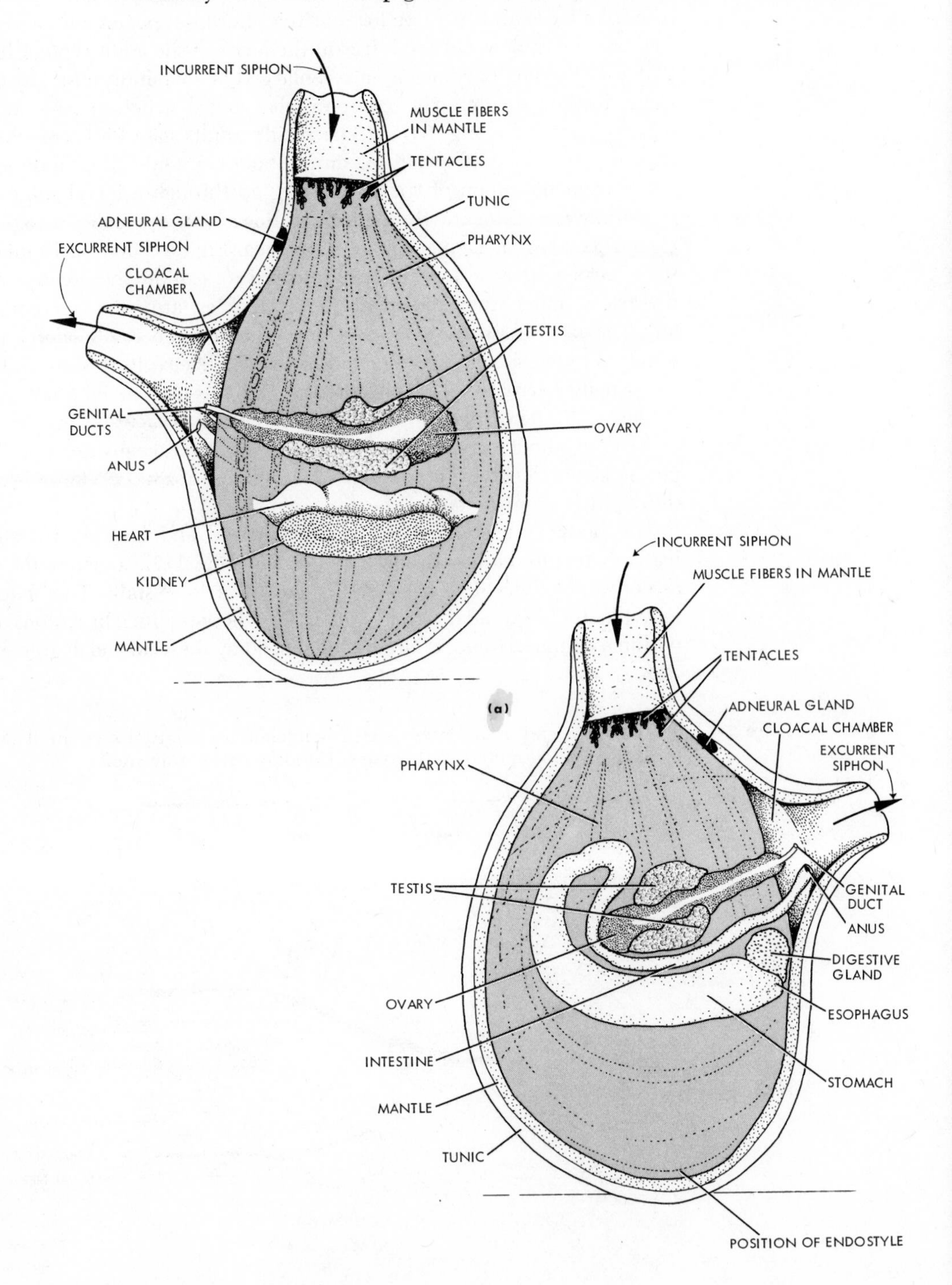

pumps it into and out of the same vessel. The blood of ascidiaceans has many interesting features and differs markedly from that of vertebrates, most notably in salt content and cell types. Reproduction in most ascidiaceans occurs either sexually or by budding. Without exception they are hermaphroditic, but apparently because of the different temporal activity patterns of the male and female gonads in the same individual only cross-fertilization occurs.

Figure 2-5 Representatives of the three major classes of tunicates. (a) A solitary ascidiacean-left and right lateral views. (b)–(c) A colonial ascidiacean—(b) is an individual removed from the colony shown in (c). (d) A pelagic thaliacean. (e) A pelagic larvacean.

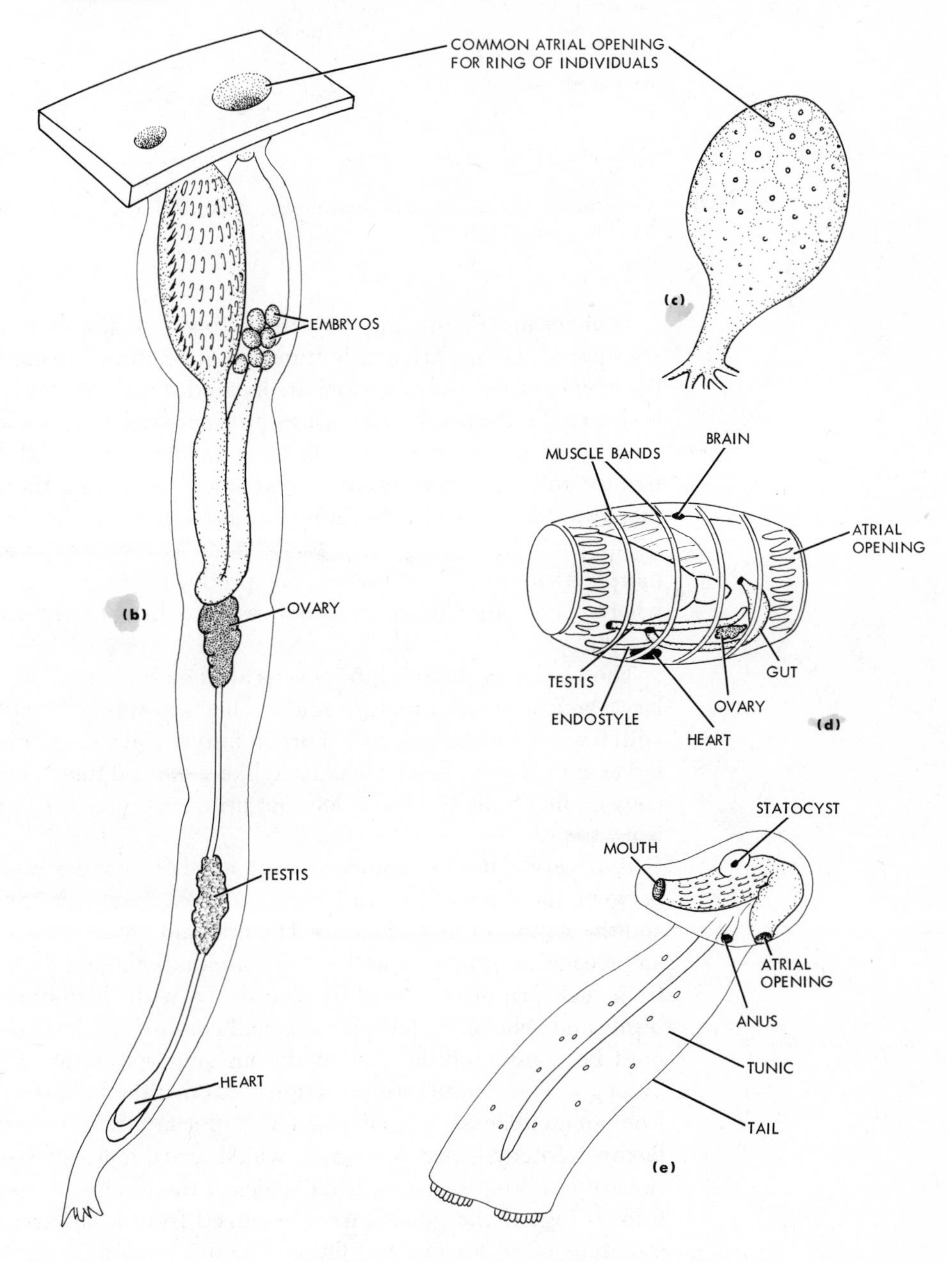

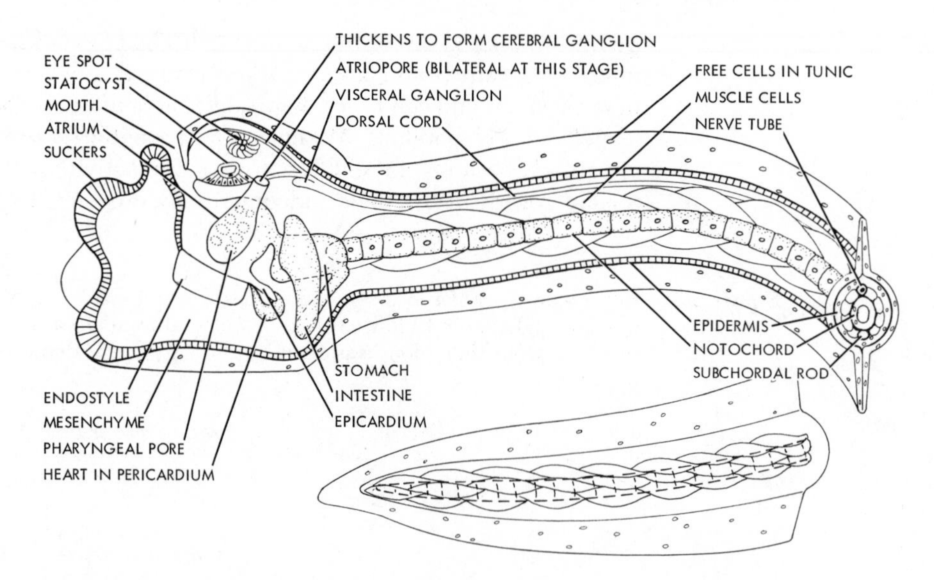

Figure 2-6 A summary of the internal anatomy of a larval ascidiacean (genus *Clavelina*). (Modified after Jollie.)

Thaliaceans (Figure 2-5) with the following major exceptions are similar to sea squirts: (1) they are nearly transparent; (2) they are barrel-shaped, and the incurrent and excurrent siphons are located at opposite ends of the longitudinal body axis; (3) the body wall encases six to ten bands of muscle that contract and thereby jet-propel the organism through the water; and (4) reproduction is usually asexual. Some sexual reproduction exists, and these animals are hermaphroditic. Probably the most distinctive feature of the thaliaceans is their complicated life cycle. For example, some species are dimorphic—that is, one morph is solitary, has no gonads, and produces a chain of individuals asexually, whereas the other morph reproduces sexually and gives rise to the asexual morph.

The Larvacea (Figure 2-5) possess all the usual chordate characters both as larvae and as sexually mature adults. They are minute and free-swimming, and with few exceptions possess a short endostyle and a single pair of stigmata. The larval animal may shed its thaliaceanlike tunic and then secrete another. When encased in a tunic the larva does not utilize its elongate tail for propulsion; it is nectonic.

Two very different points of view are held on the phyletic relationships between the three classes of tunicates. These views are based on morphology and the nature of reproduction. The first and oldest interpretation is that the larvaceans are primitive and that the ascidiaceans and thaliaceans are derived from such primitive forms. In accordance with this point of view, both the sessile and pelagic nonlarvaform sexually mature adult stages of the two classes must be considered derived conditions acquired within the radiation of the Tunicata. The second interpretation holds that the Larvacea are derived either from an ascidiacean or thaliacean ancestor and that neoteny was a significant factor. Neoteny is that process by which sexual maturity becomes fixed evolutionarily in progressively earlier stages of the life history until the usual adult form is lost. If the gonads were removed from a larvacean adult, all of the essentials of an ascidiacean tadpole would remain.

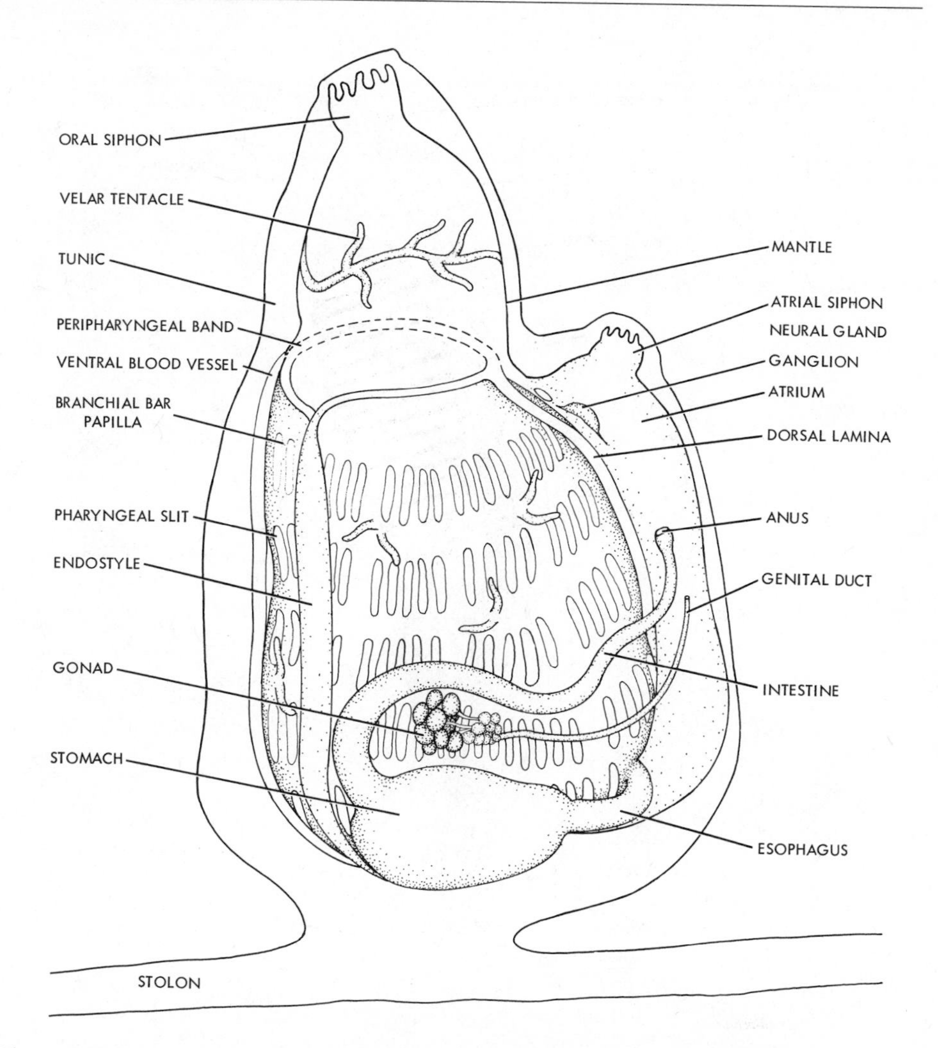

Figure 2-7 A summary of the internal anatomy of an adult ascidiacean (genus *Perophora*) shown in lateral view. This form is colonial. (Modified after Crouch.)

Cephalochordata

All adult cephalochordates are small, five to one hundred millimeters, compressed and elongated fishlike marine organisms (Figure 2-8). They possess all the obvious attributes diagnostic of the phylum Chordata (see p. 30), and similarities to the vertebrates are most obvious in this group. However, cephalochordates are conspicuously exceptional in the developmental asymmetry found in many systems, some of which remain in the adult. Although the group consists of a very homogeneous assemblage of approximately twenty-five living species distributed among three closely related families, it almost certainly had diverged by the Precambrian period from the line leading to vertebrates. Probably the best-known species biologically is *Branchiostoma lanceolatum*, often referred to by the common name amphioxus.

Most cephalochordate species are found in shallow marine waters in many regions of the world, and in some they are very abundant. Poor fin development makes most species relatively ineffective swimmers; they spend most of the time buried in the substrate with only their anterior end exposed. No true paired fins are present; cartilagelike material supports the dorsal fin, and the

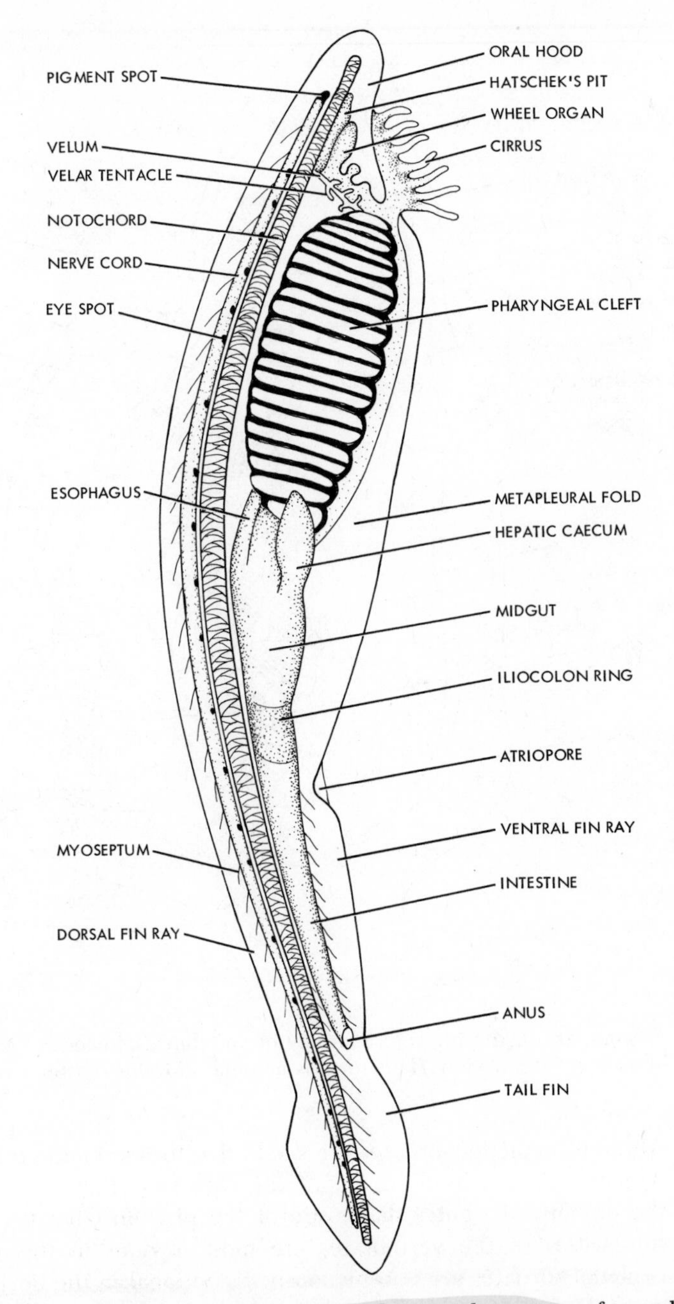

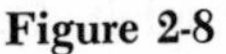
Figure 2-8 A summary of the internal and external anatomy of a cephalochordate (genus *Branchiostoma*) shown in sagittal section. (Modified after Crouch.)

end of the animal terminates in a single caudal fin resembling superficially that of many fish. Cephalochordates are covered with an acellular cuticle secreted by the underlying epidermis. The epidermis is only one cell layer in thickness. Beneath this lies gelatinous mesodermal tissue. The endoskeleton of the adult consists of a notochord extending the entire length of the animal, U-shaped cartilagelike rods in the tissue surrounding the gill slits, jointed chitinlike rods inside the oral projections or cirri, and fin rays, each within a fluid-filled cavity, in the single dorsal and paired ventrolateral folds of skin (metapleural folds) that occur along most of the length of the body (Figure 2-9). The coelom is much reduced, with spaces persisting along the dorsal wall of the pharynx and in the metapleural folds. These cavities develop from mesoderm, and they are typically enterocoelic in their mode of development. Dorsal to the notochord is a hollow nerve cord, enlarged anteriorly into a brainlike vesicle, and with paired dorsal and ventral nerves arranged segmentally along its length. Dorsal

Figure 2-9 A summary of the internal and external anatomy of a cephalochordate (genus *Branchiostoma*) shown in cross-section through the pharyngeal region. (Modified after Crouch.)

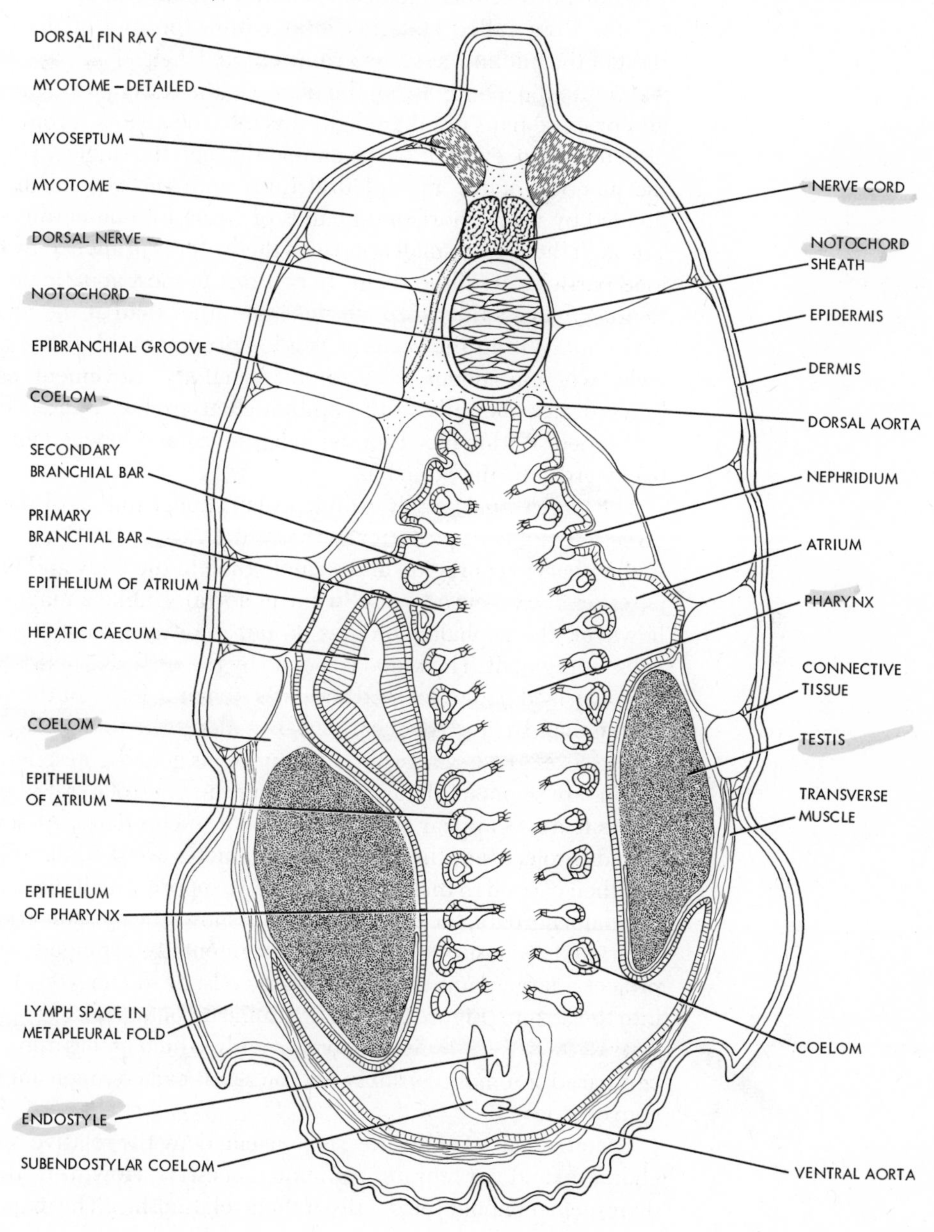

and ventral root ganglia are absent. Sense organs corresponding to the olfactory and lateral line systems of vertebrates appear to be lacking, as is also a trace of any type of eyelike structure. Photosensory cells are present in the lateral walls of the hollow nerve cord.

The body musculature consists of a segmentally arranged series of <-shaped blocks or myomeres with the apex of each directed anteriorly. An alternate contraction of the myomeres produces the undulatory swimming motion. The digestive tract is a simple type. Within the oral funnel is a mouth opening almost completely surrounded by stiffened cirri that, in turn, are covered with sensory papillae. Posterior to the straight, elongated pharynx, the gut is a simple tube with little division into chambers. The diverticulum from the anterior portion was once considered a homologue of the vertebrate liver. A ciliated fossa called Hatschek's pit within the roof of the oral funnel to the right of the midline, as well as some ciliated "wheel" organs, direct a current of water into the mouth. From the pharynx, the water passes outward through the fifty or more pairs of ciliated gill slits into the atrium. From there, the water is directed to the exterior of the animal through the single atriopore located near the junction of the metapleural folds with the ventral fin. Each gill slit is divided by nearly horizontal bands of tissue into numerous smaller openings. Through these very small apertures the water is propelled by ciliary action and food particles are filtered out. In contrast to most aquatic vertebrates, respiration occurs generally through the skin rather than at the level of the pharyngeal bars. Within the pharynx is a longitudinal midventral groove, the endostyle, wherein mucus is secreted. As ciliary movement carries the mucus upward past the gills to the epibranchial groove, filtered food particles are entrapped. Within the epibranchial groove, food-ladened mucus is transported backward into the stomach.

The large blood vessels exhibit a plan similar to that of the vertebrates. The direction of flow is posterior in the dorsal vessel (dorsal aorta), which is paired in the pharyngeal region and a single tube in the body and tail, and the flow is anterior in the ventral vessel (ventral aorta). Unlike a majority of vertebrates, however, the cephalochordates do not appear to have any capillaries, blood cells, or pigment. The blood is moved by the pumping action of a large series of muscularized bodies located along the major arteries of the pharyngeal region rather than by a single heart located along the ventral aorta.

The sexes are separate and the numerous gonads, mesodermal in embryonic origin, are segmentally arranged along the ventrolateral margins of the body. These features appear to be exceptional in chordates. At sexual maturity the gonads project into the atrium. The gametes are shed directly into the atrium and then carried to the outside via the atriopore. Of all the organ systems in the Cephalochordata, the excretory system shows the most divergence. Ectodermal in origin, it consists of a series of segmentally arranged vesicles into which project a large number of flagellated cells, or solenocytes. Each vesicle drains into the atrium (Figure 2-10). The similarity of morphology and function of this system to that of annelid worms, with which it has most frequently been compared, is almost certainly not the result of a common ancestry but rather of convergence.

Given so many similarities, no one can deny the relative recency of cephalochordate and vertebrate common ancestry. However, there is still some disagreement about the details of their relationship. The majority opinion holds that the vertebrate line was derived from an ancestor similar to cephalo-

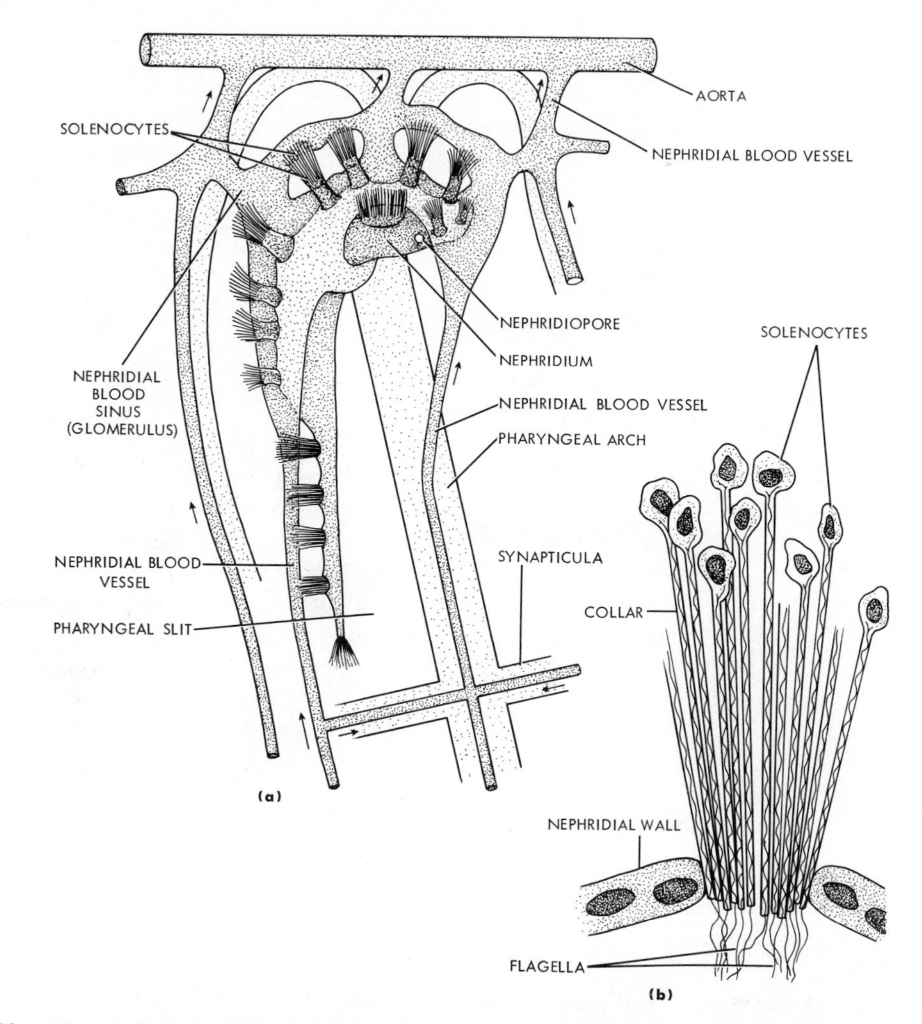

Figure 2-10 The nephridium and associated structures in a cephalochordate (genus *Branchiostoma*): (a) general view of the nephridium; (b) a detailed view of solenocytes projecting through the nephridial wall. (Modified after Dawydoff.)

chordates, and a minority believe that cephalochordates are "degenerate" vertebrates. The latter point of view is predicated on the morphological and behavioral similarity of adult cephalochordates and the ammocoete larva of lampreys. This minority reasons that the larval stage of a primitive vertebrate became sexually mature (neoteny) and the typical adult stage disappeared to give rise to the Cephalochordata. The arguments against this interpretation almost always rest on the presence of the atrium and on numerous differences in the excretory, circulatory, and reproductive systems. Here it is argued that if cephalochordates were degenerate vertebrates one should find primitive states of some of those characters among vertebrates.

Ancestry and Evolution of Chordates

The three most prominent theories concerning the evolutionary origin of chordate animals are based on annelids, spider and scorpionlike arthropods (Arachnida), or echinoderms as ancestors. Annelids have been considered because they are similar to vertebrates in bilateral symmetry, segmentation,

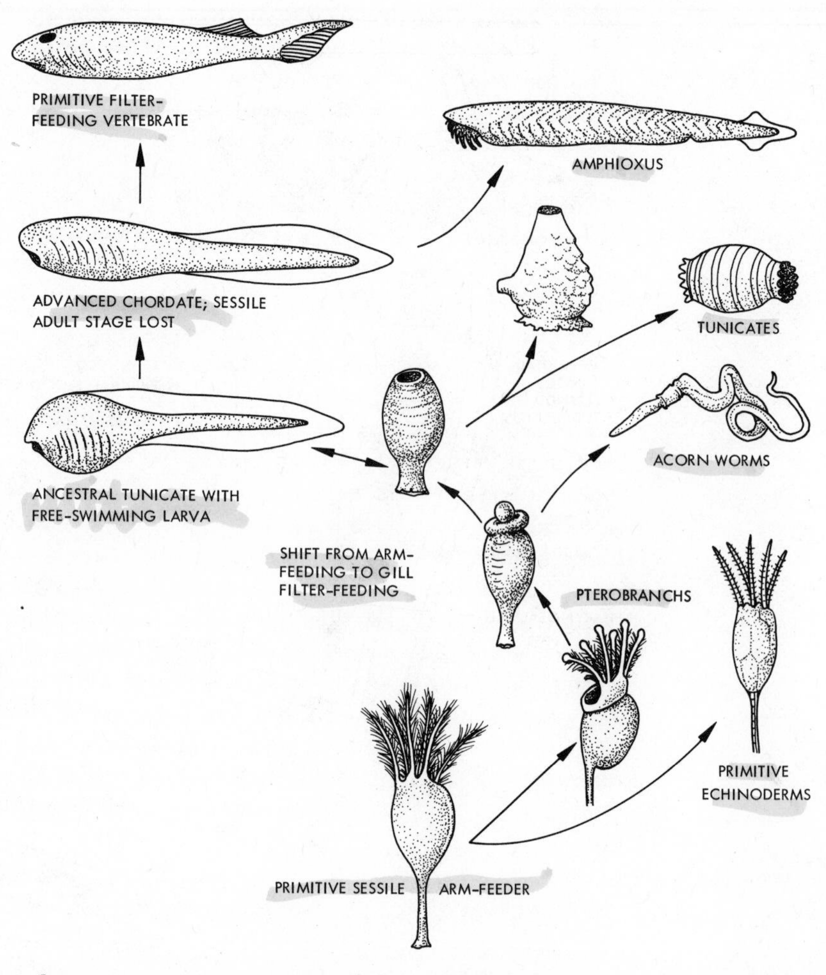

Figure 2-11 A diagrammatic representation of the probable major stages of evolution from echinoderms to vertebrates. (Modified after Romer.)

and the presence of an enlarged "brainlike" structure at the anterior end of the dorsal longitudinal nerve cord. However, the following points have been used most effectively to remove annelids from further consideration as a vertebrate ancestor: (1) their segmentation is complete; (2) the large longitudinal nerve cord is ventral in position; and (3) the mesoderm formation is schizocoelous. These same arguments, both for and against, also are used to support the theory that chordates evolved from arachnids.

The hypothesis of a close phylogenetic relationship between chordates and echinoderms is widely accepted today. This relationship is inferred from the similarity of tunicates to hemichordates and, in turn, of hemichordates to echinoderms. The similarity of the larvae of hemichordates and echinoderms and the detailed similarity of the development of certain of their systems seem to document this interpretation adequately. Fundamental to this argument is the recognition that primitive echnoderms are sessile forms, some of which are firmly anchored to the substrate. It is precisely this kind of echinoderm that possesses the series of ciliated tentacles (lophophore) that surround the mouth and that also characterize the pterobranch hemichordates. This thesis necessarily requires that we accept chordate ancestors as being sedentary, ciliary-tentacle feeding forms, from which evolved sedentary species with a gill-filter feeding system; these first chordates, in turn, gave rise to mobile organisms that actively sought their food. Neoteny has been hypothesized (Figure 2-11) as the process through which the sedentary adult prevertebrate stage was lost and the reproductively mature larvaform vertebrate ancestor evolved. The hypothesis gains some plausibility from the well-documented occurrence of neoteny in the amphioxides type of cephalochordate and in the tunicates as well.

From the Cambrian to Devonian period a little-known group of species existed, the class Stylophora, that are very similar to echinoderms, hemichordates, and chordates. At least one recent publication [1] suggests that the stylophores be included in the Chordata as the subphylum Calcichordata, rather than in its customary position in the Echinodermata. The stylophores are characterized by an external skeleton of calcite plates, a body (theca), and a strongly flexed tail (stem) (Figure 2-12). A large mouth is located at the anterior end of the theca, and branchial slits and an anus are also present; all exhibit some degree of asymmetry. The theca includes a buccal cavity, a pharynx, and coelomic cavities. Superficially, the stem of stylophores is similar to the stock of crinoid echinoderms and in detail it is believed to include primitive states of the chordate notochord, dorsal nerve cord, and segmented blocks of muscle and nerves with paired segmental ganglia. It seems that stylophores were bottom dwellers, although a "swimming" mode of locomotion has been postulated. They are believed to have filtered food out of suspension or directly from the substrate. These few generalizations suggest numerous intermediate conditions between typical echinoderms and chordates, emphasizing the need for further study of this group.

Figure 2-12 A summary of the external anatomy of a calcichordate (genus *Mitrocystella*): (a) dorsal view; (b) lateral view; (c) ventral view. (Modified after Jeffries.)

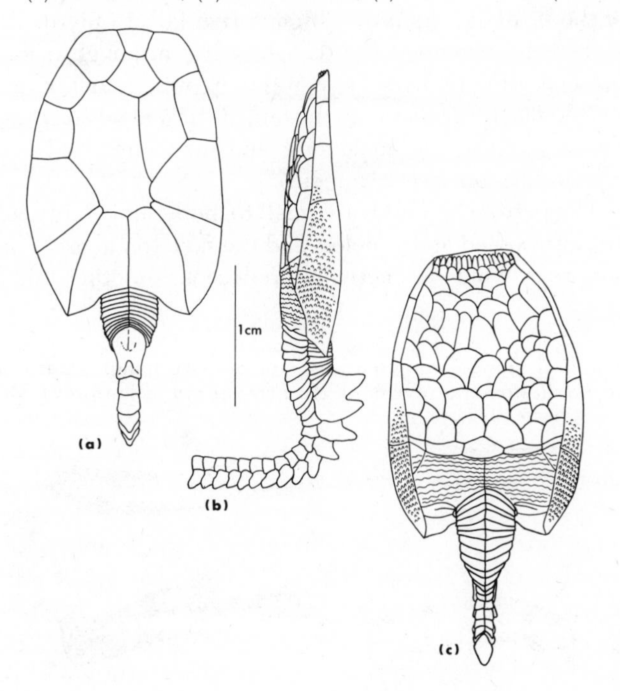

Stem Vertebrates (Agnatha)

The lampreys, suborder Petromyzontoidei, and the hagfish, suborder Myxinoidea, are the only living species of the class Agnatha (Figure 2-13). In addition to these two, there are at least four extinct orders of jawless fish, collectively called ostracoderms, that clearly document an extensive and diverse adaptive radiation of early vertebrates. Both the living and fossil agnathans, as the name implies, lack jaws. Moreover, all living species and most fossil forms lack paired appendages.

The ostracoderms are geologically the oldest (Ordovician-Devonian) and structurally the most primitive vertebrates known. Superficially, ostracoderms are unlike lampreys and hagfish (Figure 2-14), but the similarity of some of the more detailed features of their morphology indicates that they are almost certainly ancestral to them and, therefore, nearer the stem of vertebrate phylogeny. Ostracoderms were relatively small fish, less than one foot long usually, and were restricted largely, if not exclusively, to fresh-water habitats such as streams and ponds. Most of the species had a single nostril located high up on the head, and most lacked paired appendages; paired spines or flaplike appendages extending from the enlarged head region were present in others. Apparently, ostracoderms were not predators. Their large gill chamber suggests a filter-feeding mode of life. Some with flattened heads and body regions and compressed tails were almost certainly bottom dwellers. Others were compressed throughout their length and were probably active swimmers. All ostracoderms were covered with a dermal bony armor that varied from large plates to small scales, and in some the endoskeleton was bony as well. It has been hypothesized by some researchers that large predaceous aquatic arthropods were the selective force responsible for the evolution of heavy armor so early in the history of vertebrates.

The ostracoderms are usually divided into two groups, the Monorhina and Diplorhina. The latter subclass is distinguished from the former by the following characteristics: (1) the paired eyes are farther apart and located on the side of the head; (2) the pineal organ often fails to pierce the top of the head; (3) there is no opening for a dorsal nostril, although evidence now suggests the presence of paired nasal sacs at the tip of the snout or on its ventral surface; (4) the median fins, other than the tail, are absent, although spines protruding from the armor may be analogous; and (5) an internal bony skeleton is absent (apparently only cartilaginous).

The petromyzontoids are small to moderately large eel-like organisms. They are soft-bodied and scaleless and the skeleton consists only of cartilage. Adult lampreys, which are actively predaceous on other fish, are primarily marine.

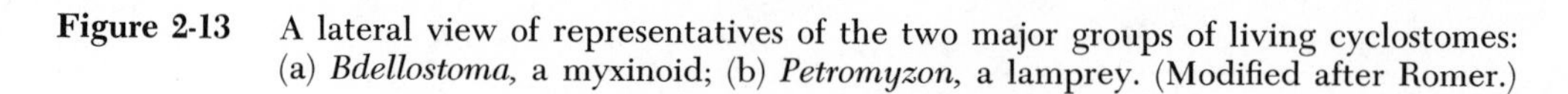

Figure 2-13 A lateral view of representatives of the two major groups of living cyclostomes: (a) *Bdellostoma,* a myxinoid; (b) *Petromyzon,* a lamprey. (Modified after Romer.)

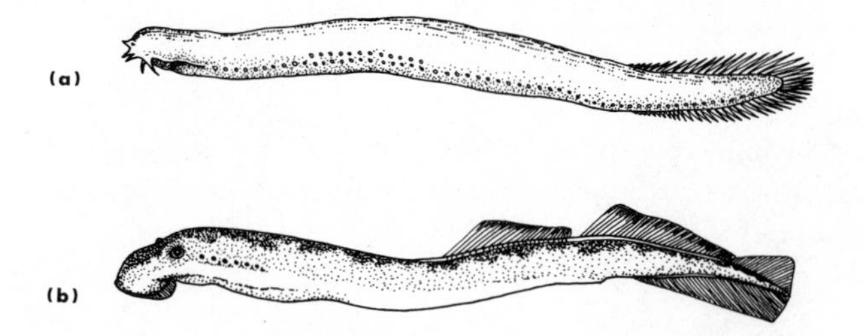

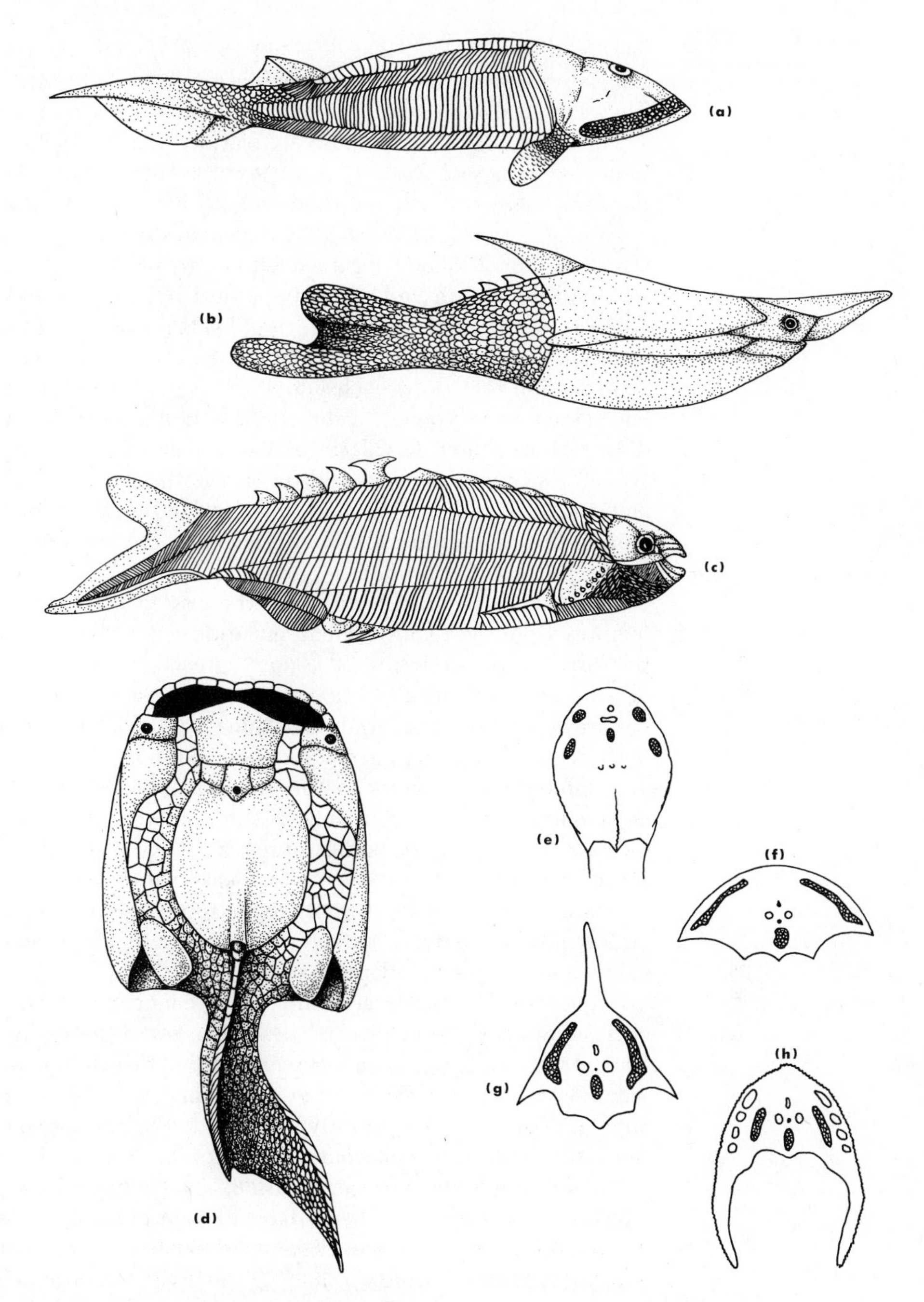

Figure 2-14 Representative ostracoderms: (a)–(c) lateral views of three distinct forms; (d) a dorsal view of another; (e)–(h) outline drawings of the cephalothoracic region of other species. (Modified after Romer.)

Some lampreys spawn in fresh water, others are landlocked. The funnel-shaped mouth is lined with circlets and rows of cornified teeth; the tongue also is covered with horny denticles. The buccal funnel forms a cup by which the lamprey attaches to its prey, and the rasping tongue is used to penetrate the host's skin. Lampreys possess a single nostril, more appropriately called a nasohypophyseal duct, that does not extend completely through to the pharyngeal region. Behind the tiny eyes is located a short series of gill openings. They are small and nearly circular and each possesses a leaflike valve.

The slimy hagfish are exclusively marine. They differ from lampreys in the following principal ways: (1) they are scavengers; with their horny tongues they burrow into the flesh of dead or dying fish; (2) the single nasohypophyseal opening is at the tip of the snout and connects with the pharynx; (3) the mouth is neither funnel-shaped nor lined with horny teeth; (4) three pairs of tentacles surround the mouth and nasohypophyseal opening; (5) the eyes may not be distinct; (6) there are five to fifteen gill slits per side that may be united by a common duct to the exterior; and (7) there are numerous slime glands in a ventrolateral row along each side of the body. Furthermore, hagfish eggs are laid at sea and the young develop directly, unlike some lampreys, which have a distinct fresh-water larval stage and a marked metamorphosis. Lampreys typically spawn in streams and the larvae that develop, often referred to as ammocoetes, remain in the mud and gravel of the stream bed for several years. During this period they are filter feeders; in this and other respects they are similar to cephalochordates.

The absence of fossil hagfish and the existence of only a single species of lamprey from the Pennsylvanian makes it very difficult to resolve their relationships to ostracoderms. It is now agreed generally that cyclostomes are derived from ostracoderm ancestors, and most researchers also agree that lampreys were derived from monorhine ostracoderms. The origin of hagfish, however, is still being argued. One school of thought holds that the cyclostomes are a monophyletic group; another thinks that hagfish evolved from diplorhine ostracoderms. If the latter view is substantiated, then the use of the order Cyclostomata to include both lampreys and hagfish would be in violation of the rule of monophyly. In any event, it appears that the absence of a bony skeleton in lampreys and hagfish, the only living representatives of the jawless class of vertebrates, is a derived character state, as are the predaceous and scavenging habits and all related adaptations.

A question of considerable interest in connection with phylogeny of the cyclostomes and ostracoderms is whether vertebrates originated in a fresh-water or marine environment. It seems certain that the ancestors of vertebrates were marine because the main groups of invertebrates are marine, as are the Hemichordata and the primitive Chordata, the enteropneusts, pterobranchs, tunicates, and cephalochordates. Of interest, then, is the question whether primitive vertebrates too were marine. The study of the kinds of deposits in which the earliest known fossils have been found coupled with the structure and physiology of the excretory system of the living groups has led some to the conclusion that the origin was in fresh water and others to believe that it was in a marine environment. The issue may be resolved in favor of brackish water as the original habitat of the earliest vertebrates.

An Early Adaptive Radiation (Placodermi)

Placoderms are a highly varied group of fish that appeared at the beginning of the Devonian period, during the peak of ostracoderm radiation, and became extinct in the Mississippian. The early portion of placoderm evolution appears to have taken place in fresh water, with derived radiation in the marine environment. Adult placoderms varied in size from a few inches to more than

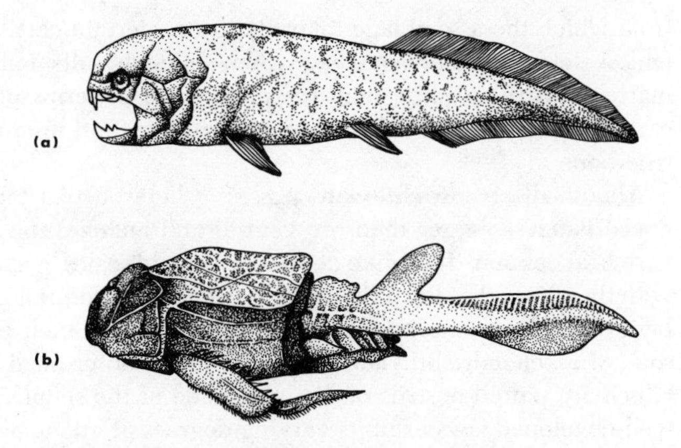

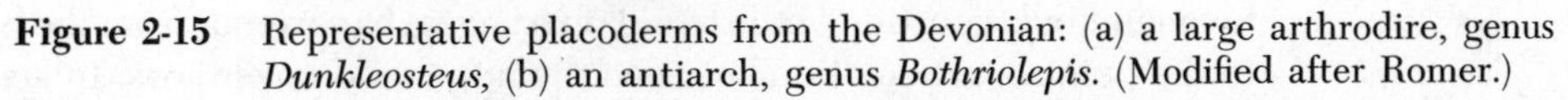

Figure 2-15 Representative placoderms from the Devonian: (a) a large arthrodire, genus *Dunkleosteus,* (b) an antiarch, genus *Bothriolepis.* (Modified after Romer.)

thirty feet in total length. Typically placoderms are flattened, and they were probably bottom-dwelling species that may have eaten molluscs and bottom vegetation (Figure 2-15). A few streamlined species are known, and they are believed to be pelagic predators. All placoderms appear to have a distinct point of articulation between the cranial and trunk sets of bones. These armorlike coverings, as well as the more caudally located individual scales, consist of three histologically different layers: (1) a basal layer of laminar bone; (2) a middle layer of trabecular bone; and (3) a superficial layer of laminar bone, which, in more primitive species, forms a dentinal tissue with enclosed cell spaces. The texture of the surface of the armor is complicated by tubercles, ridges, and grooves. Unlike the ostracoderms, placoderms possessed upper and lower jaws. It seems that at least part of the upper jaw was fused to the skull and that no species possessed true teeth on the jaws. All placoderms are believed to have possessed paired appendages. Some of the fins consisted of a large spine followed by a fleshy web of skin; other types included jointed solid spines without a connecting web, hollow spines fixed to the thoracic armor, or fleshy fins without a leading spine. Also worth emphasizing is the fact that diverticula from the pharynx have been observed in some placoderms and they are interpreted as the forerunners of lungs.

The common ancestry of placoderms with jawless and other jawed fishes, such as the acanthodians, remains a hotly debated subject. The absence of fresh-water Silurian deposits may forever keep us from observing some of the transitional stages. Similarly, the specific relationship between placoderms and chondrichthyian fish is difficult to interpret. Not only is the placoderm ancestor for the latter group still being debated, but mounting evidence indicates that it was not a single stock.

A Major Evolutionary Side Branch (Chondrichthyes)

In the middle Devonian fossil record appeared what has become a highly differentiated lineage of fish, the Chondrichthyes. Many representatives of the group still exist today; in the past, however, they were far more numerous, with their greatest radiation centered in the Carboniferous and Permian. Bone is unknown in all living and most fossil species. The presence of only a cartilaginous endoskeleton is the most striking characteristic of the group, and the one from which the assemblage takes its name. Certain cartilaginous parts of the endoskeleton become calcified when calcium salts infiltrate the cartilage matrix, and these parts may resemble bone in terms of firmness. Calcified cartilage still lacks the developmental history and microanatomical form of true bone.

Almost all chondrichthyians possess a heterocercal tail (Figure 2-16); the dorsal flange is larger than the ventral and includes the tip of the upturned vertebral column. Toothlike denticles, placoid scales, present in the skin along with the spines that form the leading edge of the dorsal fins of some species are believed to be remnants of the armor that encased their placoderm predecessors. Most chondrichthyians possess ventral subterminal mouths, in front of which are paired nostrils on the underside of the snout. The upper ramus of well-developed jaws exhibits varying degrees of attachment to the skull. The hyomandibular portion of the hyoid pharyngeal bar extends from the brain case to the point of articulation between the upper and lower jaws. In accordance with the position of the hyomandibular element, the gill slit that precedes it is

Figure 2-16 Representative chondrichthyian fish: (a) an extinct elasmobranch, genus *Cladoselache;* (b) a living selachian, genus *Mustelus;* (c) a living batoid, genus *Dasybatus;* (d) a living chimaera, genus *Chimaera.* (Modified after Romer.)

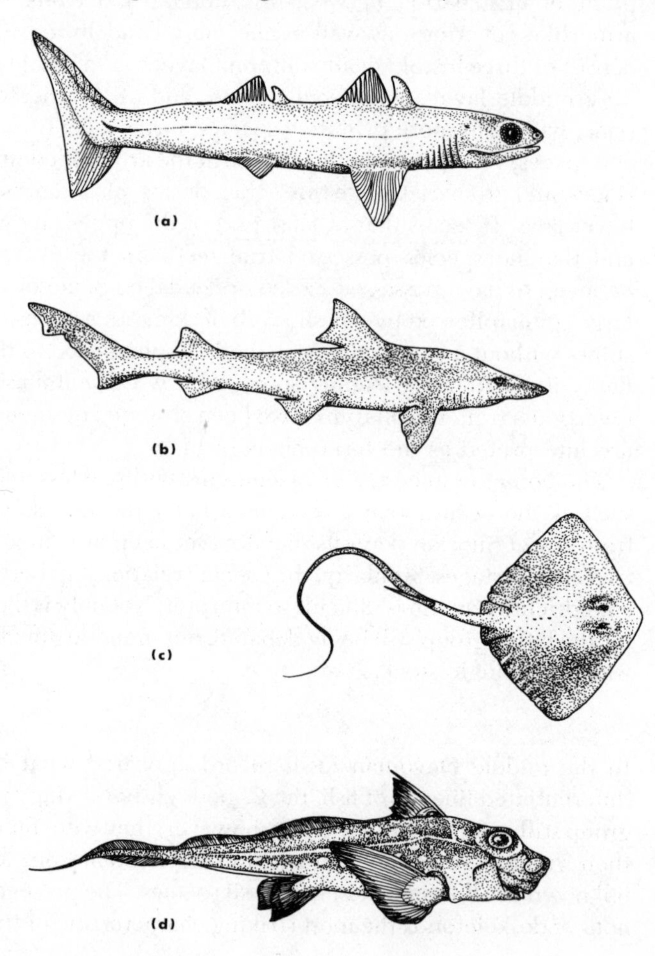

reduced to a very small dorsally located aperture, the spiracle. Typical gill slits are located posterior to the spiracle, and lungs and swim bladders are absent. The male claspers are characteristic of the group. These modified inner portions of the pelvic fins are intromittant organs that accomplish the act of internal fertilization that typifies the assemblage.

Within the class Chondrichthyes are two evolutionary lineages that have living representatives readily distinguished from each other: the subclass Elasmobranchii, the sharks, skates, and rays; and the subclass Holocephali, the chimaeras or ratfish (Figure 2-16). In the latter subclass, the external gill slits are covered by a fleshy operculum. Sharks first appeared in the latter part of the Devonian. Most sharks are predaceous, and with a few tropical exceptions are marine. The upper jaws are not attached directly to the skull and, typically, there are five to seven pairs of gill slits in addition to the spiracle. Some species lay large, heavily yolked eggs enclosed in a leathery egg case; other species bear living young.

During the Mesozoic, skates and rays evolved from the more primitive sharks. Unlike the sharks, the gill slits, other than the spiracle located behind the eye, are on the ventral surface. Typically, the tail, pelvic, and dorsal fins are reduced in size, and locomotion occurs by the undulating movement of the greatly enlarged pectoral fins. Skates and rays developed a bottom-dwelling existence and a diet largely of molluscs.

The chimaeras are a special group of marine cartilaginous fish and almost all are inhabitants of the deep sea. As their teeth are modified into large plates, it is not surprising that they too are mainly mollusc feeders. The upper jaw is solidly fused to the brain case, the well-developed notochord persists into adult life as the main axial body support, the vertebrae are poorly developed, and few placoid scales are present in the adult skin. There is some evidence that chimaeras had their origin among the predecessors of modern sharks.

The exact derivation of chondrichthyians from the placoderms is still disputed; however, that they are derived from that group of bony fish is no longer questioned. Given this relationship it must follow that the absence of bone in chondrichthyes is a result of the reduction from the primitive bony state of placoderms; it does not represent an evolutionary stage leading to the placoderm condition as once was thought. Although the fossil record of chondrichthyians is poor because of their reduction bone, it now seems certain that they diverged earlier than did most osteichthyians and that the chondrichthyian line, in the sense that it did not give rise to a subsequent major adaptive radiation, was a side branch.

Diversity of Form and Function in the Aquatic Environment (Osteichthyes)

The fish included in the class Osteichthyes first appeared in fresh-water deposits of late Silurian age. Since their origin, bony fish have continued to radiate in both fresh-water and marine environments, with the major expansion taking place during the Cretaceous. The marine environment does not appear to have been invaded until the Mesozoic. The more primitive osteichthyians and most of the derived forms have a bony endoskeleton. Typically, the operculum is bony and covers a gill chamber of not more than five gill slits. The operculum is supported by a series of bones that, in turn, form a hinge with the cranium. Lungs appear to have been present in many primitive members of the class. In at least one major phyletic line, lungs have become highly modified in structure and function as a hydrostatic organ, the swim bladder. It is now generally recognized that the greatest selective advantage for lungs was realized in fresh water where considerable seasonal drought occurred. Alternatively the selective advantage of a swim bladder exists in the marine environment where the problem of buoyancy is greater.

Even the broadest outlines of bony fish phylogeny cannot be agreed on by a majority of fish evolutionists. In this book an early phyletic divergence in the Silurian is recognized by the reference to the Acanthodii (Figure 2-17) and in the Devonian by the reference to the subclasses Actinopterygii and Sacropterygii (often called Choanichthyes).

The acanthodians, or spiny sharks as they are occasionally called, are the first jawed vertebrates (gnathostomes) to appear in the fossil record. The lineage originated in the Silurian, diversified in the Devonian, and became extinct in the beginning of the Permian. Compared to the contemporaneous ostracoderms and placoderms, acanthodians were a structurally conservative group. In general, they were small and sharklike in body form and tail shape, large headed, and had prominent spines on the anterior edge of their median and paired fins. They were distinctive fishes in possessing a large, hinged lower jaw, subterminal mouth, very large eyes located far forward on the side of the head, and true teeth in most species. The toothless forms were plankton strainers, whereas the toothed species were predaceous on large swimming fish and invertebrates. In both cases the gape of the mouth was considerable. Acanthodians did not possess extensive armor. The operculum was covered by small bony plates and the head and body by regularly arranged close-fitting scales. In addition to their jaws, this subclass is well known for its ventrolateral row of paired fins. It is most unfortunate that the relationships of acanthodians to ostracoderms and other jawed fishes is not well understood because therein lie the details of two of the most significant evolutionary trends in chordate history: the origin of jaws and paired fins. At least one prominent school of paleontologists has concluded that acanthodians evolved from pelagic ostracoderms and that they, in turn, gave rise to placoderms and other osteichthyians.

It was only with the appearance of jaws that a truly predatory vertebrate could be realized—for the first time food of more than microscopic size was readily ingested. One of the anterior-most pharyngeal bars (the mandibular arch) is believed to have been modified into the jaws while the other bars remained relatively unmodified as typical gill-supporting structures. Most of the evolutionary steps have been inferred from the morphology and development of living species of fish. The large number of paired fins found in primitive bony fish has been inferred to be the remnants of the breakup of an ancestral continuous lateroventral fold of skin similar to that seen in the metapleural folds of the Cephalochordata.

The actinopterygians, or ray fins as they are commonly called, are not on the main phyletic branch that led to tetrapods. They are believed to have evolved

Figure 2-17 A primitive acanthodian, genus *Climatius*, from the Devonian. The Acanthodii are tentatively placed in the Class Osteichthyes. (Modified after Romer.)

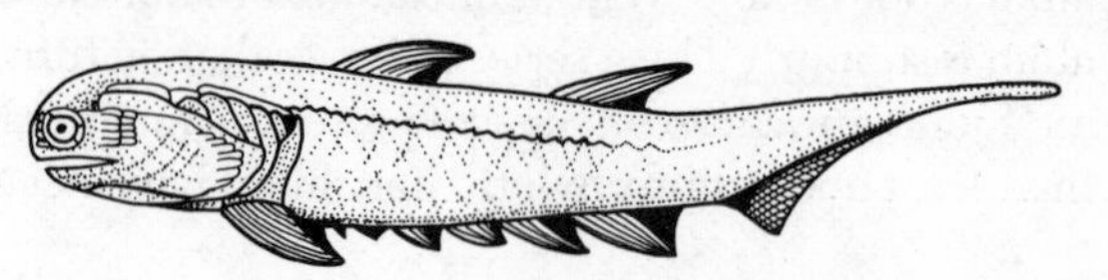

from either acanthodians or an ancient sarcopterygian ancestor. The ray fins became the dominant group in the Carboniferous and have remained dominant ever since; there are an estimated thirty thousand living species. The group is typified by the absence of internal nostrils, and with few exceptions its members do not possess fleshy-lobed, paired fins. Both characteristics distinguish them from the Sarcopterygii. The fins of actinopterygians are almost always membranous and are supported by horny or bony rays that may be jointed and branching and originate at the point from which the fin itself rises from the body. Two main lines of evolution are delimited within the ray-fin line: the infraclasses Chondrostei (Figure 2-18) and Neopterygii (Figures 2-19 and 2-20). The chondrosteans are the most primitive; they were the abundant ray-finned fish from the Devonian to near the end of the Mesozoic. The neopterygian line evolved from the chondrosteans and became dominant in the middle of the Mesozoic. The Teleostei diverged from the Halecomorphi, an early radiation of neopterygians in the Triassic. Since the late Mesozoic the former assemblage has remained the dominant group. The following list describes the general evolutionary trends that typify the major successive stages of actinopterygian evolution.

Chondrostei	*Ginglymodi*	*Halecostomi*
Relatively little bone, particularly in more recent species	Moderate amount of bone; internal skeleton partly cartilaginous	Completely ossified skeletons for almost all species
Spiracle present in most species	Spiracle present in only a few species	Spiracle absent
A strongly developed heterocercal tail in most species	Tend to lose upturned sharklike tail	Symmetrical tail in almost all species, at least superficially
At least primitively, scales thick and rhombic, covered with an enamel-like layer called ganoin	Tendency of scales to lose their ganoin covering and become thinner; mostly rhombic in shape	Scales thin, round in shape; ganoin layer lost
Maxilla attached to cheek	Maxilla not attached to cheek; jaws shortened eventually	Maxilla free, functions in protruding mouth; as a result, cheek region becomes open
Pelvic fins usually located posteriorly	Posterior location of pelvic fins usually	Location of pelvic fins often relatively far anteriorly
Lungs in most species; swim bladder in only a few species	Lungs absent; swim bladder present	Lungs absent; swim bladder present
Fresh water	Primitively fresh-water, extensive marine radiation in Mesozoic; only fresh-water survivors	Numerous fresh-water forms; however, marine radiation more extensive
Only sturgeons, paddlefish, and bichir and related forms extant (Figure 2-18)	Only gars extant (Figure 2-19)	Dominant group of living fish (Figures 2-19 and 2-20)

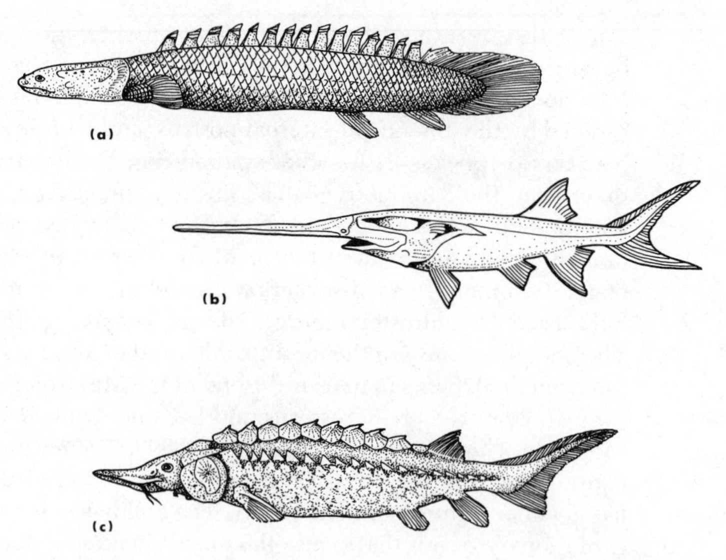

Figure 2-18 Representatives of the three major groups of living chondrosteans: (a) the African bichir, genus *Polypterus;* (b) the North American paddlefish, genus *Polyodon;* (c) a North American sturgeon, genus *Scaphirhynchus.* (Modified after Romer.)

One of the extant chondrosteans of Africa, the bichir (*Polypterus*), is of particular significance because of its fleshy-lobed, paired fins, a well-ossified internal skeleton, typical lungs, and primitive ganoid scales. These features strongly suggest that the bichir is the most primitive living descendent of the ancient actinopterygians.

The progenitors of the subclass Sarcopterygii were probably acanthodians. A knowledge of the constitution of this phyletic line is critical to the understanding of vertebrate phylogeny because the ancestors of the first land dwellers are found there. Sarcopterygians are typified (with few exceptions) by internal nostrils; fleshy-lobed, paired fins; and heavy scales that consist of a bony base covered by a thick layer of cosmine (a dentinelike substance) and spongy bone. The sarcopterygians are almost always divided into two taxa, the

Figure 2-19 Representative living neopterygians: (a) the gar, genus *Lepisosteus;* (b) the bowfin, genus *Amia.* (Modified after Romer.)

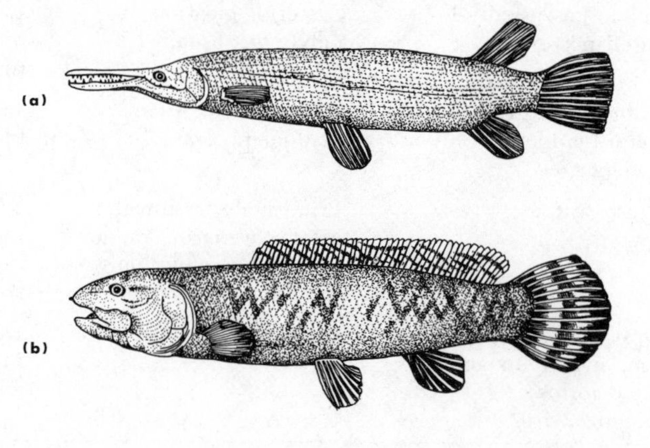

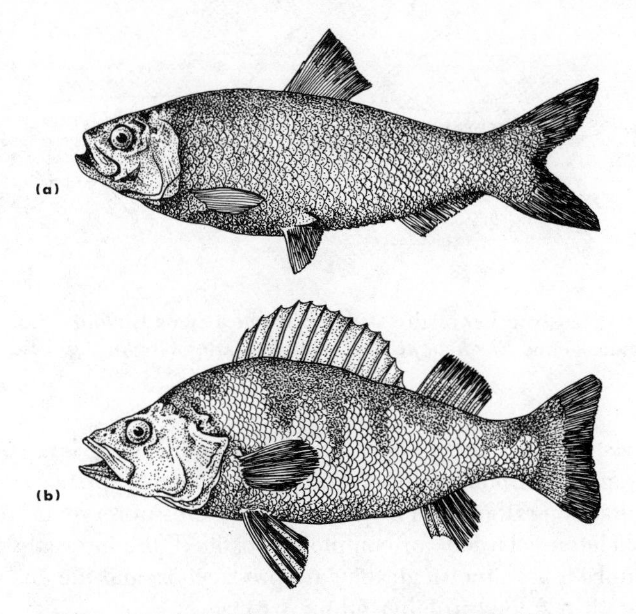

Figure 2-20 Two living teleostean fish: (a) a more primitive herring, genus *Clupea;* (b) a more highly modified perch, genus *Perca.* (Modified after Romer.)

orders Crossopterygii and Dipnoi, both of which appeared as fossils in the late Silurian-lower Devonian (Figure 2-21).

Most dipnoans, or lungfish as they are commonly referred to, differ from the crossopterygians in these ways: absence of a spiracle, reduced ossification of the skeleton, symmetry of the appendicular skeleton (consisting of a central series of elements in each appendage from which radiate, both anteriorly and

Figure 2-21 Representative choanichthyian fish: (a) the only living genus of crossopterygian, *Latimeria;* (b) a living dipnoan from Austrialia, genus *Epiceratodus.* (Modified after Romer.)

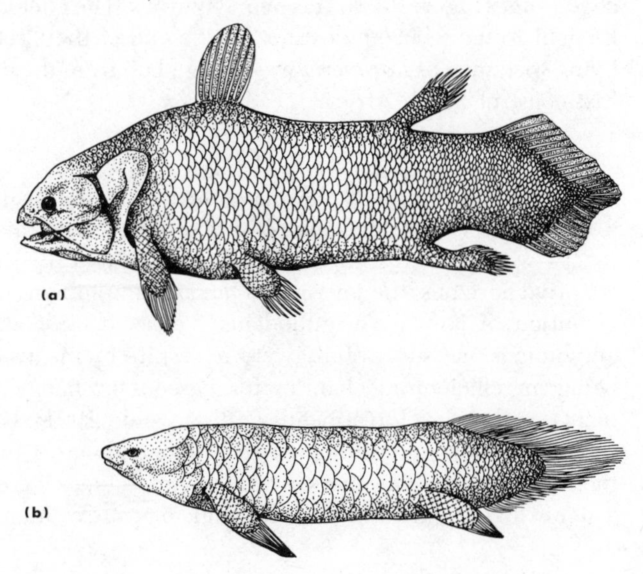

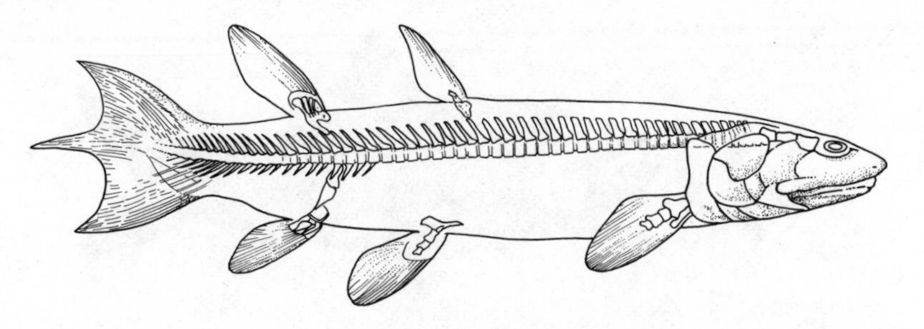

Figure 2-22 A reconstruction of the skeleton of the Upper Devonian rhipidistian crossopterygian, genus *Eusthenopteron*. (Modified after Gregory and Raven.)

posteriorly, smaller bones), and teeth modified into large plates for eating plant material. Among the living dipnoans the nares appear to be used primarily for olfaction rather than respiration. Most crossopterygians possess a spiracle; the skeleton is largely, or completely, ossified; the internal skeleton of the paired limbs is asymmetrical; they are predaceous; and the enamel of the individual teeth is folded in labyrinthine patterns.

The dipnoans resemble the amphibians, the most primitive terrestrial vertebrates, in several respects: the lung is supplied by a branch of the sixth aortic arch, the atrium of the heart is partially divided into two chambers, the mode of development is comparable, and external gills are present in their larvae. These few similarities caused early evolutionists to propose that lungfish were ancestral to terrestrial vertebrates. The similarities between crossopterygians and amphibians, more specifically such forms as the rhipidistian crossopterygians *Eusthenopteron* (Figure 2-22) and *Osteolepis*, are now believed to be so numerous and so detailed that few can argue against their relatively close common ancestry. The rhipidistians, which were primarily fresh-water species, became extinct in the early part of the Permian, but not before they gave rise to a terrestrial lineage. Today crossopterygians are represented only by a single living marine species, the coelocanth, *Latimeria chalumnae* (Figure 2-21). It is ovoviviparous. The coelocanth phyletic line was thought to have become extinct by the end of the Cretaceous, but in 1939 a living specimen of *Latimeria* was dredged up from the depths of the sea off the east coast of South Africa.

Invasion of the Land (Amphibia)

Much of the evolution of the anatomical and functional features necessary for a terrestrial vertebrate—and one completely free of the aquatic environment—was realized fully by the class Amphibia. Because it opened up new adaptive horizons, the invasion of land by amphibians has been likened to the evolution of jaws by acanthodians. The two most obvious challenges that amphibians met successfully were to respire by means of lungs—the gills not being an efficient mechanism for gaseous exchange in an aerial environment—and to develop appendages that would lift the body off the ground and permit a walking gait rather than a fishlike undulatory movement. The limbs of those crossopterygians that gave rise to amphibians were already well suited to a semiwalking gait and posture. And, of course, lungs had been secondary respiratory organs since the time of placoderms. The limbs of crossopterygians were very effective for moving over the debris of a swamp environment, certainly much more so than the ray fins of actinopterygians, which are relatively easily torn.

The end of the Devonian was a period of great seasonal droughts, and it is assumed that the lungs of crossopterygians allowed them to move from one pool to another as each dried up; their fleshy-lobed, paired fins aided them in their movement on land. Furthermore, the earliest amphibians were little more than fish in that they too encountered dry land only in their journeys between pools. In the search for those selective forces responsible for the evolution of terrestrialism, the then unexploited terrestrial food supply of insects should not be overlooked. It is important to emphasize that amphibians today are still not completely free of the aquatic environment. Most species have retained the thin, slimy, fishlike skin and must return to the water to reproduce.

The Amphibia differ from their crossopterygian ancestors in the reduction and loss of many of the opercular bones. They differ from their reptilian derivatives in not possessing epidermal scales; amnionic and chorionic membranes around the embryo; true claws; and, generally, a corneal layer in the epidermis. In addition, almost all species have single cervical and sacral vertebrae, and most adult amphibians lack external gills and gill slits. The extant species, at least, have a highly glandular skin and are incapable of involuntarily regulating their body temperature within narrow limits. Certain aquatic forms retain a lateral line system very much like their crossopterygian forebears. Almost all extant amphibians require an aquatic, or at least moist, habitat during some part of their life history. This environmental constraint is most often realized during the courting season and the period of egg laying as there are no extraembryonic living protective membranes to prevent dessication of the developing embryo. Most extant amphibians have a free-living stage, either typically larvaform or tadpolelike, and both types respire by means of gills. The tadpole differs from a larva in that its metamorphic changes to a subadult are considerably more drastic. The larva resembles the adult of the species in general body form and proportions and possesses typical adult appendages and true teeth on both jaws. It is not unusual for the larva to become sexually mature. The tadpole, on the other hand, is dramatically unlike the adult in that the head and body are not distinct; it possesses a long tail (often with a median fin), a horny beak, and rows of cornified denticles. At the time of metamorphosis the gills disappear and lungs develop, the tail is reabsorbed, and the legs appear.

Three major assemblages of amphibians are usually recognized: the subclasses Labyrinthodontia, Lepospondyli, and Lissamphibia (Figures 2-23 to 2-26). Only the latter group possesses living species and they occur in each of three orders: Anura, Urodela, and Apoda. The labyrinthodonts were dominant during the late Paleozoic and early Mesozoic periods. The labyrinthodont radiation had died out by the end of the Triassic, but not before it gave rise to the class Reptilia in the Carboniferous. Labyrinthodonts were large as adults, differing from their immediate crossopterygian ancestors in only a few attributes: their median fins were greatly reduced or absent and they possessed short, sturdy legs; large heads; and long tails. Labyrinthodonts, at least primitively, possessed archlike vertebrae very similar to those of crossopterygians.

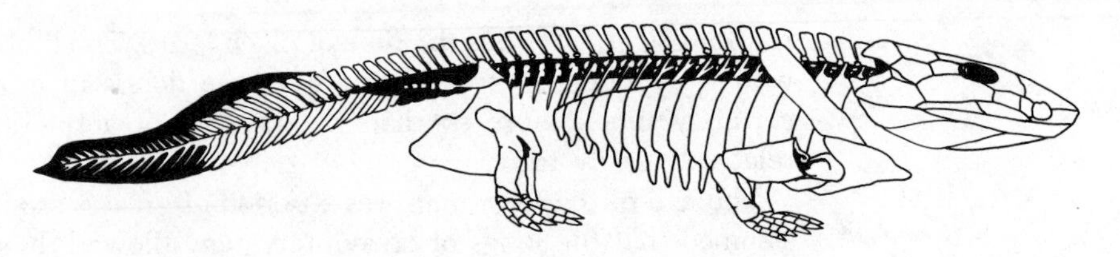

Figure 2-23 A reconstruction of the skeleton of a Devonian amphibian, genus *Ichthyostega*. (Modified after Jarvik.)

The ichthyostegid group of labyrinthodonts, the most primitive of amphibians (Figure 2-23), made its appearance in the middle Devonian. Although they possessed fishlike tails with fin rays and the external nostril was located far down on each side of the head near the margin of the upper jaw—very much like their crossopterygian progenitors—they had five-toed appendages, the hyomandibular bone had become the innermost middle-ear ossicle, and a distinct otic notch was present in the posterior portion of the skull. The otic notch probably held the tympanic membrane. Both the ear ossicle and the tympanic membrane were almost certainly used in the reception of sound and its subsequent transmission to the central nervous system. The lepospondyls appear to have evolved from the labyrinthodonts. Typically, they were of smaller adult size and differed most conspicuously in their spool-shaped vertebral centra.

Anurans, the frogs and toads (Figure 2-24), occupy a wide variety of habitats from deserts to fresh water. Their greatest diversity occurs in the tropics, although they range well into the temperate regions of both hemispheres. They exhibit a tadpole type of life history, and as adults most of their morphological and functional characteristics relate to their saltatory (jumping) mode of locomotion. The evolution of saltation is correlated with a shorter trunk with a reduction in the number of vertebrae (as few as five are known), lack of definition of the neck region, reduction of the skull and ribs, lengthening of the hindlimbs (an extra joint occurs at the level of the ankle), and absence of a tail; the posterior vertebrae are fused into a single spikelike bone, the urostyle, located between the two rami, or ilia, of the pelvic girdle. The hind feet of anurans may be fully webbed, functioning in the aquatic environment as major

Figure 2-24 An anuran amphibian. (Modified after Noble.)

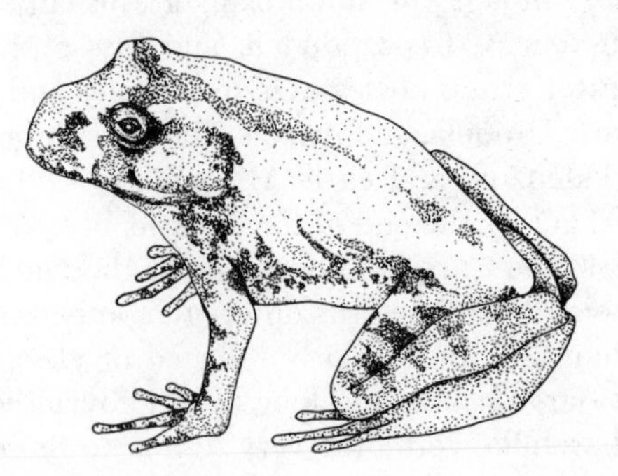

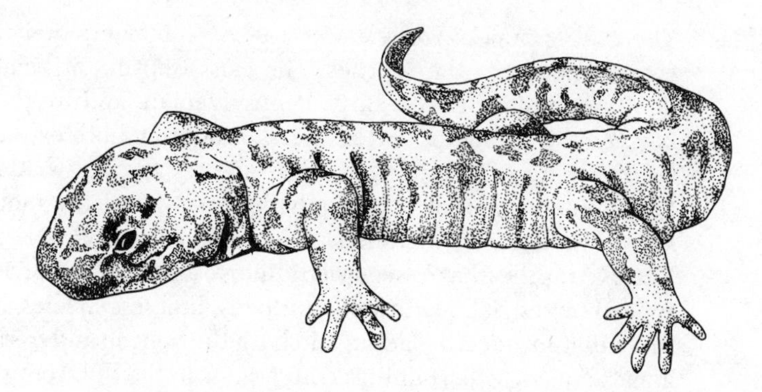

Figure 2-25 A North American salamander, genus *Dicamptodon*. (Modified after Noble.)

propulsive organs. Most species produce sounds by passing air over vocal cords; they receive airborne sounds through the tympanic membrane.

The Urodela, or salamanders as they are most often referred to (Figure 2-25), are almost always found in moist habitats. They are widely distributed in the northern temperate regions and throughout the New World tropics. In general appearance they are most like primitive tetrapods. Median fins are usually absent; the fore-and hindlimbs are nearly equal in length; the tail is well developed; and the head, neck, and trunk regions are well differentiated. There is considerable reduction of dermal bone and a general trend toward retention of embryonic cartilage in the skull and, in particular, in the pectoral girdle. Vocal cords and tympanic membranes are absent. In the large group of lungless salamanders, integumentary respiration is the primary method of gaseous exchange.

The Apodans, or caecilians, are believed to be more closely related to salamanders than to anurans (Figure 2-26). They are wormlike burrowers; however, a few species are completely aquatic. Both limbs and girdles have been lost, the tail has been greatly reduced, the anus is almost terminal, vocal cords are absent, and the eyes are usually greatly reduced in size. Caecilians have retained typical fishlike dermal scales embedded in the skin. As adults they may reach a maximum total length of about two feet. They are circum-tropical in geographic distribution.

Figure 2-26 An apodan amphibian shown encircling a clutch of eggs. (Modified after Smith.)

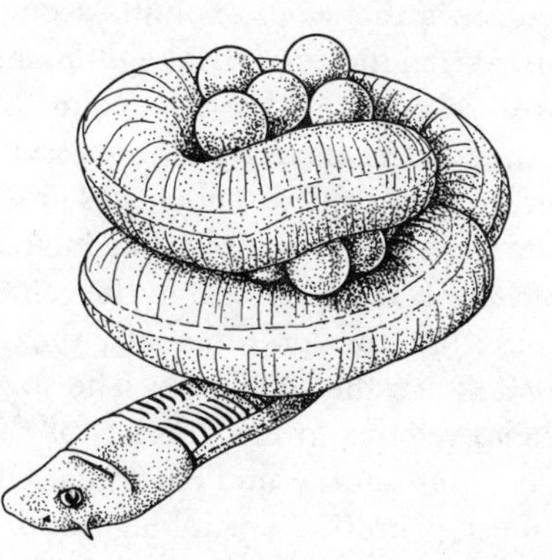

The Amniote Egg (Reptilia)

The first group of vertebrates to become completely free of the aquatic environment was the reptiles. The class Reptilia originated from labyrinthodont amphibians in the early Pennsylvanian, and by the Mesozoic they had become the dominant vertebrate. Considering both extinct and extant species, the diversity of reptiles has been greater than any other group in tropical environments, and at least equivalent to that of the more recent and better-known radiation of mammals. During some period in their long phylogenetic history, reptiles have occupied almost all possible environments. There are typical terrestrial quadrupeds, bidpeds, limbless species, burrowing forms, and those that invaded the aerial and both the marine and fresh-water habitats. The aquatic invasion certainly is convergent in the different reptile lineages and is not the primitive state. Reptiles are characterized by epidermal scales; dry skins, owing to the absence of mucous glands; true claws; and the inability to regulate involuntarily their body temperature within narrow limits. A few species of reptiles give birth to living young, but the majority lay heavily yolked, terrestrial eggs enclosed within protective layers of albumin, shell membranes, and shell. The yolk serves as a large endogenous food supply. Very early in development, the yolk becomes enclosed within a cellular sac—the yolk sac—and the embryo itself is surrounded by several embryonic membranes, amnion, chorion, and allantois, which are discarded at hatching when the young reptile leaves the shell. These membranes, present in the embryonic stages of birds and mammals, are first found among the reptiles. Albumin also is consumed during incubation. Unlike most amphibians, reptiles have no metamorphosis, and even during very early stages of development the young do not breathe by means of gills.

Laying an egg on land requires certain basic conditions, all of which appear to have been met by the so-called amniote egg. The leathery, or lime-impregnated, shell provides protection from physical and chemical shock and dessication and is sufficiently porous to allow exchange of oxygen and carbon dioxide. Internal fertilization is necessary because the laid egg is enclosed by a shell. The amnion surrounds the developing young, enclosing it in a fluid-filled cavity, and the amniotic fluid secreted mainly by cells of the amnion prevents dessication. The allantois is a saclike development from the embryo's hind gut. It increases in size with the growth of the embryo, comes into contact with the chorion, and with the chorion serves both as a repository for nitrogenous metabolic wastes and as a respiratory surface. The extraembryonic layer from which the amnion is formed gives rise to the chorion. (See Chapter 3.) This thin cellular membrane encloses the whole complex: embryo, amnion, yolk sac, albumin, and allantois. When did terrestrial egg-laying originate? What were the selective pressures responsible for its evolution? These are two very important questions that as yet have not been answered satisfactorily.

At present there are recognized six subclasses and about sixteen orders of reptiles. Of the later taxonomic level, only four include living species (Figure 2-27). The number of fenestrae (openings) in the temporal region of the skull and the specific bones that border them are the sets of characters most commonly used to distinguish the subclasses. The subclass Anapsida does not possess a temporal fenestra; this group consists of three orders: the Cotylosauria, or stem reptiles; Mesosauria; and Chelonia. Of all known reptiles, the cotylosaurs are the most primitive assemblage and are considered to have evolved directly from Pennsylvanian labyrinthodonts. These short-legged

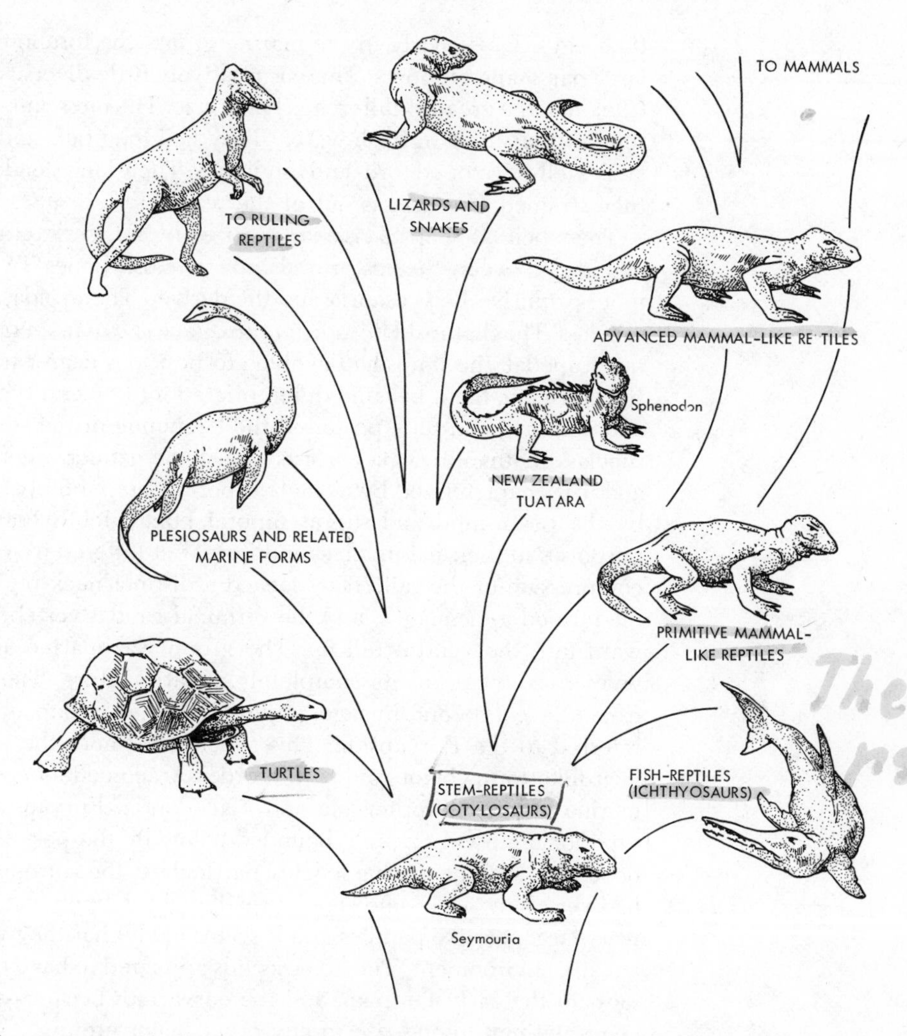

Figure 2-27 A suggested phylogenetic history of the major groups of reptiles. (Modified after Romer.)

animals, with their massive skulls, appear to have been terrestrial; therefore, one can only assume that the aquatic to terrestrial transition period preceded their appearance. Before they became extinct in the late Triassic, cotylosaurs gave rise to most of the subsequent radiations, such as turtles, mesosaurs, pelycosaurs, eosuchians, and thecodonts. The chelonians, or turtles as they are usually called, first appear in Triassic deposits, and look very much like modern forms. The group is characterized by a protective shell of enlarged dermal bony plates and overlying horny epidermal scales, constituting the dorsal carapace and ventral plastron. With the evolution of the shell came the loss of many of the muscles of the body wall, and in most species the ribs and trunk vertebrae are solidly fused to the carapace. Except for the most primitive forms, all of which are extinct, turtles lack teeth. All species lay eggs, and the males possess a single median copulatory organ; the anal opening is often a longitudinal slit. There are terrestrial, semiaquatic, and fully aquatic marine and fresh-water forms. Many of the aquatic species come on land only to lay their eggs. Particularly in the marine group, the forelimbs are modified into large oar-shaped flippers. There is relatively little diversity of form within the Chelonia, the greatest difference being size. The mesosaurs were slender, small reptiles that lived in fresh water. They had long tails and snouts and moderately well-developed fore- and hindlimbs. Their long slender teeth suggest that they strained crustaceans out of the water.

The subclass Synapsida is characterized by a single lateral temporal opening, bounded above by the postorbital and squamosal bones. Two extinct orders, the more primitive Pelycosauria and the derived Therapsida, are included in this subclass. The therapsid lineage, in turn, gave rise to mammals. Therapsids were quadrupedal, the limbs had evolved to positions nearer to the median line of the body, the teeth became differentiated into several types (molars, canines, and incisors), and they possessed the beginning of the secondary palate. The subclass Ichthyopterygia consists of a single extinct order, the Ichthyosauria, and it is characterized by a single superior temporal opening bounded below by the postfrontal and supratemporal bones. Ichthyosaurs were similar to porpoises in their habits, size, and form, but differed from them in the lateral compression of the tail. They lacked a distinct neck region; had short, paddle-formed appendages; and the terminal caudal vertebrae projected downward into the ventral tail fin. The group was marine and almost certainly viviparous, owing to its completely aquatic nature. They probably evolved from the cotylosaur lineage. A highly varied group of extinct lineages is referred to the Euryapsida. This subclass includes the primitive, terrestrial lizardlike forms belonging to the order Araeoscelidia and the more familiar marine orders Sauropterygia and Placodontia. Euryapsids all share a single superior temporal opening bounded below by the postorbital and squamosal bones. Some of the marine species, particularly the sauropterygian pleisiosaurs, had very long necks and short, broad, and relatively flat bodies. Their limbs were large, oarlike paddles and it seems likely that they were confined to the aquatic environment. The araeoscelids appeared to have evolved from cotylosaurs in the early Permian, and the euryapsids became extinct in the Cretaceous without giving rise to any other major group.

The subclasses Archosauria and Lepidosauria are typified by two temporal openings separated by the postorbital and squamosal bones. The Archosauria includes five orders: Thecondontia, the "ruling reptiles" of the Mesozoic; Crocodilia, the alligators, crocodiles, and relatives; Pterosauria, the "flying reptiles"; Saurischia, the reptilelike dinosaurs; and Ornithischia, the birdlike dinosaurs. Thecodonts were mostly small, slender, predaceous, and largely bipedal in their locomotion. It was the thecodont stock that very likely gave rise directly and independently to almost all the other archosaurian orders. The Crocodilia are characterized by very large cornified epidermal scales; those on the dorsal side contain plates of dermal bone. They have a single median copulatory organ, a bony (hard) palate separating the mouth from the nasal passages, and a longitudinally oriented anal slit. It appears that at least some of the modern crocodilians possess a completely divided ventricle in the heart. Crocodilians are predatory, equipped with large teeth set in sockets, and are all semiaquatic. There are only a few living species, all of which are relatively similar in general body shape. The Crocodilia are usually considered to have evolved from a saurischian ancestor; however, recent work suggests an independent origin from the thecodont lineage. Pterosaurs probably glided and soared, but it seems unlikely that they actually flew. The largest had wing spreads of more than twenty-five feet. They left no descendants. It is recognized that they did not give rise to either birds or bats, as was once commonly thought. The pterosaur wing was a leathery web stretched between an elongated finger, the bones of the forelimb and the body wall. The bones were light and compact, some species were toothless, and all appeared to have had small, poorly developed feet. Saurischian dinosaurs possessed a typical reptilian pelvis, in contrast to the ornithischian dinosaurs with a birdlike pelvis. The Saurischia included both primitive bipedal and quadrupedal species. The primitive bipedal forms were either carnivorous or herbivorous. The largest of all terrestrial carnivores, *Tyrannosaurus,* belongs to the Saurischia, as does the largest terrestrial animal known, *Brachiosaurus. Brachiosaurus* and related species probably spent most of their lives grazing in shallow lakes. This supposition is based on the fact that only water could buoy up the tremendous weight. The Ornithischia also included both bipedal and quadrupedal species, all of which were herbivorous. Both groups of dinosaurs are believed to have been endothermic. Until recently, saurischian and ornithischian dinosaurs were considered to have evolved independently from the Thecodontia, however at least some recent workers have concluded that they are a monophyletic derivation of that subclass. Why did dinosaurs become extinct so abruptly near the end of the Cretaceous? This, again, is a major question that still remains to be answered satisfactorily. Suggestions have included the drying up of swamps and the decline of essential food resources.

The Lepidosauria includes the primitive extinct order Eosuchia, and the orders Squamata and Rhynchocephalia, both of which have living representatives. The rhynchocephalians had an extensive radiation that existed as far back as the Permian. Today the order is represented by only a single species, *Sphenodon punctatus,* that is restricted to New Zealand. The modern form looks like a lizard, although unlike that group it lacks a copulatory organ. Its anal opening is a transverse slit, and the well-developed pineal eye contains both a lens and a retina. The Squamata evolved in the early Mesozoic from an eosuchian ancestor and radiated in the Cenozoic. The group is characterized by paired copulatory organs (hemipenes) in males and a transversely oriented anal slit. Squamates are the most numerous of living reptiles. Like their reptilian ancestors, their greatest diversity is in the tropics. The suborders Lacertilia (lizards) and Ophidia (snakes) are the most familiar groups. Typically, lizards have legs, eyelids, ear openings, and a relatively immovable quadrate; the halves of the lower jaw are firmly united, the teeth are not set in sockets, and a single temporal arch is present. Snakes, conversely, lack legs, eyelids, and ear openings; the quadrate is highly movable, the lower jaws are loosely united by a ligament that permits the swallowing of very large prey, and temporal arches are absent. Most of the differences between the two suborders reflect an early ophidian evolutionary history of burrowing. As evidenced by the vestiges of the hindlimbs and pelvic girdles found in some species, the Ophidia almost certainly evolved from a quadruped, probably a primitive lacertilian.

Feathers and Sustained Flight (Aves)

It now seems reasonably certain that the class Aves evolved from an archosaurian ancestor and—more specifically—from the saurischian line. Although it also seems certain that the archosaurian pterosaurs were capable of soaring and gliding short distances, as do some squamates and a few modern frogs, birds

Figure 2-28 A skeleton with impressions of feathers of the earliest bird known, genus *Archaeopteryx*. (Modified after Colbert.)

appear to have been the first animals to evolve self-sustained flight. Birds are characterized most obviously in terms of their feathers and wings, which are highly modified pectoral appendages. The wing and tail feathers greatly increase the surface area of the animal, adding relatively little weight, and thus offer a great deal of resistance to the air when the animal is in flight. Birds have lightweight bones, many of which are hollow and contain air sacs projecting from the enlarged lungs. Birds are necessarily bipedal, owing to their pectoral appendage modifications for flight, and the enlarged sternum provides increased surface area for pectoral muscle attachment. They usually have large, well-developed eyes; horny scales on legs and feet; and only a left ovary and oviduct. Their body temperature is high, and their ability to regulate it involuntarily within narrow limits, together with the modifications of the circulatory system, aid the high metabolic rate necessary to supply the energy for sustained flight. As a rule, soaring birds have wings with a large surface area; fast-flying birds have small, short, rapidly beating wings; and flightless birds are wingless or have wings that are greatly reduced in structure and size. Because they are capable of sustained flight, many birds can migrate great distances annually, and many species have the ability to home—that is, to find their way back to their home area over short to very long distances. Birds are peculiar in their possession of a relatively short intestine and the absence of a urinary bladder. Teeth are absent in modern forms; their function has been replaced by a horny derivative of the skin that forms the beak, or bill, and that has been modified in numerous ways for different food habits. Birds usually exhibit a very elaborate social behavior. Eggs that are laid on the ground or in a nest must be incubated by the parent. Not all birds are small; the adult of at least one species now extinct on Madagascar was ten feet tall and weighed eight hundred pounds. As a group, birds are relatively invariant compared with other classes of vertebrates. There are numerous orders, most of which differ in relatively few characteristics.

Usually birds are divided into two subclasses, Archaeornithes and Neornithes. The former subclass is known only from two species (about four individuals) from the late Jurassic. These species were about the size of a crow (Figure 2-28). Their teeth, set in sockets, were present in both jaws; the wings had three-clawed fingers; the tail was very long and had typical feathers projecting outward along the margins; and abdominal ribs (gastralia) were present. The fact that the sternum was very small in these primitive forms strongly suggest that sustained flight had not yet evolved. Like modern birds, the feathered wing contained at least parts of three digits, and the hindfeet had four toes. Neornithean birds, in contrast to their archaeornithean ancestor, possess a short tail, of thirteen or fewer compressed vertebrae, from which the tail feathers radiate. The wing bones are reduced in number and with very few exceptions there are no clawed digits on the wings. This subclass encompasses all modern birds, with about 8,500 species recognized currently, and a primitive group of toothed forms that are known from marine Cretaceous deposits of North America. Teeth appear to have been lost in all neornithean phyletic lines by the Mesozoic.

The Evolution of Higher Levels of Integration in the Central Nervous System (Mammalia)

The class Mammalia evolved from a therapsid reptile stock in the Triassic, radiated widely in the Cenozoic, and today is the most diverse class of terrestrial vertebrates. All mammals have at least some hair and suckle their young; these two features are unique to them. A few species have retained the reptilian epidermal scale in certain hairless areas on the body and appendages. All mammals, other than the egg-laying monotremes, have an intrauterine parent-embryo connection of some type called a placenta, which serves to nourish the young and to remove its waste products before birth. Although the placenta is a highly variable structure interspecifically, it can be characterized for the entire class as a functional intermediary between the blood vascular system of the young and that of the mother. The paired sets of mammary glands are modified skin glands, which furnish nourishment until the young can fend for themselves. When first born, mammals are relatively helpless, compared with other vertebrate offspring, and the young require the parent for nutrition, protection, and some education. Mammals can be characterized further, with few exceptions, as having (1) a four-chambered heart; (2) a single aorta (the left fourth); (3) three small bones (the malleus, incus, and stapes) in the middle-ear cavity that transmit the airborne sounds by mechanical vibration to the inner ear; (4) a temporal cranial element that represents a fused series of reptilian skull bones; (5) a single bone, the dentary, in the lower jaw on each side of the mandibular symphysis; (6) a muscular diaphragm separating the thoracic and abdominal cavities; (7) sweat glands; (8) urogenital and intestinal systems that open to the outside separately, rather than empty their products into a common cavity; (9) heterodont dentition; (10) two successive sets of teeth, the "milk" teeth being replaced by a permanent set; (11) marrow cavities in the long bones; (12) circular, biconcave, and enucleate red blood cells; (13) a well-developed sound-collecting lobe, the pinna, accessory to the outer-ear cavity; (14) a larynx that produces a wide range of sound; (15) an extensive development of the cerebral cortex; (16) a double occipital condyle similar to amphibians and their most immediate ancestors, the synapsid reptiles; and (17) the physiological capability of involuntarily regulating their body temperature within narrow limits—that is, they are homeothermic. Homeothermism has been responsible almost certainly for the ability of mammals to invade successfully those regions where the climate is very extreme and fluctuating. Mammals are a diverse group, particularly in their patterns of locomotion and their teeth. Both sets of characters are often used to exemplify adaptive radiation. About thirty-two orders of mammals are recognized—fourteen of which are extinct—and they are divided among three subclasses. Only two of the three subclasses, Prototheria and Theria, include living species.

The order Monotremata contains the only living prototherians. More specifically, the extant species are the duckbill platypus and spiny anteater, and both are restricted to the Australian region today (Figure 2-29). Monotremes are truly exceptional mammals in that they lay large reptilelike eggs that are incubated in a nest or in a temporary abdominal pouch. The mammary glands are poorly developed, there are no nipples, and the milk simply accumulates on tufts of hair. Monotremes have a number of features that are obviously primitive: (1) a cloaca is present and receives both digestive and reproductive products; (2) the pectoral girdle includes bones that are lacking in other mammals; (3) a ventral mesentery is present and extends the length of the abdominal cavity; (4) a functional abdominal vein has been identified; (5) the testes are abdominal; (6) the pinna is very small or absent; (7) the malleus and incus are very large and resemble the articular and quadrate bones of reptiles, from which they were modified; (8) the fiber tract that connects the two cerebral hemispheres in higher mammals, the corpus callosum, is absent; and (9) they exhibit little ability to maintain a constant body temperature. On the other hand, certain characteristics of monotremes, such as the absence or reduction of their teeth, are obviously derived features. It has been reemphasized recently that monotremes may have evolved from a therapsid stock very different from the one that gave rise to the therians. Early fossils are only now being discovered in Australia, and the question of the phyletic relationships of monotremes may yet become clearer.

The subclass Theria, in contrast to the Prototheria, does not lay eggs. They bear living young, and their mammary glands are supplied with well-formed nipples. The Marsupialia is the only order in the therian infraclass Metatheria with living representatives (Figure 2-29). Fossil marsupials are known as far back as the middle Cretaceous, and it is certain that the group was once

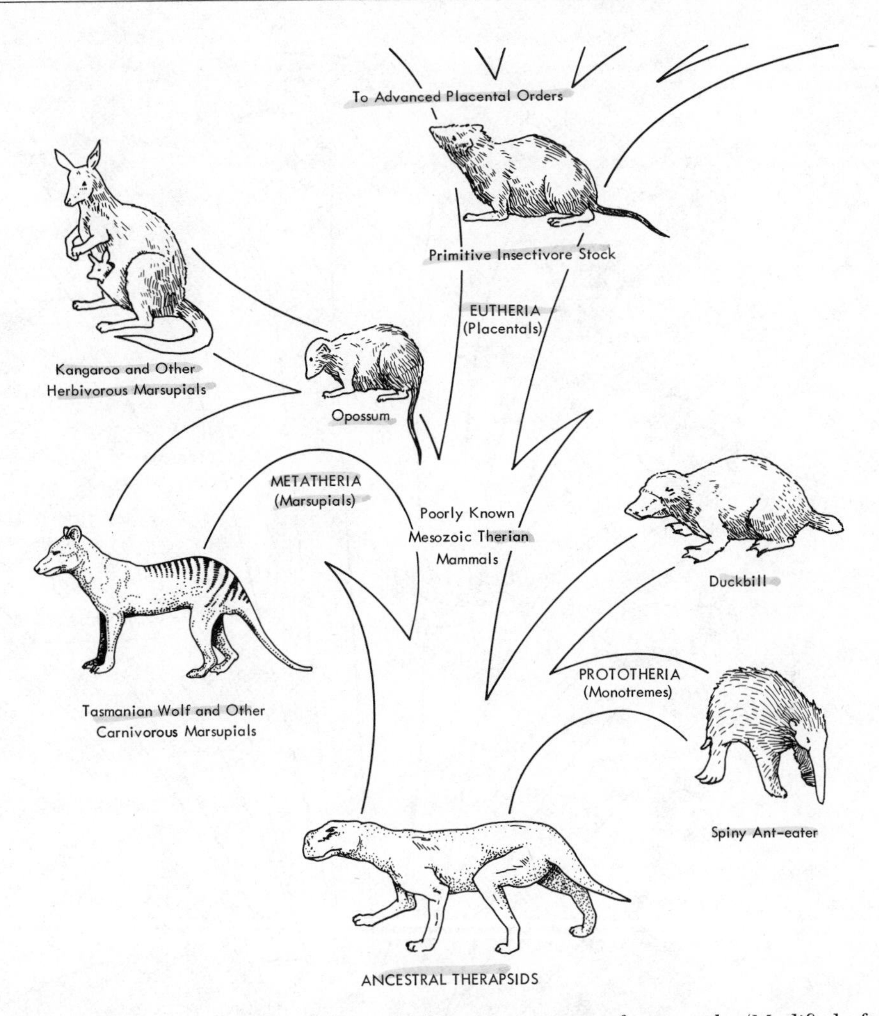

Figure 2-29 A suggested phylogenetic history of the major groups of mammals. (Modified after Romer.)

widespread throughout the world. Today marsupials are restricted to the Australian region, with the exception of two genera in the New World. A well-developed abdominal pouch, or marsupium, is present in all species, with a few exceptions where it seems to have been lost. The marsupium opens either anteriorly or posteriorly, and its walls are supported by two slender bones (called epipubics) that are attached to the pelvic girdle. The newborn marsupial is poorly developed, and it undergoes a second period of maturation in the parent's pouch. Once the young animal has made its way inside the marsupium, it becomes firmly attached to one of the nipples located there, and it completes its development. The female marsupial produces eggs with very little yolk. Following fertilization the young embryo comes into contact with the parent's uterine wall to form a poorly developed placenta of the "yolk-sac" type. Marsupials are convergent with many kinds of eutherian mammals in many different ways. There are, or at least were, large grazers, wolf- and catlike carnivores, fossorial molelike species, rabbit and bear forms, as well as one group that glides from tree to tree and looks much like a "flying" squirrel. The

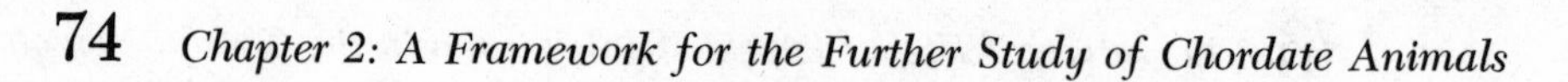

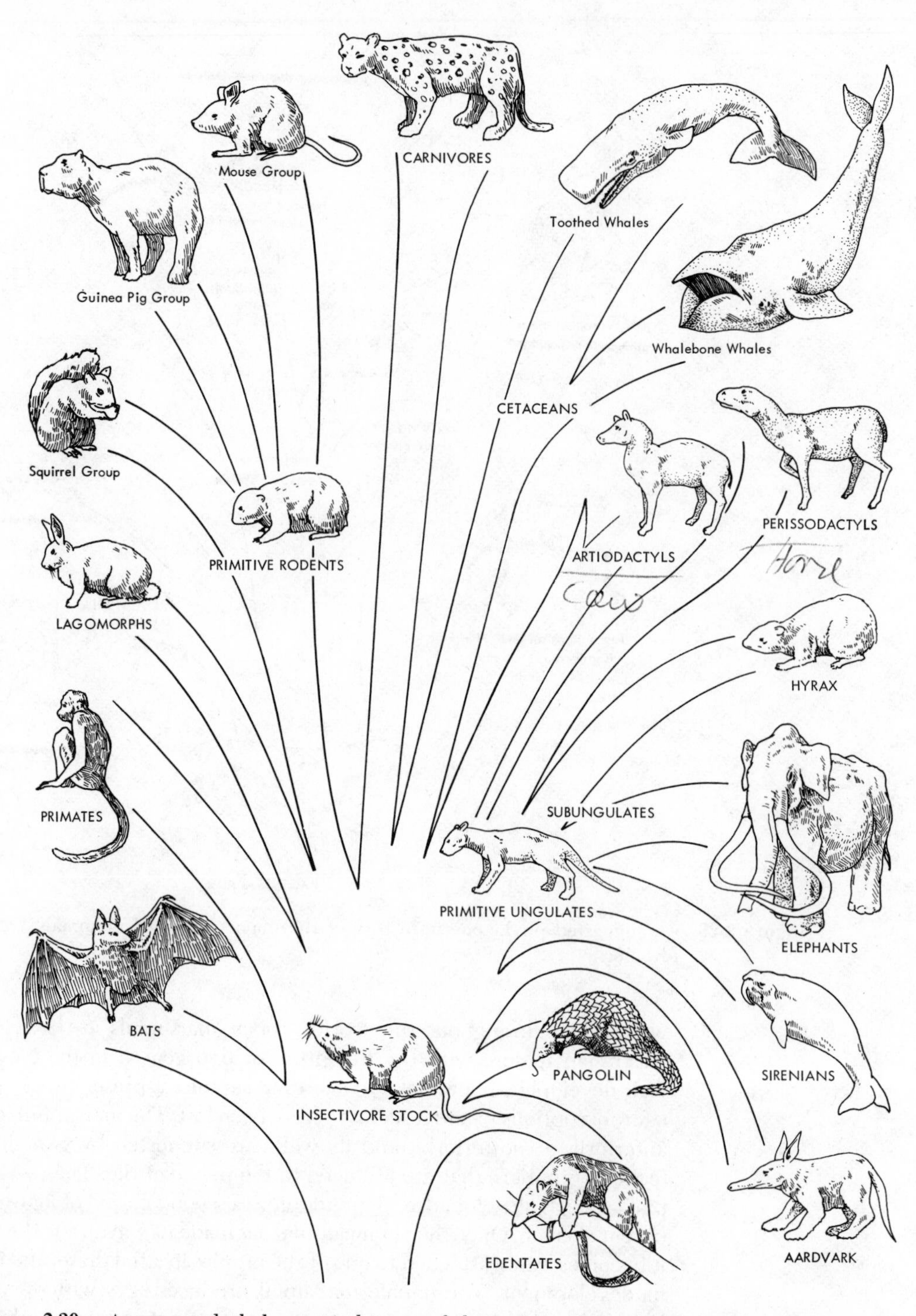

Figure 2-30 A suggested phylogenetic history of the major orders of eutherian mammals. (Modified after Romer.)

marsupial radiation is believed to have declined as a result of competition with eutherian mammals.

The infraclass Eutheria, or so-called "true" or placental mammals, encompasses about twenty-five orders (Figure 2-30). The earliest known members of fifteen living orders can be traced back to the Eocene or Paleocene, with the exception of the insectivores and primates that are known from the Cretaceous. In contrast to the marsupials, eutherians possess a well-developed placenta, most often referred to as the chorioallantoic type; the yolk sac is vestigial and does not contact the wall of the uterus. Of course, eutherians also differ in the absence of a marsupium and epipubic bones. In the Eutheria, the young are relatively well developed when born; however, there is still considerable postparturition care in many species. Unfortunately, many of the primitive orders of eutherians are now extinct; however, the living members of the order Insectivora do provide some clues to the ancient relationships between marsupials and eutherians. Insectivores have a shallow cloaca; flat feet; five toes; smooth cerebral hemispheres; small, sharp, pointed teeth; and some have permanent abdominal testes. Of a number of general evolutionary trends delimited in eutherians, the following appear to be the most conspicuous: (1) increase in tooth size; (2) reduction in tooth number, (3) reduction of toes, and (4) an increase in the size of the forebrain and complexity of the cerebral cortex. Much of the eutherian diversity is best correlated with the adaptability of their dentition and limbs and feet. It is also probable that the increase in number and complexity of the integrative pathways and centers in the central nervous system, particularly the cerebral cortex, is related to their successful diversification.

References for Chapter 2

[1] BARRINGTON, E. J. W. *The Biology of Hemichordata and Protochordata.* San Francisco: W. H. Freeman & Co., Publishers 1965. This is a concise review of the biology and evolutionary relationships of hemichordates and primitive chordates.

[2] CARTER, G. S. *Structure and Habit in Vertebrate Evolution.* Seattle: University of Washington Press, 1967. Carter's work was intended to be a complete textbook of vertebrate zoology. Its phylogenetic framework is based on slightly outdated information; however, the facts of organismal biology are still very useful.

[3] GANS, C. *Biomechanics, An Approach to Vertebrate Biology.* Philadelphia: J. B. Lippincott Co., 1974.

[4] GREENWOOD, P. H., R. S. MILES, and C. PATTERSON, eds. *Interrelationships of Fishes.* New York: Academic Press, Inc., 1973.

[5] JEFFERIES, R. P. S. "The Subphylum Calcichordata (Jefferies 1967), Primitive Fossil Chordates with Echinoderm Affinities" Bulletin of the British Museum (National History), *Geology,* **16**(1968), 243–339. This is the first comprehensive review of the relationships of the extinct calcichordates.

[6] ROMER, A. S. *Vertebrate Paleontology.* 3rd ed. Chicago: University of Chicago Press, 1966. This is an accurate summary of our knowledge of the fossil history of vertebrates.

[7] YOUNG, J. Z. *The Life of Vertebrates.* 2nd ed. Oxford: Clarendon Press, 1962. This is a lengthy account of structure and function in vertebrates.

3

Changing Patterns of Structural Organization During Early Development

Introduction A discussion of the early development of vertebrate embryos is included at this point in order to lay a foundation for an analysis of the origins of the structural and functional patterns in adult chordates. At any developmental stage there is a correlated, although transitional, organization between what precedes and what follows. This becomes increasingly complex with age, as the genotype progressively sends forth its coded messages. The brief survey here will include the structure that is first identifiable as presumptive areas and the symmetry within the fertilized egg. This initial structure becomes cellular as the embryo progresses through cell division (cleavage). It is rearranged definitively during gastrulation, when masses of cells move to new and final locations within the young embryo; the embryonic germ layers of ectoderm, endoderm, and chordamesoderm become identifiable; and determinative (inductive) factors come into play. Finally, the embryo attains visible adult and species organization during neurulation, with the progressive appearance of primordia of future organs and organ systems from the embryonic germ layers. From these early beginnings, the basic structural pattern of the adult organism gradually unfolds, a pattern that will undergo subsequent modification only with the progress of growth and differentiation in size, shape, structure, and function of the various components.

The course of growth and differentiation will be followed here as far as seems necessary to clarify the base lines of definitive structure. Ontogeny of the organs and organ systems will be taken up where relevant in the chapters that follow.

A knowledge of development has contributed immeasurably to our present understanding of organic structure and its evolution, as well as to the relationships existing between specific parts, e.g., between the male reproductive and excretory systems, sex intergrades, and so on. A knowledge of embryology gives more meaning to structure, just as the study of evolution has contributed to comprehension of the bases of vertebrate variability and homologies. Reference should be made to the facts of development in order to interpret the presence of modified or apparently different structures and teratologies, as well as to show the origin(s) of others.

Animals chosen for this introduction are specimens that have found wide usage among investigators interested in development at any stage, largely owing to their availability, hardiness, acclimation to laboratory environment, reproductive behavior, and manipulative suitability. A number of references dealing with various aspects of vertebrate embryology are listed in the bibliography at the end of this chapter.

In its broadest perspective, development can be viewed as covering the whole life span of an organism from fertilization to old age and death; possibly, in this definition, we should also include the germ cells whose union initiates the new organism in most animal species (Table 3-1). Although both progressive and regressive aspects may, thus, be included and the term is synonymous with change, it is commonly and more restrictively applied to the following: continuously progressive phenomena seen in the embryo, fetus, and/or the growing organism; replacement or regenerative phenomena in wound healing, limb regeneration, and replacement of blood cells or germ cells; growth and differentiation of blood vessels, kidneys, skin, and so on. Development has even been applied to the degenerative, or regressive, processes in aging organs, as in the case of the progressive disappearance of optic nerve fibers, neurons, islets of Langerhans, renal glomeruli, and skin elasticity, along with other phenomena characteristic of aging.

Development takes into account the whole gamut of physiological processes: replication, substitution, growth, interaction, conversion, degeneration, regeneration, differentiation, and aging—including metamorphosis (amphibians),

Figure 3-1 Spermatozoa of the bull. (Courtesy General Biological Supply Co.)

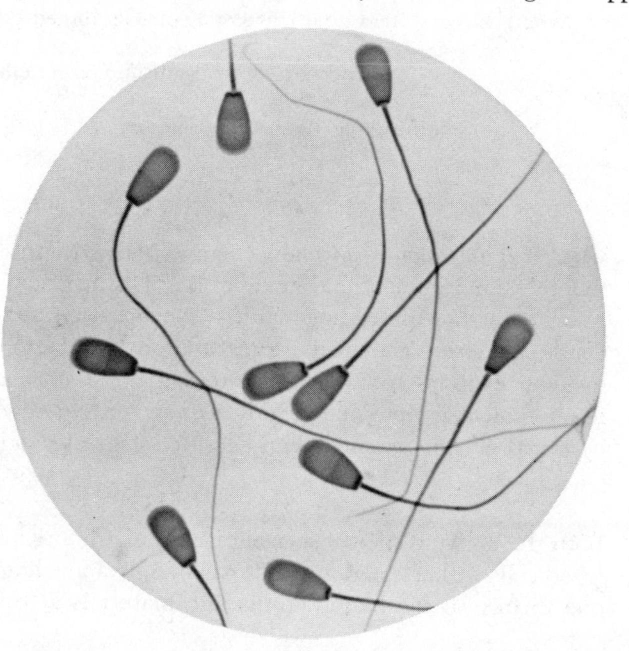

Table 3-1 Course of Progressive Development of an Organism. Each Stage Is Precisely Structured.

Female	*Male*
Oogonia (ovary)	Spermatogonia (testis)
Oogenesis	Spermatogenesis
Egg	Sperm

Fertilization: Conception. Oocyte is activated; paternal genotype added; shifts within the zygote cytoplasm.

Zygote: Genetic endowment from each parent.

Cleavage: Cell increase in number. Presumptive organization made cellular; no increase in embryonic mass.

Blastula: Hollow ball of cells.

Gastrulation: Cell movements and inductions. Germ layers appear. Cells shift to new locations in the embryo and cellular fates become determined.

Gastrula: A new structural pattern. Endoderm, ectoderm, and chordamesoderm visible.

Neurulation: Central nervous system develops in the neural plate.

Neurula: Other organ primordia appear from the germ layers.

Differentiation and Growth: Structure and function of tissues and organ systems. Body takes shape.

Embryo: Culminates in hatching or birth.

Postnatal Development: Continued growth and differentiation.

Sexually Mature Organism: Germ cells for the next generation.

Aging: Regressive changes.

Death

(Cycle Is Completed)

when striking changes normally occur in body structure as the organism adjusts to a new environment and mode of life. Disturbance of developmental processes at any stage preceding maturity may result in abnormalities, or even early death. Numerous chemicals (drugs) and physical factors (X rays) to which man may be exposed today for one reason or another are well known to be teratogenic in nonphysiologic doses. Even vitamin A, when taken in large doses by the pregnant mammal, can cause cleft palate and spina bifida.

Gamete Level of Organization; Sperm Both the egg and sperm (spermatozoan) of any organism are precisely structured cells with a functional life span of only a few hours unless they unite; the one to furnish the basic cytoplasmic materials at this level of organization along with the female genetic material, the other to activate the egg and contribute the paternal inheritance. Mammalian sperm possess certain characteristics, such as a short period of sustained independent movement, that, in association with transport and maturation mechanisms within the male and female ducts, enable them to reach the oviductal ampulla where union with the egg (ovum, oocyte) takes place. Also present are the enzymes facilitating sperm penetration of the cellular layers (cumulus and corona), membrane (zona pellucida), and egg itself to initiate development of the new individual. Attraction between sperm and egg may be a function of a chemical signal emitted by the egg (clearly demonstrated for many invertebrate eggs, but not for mammals) to which the sperm responds through properties of its plasma membrane which may involve the presence of specific receptors and materials adsorbed by the sperm from genital secretions (*coating antigens*). Quick penetration of the cell layers and membrane around the mammalian egg is the function of the sperm acrosome and its enzymes; the source of energy for motility is the mitochondria of the middle piece; and motility is provided through the sliding action of the microtubular structure of the flagellum.

The groundwork of sperm structure is laid down during the first phase of spermatogenesis, which takes place within the testicular cysts (fish, amphibians) or in the walls of the seminiferous tubules (amniotes: reptiles, birds, mammals), subsequently culminating in cell-like, nonmotile spermatids, each containing the haploid (half) number of chromosomes characteristic of the species (Figure 3-2). Reduction in the number of chromosomes is accomplished during this process by two meiotic cell divisions, during which the chromosomes of the spermatocyte replicate only once. In turn, the spermatid metamorphoses, or transforms, into the functionally motile sperm composed of head, neck, middle piece, and flagellum (tail). After a period of maturation in the epididymis, and even capacitation in the uterus in some species (rabbit), the sperm has the ability to fertilize an egg. This final phase of the metamorphic process may take place in water (fish, amphibians) or in some portion of the reproductive system (mammals). Interest in the development, structure, and aging of these cells; their transportation to the site of fertilization; maturation (capacitation); and mechanism of movement is shown by current research.

Morphology of a sperm is illustrated in Figures 3-1 to 3-4. Sperm of different species exhibit great variability in shape and size, particularly of the head, but the basic structural plan is remarkably similar throughout the chordates and among some of the invertebrates (echinoderms). Certain types may possess accessory structures, such as the vibratile membrane on the flagellum of *Amphiuma* sperm (an amphibian). Length varies from around twenty microns in the crocodile to two millimeters in *Balanoglossus* (a hemichordate). Throughout the chordates, sperm are of the flagellated (tadpolelike) type; ameboid types are encountered among the invertebrates (e.g., the nematode, *Ascaris*). The crab, *Libinia*, possesses three flagella-like components.

Within the sperm head is the nucleus, which in Figures 3-2 to 3-4 has been stained with basic dye preparations and is electrondense, absorbing strongly in the ultraviolet. Capping the nucleus, the acrosome with its enzymes, derived from the Golgi body of the spermatid during spermatogenesis (sperm development), has a conspicuous role in the fertilization reaction. The remainder of the nucleus is enclosed within the laminar argentophilic postnuclear cap. Its enzymes include hyaluronidase, a corona-penetrating enzyme, and an egg

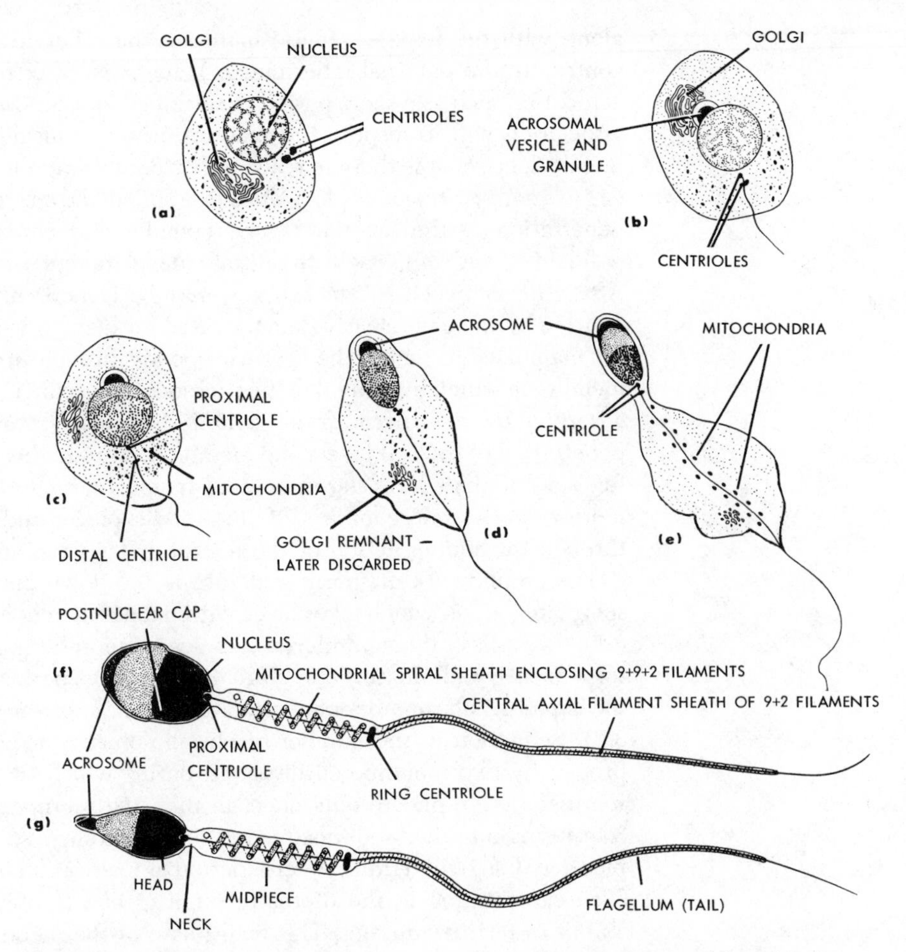

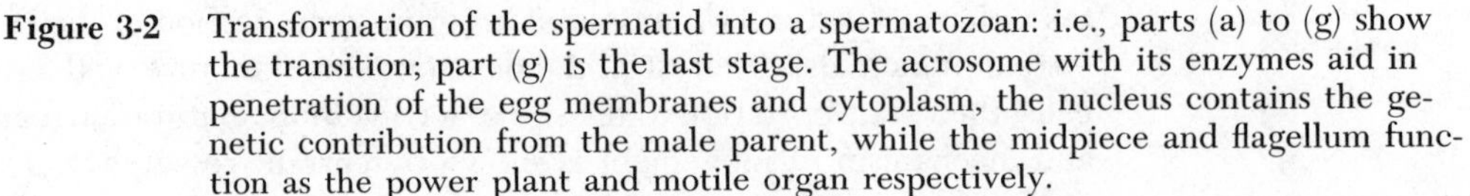

Figure 3-2 Transformation of the spermatid into a spermatozoan: i.e., parts (a) to (g) show the transition; part (g) is the last stage. The acrosome with its enzymes aid in penetration of the egg membranes and cytoplasm, the nucleus contains the genetic contribution from the male parent, while the midpiece and flagellum function as the power plant and motile organ respectively.

membrane lysine, trypsinlike in its action on the zone pellucida. An acrosomal space, or perforatorium (extra large in the rat), separates acrosome and nucleus. This fibrous, or electron-opaque perforatorium, is present in most species, but exhibits great variability in size and shape: large in rodents and hamster, very small in boar, bull, dog, guinea pig, and rabbit, and reportedly composed of scleroprotein in the toad.

The remainder of the sperm structure is involved with motility. In the middle piece and flagellum reside the energy source (mitochondria) and mechanism (microtubules) to aid in bringing sperm to the egg, along with the churning movements, ciliary actions, and secretions of the oviduct. Mitochondria comprising the spiral elements of the middle piece lie just beneath the sperm plasma membrane. Within the circle of mitochondria, the microtubules are arranged in the typical ciliary pattern of two central tubules ringed by nine doublets (9–2). Additionally, the middle piece, in contrast to the flagellum, has a ring of coarse tubules between the doublets and the mitochondria (9–9–2).

Centrioles are present in the neck portion posterior to the nucleus and at the flagellar end of the middle piece.

Current methods of selective staining, pulse labeling, differential centrifugation, electron microscopy, as well as immunological and biochemical analyses are contributing rapidly to our knowledge of sperm structure in relation to function. Sperm can be frozen in suitable media and stored in sperm banks for artificial insemination for a period of time without diminishing their capacity for fertilization (bull, human). In fact, human sperm banks are proving an effective source for the treatment of infertility, and even as a reserve storage place for sperm when planning on vasectomy or sterilization. Artificial insemination is now widely practiced in breeding some domestic animals, especially dairy cattle, eliminating the otherwise useless and food-consuming male. As a result of the current human population explosion, much research is in progress into the biology of male reproduction because no effective, acceptable male contraceptive is known that does not have disturbing side effects. It is now

Figure 3-3 Diagram of the tentative structure of the human sperm: (a) longitudinal section through the head at right angles to the page; (b) section through the middle piece; (c) section through the flagellum or tail.

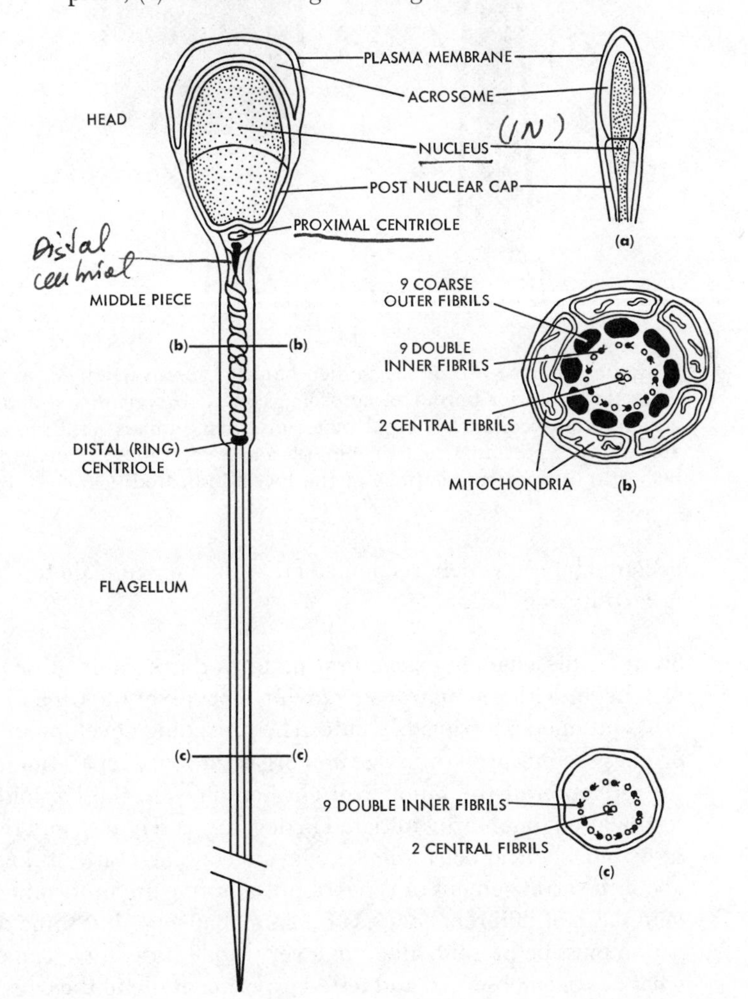

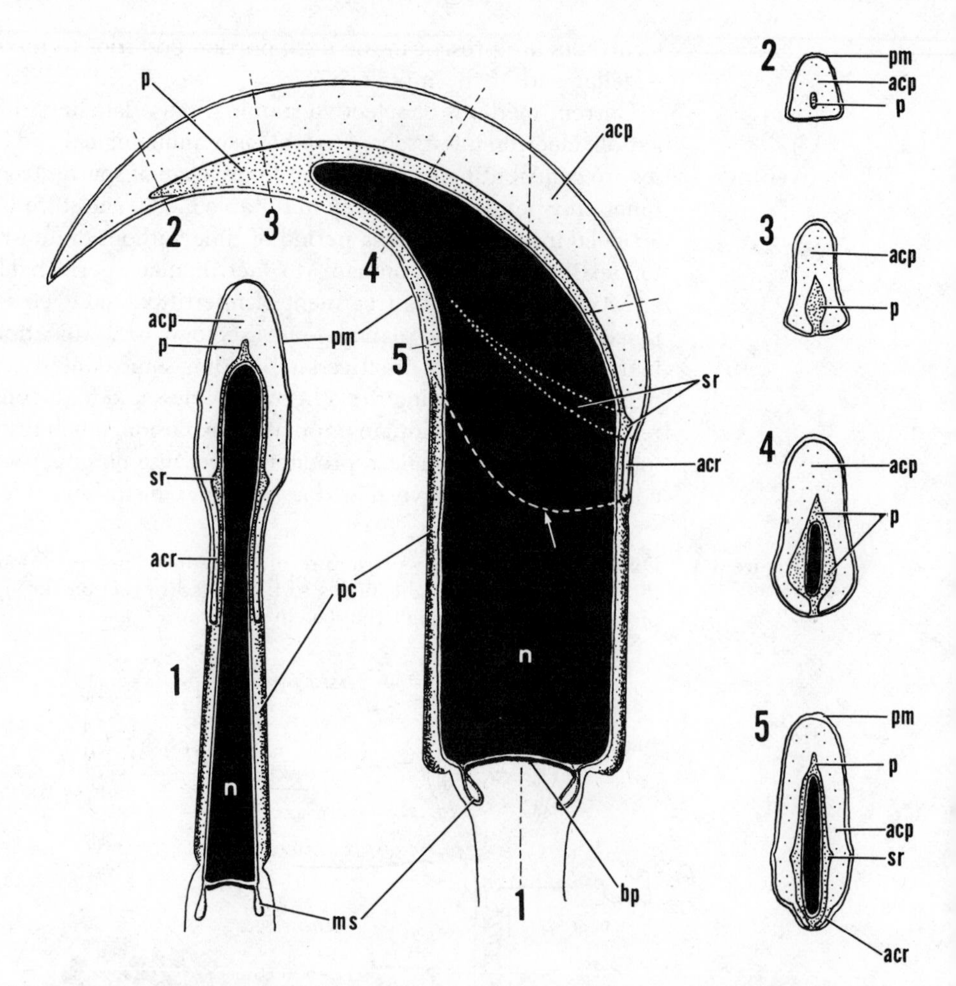

Figure 3-4 Diagrams of structure of the golden hamster sperm head. An arrow (white) indicates the posterior border of acrosome collar. *Abbreviations:* acp, acrosome cap; acr, acrosome collar; bp, basal plate; ms, membranous scroll; n, nucleus; p, perforatorium; pc, post-nuclear cap; pm, plasma membrane; sr, subacrosomal ring. Numbers 1 to 5 represent sections at the levels indicated (Courtesy R. Yanagimachi).

becoming more widely recognized that the human male too has a role to play in fertility regulation.

Egg at Ovulation

Study of the changing structural patterns during animal development might well begin with the mature egg (ovum, oocyte) at the time of ovulation, when it has attained a fertilizable state. The preceding developmental period of the egg has seen its growth in size and organization, preparation during oogenesis for reduction of the number of chromosomes to the haploid condition after fertilization (but hardly initiated in dog, fox, and horse), and the differentiation of its biochemical constituents (enzyme content). Little is known at this time about the arrangement or types of presumptive organ-forming materials in the cytoplasm of different species of mammalian eggs, but some degree of organization must be present, albeit in a very labile state, if we can extrapolate from what has been observed and tested experimentally in the eggs of tunicates and amphibians.

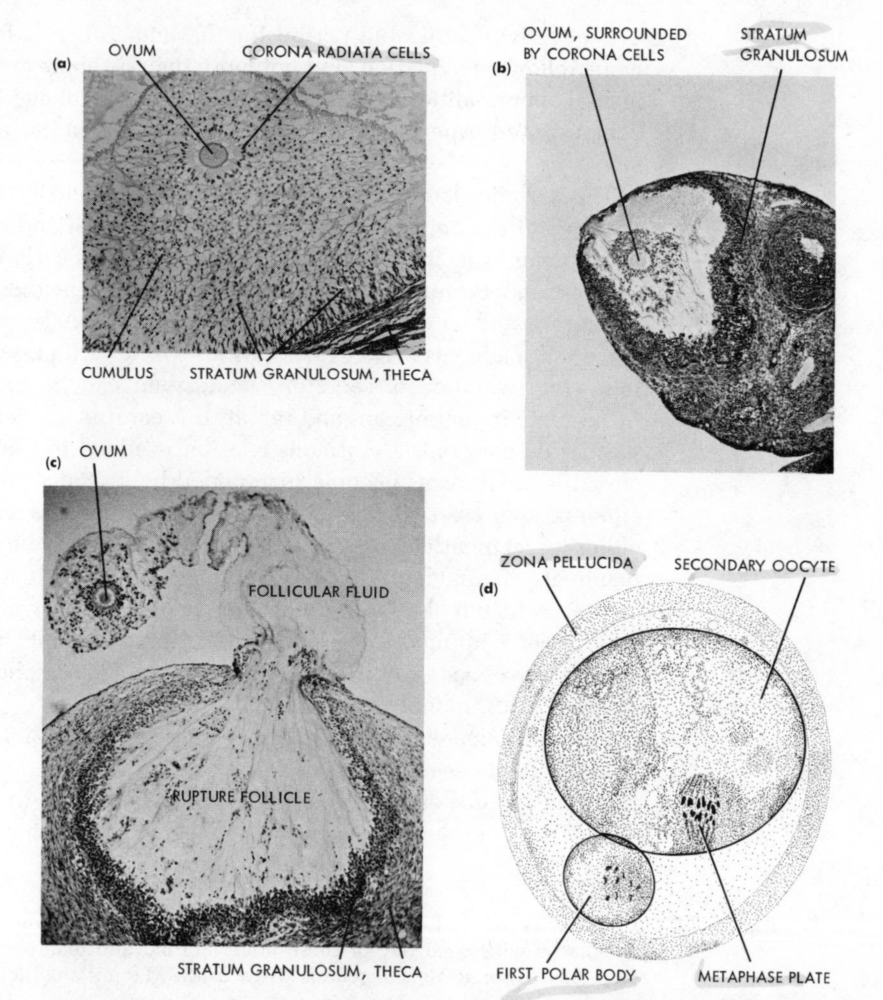

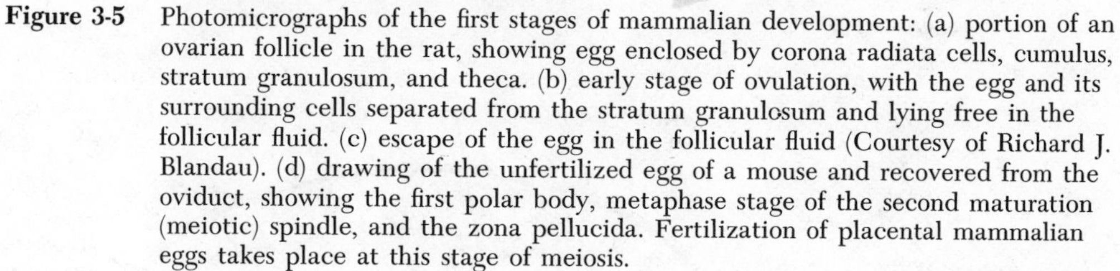

Figure 3-5 Photomicrographs of the first stages of mammalian development: (a) portion of an ovarian follicle in the rat, showing egg enclosed by corona radiata cells, cumulus, stratum granulosum, and theca. (b) early stage of ovulation, with the egg and its surrounding cells separated from the stratum granulosum and lying free in the follicular fluid. (c) escape of the egg in the follicular fluid (Courtesy of Richard J. Blandau). (d) drawing of the unfertilized egg of a mouse and recovered from the oviduct, showing the first polar body, metaphase stage of the second maturation (meiotic) spindle, and the zona pellucida. Fertilization of placental mammalian eggs takes place at this stage of meiosis.

Fertilizable eggs of any animal are complex, structured cells that have advanced so considerably at the subcellular level of organization that they can give rise to a new organism when suitably provoked. The eggs have attained a state of unstable equilibrium that requires only some kind of stimulus to induce resumption of development, usually entrance of a sperm, although in many species eggs may be parthenogenetically activated by one or more methods (amphibians, mouse, rabbit). The stimulus starts the potential organism on its long, irreversible, and progressively complicated pathway of development. A specialized cell with all of its potentialities for producing an organism has been abruptly brought to a halt at the time of ovulation; early maturity or aging of some type has been attained; and the cell will die unless activation occurs soon.

The effective fertilizable period for the human egg is fewer than thirty-six hours following ovulation, and not more than six hours in the rabbit. Developmental abnormalities may result from fertilization of aged eggs. This has been demonstrated experimentally among lower vertebrates and inferred for humans.

Types of chordate eggs, illustrated in Figures 3-5 to 3-8, may be classified on the basis of the amount of contained food material and on whether they are determinate (specifically organized in structure as early as the fertilized egg stage) or indeterminate. All types are intimately enclosed by a living plasma membrane and an external vitelline or primary membrane (zona pellucida in placental mammals). As eggs of various species are passed down the oviduct following ovulation, a secretory acquisition can be made to the primary membrane. In amphibians and rabbit, this consists of a jellylike protein-polysaccharide material; a chitinous chorion is added to fish eggs. Cleidoic (enclosed) amniote eggs become surrounded by an inner nutrient bacteriostatic albumin and, external to this, by an inorganic fibrous shell membrane, plus either parchmentlike or calcareous shells (reptiles, birds, and monotreme mammals). Throughout the chordates, eggs vary greatly in their nutritive yolk content, which is also important in influencing the cleavage pattern. The term *telolecithal* is applied to eggs containing large amounts of this inert material more or less concentrated in the lower or vegetal hemisphere (fish, amphibians, reptiles, birds, monotreme mammals).

For convenience, telolecithal eggs can be classified into two groups: (1) In macrolecithal eggs, the large amount of yolk does not divide, and the embryo forms on one side of the yolk from a cytoplasmic cap—the blastoderm (blastodisc). The yolk becomes enclosed in a cellular yolk sac (as in many elasmo-

Figure 3-6 Unfertilized eggs of frog and hen, showing the primary and secondary membranes associated with each one of these microlecithal and macrolecithal eggs. (From Saunders' *Animal Morphogenesis.* New York: Macmillan, 1966.)

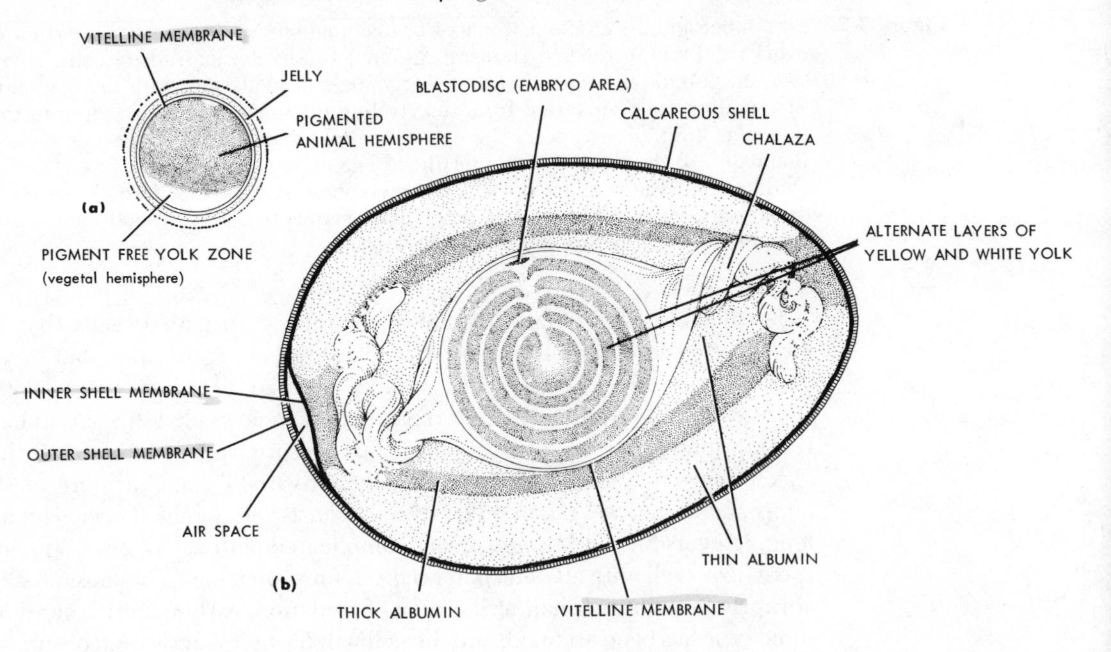

(a)

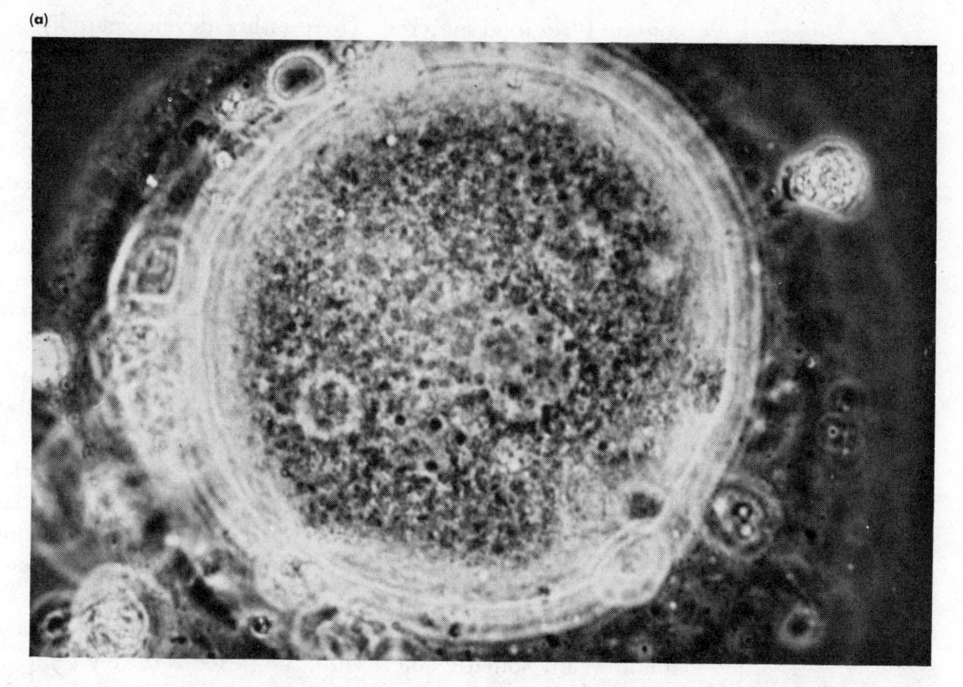

Figure 3-7 (a) Phase contrast photomicrograph of the fertilized human ovum at the pronuclear stage. The zona pellucida membrane encloses the egg. (Courtesy of Stanley R. Glasser, Vanderbilt Medical School. Photographed by the Vanderbilt Medical Center. The specimen was discovered by T. H. Clewe, M.D.)

(b)

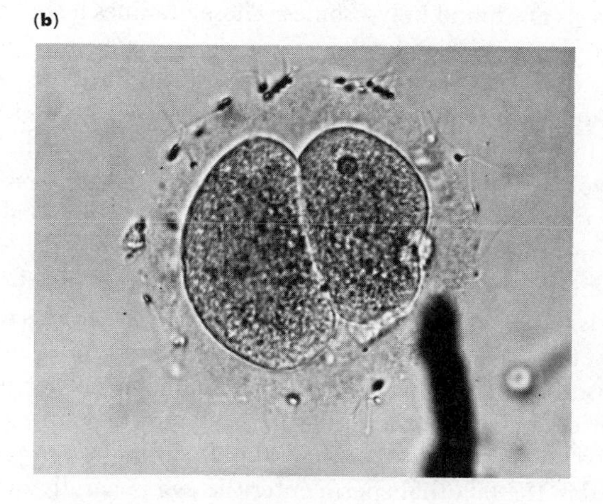

(b) Human two-cell stage, with sperm. (Courtesy of John Rock.)

(c)

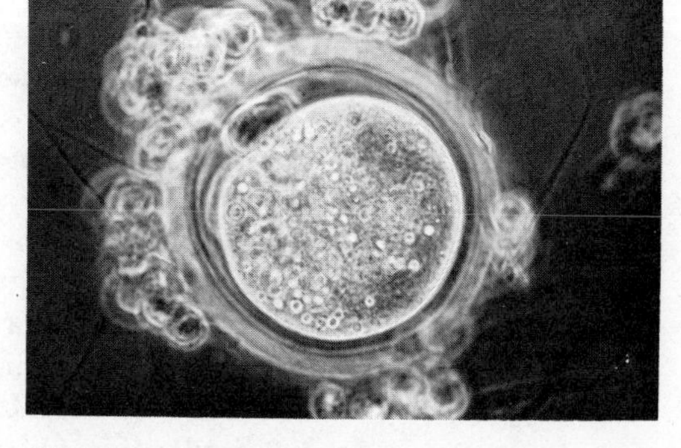

(c) Guinea pig egg with first polar body, unfertilized, enclosed by the zona pellucida.

branchs, most teleost fish, reptiles, and birds). (2) In microlecithal (mesolecithal) eggs, there is relatively less yolk that divides into cells (amphibians, sturgeon). The telolecithal type is widespread among primitive living vertebrates, and it may have been the ancestral condition.

Isolecithal (oligolecithal) eggs of primitive chordates, marsupials, and placental mammals (Figure 3-7), the smallest of the three types of chordate eggs, contain little food material. These either develop rapidly to a feeding larval stage (protochordate species), or the early embryo (mammal) soon acquires a nutritive connection with the mother via the placenta. Although in most respects mammalian development is distinctive, it has long been recognized that mammals have probably descended from reptilianlike ancestors that possessed heavily yolked eggs. One reason for this belief is that the early human embryo, lacking any yolk, possesses for a time a vestigial yolk sac.

When chordate eggs are mature and ready for fertilization, they are at some stage of the oogenetic (meiotic) cycle—generally at the metaphase stage of the second of the two divisions (Figure 3-5). The first polar body has been given off and division for the second has been initiated in the secondary oocyte. At this time the structural organization shows a cell rich in RNA, as demonstrated by numerous cytochemical studies. Eggs stain deeply basophilic with basic dyes (e.g., egg of the placental mammal), and the stainable material can be removed with the enzyme ribonuclease. The several RNAs differ relatively in eggs of different species, with ribosomal RNA particularly abundant in amphibian eggs. Little is known about the origin or pattern of aggregation of this material or its distribution within the cell. Cytoplasmic DNA has been reported. Yolk is present as granules of lipid, protein, lipoprotein, and fat.

Within the egg cytoplasm, in addition to yolk, are most of the organelles found in any somatic cell (Figure 3-8), except for the temporary absence of a nucleolus, such as ribosomes, lysosomes, pigment granules, endoplasmic reticulum, mitochondria, and Golgi apparatus. In contrast to somatic cells, the scattered endoplasmic reticulum is poorly formed, with varying numbers of ribosomes lying free or attached to it. Granules in the outer cortical cytoplasm of the egg beneath the plasma membrane contain the same enzymes, with similar staining reactions, as are found in lysosomes. These granules have been described as submicroscopic, saclike organelles about one-quarter micron in diameter, containing enzymes that break down carbohydrates, proteins, and fats. All this organization has been affected during the ovarian growth of the egg in its development from an oogonium of germinal epithelial origin.

The concentration of yolk in the lower hemisphere of telolecithal eggs points to a visible polar organization: the yolk-free upper cytoplasmic region is called the animal hemisphere and the yolk-laden lower region is the vegetal hemisphere. An imaginary line drawn through the centers of these hemispheres delineates the central axis of the cell. Internal organization relates to this axis; it may be radially symmetrical like spokes of a wheel from the hub, as in the frog's egg, or bilateral to either side, as in the hen's egg. Symmetry may change upon fertilization. Other evidence of structural polarity in any type of egg is seen in the location of the nucleus toward the animal pole, formation of polar bodies in that region, and in the fact that sperm enter the egg generally in the animal hemisphere. In the heavily yolked eggs of amniotes (reptiles, birds, monotreme mammals), cleavage is restricted to the yolk-free blastoderm.

Within the cytoplasm of the egg there is some additional type of cytoplasmic organization of ground substances, the precursors of the embryonic germ layers of ectoderm, endoderm, and chordamesoderm; a pattern of organization that will change first after fertilization and later in response to morphogenetic movements and inductive actions during gastrulation. As development progresses, this cytoplasmic organization will be controlled by genes from the

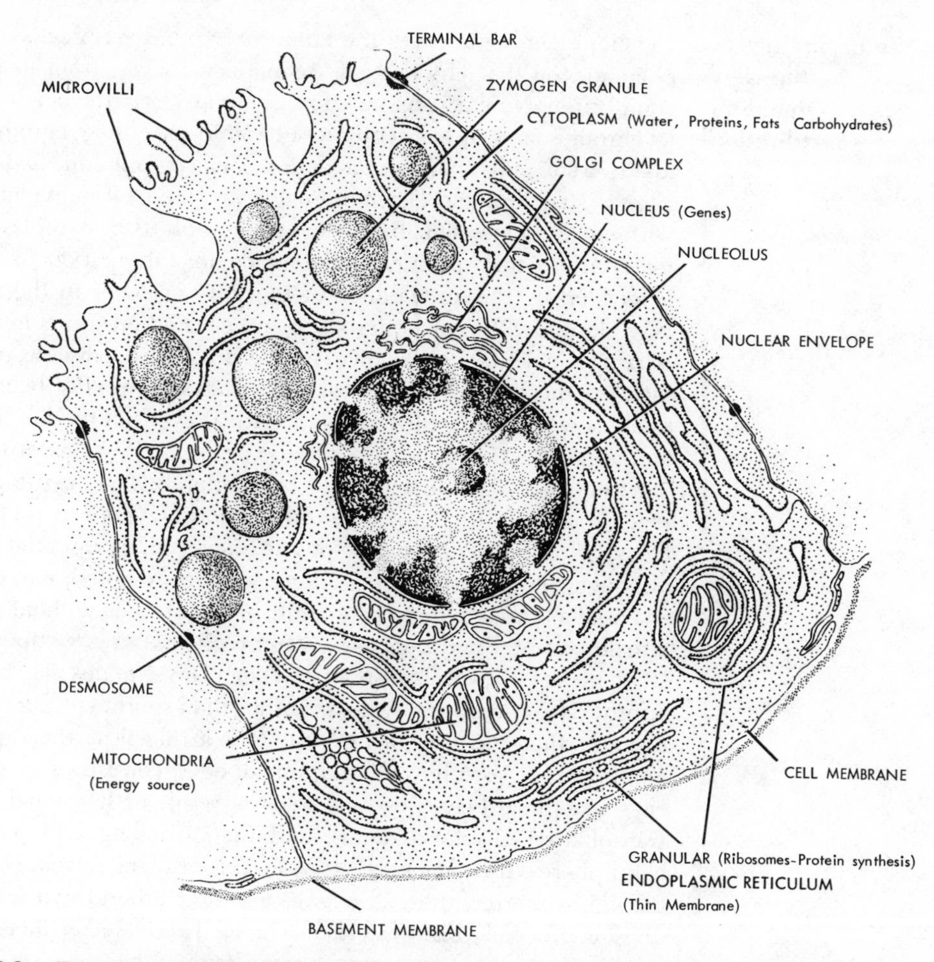

Figure 3-8 Diagrammatic drawing of a secretory-type body cell of the adult organism showing the several types of organelles found in somatic cells. A number of the organelles may be found in the germ cells. (After C. H. Waddington.)

parents. Patterns of these regional precursor areas, established genetically during the period of embryonic growth, vary slightly in different types of eggs, but in most cases they are not irreversibly determined at this time. Certain areas of the egg may be marked by the presence of several colors of pigment (tunicates, amphibians) that identify presumptive germ layer precursors; these will later change position within the young embryo. It must be kept in mind, however, that pigment serves only as a convenient marker, bearing no structural or functional relationship to the formative materials that characterize the different presumptive areas. Wide variation exists in the size of chordate eggs from placental mammals (including human)—about 0.1- to 0.2-millimeters in diameter without the zona pellucida, or just at the limit of visibility for the human eye, to the markedly telolecithal ostrich egg with its large nutritive yolk content. There is slight relationship between the size of the mammalian egg and the adult animal from which it comes. In any case, from the egg emerges the entire organism for the next generation.

The Beginnings of the New Organism: Fertilization

Development begins with the union of haploid germ cells, egg and sperm, a phenomenon that jars the egg (secondary oocyte) from its state of unstable equilibrium. Now certain new forces are at work: the egg has lost one half of its chromosomes; it has acquired a haploid set of new chromosomes from the male parent; and some reorganization of egg cytoplasmic materials occurs. The genetic structure and chromosomal sex of the new individual are established with equal endowment from each parent. Shifts of cytoplasmic presumptive materials to new positions take place in anticipation of development, as evidenced by the appearance of the gray crescent in the frog's egg where gastrulation will later be initiated and colored crescents in the tunicate egg. Barring accidental disturbance (such as drug intake, disease, or unusual genetic factors), the basis of structural organization normally becomes irreversibly established. Thus, fertilization is a critical event in the life history as the fate and appearance of a new individual are sealed by this process.

After sperm entrance, the second polar body is extruded as the second meiotic division is completed, and the ootid (zygote) is prepared for the first cleavage division. The zygote possesses everything essential for development, although the unfertilized egg alone, as already noted, may contain sufficient essential material to form a complete organism even without contribution from a male parent. Sperm entrance can be likened to any stimulating agent that triggers the basic mechanisms in the egg. This stimulus may be furnished under appropriate conditions by means other than sperm entrance, e.g., by artificial means, either chemical or physical. One artificial parthenogenetic technique, employed successfully with amphibian eggs, consists of spreading blood over eggs that have been newly shed as a result of hormonal stimulation from transplanted pituitary glands, and then pricking each egg in the animal hemisphere with a fine needle. When this pricking introduces a blood cell into the egg's interior, the cell acts as a center around which the first cleavage division proceeds. Fish eggs also have been activated artificially by a variety of techniques, and parthenogenetic development has been reported for chicken, turkey, mouse, and rabbit.

Gametogenesis is another critical process in reproduction, contributing to the formation of the total genotype of the fertilized egg as chromosomes are segregated in a random manner during spermatogenesis, as well as between oocyte and the polar bodies. Errors of gametogenesis include the production of multinuclear and giant gametes, sperm with two heads or tails, and chromosomal anomalies resulting from scattering or lagging chromosomes on the meiotic spindle leading to subnuclei. Chromosomal nondisjunction after pairing and replication, reciprocal translocations between members of chromosomal pairs, and polar body inhibition are other types among the many that have been reported. Examples include Klinefeltor syndrome; XXY or XXX sex chromosome complement in the fertilized egg; Turner syndrome, XO; and Down syndrome, trisomy 21 (mongolism). Among other anomalies of fertilization are polyspermy, failure of pronuclear development, and supernumerary pronuclei. Aged eggs are prone to maldevelopment. Although much study in past decades has been devoted to these two periods, it is surprising how little significant basic information exists about either one.

Pattern Changes at Fertilization

The earliest of the visible morphological changes taking place after sperm entrance is the breakdown of the cortical vesicles, liberating vacuolar material containing sulfonated glycoprotein that is added to both vitelline and hyaline plasma membranes around the egg. Subsequent elevation of the combined vitelline-secretion membrane, now termed the fertilization membrane, leaves the perivitelline space between it and the plasma membrane. The second polar body is pinched off, resulting in an egg with half of the maternal genetic material. In mammals, the perivitelline space appears more as a result of a slight shrinkage of the egg. Sperm and egg nuclear material subsequently form two haploid pronuclei that either unite or come to lie in close proximity, constituting the zygote (diploid) nucleus preceding the first cleavage (mitotic) division.

Streaming movements within the egg may cause extensive reorientation of internal symmetries and ooplasmic substances, including pigmentary markings. In amphibians, there is a withdrawal of melanin pigment from a region opposite the sperm entrance point that creates the gray crescent area in which gastrulation will be initiated later, marking the change from radial to bilateral symmetry in the arrangement of the internal presumptive materials. In the egg of the tunicate *Styela*, colored pigments in the cortex appear to mark presumptive materials that later turn up in the embryonic germ layers (Figure 3-15). Also, future dorsal and ventral sides of the embryo may now be identified, as well as bilateral symmetry. These eggs are examples of the determinative type of localization of organ-forming substances. In contrast, the mammalian egg is more labile, as will be discussed subsequently. Underlying biochemical events taking place at this time are beyond the scope of this chapter, such as protein and nucleic acid synthesis as well as the necessary enzyme systems, unmasking of ribosomes and messenger RNA, activity of the lysosomes, induction of blocks to polyspermy, emergence of the chromosomes, and so on.

Fertilization establishes the diploid chromosomal state and also at this time among many vertebrates the chromosomal sex of the new individual, with either the presence of the male-producing combination of the sex chromosomes XY or XO, or the female-producing combination. In mammals, all mature eggs contain an X chromosome, and the female is said to be homogametic, whereas sperm contain either an X or Y chromosome and are heterogametic. This is not the case throughout the vertebrate group; among fish, amphibians, and birds the female is heterogametic; hence, chromosomal sex is determined by the mature egg.

During the last six decades, efforts have been made to separate X- and Y-bearing sperm, thus controlling the sex of offspring by artificial insemination, which would be a boon for the cattle industry. Although there have been claims of success with certain methods and species, nothing practical has been developed. Research in the area of fertilization and egg and sperm development continues to be active, as an examination of current scientific literature will show, particularly in efforts to reproduce these processes *in vitro*, and in searching for mechanisms to interrupt these processes *in vivo* with the idea in mind of a new male or female contraceptive.

Cleavage Patterns

Cleavage represents more than mere cellular multiplication, as evidenced by the arrangement of blastomeres with respect to each other in forming the different cellular patterns and symmetries seen in early embryos (radial, bilateral). It must be assumed that there is a genetically controlled coordinating mechanism that determines the timing and orientation of each mitotic division to produce these patterns. Furthermore, cleavage converts the presumptive areas of the fertilized egg into a cellular structure that later facilitates the migration of cells during gastrulation when surface cells move into the interior of the embryo, as in the case of presumptive endoderm and chordamesoderm. Cell patterns during the early part of the cleavage period may be experimentally deranged or deleted, or two or more cleaving embryos may be combined, without disturbing normal development in many species—a fact that confirms the labile or relatively undetermined character of such early arrangements in the egg. In other species, however, any manipulation of the embryo results in developmental deficiencies or abnormalities (determinate cleavage). Evidence from nuclear transplants into enucleated eggs (amphibians), separation of blastomeres in eggs with holoblastic division, constriction or pressure of the zygote that results in irregular nuclear distribution (amphibians), and destruction of nuclei point to the genetic equivalence of the gynome of the blastomeres and to their lack of appreciable change during this period. There is no change during cleavage in the original cytoplasmic structural pattern of the zygote, nor increase in total size of the embryo. The first evidence of any critical change within nuclei has been found in nuclei of blastulae (amphibians).

Cleavage patterns appear to be correlated with the developmental patterns of embryos and also with amount and distribution of nutrient yolk that influence the rate and pattern of cleavage in heavily yolked eggs. Isolecithal eggs are characterized by total (holoblastic) and practically equal cell division that results in slight differences in blastomere size between animal and vegetal hemispheres (tunicate, rat, human). Among placental mammals, the cleavage rate and cellular arrangement of the colorless cells are slightly irregular after the first zygote division (Figure 3-9), and cleavage continues as the young embryo is moved down the oviduct by ciliary action and tubal peristalsis. After about three days, the embryo is passed into the uterus as a compact mass of cells, the morula (twenty cells, mouse; around one hundred cells, rabbit).

In amphioxus, whose determinate type of development has long been known, bilateral symmetry is established before fertilization. First cleavage, which passes through the center of this egg, produces two cells destined to become the left and right sides of the embryo. The second cleavage occurs at right angles to the first, and the third cuts the embryo at right angles to the first two and slightly above the equator. This mass of eight cells enclosing a small

Figure 3-9 (a and b) Sections of the ovary of the rat showing (1) developing Graafian follicle with egg; (2) egg enclosed by corona radiata cells and separated from the follicular stratum granulosum, previous to rupture of the follicle during the process of ovulation; (3) early stage in the transition of the ruptured Graafian follicle to a corpus luteum; (4) corpus luteum of pregnancy. (Courtesy of Dr. Richard J. Blandau, University of Washington School of Medicine.)
(1) Portion of a section of the ovary of the golden hamster, *Cricetus auratus,* at 1 a.m. of day 2 of the estrous cycle, showing ovulation of an egg from a Graafian follicle. (After M. C. Ward.) (2)-(4) Cleavage in a hanging drop culture of a two-cell stage of the rhesus monkey recovered about thirty hours after ovulation. (After W. H. Lewis and C. G. Hartman.) First cleavage division in placental mammals is initiated within twenty-four hours following sperm penetration.

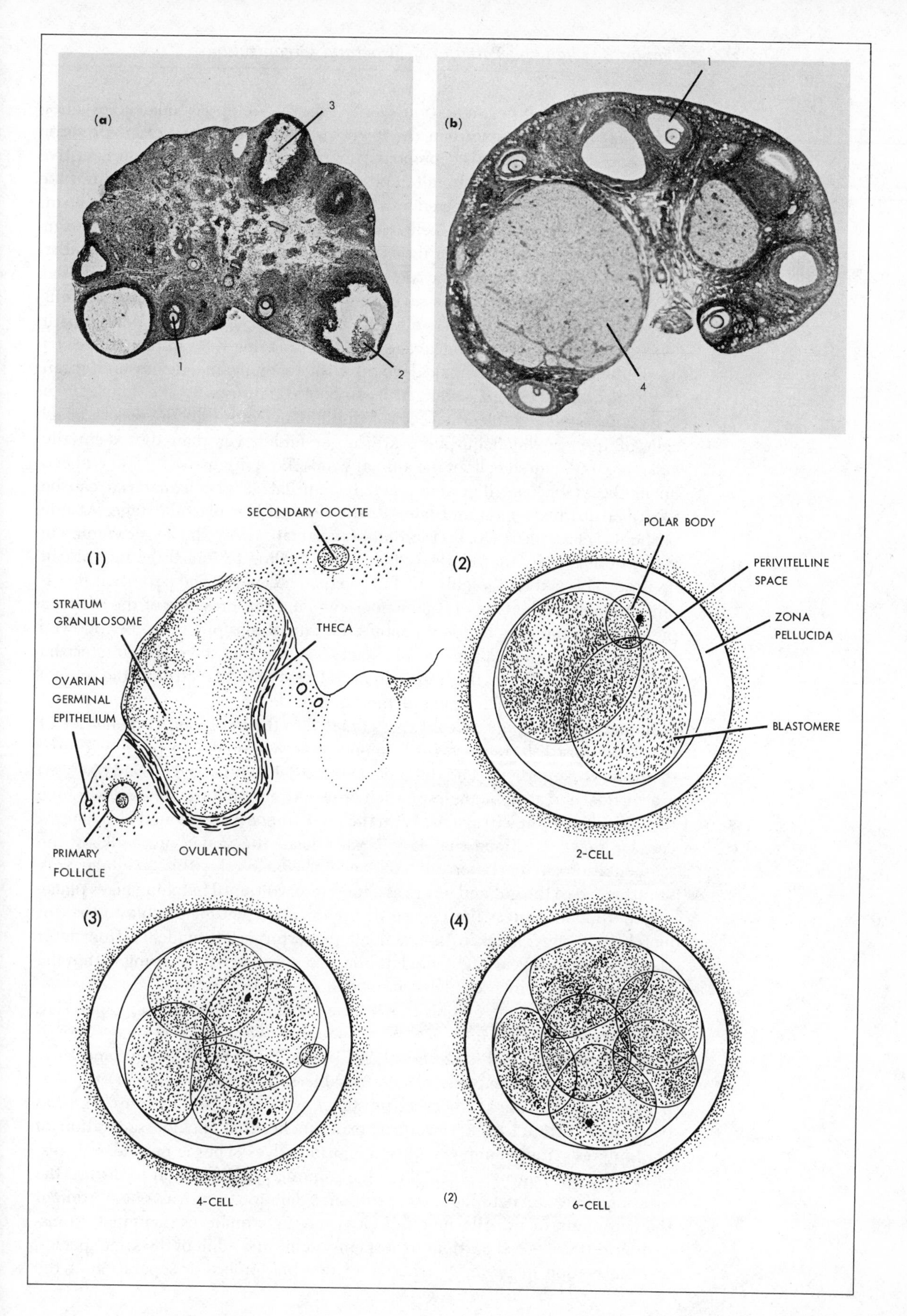
(a)
3
1
2
(b)
1
4
SECONDARY OOCYTE
(1)
STRATUM
GRANULOSOME
THECA
OVARIAN
GERMINAL
EPITHELIUM
PRIMARY
FOLLICLE
OVULATION
POLAR BODY
(2)
PERIVITELLINE
SPACE
ZONA
PELLUCIDA
BLASTOMERE
2-CELL
(3)
(4)
4-CELL
(2)
6-CELL

central cavity, the segmentation cavity, shows an upper quartet of cells somewhat smaller in size than the lower quartet where the yolk is present. Cleavage continues in this geometric progression. The next division is in the vertical plane of each blastomere and results in sixteen cells arranged in two tiers of eight cells each, followed by a horizontal division with the formation of four rows each with eight cells. After the sixty-four-cell stage, cleavage becomes irregular. Cells from these different rows have been traced into the embryonic germ layers, and the removal of cells at any cleavage stage provokes abnormal development whose severity is conditioned by the number of cells removed at any stage. If the egg of the tunicate *Styela*, another determinate type, is centrifuged before cleavage begins, displacing the substances marked by pigments, an embryo may develop with recognizable tissues and organ rudiments, but these are arranged in a chaotic pattern.

By contrast, in microlecithal eggs of amphibians (some fish), the yolk-ladened cells of the vegetal hemisphere are larger and fewer than the essentially yolk-free, pigmented cells of the animal hemisphere (Figure 3-10) and will end up in the embryonic digestive tract. Primordial germ-cell material can be identified in this region and later in the floor of the blastula stage. At any particular cleavage stage, division rate is correlated with the larger volume of yolk in this region, the relatively inert yolk tending to retard the rate of cell division. This is total (holoblastic) but unequal cleavage. The pattern at first is quite regular in the frog or salamander egg, and the presence of the melanin pigment and the gray crescent enables the fate of blastomeres to be followed through gastrulation to the neurula stage. The blastula stage is characterized by a wall several cells thick and a central cavity displaced into the animal hemisphere from which much of the embryo develops.

In macrolecithal eggs, cleavage is confined to the blastoderm on one side of the yolk (most fish, reptiles, birds, monotreme mammals)—a pattern termed meroblastic (superficial) and discoidal (Figure 3-11). First and second divisions are vertical and at right angles to each other. After about the sixteen-cell stage the divisions become irregular. With the appearance of horizontal cleavages, a cellular sheet is formed that is separated from the underlying yolk by the segmentation cavity. Later, the yolk becomes enclosed within a cellular yolk sac attached to the gut and serves as a food reservoir until hatching takes place.

Experiments to test the degree of organization within the blastoderm of meroblastic eggs (chicken) are difficult to perform. All evidence from later cleavage stages reveals right and left bilateral symmetry developing within the blastomeres, as shown by deletion or cell destruction studies.

As pointed out here, organization of the presumptive cytoplasmic materials is less rigidly specified in eggs of some species at the onset of cleavage than in others (mammal, amphibians, chicken). In such instances, the eggs and early cleavage stages are capable of considerable self-regulation (repair) to compensate for any loss of either egg substances or later of blastomeres. This has been demonstrated by a variety of experiments that include separation of blastomeres, vital staining (with noninjurious dyes such as neutral red, Bismarck brown, Janus green B, Nile blue sulfate), centrifugation to disturb the location of certain cytoplasmic constituents, deletion (removal) of a portion of the cytoplasm or of cells, union of cleavage-stage embryos (mammal), transplantation of selected portions to host embryonic and adult of the same species, and cultivation *in vitro*. Destruction of one blastomere or separation at the

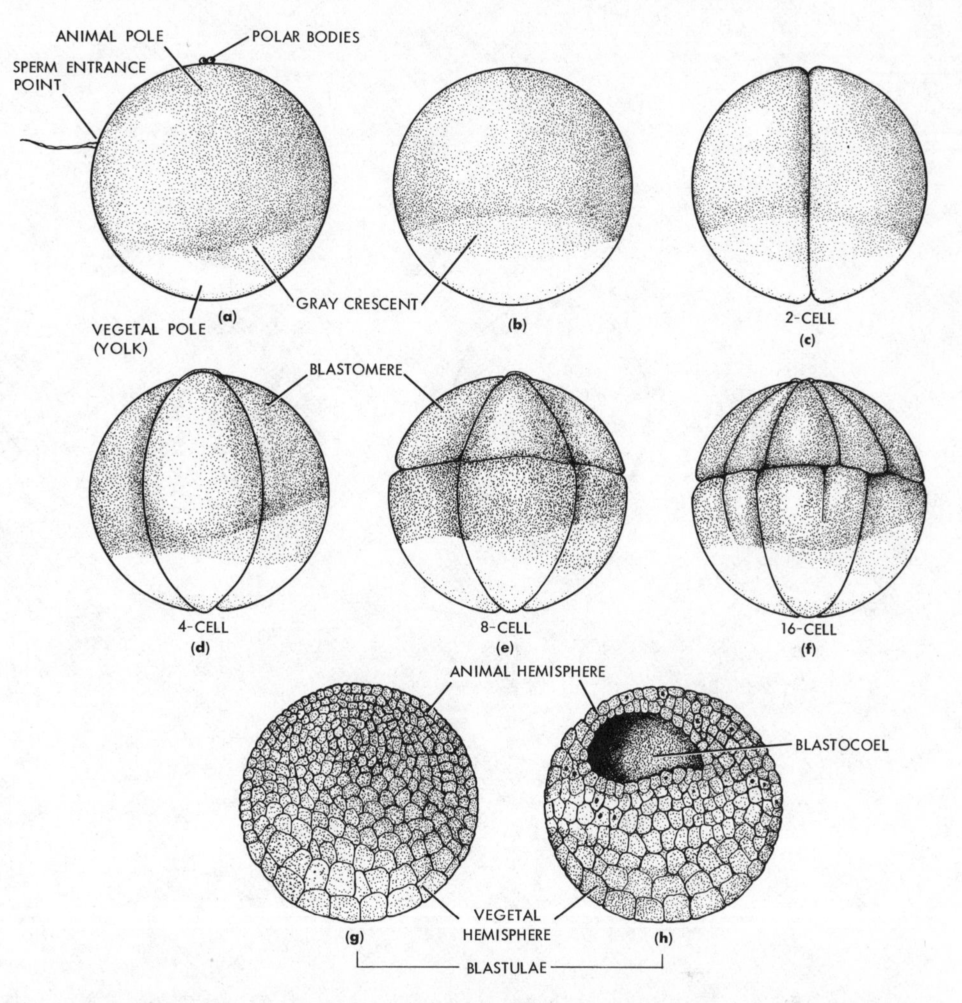

Figure 3-10 Cleavage stages and blastula of the frog. Cleavage is holoblastic but unequal, beginning within an hour following sperm penetration, and the blastula is an unequal coeloblastula. Fertilization and jelly membranes are not shown. Gastrulation will be initiated within the gray crescent area. This type of cleavage is also found in the sturgeon. (From Waddington's *Principles of Development and Differentiation.* New York: Macmillan, 1966.)

two-cell stage (mouse, rat), or even later (rabbit), does not hinder typical development of the survivors as tested by culture *in vitro* or transplantation. Such may be the origin of identical or Siamese twins in the human, or even by other types of separation in older embryos. Regulation of cell size to that of the adult takes place in these cases sometime later in development. It has been reported that even seven cells of the eight-cell rabbit embryo can be removed without interfering with the formation of an embryo from the eighth blastomere. Obviously, no segregation of vital ooplasmic materials has occurred during the first three cleavage divisions in this animal. These and other studies point to a more developmentally labile or undetermined type of organizational pattern of the cytoplasmic structure in such "regulative" eggs, and cleavage is said to be indeterminate.

A special instance of the lability of mammalian blastomeres during early

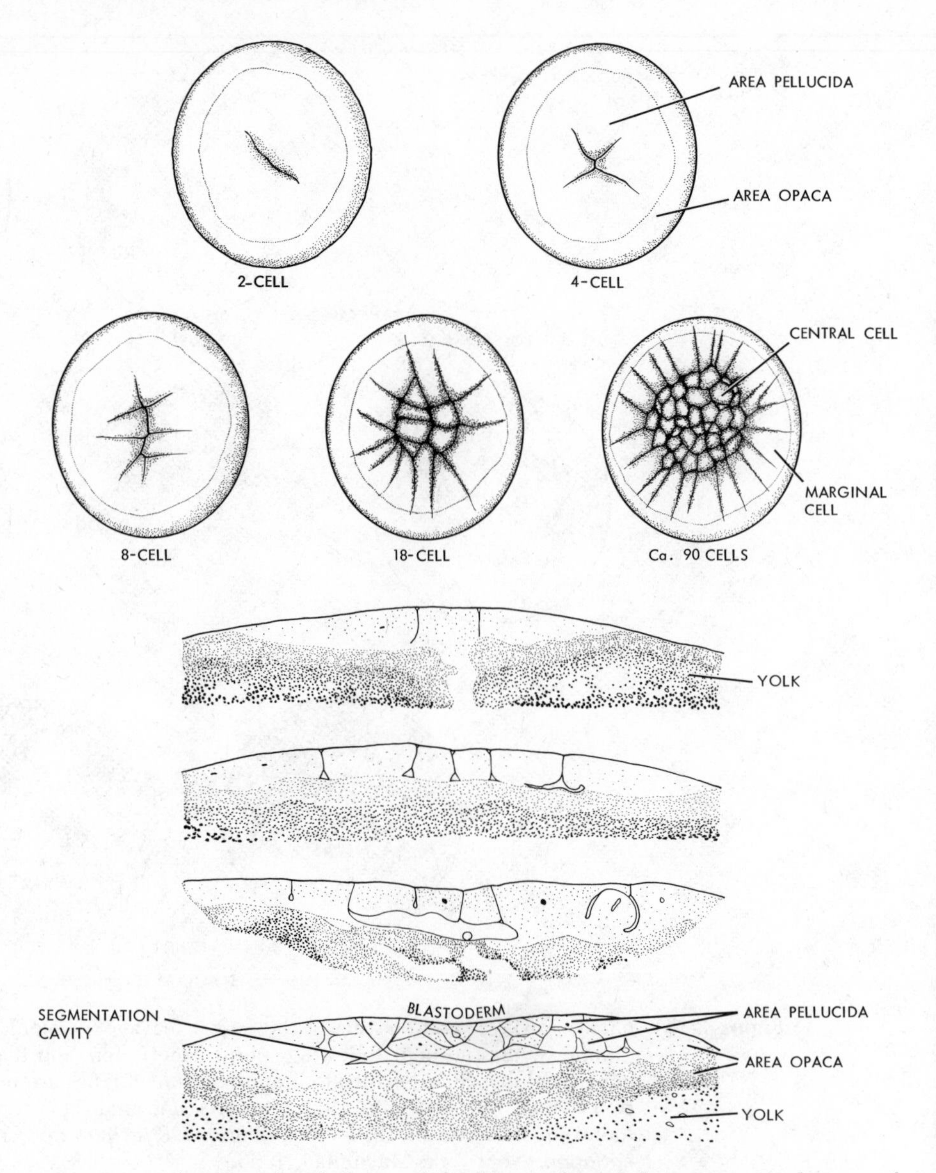

Figure 3-11 Cleavage in the macrolecithal egg of the hen, surface and cross-sectional views of the blastoderm. (After Patterson, 1910.) Through the 32-cell stage the cleavage furrows are all perpendicular to the blastodermic surface. Yolk mass is not shown.

cleavage has been demonstrated by Drs. Beatrice Mintz and K. Silvers [12] who, in their research on immunological tolerance (Figure 3-12), studied "multimice" produced from pairs of conjoined cleavage-stage embryos of the same or different histocompatible genotypes. These individuals of multi-embryonic origin are permanently tolerant of cells of both original parents, but they display a normal and specific immune response to the introduction of a

Figure 3-12 Following removal of the fertilization membrane the cleavage stages of different ova may be combined to form a single embryo, a "multi" or allophenic mouse. When transplanted to the uterus of a stimulated female mouse such cell combinations have given rise to viable young, demonstrating the organizational lability of mammalian cleavage stages. (Courtesy of Dr. Beatrice Mintz.)

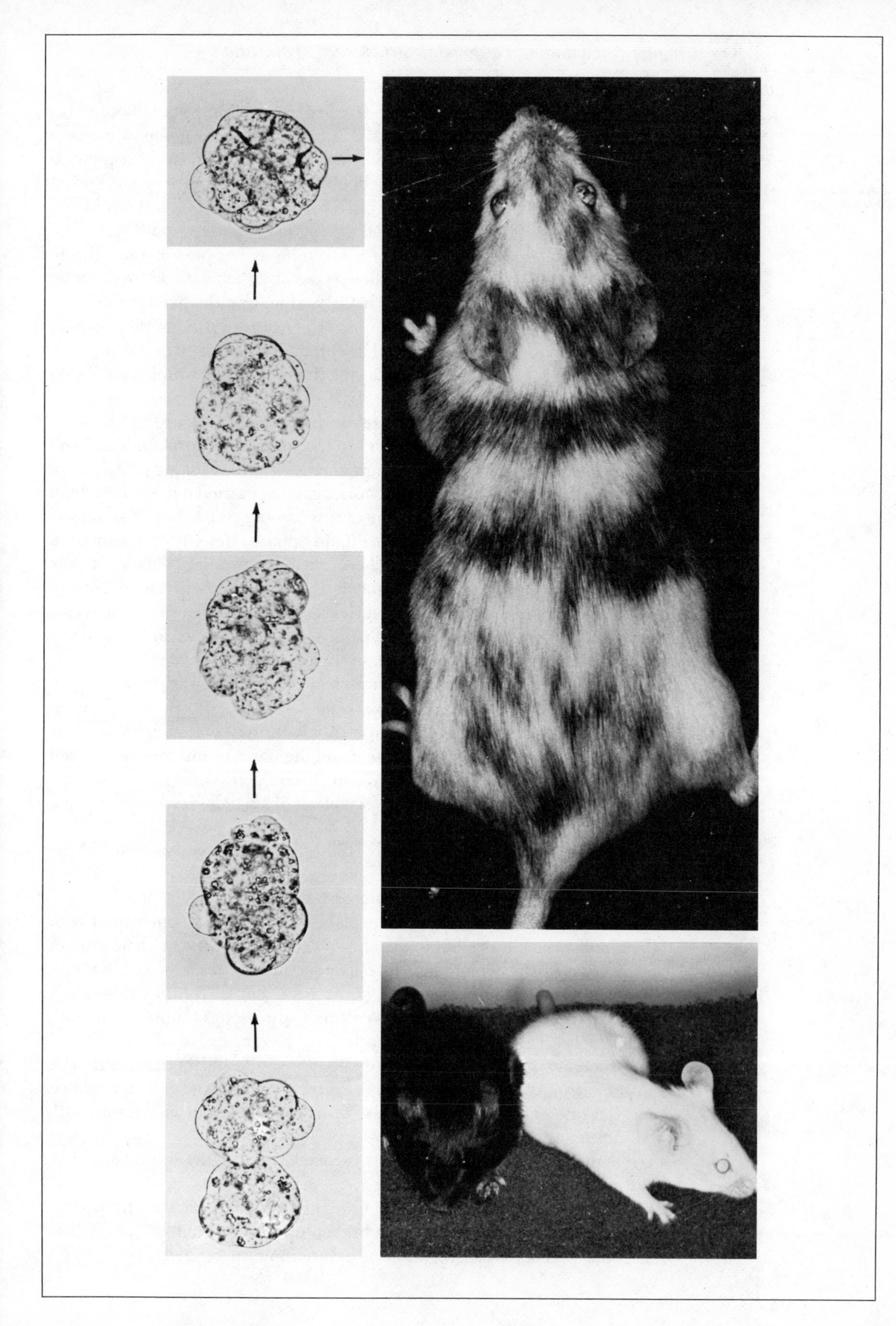

foreign antigen. The animals are formed by first assembling all the blastomeres from two (or more) genetically distinctive embryos into one composite group *in vitro*; later, this group is transferred surgically to the uterus of a pseudopregnant female. Here, regulation from double to single embryos occurs during implantation. Normal development to birth frequently follows, and Mintz and Silvers have obtained healthy adults comprising many pairs of genotypes. The name *allophenic* has been applied to such animals because of the orderly coexistence of cells with different phenotypes ascribable to known allelic genotypic differences. An embryo may have had four or more parents. It would be interesting to know more about immunologic properties as they appear during vertebrate development as well as cell interactions and factors responsible for spotting of body color. Studies of this and related topics are being actively pursued with this material.

Control of cleavage probably operates at the cortical level of egg organization. Changes in the cortex are believed to initiate furrow formation and cell division, with subcortical synthesis of new cortical material taking place before a cleavage furrow forms. Much of this work has been carried out on amphibian eggs. Research has determined that the cortex has the properties of an elastic solid, the egg does not behave like a liquid sphere, the cortex transmits a surface change at fertilization, and there occurs a passive behavior at the cleavage furrow where the cortex appears to buckle inward. These findings have dispelled a number of earlier theories as to the mechanism of cleavage, e.g., the "surface tension theory" of Spek. Attempts to slow down cleavage with mitotic inhibitors have not been enlightening.

In summary, cleavage is more than mere cellular multiplication.

1. Cells ultimately have a definite and predetermined pattern in relation to each other and the embryonic parts to which they will contribute their progeny. According to this, cleavage and the onset of differentiation (cell specialization) occur concurrently and interrelatedly.
2. Cleavage is characterized by the formation of smaller-sized cells, termed blastomeres.
3. There is little or no increase in the total mass of the embryo during the period of cleavage.
4. No rest periods occur between successive cell divisions.
5. Synthesis of nuclear material takes place at a rapid rate, correlated with the number of cells present in the embryo at any one time, with no growth in the size of the daughter cells between successive cleavage divisions.
6. The egg is the largest cell in the body. An eventual return to the adult nucleocytoplasmic ratio occurs as cell size approaches that of adult cells, which is critical for genetic control.
7. The fundamental basis of cell heredity is manifested because cleavage patterns are genetically determined. Patterns of cleavage can be modified and the rate changed by certain extrinsic forces without interfering with the normalcy of the embryo.
8. The amount and distribution of yolk is correlated with the several cleavage patterns.
9. During cleavage the *organ-forming areas* become separated from each other more distinctly and also are broken up into cell units. This emphasizes the precocious segregation of cytoplasmic materials.

10. Cellular architecture of the embryo is initiated, to be subsequently acted on by the processes of gastrulation.
11. Synthetic mechanisms are chiefly those related to mitotic activity. As cleavage progresses, the segmentation cavity appears among the cells in many animal eggs, and soon afterward the embryo attains the blastula stage of development.

Blastula Stage of Development

Attainment of the blastula stage marks the end phase of the period of cleavage. In most chordates, the embryo at this time consists of either a single- (amphioxus, placental mammals) or multicell layer (amphibians) in the form of a hollow sphere or a flat sheet upon the surface of a mass of undivided yolk (most fish, reptiles, birds, monotreme mammals). Differences in the size of cells, pigmentation, and amount of yolk in different parts of blastulae continue to emphasize the specific areas destined to form the various embryonic germ layers from which the tissues and organs of the adult organism will develop. The blastula stage is reached at two days following fertilization in the frog; late during the first day in the incubated hen's egg; and by three to four days in

Figure 3-13 Blastulae. (a) Section of the blastula of *Amphioxus,* location of the germ layer material of ectoderm, endoderm, and mesoderm (after Conklin). (b) Photomicrograph of a section of a frog blastula. Germ cell material has been located among the cells in the floor of the blastocoel. (Courtesy of the General Biological Supply House, Inc.). (c) Section of the blastula of a bird.

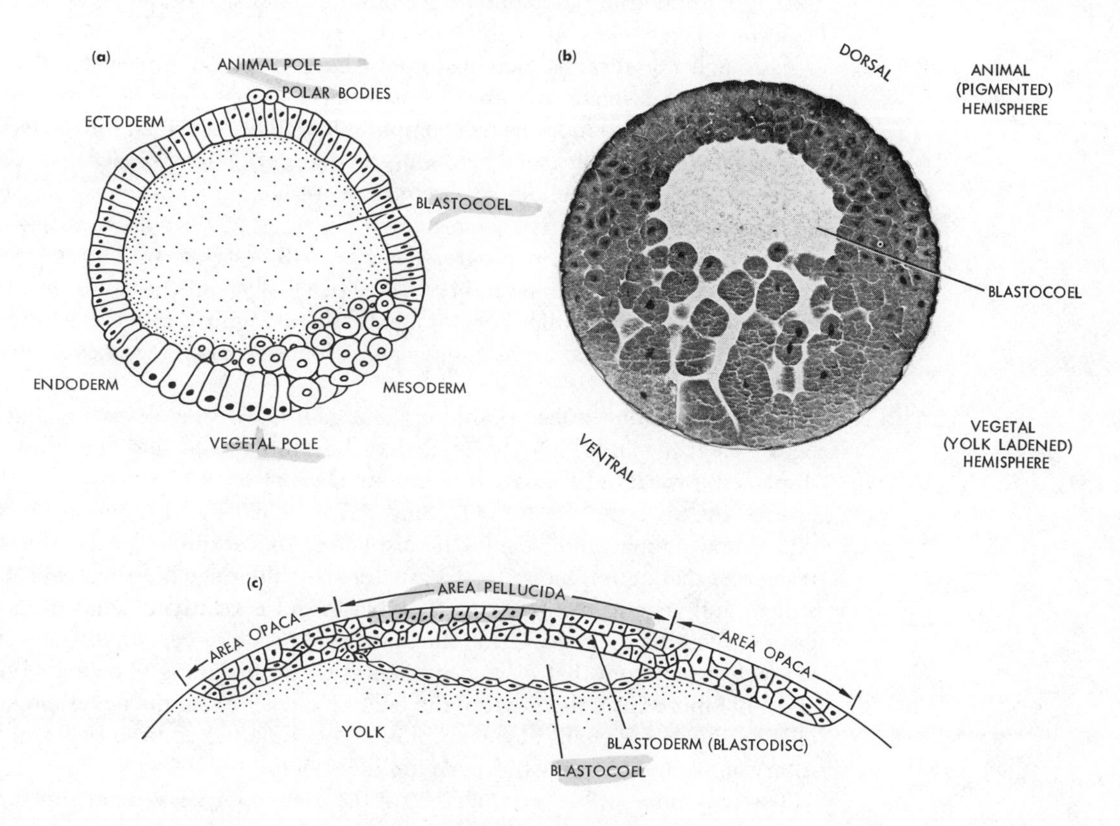

placental mammals when the embryo is entering the uterine cavity. The coeloblastula of an isolecithal egg, like that of amphioxus, is a hollow sphere with a wall one-cell thick, enclosing a cavity, the blastocoel, that contains a chemically complex fluid of cellular derivation (Figure 3-13). It is still surrounded by the fertilization membrane, or zona pellucida in mammals. Polarity continues to be indicated by the presence of smaller-sized blastomeres—micromeres—in the animal hemisphere and larger cells—macromeres—in the vegetal hemisphere. "Fate" maps of the distribution of presumptive formative materials of blastulae among chordates, illustrated in Figure 3-15, show the position of the presumptive ectoderm, endoderm, and chordamesoderm. Knowledge of this type has markedly aided the interpretation of what takes place during gastrulation. The basic structural plan observed in the fertilized egg has not undergone much if any change.

Another example of coeloblastula of the equal type is the mammalian blastocyst. It too is a spherical (human, rabbit) or elongated (pig, sheep) hollow sac consisting of a wall one cell in thickness; the trophoblast (important in formation of the placenta); and an inner cell mass attached to one side of the trophoblast, from which the embryo, amnion, allantois, vestigial yolk sac, and part of the chorionic mesoderm will arise. At five to six days after mating, the spherical blastocyst of the rabbit, like that of the human, has attained considerable size and can no longer be flushed from the uterine tube. Rabbit blastocysts of this age can be obtained easily by gently tearing open the uterine tube from one end to the other while the tube lies immersed in mammalian physiologic solution. Implantation into the uterine wall has not yet been initiated. Endoderm is delaminating from the under side of the inner cell mass to form a layer one cell in thickness (Figure 3-14).

Although not directly demonstrated, there is reason to suspect that the pattern organization of presumptive materials within the inner cell mass at this time is bilateral and more or less comparable to that described for the earliest blastoderm of the chick embryo. One to two days later *deciduation* (disappearance of the zona pellucida) occurs, and uterine implantation is initiated at a time when the definitive primitive streak stage of development has been reached, gastrulation is in progress, and a small vestigal yolk sac is present. Implantation of the human embryo takes place at about the same time after fertilization and at a similar developmental stage (Figure 3-29). Descriptions of other types of mammalian blastulae can be found in the references at the end of this chapter.

Unequal coeloblastulae characterize organisms with holoblastic, but unequal, cleavage in which the wall may be mutilayered and the blastocoel displaced toward the animal pole. Among chordates the blastulae of amphibians are of this type (Figure 3-13) and show a clear-cut distinction between pigmented animal and vegetal hemispheres. In certain respects, this type resembles that of amphioxus and a greater size difference between cells in the animal and vegetal hemispheres. The blastocoel is relatively small. Analyses have shown that the fluid within the blastocoel possesses certain distinguishing characteristics from that of environmental fluids, including chemical content and a pH more basic than that of the enclosing cells. With the development of cilia on the surface cells of the amphibian blastula, the embryo begins a slow rotary movement within the perivitelline space.

The only perceptible organization of the amphibian blastula at this time is

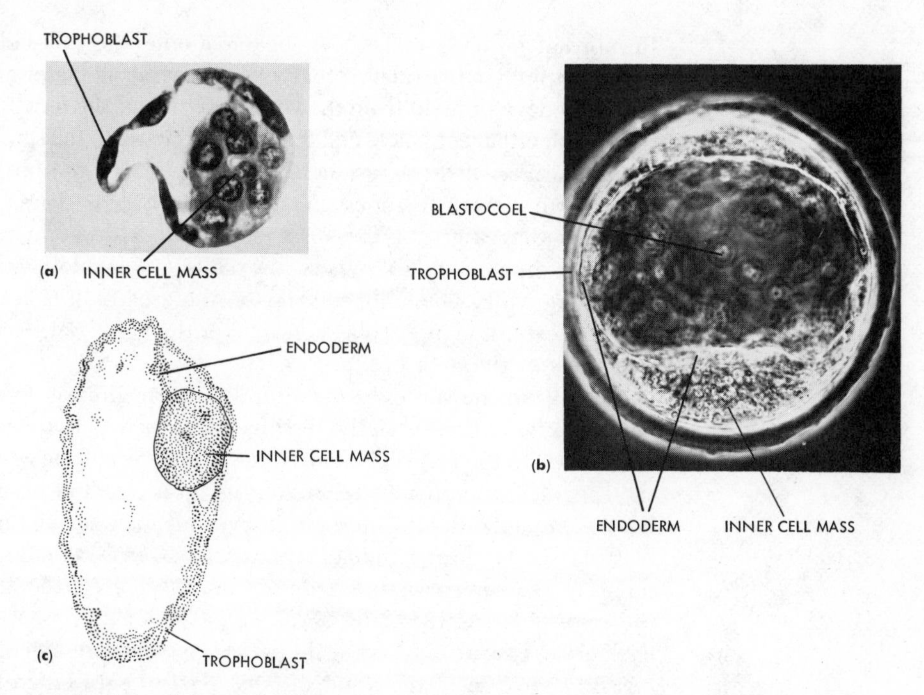

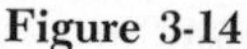
Figure 3-14 (a) Section of the early blastocyst of the guinea pig, six days following fertilization. (b) Living blastocyst of the guinea pig, enclosed by the zona pellucida; endoderm lines the blastocoel. (c) Section of a sheep blastocyst, 10 days and 22 hours of age. (Courtesy of Richard J. Blandau, and of W. W. Green and L. M. Winters.)

reflected in the rapid multiplication of the small cells in the gray crescent region and in the initiation of cell growth in the margin of the *pigmented region* along with its downward movement over the yolk-laden cells of the vegetal hemisphere (*process of epiboly*). This region of overgrowth, the germ ring, will continue to spread downward until all the yolk-laden cells eventually are enclosed by pigmented cells. Indirect evidence of structural organization in the blastulae of different species of amphibians has been obtained by staining the surface cells and following their subsequent movements during gastrulation. This, along with deletion and transplantation studies, has demonstrated a definite arrangement of organ-forming materials in areas in the wall, including the gray crescent area, which mirrors the structural pattern of the fertilized egg (Figure 3-15).

Although such fate maps may differ somewhat in pattern between different genera of amphibians, the basic arrangements are comparable. Some of the cellular areas, such as the presumptive mesoderm and notochord, will later move to the interior during gastrulation; others, such as the presumptive pigmented neural ectoderm, will spread over the dorsal surface of the embryo. It should be reemphasized that although these areas will definitely appear in certain embryonic tissues in normal development, they are far from being irreversibly destined as to types of tissue into which they will differentiate. Neural ectoderm normally ends up in the nervous system, but at this stage its fate is labile, or loosely determined, and can be channeled easily into other types of tissues through appropriate inductor (evocator or releasing) influences.

Primordial germ plasm has been located among the yolk-laden cells in the floor of the amphibian blastula. Actually, this generative material has been reported earlier in development from the vegetal region of the fertilized amphibian egg.

Macrolecithal eggs have cell division restricted to the small blastodermic cap on the surface of the large inert mass of yolk. In the bird, for example, the blastoderm shows two zones: a central lighter-colored area pellucida, beneath which is the cavity of the blastocoel, and a darker peripheral band, the periblast or area opaca, where the cells are in close association with the underlying yolk. The embryo forms in the area pellucida while a subsequent outward extension of the blastoderm over the yolk continues by growth within the zones of the area opaca.

Thus far in the survey of structural patterns during development we have noted the types present in the fertilized egg, as well as the ways in which these are related to the respective axes and symmetries of the egg. During cleavage, the specific components are broken up into a cellular organization without much reorganization or increase in over-all mass of the embryo. Even at the blastula stage, these original structural patterns remain approximately the same. In the next period of gastrulation, there are striking alterations within the areas of formative materials as well as in location, shape, and neighbors of each area. Toward the end of this phase, the earliest signs of the basic characteristic structure of the adult organism will begin emerging from the germ layers in the form of future tissue and organ primordia. Additional primordia will continue to appear and undergo differentiation in structure and function as the body of the embryo, developing in an orderly fashion, takes shape. There will be no subsequent relocation of the structural units.

Gastrulation: The Beginning of Differentiation

Gastrulation is another critical time in early embryonic development. It is characterized by profound morphogenetic cell movements accompanied by tissue interactions of a determinative (restrictive) or specializing nature. Labile potency is changed to a specific fate that is irrevocable and unchangeable. As development progresses this will occur sooner or later to other tissues and organ primordia. At this time, three embryonic germ layers make their appearance: ectoderm, endoderm, and chordamesoderm. In the late blastula stage, the surfaces of cells in the presumptive areas begin exhibiting intense activity both individually and in relation to one another. This activity, which continues in cells as they move to other locations, is not clearly understood. Certain surface cells are, thus, brought into the interior of the embryo (chordamesoderm), whereas others spread out to occupy the areas just vacated

Figure 3-15 "Fate" maps of the presumptive tissue-forming areas in early stages of chordate embryos. (a) Areas of embryonic tissues in 2- and 4-cell stages of the tunicate, *Styela.* (b) Teleost blastoderms: (1) *Fundulus,* (2) trout. The undivided yolk is not shown (after Oppenheimer, 1947; Pasteels, 1936). (c) Early frog gastrula. Dorsal lip of the blastopore (future posterior end of the embryo) is used as a point of reference. (d) Chick blastoderm previous to appearance of the primitive streak (after Waddington). (e) Bilateral location of several organ-forming areas in the chick blastoderm at the head process stage (after Willier and Rawles, 1935; Rawles, 1943). Germ layer materials are assuming final positions as cell movement takes place through the primitive streak (see Figure 3-22).

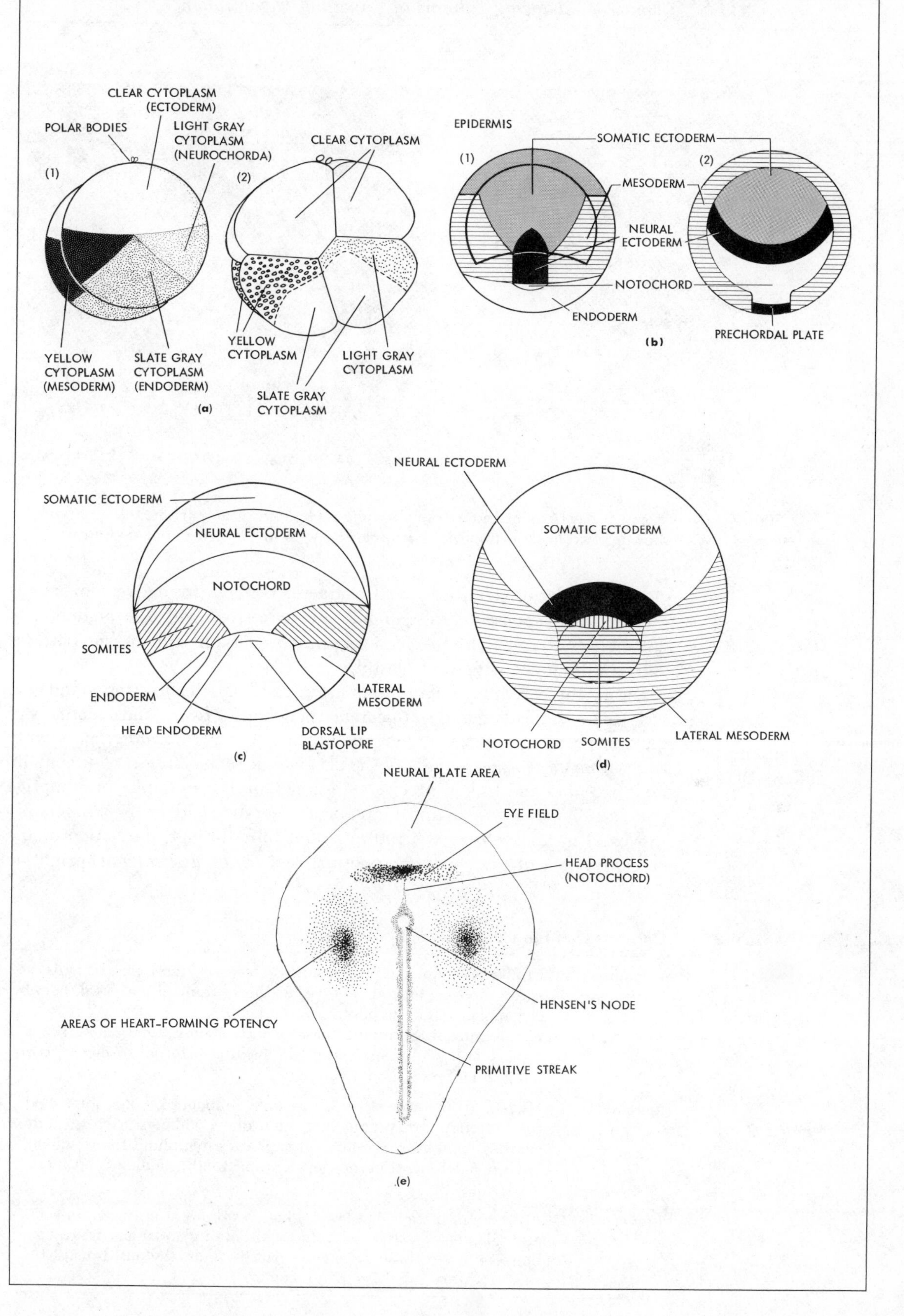
CLEAR CYTOPLASM
(ECTODERM)
POLAR BODIES
LIGHT GRAY
CYTOPLASM
(NEUROCHORDA)
(1)
(2)
CLEAR CYTOPLASM
YELLOW
CYTOPLASM
(MESODERM)
SLATE GRAY
CYTOPLASM
(ENDODERM)
YELLOW
CYTOPLASM
LIGHT GRAY
CYTOPLASM
SLATE GRAY
CYTOPLASM
(a)
EPIDERMIS
SOMATIC ECTODERM
(1)
(2)
MESODERM
NEURAL
ECTODERM
NOTOCHORD
ENDODERM
PRECHORDAL PLATE
(b)
SOMATIC ECTODERM
NEURAL ECTODERM
NOTOCHORD
SOMITES
ENDODERM
LATERAL
MESODERM
HEAD ENDODERM
DORSAL LIP
BLASTOPORE
(c)
NEURAL ECTODERM
SOMATIC ECTODERM
NOTOCHORD
SOMITES
LATERAL MESODERM
(d)
NEURAL PLATE AREA
EYE FIELD
HEAD PROCESS
(NOTOCHORD)
HENSEN'S NODE
AREAS OF HEART-FORMING POTENCY
PRIMITIVE STREAK
(e)

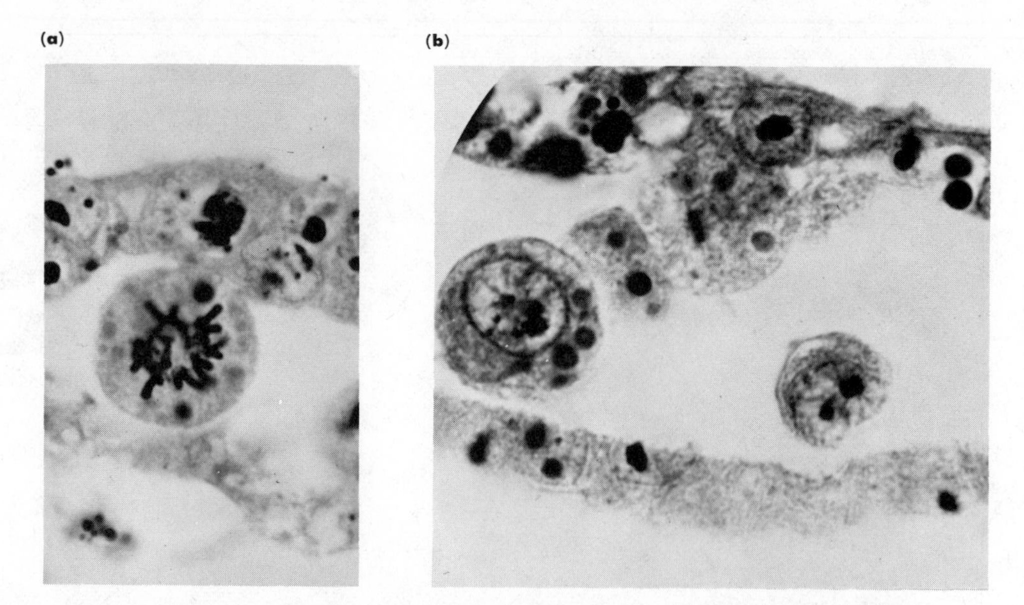

Figure 3-16 (a and b) Sections of the extraembryonic endoderm of an early chick embryo, showing two large primordial germ cells; one is in the process of dividing.

(neural ectoderm). The order and character of these formative movements executed in response to the *gynome* are correlated with the amount and distribution of yolk. The mechanics appear essentially similar, but patterns differ in the several types of blastulae.

During this early phase of differentiation, cells take the first step in the long series of events leading to the appearance of definitive tissues and eventually of the association of tissues in organs and organ systems. It was noted previously that within any formative area of the blastula the cells were developmentally multipotential and lacked precise determination. For example, presumptive muscle-cell mesoderm in the amphibian embryo could be experimentally provoked to join cells of the future neural tube through the action of the *evocator* effect of the chordamesodermal roof of the *archenteron* (primitive

Table 3-2 Derivatives of the Germ Layers in Vertebrates.

Ectoderm	Epidermis, hair, cutaneous sensory receptors and glands, nails, claws, hooves, lens of the eye, feathers, brain, spinal cord, nerves, chromaffin tissue, hypophysis. Epithelium of the mouth, nasal cavity, sinuses, anal canal, sense organs, oral glands, enamel organ, meatus, external surface of tympanic membrane, body of tongue.
Mesoderm	Muscle, connective tissues, cartilage, notochord, bone, joint cavities, adrenal cortex, gonads, genital ducts, kidneys, ureters, body cavities, blood vessels, heart, lymphatics, lymphoid tissue, blood, bone marrow, dentine and cementum (tooth), malleus and incus (ear bones).
Endoderm	Epithelium of the bladder, vagina, vestibule, urethra and associated glands, digestive tube and associated glands, larynx, trachea, lungs, pharynx, base of tongue, parathyroids, thymus, tonsils, thyroid, auditory tube, tympanic cavity.

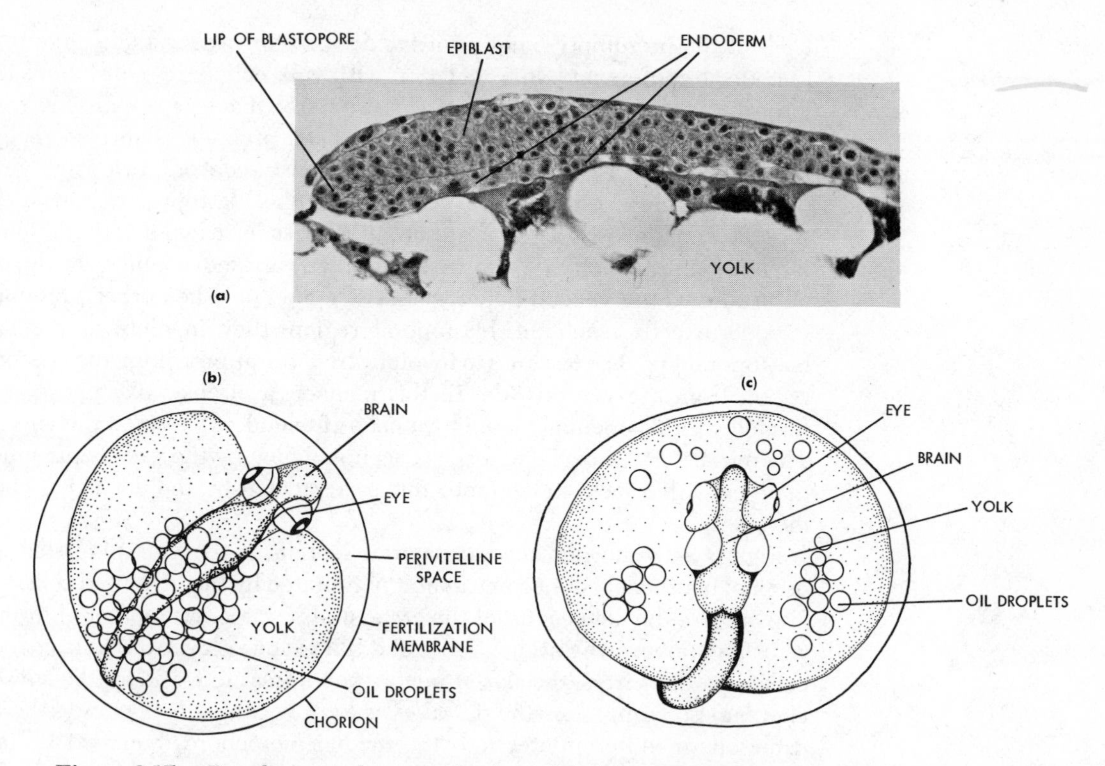

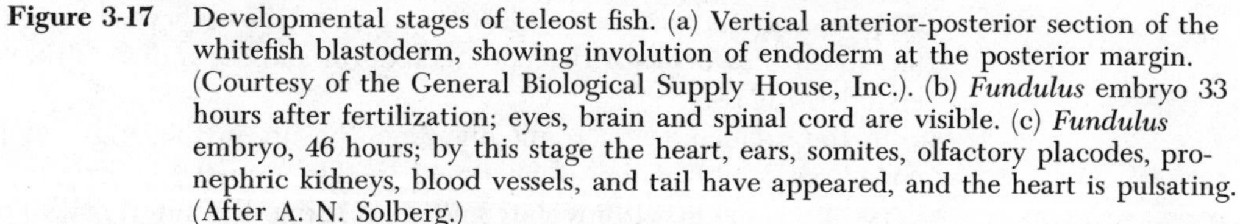

Figure 3-17 Developmental stages of teleost fish. (a) Vertical anterior-posterior section of the whitefish blastoderm, showing involution of endoderm at the posterior margin. (Courtesy of the General Biological Supply House, Inc.). (b) *Fundulus* embryo 33 hours after fertilization; eyes, brain and spinal cord are visible. (c) *Fundulus* embryo, 46 hours; by this stage the heart, ears, somites, olfactory placodes, pronephric kidneys, blood vessels, and tail have appeared, and the heart is pulsating. (After A. N. Solberg.)

gut). Fates now become fixed, and cells of a developing tissue in the future will give rise only to cells characteristic of that particular tissue or its derivatives. In other words, the developmental potentiality of a cell within the new location of a presumptive area gradually becomes restricted by new environmental influences until it attains a nonmodifiable condition. Not all presumptive areas become restricted during the period of gastrulation. Other evidences of inductive action can be seen later in development, as in the determination of the lens of the eye from body ectoderm through inductive action of the optic cup, or of the cornea of the eyeball through influences from the combined optic cup and lens. Few organs in the body have been found to complete differentiation in structure and function without the involvement of the inductor (organizer) phenomenon in the course of their early developmental history.

Gastrulation and later developmental phenomena will now be discussed briefly for eggs containing large amounts of yolk and for the special egg of the placental mammal. For other types of eggs, reference should be made to current texts on embryology.

Frog Embryo Gastrulation

An initial step in gastrulation in the microlecithal frog's egg is the formation of the dorsal blastoporal lip within the gray crescent area about twenty-nine hours after fertilization. It will be recalled that the crescent originally appeared opposite the sperm entrance point during fertilization of the ovum.

Cells of the presumptive notochord and somatic mesoderm converge toward the dorsal midline and, in association with yolk-endoderm cells, move (involute) into the internal blastocoel (Figure 3-18). Factors responsible for the activity at this particular place have not been precisely identified. The new cavity taking form in this manner is the gastrocoel, or primitive gut; its opening to the exterior is the beginning of the blastopore. Elsewhere, the superficial pigmented layer of presumptive mesoderm and ectoderm, showing intensive mitotic activity, continues its downward movement over the endoderm toward the vegetal pole (process of epiboly). When other presumptive mesoderm cells reach the blastoporal region, they involute at the lateral blastoporal lips. The last ones to involute are those approaching the blastoporal region from the ventral side. In this manner, a circular blastopore is soon formed with its opening closed by a mass of endoderm, termed the yolk plug. The dorsal surface of the embryo is occupied now by the presumptive neural plate cells that have moved into the area vacated by the departing chordamesoderm.

Figure 3-18 shows a median sagittal section of the yolk-plug stage. The original blastocoel has become almost obliterated by the involuted cells, which also enclose the new cavity of the archenteron. Endoderm material forms the walls and floor of the archenteron, and the roof is a layer of inner endoderm cells separated from the dorsal neural plate area above it by the involuted chordamesoderm. Soon, the dorsal endoderm becomes distinct, and the chordamesoderm differentiates into the median notochord, bounded by lateral strips of somatic mesoderm. Elsewhere around the embryo, a layer of somatic mesoderm lies beneath the body ectoderm and separates it from the large endoderm mass.

In the salamander, a slight difference occurs in the origin of the dorsal endodermal roof of the archenteron. In this form the endoderm above the dorsal lip is the first to involute and gives rise to the anterior wall of the gut. The endoderm is followed by chordamesoderm, which soon forms into distinct notochord and somatic mesoderm as the temporary roof of the archenteron. Shortly thereafter, the side endodermal walls of the archenteron extend toward the median line beneath the chordamesoderm and unite to form the endodermal roof.

Many years ago, Hans Spemann and his students showed that the dorsal blastoporal lip and dorsal archenteric roof are causative factors in the differentiation of the neural ectoderm from its presumptive competent predecessor, and they gave the term *primary inductor* (organizer) to this material. When transplanted beneath indifferent ectoderm of another embryo of similar age, either material induced and organized a secondary embryonic axis containing elements that originated from the competent ectoderm as well as those differentiating from the transplant itself (Figure 3-19). As chordamesoderm cells pass through the lip during the course of normal development, they become endowed with organizing capacity. The chordamesodermal layer becomes the inductor of the central neural tube from the presumptive neural ectoderm above it and, thus, determines the fate of this tissue. Actually, it was demonstrated that the chordamesodermal roof possesses regional evocating capacities; transplantation of the anterior part induces brain and head structures, whereas the posterior part initiates trunk and tail structures.

By the end of the gastrulation period in amphibian development, the major

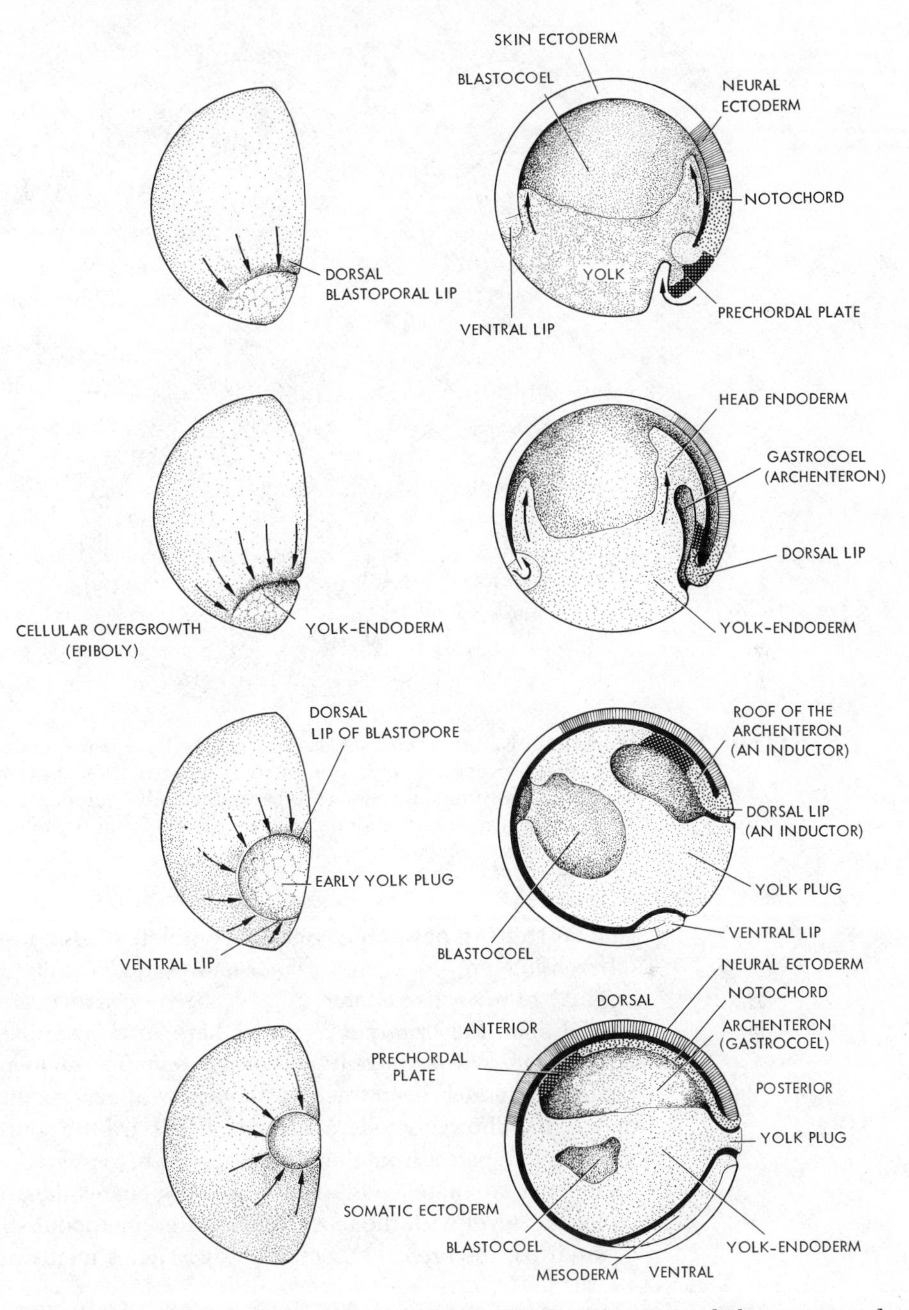

Figure 3-18 The course of gastrulation in the embryo of the frog shown in whole mounts and cross-sections. The structural pattern changes markedly as areas of presumptive organ-forming materials move to other positions either on the surface or to the interior of the embryo. The first cellular interactions, known as inductions, have been identified during this phase of development. (From Saunders' *Animal Morphogenesis,* New York: Macmillan, 1966.)

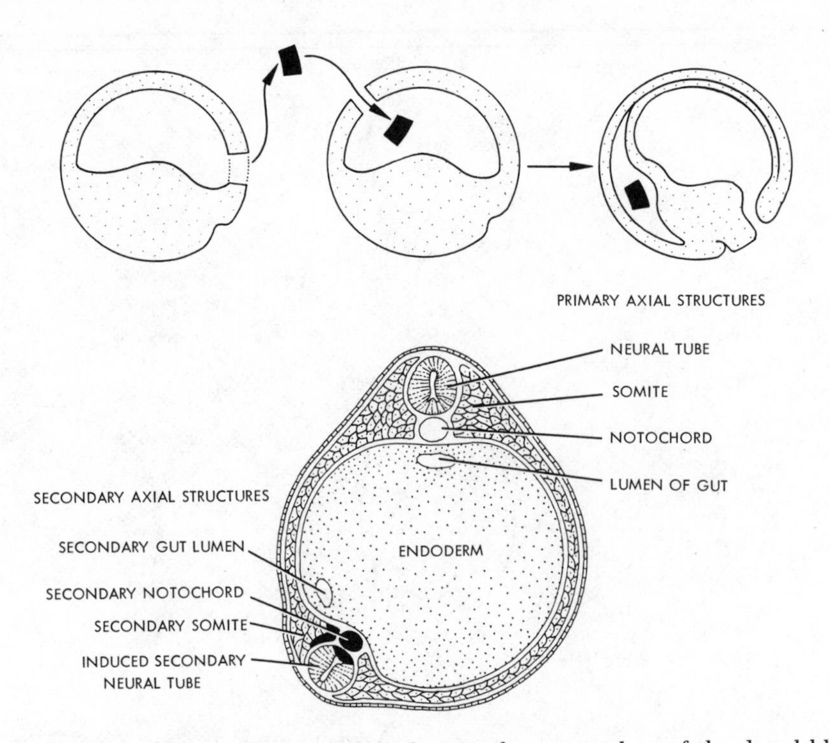

Figure 3-19 Induction of a secondary set of axial organs by a transplant of the dorsal blastoporal lip (prospective archenteron roof). Gastrulation movements relegate the graft to a position subjacent to prospective belly ectoderm. Both ectodermal and mesodermal organs are induced by the graft. (From Saunders' *Animal Morphogenesis*. New York: Macmillan, 1968.)

cellular shifts in position have been completed. Also, instead of being able to differentiate into a variety of structures under suitable stimulation, some original presumptive materials have become irrevocably determined along special lines. For example, as shown here, after gastrulation the presumptive neural plate will form only nervous system. The influence of this particular determination has been traced by a variety of experimental techniques to the underlying chordamesodermal layer. What are the source and nature of the stimulus that passes from the chordamesoderm to influence the prospective fate of the neural plate cells? The search for an explanation has reached the molecular level with the most sophisticated methodologies, but after decades of intensive research the question still remains unanswered.

Neurulation About the time the blastopore finally closes and the yolk plug becomes covered, two longitudinal neural folds, separated by a groove and connected by an anterior transverse fold, appear in the neural plate ectoderm. These folds proceed to arch dorsally and unite, in this manner cutting off an internal neural tube that shows initial division into future brain and spinal cord regions. Three parts of the embryonic brain soon become distinguishable: fore-, mid-, and hindbrain regions, designated respectively as the *prosencephalon, mesencephalon,* and *rhombencephalon.* Some of the main derivatives of these regions are briefly summarized in Table 3-3. The nervous system will be discussed in Chapter 10.

Mention should be made of the neural crest cords that arise from the neural

Table 3-3

Prosencephalon	Telencephalon	Cerebral hemispheres, corpus striatum Lateral ventricles (1 and 2) Olfactory lobes
	Diencephalon	Optic cups, thalamus (epi- and hypo-) Third ventricle, anterior tela choroidea Neurohypophysis and stalk, pineal body
Mesencephalon	Mesencephalon	Optic lobes, tectum, tegmentum Portion of the aqueduct of Sylvius (iter)
Rhombencephalon	Metencephalon	Cerebellum, pons Portion of the aqueduct
	Myelencephalon	Medulla oblongata, pyramid Fourth ventricle, choroid plexus Motor components of certain cranial nerves (nerves 5–12)

plate as the neural folds close. Fate of the neural crest cells has been traced by deletion, transplantation, and staining methods. If these cords are experimentally removed, the organism will lack the medulla of the adrenal gland, ganglion cells of spinal and certain cranial nerves, the autonomic system, pigment cells or *melanocytes,* cells ensheathing nerve fibers, visceral cartilage, parts of tooth germs, meninges of cord and brain, and so on.

Other structures make their appearance during neurulation as seen in Figures 3-20 and 3-21, as the embryo undergoes a change in shape with the appearance of the head and early tail bud. The notochord becomes distinct from the mesoderm as a median cylindrical rod. It is now bordered on each side by somites, forming in the somatic mesoderm, from which muscle cells, dermis of the skin, and units of the vertebral column will develop later. Ventral to the somite band on each side of the body is a narrow strip of somatic mesoderm, the *nephrotome,* from which will develop the elements of the kidneys and their ducts. Below the nephrotomal strip the lateral mesoderm consists of two

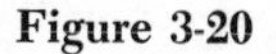

Figure 3-20 Three stages of neurulation (neural tube formation) in the frog: early (a,a′); middle (b,b′); and late (c,c′), with the neural folds about to fuse.

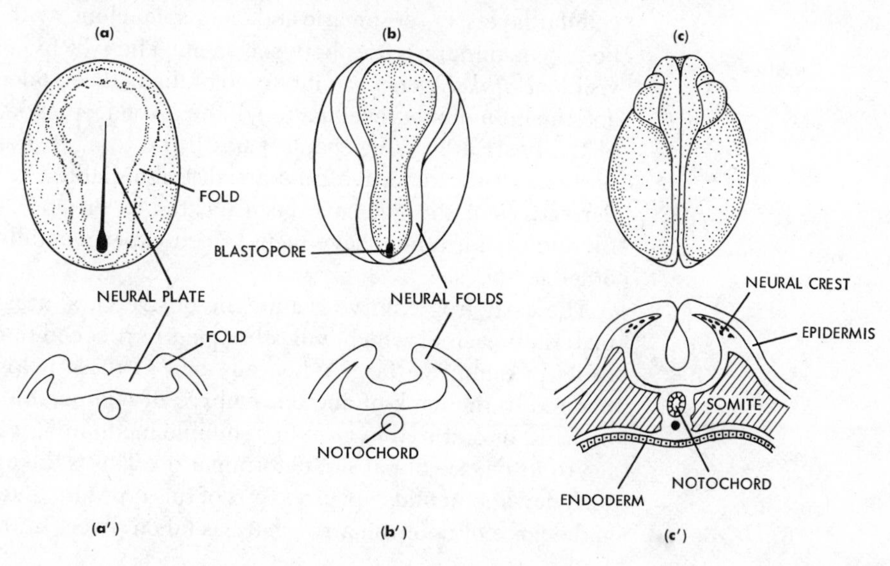

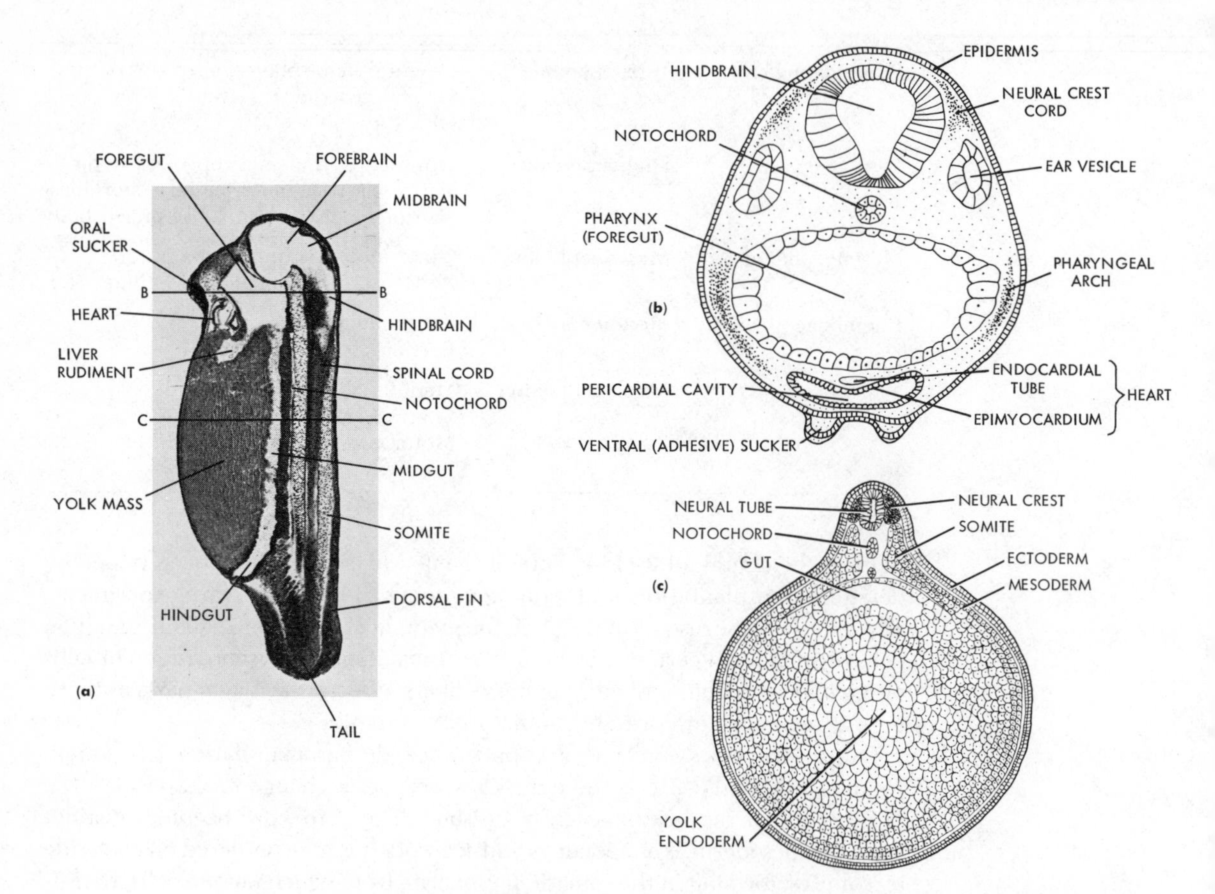

Figure 3-21 (a) Photograph of a sagittal section of a 4-mm frog embryo at the tail-bud stage. Developing brain and spinal cord are visible in the dorsal region of the body. Ventral to these are the notochord, gut and darkly stained yolk mass. Anteriorly, the heart and foregut lie ventral to the brain. At this and earlier stages numerous other structures are present only as "fields" or undifferentiated areas. (Courtesy General Biological Supply House, Inc.) (b) Cross-section through the foregut and ear region at level "B". (c) Cross-section through the trunk region at level "C".

cellular layers, outer somatic and inner splanchnic, with the primordium of the body (coelomic) cavities between them. The two hypophyseal primordia are visible: *Rathke's pouch*, an ingrowth of the body ectoderm ventral to the brain, and the infundibular evagination of the prosencephalon. The paired rudiments of the heart are taking shape beneath the foregut. Fore-, mid-, and hindgut portions of the digestive tube are distinguishable. At this time, subordinate derivatives of the gut have been located as primordia: liver, bile duct, and thyroid gland; others have been traced experimentally by vital staining and other methods.

The early frog embryo is a mosaic of structural areas or "fields" of specific materials, each of which will subsequently proceed to differentiate according to its presumptive fate. When any one of these fields is isolated and transplanted to the flank of another embryo of comparable age or stage of development, or cultured *in vitro* in a suitable medium, it will differentiate according to its degree of pattern determination. This is the usual method for testing such developmental capacity. Tests of this type also give evidence of the nature and degree of determination that has taken place. In this new environment, a field will not be subjected to additional inductive influences and, hence, will differentiate independently of other structures on the basis of its inherent organization. It is said to show a capacity for independent self-differentiation that will become more complete with age.

Not all structures of the organism are determined at this time, nor are fields capable of differentiating into all parts peculiar to the mature organ. Determination of some organs takes place later, and even that within a field may occur slowly. An early field may give rise to an organ that is deficient or deformed. With advancing age, a field progressively differentiates a more normal structure. The field concept of organization is easily applied to the embryos of fish and chick although these organisms are not so readily obtainable and the techniques of transplantation are more difficult. Little is known about reptilian and mammalian embryos of comparable stages of development, but there is reason to believe that a similar type of organization is present.

A number of facts have been ascertained about fields, as illustrated by the ear and limb fields. (1) Any field is larger in area than that portion that will later become incorporated into the organ, and any major part of an ear field will differentiate an internal ear. Several internal ears can be obtained from one ear field, which indicates the labile nature of the field at this time. A transplant from the center of the field gives a more complete ear than one from the periphery; there must then be an existing gradient of internal organization. The gradient concept previously has been referred to in connection with the fertilized egg, cleavage, and blastula stages of development. (2) As early stages of any field may show some overlapping with neighboring fields, transplants from peripheral regions usually contain more than one organ. (3) Transplants combining two similar fields may form one giant organ; this result demonstrates a capacity for regulation on the part of the fields. Another type of regulative ability during morphogenesis is seen in the capacity of a deficient or partial field to form a complete internal ear. These examples of regulative ability to compensate for loss of a part of itself or to combine two fields forming one ear are reminiscent of a similar capacity exhibited by fertilized egg or cleavage stages of certain indeterminate embryonic types. (4) Gradually, the cells lose their lability and for a time may form different subparts of a field as progressive differentiation takes place. Even the major axes of an organ, such as the limb, become fixed in a characteristic order. The anterior-posterior axis becomes fixed first as shown by transplantation of a rotated field, followed shortly thereafter by the dorsoventral axis. The last to be determined is the medial-lateral axis. Within amphibian limbs, determination may never become absolutely fixed, as shown by the regenerative capacity of limbs of the newt or of the aquatic salamander following amputation.

Tail-bud Stage

By the following tail-bud stage (Figure 3-21) organization of certain primordia has reached a stage of differentiation when further development can no longer be modified experimentally. Now the embryo is becoming structurally organized in the pattern of the tadpole. The segmental arrangement of structures is seen in the somites, primitive kidney tubules, neural crest derivatives, and vascular system. The tail bud itself is a mass of undifferentiated chordamesodermal material that was the last to involute at the dorsal lip of the blastopore. The heart develops at the tail-bud stage from lateral mesoderm beneath the pharyngeal region. Blood islands and the first blood cells have appeared in the vicinity of the liver and at the base of the tail.

Numerous and radical changes will take place later during the course of metamorphosis from the tadpole to the adult structural pattern when certain organs will disappear and others become modified for the new life of the land-dwelling animal. Discussion of these cannot be taken up here; for further details reference should be made to a textbook of vertebrate embryology.

Chick Embryo Two steps mark the course of gastrulation in the chick embryo. During the first, which is initiated before the egg is laid, the primitive endodermal layer forms by a splitting or proliferation (*delamination*) of cells into the blastocoel from the thickened posterior portion of the area pellucida. As these cells accumulate, they spread outward with other contributions from the blastoderm to form an inner cell layer beneath the blastoderm. This new layer includes the endodermal material and is called the *hypoblast*, whereas the original layer,

Figure 3-22 (a) Prospective organ-forming areas in the definitive primitive streak stage of the chick. Superficial areas are shown in the left side, invaginated material in the right. (After Rudnick.) (b) Diagrammatic cross-section of the primitive streak to illustrate movement of epiblast cells into the streak and their passage outward to contribute to the mesodermal and endodermal layers during gastrulation. (From Waddington's *Principles of Development and Differentiation*. New York: Macmillan, 1966.)

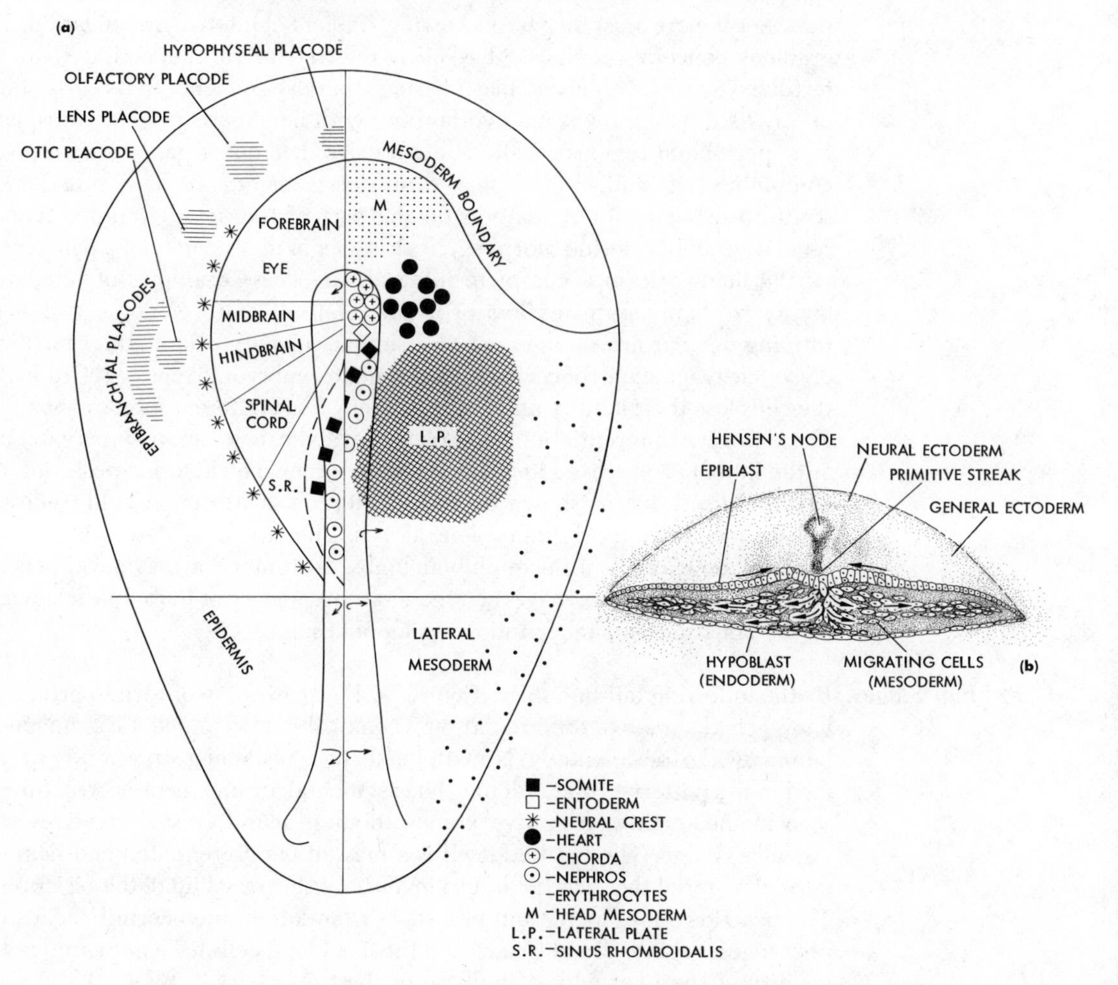

now referred to as the epiblast, contains the chordamesoderm and ectoderm. Long ago, C. H. Waddington showed that this endodermal sheet is the inductor of the primitive streak within the epiblast dorsal to it. The epiblast continues to contribute cells to the hypoblast for some time. In placental mammals, the endoderm originates in the same manner, i.e., by delamination of cells from the inner cell mass.

In the second step of chick gastrulation, the cells in the epiblast (i.e., those containing presumptive material of the mesodermal somites, heart, and lateral and extraembryonic mesoderm) migrate posteriorly and toward the anterio-posterior midline of the epiblast. Here they accumulate and invaginate to form a structure known as the primitive streak (Figures 3-22 and 3-23). These cells

Figure 3-23 Early developmental stages of the chick: (a) Pre-streak blastoderm, 4–6 hours of incubation. (b) Beginning primitive streak, 7–9 hours. Courtesy of Nelson T. Spratt, Jr. (c) Definitive primitive streak stage, circa 14–16 hours. (d) Early neural folds, showing the head process and head fold, circa 18–19 hours. (e) Pre-somite embryo, circa 19 hours. (f) 4-somite stage, circa 25 hours of incubation. Courtesy of Viktor Hamburger and Howard Hamilton.

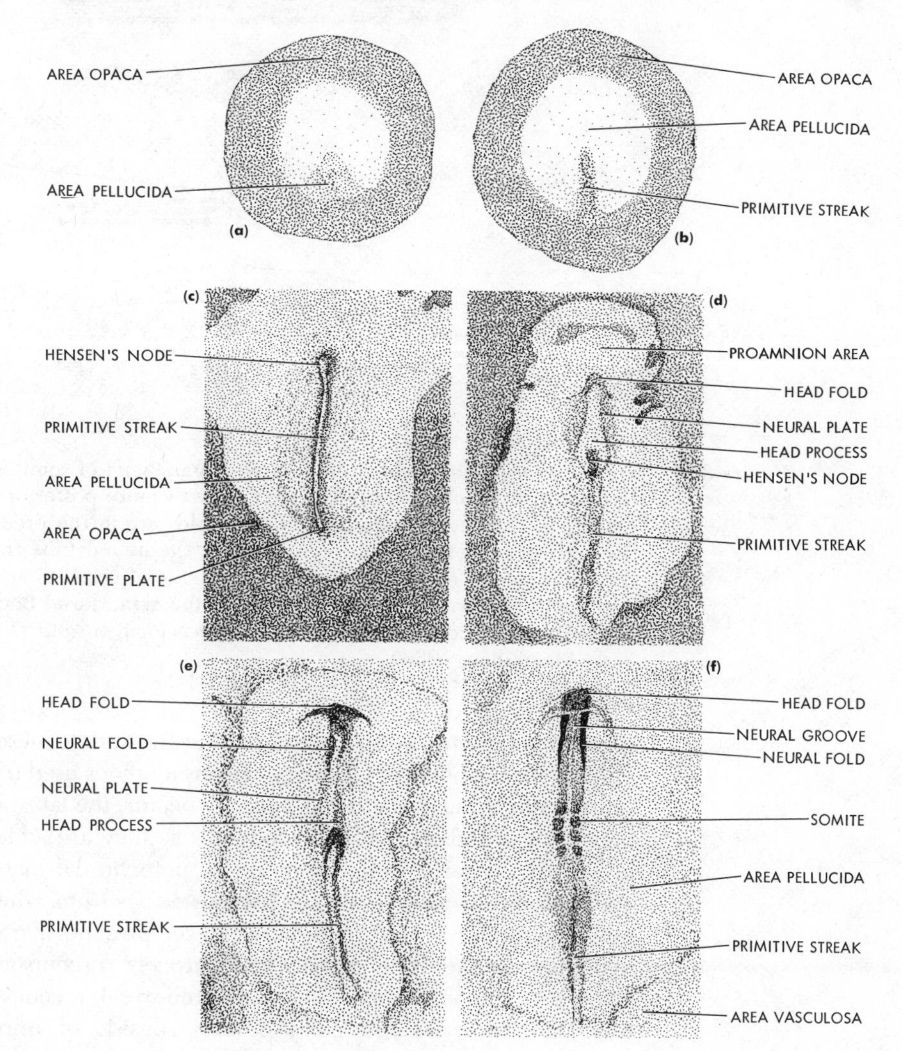

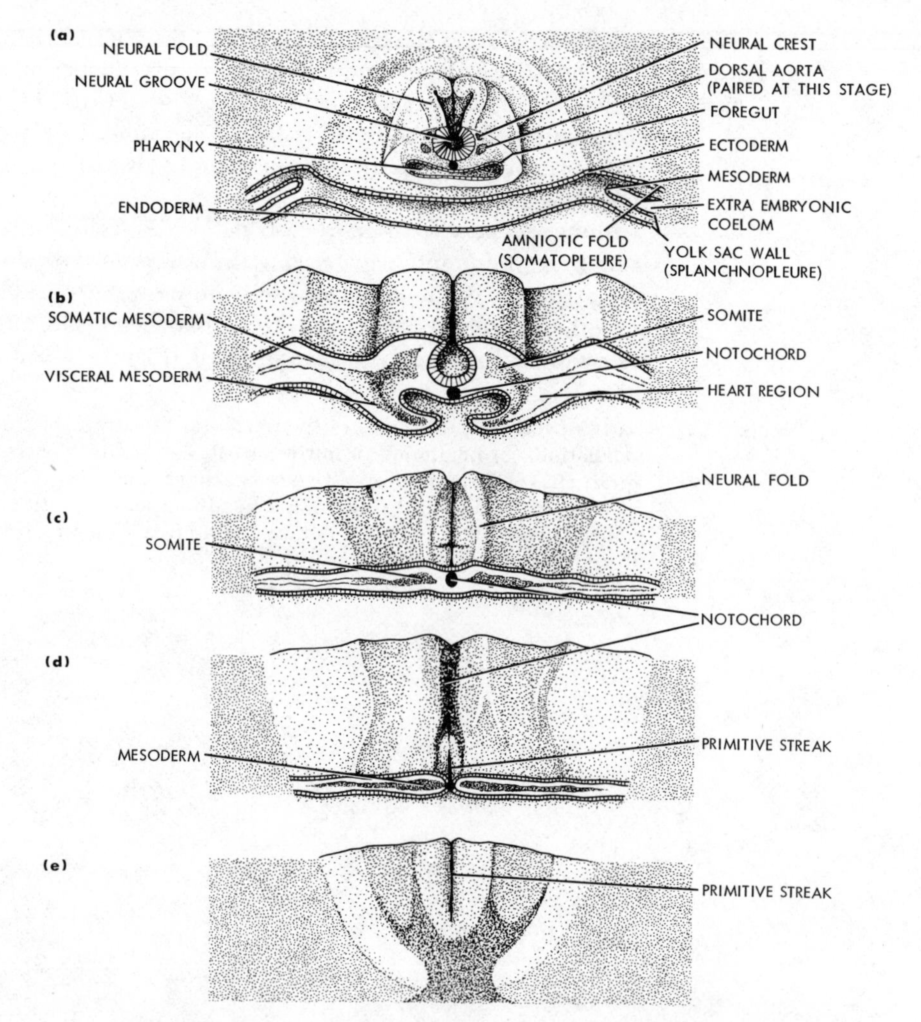

Figure 3-24 Structure of an early chick embryo with four pairs of somites. The embryo begins to form in front of the primitive streak, with more posterior parts taking shape as the streak recedes and shortens. Neural folds are in the process of fusing about the level of the future hind brain to form the neural tube from which will arise the brain, spinal cord, and motor components of cerebral and spinal nerves. Somite material gives rise to the dermis of the skin, dorsal body muscles, and vertebrae. (From Waddington's *Principles of Development and Differentiation.* New York: Macmillan, 1966.)

spread laterally toward the area opaca as the middle mesoderm layer between epiblast and hypoblast. Of the numerous methods used to show that invagination of cells through the streak actually occurs, the latest has been to trace the migration of radioactively labeled cells as they are added to the underlying layers. At the same time, presumptive notochordal material moves into the anterior end of the streak at Hensen's node, from which it extends in the midline anterior to the streak between the presumptive neural ectoderm and hypoblast in the form of the head process (precursor of the notochord). Gradually, the notochord elongates as the streak recedes in a posterior direction. The three-layered embryo now consists of upper epiblast, middle chordamesoderm, and lower hypoblast layers. Cells that do not invaginate through the primitive streak and notochord remain in the epiblast as the ectoderm. In rabbit and man the streak is believed to form in a manner comparable to that of the chick; there is no contrary evidence.

A fate map showing the locations of presumptive organ-forming materials at the maximum (definitive) primitive streak stage is shown in Figure 3-22.

Culture *in vitro* is a technique widely used by investigators. The course of cellular movement during gastrulation also has been followed by sprinkling finely divided carbon on blastoderms cultured on the surface of plasma clots or on artificial culture media, both media containing embryo extract as a food source. Marking cells with vital dyes has been employed successfully with both living blastoderms *in situ* and transplants to the chorioallantoic membrane as well as to the embryonic coelomic cavity to determine the locations, extent, and organization of the presumptive organ-forming areas.

The primitive streak of the chick embryo has been likened to the blastoporal lips of the amphibian gastrula. Both possess inductor capacity. Cellular migrations toward the midline occur in both forms, and similar presumptive materials invaginate (Figures 3-18 and 3-22). Furthermore, in both chick and amphibians, the presumptive areas have greater potency than their presumptive values and are still labile at this time. If the chick preprimitive streak blastoderm is cut transversely into equal anterior and posterior halves, each will form a complete embryo. This could be one of the methods of the origin of monozygotic twins; another may involve the separation of cells during some stage of cleavage. Transplants of the anterior part of the chick primitive streak will induce neural tubes in indifferent ectoderm as well as differentiate certain organ structures such as the heart and kidney. This is another example of the interaction of tissues by induction.

Early Development

As the area pellucida of the chick embryo elongates, the zone of cell invagination (streak) appears to recede posteriorly. This involves an initial displacement followed by a progressive shortening of the streak itself as the last of the invaginating epiblast cells destined to change positions passes through it and the embryo begins to take form anteriorly (Figure 3-23). These cells also are added to the internal layer of mesoderm and notochord. Structures of the embryonic axis have begun to make their appearance in an anteroposterior order. Stages of early development of the chick embryo are shown in Figure 3-23. At early stages, some remnant of the primitive streak can still be found at the extreme posterior end of the embryo, showing that the process of gastrulation in this form overlaps early structural development.

Within seventy-two hours of incubation, the beginnings of many adult organs have made their initial appearance from the germ layers (Figure 3-25). From the neural ectoderm have arisen the primordia of the spinal cord and brain, following the fusion of lateral folds in the neural plate, the neural crest cords from which many diversified structures will later differentiate, the eyes, and certain cranial ganglia. Body ectoderm has formed the lens and the beginning of the cornea of the eye as a result of induction by the optic cup and lens itself, the otic vesicle (otocyst, primordium of the internal ear) by inductive action of the hindbrain, and the beginning of the mouth and cloacal cavities. Cephalic, caudal, and lateral body folds have given shape to the embryo and formed the foregut and hindgut from the endodermal sheet. Derivatives of the foregut include primordia of the visceral pouches, thyroid gland, lungs, trachea, thymus and parathyroid glands, liver, pancreas, esophagus, and stomach-intestine. The allantois is an outgrowth of the hindgut. Derivatives of the mesoderm at this stage are heart, early pro- and mesonephric tubules, somites, and certain arteries and veins with their connections. Mesoderm in the form of mesenchyme has been induced to aggregate around certain primordial structures such as optic cup, otocyst, neural tube, and others. In addition to subordinate parts already initiated, many other structures will subsequently take form. At this stage, the primordia of the organs just mentioned have all attained a state of development in which they are capable of independent self-differentiation when transplanted to a suitable site or cultured *in vitro*. The development of the structure and function of organs arising from such primordia is a matter discussed in subsequent chapters.

At this developmental stage, the anterior half of the embryo is enclosed by inner amniotic and outer chorionic membranes that have had their origins in the fusion of folds of the somatopleure, or extraembryonic ectoderm-mesoderm layer, lateral and anterior to the embryo (Figure 3-27). The extraembryonic splanchnopleure or mesoendodermal layer now partially encloses the yolk in a yolk sac. As an evagination of the hindgut, the allantoic sac is a small projection into the extraembryonic coelom between the chorion and yolk-sac wall; later, this will expand into the extraembryonic coelomic cavity. There is active circulation of blood at this time; the heart has been beating since the twenty-eight-hour stage. In older embryos (Figure 3-26), food will be obtained from yolk and albumin by way of the vitelline circulation; nitrogenous wastes of metabolism will be collected by the kidneys to be stored as insoluble uric acid in the allantois. Exchange of oxygen and carbon dioxide will take place between the blood and external air through the fused chorio-allantoic membrane via the allantoic circulation (Figure 3-27).

The Mammalian Embryo

Early development of a placental mammalian embryo, e.g., rabbit or man—follows the same general pattern described for the chick. The exceptional features include the various ways in which the extraembryonic membranes form, additional roles of the allantois and/or its blood vessels in excretion, nutrition and gaseous exchange via the placenta, and the relationship of the chorion and uterine endometrium in the placenta. The isolecithal mammalian egg, which contains little food material, must secure it from external sources. During cleavage and development of the blastocyst, the luminal fluids of oviduct and uterus probably constitute the earliest food sources; however, seven or eight days following fertilization, the blastocyst undergoes nidation and initiates invasion of the uterine wall. For a time the debris or histotroph resulting from this implantation process may be engulfed by cellular (trophoblastic) phagocytic action of chorionic cells, but the invading embryo soon establishes a functional relationship with the mother's blood via chorionic villi

Figure 3-25 Chick embryos showing completion of the process of neurulation, beginning of body formation and flexion (see also Figure 3-23), and appearance of organ primordia. (a) Seven pairs of somites, 29 to 33 hours (stages after Hamburger and Hamilton). (b) Ten somites, 33 to 38 hours. (c) 51 to 56 hours. (d) 62 to 75 hours. (Courtesy of the General Biological Supply House, Inc.)

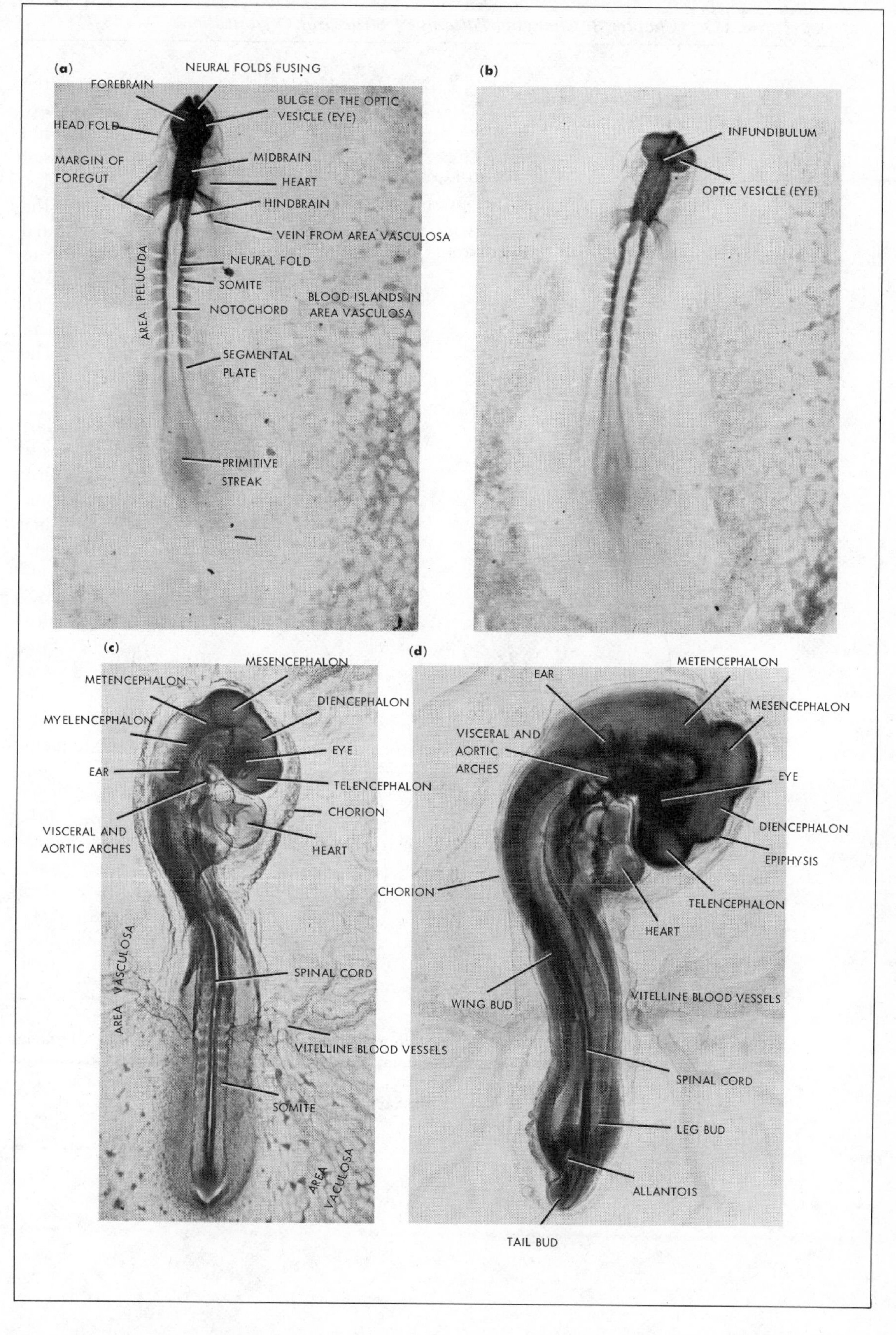
(a)
NEURAL FOLDS FUSING
FOREBRAIN
BULGE OF THE OPTIC VESICLE (EYE)
HEAD FOLD
MARGIN OF FOREGUT
MIDBRAIN
HEART
HINDBRAIN
VEIN FROM AREA VASCULOSA
NEURAL FOLD
AREA PELUCIDA
SOMITE
BLOOD ISLANDS IN AREA VASCULOSA
NOTOCHORD
SEGMENTAL PLATE
PRIMITIVE STREAK
(b)
INFUNDIBULUM
OPTIC VESICLE (EYE)
(c)
MESENCEPHALON
METENCEPHALON
DIENCEPHALON
MYELENCEPHALON
EYE
EAR
TELENCEPHALON
CHORION
VISCERAL AND AORTIC ARCHES
HEART
AREA VASCULOSA
SPINAL CORD
VITELLINE BLOOD VESSELS
SOMITE
AREA VACULOSA
(d)
METENCEPHALON
EAR
MESENCEPHALON
VISCERAL AND AORTIC ARCHES
EYE
DIENCEPHALON
EPIPHYSIS
CHORION
TELENCEPHALON
HEART
WING BUD
VITELLINE BLOOD VESSELS
SPINAL CORD
LEG BUD
ALLANTOIS
TAIL BUD

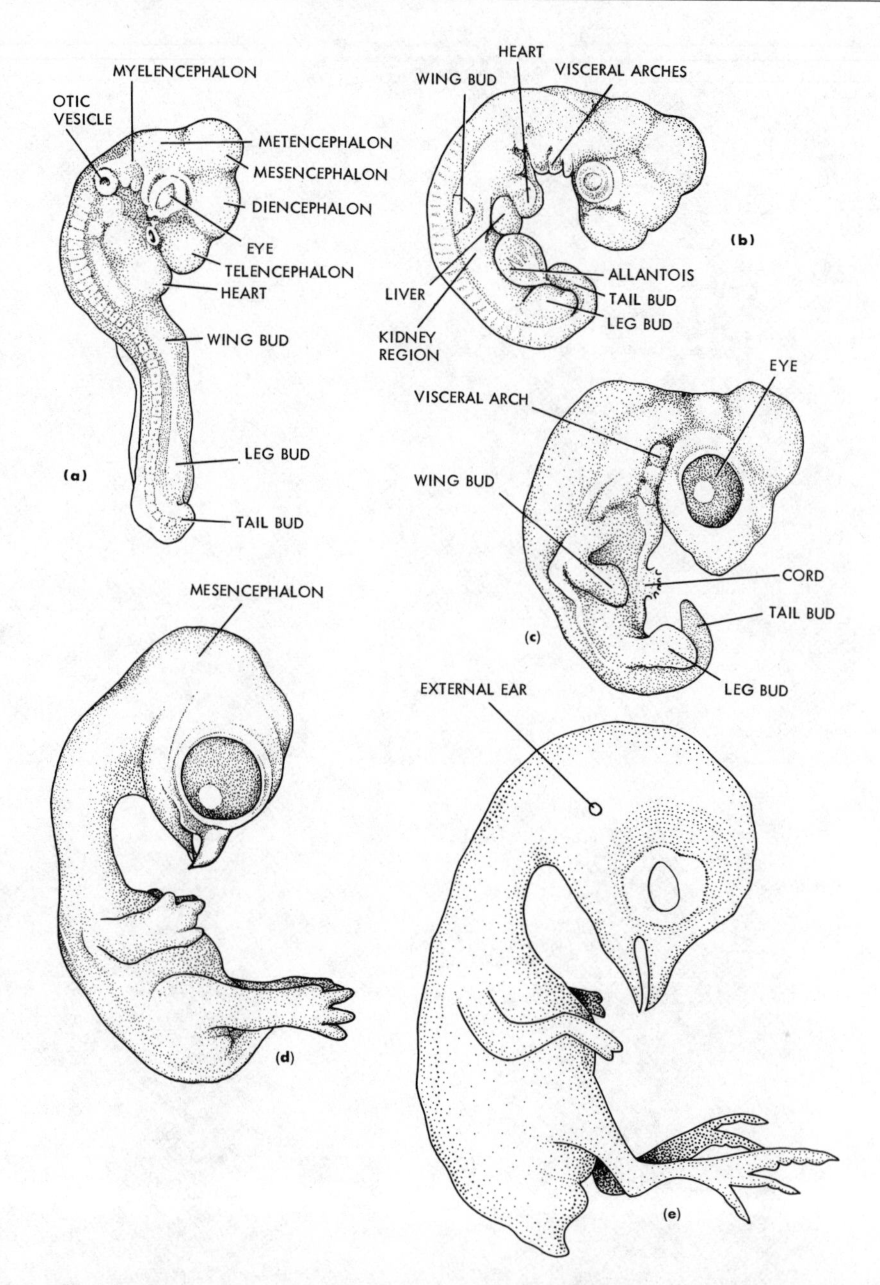

Figure 3-26 Chick embryos, illustrating development of body form. By circa 10 days the avian features are clearly recognizable. (a) circa 65 to 69 hours. (b) circa 3.5 days. (c) circa 5 days. (d) circa 7.5 to 8 days. (e) circa 10 days. (Stages after Hamburger and Hamilton. Figures a–d from Saunders' *Animal Morphogenesis.* New York: Macmillan, 1966)

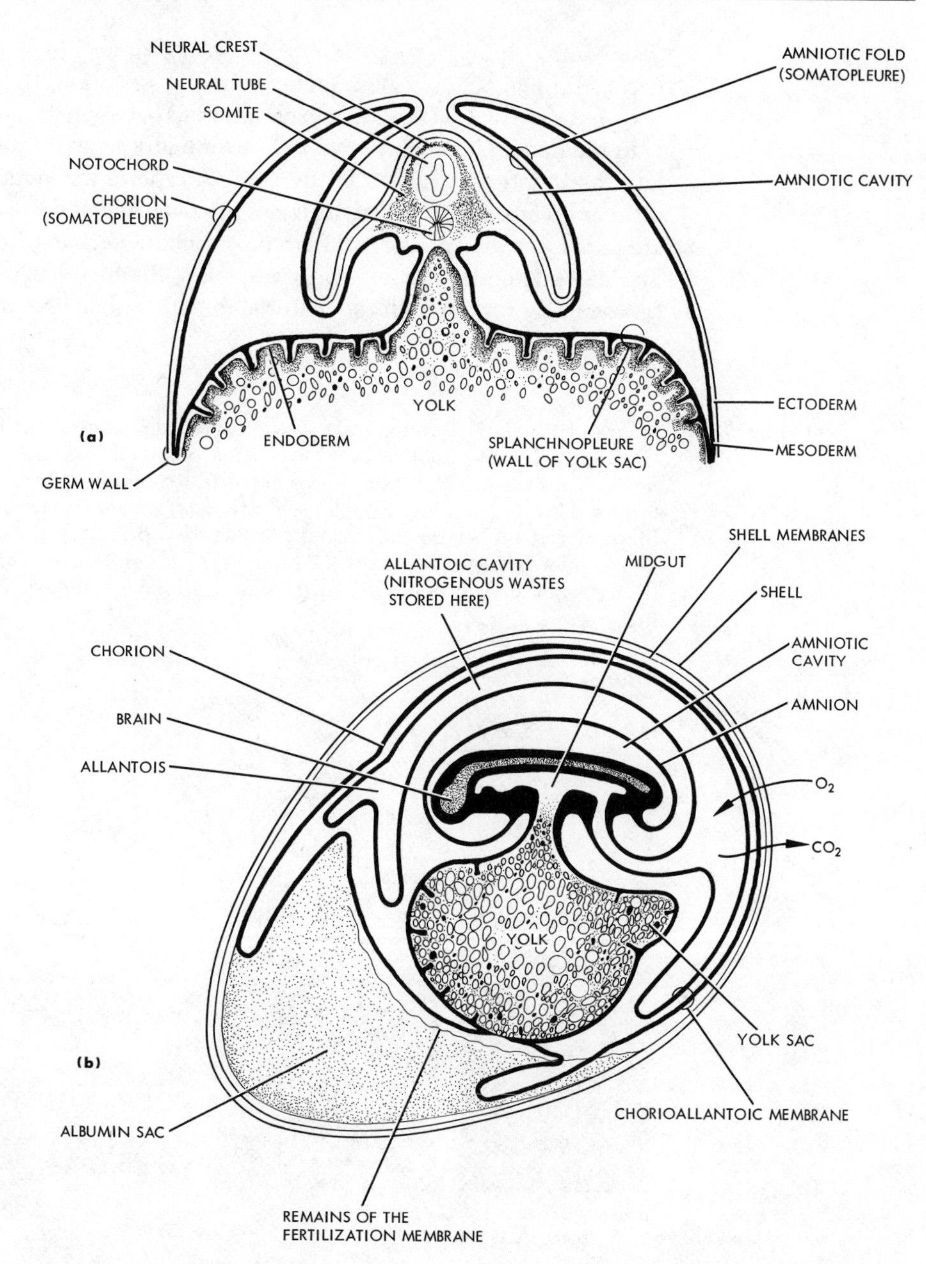

Figure 3-27 Formation of the extraembryonic membranes (amnion and chorion) and appendages (allantois and yolk sac) of the chick embryo. Oxygen and carbon dioxide are exchanged with the outside air via the circulation in the wall of the allantois. Nitrogenous waste products of metabolism in the form of insoluble uric acid is stored in the allantois, an evagination of the hind gut; the amnion and chorion arise as folds of the embryonic somatopleure and serve a protective function; the wall of the yolk sac is an extension of the splanchnopleure. (After Kent's *Comparative Anatomy of the Vertebrates.* The C. V. Mosby Company, 1965.)

that contain branches of the embryonic allantoic blood vessels. From this period throughout the subsequent course of gestation, the needs of the embryo-fetus are met by the passage of materials between the two vascular tissues.

In the human, a few layers of cells eventually separate the blood streams of fetus and mother, whereas in the contact type of placenta found in the pig, little or no erosion of maternal tissues has occurred and the two blood systems are separated by all the cellular layers characterizing both chorionic and uterine endometrial structures. In any case, the blood of the embryo is at all times in equilibrium with the maternal blood, a phenomenon that is regulated

Figure 3-28 Stages of the early development of the rabbit, an induced ovulator: (a) ovulated tubal egg, showing nucleus and outer zona pellucida membrane; (b) surface view of the thicker inner cell mass surrounded by trophoblast from a blastocyst stage, 6 days following mating; (c) whole blastocysts, showing denser inner cell mass; (d) blastocyst at higher magnification; (e) definitive primitive streak stage, 6 days, 23 hours; (f) neural plate, 7 days, 15 hours; (g) presomite embryo, 7 days, 21 hours; (h) two-somite, 7 days, 23 hours; (i) neural fold stage, 8 days, 5 hours after mating.

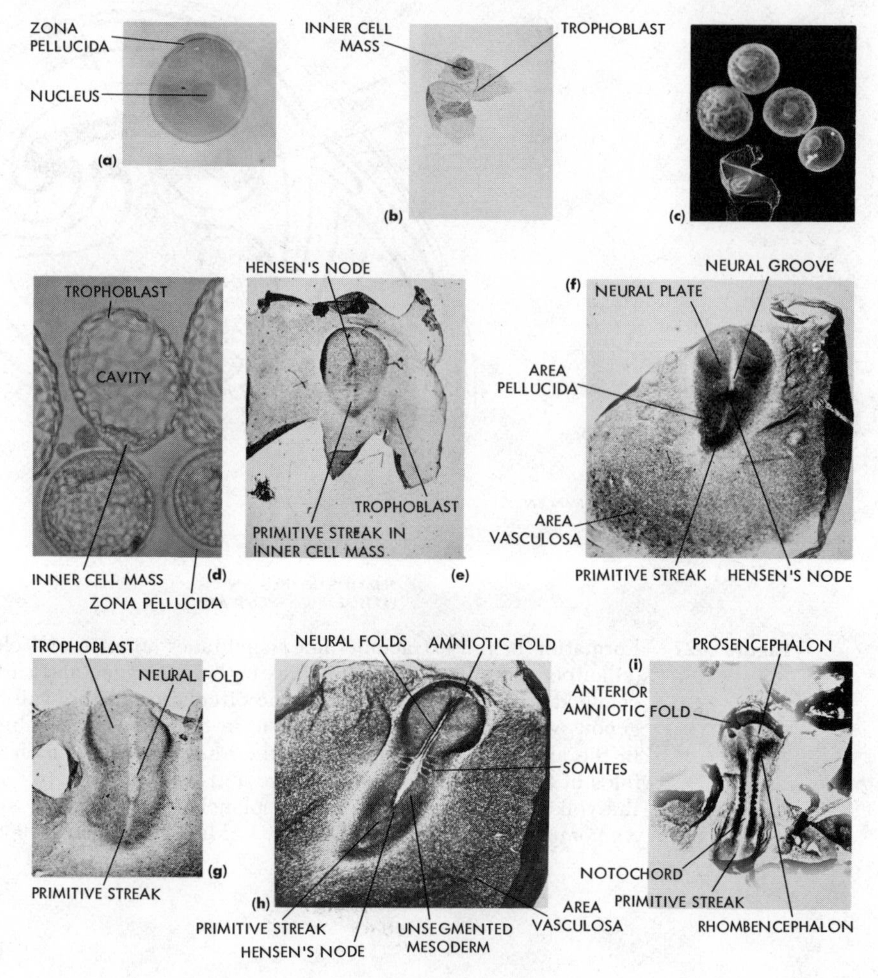

by the permeability of cell membranes and other factors. The placenta has additional functions; e.g., it becomes an endocrine gland in the human, secreting those corpora luteal-like hormones that maintain pregnancy. Many substances pass more or less freely across the placenta, whereas others are stopped by this screening barrier.

The developmental capacity of early mammalian embryos cultured *in vitro* has long been of interest. It is known that the inner cell mass of the rabbit will grow for a time on the surface of chicken plasma clots containing rabbit embryo extract as a food source, or on food-enriched artificial media. Recently, the growth of blastocysts has been reported to take place in both artificial media and serum. Mammalian eggs have been fertilized *in vitro,* and the resulting development similarly was followed in the culture medium to the blastocyst stage (mouse, rat). When transplanted to the uterus of a hormone-activated foster mother, such blastocysts have become implanted. Transplants have been made to the anterior chamber of the adult eye in chicken, rat, and rabbit. In placental mammals, transplants also have been made to the omental bursa and beneath the kidney capsule where the foreign material becomes incorporated and vascularized as a graft. Such transplants are not rejected at once by host tissue because the embryo has not developed an immune system at this time.

Fertilization of the human egg occurs in the upper portion (ampulla) of the oviduct following ovulation. During its passage down this tube cleavage takes place, and when the embryo reaches the uterus three days later it has attained the late morula stage of development. During the following four days changes occur that convert the young embryo from a more or less compact group of cells to the blastocyst. Fluid accumulates within the cellular mass, and by the one-hundred-cell stage the embryo consists of a fluid-filled vesicle with a wall or trophoblast one-cell thick and a small mass of cells, the inner cell mass, attached to the inner surface of the trophoblast (Figure 3-29).

Near the end of this free uterine period, the inner cell mass has been divided into an upper ectodermal portion, the ectoderm layer of the amnion, and a lower embryonic disc from which the embryo will develop. Separating the two is a fluid-filled space, the progenitor of the amniotic cavity. Endoderm is proliferating from the posterior lower surface of the embryonic disc; chorda-mesoderm is invaginating through the primitive streak in a manner comparable to the chick embryo (Figure 3-22); and the mesoderm has split peripherally, with one layer (somatic) growing over the amniotic ectoderm and the other over the endoderm to form the vestigial yolk sac and primitive gut. Meanwhile, additional mesoderm has formed over the inner surface of the trophoblast. Once attached to the uterine endometrium, the trophoblast cells undergo rapid multiplication and in the first stages of implantation destroy and invade the uterine endometrium. By the ninth day, the embryo is completely embedded in the endometrium, cavities are appearing in the mass of trophoblast cells in the initial stages of placentation, and embryonic structures are beginning to form in front of the primitive streak. The amnion has expanded to enclose the embryo in a sea of amniotic fluid, and the whole is enclosed within the mesectodermal trophoblast. Space does not permit further description of human origins, but additional information can be found in the following references.

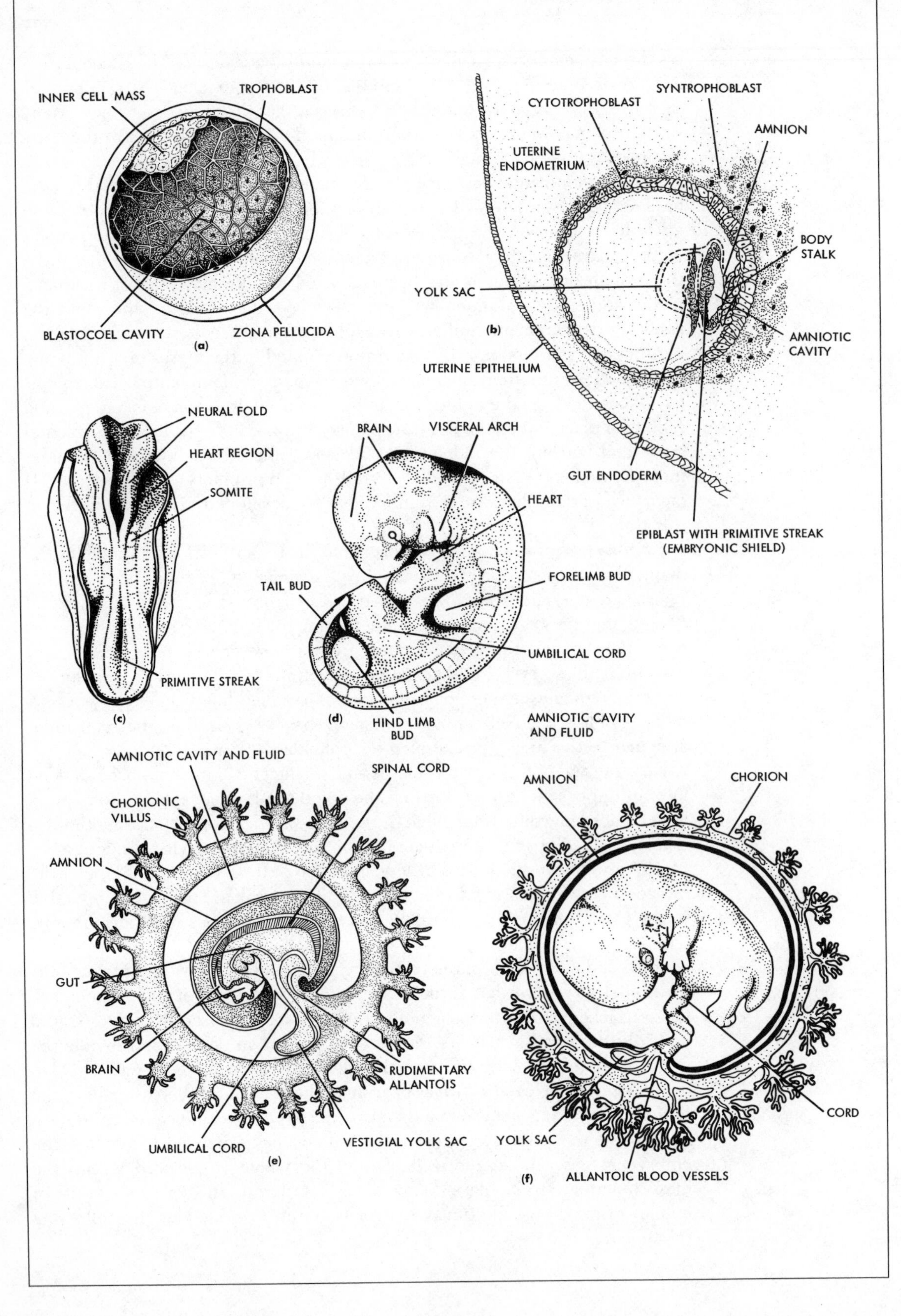
INNER CELL MASS
TROPHOBLAST
BLASTOCOEL CAVITY
ZONA PELLUCIDA
(a)
SYNTROPHOBLAST
CYTOTROPHOBLAST
AMNION
UTERINE ENDOMETRIUM
BODY STALK
YOLK SAC
(b)
AMNIOTIC CAVITY
UTERINE EPITHELIUM
GUT ENDODERM
EPIBLAST WITH PRIMITIVE STREAK (EMBRYONIC SHIELD)
NEURAL FOLD
HEART REGION
SOMITE
PRIMITIVE STREAK
(c)
BRAIN
VISCERAL ARCH
HEART
FORELIMB BUD
TAIL BUD
UMBILICAL CORD
(d)
HIND LIMB BUD
AMNIOTIC CAVITY AND FLUID
SPINAL CORD
CHORIONIC VILLUS
AMNION
GUT
BRAIN
RUDIMENTARY ALLANTOIS
UMBILICAL CORD
VESTIGIAL YOLK SAC
(e)
AMNIOTIC CAVITY AND FLUID
AMNION
CHORION
CORD
YOLK SAC
(f)
ALLANTOIC BLOOD VESSELS

Figure 3-29 Early stages of human embryos with enclosing membranes. (a) Blastocyst, comprising inner cell mass, trophoblast, and blastocoel cavity (after Streeter). (b) Implanted embryo, showing differentiation of the trophoblast into syntrophoblast and cytotrophoblast during implantation; formation of the embryonic shield and amniotic ectoderm by cavitation within the inner cell mass, and development of the yolk sac and mesoderm. (c) Embryo of six pairs of somites and in the process of neurulation. (d) Thirty-two-day embryo containing a number of organ primordia. (e) The embryo floats in amniotic fluid, and is attached by the umbilical cord to the developing placenta marked by the numerous villi; within the cord are the vestigial allantois and yolk sac with their blood vessels. (f) Embryo of forty-one days following fertilization. (From Saunders' *Animal Morphogenesis*, New York: Macmillan, 1966.)

References for Chapter 3

[1] Arey, L. B. *Developmental Anatomy.* 7th ed. Philadelphia: W. B. Saunders Company, 1965. This is a textbook and laboratory manual of human embryology, including teratologies, reproductive cycles, formation and function of the placenta.

[2] Austin, C. R. *Fertilization.* Englewood Cliffs, N.J.: Prentice-Hall, Inc., 1965. (Extensive bibliography.) This book is for students and research workers who seek a general knowledge of fertilization in its comparative aspects, as well as of the various cytological, physiological, and behavioral mechanisms concerned with the union of the gametes.

[3] Balinsky, B. I. *An Introduction to Embryology.* 4th ed. Philadelphia: W. B. Saunders Company, 1975. Embryology is presented as a single science in which the descriptive morphological and experimental physiological approaches are integrated. The subject is interpreted in a comparative sense, as the science dealing with ontogenetic development of animals. Included are such topics as postembryonic development, regeneration, metamorphosis, sexual reproduction, and gene control. (Good bibliography.)

[4] Barth, L. G. *Embryology.* New York: Holt, Rinehart & Winston, Inc., 1953. The author combines a clear, uncomplicated discussion of the basic principles of embryology with a survey of the anatomy of selected embryos. A lucid, readable discussion.

[5] Bell, Eugene. *Molecular and Cellular Aspects of Development.* New York: Harper and Row, Publishers, 1965. The author focuses on some problems of developmental biology through the presentation of key papers in fourteen different, but closely allied, areas: serial transplants of embryonic nuclei, differentiation of cells *in vitro,* reconstruction of tissues by dissociated cells, directed movements and selective adhesion of cells, embryonic potency of embryoid bodies, chromosomes, and cytodifferentiation.

[6] Deuchar, E. M. *Biochemical Aspects of Amphibian Development.* London: Methuen and Co., Ltd., 1966. This is a chemical approach to the understanding of the problems of amphibian development. It is recommended for the more advanced student.

[7] Ebert, J. E. *Interacting Systems in Development.* New York: Holt, Rinehart & Winston, Inc., 1965. The author discusses the interactions that occur between cells at several critical periods of development: for example, fertilization and gastrulation. Later developmental phenomena also are discussed.

[8] Gurdon, J. B. "Transplanted Nuclei and Cell Differentiation." *Scientific*

American, **219**(1968), 24–35. The nucleus taken from early embryonic cells, up to the blastula stage, and transplanted into an enucleated recipient egg, is usually capable of causing complete development. Such experiments aid the study of the interaction of nucleus and cytoplasm in the development of structure.

[9] JACOBSON, A. G. "Inductive Processes in Embryonic Development." *Science*, **152**(1966), 25–34. Most organs form as a result of gradual cumulative effects of interactions between embryonic tissues. The mechanisms of cell differentiation present some of the more provocative problems of biology; in animal embryos one such mechanism is embryonic induction.

[10] MASCONA, A. A., and A. Monroy, eds. *Current Topics in Developmental Biology*. New York: Academic Press, Inc., 1966. This is a series of thought-provoking papers dealing with current topics in developmental biology.

[11] METZ, C. B., and A. Monroy, eds. *Fertilization, Comparative Morphology, Biochemistry and Immunology*. New York: Academic Press, Inc., 1967. This is an excellent account of the process of fertilization based on data from several disciplines.

[12] MINTZ, B., and W. K. Silvers. "Intrinsic Immunological Tolerance in Alophenic Mice." *Science*, **158**(1967), 1484–1487. This is an illustration of the lability of mammalian cleavage cells, mice of multiembryonic origin and the problem of immunity.

[13] NELSEN, Olin E. *Comparative Embryology of the Vertebrates*. New York: Blakiston, 1953. This is a broad, comprehensive, and descriptive presentation of the morphology of vertebrate development in a wide variety of species. (Profusely illustrated; references at the ends of the chapters.)

[14] SAUNDERS, J. W., Jr. *Animal Morphogenesis*. New York: Macmillan Publishing Company, Inc., 1966. The author deals with the way form and function of the organism develop from the fertilized egg. (For beginning students.)

[15] SPRATT, N. T., Jr. *Introduction to Cell Differentiation*. New York: Van Nostrand Reinhold Company, 1964. The author provides information for beginning students on the principles of the differentiation of cells that occurs during the development of an individual—a presentation of a point of view.

[16] WADDINGTON, C. H. *Principles of Development and Differentiation*. New York: Macmillan Publishing Company, Inc., 1966. The author mainly is concerned with principles. The book contains little discussion of the changes in anatomy that go on within various types of embryos. (For the beginning student.)

[17] WESSELS, N. K., and W. J. Rutter. "Phases of Cell Differentiation." *Scientific America*, **220**(1969), 36–44. By cultivating embryonic tissue in the laboratory, the specialization of cells can be studied in the mammalian pancreas. Three regulatory phases seem to occur, each of which leads to a new stage of organizational differentiation.

[18] WILLIAMS, M. A., and L. D. Smith. "Ultrastructure of the 'Terminal Plasm' During Maturation and Early Cleavage in *Rana pipiens*." *Developmental Biology*, **25**(1971), 568–580. Ultrastructural analysis of the germinal plasm region (vegetal pole region) of *Rana pipiens* eggs prior to fertilization and to the sixteen-cell stage is made.

4

The Integumentary System

Introduction One of the body's organ systems that contributes to the homeostatic steady state is the skin and its derivatives, known collectively as the integumentary system. This system envelops the entire body and lines oral and rectal cavities; thus, it serves uniquely as an effective barrier between the internal environment of cells and organs bathed in tissue fluids and the external, unpredictable environment of air or water. It is the chief organ of human identification as well as sexual attraction. Undergoing continual adjustment and often profound changes in response to stimuli, this versatile, dynamic organ is more effective in maintaining constancy in some animals (terrestrial reptiles, birds, mammals) than in others (aquatic cyclostomes, fish, amphibians). Aquatic forms are better protected from rapid fluctuations in temperature and humidity, skin abrasion, radiation, and other disturbing environmental factors and are correspondingly deficient in the compensatory mechanisms required by land-dwelling vertebrates to meet sudden changes. The capacity of the skin to adjust unconsciously is provided by the negative feedback, or biological control, principle; a change in one direction brings forth a compensatory reaction in the opposite direction. The skin has numerous vital functions, such as thermal regulation, that are under the control of self-regulatory autonomic processes.

Probably no organ has received less attention than the skin, and consequently a host of misconceptions exists about it. Knowledge has evolved as the role of the integument in homeostasis has become better understood. Robert F. Rushmer and his colleagues at the University of Washington have called attention to the opportunities for effective collaboration on problems related to biological structures and organisms by using the skin to show the diversity of problems that can be studied by investigators with widely different but related scientific backgrounds [24]. They selected the skin for many reasons: it is obviously accessible to attack; everyone is familiar with many of its characteristics; samples are at hand for direct examination or quantitative measurements; and it has both living and nonliving components susceptible to physical and chemical measurements. The cells of the skin have a chemical profile that is distinct to this organ; for example, the study of the unique lipids, secreted primarily by the sebaceous and sweat glands and the stratum corneum cells as a class of chemical constituents of the skin, offers an opportunity to investigate the functional specialization of this tissue and the roles of the lipids [21]. A remarkably complex organ, it consists of many different tissues, including blood vessels, glands, nerves, sensory receptors, smooth muscle, connective tissue, and fat (Figure 4-1). Diseases and lesions of the skin generally are identified primarily on the basis of their appearance. Comprehensive descriptions of the biological, physical, and chemical properties of this highly specialized, multifunctional organ are in demand in order to clarify its function in disease and in health.

A wide range of methods and concepts is being applied to the study of the integumentary system. Techniques include those dealing with cytodifferentiation, biochemical biophysical properties, immunological responses, morphogenetic potentials, phenotypic characteristics, and functions of its components. A crucial problem has been the basic mechanisms by which cells of the same genotype eventually develop into different phenotypes. Transplantation of epidermis and dermis, separated by trypsin and united in different combinations, has been an effective mechanism for the study of the principles of morphogenesis and physiology. The manner in which genes determine skin and hair color, the functional importance of differences arising between epidermis from the different regions of the body, the role of the dermis, regeneration, aging phenomena, the ability of the skin to withstand certain treatments such as temperature changes and stress, evaluation of tissue viability, and other topics have been studied by transplantation techniques. Grafts are of many types: isolated nuclei, single cells, disorganized masses of cells, solid tissues, combinations of tissues, whole and even nonviable integumentary tissues. The incompatibility of homografts has been analyzed by this technique, as well as the means to combat it. In certain analytical studies, culture *in vitro* has been productive, but this method is more complicated than grafting. Laboratory-grown cell suspensions have yielded interesting results. Biochemical methods are contributing data on metabolic pathways within the skin along with the enzymes involved, melanogenesis, vitamin D synthesis, and keratinization. Physical studies are yielding information on the flow of blood through the dermal blood vessels under different conditions.

Man, in his efforts to meet the many problems that afflict the skin, has utilized the techniques of various scientific disciplines. The intense direct heat of the laser light rays is just beginning to find biological application. Formerly untreatable birthmarks, formed by purplish blood vessels adjacent to the skin and extending into the dermis, are now being erased by these rays. The rays also are reported to remove warts, moles, and tattoo marks. Both warts and cancer are examples of uncontrolled growth. Warts have been shown to be nonmalignant tumors caused by a virus, and although the vital etiology of certain kinds of cancer has not yet been demonstrated clearly, it is suspected from studies on lower animals. The continuing function of skin transplants between individuals depends on the development of some "tolerance" between the graft and the recipient, along with the use of immunosuppressive therapy

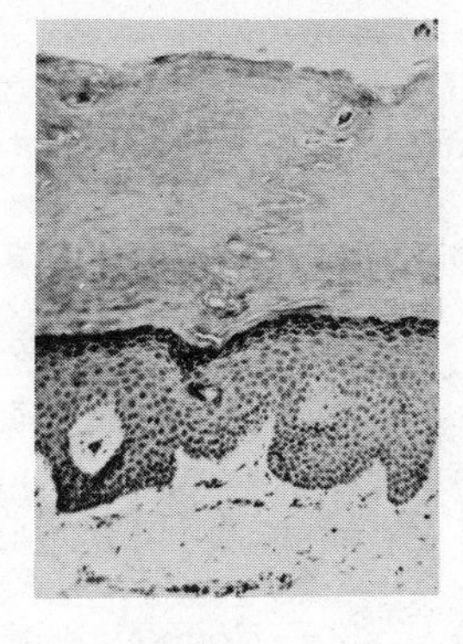

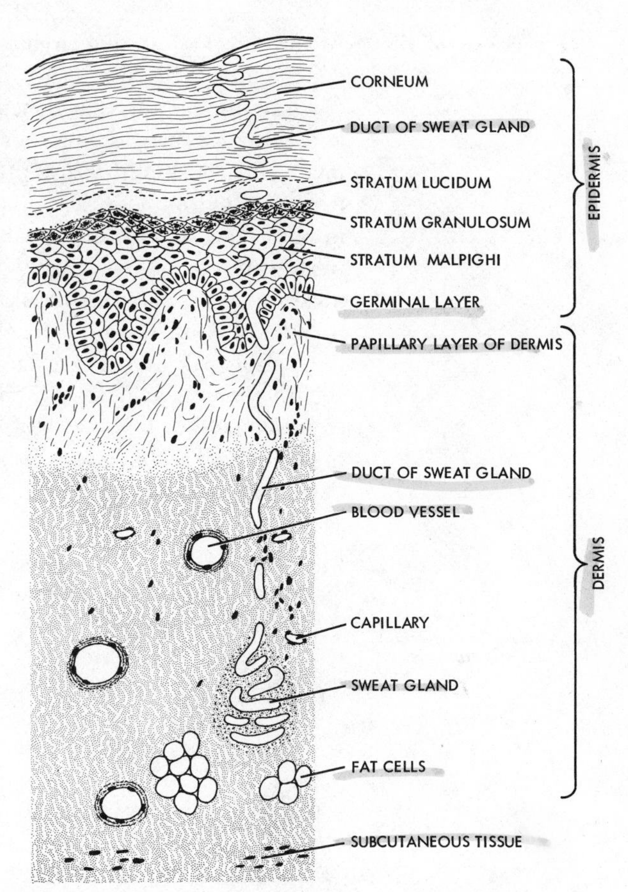

Figure 4-1 Diagram with accompanying photomicrograph of a vertical section of thick human skin from the palm, stained with hematoxylin and eosin, with the principal layers indicated. Note the path of the thickened corneum of flattened, anuclear, keratinized cell remnants. Below this in order are: (1) the homogeneous-appearing stratum lucidum, a nonstainable hyaline layer characteristic of such thick skin as that on the palm and sole, (2) the stratum granulosum or granular layer of flattened cells containing large melanin granules, (3) stratum spinosum or prickle (spinous) layer of variable thickness and changing appearance as successive generations of cells are moved toward the surface of the skin, and (4) the basal germinative layer of columnar cells and the cell bodies of melanocytes. Mitosis takes place within this layer. Below is the thick dermis showing the vascular connective tissue and the secretory portion of a sweat gland. (Photograph by courtesy of Dr. William Montagna, Oregon Regional Primate Research Center, University of Washington, and Academic Press, New York.)

to encourage adaptability. When major incompatibility is present, the graft is rigorously rejected. This area of study has received much attention in recent years and will receive increasingly more as we seek to replace lost or diseased parts of the body in animals that have lost all powers of regeneration except wound healing. Along with methods of matching graft and recipient, massive steroid therapy has been employed. Another skin problem, hyperplasia, is provoked by mite bites and X rays.

Functions of the Skin

At any level of animal organization, from protozoa to man, organisms are dependent on the properties of this uni- or multilayered, never stationary, covering membrane that constitutes an unbroken continuous barrier for the body. In one way or another any derivatives, if present, are also vitally involved. The skin furnishes protection against physical and chemical attacks from the environment; it is pliable, yet resistant to fracture. Any loss of water in both aquatic and terrestrial organisms, or too great an influx of water in fresh-water forms, is largely controlled. Although skin resists organismic invasion, some parasites (hookworm) have little difficulty in penetrating the thickest of skins. Metallic nickel, the oleoresins of poison ivy, and secretions of the poison oak and sumac enter the skin easily. The surface of the skin is slightly on the acid side of neutrality. Communication with the external environment occurs by way of the skin—within it are located the major sense organs as well as the lesser sensory receptors of pain, temperature, and pressure that alert the organism (Figure 4-2).

Figure 4-2 Cutaneous encapsulated sensory receptors: (a) Meissner's corpuscle from the dermal papilla; (b) Pacinian corpuscle; (c) end bulb of Krause; (d) Meissner's corpuscle.

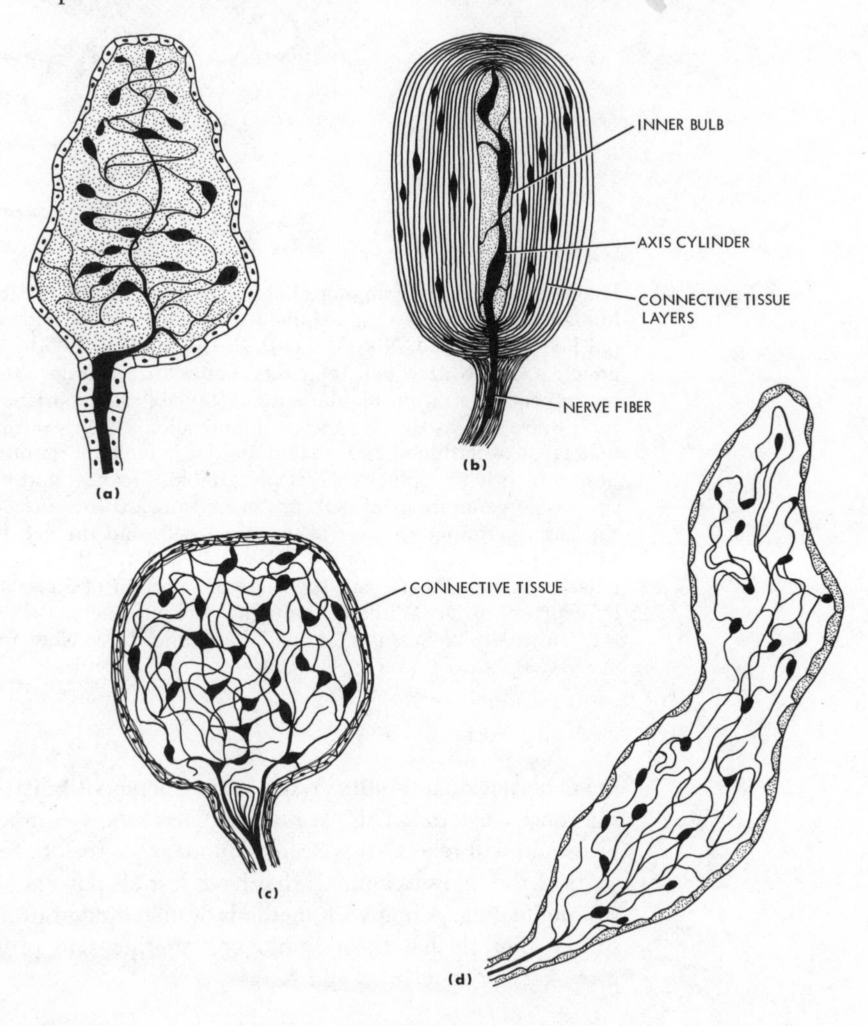

Through the reaction of pigment cells and changes in corneal thickness, the skin offers protection against injurious amounts of light. This organ, along with hair, feathers, or thick scales, insulates the body against sudden loss or gain of heat. It is a nonconducting retainer that includes elements assisting in body-temperature regulation. Heat losses and gains are always in delicate balance. The metabolic rate—the heat produced in the body over a period of time—is balanced against the rate of evaporation, rate of loss or gain due to radiation or convection as wind passes over the body, and the rate of storage of heat in the body. This metabolic rate usually is measured by rate of oxygen consumption. The heat generated per hour varies from around two-hundred British Thermal Units (BTU) in resting women (270 in men) to 1,100 BTUs walking at four miles per hour (men 1,500), or 1,000 to 1,400 BTUs dancing. In the latter, 0.9-pound (1 pint) of water may be lost by the average person.

Among other functions of the skin is the control of blood pressure and blood flow. Through moist skin membranes or glandular structures it has an active physiological part in absorption or elimination of materials. Elements for sex distinction (pheromones) and sexual attraction through coloration are also present. Coloration is a conspicuous property of the skin and its derivatives. Through matching or contrasting color patterns animals can blend with their backgrounds or stand out prominently (skunk). The human skin follows a characteristic life cycle from fetal to old age; epidermal proliferation shows an endogenous or circadian rhythm of cell activity.

Along with underlying muscles and bone structure, the skin, aided by fat distribution, is responsible for shaping the face and body contours. Human organisms have long been identified by the pattern of ridges and sulci on the finger pads (Figure 4-6). No two sets of whorls and loops are similar, except possibly those of monozygotic twins, and females have distinctly different patterns from males. Such patterns are established during the third and fourth fetal months. Human skin has a distinct odor that is peculiar for each individual—a chemical fingerprint.

A dynamic organ that continually undergoes growth, differentiation, and replacement in its reaction to external stimuli, the skin reflects the physiologic state of the body's internal environment. Wound healing, and in lower organisms total regeneration of certain parts, may take place. Man loses cells from the surface of the skin, and lower organisms can molt whole sheets of skin (salamander, snake). Food in the form of fat or blubber (whale) is stored in the deeper subcutaneous layer where it takes the form of pads or cushions of adipose tissue, serving also for insulation.

Skin produces numerous and different end products: e.g., secretions (sebum—an oily, waxy substance) and bony or horny structures. Irritating and even poisonous substances serve many animals as a defense against predators. Most of the complex biological syntheses occurring in the skin are involved in the production of secretions—e.g., sweat, sebum, milk (Figure 4-3), and proteinaceous keratin. The epidermis is also an effective barrier against the penetration of a wide variety of substances, the absorption of which depends on their physical and chemical properties as well as on the normal and abnormal state of the skin. Studies of cutaneous permeability, unlike transport investigations in other cell systems and biological membranes, are only beginning to yield useful information.

No other single organ of the vertebrate body has so large a total mass or

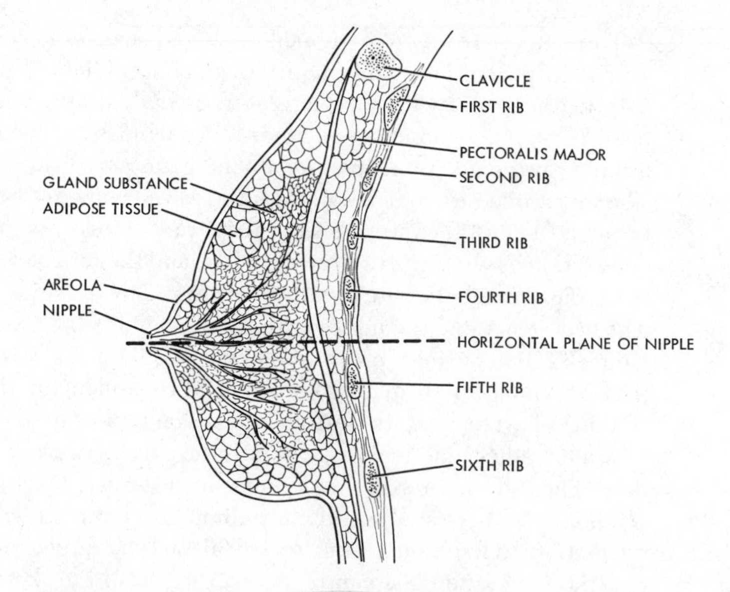

Figure 4-3 Another type of integumentary gland is the mammary gland. Diagrammatic section of the structure of the human mammary gland.

weight; with the exception of the lungs, none can compare with its area of free surface. For example, in an average sized man thirty-three years old, the weight of the skin with subcutaneous fat removed has been estimated to be between ten and eleven pounds, and that of a female between seven and eight pounds, with an area of approximately five to six square feet. Most of the skin consists of the dermis, with a relatively inactive metabolism, as shown by its small oxygen consumption compared to the epidermis [1]. In rabbits the weight of the skin may constitute 13 to 17 per cent of the total body weight, with some variation existing between sexes, among species, and during different seasons of the year. This percentage is reported to be even greater among reptiles. The skin, which forms about 5 per cent of the body weight in the smallest cat fetus, increases rapidly initially and then more slowly until a maximum of a little more than 20 per cent is attained at birth. A postnatal decrease is shown by the newborn cat, whereas the adult ratio is about 13 per cent.

The Skin of Man: The Epidermis

Human skin, like that of all vertebrates, has a dual origin. The epidermis arises from the surface layer of cells in the embryo, known as the nonnervous ectoderm. At first it is only a single layer of cells and constitutes the early periderm. This primary "skin" is characterized by the swollen appearance of the cells and their peculiar staining qualities, which indicate particular chemical properties. Shortly thereafter, mitotic activity accelerates and the dermis begins gradually to differentiate from the underlying mesoderm. By the third month of fetal life, the epidermis shows approximately five layers of cells, one of which is a basal layer, covered on its free surface with the single-layered epitrichium. At five months, hairs make their appearance and the fetus is soon almost completely covered with these epidermal derivatives. At this time the epitrichium, except for the palms and soles, is shed from the body; the epidermis and dermis continue to show mitotic activity and structural differentiation. Meanwhile, a subcutaneous fatty layer, also of mesodermal origin, has made an appearance between the dermis and the underlying connective tissue fascia covering the muscles.

Human adult epidermis (Figures 4-1, 4-4, and 4-5) includes a number of regions. Essentially cellular in character with little intercellular material and no vascularization, it has (1) a single-cell, basal, germinative layer from which successive generations of epidermal cells are produced, more distinct than in lower vertebrates; (2) a thick, living Malpighian layer of several zones; and (3) an outer, specialized, transparent, corneal layer of variable thickness. This stratum corneum of dead, flattened, keratinized cellular remains is largely responsible for the impermeability of the skin. Following its removal or chemical alteration, the remainder of the skin becomes freely permeable to water and dissolved substances.

Cells produced by mitosis in the germinative layer are pushed through the spinosum and granulosum zones of the Malpighian layer by successive generations of new cells toward the free surface of the epidermis. Each new generation is discarded from the surface of the overlying corneal stratum in flattened squames of dead cells. During this journey the cells become progressively modified in shape and composition. When they reach the corneal region, all nuclei are lost and the cytoplasm has been replaced by horny keratin (Figure 4-5). The process of keratinization, which has intrigued biologists for many years, is still not clearly understood. A point of interest has been why some of the progenitors of the epidermal cells retain mitotic capacity and remain in the germinative layer to give rise to future generations of epidermal cells and others lose this capacity and undergo these processes of irreversible differentiation (keratinization).

The human skin continues to change throughout life, as is obvious when contrasting the skin of a young child with that of an adult or aged person. It

Figure 4-4 Section of the epidermis from the sole at higher magnification. As the cells of the stratum Malpighi are pushed toward the surface of the skin, they become progressively larger and stainable granules appear in the cytoplasm; such cells constitute the stratum granulosum. The cells also become markedly flattened. The photograph emphasizes the characteristics of the cellular constituents of the several strata. Although nonliving in the stratum corneum, the cell remnants in "pressure areas" of the epidermis retain some cellular appearance, are firmly cemented together, and are not easily sloughed. (Courtesy of Dr. William Montagna, Oregon Regional Primate Research Center, and Academic Press, New York.)

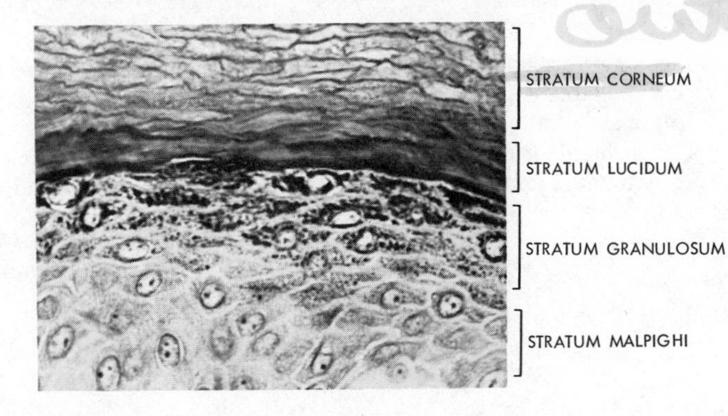

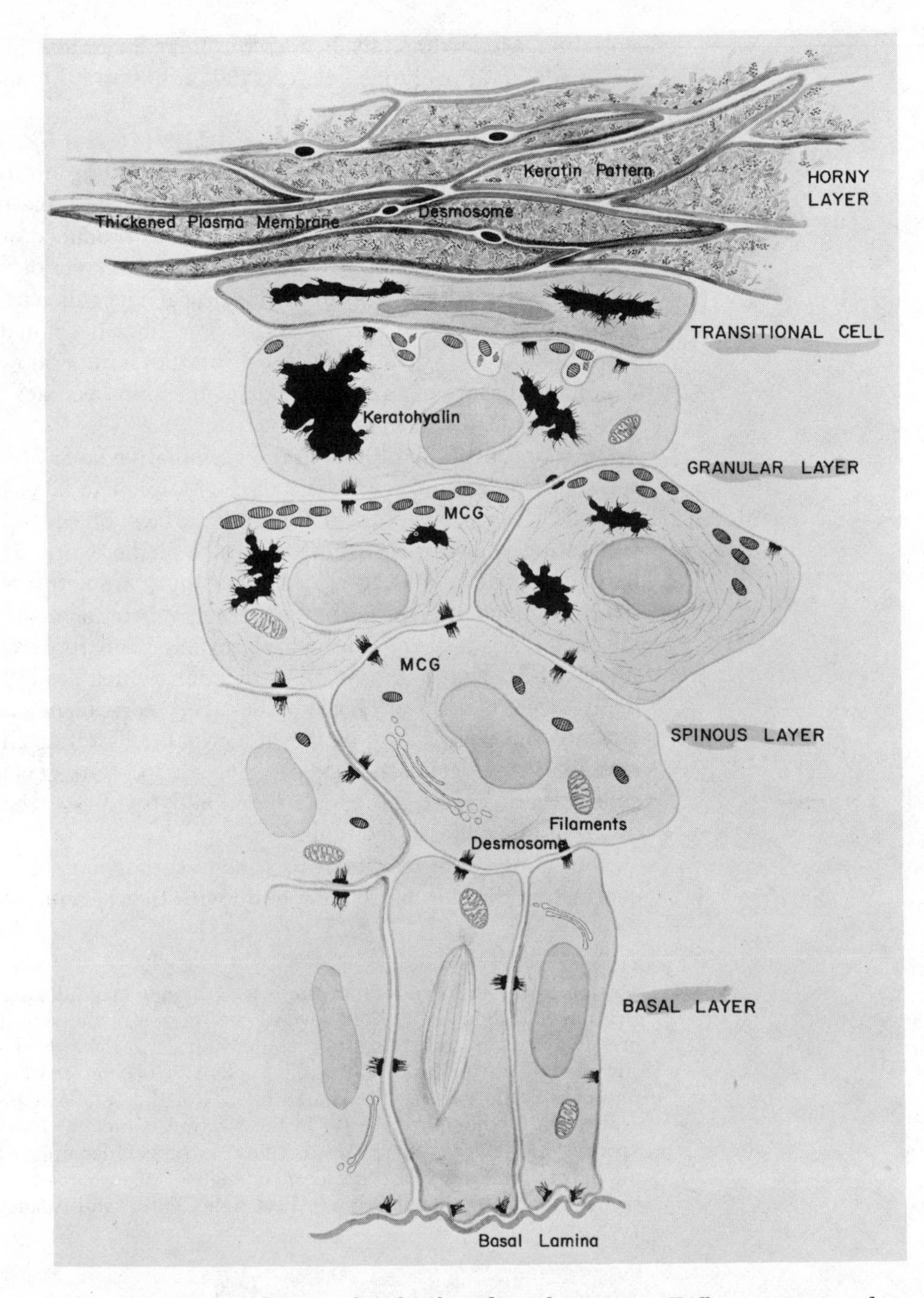

Figure 4-5 Schematic diagram showing details of epidermal structure. Differentiation products—filaments, membrane-coating granules (MCG), keratohyalin, and thickened plasma membrane—are shown in the different layers. Fully cornified cells contain much filament-matrix and exhibit the "keratin pattern." (Courtesy of William Montagna, Oregon Regional Primate Research Center, and Academic Press, New York.)

Figure 4-6 Scanning electron micrograph of the intact surface of a human finger tip, showing loose surface squames, orifices of two sweat glands, and several sulci and gyri. (Courtesy of W. H. Fahrenbach, Oregon Regional Primate Research Center.)

is thinnest over the eyeball, where it is also transparent, and thickest on the palms and soles even in the early fetus. Dorsal and extensor surfaces of the body have thicker skin than ventral and flexor surfaces; skin is also thicker and less supple in men than in women. Corns, callouses, and other local thickenings form in response to excessive friction or pressure, and freckles are a melanophore reaction to sunlight. With age the skin becomes thinner, discolored, and increasingly wrinkled, while chemical and structural changes occur throughout its several regions. Thickness of skin declines by about one half between the ages of twenty and eighty, due to structural changes in the collagen accounting for 750 per cent of the dry weight of skin. Wrinkles formed in smiling or by knitting the brow and the weathered skin of the aged result from the pull of muscles together with the loss of both elasticity and subcutaneous fat. The capacity of the skin to be stretched depends on its thickness, the presence of folds, elasticity and the composition of the dermis, firmness of attachment, and age; the greatest capacity is in skin from the abdomen.

The surface of the human epidermis is both smooth and heavily marked with ridges and furrows. It is characterized by the exit point of hairs, pores of sebaceous and sweat glands, and by wrinkles. Some areas (palms, soles) are hairless, others are not. The thickness and surface of the corneum, together with the amount and nature of the cutaneous secretions, make the epidermis rough or smooth, dry or moist. Tensil strength and resiliency vary. On the whole, inherited patterns in the skin, which are genetically established and determined by the orientation of dermal and other factors, make their appearance early in fetal life. The patterns of folds, ridges, creases, and flexor lines finally established remain unchanged throughout life and can be altered only by damage to the underlying dermis. The epidermis varies from site to site in the human body, as shown by differences in keratinization and morphology of surface patterns.

The epidermis and dermis of whole skin can be cleanly separated by maceration in acetic acid, brief treatment with trypsin or calcium chloride, heating to sixty degrees centigrade, digestion of thin shavings, treatment with weak ammonia solution, or use of 2N sodium bromide (Figure 4-7). The enzyme method, which leaves the cells alive, permits grafting, culture *in vitro*, supravital staining, or enzymical study.

Among lower vertebrates, the surface at the dermoepidermal junction is more or less flat, whereas in reptiles, birds, and mammals (man) epidermal cones or ridges of different sizes project downward into the dermis (Figure 4-7). Enclosed between them are dermal papillae that may contain capillary knots or encapsulated sensory receptors. The characteristic regional differences in the architecture of the underside of the epidermis can be correlated with the pattern of hair and sweat gland arrangements [20]. With advancing age these folds tend to flatten out. Such folds may provide for increased cell proliferation by increasing the area of the basal layer. The skin of the guinea pig, mouse, rat, and rabbit has no epidermal ridges, and the dermoepidermal junction is less intimate; consequently, as cell proliferation occurs vertically, the horny cells tend to be arranged in tiers. From the basal layer of epidermal cells, delicate

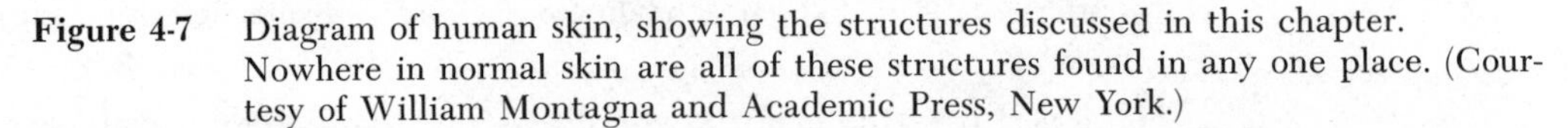

Figure 4-7 Diagram of human skin, showing the structures discussed in this chapter. Nowhere in normal skin are all of these structures found in any one place. (Courtesy of William Montagna and Academic Press, New York.)

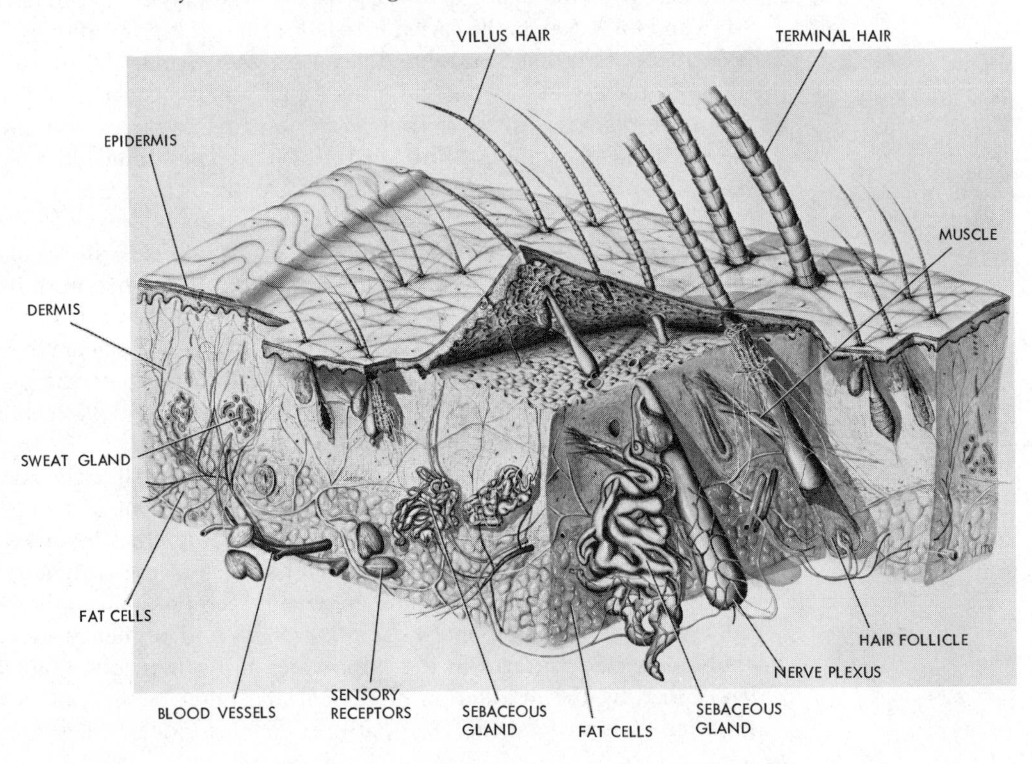

protoplasmic processes extend into the dermis to provide a closer union between the two layers. Anchoring roles have been assigned to all the connective tissue fibers: collaginous, elastic, and reticular.

The human epidermis, except for the palm and sole at one extreme and the cornea of the eye and the wall of the scrotal sac at the other, is remarkably constant in thickness; regional variations are attributable chiefly to the dermis. In general, it is around sixty to one hundred microns thick, whereas the corneum of the palm and sole alone may reach one hundred to six hundred microns [20, 24]. In the mouse and hamster the stratum Malpighi is two to three cells thick, the spinous and granular layers are not distinguishable, and the stratum corneum varies between five and ten layers of cells. Larger mammals possess a thicker epidermis in which layering is more distinct. Fur-bearing mammals show a thin epidermis that consists of little more than the Malpighian layer covered by a thin corneum. As no blood vessels penetrate the epidermis, all the metabolic requirements must be met solely by diffusion from the dermal blood vessels. Free nerve ends, branched or unbranched, may end in it or be associated with particular epithelial cells. Some of the changes that the epidermal cells undergo during their progression through the several layers can be correlated with the thickness of the tissue layers through which this diffusion occurs.

Supplying fresh epidermal cells, which are continually replenished, involves a complicated process of differentiation during which the cells become completely changed (Figure 4-5). For example, ultrastructural fibrils or filaments are arranged in bundles attached to desmosomes, and finer filaments associated with microvilli are unattached. These may be supportive in function, or those in the vicinity of the plasma membrane may be contractile action filaments. Various types of junctions occur between epidermal cells, the most common being the desmosomes (intercellular bridges, macula adhaerens), which account for the prickle appearance of the cells in the spinous layer. Fused membrane junctions—the zonula accludens—occur between cells showing active transport and between superficial cells. The keratohyaline granules develop in cells that lie within the zone of the granulosum. Keratohyaline is a complex of calcium, phospholipid, and protein that stains dark blue with haematoxylin. Wherever the stratum lucidum is present (sole), the substance eleidin develops. Another structural change is the marked thickening of cell membranes. The material within these membranes is keratin, a substance that is resistant to many factors including pH, high and low temperatures, and enzymatic digestion. For more information on this subject, consult the references at the end of this chapter.

The stratum corneum is an interesting layer. Transparent, colorless, generally thin except in areas of the body subject to friction and pressure, it can be stripped off in layers by cellulose adhesive tape. Cells are joined together by attachment plaques. The stratum corneum has a function in the reaction of skin to ultraviolet rays (sun tanning), contributes to skin color, and is chiefly responsible for protecting the body against diffusion of substances. Transfer of materials probably occurs in the liquid phase and is brought about by a concentration gradient within the layer itself. The corneum is far more hygroscopic than other keratinized materials, such as hair or horn. Dry skin (corneum) has a high electrical impedance to the flow of electric current and, thus, provides some protection. For further details on the corneum and skin as an electrical, thermal, and radiation barrier and protector and as a potential focus for multidisciplinary research, see Rushmer et al. [24]

KERATINIZATION IN THE EPIDERMIS. Keratinization is a process peculiar to the epidermis of amniotes, although to a lesser extent it may also take place in the skin of certain land-dwelling amphibians. Keratin is a high-molecular-weight protein with a complicated structural development. As the differentiating cells are crowded through the several zones of the granulosum and lucidum and outward into the corneum toward the surface of the epidermis, each undergoes a profound transformation from a nucleated, metabolizing unit to one that is dead, markedly flattened, nonnucleated, and keratinized (Figures 4-4 and 4-5). The first of these changes occurs in the prickle-cell layer and involves an increase in size and the initial formation of structural proteins, enzymes, and keratin precursor polypeptides. Subsequently, there occurs a flattening of cells that are still a part of the Malpighian layer. Further flattening takes place in the granulosum, and the nucleus is now pycnotic or fragmented, the cell membrane is thicker, and fine keratohyalin granules have become a conspicuous feature of the cytoplasm. The functional role of these is not known, although much work has resulted in various interpretations of the data [19]. This type of keratinization occurs in epidermal cells, internal root sheaths of hair follicles, and medulla of the hair. Only the thickest portions of human skin contain the lucidum type of keratin (sole of the foot, finger pad) as do the cortex and cuticle of the hair. A different chemical compound (eleidin), possibly a derivative of keratohyalin, is visible in the semitransparent, nonstainable stratum lucidum, which appears to be homogeneous. The skin pigment (melanin) is present in cellular melanocytes of the Malpighian layer. W. Montagna has likened the entire cutaneous structure to a huge glandular system with keratin, gland secretions, tooth enamel, and color pigments viewed as epidermal secretions.

Keratinization in scales on the tails of marsupial, rat, and mouse is essentially similar to that of reptiles and birds, and no granular layer appears. Lizards and snakes have cycles of keratinization. Sea lions and seals possess a thick phospholipid-rich horny layer that appears scalelike and is termed a hyperkeratotic condition. This is also the case in the skins of elephant, hippopotamus, and the armor-plated rhinoceros. Other regions of the human body undergoing keratinization, some in response to hormones (vagina), include the tongue and external auditory meatus. Differences in degree of keratinization are seen in hairy sites, palm and sole, and other regions of the body.

DESQUAMATION. Completely keratinized cells are finally lost by desquamation from the skin surface (Figure 4-6). It has been estimated that a seventy-year-old man, thus, has lost some forty pounds of cornified cells during his lifetime. This is another example, along with blood and sperm cells, of the amazing production of cells in several regions of the body to replace those lost during activity. In this case, it is a protective feature characteristic of land-dwelling vertebrates. Desquamation is a continuous process in hairy skin, but in pelted mammals the cells are trapped until the hair is molted. In the elephant seal, the horny layer is sloughed yearly as an intact sheet containing the molted hair. A regulatory mechanism controlling desquamation in man has been described. Normally, an equilibrium, conditioned on a sensitive feedback mechanism, exists between cell proliferation in the basal layer of the epidermis and cell loss, whereby the thickness of the epidermis changes very slightly. Under environmental stress, this equilibrium becomes modified and the epidermal layer, especially its corneal component, thickens markedly. Actually, there is considerable variation in the equilibrium of this relationship, as indicated by skin thickness in different regions of the body. Thyroid hormones accelerate, whereas adrenocortical steroids retard, both cell proliferation and desquamation. A local, tissue-specific hormone has been described as inhibiting mitosis and, thereby, controlling the balance of cell proliferation and cell loss. When hormone concentration is low, cells produce the proteins that promote mitosis; when it is high, cell differentiation is favored.

Molting in lower vertebrates occurs periodically and can be specifically controlled. In mammals, cornified cells flake off in small aggregates from the free epidermal surface (human dandruff in hairy sites) because a change, different from that in the reptilian molt, takes place in the intercellular pathway of the horny layer where fused cellular junctions change in character so that dissolved substances cannot pass. Junctions break, possibly by enzymic hydrolysis, and small lysomal bodies pass through plasma membranes into the intercellular spaces and release destructive enzymes. Investigators have speculated on whether this change is simply a physical one of drying or a chemical decomposition of the mucolipoprotein. Conversely, several suggestions have been advanced to account for the accumulation of cells on the palm and sole in response to environmental stress. One of the more interesting speculations suggests a change in the differentiation process affecting the intercellular cement at least, and a modification of the hormonal feedback pattern normally accounting for the maintenance of balance between sloughing and proliferation in the basal layer.

GROWTH OF THE EPIDERMIS. In vertebrates, mitosis is confined mainly to the stratum germinativum of the epidermis, although it also occurs to a lesser degree throughout the Malpighian layer of amphibian and fish skins characterized by little or no keratinization. The rate of mitosis is variable. Hair growth occurs in cycles: during the early phase of each cycle the surrounding epidermis is increased in thickness, and then the process subsides. Where single waves of hair growth occur, neighboring follicles are in phase as well as related cyclical changes in epidermal mitosis. However, in humans, the neighboring follicles are in different stages of mitotic activity, so epidermal cycles are not seen. A thick, keratinized epidermis in the rabbit can be regenerated within two weeks from cells that are "seeded" on an epidermally denuded area. A host of different factors affects mitosis, including loss of corneal cells, shock or stress, temperature, muscular exercise, level of certain hormones, age of animal, stage of estrous cycle, and blood supply as the source of raw materials. Epidermal growth is cyclic and exhibits a diurnal circadian rhythm along with a less defined estrous rhythm. The period of mitotic frequency in the nocturnal mouse and rat epidermis is reported to be greatest around noon to 1 P.M. and slightest around 10 P.M. The number of dividing cells is more than double that at night. In man, the mitotic cycle is highest in the night hours. Under typical conditions, human epidermis can duplicate itself in about twenty-seven days.

Growth has been described as the accretionary type in which the semi-specialized cells of the basal generative layer retain the ability to divide mitotically and, hence, serve as a progenitor pool for new generations of cells. Cells of other layers have lost this capability. Exceptions can be found occasionally in noncornified skins of anamniotes. Clearly, the generative cells are multipotential as to these different properties, but their capacity to undergo division must be masked in some fashion because it occurs only while in the environment of the basal zone. When a daughter cell in a mitotic division appears outside the generative area, it suddenly acquires morphogenetic and physiological properties of functionally differentiating cells. However, because the basal cells give rise only to units of the epidermis, they have attained a certain level of functional specialization.

A slightly more advanced stage of differentiation of the generative cell is seen in the papillae of hairs and feathers; the daughter cells differentiate into localized cornified structures showing various specializations, but not into generalized epidermal cells. Subsequently, the process of keratinization is essentially similar, but the cells remain adherent through some peculiar property of the intercellular material. Other examples include specialized epidermal derivatives such as the nail, claw, hoof, horns of several types, and epidermal scales. There are instances in which highly specialized epidermal cells divide to replace lost or destroyed parts—for example, the mammary gland. In this specialized type of alveolar sweat gland of the apocrine type, growth occurs in the secretory cells after some protoplasm is lost during milk secretion. Other examples include sebaceous glands and the mucous gland of amphibian skin.

WOUND HEALING. Regeneration among the warm-blooded amniotes is limited to wound healing, in which the skin is involved. Much attention in the form of voluminous literature from both basic and practical standpoints has been given to this phenomenon, particularly, for obvious reasons, in man. During epithelialization and reunion of the connective tissue components, which are generally the principal processes, the adjacent epithelium migrates inward to cover the wound. Little "granulation" tissue forms if the wound has previously been sutured. Conversely, when there is an open wound to be covered, marked inflammation and tissue destruction may first occur, accompanied by initial proliferation in tissues affected by the injury. As the proliferating epithelium migrates to cover the wound, it moves beneath the fibrin clot to unite eventually with the dermal connective tissue; cell division is limited primarily to marginal cells. At first, the wound epithelium thickens; buds from this area grow inward only to be pinched off later as the thickening regresses. This type of skin reaction has been likened to that occurring in the healing of the adult amphibian limb following amputation. Phosphatases and proteolytic enzymes with unknown functions have been found in both amphibians and mammals. Wound healing, more rapid in the young than in adults, is, unlike amphibian regeneration, influenced by nutritional deficiencies (proteins, vitamin C). Investigators have given attention to the effects of hormones and X radiation.

In certain portions of the mammalian body, large wounds become reduced in size within a few days as a result of the activity of the encircling connective tissue under the influence of integumental or dermal muscles (panniculus carnosus) attached to the skin. These are the muscles by which a horse twitches its skin, a dog shakes its wet coat, the armadillo and European hedgehog roll themselves into a ball, a snake moves by traction on the ground with its ventral epidermal scales, and birds fluff their feathers in regulating body temperature. Other dermal muscles, e.g., the orbicularis—are responsible for the expression of human emotions, such as wrinkling the forehead and other skin movements of the neck and head.

Dermis Beneath the epidermis is the second and thicker layer of the skin, the tough dermis or corium of collagenous connective tissue (Figures 4-1 and 4-7). Although it also is elastic, the connective tissue can be stretched only slightly. In mammals such as horse, hippopotamus, and walrus, dermis in certain body parts has little elasticity. Structurally different in every way, it consists largely of collagenous and intertwined elastic-felted fibers enmeshed in the gel-like matrix. These give it the characteristic tensile strength. Leather, which is tanned collagen, has even been made from human dermis. Small blood vessels and nerves penetrate this region; hair follicles and sebaceous and sweat glands are located here; and sensory organs of tactile and temperature functions are abundant. Bundles of smooth muscle fibers—the arrectores pilorum muscles of ectodermal origin—are attached to hair follicles. Smooth muscles fibers constitute a relatively continuous layer, the tunica dartos, in the dermis of the scrotum and penis. Such fibers are also abundant in the nipple and aureola of the breast (Figure 4-3) as well as in the perineal and circumanal regions. Living dermis is a resilient fibrous gel that changes markedly with age, as is shown by sagging and wrinkled skin, as elasticity is lost. This layer is the distinctive part of vertebrate skin; it is absent as such from invertebrates.

Figure 4-8 Scanning electron micrograph of the underside of split skin from the chest, showing sebaceous glands and the honeycomb appearance of the dermoepidermal junction. (Courtesy W. H. Fahrenbach.)

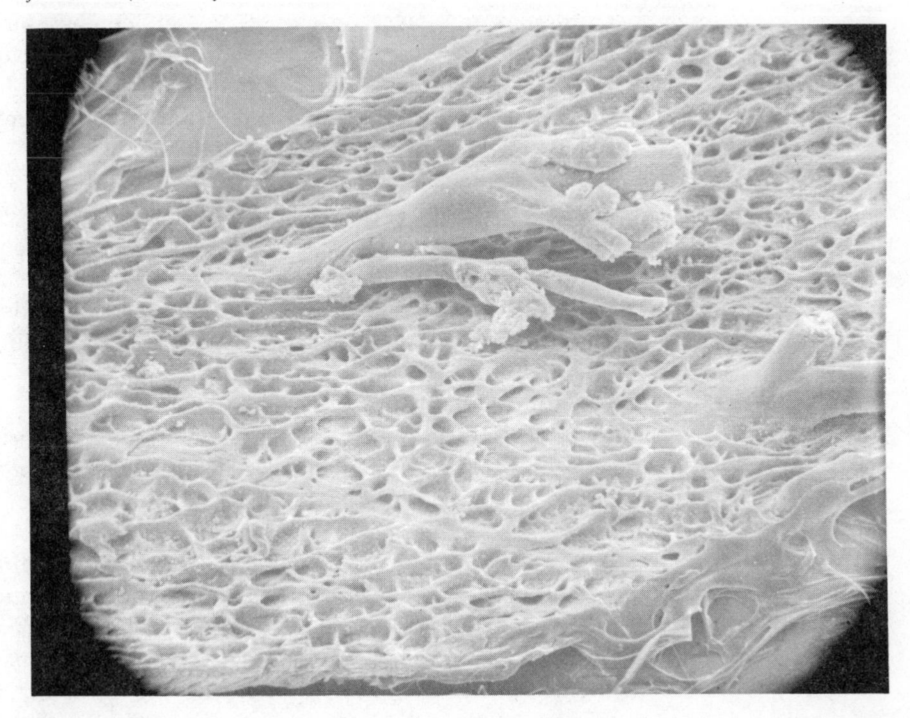

Predominantly of mesodermal origin from the dermatome and lateral somatic mesoderm of the embryo, the thickness of the dermis varies in different regions of the mammalian body, being thickest on dorsal surfaces as well as on palms and soles of primates. Its inner surface is not clearly marked but grades off imperceptibly into the fatty subcutaneous layer. Two parts make up the dermis. A superficial papillary region contains a number of specialized elements: reticular or possibly precollagenous material, collagenous and elastic protein fibers (elastin is lacking in cyclostomes), abundant capillary and lymphatic vessels, nerve fibers and sensory structures, and deposits of glycogen and adipose tissue along with sebaceous and sweat glands. These are all embedded in the matrix. Several types of cells occur: fibrocytes, which secrete collagen and matrix materials; phagocytic macrophages (histocytes); mast cells; and pigment cells. Also, various kinds of dermal (scales) and epidermal derivatives (hair or feathers, glands of mucous, serous or mammary types) occur among the several vertebrate classes. Teeth are laid down in this zone and the basal lamina. In mammals, the surface adjacent to the epidermis may be molded around the ridges and valleys of the epidermis in the form of dermal papillae. This layer also contributes to the connective tissue sheaths, excapsulating hair and feather follicles, as well as to the ducts of glands projecting through the papillary layer. Reticular fibers constitute, or are a major component of, the basement membrane (basal lamina) underneath the epidermis.

Sebaceous glands occur in most regions of the human body. There may be as many as nine hundred glands per centimeter on the scalp and face; there may be fewer than fifty on the forearm [21]; and they may be absent entirely from the palm and sole. Eccrine sweat glands function as thermoregulators, although the role of the apocrine glands is not clear. The former are widely and uniformly distributed over the body, but the apocrine type is found chiefly in the axilla, pubic area, areola, and nipple. Specialized types of sebaceous glands occur in man and lower vertebrates at skin openings, such as the Meibomian glands on the eyelids, cerumenous in the external air canals, the nostrils, in the mouth (Fordyce's glands), in the nipples (Montgomery's glands); also glands are found in the lips, prepuce, and anogenital regions (scent glands).

The deeper fibrous reticular region of the mammalian dermis contains, in addition to many elements of the papillary layer, bundles of coarse, branching collagenous fibers arranged horizontally at angles to each other and parallel to the epidermis. Some fibers extend downward into the framework of the underlying subcutaneous or adipose tissue, probably serving as anchors for the skin. Between the collagenous fibers is abundant elastic tissue, that sheaths any integumentary derivatives lying wholly or in part within this reticular layer. Much has been written about the histochemistry, origin, and structure of any one of the cellular units.

Evidence indicates that the ground substance has several physiological functions. Probably, the reticular fibers are "precollagenous" elements, whereas fibroblasts are the progenitors of connective tissue cells such as histocytes and mast cells and, in fact, of all other types of cells found in the dermis. Mast cells contain cytoplasmic granules and stain metachromatically with toluidine blue. These granules can be demonstrated by a number of techniques, with certain differences occurring among species. Mast cells are reported to increase in various itching skin diseases and may contain histamine, as in the case of mast cell tumors. The degree of granulation responds to

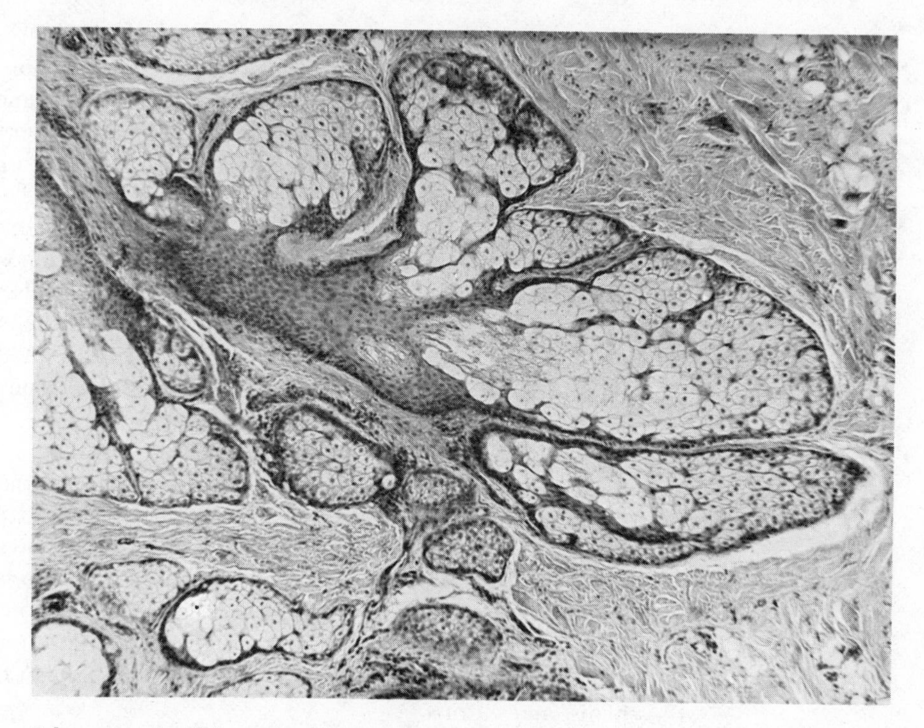

Figure 4-9 Sebaceous gland from the frontal area of the scalp. (Courtesy of William Montagna and the Academic Press, New York.)

trauma, mechanical stimuli, thyroxine deficiency or excess, and to other hormones.

The structure, distribution, and physiology of cutaneous nerves and sensory organs have been studied by a variety of techniques (Figure 4-7). Several plexi have been investigated; a deep cutaneous plexus is present in the panniculus adiposus with fibers extending to a superficial subcutaneous nerve plexus beneath the dermoepidermal boundary. Other nerve meshes have been reported, indicating that the dermis is well supplied with interconnected nerves and nerve plexi. Until recently, it was thought that many morphologically different types of sensory organs existed in the dermis, each of which had a specific physiological role, and studies were devoted to the identification of these roles. This view appears to be changing, with similarities being reported in the origin and histochemical and functional properties of the cutaneous receptors, as well as between species. Of the so-called sensory end organs, the following are reported to be morphologically identifiable: Meissner and Vater-Pacine corpuscles, hederiform endings, mucocutaneous end organs, and the dermal and hair follicular nerve networks. Cutaneous sensory nerves are distributed in regions where hair follicles are abundant; whereas in locations of sparse follicles, the networks are profuse. Other dermal structures, such as sweat and sebaceous glands and smooth muscle fibers, affect the pattern of nerve nets.

Little can be said in the space available about the complex types of patterns in the cutaneous blood vessels (Figure 4-7). The pattern is affected by many factors, ranging from thickness of the dermis and its relation to the epidermis to types and distribution of cutaneous appendages and the relationship of the skin to underlying structures. Throughout the dermis are large and small networks of anastomosing arterioles and capillaries with minor networks associated particularly with hair follicles and sweat glands. Numerous capillary loops extend into the dermal papillary ridges, chiefly in friction areas. Associated with the networks of arterioles are venous plexi, with preferential channels between the arterioles and venules, surrounded by one layer of smooth muscle. There are also direct shunts that permit blood to bypass the capillaries, particularly by way of large arteriovenous anastomoses under vasomotor control in the fingers and toes, pinnae of the ears, and elsewhere. These factors, together with several types of vascular shunts and the capacity of venules to contract and dilate, provide the mechanism of this highly versatile organ for cutaneous regulation of heat and blood pressure, respiration among the amphibians, and other functions.

The Subcutaneous Layer

Beneath the dermis, and considered by some as the third component of the skin, lies the subcutaneous transitional zone characterized by the presence of abundant fatty tissue (panniculus adiposus), vascular and nervous elements, and fibrous attachment with the dense connective tissue fascia covering the deeper lying muscles. Owing to its elasticity, mammalian skin can be pinched into folds, but when released it returns to its natural position, an indication of the firmness of this attachment, particularly in the human face, palm, sole, glans penis, and clitoris.

The boundary between subcutaneous tissue and dermis is less clearly demarcated than between the dermis and epidermis due to accumulation of adipose tissue. This layer varies widely in thickness in different parts of the body, among individuals, and even among species. Fat is abundant in some mammals (seal, whale, walrus). Among the factors affecting the deposition of fat in man are the types and abundance of food, alcohol, drugs, neuroendocrine disturbances, genetic inheritance, stress, and bodily activity. The brown fat of true hibernators has a vital function in arousal from the state of hibernation. Another distinctive characteristic is the loose organization of the felted reticular fiber network that permits increased freedom of motion in the underlying muscles. Here the vascular network similarly is extensive and capable of holding a large fraction of the total amount of blood in the body.

Comparative Anatomy of the Skin

Protovertebrates

It is interesting that the transparent skin of the primitive chordate *Branchiostoma lanceolatum*, amphioxus, foreshadows the bilaminar structure of the vertebrate skin. No recognizable predecessor of this type of organization is found in invertebrates. It is known that amphioxus possesses a number of other anatomical features characteristic of vertebrates. The epidermis includes a single layer of columnar epithelial cells that is characterized by an abundance of tonofilaments and a border of microvilli, among which are sensory cells, cuticularcytes, and goblet-shaped, unicellular glandular cells (Figure 4-10). Beneath this lies a thin, gelatinous, connective tissue dermis overlying the muscles. In addition, the free epidermal surface is protected by a thin fibroprotein cuticle, but in the larval stage it is ciliated.

Another interesting type of protovertebrate skin is found in the tunicate, *Molgula manhattensis*. The tunic, a complicated structure, exhibits considerable power over regeneration. The cellular layer, resembling the hypodermis or

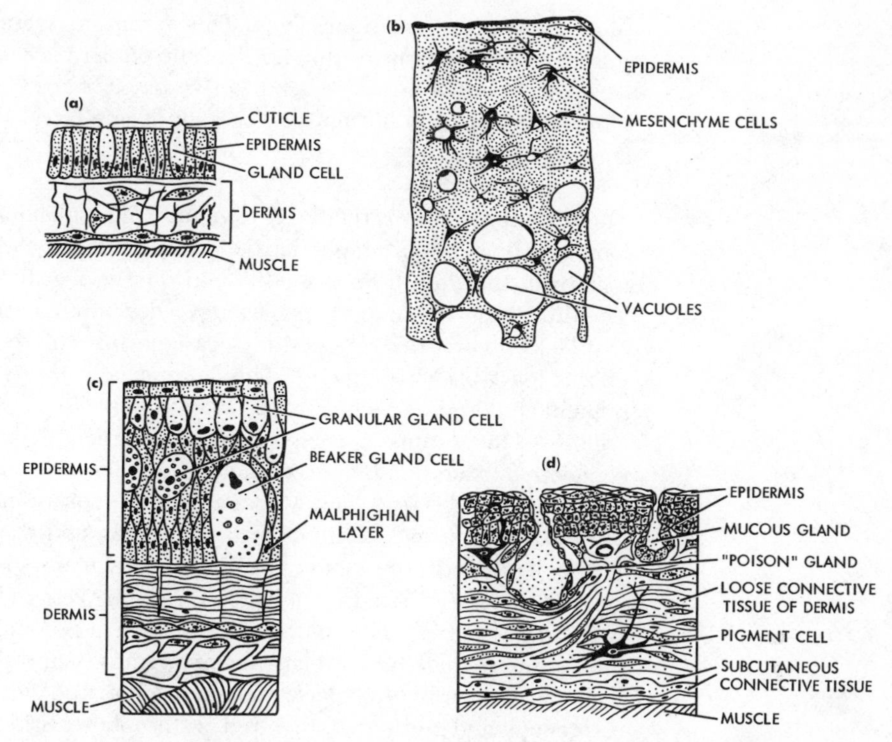

Figure 4-10 (a) A section through the integument of amphioxus (after Haller); (b) Section through the mantle of a tunicate, *Phallusia.* The wandering mesenchyme cells secrete the intercellular *tunicin* (after O. Hertwig); (c) Diagrammatic section through the integument of a lamprey eel, *Petromyzon* (after Haller); (d) Section through the integument of a frog (after Haller).

skin of invertebrates, secretes a thick, tough, protective mantle that contains tunicin and also scattered cells and fibers (Figure 4-10). Tunicin appears to be chemically related to cellulose, a substance not found elsewhere among the chordates nor outside the plant kingdom. Stellate mesenchyme cells, fibers, blood vessels, and nerve fibers invade the mantle.

Cyclostomes Cyclostomes, Agnatha, the cladistically oldest of living vertebrates (the jawless, limbless lampreys and marine hagfish), possess a highly glandular skin (Figure 4-10). Unicellular beaker and granular glands occur among the thick layer of epidermal cells, and mitosis is restricted to the basal layers. These glands secrete a thin cuticle on the surface of the skin. Unlike the cells of higher vertebrate skin, little change occurs in the shape of cells in the Malpighian layer. However, those in the outermost layer possess a border of microvilli and may appear striated. Epidermal cornification results in horny teeth within the buccal cavity and on the rasping tongue. A peculiarity of the hagfish is the presence of a single row of conspicuously large multicellular epidermal slime glands that extend on either side of the elongated body and open exteriorly by large pores. The animal is capable of secreting a profuse mucous slime when stimulated. It has been estimated that an irritated, average-sized hagfish may produce as much as six or seven hundred cubic centimeters of watery mucus at one time. Some species, when in the resting state, are reported to secrete a slime cocoon around themselves. The dermis in these descendants of the primitive ostracoderms is thinner than the epidermis and consists of an interwoven network of nonelastic connective tissue fibers. Control of the melanophores is exclusively hormonal. The ammocoete larval skin resembles that of amphioxus.

Fish The scaly skin of fish is relatively thin and highly glandular. In contrast to the loosely fitting skin of amphibians and birds and the elastic, wrinkled skin of mammals, that in fish fits smoothly and tightly over the underlying muscles. The thin epidermis in most species never develops a corneal layer and generally lacks color cells, except for occasional irregularly distributed melanophores (dark-brown pigment). This is especially the case in shark (elasmobranch) epidermis, which is darker than in most fish. There are generally two regions in the dermis of a teleost fish: a superficial layer of compact, fibrous connective tissue tied by perpendicular bundles of connective tissue fibers to the surface of the scale and a deeper portion consisting of loose connective tissue, with its typical content of cells, nerves, and capillaries.

The depth and variety of colors exhibited by many species of coral reef fish (among these are the parrot fish [6,13]) of the West Indies and Bermuda are particularly striking. The color cells, all of which occur in the upper portion of the dermis, include brown-black melanophores with short or elongated pseudopodia with brown to black melanin; yellow-red xanthophores containing carotenoids and pterines; orange-red erythrophores; iridocytes or guanophores containing crystals of nitrogenous guanin arranged in a network; noncellular translucent bodies with a blue pigment (or blue pigment diffused through the tissue); opalescent bodies; and rodlike doubly refractive bodies resembling guanin crystals. Reflection from iridocytes produces the silvery color. Blue pigment is relatively rare among animals—in most insects, fish, and birds it results from the refraction of light. The blue head (*Thalassoma bifasciatum*), which shows one of the most brilliant blues among tropical coral reef fish, has no blue pigment. Its color appears to emanate from regions of the skin supplied with melanophores overlaid by clusters of iridocytes that give an interference effect. Chemical analysis of the blue pigment in a Mediterranean wrasse, *Crenilabrus pavo*, has revealed the presence of a carotenoid albumin.

The extraordinary complex of color-producing tissues found among teleost fish provides an ample structural basis for color changes. In the several types of chromatophores, the concentration and dispersal of pigment, presenting smaller or larger light absorptive surfaces and concealing or revealing the refractive iridocytes and the blue pigment, are believed to be the mechanisms involved. These pigments, together with the physical structure of the skin plus the transparency of the epidermis, all contribute to the colors that make fish so attractive. This color change is produced by contractile fibers within the chromatophores. An expanded cell with pigment distributed into the cellular processes results in a deeper or darker color. Certain fish are able to discriminate between illuminated bodies that reflect light of equal brightness but of different wave lengths. Melanophores are the melanin-bearing chromatophores largely responsible for the adaptive color changes in cold-blooded vertebrates. In most teleosts these are controlled by two sets of autonomic nerve fibers: melanin-aggregating (adrenergic) and melanin-dispersing (cholinergic), with several neurons to each melanophore. Some bottom-dwelling flatfish (plaice) can change color in a matter of seconds to blend into the background. In elasmobranchs, pigment-clumping nerves are present in dogfish (*Mustelus, Squalus*), whereas in *Scyllium* and *Raia* (skate) only hormonal control exists. A melanin-clumping hormone has been reported in *Raia* and *Scyllium*.

Among deep-sea teleost fish of the suborders Iniomi and Stomatioidae, certain sharks, and the West Coast estuarine midshipmen, *Porichthys* (a relative of the toadfish), to mention a few, skin glands have become modified to form light-producing organs, or photophores [23,29]. These have definite arrangements in patterns characteristic of the species and may serve the general functions of warning, defense, and attraction. The "cold" light produced does not project any distance. Generally, the organ consists of a glandular portion that produces a luminous secretion and is bounded on one side by a translucent lens of modified cells and on the other by a concave layer of flattened cells, called the reflector, plus a single-cell layer of pigment cells (Figure 4-11). In the shark, *Spinax*, the photophore develops by modification of cells in the basal germinative layer of the epidermis. These gland cells form a cup that lies beneath the epidermis and, in association with other enlarged cells, forms a

Figure 4-11 Light organs of fish: (a) the teleost, *Cyclothone* (after Brauer); (b) the teleost, *Porichthys* (after Greene).

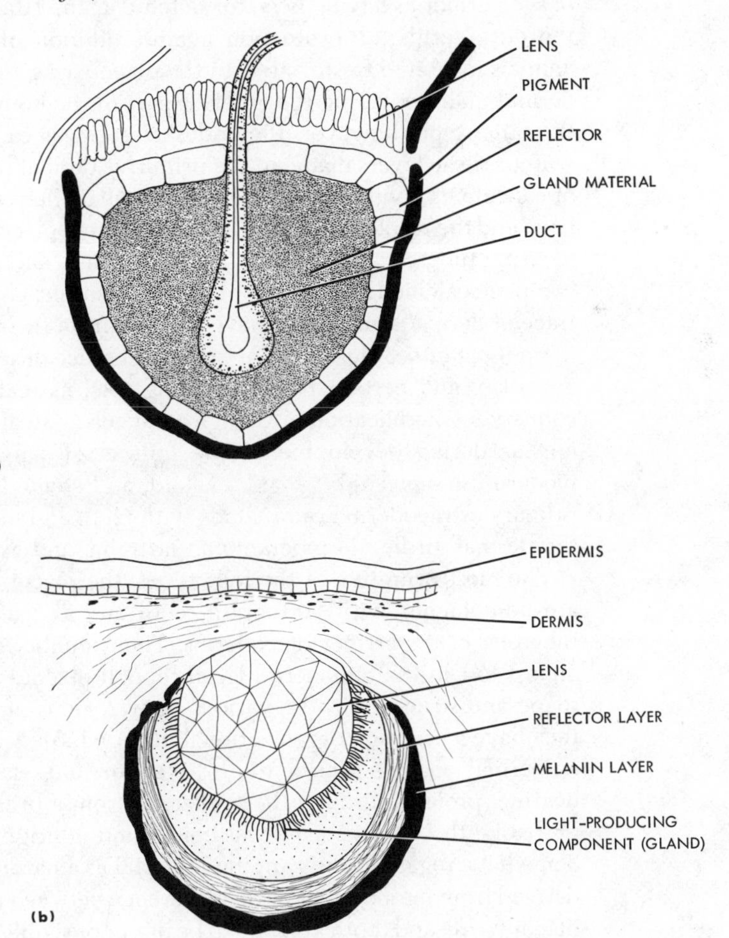

lens. Illumination arises through the oxidation of lucifern, a phosphorus-containing compound, by the enzyme luciferase. In some types of photophores, luminous bacteria appear to be the source of illumination.

Free nerve endings and touch receptors, abundant in the epidermis and dermis, have multiple functions. They serve for thigmotactic orientation (mechanical contact and currents) and avoidance and also may aid in discriminating between palatable and nonpalatable food. Certain avoidance reactions are mediated by the integumentary lateral line system. The common catfish possesses taste buds even in the tail skin. In fish and larval amphibians, the lateral line system consists of either open or closed tubes in the skin of the head and extends tailward within the intersomitic septum.

A dermal armor (or remnants of one) can be found in various guises throughout the classes of vertebrates both living and extinct. The oldest and most primitive of fossil vertebrates, the small, sedentary, bottom-dwelling, nonpredaceous, fresh-water ostracoderms, found in Silurian and Devonian deposits, were heavily armored with dermal plates. Placoderms, the choanichthyes, and other primitive types also were armored, but in not one of the remains can an epidermis be found. Probably little typical dermis, as we know it, was present in the skin. Remnants can be related to dermal structures present among living vertebrates. Students have speculated about the function of such armor as having been for defense against the gigantic water scorpions, the eurypterids, or protection against dilution of body fluids because the animals inhabited fresh water. In the placoderms, the armor consisted of bony dermal plates covering the anterior part of the body and free caudal scales of the same type; however, the endoskeleton was cartilaginous.

Four tissue layers made up the primitive dermal plates: a central thick layer of cancellous bone, with concentric lamellae, between two layers of lamellar bone and the whole topped with dentine. Some placoderms showed a reduction (dentine tubercles) or absence of the dentine, and in the arthrodires a hard, compact, calcified layer had replaced the dentine. Cyclostomes, which lack any trace of dermal armor, may have evolved from the armorless Silurian ostracoderm, *Jamoytius*. Most believe that the absence of scales in cyclostomes, many Siluridae, and certain bottom-feeding teleosts (catfish) must be taken as a convergent modification. Although adult eels are scaleless, temporary remnants appear during development. The four chief types of scales found among modern fish—placoid, ganoid, cycloid, and ctenoid—have evolved from this primary ostracoderm armor, along with teeth and such skeletal components as the dermal girdle, dermocranium, gastralia, and osteoderms [23,25,27,29].

The most primitive of the four types, the placoid scales of elasmobranchs, are homologous with vertebrate teeth and also are related to the dentine tubercles of the extinct placoderms. Their number is reduced in the rays and almost lost in the chimaeras. Even though placoid scales differ somewhat in shape and structure in the various body regions of the dogfish, for example, they have a common organizational pattern [25]. A tapering spine, capped by epidermal enamel (like a tooth) over an inner layer of hard mesodermal dentine, projects through the epidermis from a broad basal plate of dentine; probably, this type of scale was compound in origin (Figures 4-12 and 4-13). Some have regarded the capping material as a harder vitrodentine that also is derived from mesoderm, but the difference between the dentine and enamel of placoid scale and tooth appears to be of minor significance. Obvious transitions

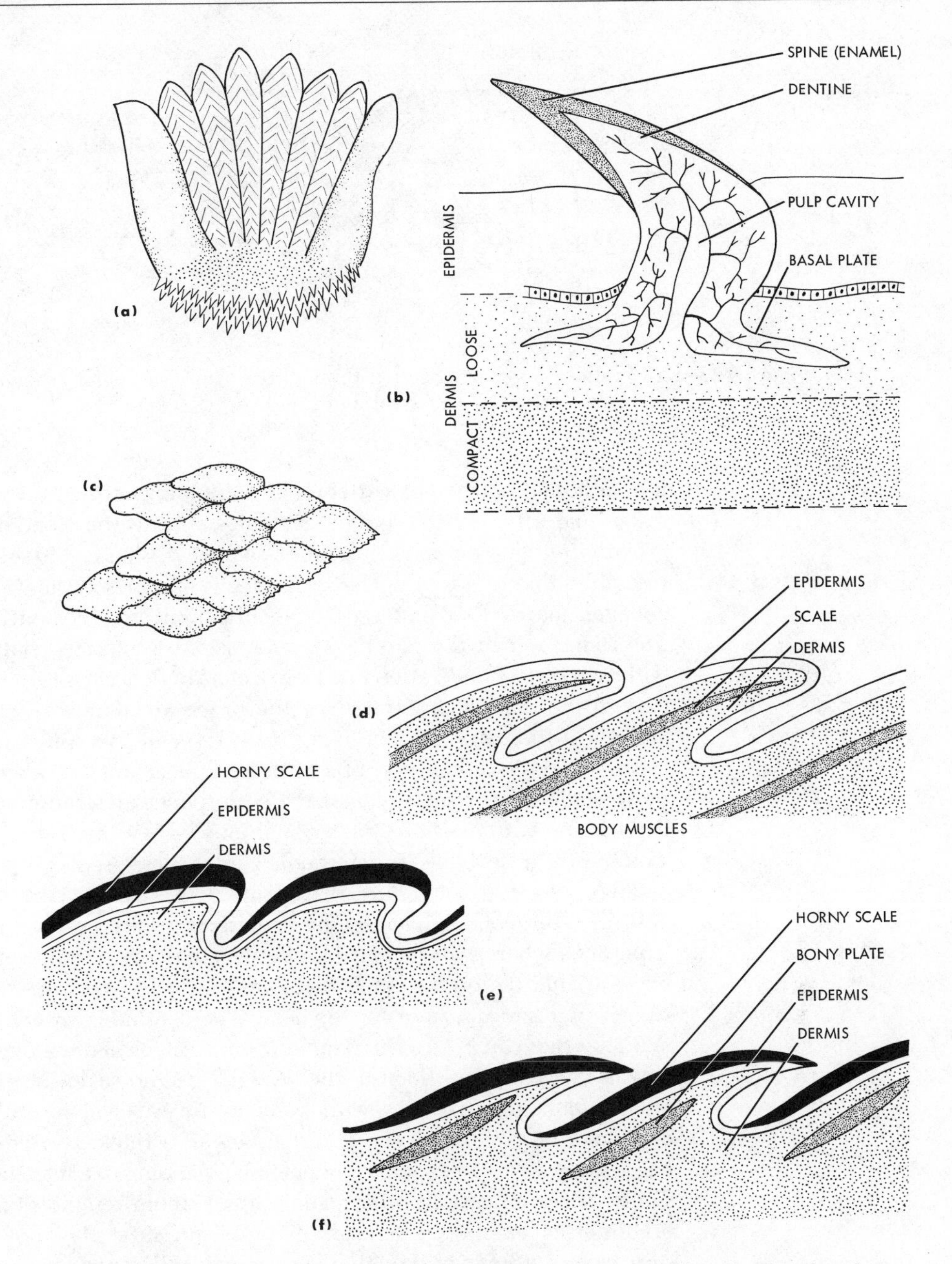

Figure 4-12 Types of vertebrate scales: (a) ctenoid scale of a bony fish; (b) section of a placoid scale; (c) ganoid scales of the garpike; (d) position of scales in the skin of a teleost fish; (e) horny epidermal scales of the snake; (f) diagram of a vertical section of lizard skin showing bony dermal plates beneath horny epidermal scales.

from scales to rows of teeth on the elasmobranch jaw lends additional evidence for the evolution of teeth from this type of scale [23,27]. The scales account for the rough surface of shark leather (shagreen).

New placoid scales, unlike other types of scales, continue to be formed throughout life, whereas fully formed scales cease growth and may eventually become worn or torn out (Figure 4-13). Scale addition is a necessary feature as

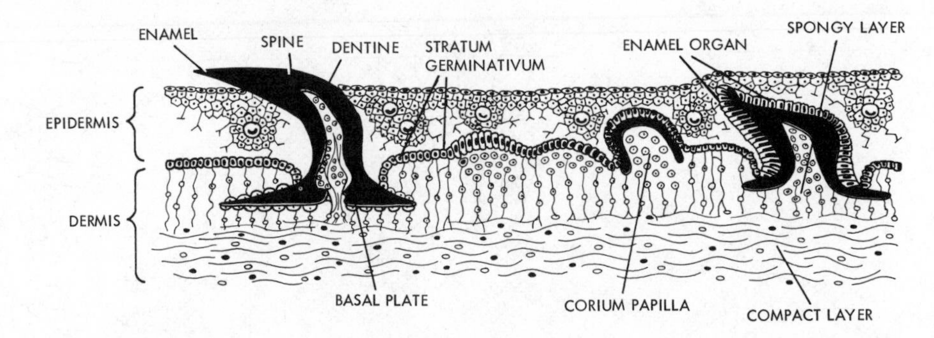

Figure 4-13 Diagram of a vertical section through the skin of a dogfish, showing development of placoid scales. (After Neal and Rand.)

fish, although not higher vertebrates, continue to grow in size after reaching sexual maturity. Other types of fish scales, such as the ctenoid, show periodic growth rings that determine the age of a teleost fish. Placoid scale development resembles that of a tooth. The surface cells of a mesodermal papilla secrete a collagen matrix that undergoes crystallization (hydroxyapatite plus calcite). The remainder of the papilla becomes soft, vascularized pulp well supplied with nerves. Crystallization produces a cone of dentine with a broad anchoring base in the dermis penetrated by the processes of the secreting cells. With growth of the cone, the overlying basal layer of the epidermis—the enamel organ—deposits a thin layer of hard crystalline enamel. As a result of these two processes, the growing point of the scale penetrates the epidermis. Enamel deposition continues until the scale attains its definitive size. Dentine differs structurally from bone in being noncellular, nonvascular, and lacking nerve endings.

In the sturgeon, *Acipenser,* the ganoid scales are large, isolated, platelike structures; whereas in the garpike Lepidosteus and certain related forms, the scales are hard, rhombic plates that fit edge to edge to form a complete armor. Larger plates form armor for the head. The dentine layer consists of separate denticles that can pierce the skin; new ones are formed with age. In *Amia,* the bowfin or Mississippi dogfish, the vestigial ganoid scales on the head consist only of collaginous plates with bony particles. Cycloid scales are present elsewhere on the body. The outer surface of a thick, bony ganoid scale is a hard, shiny layer of calcified noncellular ganoin that may be lacking from similar scales in certain other ganoid fish. Ganoin resembles enamel, but the origins of the two differ. Beneath this outer surface is a layer of dentine that, in turn, covers either a middle thin layer of cancellous bone over a lower layer of lamellar bone (isopedine) or just a layer of lamellar bone (*Polypterus*). Growth occurs in such a scale by the deposition throughout life of layers of ganoin and of lamellar bone on the underside and laterally. In embryonic development, the scale originates as a bony plate in the dermis.

Further evolutionary reduction from the dermal scales of ancient placoderms is seen in cycloid (cod) and ctenoid (perch) scales of modern teleosts, with only slight differences (Figure 4-12). Both are thin layers of calcified collagenous matrix embedded in pockets of dermis covered by epidermis that overlap each other like the shingles of a roof. In the deeper dermis the collaginous fibers are arranged more or less parallel; some teleosts (herring) store fat in the subcutaneous tissue. The more primitive cycloid typ elongated with a smooth, inserted edge and thicker center; in contrast, the ctenoid is rounder, having a conspicuous serrated inserted edge and tiny spines (tines) on the surface of the exposed section. *Amia,* bony dipnoans, and primitive teleosts (cod) have cycloid scales; most teleosts have ctenoid, although intermediate types have been reported. Among flounders, ctenoid scales are found on the dorsal side of the body with cycloid on the ventral. Bony spines occur in the porcupine fish and dermal dentine denticles in the giant sunfish.

Each scale has a base of collagen fibers on which hydroxyapatite and carbonate are deposited. Pigment cells of several different type are present. Because both types of scales can be transplanted successfully between fish, they have been employed in genetic and physiological studies and also in studies of color patterns [13]. The outer layer is bony, and the inner surface fibrillar, with much calcareous material. Because growth continues throughout life, and in temperate climates is conditioned by the available food supply, the addition of new material to a scale by the scleroblastic cells is seasonally interrupted. The resulting rings of growth may mirror the age of a fish rather closely. Embryonically, a scale is formed between two layers of cells—one possibly epidermal—and additions continue to be made. In the pipefish, *Syngnathus,* and its relative the seahorse, *Hippocampus,* the dermal scales are united into a compact case.

A similar line of scale evolution took place among the lungfish (Dipnoi). Ancient forms possessed dermal cosmoid scales consisting of an inner layer of lamellar bone (isopedine), above this in successive layers were cancellous bone, cosminelike dentine, and vitrodentine; the whole grew by the addition of lamellar bone. Recently discovered living *Latimeria* (Crossopterygii) shows a reduction of the dentine layer. Living lungfish have a scale reduced to thin calcified and fibrous layers.

Enlarged dermal spines occur commonly in fish at the anterior borders of fins—in some species, calcified (stickleback). Spines also may serve as support for the dorsal, ventral, and even the anal fins. Fin rays may be cartilaginous (lamprey, hagfish, dogfish), bony, fibrous, horny, paired, jointed, or nonjointed.

The thin, abundantly innervated, well-vascularized teleost skin also serves as an accessory respiratory organ, and the capillary plexus associated with each scale is supplied with its own arteriole and venule. An example is the turret-eyed mudskipper of the coastal swamps in Old World tropics that spends much of its time on land, but when at rest keeps its tail submerged in the water. Mucous-secreting and sensory cells are found in the teleost epidermis, along with cells that produce the fragile cuticle described for many fish skins that is reported to be ecdysed periodically. Like amphibians, fish epidermis shows no particular regional cellular specialization; there is no exclusive basal germinative layer.

A special sensory receptor system not found among land vertebrates is the lateral line organs peculiar to fish and aquatic larval amphibians. The receptors consist of clusters of sensory and supporting cells, termed neuromasts, scattered in the epidermis (pit organs) but particularly located along a complex pattern of canals on the head and extending caudally along each side of the body between the epaxial and hypaxial muscle masses. These canals open by pores at intervals along their length. They are open grooves in some cartilaginous and bony fish, a condition found among fossil amphibians. In living amphibians, neuromasts generally are scattered in the epidermis, but with some linear arrangement.

The structural organization of a neuromast resembles that of a taste bud. Sensory cells have hairlike processes extending into an overlying gelatinous mass, termed a cupula, that is secreted by these cells. They probably react to hydrostatic pressures that move the cupula and, thus, cause a bending of the cellular processes. Neuromasts originate as thickenings of the head ectoderm of the embryo lateral to the brain and become supplied by the facial, glossopharyngeal, and vagal cranial nerves. Probably homologous to neuromasts, but more sensitive to temperature fluctuations of the water, are the *Ampullae of Lorenzi* found on the head of elasmobranch fish, which consist of clumps of jelly-filled, vase-shaped minute tubes supplied by the facial nerve.

Amphibians

The epidermis of adult amphibians is seldom more than a few cells in thickness, and although mitoses are largely confined to the basal layer, dividing cells occasionally are found in the upper layers. In terrestrial species, such as the frog, it is generally five to eight cells thick and rests on a basement membrane of collaginous fibers. Stellate melanophores are present (Figures 4-10 and 4-14). A single layer of flattened, keratinized nucleated cells constitutes the outer stratum corneum, but in some species, such as the giant salamander of Japan, *Cryptobranchus*, the keratinized corneum is several layers in thickness. Beneath this is the broad layer of the stratum Malpighi; the basal layer shows no particular specialization. Nonkeratinized epidermal cells are joined by simple desmosomes, but membranes of the superficial cells are fused in watertight junctions, the *zonulae occludens*.

Desquamation occurs in the frog as unicellular sheets. Adult toads slough skin about once a week in hot weather, less frequently in cool temperatures. In this process the horny layer splits along a dehiscense line on the back, and the toad aids removal by puffing up its body, rubbing against any hard surface, and by other body movements that reveal the new, shiny keratinized layer underneath. The discarded tissue is usually eaten. Desquamation is hormonally controlled as hypophysectomy of the toad inhibits sloughing and the keratinized cells merely continue to pile up.

Figure 4-14 Vertical section of the frog's skin through a dermal plica. Above the large granular (poison) glands lie the smaller mucous glands, and a layer of dermal melanophores.

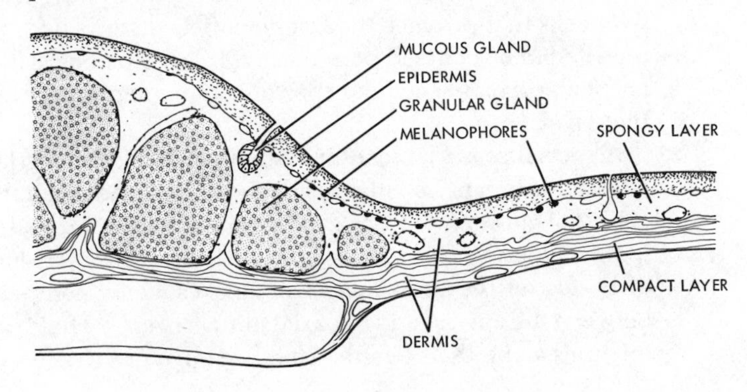

Aquatic species and amphibian larvae in general lack keratinized skin structures and are more glandular. Embryonic epidermis is ciliated and is only one cell thick. In *Necturus*, a nonmetamorphosing urodele, the late larval condition of the skin is retained. Large, club-shaped cells are abundant in the stratum Malpighi, and small goblet mucous cells occur on the head. A thin cuticle covers the surface of the skin, as in teleost fish. Neuromasts similar to those of fish, but not in as precise organization, are found in larvae and adults. In tadpoles and neotenous urodeles the epidermis is not keratinized; a cuticle is present. Evidence of cornification, however, is seen in the several rows of minute horny "teeth" on the lips of tadpoles (Figure 4-15), horny beaks in certain species of tadpoles, and the claws on the African clawed toad, *Xenopus*. In spite of evidences of keratinization among amphibians, the skin is a major accessory respiratory organ with an abundant vascularization and other features peculiar to this function.

The dermis of the skin of *Rana pipiens*, the common leopard frog of biology courses, includes basically two cellular components: outer spongy and inner compact layers (Figure 4-10). The stratum spongiosum of loose aerolar connective tissue, with collaginous and interlacing elastic fibers along with cells of several types (fibrocytes, macrophages), encloses the secretory portions of large multicellular mucous and granular (poison) glands. Both are of ectodermal origin and of simple alveolar type, with short ducts extending through the epidermis to the surface of the skin. A sheet of stellate melanophores lies just beneath the basement membrane of the epidermis.

The deeper layer of the dermis includes compactly arranged collaginous fibers and occasional bundles of smooth muscle fibers that extend toward the epidermis. Mucous glands are lined with tall columnar epithelial cells that stain with the dye hematoxylin. Slimy mucus helps maintain a moist skin, protects against friction and irritants, plays a role in desquamation, and contributes to the cocoon that some species, particularly those inhibiting dry environments, secrete around themselves during estivation. The whitish, acrid fluid of the granular glands is distasteful to some enemies and may be poisonous in certain species of toads and tropical frogs [9]. Granular glands have a restricted distribution in the skin, being especially abundant in the ridges of the epidermal plicae along the dorsal lateral margins of the frog's back. Structurally, each consists of a single layer of flattened cells lining a central cavity that is packed

Figure 4-15 (a) buccal funnel of a lamprey lined with horny "teeth"; (b) horny beak and "teeth" of a tadpole. (Courtesy of Theodore H. Eaton, University of Kansas, and Harper & Row, Publishers.)

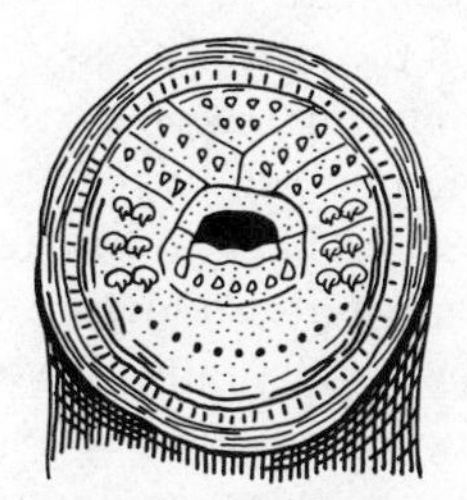

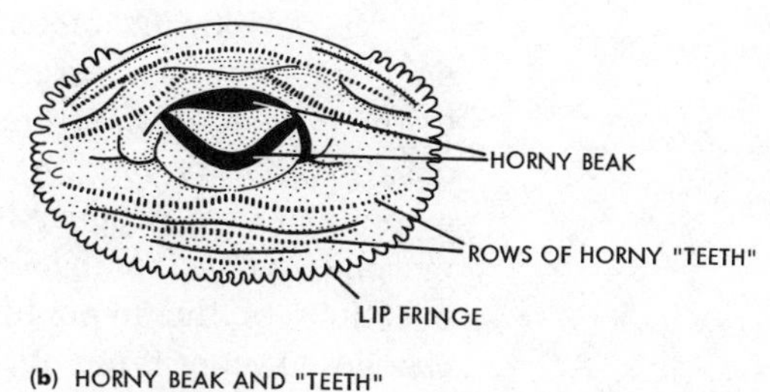

with granular material staining with any cytoplasmic dye (e.g., eosin). Poison glands of *Necturus* have an outer sheath of smooth muscle, and the whole is innervated. Discharge of the secretion follows reflex contraction of the muscle layer. Fossil amphibians, *Stegocephalia*, possessed bony scutes in the dermis, but modern amphibians generally lack any dermal armor. Exceptions include the legless caecilians, in which small bands of dermal scutes are embedded in the skin, and the head processes of the horned toad.

Amphibian coloration is highly developed, as in fish. Color cells include branching melanophores in both epidermis and dermis, and numerous xanthophores and iridophores in the dermis. Melatonin of pineal origin is the most potent agent to provoke contractions of the melanophores in fish and amphibians. At a concentration of 10^{-13} gram per milliliter, the skin of these animals quickly blanches. The pineal gland also contains the enzymes that produce melatonin. The melanocyte-stimulating hormone (MSH) of the pituitary's intermediate lobe produces quick darkening of the skin through melanin dispersion in the melanophores; thus, the animal darkens as the responding iridophores also contract their guanine crystals to the region of the cell nucleus. Temperature and humidity are correlative factors. MSH is active in melanophore control in those fish and reptiles that can change color and increases epidermal pigmentation in mammals.

Removal of MSH from the animal's blood stream by hypophysectomy causes the withdrawal of melanin pigment from the cellular processes and expansion of the iridophores, and the animal turns a silvery color. Strong light stimulation via cholinergic nerves, as on a bright background, reduces the release of MSH and clumping of the pigment results, as well as movement of guanine into the processes of the iridophores. A reflecting surface is formed and the animal becomes lighter. As a result, the melanophores appear contracted and the iridophores expanded. The reverse of this process occurs in the absence of light, and the animal darkens. Because of hormonal control, such color changes require minutes to hours. Xanthophores contain carotenoids and act as yellow filters in reflecting light. These cells constitute the orange spots on the spotted fire salamander. By diffraction, the guanophores produce the blue green. Yellow lipophores overlie the guanophores and filter out the blue. Their disarrangement produces yellow. Red color results from pigment in the lipophores.

Two metamorphic periods characterize the life cycle of the spotted newt, *Notophthalmus viridescens.* During each of these periods profound changes, hormonally controlled, cause the skin structure and other organs to adapt to the new environment. In each of these periods the skin has a characteristic structure and color. In the first metamorphic period the aquatic larval type of skin (thin, noncornified, highly glandular, living) undergoes transformation to the type of skin characteristic of the red eft stage—a thickened, cornified, nonglandular, heavily granular epidermis adapted to life on dry land. After two to three years of land life, the animal undergoes a second metamorphosis, during which the reddish skin changes back to the distinctly aquatic greenish amphibian type peculiar to the adult reproductive stage. The second metamorphic change in skin has been experimentally provoked within three days by injections of the hypophyseal hormone prolactin [14]. Metamorphic skin changes of other types of amphibian larvae can be invoked with injections of thyroid hormone [11].

Reptiles Much of the success of reptiles in becoming adapted to complete land life resulted from the evolution of protective mechanisms such as a type of membrane-enclosed egg that need not be deposited in water, a more advanced type of kidney (the metanephros), and a thickened, keratinized epidermis relatively impervious to water loss. Land-dwelling higher vertebrates (reptiles, birds, mammals) have lost most skin glands, man excepted, during their evolution to total land life and have developed additional means of protection in the shape of horny scales with flexible (snake) or rigid hinge connections (turtle), feathers, or hair. In spite of all these factors favoring water conservation, the reptilian skin can suffer water loss as reported for lizards, snakes, and turtles of arid climates, but at a reduced rate in comparison to amphibians. Such animals, including desert amphibians, have been shown to survive arid conditions with little or no water intake because they have been able to absorb moisture from the walls of burrows and from water metabolized from food (metabolic water).

The few glands in reptiles have special functions related to reproductive behavior and protection against aggressors. In the crocodilia, one pair of musk glands is located on the ventral side of the lower jaw and another within the cloacal opening in both sexes. A row of skin glands extends down either side of the back between rows of dermal plates, the secretion of which has a nauseous odor. Musk glands in turtles are found along the jaw and between the carapace and plastron. On the ventral side of the thigh of most lizards lie small pits of epidermal origin, the so-called femoral and preanal glands, vestigial in some females, whose viscous yellow-colored secretion may act as a sexual stimulation during copulation.

The thin reptilian epidermis includes a basal layer where cell multiplication takes place, a middle layer of prickle cells, and an outer thick stratum corneum that may be shed periodically during ecdysis (snakes, lizards) or retained as growth occurs in the form of marginal rings on each scale in most turtles (dorsal carapace) and alligators. The latter results in a gradual increase in thickness of the scale as well as in its peculiar shape. Between the overlapping scales of lizards and snakes, for example, a thinner, more flexible epidermis, aided by rows of abdominal keratoid plates, permits movement (Figure 4-12). Those of turtles and the scutes of alligators are firmly attached to each other. Scales are modified for particular functions, such as over the tips of digits where they act as suction pads, on the abdomen of the snake and the foot pads of crocodiles for aid in motion. Also, the protective spurs (lizard), beaks, and claws are corneal derivatives. Scale patterns and numbers are constant for each species but may be influenced by temperature. Commercial tortoise shell is the carapace of the marine hawkbill turtle. An exception among turtles is the jellyfish-eating, marine leatherback. This giant of the seas lacks the carapace and instead has a leathery exterior with seven bony ridges extending the length of the back and five along the belly. Imbedded in the skin is a mosaic of small, flat bones forming a hard inner shell that, together with a layer of oily cartilage as much as two inches in thickness, substitutes for the typical reptilian carapace and plastron.

Mitosis ceases periodically in the basal epidermal layer of a snake, for example, and the corneum simultaneously becomes dry and hardened. On resumption of mitotic activity, a new layer of corneum accumulates beneath the old hardened layer, which is subsequently sloughed off (molted) in its entirety. In turn, the soft, vulnerable new corneum quickly becomes keratinized. Snakes usually turn the old skin inside-out as it is discarded; remnants on the tip of the rattlesnake's tail, the so-called rattles, represent previous molts. At each successive molt, possibly several times a year, a new, larger rattle is added as the animal grows in size. Lizards are able to cast off the whole tail, or a piece of it, when it is grasped by a predator, regeneration subsequently taking place from the stump.

Reptilian dermis shows two layers, superficial and deep, varying in thickness among the species but otherwise of typical vertebrate structure of elastin and fibroblasts. Dermal bones are represented by the rod-shaped abdominal gastralia of crocodiles and the tuatara, and bony plates in crocodiles and turtles. The dermo-epidermal junction is generally flat. Leather has been made from those types characterized by large collaginous fiber components and also from whole skin (lizard). Man's quest for exotic materials and his disturbance and pollution of natural environments, as the human population continues to increase rapidly, are factors leading to the eventual extinction of reptilian as well as many other vertebrate species.

Melanophores and melanocytes are found in the Malpighian layer and superficial region of the dermis. Dermal melanophores under hormonal control (MSH) account for most of the color changes in lizards (*Anolis*) but are under sympathetic nervous control in chameleons. Adrenalin may play a role in the clumping mechanism in certain lizards. The skin is well innervated by nerve plexi in the dermis, from which sensory nerves extend into the dermis and epidermis. These endings may be knoblike or encapsulated by cells (lizard) of possible epidermal origin, and they have been described as precursors of dermal sensory structures of birds and mammals. Special sensory structures are the sensory pits, tactile prototrichs, and innervated scale spines of certain species.

Birds

The thin, delicate, reptilianlike, loosely attached avian skin, well protected by coverings of feathers, scales, and horny beak, is particularly adapted to the free movement of the underlying muscles. Although the epidermis is only a few cells thick in areas covered by plumage, it is thicker elsewhere, as on the legs and toe pads (chicken). Thin corneal scales are present on legs and feet. Except for the uropygial or preen gland on the dorsum of the rudimentary tail in most birds (absent entirely in emu, cassowary, and ostrich), the skin is almost devoid of glands. A simple branched structure of the saccular type with a single duct and separated by a septum into two parts, this gland produces an oily secretion used to preen the feathers. The holocrine secretion may include cholesterol and is rich in hydrophobic lipids; without it most aquatic birds cannot float. Even though pigment is present in feathers, scales, and beak, the skin itself will not tan. Melanocytes of the mammalian type are found in feathers, but iridescent colors are attributed to interference spectra.

The thin, well-vascularized dermis of birds, composed primarily of irregularly interlacing fibers, is rich in several types of multicellular and unencapsulated sensory receptors and nerve plexi that are lacking in the epidermis. Smooth arrector muscles attached to the feather follicles raise and fluff the feathers. Modified sebaceous glands near the ear are found in a few birds. Dermal ossifications are entirely lacking. The ostrich skin possesses sufficient strength to be used as leather. Deposits of subcutaneous fat are found in aquatic species and in the domestic chicken.

Keratinized Derivatives

Beaks of thickened, keratinized, compact epidermal cells are common among higher vertebrates of the amniote group. When present, teeth are lacking in the adult animal, although tooth germs may make a transitory appearance in the monotremes and certain birds (parrots) during development. Many variable shapes of beaks are found, especially among birds (Figure 4-16). The duck's soft beak is only thinly keratinized and is well supplied with nerve endings from the dermis. Experimental studies of beak morphogenesis have shown that this structure is specifically induced in the head ectoderm of the embryo by the corresponding underlying specific mesenchyme.

Figure 4-16 Bills of vertebrates to illustrate variations in form and functional adaptation. Fruit and/or seed eating: parrot, hornbill, toucan, grebe, helmet-bird, cardinal, grosbeak; mud straining: duck, flamingo, spoonbill; fish eating: heron, auk; insect catching: whippoorwill; crustacean catching: petrel; flesh tearing: hawk, eagle; scavenger: gull; wood chiseling: hornbill, woodpecker (climbing). Any type of bill consists of elongations of nasal, maxillary and premaxillary bones of the upper jaw and the dentary bone of the lower jaw, all covered by heavily cornified epidermis. Turtles and monotreme mammals (platypus) possess bills. (From Kent, George C., *Comparative Anatomy of the Vertebrates*, 3rd ed., St. Louis, 1973, The C. V. Mosby Co.; and from Atwood, William Henry, *Comparative Anatomy*, 3rd ed., St. Louis, 1973, The C. V. Mosby Co.)

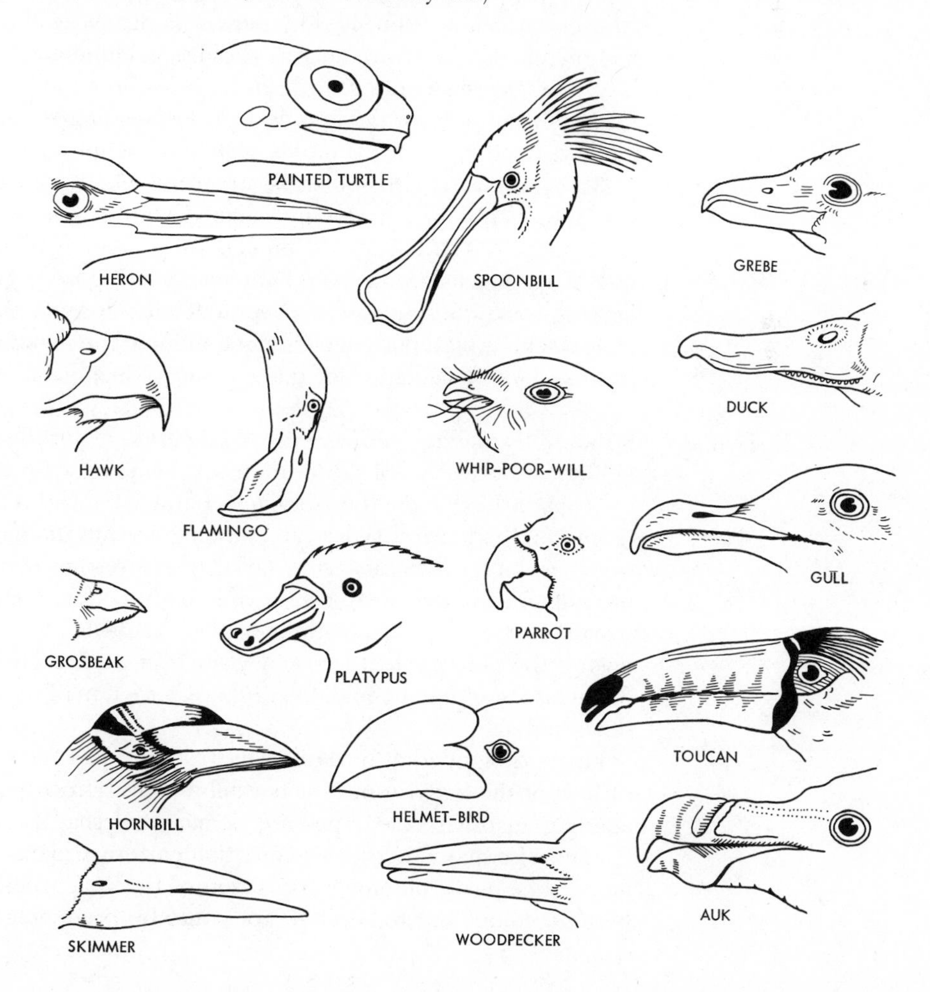

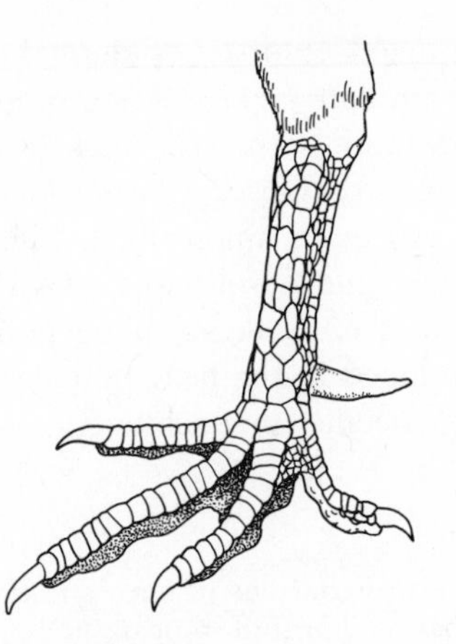

Figure 4-17 Horny scales, claws, and spur on the leg of the fighting cock.

Epidermal scales are most often thought to be peculiar to reptiles, but in one form or another they also are found on the legs of birds (Figure 4-8), the tails and legs of rats, the bodies of armadillos, and the feet of kangaroos. They constitute the hard armor of the scaly anteater, or pangolin, on which they form overlapping plates. As in reptiles, they develop as rigid masses of the stratum corneum, connected by narrow zones of flexible epidermis. In birds and mammals, the scales enlarge during development but are never shed; if lost, they are regenerated. In birds with feathered legs (owl, ptarmigan), the feathers emerge from the apexes of scales, lending support to the belief that feathers evolved from primitive, scalelike beginnings.

Horny spurs are hard epidermal structures with bony cores that occur on the legs and even wings of certain species (tropical jacanas and screamers, cock pheasant). Such structures are found more commonly in males of the species as one of the secondary sexual characteristics (Figure 4-17). In the rooster or fighting cock, the spur is employed in defense or as an offensive weapon. The male duckbill platypus (an egg-laying monotreme mammal) has a spur on the inner surface of each hind leg that contains a duct leading from a poison gland.

Feathers

Structurally and developmentally, feathers are by far the most intricate of any epidermal derivative of the stratum corneum. They occur in specific areas or tracts of the bird's skin, the pterylae, separated by featherless zones, the apteria (Figure 4-18). Several types can be recognized in the chicken: flight feathers (remiges) on the wing margins; tail feathers (retrices) on the vestigial tail, or uropygium; contour feathers on the body surface that contribute to the streamlined shape; and inconspicuous filoplumes, or "hair" feathers, distributed among the contour feathers and visible on a plucked bird. By fluffing the feathers in response to cold, the bird creates a film of insulating air next to the skin's surface.

Feathers of an adult bird differ in structure with the several tracts over the surface of the body; some tracts exhibit transitions from one feather type to another, in others the types are demarcated sharply. Within these tracts, feathers also may differ in pigmentation pattern, order of embryonic development, time of development, or season of the year when the feather follicles become competent to react to hormones by producing feathers of different

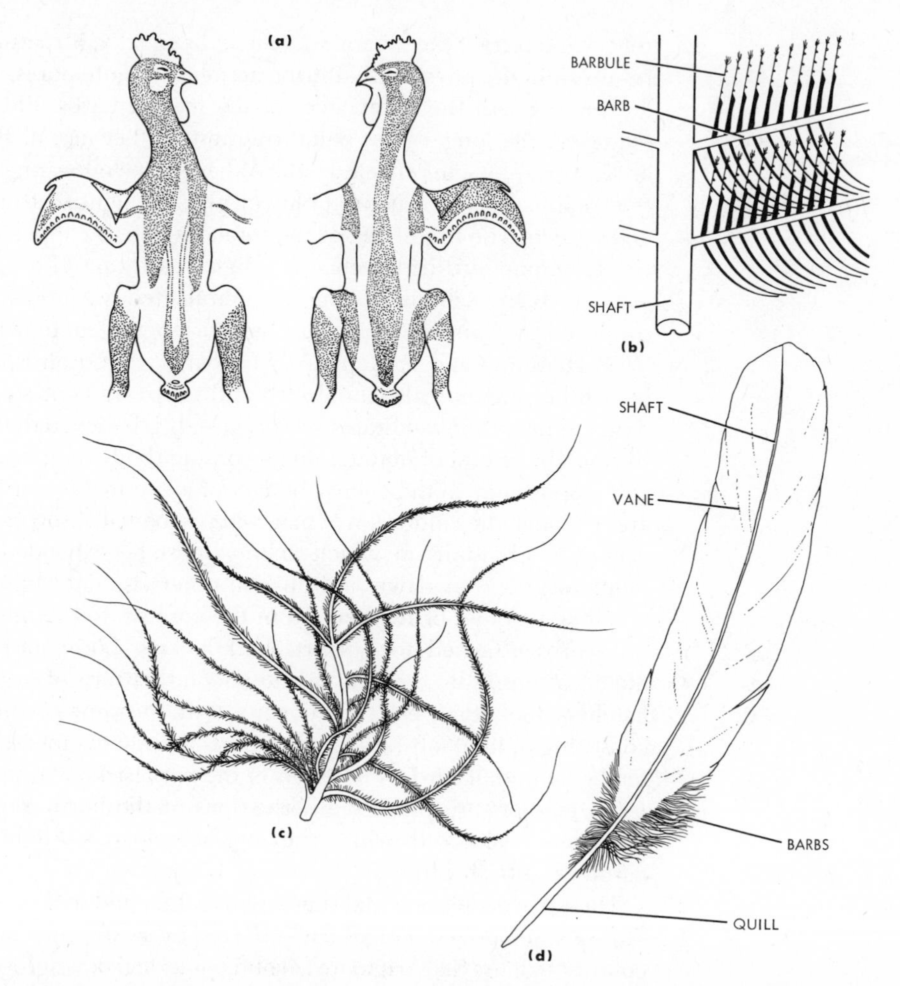

Figure 4-18 Feather structure: (a) pterylae, or feather tracts; (b) details of flight feather structure; (c) powder-down feather; (d) quill feather.

colors. During the molting process, a characteristic order occurs in which feathers in different tracts are lost, apparently in the reverse order of their development. Feather follicles usually are arranged in rows, although in certain areas, such as under the wings, around the cloacal opening, and the flanks, feathers may be arranged in a random fashion. Probably few feather follicles are added after hatching. The overlapping scales on the legs also appear to have a regular pattern, although this is reported to be a less orderly development than that of feathers.

Feather color variation results from either one or both of two mechanisms: physical structure and the presence of chemical pigments. White, blue, and iridescent colors are structurally derived, whereas red, yellow, brown, and black are of pigment origin. Brown- and black-colored feathers are caused by melanin taken up from a reservoir of melanocytes in the germinal region of the follicle. Red and orange result from the presence of carotinoids in the keratinized cells of the feather (scarlet ibis), and like the pink color of the flamingo are diet related. There is no pigment in a white feather; the presence of air in cells and the polygonal shape of the barbule cells simply break up the light and reflect and refract the several wave lengths equally. Structural, or Tyndall, blue results from the physical scattering action of minute pores, or particles, in cells located beneath the outer layer of the feather barbs. Blue wave lengths are scattered, the longer ones being transmitted; because of this scattering, such areas appear blue in reflected light. When some yellow pigment also is present, the combination of structural blue or physical light scattering and the yellow gives green. The bright red of the turaco is due to copper-containing porphyrin. Hummingbird iridescence is a light-scattering effect [29].

Many birds exhibit strikingly complicated variations of color patterns produced by contributions from the overlapping contour feathers. Studies by W. J. Hamilton and F. Heppner with dark-colored birds have indicated that homeothermic animals can absorb and utilize radiant solar energy and that dark pigmentation facilitates this process. It is believed that color, by reducing the metabolic cost of maintaining a constant body temperature, is of considerable importance in the energy budget of an animal in nature. The evidence is applicable in the coloration of man when maximal absorption of solar radiation occurs in situations in which energy must be expended to maintain body temperature, as at dawn and dusk in otherwise hot climates.

Different types of feathers, from the contour to the filoplume with its few soft barbs attached to a short shaft lacking both quill and vane, are all modifications of a common developmental pattern of organization. A newly hatched chick (Figure 4-19) is covered with the same nestling down (plumules) consisting of fine, soft barbs and a short quill that is found among the contour feathers of adult birds. First seen in the earliest known fossil bird, *Archaeopteryx*, feathers are primitive derivations of the horny reptilian scale, as are claws and beaks, with which they are homologous. Flightless birds have free barbules (ostrich plumes).

The early developmental stages of a feather and scale are quite similar; both show a vascularized dermal core covered by a thickened layer of epidermis, in contrast to hairs that originate as solid epidermal downgrowths into the dermis. As the feather germ lengthens into a cylindrical cone, it grows down into the dermis forming a pit around its base. The base, or germinal collar, is the growing portion that continues forming the follicle as well as deepening the pit in which the follicle lies. Concomitantly, the epidermal component of the follicle grows outward as it differentiates into the keratinized quill from the epidermal prickle-cell layer, enclosing the dermal pulp and the outer feather sheath from the corneum. When fully formed, the sheath splits and, as the feather emerges, the barbs and barbules become separated and spread flat forming the vane. By the time development ceases, the downward growth of the germinal collar cells shuts off the blood supply to the pulp, which then dries up to form the pith, leaving a cavity with openings—the umbilia—at either end. Contour feathers grow from the same follicles as do plumules, with which they are actually continuous.

Developmental patterns in all kinds of feathers are differentiations from a common basic type. On the fifth day of incubation, the prospective epidermis of the chick embryo is a flat, unspecialized layer two cells thick; in the basal layer, the cells are cuboidal, and they are flattened in the outer layer and less numerous. After six days of incubation, the epidermis shows beginnings of the adult condition: a basal, or germinative, layer of columnar cells covered by a superficial periderm of flattened cells. The seventh day shows the first indica-

Figure 4-19 Nestling down feathers on a newly hatched chick. (Courtesy Victor Hamburger, Washington University.)

tion of feather germ formation along the mid-dorsal line of the back in the form of small, protruding aggregations of dermal cells covered by thickening epidermis. Later, new feather primordia form successive rows on each side of the original row. The collar, or basal region of the feather germ, sinks into the dermis and is the growth, or proliferative, zone of the developing feather. At the same time, the overlying thickened epidermis begins to grow outward. As differentiation proceeds, it will give rise to all parts of the feather. By eight days of embryonic development, the more advanced feather differentiations have attained the form of cylindrical papillae protruding above the skin's surface. Keratin can be detected at fourteen days; the primordium is birefringent at twelve days.

A replacement feather for the next molt develops by the formation of a new feather papilla beneath, and for a time it is attached to the proximal end of the old calamus. The old feather is shed when the sheath of the replacing feather to which it is attached is discarded. Not all feathers of a bird are usually molted at once.

Formation of the epidermal component of the feather germ appears to be dependent on the presence of the underlying dermal aggregate [8,16,22]. Thus, we can speak here of an inductive interaction of epidermis and dermis. Without the formation of the dermal aggregate, brought about by localized factors at the dermal level, no differentiation of the epidermal sheath in the feather will take place. After this stage, the dermis soon loses its inductive capacity. Later, under the influence of the epidermis, dermal cells move into the epidermal sheath of the feather primordium. Thus, the epidermis plays a major role in feather development, even controlling the caudal adjustment of feather primordium by its cephalocaudal orientation. This has been demonstrated by growing combinations of epidermis and dermis *in vitro.* The tissue-culture technique also has revealed the bipotentiality of the chick embryonic skin. It is the dermis from various regions in the body that determines whether overlying epidermis remains flat and devoid of feathers, gives rise to different types of feathers, or forms scales. Although the epidermis develops according to the nature of the underlying dermis, its histological differentiation conforms with its origin: thus, the epidermis is feather tract specific. Combinations *in vitro* of dorsal epidermis, for example, with tarsometatarsus dermis give rise to thickened epidermal scales instead of feathers. It has been shown also that dermal skin cells need not be living to produce this effect, even collagen may be substituted.

Mammals Differences in epidermal keratinization occur on the palm, sole, and hairy regions of the skin. A thick, phospholipid-rich hyperkeratotic horny layer in seals and sea lions probably aids in making the skin more water resistant in these mammals. This excessive formation of keratin material is found also among large tropical mammals (elephant, hippopotamus, rhinoceros). In contrast, those with a protective pelt of hair have an epidermis only a few cells thick (mouse, sheep). Human epidermis is much thicker, possibly correlated with the absence of hair, but the horny layer remains thin. Whales show a rather poor type of keratinization (parakeratosis). Desquamation in either case occurs in flakes (most seals) rather than as individual cells in the hairy sites. In the elephant seal, the hairy layer is sloughed annually as an intact sheet containing molted hairs.

Horns An epidermal derivative found commonly, although not exclusively, among male mammals is the horn, a term loosely applied to purely epidermal structures, dermal frontal bone, or formations combining both tissues (goat, cattle, sheep, antelope). Horns have roles in reproduction, protection, and offense (Figure 4-20). In vertebrate evolution horns were first encountered in the skeletons of horned dinosaurs of the late Cretaceous period. In *Triceratops* (a three-horned dinosaur), all defensive mechanisms were located on the head. A broad frill of bone, presumably covered with skin, extended over the neck region, usually along with three horns on the head, two over the eyes, and one on the nose. Representatives of modern vertebrates with nose horns include several genera of lizards and the rhinoceros. Horns have attained their maximum specialization among ungulate placental mammals.

The strong horn of the rhinoceros consists of interwoven epidermal keratin fibers, or hairlike filaments, with a laminated structure and an irregular shape, agglutinated firmly by horny cells into a pointed mass. It lacks a bony core but is fixed over a short knob on the nasal bone; a new structure is produced if needed. African and Sumatran rhinos have two slender horns—the longer one is found anteriorly; a single short, broad horn characterizes the Indian species.

Antlers are borne generally by males of the deer family (deer, elk, moose), but both sexes of the reindeer and caribou develop them. Each antler in a sexually mature animal consists of a projection of the frontal bone of the skull. During antler development, the growing core is covered with a layer of skin

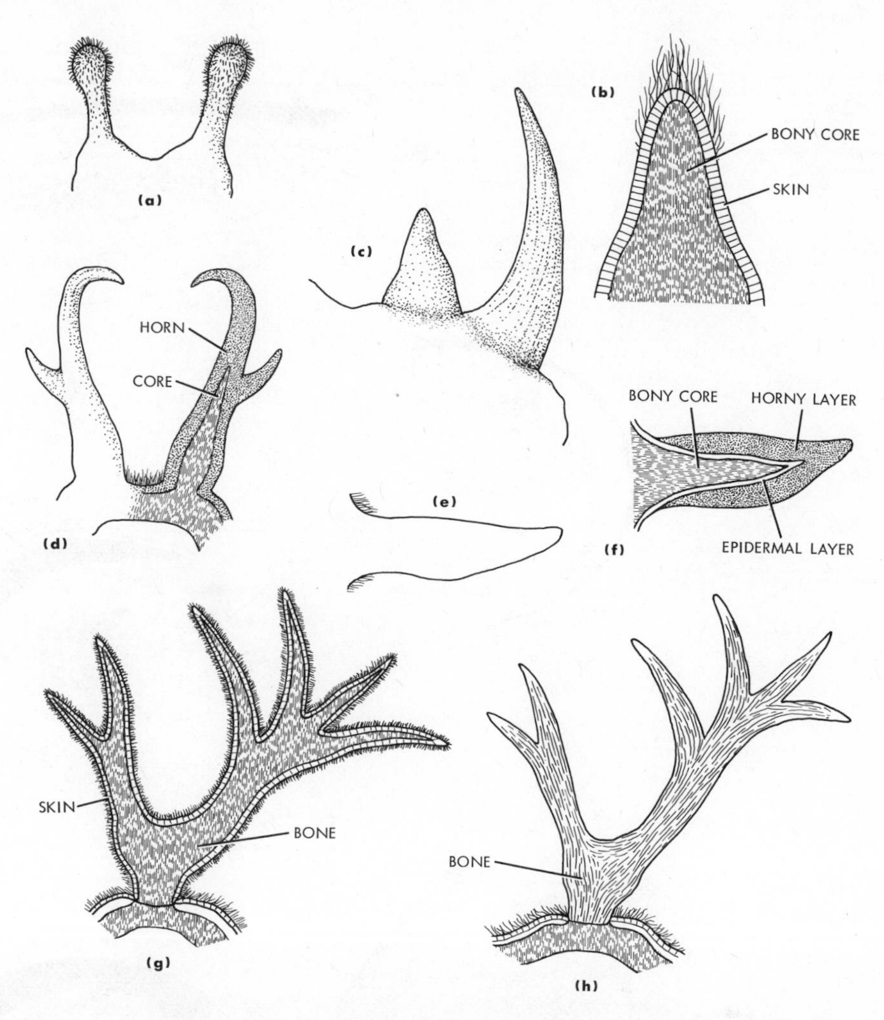

Figure 4-20 Horns and antlers: (a)-(b) "knob horn" (giraffe); (c) fiber horn (rhinoceros); (d) pronghorn antelope; (e)-(f) cow horns; (g) deer antler "in velvet"; (h) mature antler.

and the animal is "in velvet." When the antler matures, and before the onset of the mating period, this skin dries and is sloughed, leaving only the exposed bony core. Later, this antler is lost when tissue constriction at its base of attachment on the frontal bone shuts off the blood supply and weakens the connection between antler and frontal bone. This involves the breakdown and absorption of bone. Following overgrowth of the area by the surrounding skin and in hormone response to changes in light and temperature, a new antler with its covering of velvet, begins to grow. The process is repeated for each breeding season. The antler enlarges during each succeeding growth period and tends to show an increased number of prongs. This phenomenon, obviously sex-related, is controlled by the hypothalamic-pituitary-testicular system of neuroendocrine glands.

Another type of horn, the pronghorn, is found among two species of ungulates, the pronghorn antelope *Antilocapra americana,* and the Saiga antelope of Russia, *Colus tatarica.* This type consists of a permanent bony core

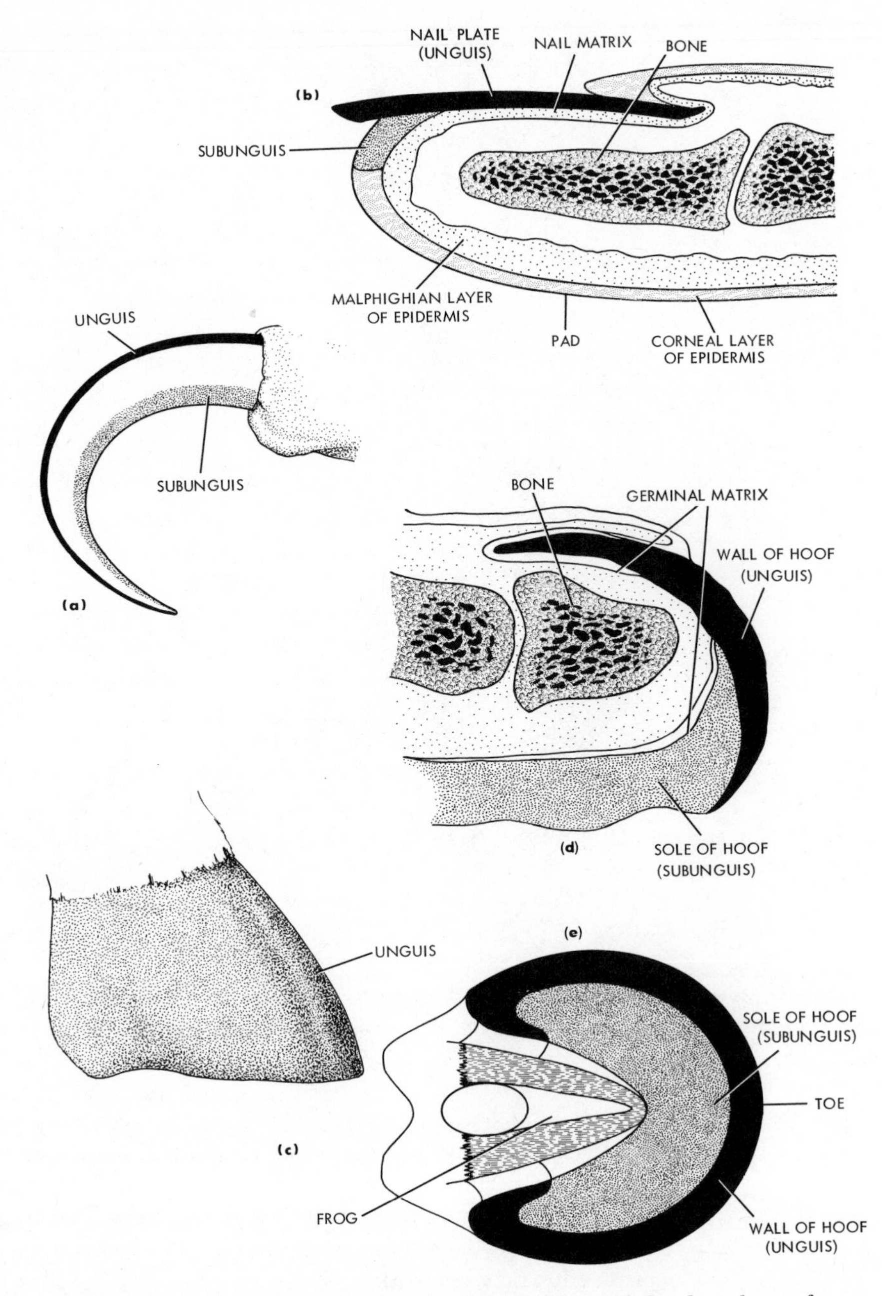

Figure 4-21 Digital tips: (a) claw of the eagle; (b) sagittal section of the digital tip of a young child; (c) side view of the hoof of a horse: (d) sagittal section of the hoof; (e) ventral view of the hoof.

encased by a sheath of horny skin shed periodically and renewed each season without loss of the core. The stubby horns of the giraffe and okapi, permanent structures, remain covered by living skin throughout life. Such horns continue growth by cell addition from the germinating layer of the skin.

Hollow horns occur in both sexes among domestic and wild cattle, goats, sheep, and antelope. Horn formation of this type is similar to the ordinary process of cell proliferation from the germinal layer of the epidermis followed by cornification of the proliferated cells. Not easily impaired, the horn accumulates thickness as a casing around a core of bone from the frontal bone of the skull. These horns are permanent, usually not branched, and may attain large size, as in the Texan steer and certain breeds of African domestic cattle.

Scales The horny scales on the dorsal side and long tail of the toothless, anteating pangolin differ in structure from those on other mammals (rat). Histologically, they resemble the human fingernail. Scales of the armadillo are less keratinized.

Nails, Claws, and Hooves These tough keratinized structures, which have a common origin and similar basic organization, serve specialized functions in different species (Figure 4-21). Claws form on the ends of digits by action of the germinative layer. First found among living vertebrates on a few amphibians, they are common in modern reptiles, birds, and certain mammals. The outer plate of the claw is the unguis hardened by keratin and calcification, possibly a modified scale, beneath which is the softer subunguis near its base. Among birds, the South American koatzin also has two movable clawed digits on its wings.

Ungulate mammals show a specialization of the claw in the form of a hoof of similar origin and structure. The unguis is the hoof proper; the subunguis functions as the underlying pad. The hoof is characterized by a basic filamentous structure but, unlike the rhinoceros horn, the filaments are cemented together with cellular, interfilamentous horn material. In addition, the hoof does not easily fray. The filamentous structure may contribute strength and rigidity to the hoof.

In primates, the claw is replaced by a nail—the nail plate, or unguis, being the conspicuous component. The unguis is a narrow band between nail plate and the epidermis of the fingertip. Cells in the nail do not progress as far as either the claw or hoof in the keratinization process, which ceases at the stratum lucidum (cf. skin) stage of differentiation when cells contain the material eleidin. Nail growth takes place at its base in the upper epidermis of the digit.

Hair Hair appeared as a new epidermal derivative early in mammal evolution, in the Triassic period according to one authority, probably between scales as tactile vibrissae (therapsids). Subsequently, as the hairs increased in number and the development of secondary follicles created the pelt of modern mammals, it became a component of the temperature-regulating mechanism of the body by trapping air next to the skin. Protection against physical injury is another function. Secondary hairs constitute the wool component and primary hairs the outer coat—a distinction that has been lost among higher primates. The bare tail of rodents serves as a mechanism for heat loss. Biologists have long speculated on the occurrence of human hairlessness and the controversy is far from over. Hair no longer affords physical and thermal protection to man.

Although given as one of the diagnostic characteristics of mammals—milk secretion by mammary glands being the other—hair is scanty or hard to find in rhinoceros, hippopotamus, elephant, armadillo, whale, and so on. Hairlessness in mice has been bred as a mutant. Absence or loss of hair and its distributional pattern on the body are secondary sexual characteristics, as the follicles are

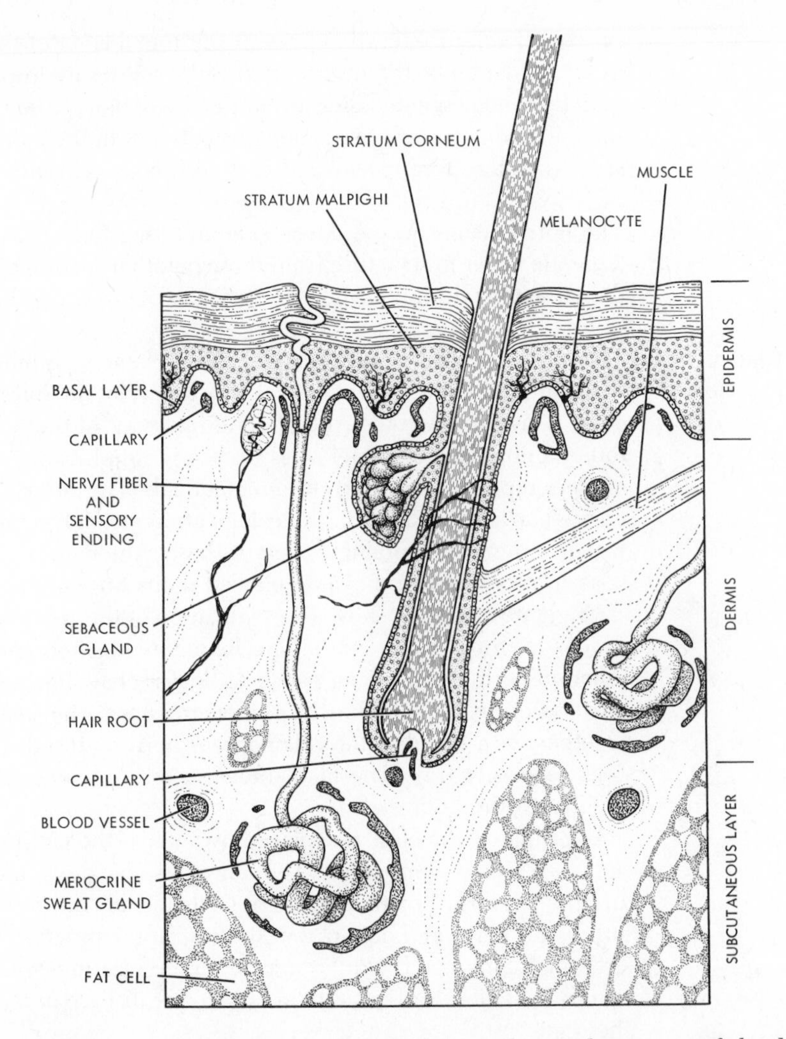

Figure 4-22 Diagram of a vertical section of the human skin to show many of the different structural elements present, especially hair, glands, nerve endings, and vascular supply.

under sex steroid hormone regulation. Terminal body hairs develop at puberty. Sexual selection by the human male has been advanced as a factor in influencing hair patterns and color. One type of baldness has been associated with an excess of male sex hormone; however, there are several different forms of baldness and the genetic factor is held to be mainly responsible [12].

The form and size of hairs among mammals exhibits numerous variations from the large, coarse spines of the porcupine and Australian anteater, *Echidna,* to the soft, fine wool of sheep and goats. Bristles on the pig are short, straight, elastic hairs, often with split ends. Fur consists of fine, soft, densely aggregated hairs with interspersed long, coarse hairs that are removed when the skins are worked into articles of commerce. Specialized types of hair include the whiskers or vibrissae of the rat and cat, the effective contact receptors associated with sensory organs; the sex-linked lion's mane; the winter fringe on the feet of the snowshoe hare; and the eyelashes and eyebrows of man. Curly hair is oval and elliptical, whereas straight hair is round and kinky hair flat. Wool hair cuticle is flat, rough, and scaly, tending to be slightly twisted—qualities that make it suitable for yarn. The late human fetus is entirely covered by abundant fine embryonic hair, the lanugo coat. Generally, this is discarded before birth along with the epidermal periderm, although occasionally infants are born with a fine crop of hair.

The distribution of hair in the human is unique among primates. Like other mammals, excepting the angora rabbit and sheep, the follicles exhibit alternate growth (mitotic) and rest periods of hair production, referred to as hair cycles. Growing (anagin) follicles show maximum energy requirement whose common source is glucose, which is utilized about twice as rapidly as by the resting (telogen) follicle. Glucose diffuses through cell membranes from the blood, becomes phosphorylated, and is subsequently utilized by way of a number of well-known metabolic enzymic pathways: e.g., pentose and TCA cycles; synthesis of glycogen, nucleic acids; ATP production; and so on. More specific information can be found among the references at the end of the chapter [1,20].

Throughout the ages, man has been concerned over his baldness (alopecia); today, information is available about the causes, development, and possible cures for different types. This may involve actual loss of hair or the replacement by a delicate fuzz resembling fetal hair. Hair follicles have been successfully transplanted from one part of the human scalp to the bald spot in an effort to correct the condition. Although there is some basis for the correlation of baldness with an excess of circulating male sex hormone [12], appearing as

Figure 4-23 The underside of the epidermis from the human male scalp, showing hair follicles with sebaceous glands. (Courtesy of W. H. Fahrenbach.)

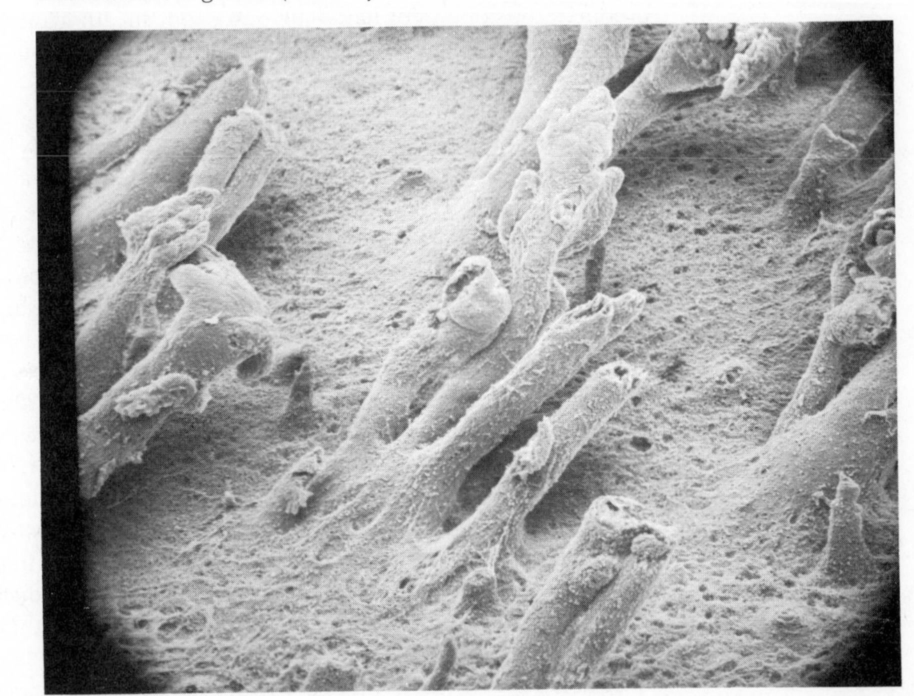

early as adolescence, there seems to be much evidence that it is gene-controlled, like other aspects of hair production. Baldness is not encountered among eunuchs, and ovarian and adrenal cortical dysfunction may induce abnormal hairiness (hirsutism), especially obvious among women. A key reaction in testosterone metabolism is the conversion of the male hormone to dihydrotestosterone as the active androgen in the hair follicles, as is the case with several other testosterone-target organs in the male (prostate gland, epididymus). Growing follicles metabolize testosterone more rapidly than resting follicles, and the role of dihydrotestosterone formation is increased.

Hair develops from a downgrowth of the epidermis. Where scales and hairs are associated in a mammalian skin, the hairs arise from the hinge regions between the scales (tails of marsupials and rodents). Essentially, a hair is a compact rod of keratinized cells with a central medullary portion enclosed by a cortex of variable thickness, the whole covered with flattened cuticular cells, on the outer surface of which is still another layer of thin, nonkeratinized chemically resistant epicuticle (Figure 4-24). Large intracellular spaces within the medulla characterize the hairs of rodents and many hooved mammals, and most coarse hairs contain such spaces (porcupine quills). Fine hairs may lack the medullary component (human), and the cortex of the human scalp hair is quite thick.

A hair originates at the base of a down-growing epidermal follicle from a group of germinative cells connected with the stratum Malpighi of the skin and enclosing a small dermal papilla. In certain cases, as the spines of the European hedgehog and vibrissae of the cat, the dermal papilla first pushes up an epidermal cap, as occurs in the development of a feather. Later, however, the epidermis grows inward and surrounds the dermal aggregate, forming a follicle. The continuous proliferation of epidermal cells from the germinal zone during the growing anagen stage, followed by their rapid keratinization, produces the column of hair. The outer root sheath arises from the outermost ring of germinal cells in the follicle bulb and is continuous with the epidermal prickle-cell layer. Progressing inward, one finds the inner-root-sheath germinal cells and the central hair-forming cells that become elongated and spindle-shaped as they enter the keratogenous zone. Biochemical changes undergone as the cells in this zone become fully keratinized have been described [29]. With the attainment of maximum hair length, the bulb cells cease to divide. During the following twenty-four-hour catagen stage in the life of a hair, autolytic breakdown of the lower part of the follicle reduces the connection between the permanent part of the follicle and the germinal cells to a shriveled cellular cord that soon disappears, leaving the club hair attached to the resting telogen-stage follicle by brushlike keratinized cells. When growth is resumed (anagen stage), the new hair grows along the side of the old club hair. This growth occurs in cycles, consisting in some species of single waves of growth that pass over the body (mouse). The factor(s) responsible for the rhythm of hair formation and quiescence in an individual follicle is not clear. In animals with seasonal changes of hair, the same follicles form the summer and winter coats. Hairs usually emerge from the skin at an angle and may point in different directions in various parts of the body. Club hairs may be retained during succeeding one or more hair generations (dog), or discarded soon after the new growths have formed (wild mammals).

Associated with the follicle are a capillary network, nerve endings (Figure

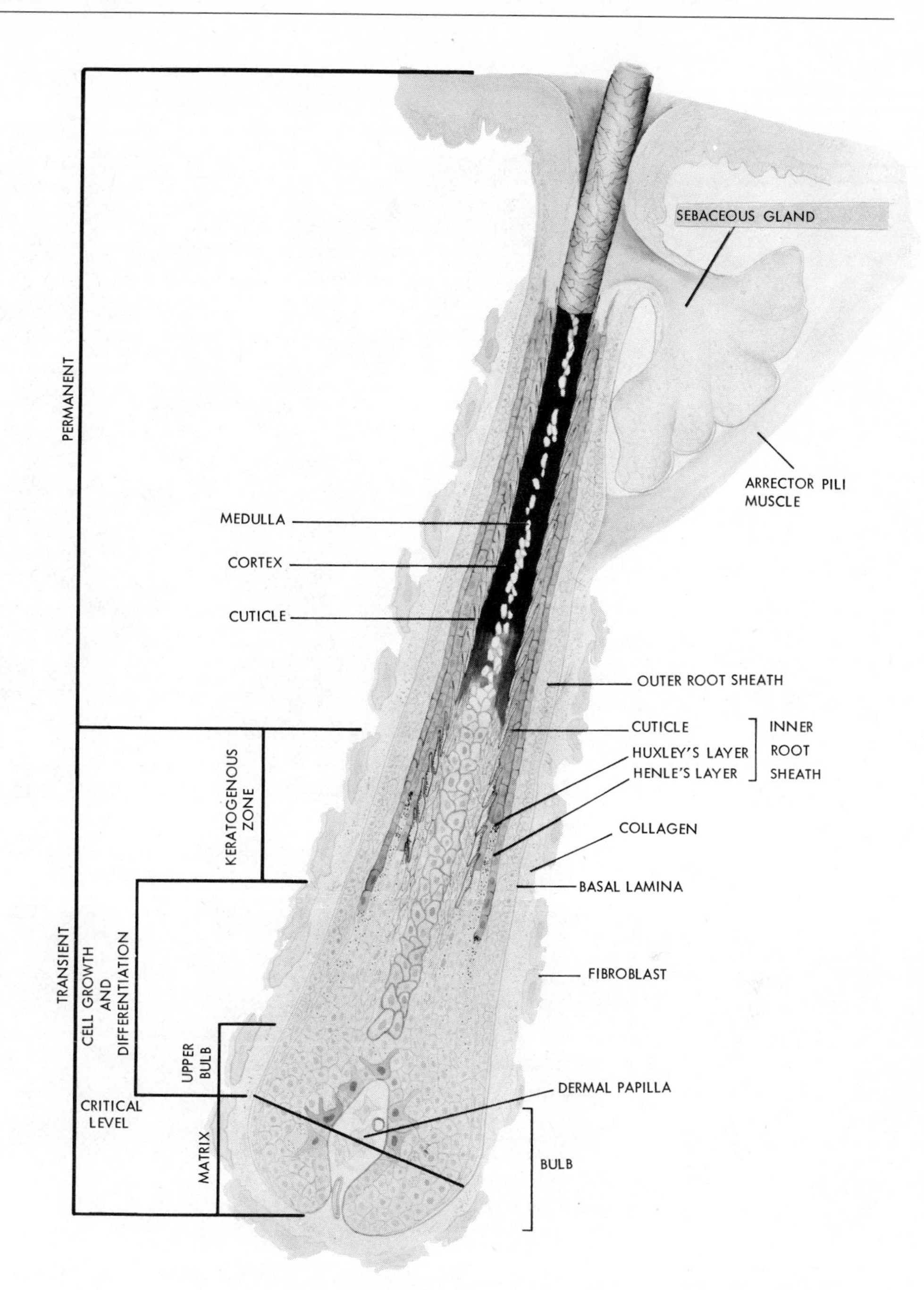

Figure 4-24 Diagram of a growing (anagen) terminal hair follicle, showing the positions of the several layers. (Courtesy of William Montagna and Academic Press, New York.)

4-25), an epidermal sebaceous gland with an opening into the follicle cavity, and an arrector pili muscle of ectodermal origin (Figure 4-30). Constriction of this smooth muscle causes movement of the hair in fright and display of anger (cat, dog) and raises the hair, increasing the amount of air trapped beneath. Each muscle is attached at one end to the hair sheath and at the other to the epidermis. On contraction, the hair is made to stand erect or the skin rough-

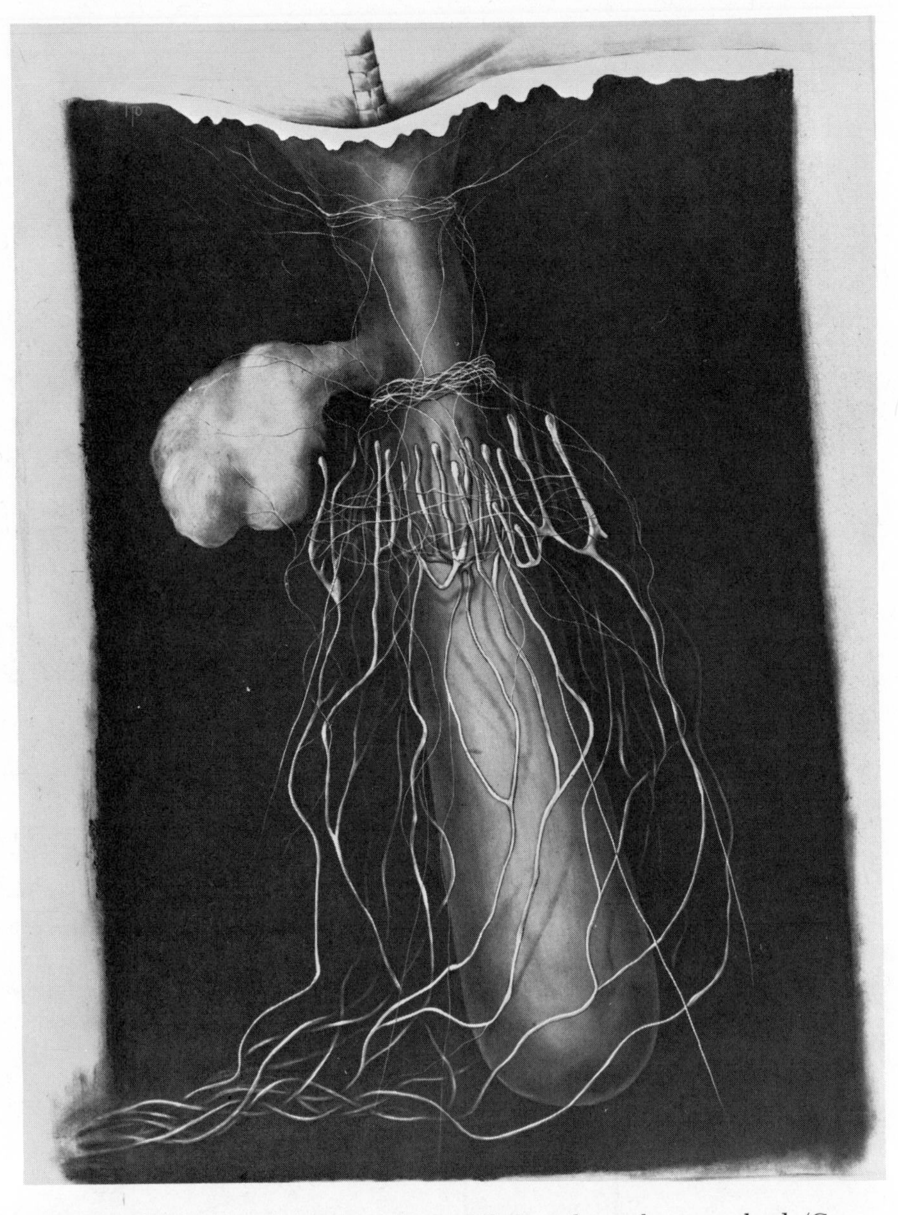

Figure 4-25 Disposition of nerves around a small hair follicle and its sebaceous gland. (Courtesy of William Montagna and Academic Press, New York.)

ened, forming a goose pimple. This may be either a heat-controlling mechanism or an emotional reaction. From the sebaceous gland comes the oily secretion (sebum) associated with the scalp. The size and function of these glands are controlled chiefly by androgenic hormones. Small in the child, they begin to enlarge in early puberty and reach full size in the adult human.

Cellular proliferation and differentiation of the hair are affected by many factors. Among certain mammals they are interrupted when hair is discarded during various seasons of the year, as when a horse sheds its winter coat in spring responding to temperature change. In other mammals, there are sea-

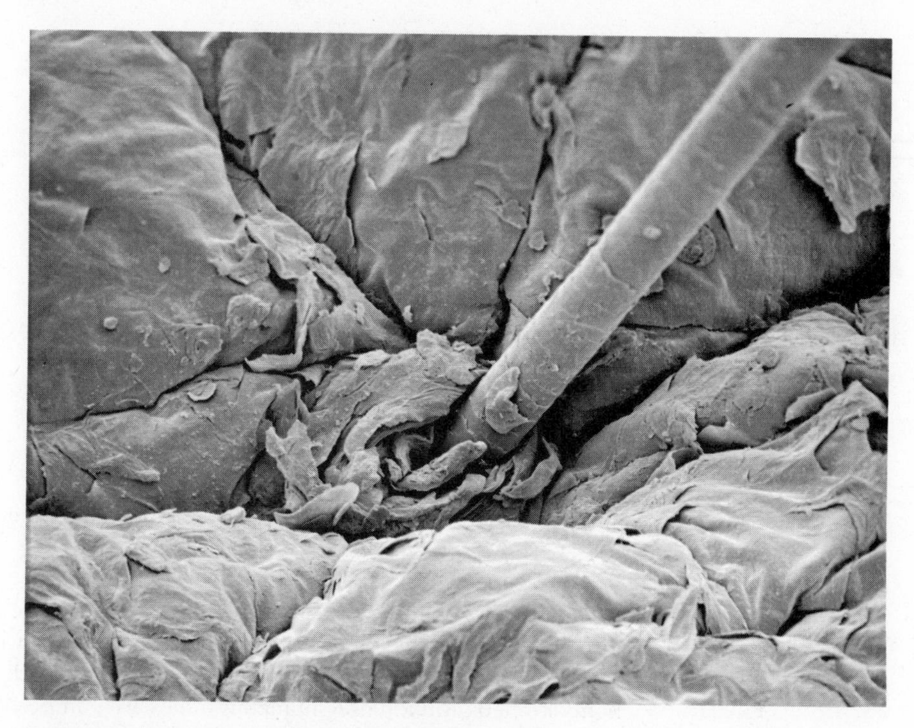

Figure 4-26 Scanning electron micrograph of the intact surface of the forearm, showing the size and loose arrangement of the squames and the emergence of a hair. (Courtesy of W. H. Fahrenbach.)

sonal changes of color, texture, and/or amount, as when the weasel's winter fur coat changes from white to a summer brown (Figure 4-27). In a number of species the young are born naked, and several pelages may develop and be replaced before the adult coat is formed. For example, in the remarkably swift cheetah, the naked cub soon develops a mantle of gray hair that is replaced later, from the same hair follicles, by the characteristic spotted coat of the adult. Hair molt (certain breeds of sheep) may occur once or twice a year in autumn and spring, or more frequently (rodents), and may be patchy (camels).

Diet is an important factor for any mammal, as are the physical, emotional, and mental states of the organism. For example, morphological and color changes may result from protein-caloric malnutrition, and fiber-growth rate (merino sheep) may be affected. Adding molybdenum to the diet of sheep arrests typical pigmentation and produces a striped fleece. A copper-deficient diet changes wool texture. Hair has many specialized uses; it protects the skin from abrasions and cuts, thermal and radiant injury, and chemical irritation. Few substances can reach the skin of a furry animal.

The color of the human skin is due primarily to two components: translucency and the presence of pigments. The former permits blood in the dermal capillaries to color the skin, as in blushing. Pigments include brown, which can darken to black, and a yellow that can modify to red. These are diffused in the tissues. Such pigments are found in varying proportions in skins of all races and even vary within a single individual of a race. Albinos lack pigment in both skin and hair. Skin color changes during a lifetime, and the female skin is more yellow than that of the opposite sex. Pigment is present in the epidermis and

Figure 4-27 Photographs of the weasel, showing the change in fur color between summer and winter seasons.

hair bulb, but rarely in the outer portion of the dermis. Not until shortly after birth are pigment granules present in any amount in the stratum Malpighi. Excessive local pigmentation produces freckles, which are more common in light-skinned individuals. Cutaneous moles are composed of local elevations of both dermis and epidermis; they may be pigmented and even involve a few hairs. When a mole is congenital and involves blood capillaries, it is known as a birthmark. Within the mammalian epidermis is a reticular system of pigment-secreting cells, the melanocytes, that constitutes a self-maintaining unit. These differentiate from melanoblasts, which in turn have their origin in the embryonic neural crest cords. From such cords, the melanoblasts migrate throughout the body and, on arrival at their destinations, differentiate into pigment-forming cells. Those in the retina of the eye, however, have their origin in the eye primordium itself. Melanocytes in the hair follicle bulb transfer melanosomes to the cells constituting the medulla and cortex. Chemical patterns of genetic origin result from chemical differences in the melanin deposited during hair growth, producing the yellow and dark-brown effects.

Hair color is produced by pigment within the intercellular spaces of the cortex: black and red in man, brown in most other forms. White is due to an absence of pigment together with the refraction of light from air spaces between the cells of the medulla. The amount of pigment determines the shade of hair color, as brown black usually masks the red yellow. Dark hair lacking red is generally blue black.

Although the neural crest origin of pigment cells among vertebrates has been well established, the mechanisms responsible for spotting patterns is not clear. For example, the white spotting in hair-color patterns, such as the pigment-free areas in piebald mice, is reported to be genetic in origin. Two alternative mechanisms have been proposed: one states that melanoblasts fail to reach the areas of the skin destined to be pigment-free because of a controlled migration defect at some early stage in embryonic development, and the other is a matter of cell antagonism exhibited by propigment cells. Appropriate spacing of chromatophores could produce color patterns as well as prevent the mingling of melanophores and xanthophores. A delayed time sequence of melanoblast

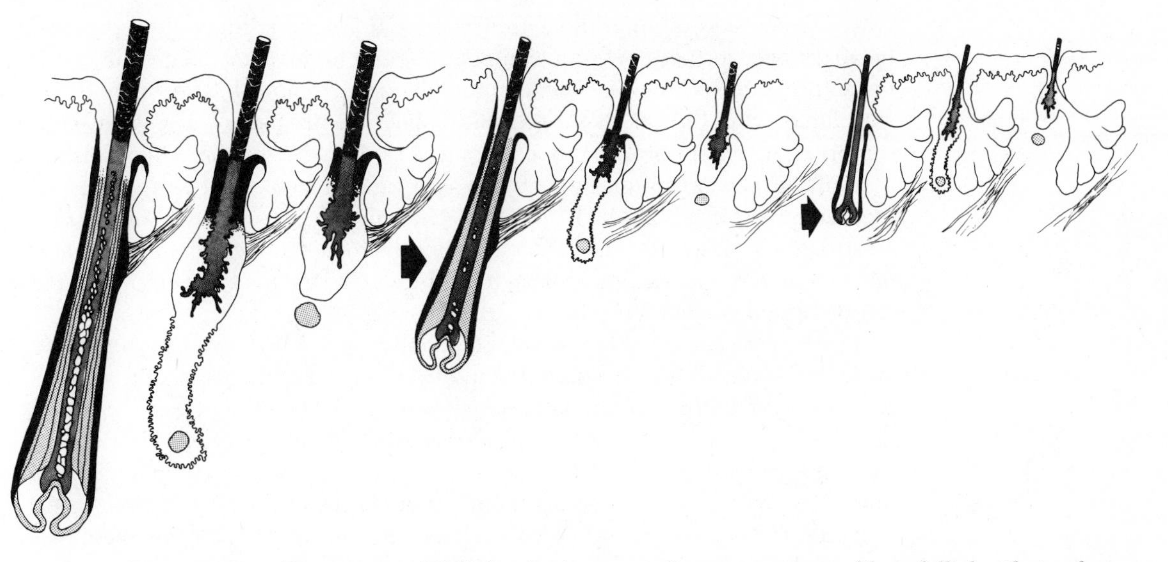

Figure 4-28 Diminution of follicle size in succeeding generations of hair follicles that culminates in the vellus type. (Courtesy of William Montagna and Academic Press, New York.)

migration has been reported. Another mechanism proposes that melanoblasts migrate into all areas of the skin but fail to differentiate in certain regions either through genetic factors within the neural crest cells or through the influence of the cellular environment. The question is still unanswered on the basis of current evidence from various genetic and experimental approaches. Something must be present in the colorless areas that inhibits melanoblast differentiation.

Numerous factors affecting integumentary colors among vertebrates include secretions of the endocrine glands (melanocyte-stimulating hypophyseal hormone—MSH, or intermedin), some steroids, and hormones (corticosteroids) that influence the corticotrophin-MSH relationship. Adrenal deficiency leads to darkening of the fur. Sunlight on the human skin produces additional melanin and also causes bleached pigment already present to darken. The darkening represents oxidation of melanin, which is present in a reduced leucoform. Melanin behaves as an oxidation-reduction indicator, the oxidized form being dark, the reduced form bleached. Actually, it is the thickening of the corneum, rather than melanin formation, that increases its screening action, and it is considered to be the principal factor in the increased resistance to sunburn. Ultraviolet radiation is a major factor in human cutaneous cancer.

Skin Reactions

The skin, a mirror of body function and internal disturbance, reflects abnormal conditions of the internal environment in specific ways. A number of diseases with pigmentary indicators are known: the paper pallor of pituitary disease, the cyanotic blue of congenital heart disease or of oxygen deficiency in the newborn, the deep tan of thyroid dysfunction, and the jaundice yellow of liver disturbance. The bronze of hemochromatosis reflects a disturbance of iron metabolism that deposits iron in the skin. Many chemicals in common therapeutic and diagnostic use, including medicinal drugs and vegetable pigments, will produce skin discoloration. In exceptional amounts, the yellow pigment betacarotene, a common skin component present in many vegetables, can make the skin yellowish. The carotenoid group color the serum and are associated with the fat of dermis and subcutaneous tissues. Excessive consumption of tomato products over a long period may color the skin reddish as a result of the accumulation of excessive amounts of another carotenoid pigment, lycopene. Clinical literature is a good source of such interesting information. Man's skin also reacts violently to a number of allergies and diseases such as measles and poison plant products. A skin test is employed for the detection of tuberculosis.

Among vertebrates, skin transplants usually survive for any length of time only between individuals characterized by a high degree of genetic similarity—those of inbred strains or monozygous twins. In other cases "foreign" tissue is rejected when the host produces an antibody reaction. [7] Positive cases of successful grafts between sibs have been found, however, under certain unusual conditions, as in small natural populations of fish that have been isolated for extended periods. This has been attributed to the achievement of a high degree of genetic uniformity through inbreeding within a restricted environment. The adaptability of skin grafts from cancer patients to healthy recipients is less easily explained. Atypical behavior of skin grafts in this case may result from the attack on the mechanisms that destroy abnormal cells. Immunity mechanisms become manifested by the human fetus around the time the lymphatic system and thymus gland make their appearance [4]. Henceforth, the body is protected against foreign cells, even from tissue and organ transplants. When genetically absent, a condition that is transmitted in the male line, the individual can survive only in a germ-free environment. Curiously, in some individuals the reaction is less intense than in others. Immunologic tolerance for a variety of antigens has been achieved in adult animals with immunosuppressive chemicals such as 6-mercaptopurine. Massive doses of tissue antigen, and of tissue antigen combined with chemical immunosuppressive agents, have been successful in making grafts more acceptable for a time (heart, kidney).

Another type of skin reaction little understood is skin cancer. Chemical and physical factors of varied types may produce at first a tiny nodule long after the inducing agent has disappeared. This nodule has been observed to grow into an epidermal cancer; destructive action takes place through uncontrolled growth and cellular-tissue modification. Some kind of change has taken place in the living epidermal cells so they become excessively stimulated to produce themselves in an unregulated manner. Such cells have lost the capacity to keratinize. Skin and mouth cancers are far more common than is realized. The changed cells are termed neoplastic because they form new tissue; the growth itself is a neoplasm. Coal-tar derivatives, excessive exposure to the ionizing radiation in sunlight, and myriad chemical and physical means can evoke tumors on the skin, which furnishes a visible culture medium. Skin cells are especially vulnerable to DNA damage provoked by immoderate exposure to sunlight.

Pharmacological Activity

Some vertebrate skins produce unpleasant or irritating secretions that may provide partial defense against predators. This is especially the case among frogs of the genus *Phyllobates* whose skins possess remarkable pharmacological activity. Often such animals of the neotropical family *Dendrobatidae* have vivid pigmented colorations. Such markings are considered to be a warning to other organisms. A number of animal groups exhibit similar warning colorations. Daly and Meyers [9] describe a Panamanian poison frog, *Dendrobates pumilio,* that shows interpopulational variation in color, degree of toxicity, size, and habits. Differences in body coloration encompass the visible spectrum from red to blue, also achromatic black and white. Wide variations in degree of toxicity are not apparently correlated with supposed warning coloration. Skin extracts yield two toxic compounds characterized as steroidal alkaloids that exhibit cardio- and neurotoxicity. Batrachotoxin may be a valuable tool for studying ion transport in electrogenic membranes. For centuries skin acretions from small, brightly colored frogs have been used by Indians of the Choco rain forest of Columbia to prepare the deadly darts for their blowguns. The species involved is the Columbian poison arrow frog, *Phyllobates aurotaenia,* that yields only fifty micrograms of a labile active venom.

Vitamin D Synthesis by the Skin

Unlike other vitamins, the "sunshine" vitamin D is synthesized in the human skin under the influence of solar rays in the ultraviolet region of the spectrum (wave length 290 to 320 millimicrons) that converts the provitamin 7-dehydrocholesterol. Vitamin D mediates the absorption of calcium from the intestine and the deposition of inorganic minerals in growing bone. Naturally, children require more than do adults. Too little results in rickets, whereas a prolonged excess may provoke chalky calcium deposits in arteries, particularly the aorta, and "stone" formation with impairment of function in the kidney. From its source of production in the skin, the vitamin is distributed in the circulatory system.

Physiologic means of regulating the vitamin D concentration in the body is apparently through control of the rate of photochemical synthesis. One square centimeter of white human skin can synthesize up to eighteen international units in three hours. W. F. Loomis [17] reasons that the rate of synthesis in the stratum granulosum is regulated by the two processes of pigmentation and keratinization of the overlying stratum corneum, which allow only regulated amounts of solar ultraviolet radiation to penetrate the outer skin layer and reach the region where vitamin D is synthesized. According to this view, different types of skin—white (depigmented, dekeratinized), yellow (mainly keratinized), and black (mainly pigmented)—are adaptations that maximize ultraviolet penetration in northern latitudes and minimize it in southern latitudes; hence, the rate of vitamin synthesis is maintained within physiologic limits (0.01 to 2.5 milligrams per day) throughout man's worldwide habitat. Evidence points to a marked correlation between skin pigmentation and equatorial latitudes. Additionally, the reversible summer pigmentation and keratinization activated by ultraviolet radiation, known as sun tan, represents a means of maintaining physiologic constant rates of synthesis despite the great seasonal variation in solar ultraviolet radiation in northern latitudes. In the tropics, man's lightly pigmented skin can synthesize a "deadly dose," up to 800,000 units, in a six-hour exposure of his whole body. Sun tanning can be overdone, as is well known when the skin blisters through the accumulation of tissue fluids following breakdown of tissue barriers. Heavily pigmented skin transmits around 3 to 36 per cent of the rays, whereas light-colored skin may allow a 53 to 72 per cent passage of the rays. Thus, a dark-skinned man in northern latitudes would be less favored and would suffer in comparison to his lighter-skinned neighbor; one exception is the Eskimo, who obtains his vitamin D from fish oils in his diet.

A deficiency of vitamin A results in atrophy of hair follicles and extensive hyperkeratosis; otherwise, the skin appears normal. Conversely, excessive amounts of this essential vitamin may cause hypertrophy of the adult epidermis, in addition to various developmental defects in fetal mammals if given to pregnant females (cleft palate, thyroid disturbances).

The skin synthesizes, primarily in the holocrine sebaceous glands and epidermis, a large number of unusual lipids (waxes, triacyl glycerols, cholesterol) that appear in the complex fatty sebum, and there is an accompanying accumulation of many unique intermediate compounds. Sebum also contains free fatty acids, squalene, phospholipids, and so on. Seal sebum contains much hydrophobic substance, and that of sheep is a source of lanolin. The uniqueness of skin lipids lies also in the large number and variety of saturated and unsaturated fatty chains that are synthesized. The skin displays many biosynthetic techniques, but certain secondary processes also take place both in the glands and on the skin surface (hydrolysis, formation of sterol esters); thus, the skin is a veritable chemical factory. Questions have been raised as to the reason(s) for the formation of so many compounds and the functional significance. One suggestion is that they constitute olfactory messages, or pheromones, for chemical communication between animals. In any case, every human being possesses a characteristic odor as distinct apparently as the whorl-loop pattern in the finger pads, by means of which the individual could be identified providing our olfactory receptors were as sensitive as those of a dog, for example (chemical fingerprint). It is well known that olfactory (chemical) communication plays a vital part in the animal role, e.g., recognition, sexual attraction, territorial marking. Functionally, such products may aid the skin itself in resisting the actions of potential pathogens by creating an environment unsuitable to their survival. Sebum flows out the hair canal, coating it and moistening the stratum corneum. This fatty secretion of the sebaceous glands is produced continuously, steroid sex hormones affect its secretion (estrogen, androgen), and there is no nervous control.

Sexual Skin

The cyclic changes that take place in the sexual skin of some species of female nonhuman primates are hormonally related. Structure of this type of skin varies with the species. It is characterized by a thickened dermis, a conspicuous edema, and an abundant blood supply. General conformation and color changes take place during the menstrual cycle. Swelling of the skin results from water deposited in the intercellular spaces and from connective tissue inhibition. Swelling and regression are associated with a rise and fall in body weight, cyclical fluctuations in daily water balance, and with concentrations of erythrocytes. For example, the red blood cell count in *Macaca nemestrina* is reported to be lowest during menstruation and highest during maximal genital edema midway in the menstrual cycle. Perineal swelling indicates estrous activity, but it is not correlated with the exact time of ovulation. Subsidence of the sexual skin follows ovariectomy. In *Papio*, perineal deturgescence coincides with the postovulatory and luteal phases of the menstrual cycle. When the estrogen level drops, wrinkles appear in the region of the anus, vagina and outer borders of the skin, and the color changes from bright red to gray-pink within a matter of a few days. (Figure 4-29).

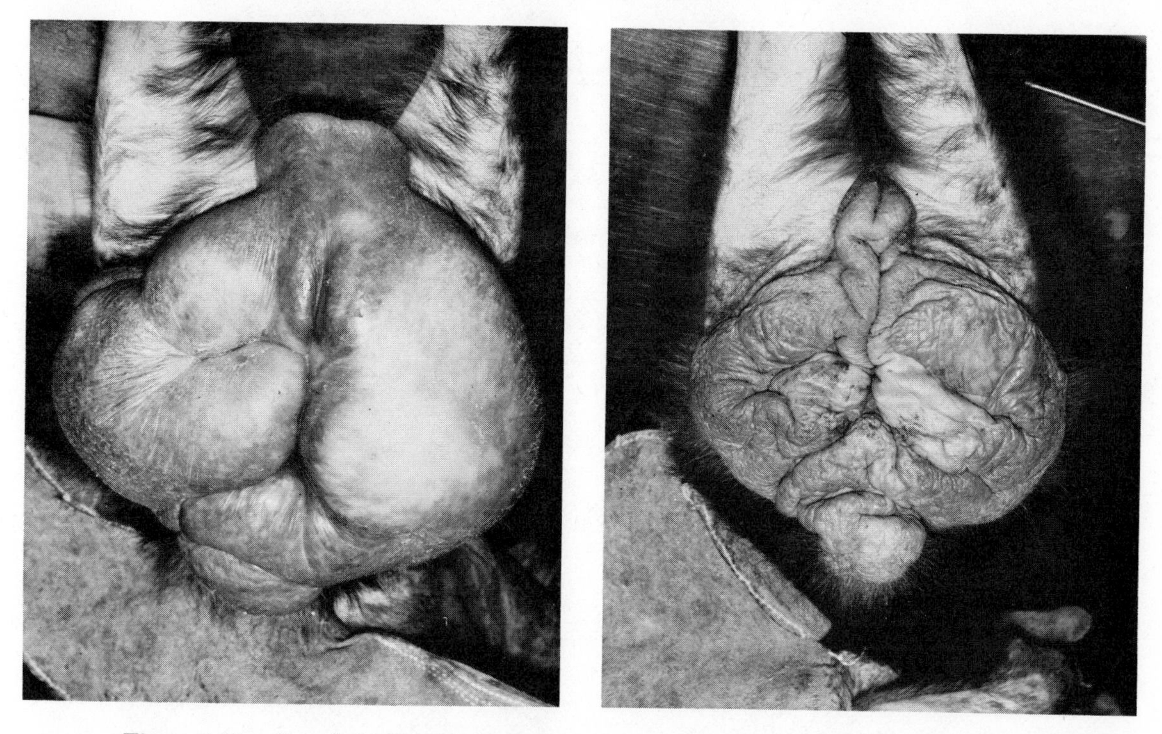

Figure 4-29 Sexual skin on the buttocks of a postovulatory and preovulatory pigtail macaque in estrus. (Courtesy of William Montagna and Academic Press, New York.)

Temperature Control

Change in the external environment is perceived by peripheral receptors in the skin that, in turn, send messages by sensory nerves to central receptors in the spinal cord and brain. Responses are relayed to smooth muscles and glands. Birds and mammals are homeotherms, maintaining a relatively constant internal temperature through a wide range of environmental temperature changes. In contrast, the body temperature of poikilothermic vertebrates varies only a few degrees from that of the environment. Hence, poikilotherms are capable of a wide span in body temperatures, as in the case of true hibernators (bat). The skin functions in other ways in body-temperature control. In man, impulses transmitted by motor nerves to the blood vessels surrounding dermal sweat glands (Figures 4-7 and 4-25) cause them to dilate and allow more blood to reach the secretory cells. By evaporation, the increased amount of secretion on the surface of the skin tends to lower the skin temperature.

The coiled sweat glands are of two types: small thermoregulatory eccrine of epidermal origin whose cells secrete continuously into a duct (merocrine secretion), and larger apocrine. Apocrine glands (thermoregulatory and scent types arise as outgrowths of hair follicles) possess storage sacs and secrete by cellular budding or disintegration into the duct (apocrine secretion). Eccrine sweat is a hypotonic solution of sodium chloride containing potassium, urea, lactic acid, little protein, and mucopolysaccharide. Some water and sodium are reabsorbed by cells of the duct. Nerve control is autonomic. Thermoregulatory apocrine secretion is watery and is expelled by the contraction of smooth muscle fibers in the storage sacs (hoofed mammals; lacking in rodents; scanty in cat and dog). Responsive to sex hormones, scent glands produce a scanty, viscous fluid, containing odorous volatile substances and mucopolysaccharides,

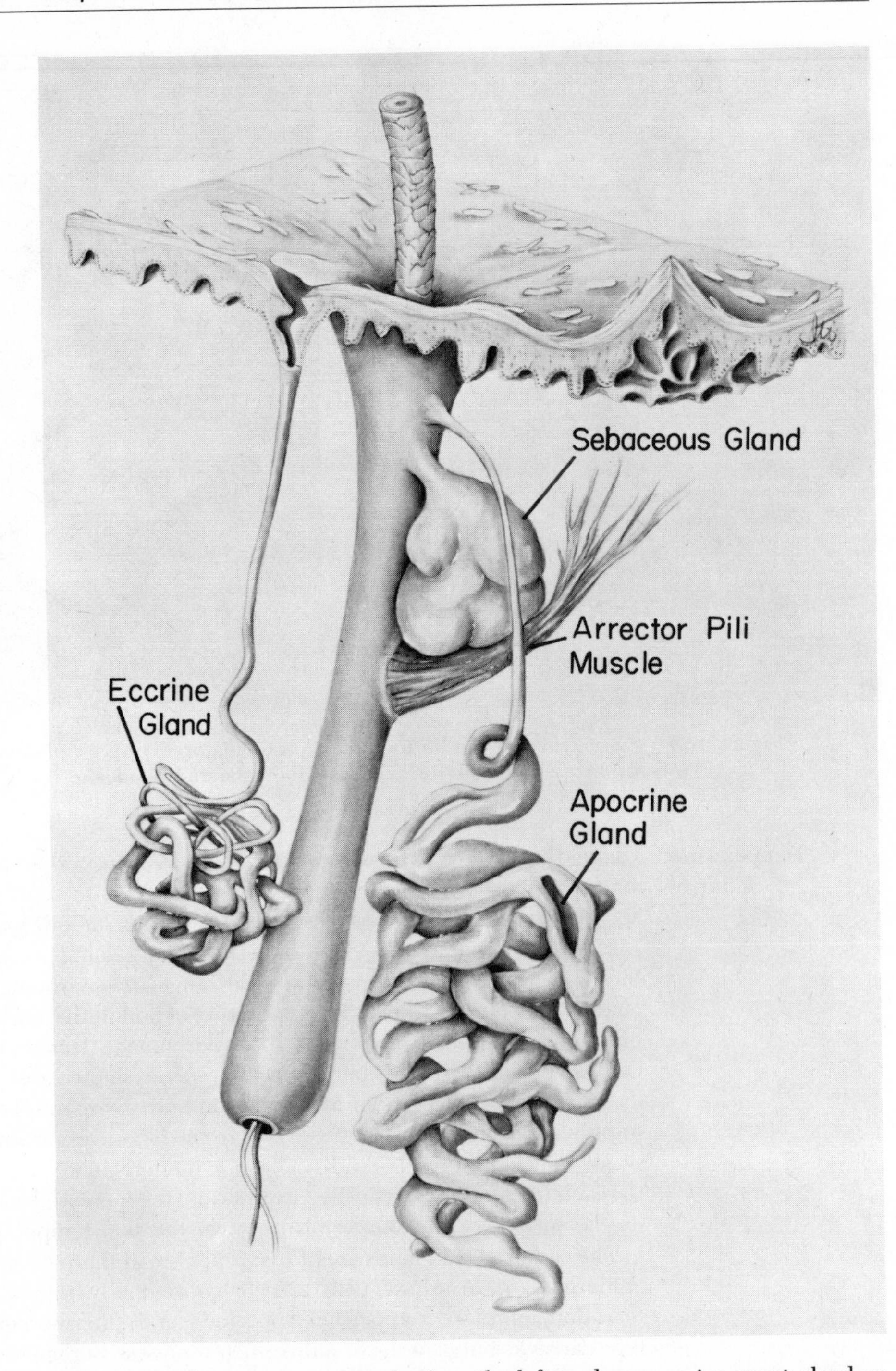

Figure 4-30 Stereogram of an eccrine sweat gland on the left, and an apocrine sweat gland on the right of a hair follicle with its sebaceous gland. Eccrine glands are simple tubular coiled structures without connections to hair follicles; function throughout life and produce a watery secretion. Apocrine glands are connected with hair follicles, begin to function at puberty, and produce a viscous secretion. (Courtesy of William Montagna and Academic Press, New York.)

which play a role in sexual behavior. They are found in urogenital skin, belly skin of the male musk deer, human axilla, and so on. The protective scent of the skunk and mink are produced by glands of the apocrine type.

Undue loss of heat is prevented by the reduction of the amount of blood that reaches the glandular tissue through the arterioles and dermal capillaries. Arterioles undergo vasoconstriction that results when the circular smooth muscles in the arteriole walls contract in response to messages from the temperature-control center in the hypothalamus of the brain. However, some heat may continue to be lost in the arms, legs, fingers, and toes. This is controlled, in turn, by a mechanism that utilizes the principle of exchange between two currents moving in opposite directions, the countercurrent heat exchanger. In the extremities, arteries, veins, and their branches lie in close proximity, permitting heat transfer from arterial to venous blood.

An interesting mechanism exists in other types of human extremities as the external ear pinna and the nose, where freezing could easily take place. Here are arteriovenous anastomoses that bypass the capillary bed and react to stimuli by expanding both in high and low temperatures, so in each case more blood reaches the skin. This system permits both cooling at high temperatures and warming at lower ones.

Additional mechanisms provide protection for homeotherms. Heat is lost from the body by radiation, conduction, and convection. Radiation in the form of electromagnetic waves is conducted from the body by contact with air or water, the movement of which carries away the heat. People living in cold climates possess a thicker epidermis. Feathers and hair are insulating devices. Contraction of the smooth muscle associated with each hair follicle causes the hairs to "stand out" from the skin, thus thickening the fur coat and causing less air movement next to the skin. In some fur-bearing mammals, coat thickness can be increased by additional hair formation or replacement. Fat deposits in the subdermal and even the dermal layers also serve as insulation.

Adaptations to Arid Environments

Unlike reptiles, amphibians never overcame their dependence on water in one way or another. Because the integument is permeable, it is unable to prevent the rapid evaporation that occurs readily as from a free surface [5]. Epidermal cells can actively transport sodium and chloride ions into the dermis, preventing leakage of body fluids. Among most amphibians the loss of fluid is inversely related to the relative humidity; at low humidity this evaporation lowers body temperature, whereas a rise in body temperature accompanies dehydration. Rate of loss is independent of the skin. Normal permeability differs between various species; certain toads and the red eft land stage of the newt lose water more slowly than frogs, as evidenced by the nature of the stratum corneum. As a living membrane, frog skin has commonly been used in permeability studies. In water, frogs take up moisture by osmosis through the skin, never by mouth, as has been demonstrated in dehydrated animals. This is an obvious advantage, as the frogs can collect necessary water from damp surfaces; in desert frogs and toads it facilitates rapid hydration from burrow walls; water also is stored in the urinary bladder.

Australian terrestrial amphibians avoid dehydration by burrowing, during which a transparent cocoon of a single cell layer of shed keratinized stratum corneum may reduce water loss. This cocoon completely surrounds the animal, even across the eyes and cloaca, although there may be tubular inserts into the external nares. Apparently, this membrane is formed by sloughing the stratum corneum as a complete unit through intercellular separation by some unknown method. Passage of water through the cocoon wall can amount to 0.22 to 0.65-milligrams of water per square centimeter per hour. Some desert-dwelling frogs have been known to spend three months or more underground making use of soil water. When the California spadefoot toad, *Scaphiopus couchii*, emerges from its burrow following a dry spell, it is covered by a hard material resembling dried skin. Lungfish estivate, during which they secrete a slime cocoon around themselves as they lie buried in the mud. Protection for many years may be gained in this manner, even after the mud has become a thoroughly dried mass on a scientist's laboratory shelf.

Water loss by evaporation may take place from the thickest cornified skins, such as of reptiles. In the desert tortoise, *Gopherus agassizi*, evaporation constitutes the major proportion of water loss. However, this loss and that by respiration are far less than in lizards and tortoises from wetter regions, as reported in 1966 by K. Schmidt-Nielsen and P. J. Bentley. This suggests that in reptiles there are adaptations of the integument associated with aridity of the environment. Birds and reptiles excrete waste nitrogen principally as uric acid, so little or no water is involved; even fecal pellets may be almost dry. The condition of the skin aids camels to sustain a far greater degree of dehydration than in the case of man. Some mammals in dry regions utilize metabolic or oxidation water produced from foodstuffs (deermouse, *Peromyscus;* kangaroo rat).

No discussion of the versatile skin would be complete without some mention of its role as a peculiar ecosystem with characteristic microscopic flora and fauna and diverse ecological niches. In a fascinating article, Mary J. Marples pictures the skin as "a kind of soil with attributes that are beneficial (temperature, food and water supplies) or harmful (slight acidity) for the organisms it supports, a discrete world where living and nonliving components, all interacting with one another, exist in equilibrium."* Some of these organisms, which are potentially lethal pathogens, cause disease only if they penetrate the deeper cutaneous layers. The flat, scalelike squames (Figures 4-6 and 4-26) of fibrous proteinaceous keratin together with secretions of various glands serve as food for the cutaneous organisms. These glandular secretions and by-products from keratinization provide free amino acids as nutrients. Carbohydrates and certain vitamins are available. Except in the extremities, little variation occurs in the warm climate of this ecosystem. Among the permanent, indigenous organisms dwelling in this environment are the follicle mite, *Demodex folliculorum*, yeasts, fungi, bacteria (staphylococci and Gram-negative bacilli), and possible viruses; all survive because they are compatible with healthy skin. Evidence shows that some of the substances on human skin prevent the growth of pathogens. The study of skin ecology is now considered essential to the control of skin and wound infections. Even the most innocent injection may provide entry for life-threatening infection [29].

References for Chapter 4

[1] ADACHI, K. "The Metabolism and Control Mechanism of Human Hair Follicles." *Current Problems in Dermatology*, **5**(1973), 37–78. Data are

*Mary J. Marples, "Life on the Human Skin," *Scientific American*, **220:** 108–115 (Jan. 1969).

presented on energy and androgen metabolism and some metabolic control mechanisms, along with speculation on the genesis of pattern baldness.

[2] Albuquerque, E. X., J. W. Daly, and B. Witkop "Batrachotoxin: Chemistry and Pharmacology." *Science,* **172**(1971), 995–1002.

[3] Axelrod, Julius "The Pineal Gland: A Neurochemical Transducer." *Science,* **184**(1974), 1341–1348. Chemical signals from nerves regulate synthesis of melatonin and convey information about internal clocks.

[4] Beer, A. E., and R. E. Billingham "The Embryo As a Transplant." *Scientific American,* **230**(1974), 36–51. Pregnancy is an unusual graft-host relationship, in that the embryo-fetus seems indifferent to the maternal immune response.

[5] Bentley, P. J. "Adaptations of Amphibia to Arid Environments." *Science,* **152**(1966), 619–623. In arid environments, amphibians have means of acquiring water for survival.

[6] Bern, H. A. "Hormones and Endocrine Glands of Fishes." *Science,* **158**(1967), 455–462. Hormonal control of coloration is explained.

[7] Billingham, R. E. "Tissue Transplantation: Scope and Prospect." *Science,* **153**(1966), 266–270. Immunological factors in connection with skin transplantations are described.

[8] Cohen, J. "Feathers and Patterns." In Abercrombie, N. and J. Brachet (eds.), *Advances in Morphogenesis.* New York Academic Press, Inc., 1966, Vol. 5. For structure and development of feathers and feather patterns, see pages 1–38.

[9] Daly, J. W., and C. W. Meyers. "Toxicity of Panamanian Poison Frogs (Dendrobates); Some Biological and Chemical Aspects." *Science,* **156**(1967), 970–973. Pharmacological activity of the skin is discussed.

[10] Daniels, F., Jr., J. C. van der Leun, and B. E. Johnson. "Sunburn." *Scientific American,* **219**(1968), 38–46. When the human skin receives an overdose of ultraviolet radiation, a complex series of events ensues.

[11] Etkin, William, and Lawrence I. Gilbert, eds. *Metamorphosis: A Problem in Developmental Biology.* New York: Appleton-Century-Crofts, 1968. A single up-to-date source that deals in an integrative way with the several aspects of metamorphosis.

[12] Giacometti, L. "A Perspective on Baldness." Publication no. **421**(1974), Oregon Regional Primate Research Center, Beaverton, Oregon. This article traces the history of some of the superstitions concerning hair and the lack of it.

[13] Goodrich, H. B., and M. Hedenburg. "The Cellular Basis of Colors in Some Bermuda Parrot Fish with Special Reference to Blue Pigment." *Journal of Morphology,* **68**(1941), 493–505.

[14] Grant, W. C., Jr., and G. Cooper, IV. "Behavioral and Integumentary Changes Associated with Induced Metamorphosis in *Diemictylus.*" *Biological Bulletin,* **129**(1965), 510–522. Changes in the newt's skin between land and aquatic phases of the life cycle are described.

[15] Hertig, B. A., M. L. Riedse, and H. S. Bilding. *Advances in Biology of Skin.* Vol. 2. New York: Pergamon Press, Inc., 1962. Advances in knowledge of the skin are presented.

[16] Kischer, C. Ward. "Fine Structure of the Down Feather During its Early Development." *Journal of Morphology,* **125**(1968), 185–204. Dermal-epidermal interactions in the development of a feather are discussed.

[17] LOOMIS, W. F. "Skin-pigment Regulation of Vitamin-D Biosynthesis." *Science*, **157**(1967), 501–506. In man vitamin D is synthesized by the skin.

[18] MODELL, WALTER "Horns and Antlers." *Scientific American*, **220**(1960), 114–122. Horns and antlers are commonly believed to be alike, but actually they are quite different: the material of horns is related to skin, that of antlers to bone. See also Goss, Richard J. *Principles of Regeneration.* New York: Academic Press, Inc., 1969.

[19] MONTAGNA, W., and W. C. LOBITZ, eds. *The Epidermis.* New York: Academic Press, Inc., 1964. This is a detailed discussion of the epidermis, its structure and function.

[20] MONTAGNA, W., and P. P. PARAKKAL *The Structure and Function of Skin.* New York: Academic Press, Inc., 1973. This is a more detailed description of the skin: its role, structure, importance, and unsolved problems; it offers a good list of references.

[21] NICOLAIDES, N. "Skin Lipids: Their Biochemical Uniqueness." *Science*, **186**(1974), 19–26. Unlike internal organs, the skin biosynthesizes and excretes unusual fat soluble substances (lipids) that may function as chemical fingerprints.

[22] RAWLES, M. E. "Tissue Interactions in Scale and Feather Development As Studied in Dermal-Epidermal Recombinations." *Journal of Embryology and Experimental Morphology*, **11**(1963), 765–789. A study of the interactions of the epidermis and dermis during development is made.

[23] ROMER, ALFRED S. *The Vertebrate Body.* 3rd ed. Philadelphia: W. B. Saunders Company, 1963. Some aspects of the comparative anatomy of the skin and its derivatives are considered.

[24] RUSHMER, R. F., K. J. K. BUETTNER, J. M. SHORT, and G. F. ODLAND. "The Skin." *Science*, **154**(1966), 343–348. Use is made of the skin to show the diversity of problems that can be approached by investigators with widely different backgrounds.

[25] SAYLES, L. F., and S. G. HERSHKOWITZ. "Placoid Scale Types and Their Distribution in *Squalus acanthias*." *Biological Bulletin*, **73**(1937), 51–66.

[26] SEARLE, A. G. *Comparative Genetics of Coat Colour in Mammals.* New York: Academic Press, Inc., London: Logos Press, 1968.

[27] SMITH, HOBART M. *Evolution of Chordate Structure.* New York: Holt, Rinehart & Winston, Inc., 1960. Evolutionary development of skin structures is discussed.

[28] ———. "The Phylogeny of Hair and Epidermal Scales." *Turtox News*, **36**(March 1958), 82–84. The evolutionary origin and development of hair and scales are discussed.

[29] SPEARMAN, R. I. C. *The Integument.* Cambridge: Cambridge University Press, 1973. This is especially valuable for a comparative account of the integument in both invertebrates and vertebrates.

5

Musculoskeletal System

Introduction: Functional Integration

The interdependence of structure and function is one of the fundamental principles in biology. The form of bones and muscles, with their preeminently mechanical function, can serve as one of the more obvious examples of this principle. It is especially by the use of bones and muscles that vertebrates are able to act on their environment by exerting forces and moving themselves around the world. Because of the intimate and mutual functional dependence between the skeletal and muscular systems, they will be discussed together here as one inseparable integration rather than as independent morphological entities. Various integrated combinations of skeletal elements and muscles function as biological machines that convert chemical energy, derived ultimately from the reaction between oxygen and food, into force and mechanical work. In this chapter, the structural components and mechanisms of biological machines that execute locomotory, feeding, and respiratory functions will be analyzed.

One of the most conspicuous influences of muscles on bones is reflected on the surface morphology of bones. The surface markings on bone faithfully delineate the shape of attached connective tissue structures; for example, fleshy muscular attachments often produce depressions called *fossae*, which vary in size and shape, that interrupt the otherwise featureless surface of bone. On the other hand, tendons usually are associated at their point of attachment with some degree of roughening, often in the form of elevations that may be elongate (crest or *crista*), pointed (*process*) or rounded (*tuberosity, tubercle*, or *trochanter*). There is some evidence that the prominence of such elevations is associated with the power of the muscles involved. Therefore, surface markings on bones furnish relevant data on the junction of bone with muscle, tendon, ligament, or articular capsule and have been used extensively by paleontologists in reconstructing vertebrate fossils. However, it is important to emphasize that even in a so-called fleshy, or direct, attachment of muscle to bone, the actual muscle fibers are not themselves adherent to the bone or its membranous covering, the periosteum. The precise mechanism by which tension is transmitted from contracting muscle fiber to bone is effected through the collagen that surrounds muscle fibers in the form of perimysium, although some details are still unknown. It is certain that, whether the junction to bone is by tendon or is fleshy, the actual agent of attachment is collagen fibers that pass into bone as Sharpey's fibers. The multitude of microscopic connective tissue ties of a fleshy attachment necessarily extend over a larger area and do not mark the bone appreciably. In contrast, where connective tissue (collagen) is concentrated in greater masses, visible markings appear on the bone surface at more localized sites. However, such localized stresses applied to bone cause, in addition to external markings, adaptive changes to occur in the internal architecture that will be discussed subsequently.

Extrinsic influences of muscles on bones such as tension and pressure are also responsible for the final form of a skeletal element. Experiments have shown that, in the absence of the normally occurring external forces, bony elements can self-differentiate only for a relatively short period. The final shaping of a bony element and its internal architecture occur only under the influence of the surrounding muscles and soft tissues. This characteristic plasticity of bone tissue, although more pronounced in younger individuals, furnishes the bony elements with a potential to undergo biologically adaptive responses to changing environmental factors.

Form, Function, and Architecture of Bone

Because the musculoskeletal system enables the organism to exert forces on its environment, the concept *force* is basic to this chapter. The concept force is implicit in Newton's three laws of motion:

1. A body remains at rest or moves at a constant velocity unless forces act on it.
2. An unbalanced force gives a body an acceleration in the direction of the force, the acceleration being proportional to the force and inversely proportional to the mass of the body.
3. If body A exerts a force on body B, body B exerts an equal and opposite force on body A. Because a force is a vector quantity, one must give force its direction as well as its size.

If a tensile or compressible force is applied to an object of length L, it will respectively increase or decrease in length by an amount of ΔL. The ratio $\Delta L/L$ is called the *strain*. *Stresses* are set up within the body to which a force is applied. Stress is measured in terms of force per unit area and can be imagined as the intensity of the forces acting on any particular part of a body.

If a bony element is loaded in some way, we can plot the strain associated with the various stresses into a stress-strain curve as shown in Figures 1(a) and (b). In the region OA of the curve $OABC$, the strain is directly proportional to the stress and the strain will return to zero if the stress is removed. Beyond the point A, however, the curve bends over because strain is no longer directly proportional to the stress. With increasing strain, the curve gets flatter and flatter until the material breaks at C. The value of CB is the ultimate stress of

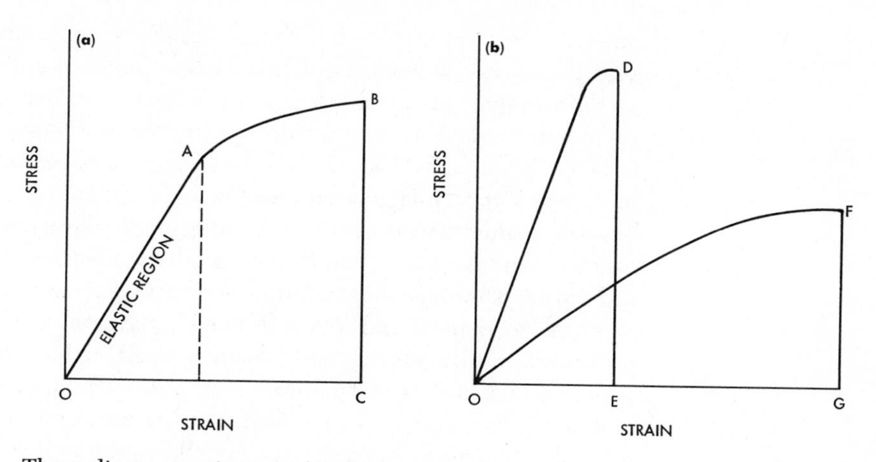

Figure 5-1 Three diagrammatic stress/strain curves.

the material, and, although at variance with the definition in engineering, can be considered as the *strength* of the bone. The area under the stress-strain curve is a measure of the amount of work per unit volume that has to be performed on the bone in order to achieve that stress. The area under *OABC* is, in effect, a measure of the amount of energy that can be absorbed before the bone fails. However, it must be emphasized that the greatest amount of energy is not necessarily absorbed by the material with the highest ultimate stress, as shown by the curves *OED* and *OFG*. *DE* is greater than *FG*, but the latter curve has a greater area under it. Energy absorption is an important adaptive feature when a bony element is loaded very quickly by impact. Because of the complexity of bone structure one cannot speak of *the* strength of the element, but can record only the approximate strength of a given type of bone when loaded in a certain way relative to its internal architecture. It is also important to know that the less the strain for a given stress, the stiffer the material is. The value of the stress divided by the strain in that part of the curve where the material is still elastic (*OA*) is called the modulus of elasticity. If the modulus is low, the material is easily deformed; if it is high, the bone is rigid. Most skeletal elements are the result of a compromise between strength (or stiffness) and lightness.

Muscles often play a major role in achieving the metabolically important goal for bones to provide sufficient strength with minimum material by preventing bones from being loaded during bending. The stabilizing influences of muscles on bones is a complex and important matter that is still being studied.

If optimal adaptive versatility is to be achieved, skeletal elements must possess proper combinations of static strength, stiffness, and resistance to impact. As we see in Figure 5-1, the area under the stress-strain curve (which is roughly proportional to the amount of energy that can be absorbed by the bone before it fractures) is larger at a given stress, the lower the modulus of elasticity. How then can bone provide the appropriate proportions of tensile strength (a low modulus of elasticity), compressive strength, and stiffness (a high modulus of elasticity)? Such a mechanical versatility has been achieved by bone because of its composite material. The mechanical properties of the composite are unlike those of either component taken alone and are not the sum or average of the two taken together. Engineers make composite materials, such as glass in resin, or tungsten in copper, that are often much stronger for their weight than are noncompounded materials. Bone is essentially an inorganic mineral laid down in an organic matrix. Basically, the mineral appears to be hydroxyapatite $Ca_{10}(PO_4)_6(OH)_2$. The organic components are fibrous proteins of the collagen type arranged in ropelike triple helices that are bound together by side groups that offer great resistance to tension, although bending can take place readily. The needle-shaped appatite crystals surround and impregnate the collagen fibrils, and there is evidence that there are bonds between the mineral and collagen. The long axes of the crystals are parallel to those of the fibrils. Varying patterns are formed by the collagen fibrils and their associated needles. In woven bone, the fibrils are tangled up more or less at random; in lamellar bone, the fibrils in any one lamella are parallel to each other, but the fibrils in successive lamellae are about at right angles to each other. With this arrangement, no crack can pass through a bone without encountering an almost infinite palisade of apatite needles separated by collagenous matrix. Because the collagen allows the needles to move relative to each other, the modulus of elasticity of bone is much lower than that of apatite, but much higher than that of collagen. The composite nature makes bone more rigid than collaginous fibers, more flexible and resistant to fracture than the mineral, and far more versatile in the kinds of loads it can withstand than either component alone.

Given the composite nature of bone, a great number of adaptive responses to changing or varying environmental demands, i.e., to mechanical stresses, can evolve rapidly in bones by regulating the appropriate proportions of their constituent elements. In most instances, bone behaves like a dynamic system controlled by feedback, as shown in Figure 5-2. An applied force causes stresses and strains in bone. Strains are the stimuli for the process of adaptive responses, or remodeling, to occur. Under normal loading, bone apposition by osteoblasts and bone resorption by osteoclasts are in balance. Greater strains will cause bone hypertrophy by a relative predominance of bone apposition. The increase of the load-bearing cross section is followed by lowering of stresses. Decreased stresses will, in turn, lead to bone resorption. As observations on astronauts demonstrate, the first reaction of bone to a decrease of stresses (i.e., weightlessness) is a rapid resorption of mineral matrix, although remineralization occurs within a short time under influence of gravitational forces. Although the responsiveness of bone to altered conditions of stress is well established, the exact morphogenetic control systems are still unknown. As is the case with many other crystalline substances, bone is piezoelectric—i.e., it generates small electric currents when deformed—and the distribution of the potential difference variations reflects the form of the deformation. It is suggested that such currents could, in some way, polarize the cells responsible for bone resorption (osteoclasts) and apposition (osteoblasts), so that the bone is remodeled in a manner that resists the predominating mechanical stresses.

Adaptation of bone to mechanical stimuli also involves the inner architecture. Mature bone is composed of two kinds of tissue, one of which is dense in texture like ivory, named compact bone; the other consists of a meshwork of thin sheets of bone, *trabeculae*, within which are intercommunicating spaces (or *cancelli*), and is called spongy (trabecular or cancellous) bone. The compact bone is always on the exterior, surrounding the spongy bone. Their relative

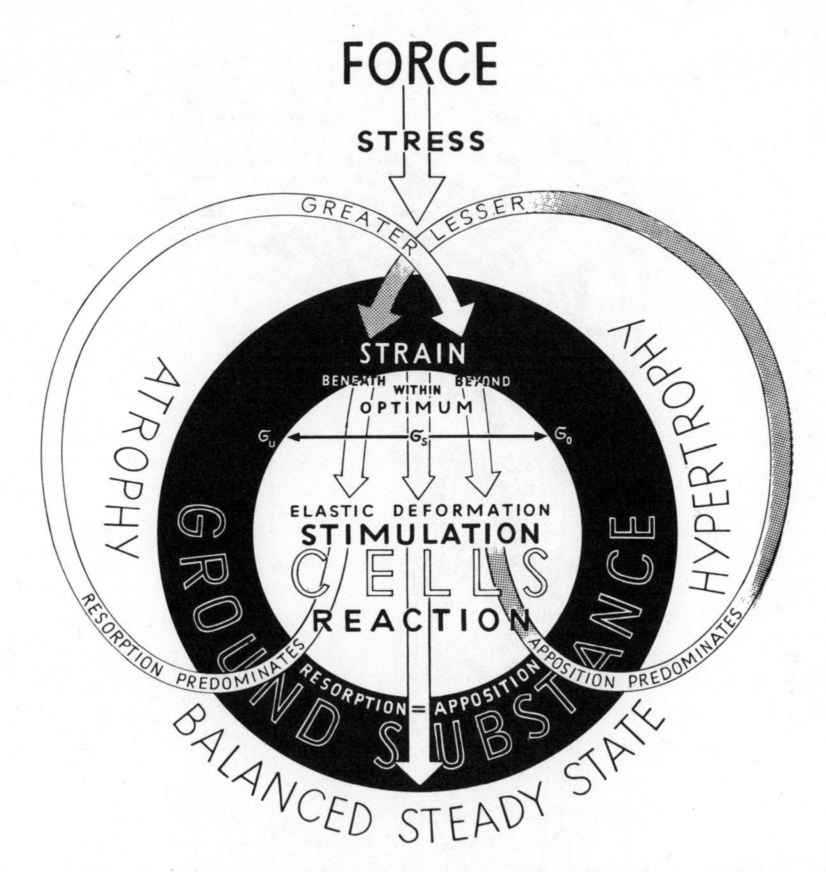

Figure 5-2 Feedback mechanisms controlling bone tissue. Intermediate stress levels maintain the status quo. Reduced or excessive stress levels result in, respectively, resorption and apposition (deposition). (From Kummer.)

quantities and architecture vary characteristically in a manner that reflects adaptations to mechanical demands.

Sections through spongy bone show trabecular patterns that correspond to the lines of forces acting on the bone. Because forces transmitted by one bone to another across a flexed joint are not parallel to the shaft of either, stress lines are across the ends of those bones. Stress lines change as loads change, so the bone responds by adopting trabecular arrangements that are a compromise of the more usual loads. Wherever localized stresses are applied to bone, as in the attachment of tendons and ligaments, a subjacent condensation and reorientation of trabeculae become evident. We must, therefore, conclude that patterns of stress and trabeculation are closely related (see Figure 5-3).

Compact bone consists mainly of a number of irregularly cylindrical units, named Haversian systems, each composed of a central Haversian canal that contains a vein, artery, and nerve and is surrounded by concentric lamellae of bone. Between these lamellae are a number of small cavities, *lacunae,* that are connected with each other and with the Haversian canal by many fine radiating canals, the canaliculi. Within the lacunae and their canaliculi are, respectively, the bone cells, osteocytes, and their fine cytoplasmic extensions. The angular intervals between the Haversian systems are occupied by interstitial bone. Encircling the inner and outer surfaces of the bone are the circumferen-

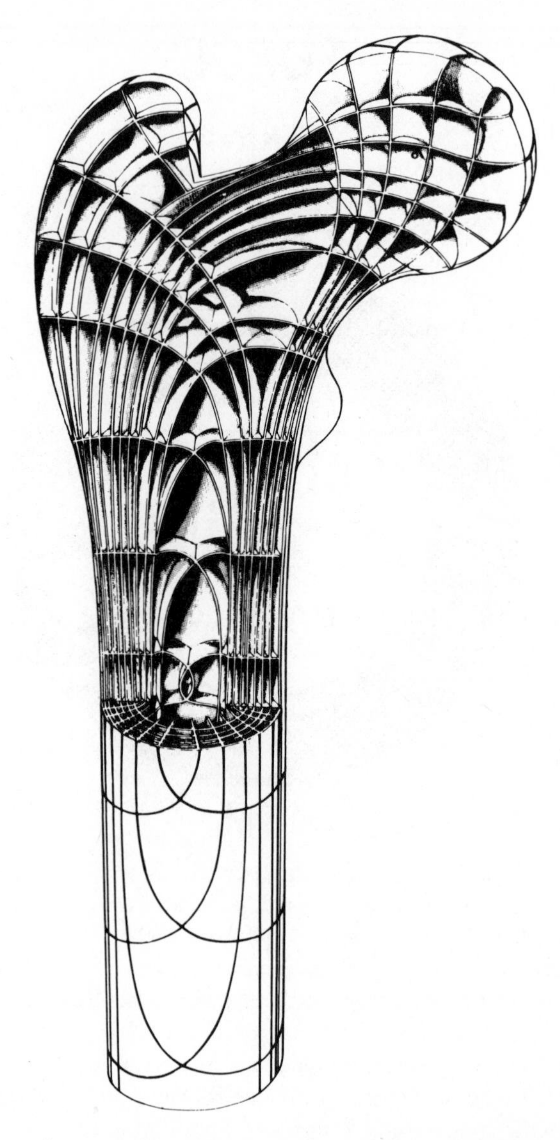

Figure 5-3 Theoretical construction of three-dimensional trajectorial system in a human femur. At the head of the femur the weight of the body is passed eccentrically directly from the pelvis to the femoral head. The vertical set of osseous bars are the compressive trajectories. Those running at right angles and more or less horizontally, are the tensile trajectories. (From Kummer.)

tial lamellae. Haversian systems are prominent in the long bones of man and some larger carnivores, but are less plentiful in herbivores, and are virtually absent in small mammals. In nonmammalian vertebrates, they appear in some reptiles and birds, but hardly anywhere else. Small cavities, notches, and channels all weaken materials by causing localized concentrations of stress. A stress concentration is depicted in Figure 5-4. In (a) the lines of force run straight from one end of the stressed material to the other. In (b) there is a very narrow crack, and the lines of force have to make their way around the tip. Although (a) and (b) have the same area at the narrowest place, in (b) there is a concentration of force at the tip of the crack so that there the stress will be concentrated. As the crack gets longer, the force needed to keep it running gets less and it accelerates, unless somehow the tip of the crack becomes blunt. When the crack runs into a cavity [Figure 5-4(c), (d), (e)] the stress concentration is reduced greatly. The lacunae in cellular bone are, therefore, thought to play a role in stopping the spread of microfractures.

In order to save on metabolic requirements, bones are functionally designed to provide sufficient strength with minimum material. Adaptive responses of bones to mechanical loading often are reflected in their gross shape. We know that the I-beam is extensively used as a girder by engineers. This beam has upper and lower bars of steel, which are stressed when loaded, and a wall to hold the bars apart. Although bones are never designed as simple I-beams, the same principle of construction explains the dumbbell-like cross sections of bones that have to resist bending in one plane. In beam-joint systems, the resistance to bending is equal to a constant times the width of the beam times the square of its height: $R = cwh^2$. If one dimension is twice the other, then the beam is twice as strong on edge as it is flat. In other words, a flat beam effectively resists bending when loaded edge on, but is weak when loaded flat side on. The long bones of the appendages of tetrapods must resist bending in many directions; hence, they cannot be flat. The cylinder is an effective shape because compressive and tensile forces can concentrate at opposite edges of the cylinder regardless of the direction of loading and can resist bending in many directions. Because resistance of a cylinder to bending varies inversely as the square of its length and because the long bones are stabilized by muscles acting at their ends, they would be most susceptible to fracture at the centers of their shafts. To protect against such bending forces, the shafts of long bones at midlengths exhibit marked hypertrophy (Figure 5-3). As animals become larger, their bones must grow at a faster rate than their bodies, i.e., allometrically. A cylinder supporting its own weight becomes relatively weaker by linear growth. In order to maintain the necessary strength, such a lengthening cylinder has to increase its circumference by a larger factor. If n expresses the

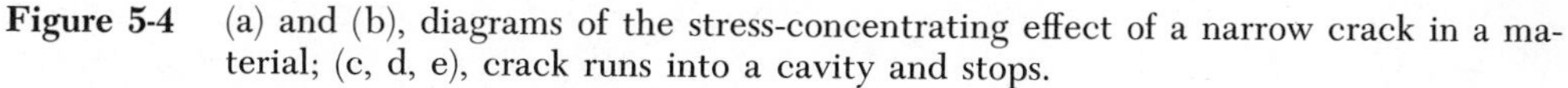
Figure 5-4 (a) and (b), diagrams of the stress-concentrating effect of a narrow crack in a material; (c, d, e), crack runs into a cavity and stops.

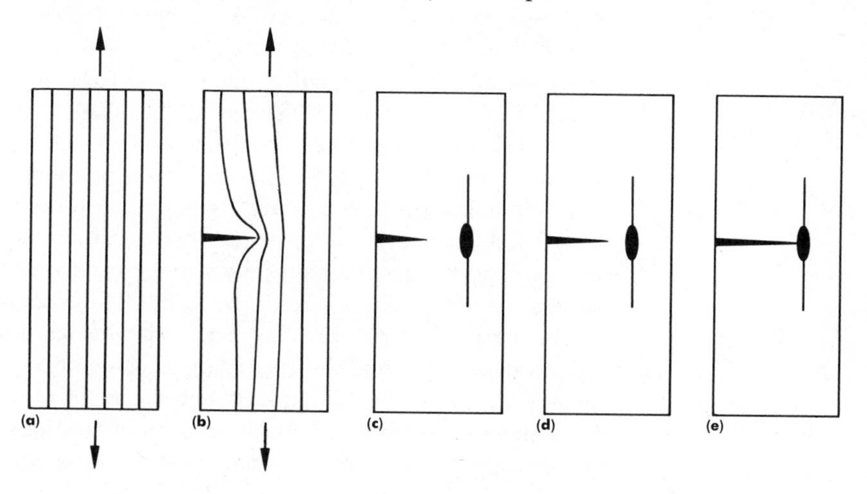

increase in length, then the diameter must increase by $n^{1.5}$. The results of this principle can be seen in all large tetrapods.

We can conclude that the adaptive processes in bone tissue are basically feedback mechanisms controlled by mechanical stresses that influence the size, shape, and composition of the organic and inorganic materials, cross-sectional configurations, and the trabecular pattern of spongy bone. In the protracted permutations possible during vertebrate evolution, the best biological solution to the mechanical demands has been arrived at, in interaction with the limiting factors of nutritional resources, available muscle power, and a compromise between size and weight.

Fine Structure and Mechanics of Contraction of Skeletal Muscle

One of the basic characteristics of living organisms is their ability to react to changes in their environment. All responses involve the utilization of metabolic energy by some sort of effector system. Contractile proteins, which can convert chemical energy into mechanical energy of motion, are present in high concentrations in muscle cells. The integration of muscle cells into distinct groups, their intimate association with the skeleton, and their coordination by the nervous system, transform them into an effector system capable of a wide range of highly complex actions. Three different types of muscle cells can be identified on the basis of structure and contractile properties: (1) *skeletal* muscle, which is attached to the bones; (2) *smooth* muscle, which surrounds such hollow chambers as the stomach and intestinal tract, the bladder, the uterus, and blood vessels; and (3) *cardiac* muscle, which constitutes the contractile tissue of the heart.

A single muscle cell, with its hundreds of nuclei, represents a muscle fiber, which in man can reach a length of thirty centimeters or more. These fibers are arranged in bundles or *fasciculi* of various sizes within the muscle (Figure 5-5). Connective tissue occupies the spaces between muscle fibers within a fasciculus, where it is known as the endomysium; each fasciculus is surrounded by a stronger connective-tissue sheath, or *perimysium;* enveloping the whole muscle is the stout *epimysium,* which is continuous with the outer perimysial septa and externally with the connective tissues of surrounding structures such as tendons and aponeuroses (flat sheets of densely arranged collagen fibers).

The cell membrane of the muscle fiber, called the sarcolemma, surrounds the cytoplasm, or *sarcoplasm,* which is divided into longitudinal threads or *myofibrils* (see Figure 5-5). Under electron microscopy each myofibril is seen to be composed of longitudinally arranged *myofilaments,* which form a regularly repeated pattern along the length of the fibril. One unit of this repeating pattern is known as a sarcomere (Figure 5-6). Each sarcomere is composed of thick and thin myofilaments. The thick ones, composed of the protein myosin, are in the middle of the sarcomere, where their orderly arrangement produces the A-band pattern. At either end of the sarcomere are the thin myofilaments composed of the protein actin. The interconnecting zone of thin myofilaments between adjoining sarcomeres is the Z-line. The H-zone is in the middle of the A-band and corresponds to the region from the ends of the thin filaments in one half of the sarcomere to the ends of the thin filaments in the other half (Figure 5-6). According to the sliding-filament theory of muscle contraction, muscle shortening results from the relative movement of thick and thin filaments past each other, rather than from a change in the length of the filaments themselves.

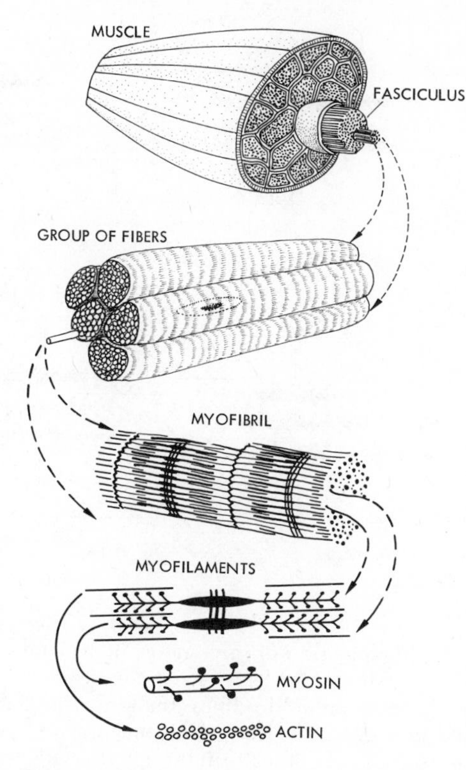

Figure 5-5 Diagram showing the various levels of organization within a skeletal muscle, from whole muscle, through fasciculi, fibers, myofibrils, myofilaments, down to molecular dimensions.

As the thick and thin filaments move past one another, the length of the H-zone changes, but the A-band remains constant. Sliding of the filaments is realized by movement of the cross bridges that exist between the thick and thin filaments. At the cross bridge, actin reacts with myosin, releasing energy and changing the configuration of the cross bridge. It has been suggested that each cross bridge must detach itself from one place on the actin and reattach itself to another site farther along, repeating the process with an action similar to that of a man pulling on a rope hand over hand.

Muscle contraction refers to the active process of generating a force parallel to the muscle fiber, causing it to shorten. The force exerted on an object is the muscle *tension,* and the force exerted on a muscle by the weight of an object is the *load.* Thus, muscle tension and load are opposing forces. In order to lift a load, the muscle tension must be greater than the load. When a muscle shortens and lifts a load, the muscle contraction is said to be isotonic (constant tension) because the load remains constant throughout the period of contraction. When shortening is inhibited by keeping the muscle at a fixed length, the generation of tension occurs at constant muscle length and is, therefore, called isometric contraction (constant length). Supporting a weight in a fixed position requires isometric contractions, whereas body movements require isotonic contractions.

A typical skeletal muscle responds to a single adequate electrical stimulus by giving a *twitch,* i.e., a brief period of contraction is followed by relaxation (Figure 5-7). After stimulation, there is a pause of a few milliseconds, the *latent period,* before the tension begins to increase in an isometric twitch. The time

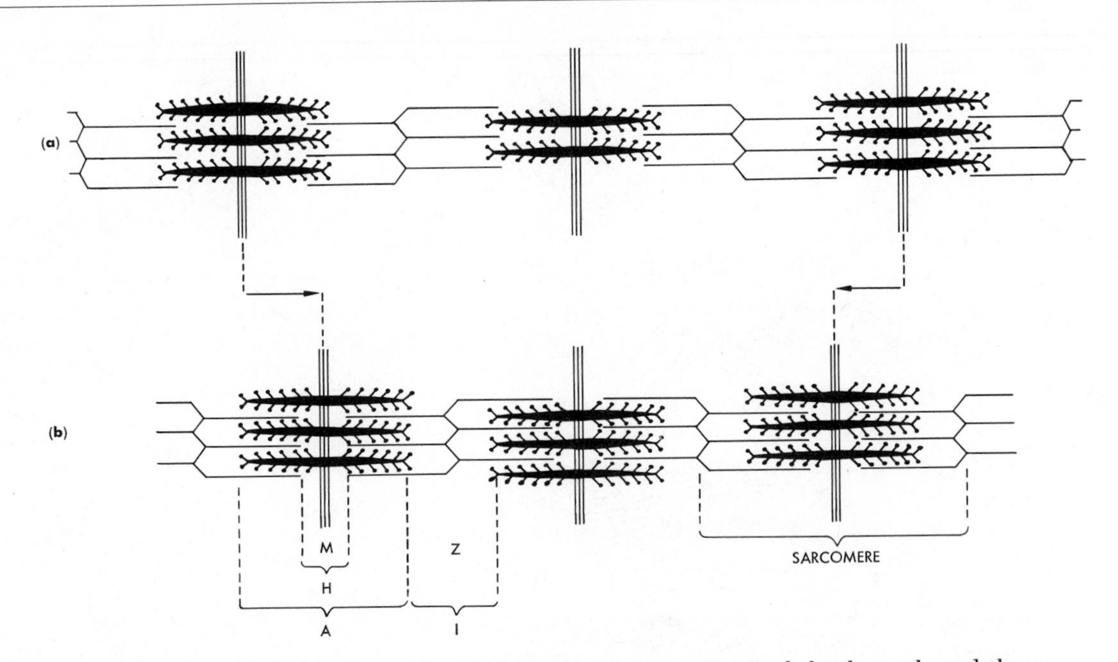

Figure 5-6 Diagram showing the organization of sarcomeres in skeletal muscle and the changes occuring during shortening. (a) resting condition. (b) contracted condition. Thin actin filaments depicted by lines, thick myosin filaments are drawn with thick lines and processes. The actin filaments are each attached at one end to a Z-band and are free at the other to interdigitate with the myosin filaments. The orderly arrangement of the thick myosin filaments produces the A-band pattern. The I-band is the region between two adjoining sarcomeres. The H-bands are the middle region of the A-band into which the thin actin filaments have not penetrated. The M-band lies across the middle of the H-band consisting of fine strands interconnecting the adjacent myosin filaments.

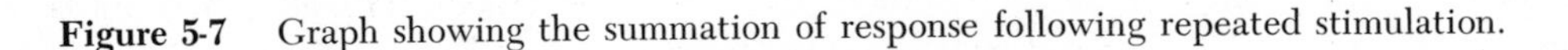

Figure 5-7 Graph showing the summation of response following repeated stimulation.

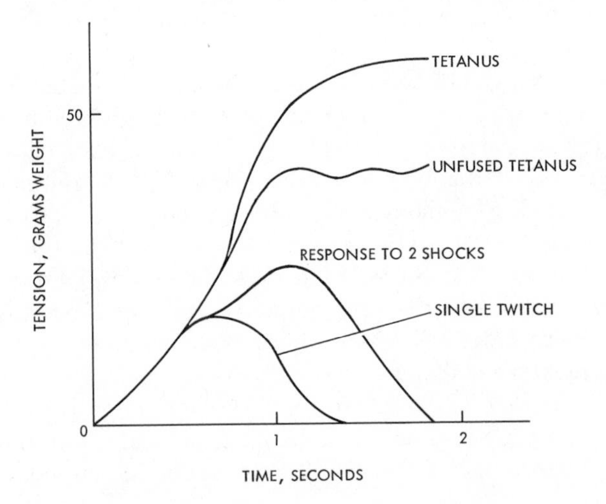

from the start of tension to the peak of tension is the *contraction time.* Not all skeletal muscles contract at the same rate. Some slower fibers have contraction times as long as one hundred milliseconds or longer, whereas faster fibers may take as few as ten milliseconds. If a second shock is given to the muscle before the response to the first has completely died away, *summation* takes place, as shown in Figure 5-7. If the stimuli are repeated at a high frequency, the result is a smooth *tetanus,* with tension maintained at a high level, as long as the train of stimuli continues, or until fatigue sets in. Such an experimental stimulation can cause the entire muscle to contract simultaneously. However, such a situation rarely occurs physiologically. Under normal physiological conditions, the number of muscle fibers contracting is controlled to regulate the total tension and shortening of the whole muscle.

The initial length of a muscle determines the amount of force it can develop. The *length-tension relationship* of a muscle is expressed in the curve of Figure 5-8. If the length at which the muscle develops maximum tension (l_o) is used as a reference point, the different muscle lengths at which isometric tension is measured can be expressed in terms of the per cent of the l_o muscle length. At a length of 60 per cent of l_o, the muscle develops no tension when stimulated. As the length of the muscle is increased, the isometric tension rises to a maximum at l_o, and further lengthening of the muscle causes a drop in tension. In the body, the relaxed length of the skeletal muscle is very nearly l_o and, thus, at optimal length for force generation.

Vertebrate skeletal muscles contain at least two basic types of muscle fiber, red and white. Because red fibers contract more slowly than white, the two are sometimes referred to, respectively, as slow and fast. Although many structural and biochemical differences exist between the two types, it is sufficient for our purposes to state that red fibers are difficult, if not impossible, to fatigue under normal physiological conditions and are mainly responsible for maintaining tension for extended periods, e.g., to run or swim a long distance or to maintain posture. The white fibers, on the other hand, are active for only short periods because they fatigue easily and are responsible for short bursts of intermittent contractions of high intensity. Most muscles are mixtures of red and white

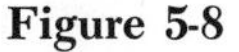

Figure 5-8 Variation in isometric tension with muscle length. (Modified from Gordon, Huxley, and Julian.)

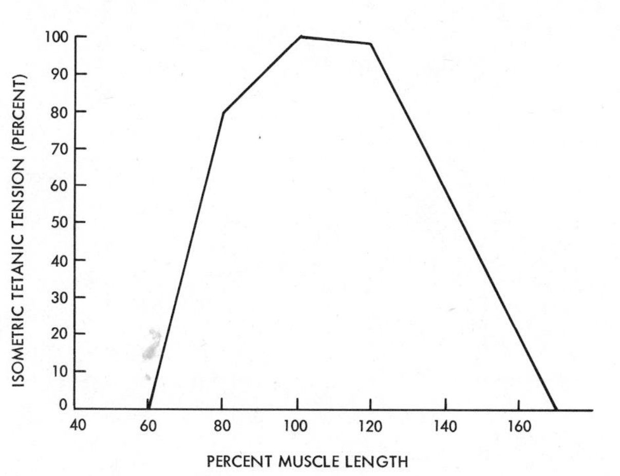

fibers. Different types of exercise have different effects on muscle development. Weight-lifting leads to hypertrophy of the white fibers, whereas long-distance running enhances the efficiency of red fibers.

Architecture of Muscles: Structural Diversity and Function

Among vertebrates there is a dramatic diversity in the size, shape, complexity, fascicular pattern, and form of attachment of skeletal muscles. Each is remarkably well adapted to provide the necessary range, direction, and force of contraction, to meet the functional requirements at the joints over which it crosses.

Muscles can be grouped according to the orientation of their fasciculi (Figure 5-9). In *parallel muscles,* the fasciculi are oriented parallel to the final direction of pull at their attachments. Parallel muscles vary from flat, short, and *quadrilateral* to long and *straplike,* the individual fibers either running almost the entire length of the muscle or traversing shorter segments between tendinous intersections distributed at intervals along the muscle. A parallel arrangement of fasciculi is frequently found in the "bellies" of *fusiform* muscles, which may, however, be quite short, converging to a tendon, sometimes of considerable length at either one or both ends.

Where the fasciculi are parallel to the line of pull at its attachments, full advantage can be taken of both the full force and complete range of excursion of the muscle. Because the maximum range of unrestrained contraction is a function of the length of its muscle fibers, straplike muscles are particularly effective in situations requiring a considerable range of action. However, the

Figure 5-9 Some representative morphological types of muscle based upon their fascicular architecture.

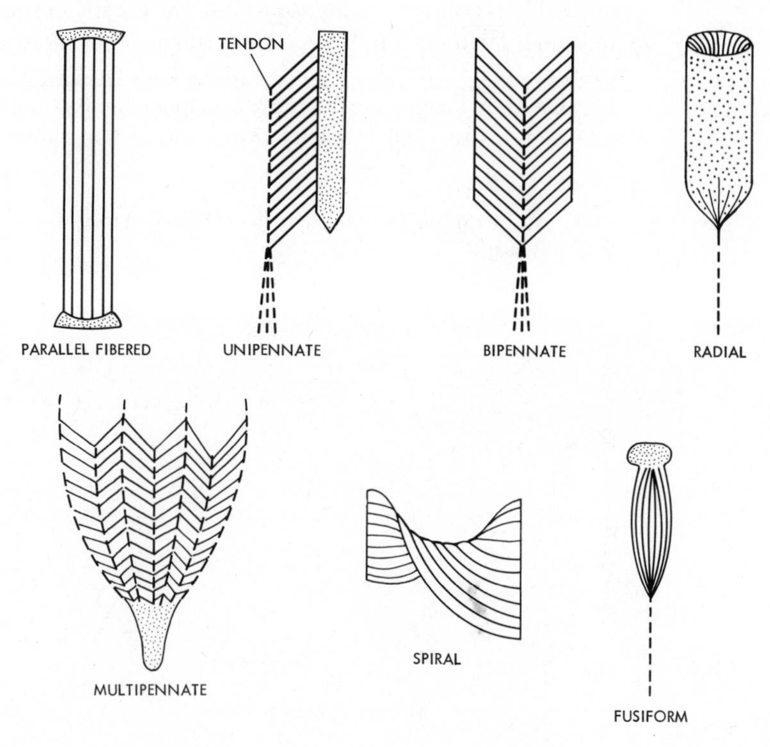

maximum force generated by a muscle is dependent on its effective mass of contractile tissue—which, in turn, reflects the number of fibers per unit of cross-sectional area, dimensions, and the nature of its fibers. Because straplike muscles often possess relatively few fibers, their actions generate moderate power.

Muscles in which the fascicular position is at an angle to the direction of exerted force are classified as *pinnate* (or pennate). Parallel pinnate muscles that connect two plane surfaces and induce them to move parallel to each other are called *unipinnate.* Two such sets of unipinnate muscles may insert on the two sides of a flat tendon in a characteristic feathered or herringbone pattern that gives the muscle the name *bipinnate.* In some situations, muscle fibers arise from around the walls of osteofascial compartments and converge obliquely toward a centrally sited tendon, forming *circumpinnate* patterns.

In all pinnate muscles, only a proportion of the force and range of action available is effective along the line of the tendon. As can be seen in Figure 5-10, the force developed by contraction of a fiber of a pinnate muscle is resolved into two components, an effective one acting along the tendon and another at 90 degrees to this; the latter does not contribute to the useful work and is counterbalanced by the opposite fiber. This loss of force establishes the fundamental disadvantage of fiber angling, which increases disproportionately with increases in the angle of pinnation. Similarly, the range of excursion is restricted and is proportional to the length of the muscle fiber times the cosine of the angle of its attachment to the tendon. The angle of pinnation between fiber and direction of tendon motion is, therefore, an important parameter of a pinnate muscle. The loss of force and range of excursion establish a disadvantage of fiber angling. What then is the selected advantage of fiber angling? The diverse sites and situations in which pinnate muscles occur make it unlikely that any single functional explanation will cover all selective influences. Pinnate muscles gain significant mechanical advantages resulting from the insertion of narrow tendons, which possess 30 to 120 times the elastic limit of that of muscles of equal cross sections, onto restricted optimal sites. Because tendinous attachments to bones can be concentrated on very restricted spaces, multiple muscle attachments on small bony elements can be realized. To create optimal lever systems, exertion sites for forces must be concentrated on restricted areas. Elongate tendons running in low friction bursae (*bursa* is a

Figure 5-10 (a) Simplified mechanical actions in a pennate muscle; (b) Diagram of mechanical properties of a parallel-fibered muscle.

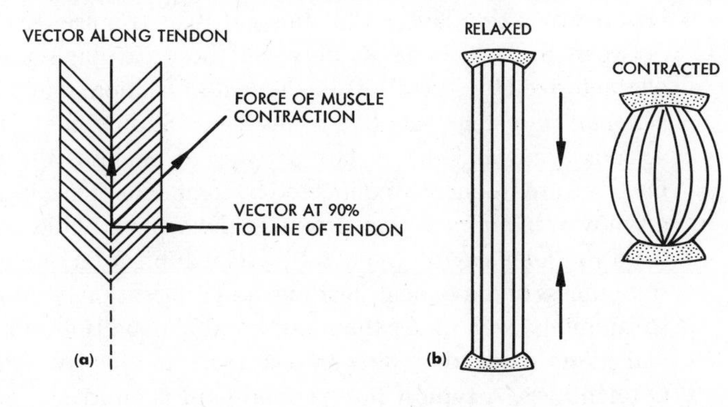

flattened sac of connective-tissue membranes filled with a fluid) can alter the direction of pull by passing through pulleys, as where tendons of digital extensors turn in front of the ankle.

Mechanical Basis of Movement: Joints and Kinematics

Vertebrate skeletal structures are only occasionally of a purely protective nature. Far more often they provide for the attachment and leverage of muscles. Together with muscles, the skeletal elements form complex lever and kinematic systems. The complex patterns of movements in the vertebrate body result from two situations: (1) because muscles only can pull, they must have an anatomical arrangement that allows them to act together in changing patterns on the complex lever system that constitutes the skeleton; (2) the neurological arrangement by which different muscles are switched on and off with fine gradations of strength and acceleration must result in controlled and highly coordinated movement.

In order to accomplish the great variety of complex movements, the skeleton must contain intervening arrangements that permit various degrees of angulation, torsion, or displacement. All vertebrates, even cartilaginous fishes, possess joints. However, not all junctions between bones allow movement. Because all bones ossify in preexisting fibrous tissue or cartilage, these two tissues inevitably appear at such primitive junctions as *sutures* and *synchondroses*. Synchondroses include the numerous temporary cartilaginous junctions between skull components and within elements of the postcranial skeleton in immature vertebrates. Many junctions between bony elements of the palate in teleost fishes remain cartilaginous throughout life. Sutures occur wherever margins or broader surfaces of bones meet, separated only by a zone of connective tissue, the sutural ligament. This is the surviving unossified part of the mesenchymatous sheet in which dermal bones develop and is the site of bone growth. When growth at a suture seizes, the bone-forming cells transform the ligament into bone, a process that is slow but leads to complete fusion of adjoining bones, i.e., *synostosis*. At sutures, the bone margins often develop spikes and recesses that interlock so firmly that bones are very difficult to separate. In some cases, two adjoining bones are bound together by an interosseous ligament, allowing a small degree of movement between the elements. Such an arrangement is called a syndesmosis, not to be confused with a *symphysis* in which a connecting pad or disc of fibrocartilage is involved. Unlike synchondroses, which are temporary junctions that disappear after the period of growth, symphyses are permanent and are concerned with movement.

In more mobile articulations of the *synovial* type, the adjoining bones are covered by a thin layer of hyaline cartilage (Figure 5-11). The actual contact between the two bones is between these cartilaginous surfaces, which are characterized by a very low coefficient of friction, that is reduced by a viscous *synovial fluid* that acts like a lubricant. Sliding contact is enhanced by the viscous synovial fluid. A fibrous capsule encloses the joint, and the bones involved are frequently connected by ligaments. The fibrous capsule is lined by a synovial membrane that produces the synovial fluid and removes materials from the joint cavity. Synovial joints exhibit a unique efficiency in terms of smoothness of movement, loading, and lubrication, without any real tendency to jamming, even under the most variable conditions. Phylogenetically, synovial joints made their first appearances in the jaw joints of the primitive Osteichthyes. A typical nonsynovial joint is multiaxial, inasmuch as it allows

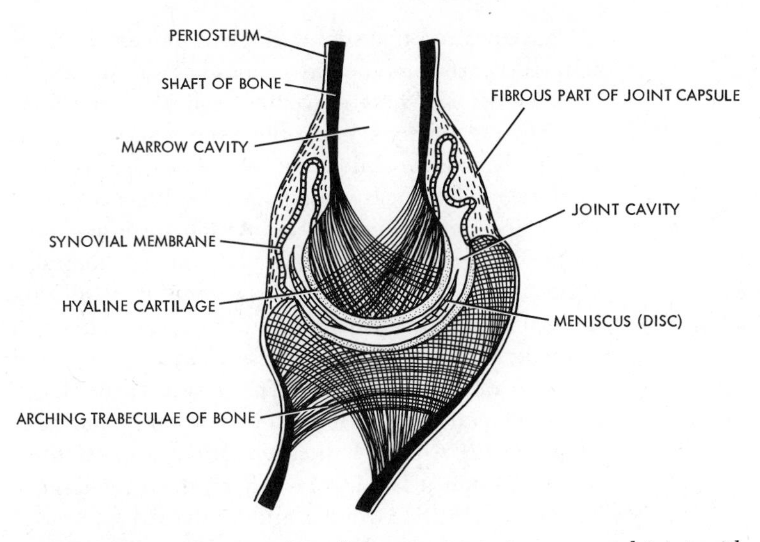

Figure 5-11 Diagrammatic representation of the structures in a synovial joint with a cartilaginous disc.

movement in all directions, however limited in range. Basically, these movements are bending, or angulation, torsion, and translation, in which one articular surface slides bodily across its partner. A multiaxial joint requires considerably more complex muscular control than an uniaxial one, in which more reliance can be put on ligaments. The emergence of a synovial joint can be correlated with a need for limitation of direction and range of movement that require refined control. It is clear that when a joint is to be maintained static in some phase of movement or stance, this can be accomplished more economically if the joint surfaces themselves, together with the disposition of adjoining ligaments, impose limitations on movement in some directions. There are, of course, biological limitations to the bulk of muscles around a joint and to the expenditure of muscular energy required to prevent unwanted movements in certain directions. It is, therefore, advantageous if unwanted movements can be transferred from muscles to the control of ligaments and to the shape of the articular surfaces.

The four major kinds of movements permitted in joints are listed here.

1. *Sliding* is the simplest kind, in which one surface slides over another without any appreciable angular or rotary movement.
2. *Angular* movement provides either a decrease or increase in the angle between adjoining bones. Within this category one can distinguish *flexion*, which is bending that often occurs around a transverse axis of a joint and results in the approximation of two morphologically ventral surfaces. The opposite movement is *extension*, or straightening. *Abduction* and *adduction* imply movement away from, and toward, the midline of the body. However, the closing and opening of jaws are often referred to, respectively, as adduction and abduction.
3. *Circumduction* is a movement circumscribing a conical space—the base of the cone is described by the distal end of the bone, and the apex is at the joint. Circumduction is in reality a combination of flexion, extension, abduction, and adduction.
4. *Rotation* is a movement of a bone around its long axis.

We have to stress that sliding, angular movements, circumduction, and rotation frequently are combined in various ways to produce an almost infinite variety of movements. Where movement is limited in extent, the reciprocal articular surfaces are of equal size, but where movement is free, the bone that is habitually more mobile exhibits the larger surface.

These movements are made possible by two fundamentally different articular surface shapes (Figure 5-12). Articular surfaces are not perfectly flat, but are parts of the surface of ovoids (egg-shaped bodies). When such surfaces are either convex in all directions or concave in all directions, they are termed male or female ovoid articular surfaces, respectively. The arbitrary subdivision of ovoid articular surfaces into hinge joints, pivot joints, and ball-and-socket joints is rather ambiguous. The second major type of articular surface is saddle-shaped because it is convex in one plane and concave at right angles (Figure 5-12). Although great variations occur in the degree of curvature from nearly flat to nearly part of a sphere, the articulations in vertebrates all can be classified as either ovoid or saddle-shaped.

Muscular action often results in a force acting around one or more joints. Because a force is a vector, it possesses both magnitude and direction. By convention we show forces graphically as straight arrows. The head of the arrow represents the direction, and the length of its shaft the magnitude. Addition and subtraction of forces can be carried out graphically or by using simple trigonometric relations. To add two forces graphically, one finds the intersection of their lines of action and there connects the forces head to tail. The line joining the free ends of the two forces is their sum, or resultant. Not only can forces be added to get resultants, they also can be resolved into components acting in various directions. For example, Figure 5-13 shows a body acted on at the point P by a force F at an angle α to the horizontal. The force can be resolved into a vertical component $F \sin \alpha$ and a horizontal component $F \cos \alpha$. A vertical force $F \sin \alpha$ and a horizontal force $F \cos \alpha$ acting simultaneously at P would have exactly the same effect as the force F. It is important to consider the tendency of forces to rotate bodies. Figure 5-13(b) represents a rigid body that is free to rotate about an axis through J, at right angles to the plane of the paper. The force F acts as shown at P and tends to

Figure 5-12 Two fundamental geometric types of articular surfaces. (a) Ovoid, which may be convex (male) or concave (female). Note that the solid body presents ovoid profiles in two planes at right angles and that the curvatures of the two may be different. (b) Saddle-shaped surfaces, which are concave-convex. In practice both types of surfaces may vary from only slightly to highly curved. Thus ovoid surfaces may be "almost flat" or "almost spherical," but the majority show intermediate grades of curvature, and much variation in change of radius from place to place. (Courtesy of Williams and Warwick, 1973, and Longman Group Ltd.)

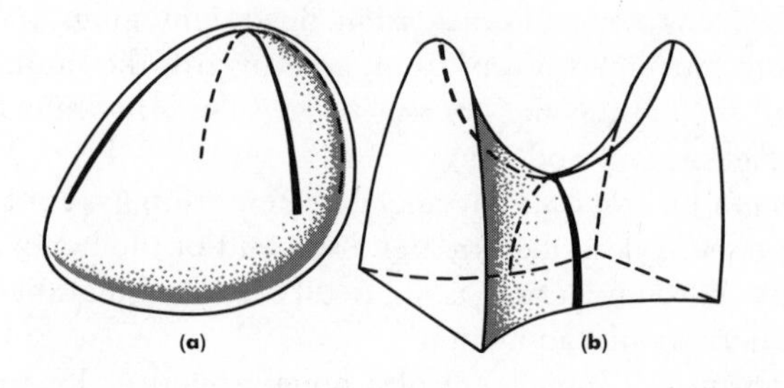

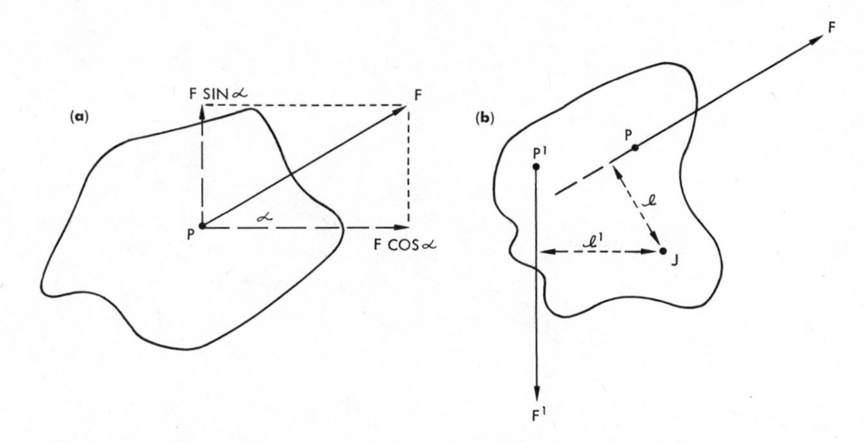

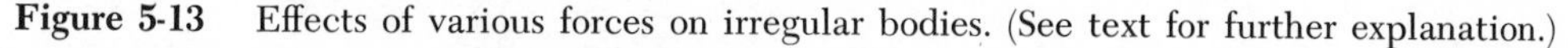
Figure 5-13 Effects of various forces on irregular bodies. (See text for further explanation.)

make the body rotate clockwise about *J*. Its tendency to do so is measured by its moment, or torque, about *J*, where l (= the moment arm) is the perpendicular distance between *J* and the line of action of the force. The force F^1 acting at P^1 tends to rotate the body in the opposite direction, i.e., anticlockwise. By convention the clockwise direction is designated as positive; therefore, we must regard the moment of F^1 about *J* as negative, i.e., $-F^1 l^1$.

All bone-muscle systems are basically machines. A machine is a mechanism that transmits force from one place to another, usually changing its magnitude in the process. We will designate any input force applied to a machine as in-force (*Fi*) and any output force derived from a machine as an out-force (*Fo*). In the vertebrate body, in-forces are applied by the pull of tendons, by gravity, and by external loads; useful out-forces are realized at the teeth, feet, digits, and elsewhere. Most locomotor and feeding mechanisms transmit forces by levers. A lever (Figure 5-14) is a rigid body that transmits forces by turning at a pivot (i.e., at a joint). Each force is spaced from the joint by a segment of the lever called a lever, or moment arm; the in-lever arm (l_i) extends from the in-force to the joint, and the out-lever arm (l_o) extends from the pivot to the out-force. As we have seen, the product of a force times its lever arm is a moment. Every functioning lever includes at least two moments, one for the in-system and one for the out-system. Thus, $M_i = F_i l_i$ and $M_o = F_o l_o$. When $F_i l_i = F_o l_o$, the system is in equilibrium. When a tetrapod is standing, the forces of all postural muscles are adjusted so that the total sum of moments is in equilibrium. It is clear that a mammal that digs with its forelimbs must produce a large out-force that is obtained by either increasing F_i and l_i or decreasing l_o because $F = F_i l_i / l_o$ A habitual runner, on the other hand, would require a high velocity (i.e., speed in a given direction). The velocity of any point on a lever is determined only by its distance from the pivot and the angular velocity, or rate of turning, of the lever as a unit. In- and out-velocities (v_i and v_o) are, therefore, also related to their respective lever arms, but in the reverse way: $v_i l_o = v_o l_i$, so $v_i = v_o l_i / l_o$, and $v_o = v_i l_o / l_i$. From these simplified equations, one can see that in a runner it is advantageous to have an increased l_o and a decreased l_i (compare (a) and (b) in Figure 5-14). In all of our considerations so far, we have omitted the action of gravity. In addition to the moments resulting from the action of all muscles acting on a bony element, one

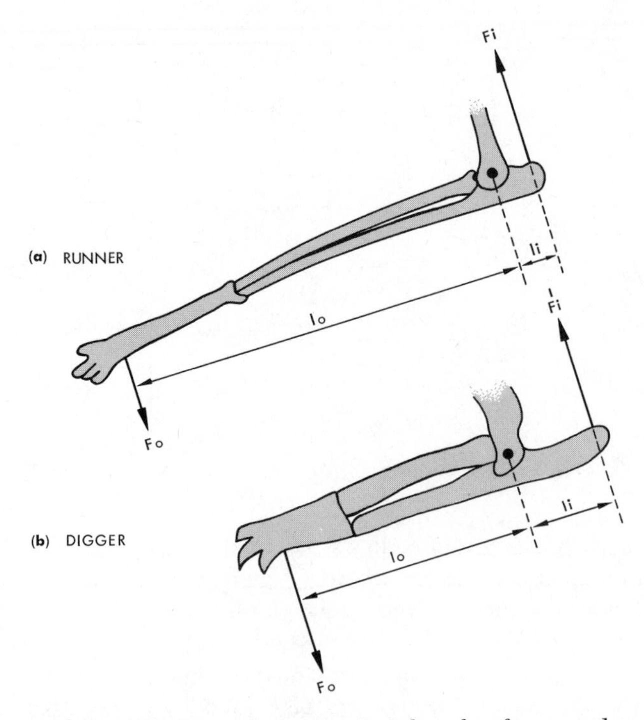

Figure 5-14 Simplified diagrams showing principles of in-forces and out-forces, and lever arms: (a) forearm of a runner and (b) forearm of a digger.

must also allow for the moment from the pull of gravity on the lever. Ideally, bone-muscle systems must be described as free-body diagrams in which all the forces acting on the bodies must be included. The only equations needed in free-body diagrams (the sums of the moments and of the vector of forces) are simple additive algebraic equations. Another complication arises from the fact that many muscles cross more than one joint. Two-joint muscles, which cross over two joints, can move either joint, both joints, or neither joint, depending on the actions of other muscles and loads. Contraction of a two-joint muscle tends to move both joints so it can provide in-forces to two lever arms simultaneously. Two-joint muscles generally have longer moment arms than one-joint muscles and can develop a greater acceleration on the bone. Another mechanical advantage for a two-joint muscle is that it can transmit tension between the end bones of the three bone series with less expenditure of energy than can the corresponding sets of one-joint muscles.

Bony elements in the feeding apparatus are often integrated into kinematic chains. A kinematic chain is an assembly of links jointed in such a fashion that if one is fixed and another moved, all other components must move in a predictable way. The parts of a kinematic chain have one degree of freedom of movement relative to each other. An example of a simple kinematic chain is the four-bar crank chain, which consists of four links joined by hinged joints whose axes are parallel (Figure 5-15). If one of the links is fixed and another moved, the movements of the rest are predictable. If the angle between any two links is given, the rest of the angles can be determined. More complicated kinematic chains can be made by joining together two or more simple ones

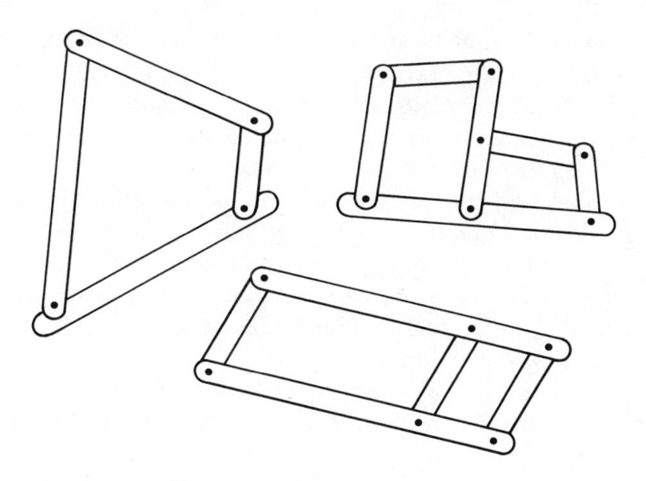

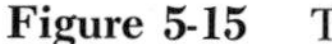

Figure 5-15 Three examples of kinematic chains.

(e.g., Figure 5-15). As we will see later, jaw mechanisms of teleost fishes, lizards, and birds can be described in terms of kinematic chains.

Locomotory Pattern and Apparatus of a Representative Teleost

The dominant mode of locomotion for fish is by waves of curvatures traveling along the body. Bending is executed against the resistance of the water. The force acting perpendicularly to the curvature of the body can be resolved into two components [(Figure 5-16(a)], a forward (*F*) and a lateral (*L*) one. The net effect is that of a series of curved planes, each displacing water posterolaterally and, consequently, producing anterolateral reaction forces [(Figure 5-16(b)]. As alternate planes face left and right, the lateral forces cancel each other; hence, the anteriorly directed components of the reaction forces are summed and induce forward propulsion of the body. However, a fish does not simply bend its trunk from side to side: the body is thrown into transverse waves that pass posteriorly along the body, increasing in amplitude as they go. The backward travel of the waves proceeds faster than the forward travel of the entire fish. As the fish travels forward, the equal and opposite lateral forces keep changing in

Figure 5-16 (a) Forces on the tail of a fish that is propelling itself in the direction of the large open arrow by sweeping its tail in the direction of the small one. The resultant force may be resolved into lateral and longitudinal components. (Courtesy of Gans and J. B. Lippincott Co.) (b) Diagrams based on a film showing the direction of forces applied to the water. The useful fraction of force increases as waves move posteriorly. (From Gray, as modified by Gans. Courtesy of J. B. Lippincott Co.)

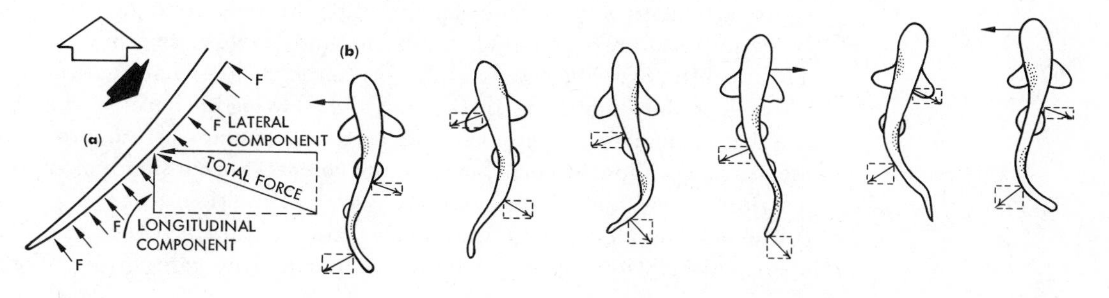

their location along its body [(Figure 5-16(b)]. Rather than simply canceling each other, they form forces creating movements that pass posteriorly and reverse their direction every time a new bend is formed. These continuously changing moments will cause some oscillations in the body of the fish. Because the degree of oscillation at a given proportion rate is inversely related to the number of bends in the body at any instant, an elongate fish capable of producing a greater number of bends would have a selective advantage over a short-bodied one. Many fish have indeed evolved elongate bodies, whereas others have developed such compensating mechanisms as stabilizing fins and lateral flattening of a stiffer body; these devices tend to reduce the lateral oscillations of the head.

In most fish, both primitive and advanced, undulations represent the major locomotory patterns. And, as we will see in subsequent chapters, lateral undulations still play a prominent role in the locomotory patterns of lower tetrapods. Given this mode of locomotion in fish, we will now examine the structural design underlying undulatory movements.

Primitively, the skeletal framework supporting the locomotory apparatus of fish is the *notochord*, which is always well developed in the vertebrate embryo. Although notochordal cells are soft and gelatinous, they are surrounded by a very firm cylindric sheath, resulting in a substantial body axis. Anteriorly, the notochord extends to the region just behind the brain; posteriorly, it reaches to the distal end of the fleshy part of the tail. The stiffness of the notochordal sheath enables the lateral body musculature, segmented into *myomeres*, to flex the body in varying degrees and ways. The notochord cannot be compressed or shortened. Recently it has been shown that the notochord in amphioxus is contractile. By modulating the degree of contraction, the stiffness of the notochord can be regulated to achieve various swimming speeds. The notochord persists as a prominent organ in the adults of the Agnatha, placoderms, pleuracanths, cladodonts, holocephali, dipnoans, and many members of the orders of the actinopterygians.

In the Selachii and more modified actinopterygians the notochord is replaced structurally and functionally by the vertebrae, which develop around it, furnishing greater strength and a far greater variety and number of muscular attachments. The principal part of a vertebra is the *centrum*, a spool-like body below the spinal cord. Extending dorsally from the centrum are two pillars that surround the spinal cord to form the *neural arch*. Above the spinal cord the neural arch forms a *neural spine*, which runs upward between the dorsal muscles of the two sides (Figure 5-18). In the tail region there is a similar hemal arch and spine that extends ventrally from the centrum to surround an artery and vein. Both the anterior and posterior articulatory surfaces of the centrum are concave; hence, this type of vertebra is called amphicoelous. Amphicoelous centra articulate only at the periphery of the intervertebral joint by means of syndesmoses, allowing limited angular motion between vertebrae. Functionally, a central axis composed of multiple amphicoelous centra will be incompressible in the longitudinal direction while allowing the necessary bending or angular movements. Therefore, the vertebral column in fish provides the mechanism to allow waves of bending movements to occur simultaneously with a series of rigid bodies in a complex system of multiple levers.

The bending movements are caused by contractions of two types of muscle fibers: red and white, which often form distinct and separate entities. When not mixed, red muscle fibers in fish are superficial (Figure 5-17) and run more or less longitudinally, whereas the white and deeper fibers are arranged in complex three-dimensional patterns in which some of them make angles of thirty degrees or more with the vertebral column. At ordinary cruising speeds, only the red muscles are active, whereas the white muscles are mainly used as an additional source of power for bursts of fast swimming. The major propulsive force in fish is generated by the axial musculature, which is segmentally arranged. Successive segments, the myomeres, run along each flank, corresponding in number to vertebrae (Figure 5-17). Between successive myomeres are strong sheets of connective tissue, the *myosepta,* in which ribs and extra intermuscular bones develop (Figure 5-18). In jawed fish, a *horizontal septum* of connective tissue divides every myomere into dorsal and ventral portions. It is in this septum that dorsal ribs develop. As a result of the formation of the horizontal septum, the axial muscles can be divided into dorsal, or *epaxial,* and ventral, or *hypaxial,* musculature.

The superficial band of red muscle fibers is roughly parallel to the long axis of the body and is active at all swimming speeds. In contrast to this relatively simple organization, the white fibers are arranged in W-formed myomeres with the upper edge turned forward (Figure 5-17). Deep to the surface, the angles of the zigzag folding become increasingly sharp. Each myomere folds backward or forward beneath its neighbors, so that any vertical cross section cuts a large number of myomeres. Folding of the myomeres serves two major functions: (1)

Figure 5-17 The axial musculature of a trout (a) consists of serial myomeres (b), each folded complexly so that their sequential contraction from head to tail causes a gradual shortening. (Courtesy of Gans and J. B. Lippincott Co.). (c) and (d), diagrammatic dorsal and lateral views, showing the course of the muscle-fiber trajectories of a teleost: obliquus externus, obliquus internus, trajectories in the epaxial bundles. (Adapted from Alexander.)

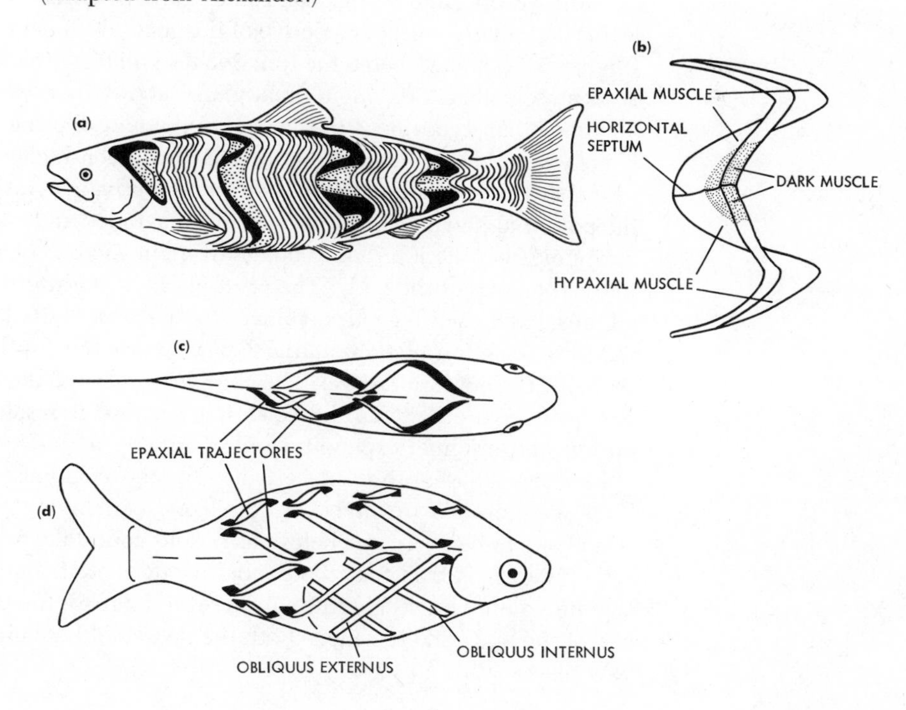

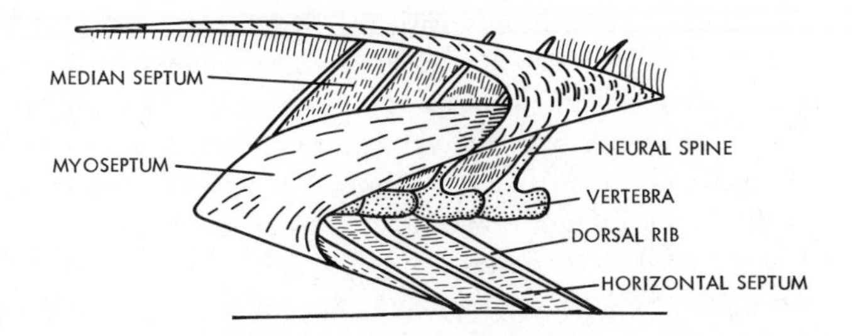

Figure 5-18 An epaxial myoseptum of a representative teleost, showing antero-posterior extent of the septum resulting from its complex deep folding. Laterodorsal view. Forces developed by epaxial myomeres will have effects on as many as ten segments. (Modified from Nursall, 1956, Proceedings Zoological Society, London.)

Any given myomere is extended cranially and caudally over a distance equal to several body segments. This arrangement enables the short, parallel muscle fibers to exert forces over a relatively great distance (Figure 5-18). The obliquity of the myosepta extends the pull over several vertebral joints. (2) As a result of any one myomere extending over several segments and overlapping several neighboring myomeres, any given vertebral joint is acted on by several myomeres at once. Multiple myomeres acting on a given vertebral joint smooth and reinforce the bending movements.

It is important to note that the muscle fibers of the myomeres in teleosts do not run longitudinally and parallel to the vertebral column. Instead, the deeper muscle fibers are arranged in complex three-dimensional patterns that make angles with the long axis of the fish (Figure 5-17). As stated before, a muscle fiber extends only from one myosept to the next. However, a line could be drawn through successive myomeres by following a fiber to its end on a myosept, continuing along a fiber starting directly opposite it on the other side of the myosept, and so on. Such lines, which we can call muscle-fiber trajectories, will give us a perspective of the general topography of the myomeres (Figure 5-17). What then is the functional significance of this structural design? If all muscle fibers ran longitudinally, the lateral fibers would have to contract far more than those near the vertebral column. During bending, the lateral fibers would have to shorten more and at a much higher rate, whereas those near the vertebral column would shorten slightly, generating only a fraction of the potential power. This situation can be avoided and made more efficient by arranging fibers at different distances from the median plane at different angles to the long axis of the body. The resulting helical pattern (Figure 5-17) allows all muscle fibers of a given myomere to contract at similar rates and to about the same fraction of their initial length when the fish bends its body, fully realizing the potential power. The complex structural design of the white fibers has been favored by natural selection because it results in optimal power output and minimal expenditure of energy.

In nearly all fish, propulsion is caused by the axial musculature acting on the axial skeleton. Fins evolved as stabilizers controlling pitch, yaw, and roll, which had to be overcome before fish could undertake more or less continuous swimming. It is true that the caudal fin does contribute importantly to the swimming of many fish because it serves to increase the thrust at the posterior end of the body and to reduce both the degree of bending of the body and the yaw of the fish.

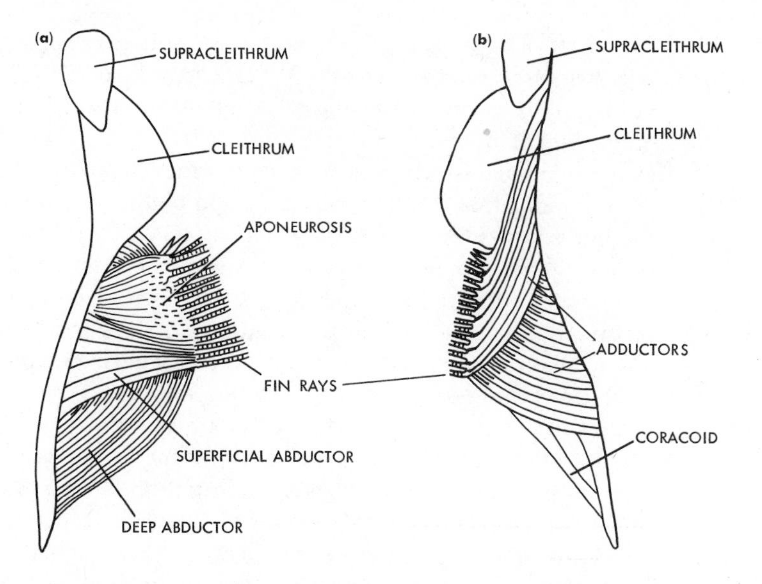

Figure 5-19 (a) Lateral view of pectoral girdle and fin musculature (ventral complex) of a perchlike fish; (b) medial view of the same showing the dorsal complex of muscles.

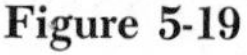

In teleosts, fins are supported mainly by integumentary fin rays (mostly of dermal origin) that occupy the entire distal, vanelike part of the fin; only the vanelike part projects from the body. These *lobeless fins* characterize the actinopterygian (ray-finned) fishes. Lobeless fins are operated by relatively simple groups of muscles; the ventral mass, occupying a more or less lateral and anterior position, is differentiated into two layers (abductor superficialis and profundus, Figure 5-19) responsible for forward and slightly ventral movements of the fin. The dorsal mass of muscle, which is also differentiated into two layers (adductor superficialis and profundus), draws the fin backward and slightly upward, generating a propulsive force.

Terrestrial Invasion by Amphibians

It is known from the fossil record that toward the end of the Devonian, the earliest amphibians developed terrestrial life habits. One of the most important architectural innovations, which enabled vertebrates to enter the new adaptive horizons on land, is the development of paired appendages that would lift the body off the ground and permit a walking gait rather than the undulatory movements so characteristic of fish. There is no doubt that the early efforts at land life were gradual. Both muscular and skeletal systems came under an entirely new set of forces when the effects of gravity became more pronounced and the support of the water was eliminated. It is not surprising that these new environmental demands produced greater evolutionary changes in vertebrate locomotion than had occurred in tens of millions of years previously. The amphibians are the first of the great group of land vertebrates, the tetrapods, characterized by the presence of four pentadactyl (five-digited) limbs. As we have seen in Chapter 2, living amphibians belong to three specialized and sharply separated orders. We will discuss the locomotory apparatus of a representative of the Urodela (newts and salamanders), with an elongate body and tail.

Locomotory Pattern of a Urodele

The pattern of terrestrial locomotion in a salamander depends on the speed with which the animal moves. When the animal is frightened, it can move quite rapidly by undulating its body violently and dragging the ventral surfaces of its body and tail over the substrate. A drastically different pattern is exhibited by undisturbed individuals moving in a deliberate manner. Although lateral undulations of the trunk produced by the axial musculature still occur, the limbs now play an important locomotory role (Figure 5-20). The body is raised off the ground, although the tip of the long tail shows a slight drag.

During undisturbed locomotion, the pattern of limb movement is characteristically diagonal: When the right forefoot is forward, the left hindfoot is also in that position, and vice versa. In every cycle one phase occurs in which all four limbs are on the ground. Starting from this position, the left forefoot is lifted and swung forward. Before this foot is set down, the opposite hindfoot is lifted and swung forward. During the swing of this right hindfoot, the left forefoot is set on the ground. Next, the right hindfoot is set on the ground, and half a cycle is completed. The body is always bent concavely on the side on which the hindlimb has been swung forward (Figure 5-20). Lateral undulations of the body cause an increase of the angle through which a limb can be displaced, resulting in a much longer stride.

Correlated with the new stresses imposed by gravity and the need to support the girdles of the limbs, the amphibian backbone has undergone many structural adaptations. In more terrestrial forms, the centra are heavily ossified and have thick laminar or cortical layers and adaptive patterns in the formation of trabeculae. Intervertebral joints are strengthened and their mobility controlled by highly specialized articulating processes, the *pre-* and *postzygapophyses* (Figure 5-21). We also can recognize the very first stages of a regional differentiation of the vertebral column to furnish increased mobility of the head (specialized cervical vertebra) and to support the pelvic girdle (sacral vertebra). The caudal vertebrae lack zygapophyses and have hemal arches. The centra are convex anteriorly, whereas their posterior aspects are concave; such *opisthocoelous* vertebrae add strength to the spine. The cervical vertebra develops a prominent process anteriorly, the *odontoid process*, articulating

Figure 5-20 (a) successive stages of the crawling newt during half of a locomotory cycle. Feet resting on the ground in successive stages are connected by lines. RF = right fore, LF = left fore, RH = right hind, and LH = left hind foot. (After Roos, 1964.) (b) movements of the forelimb during the recovery cycle. (Modified after Evans, 1946.)

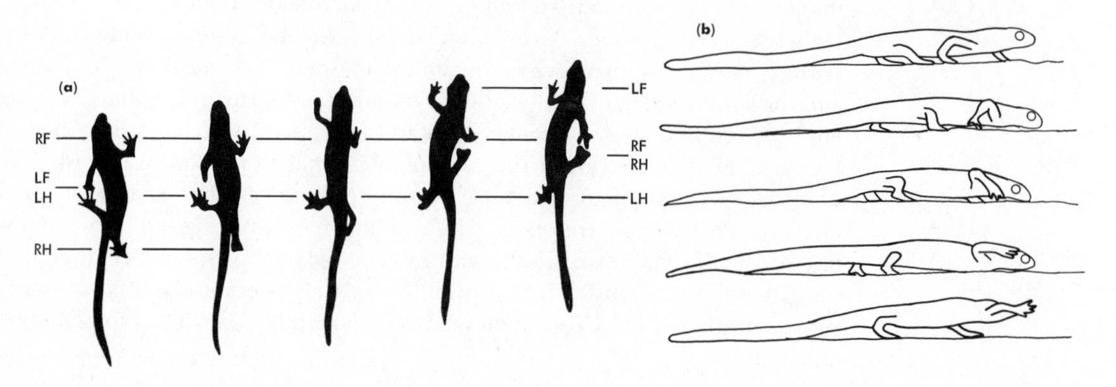

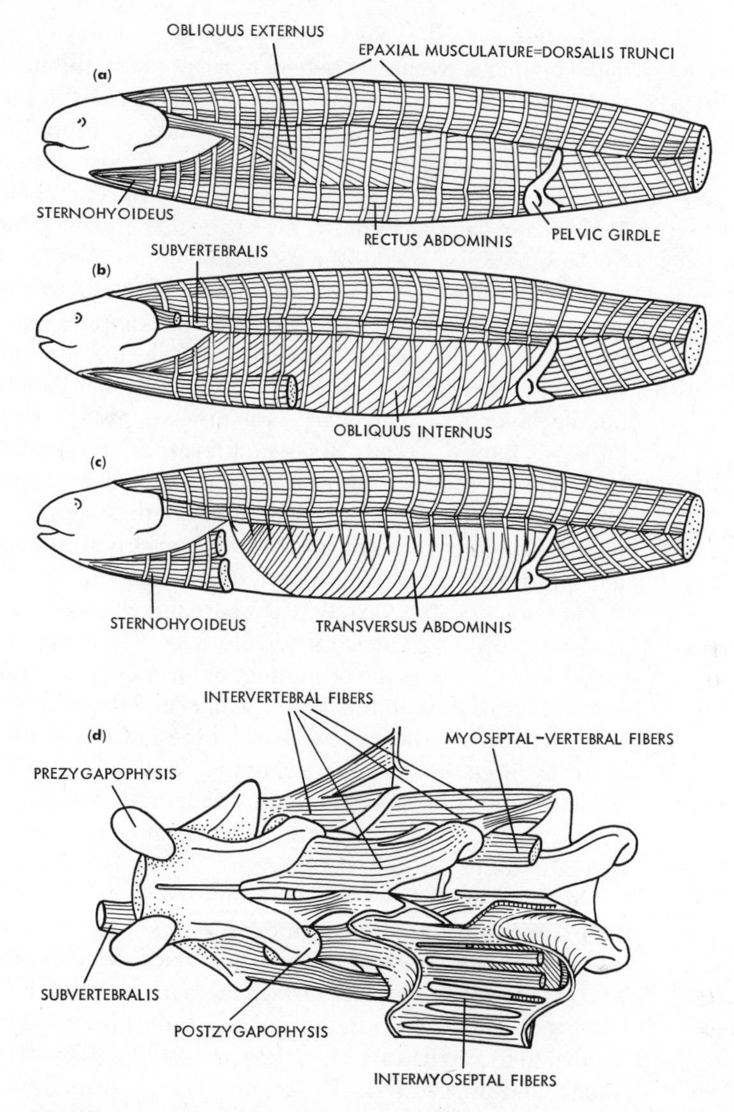

Figure 5-21 (a) Diagrammatic lateral view of the axial musculature of a salamander, showing the superficial fibers. (b) After removal of external oblique and rectus. (c) After removal of internal oblique. (d) Dorsal view of epaxial fiber tracts. (a, b, and c modified after Maurer; d after Auffenberg.)

with the foramen magnum of the skull, furnishing mobility of the head. *Transverse processes* are present on all but the cervical vertebra for support of the ribs. Ribs in salamanders are intersegmental splints of replacement bone to enhance the leverage of axial muscles. As ribs become longer, they assume a protective role for the viscera. The sacral vertebra has enlarged transverse processes for support of the pelvic girdle.

As we have seen, lateral undulations still play a salient role in salamander locomotion. Lateral bending movements are mainly caused by the action of the dorsal epaxial musculature on the vertebral column. In the salamander, the segmented epaxial muscles still can be considered a single muscle, the *dorsalis trunci* (Figure 5-21), of which many deeper fiber tracts span two or several successive vertebrae. Such an arrangement allows the dorsalis trunci to produce angular movements between vertebrae in the horizontal plane, causing the important lateral undulatory movements of the body.

With the greater emphasis on the limbs as propulsive devices, there is a trend toward reduction of the hypaxial trunk musculature in the tetrapods. The hypaxial muscle mass becomes subdivided into specialized sheets of muscles. Three major groups (Figure 5-21) can be recognized: (1) subvertebralis muscles, (2) flank muscles, and (3) the rectus abdominis muscle. Subvertebralis muscles pass from vertebra to vertebra, attaching to the ventrolateral part of each centrum and to the ventral aspect of each transverse process. Flank muscles are composed of three superimposed major sheets: the most external sheet, the *external oblique* (obliquus externus), has fibers that slant upward anteriorly; the middle layer has fibers that slant upward posteriorly, the *internal oblique* (obliquus internus); and the deepest layer, the *transverse* (transversus) muscle, has fibers in a dorsoventral direction. The diverse orientation of the three muscle sheets furnishes considerable strength to support the trunk viscera. The rectus abdominis muscle runs anteroposteriorly from shoulder to the pelvis; its left and right halves are separated by a tendon, the *linea alba.*

Functionally, the vertebral column and the axial musculature of the amphibian are being transformed into a bow-and-string construction. In salamanders we see only the beginnings of the new structural design. The vertebral column together with the dorsalis trunci and the subvertebralis muscles can be regarded as a bow bent in the dorsal direction (therefore, concave ventrally) by a string. Functioning as a string are parts of the pectoral girdle, all flank muscles, and especially the rectus abdominis and linea alba. This structural innovation allows dorsoventral bending of the column, a type of movement virtually absent in fish.

The forelimb (Figure 5-22) is supported by the *humerus* in the arm, and two bones, the *radius* and *ulna,* in the forearm. The two ends of the humerus are twisted on one another: the proximal end is essentially horizontal, the distal end tilted so that its lower surface faces forward. In the sprawling gate, such a "twisted" humerus in the horizontal plane allows the forearm to move backward and forward in a more or less parasagittal plane (Figure 5-20). Two bones, radius and ulna, are in the forearm. The proximal end of the ulna projects beyond the elbow joint as the *olecranon process.* Six or seven *carpal* bones make up the wrist (*carpus*). Two are proximal, the *radiale* and *ulnare,* on the radial and ulnar side, respectively (Figure 5-22). A small *intermedium* generally is fused to the ulnare. A single centrale lies between proximal and distal rows. Four metacarpals articulate with the distal four carpals. The first digit or *pollex* (thumb) is absent. Digits two, three, and five have two phalanges each, whereas the fourth has four.

The forelimb articulates with the pectoral girdle, which is composed of two halves. Each half has a cartilaginous ventral part with an anterior *precoracoid* and posterior *coracoid* region (Figure 5-22). A narrow bone, the *scapula,* projects dorsally. To its distal border is connected a broad *suprascapular* cartilage. The area around the *glenoid fossa,* which is a depression on the lateral border of the coracoid cartilage, also is ossified.

Specific movements occur in the shoulder and elbow joints, both of which are synovial. In the shoulder joint the humerus is allowed a wider range of movement in the horizontal plane than in the vertical plane because of the

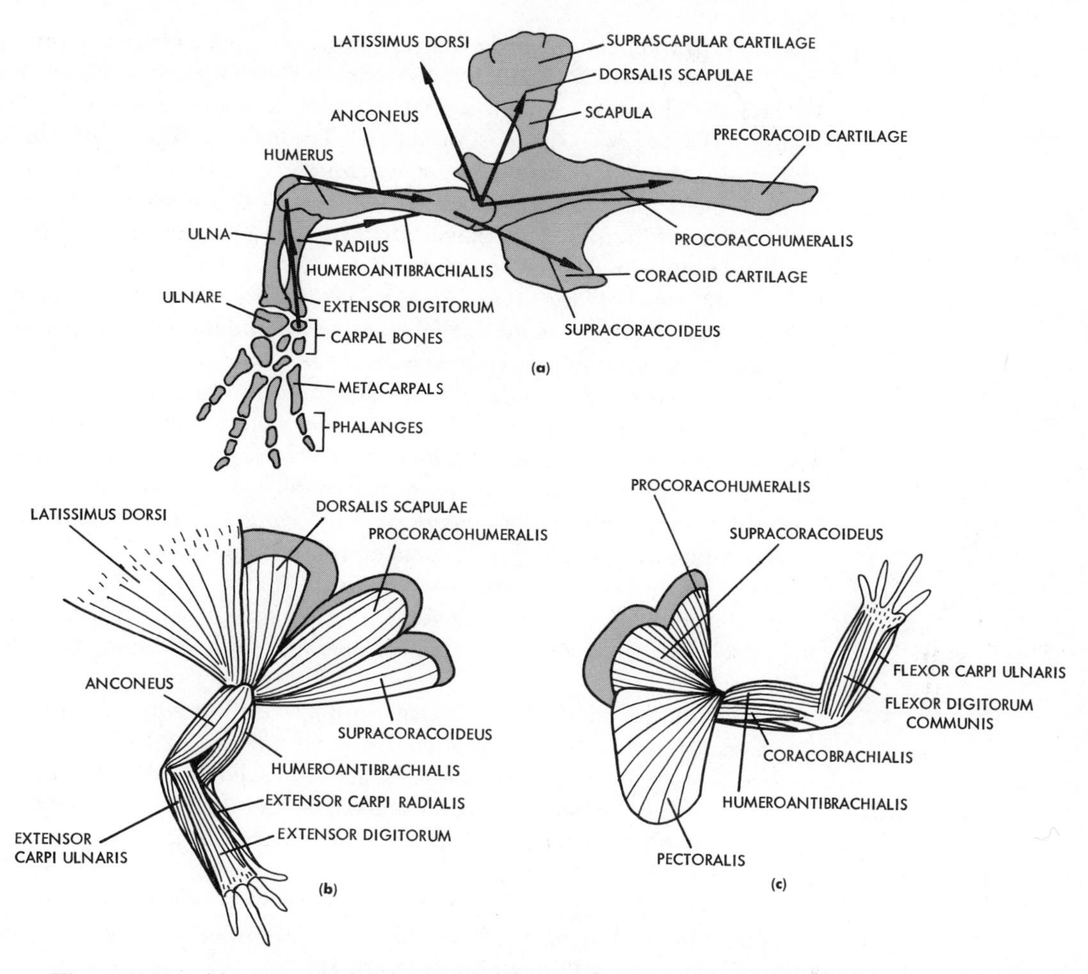

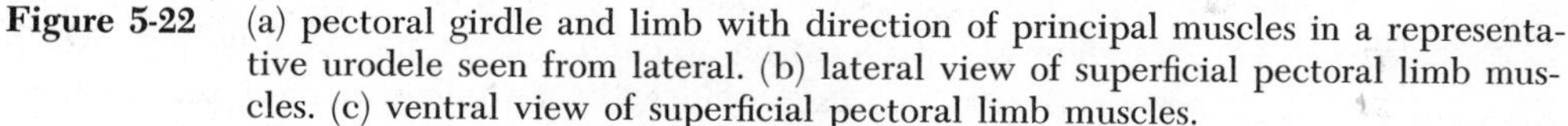

Figure 5-22 (a) pectoral girdle and limb with direction of principal muscles in a representative urodele seen from lateral. (b) lateral view of superficial pectoral limb muscles. (c) ventral view of superficial pectoral limb muscles.

elliptical shape of the head of the humerus. The elbow joint is constructed in such a way that flexion, extension, and a slight rotation of the forearm can occur. In the wrist (carpus), flexion and extension of the hand (manus) take place because of the specific shape of the articulatory surfaces between the radiale and radius and the ulnare and ulna.

Forelimb movements are caused by musculature, which is basically composed of two opposing groups comparable to those of the pectoral fins of fish. However, in tetrapods, the dorsal and ventral groups develop into complicated muscles by growth and cleavage. From the dorsal group arise the muscles on the extensor surface, whereas the ventral group differentiates into flexors on the opposite limb surface.

First, we will discuss the dorsal muscles of the pectoral or forelimb. Superficially, two dorsal muscles, the *latissimus dorsi* and *dorsalis scapulae* (Figure 5-22) attach to the humerus near its head and are responsible for much of the movement of that bone on the shoulder girdle. The dorsal surface of the humerus is covered by a stout muscle, the *anconeus* (triceps), that arises from the humerus and pectoral girdle and inserts distally on the olecranon process of the ulna (Figure 5-22). Distally, the dorsal arm musculature is continued beyond the elbow by muscles of the extensor series, of which the *extensor digitorum communis* is the most prominent. The extensor digitorum communis separates into four separate tendons, which insert on the dorsal surfaces of the digits. Other forearm extensors are the *extensor carpi ulnaris* and the *extensor carpi radialis* (Figure 5-22), arising from the humerus and inserting on the distal end of the ulna and radius, respectively.

We will now study the ventral muscles of the forelimb. On the underside of the shoulder is the *pectoralis*, the "chest muscle," and the *procoracohumeralis*, originating, respectively, from the fascia of the rectus abdominis and from the dorsal surface of the precoracoid cartilage and inserting, together with the latissimus dorsi, on the humerus near its head. Anterior to the pectoralis is a *supracoracoid muscle* extending from the coracoid cartilage to the humerus. On the ventral aspect of the humerus are two muscles, the *humeroantibrachialis*, sometimes referred to as the biceps, and the *coracobrachialis*. The humeroantibrachialis passes from the proximal end of the humerus to the proximal end of the radius; it is, therefore, easy to visualize that it tends to flex the elbow joint. The coracobrachialis originates from the coracoid cartilage to insert on the distal end of the humerus at a considerable distance from the head, or fulcrum.

In the distal part of the limb the main propulsive effort is a backward push of the forearm and hand, accomplished by the muscles of the ventral flexor surface. Consequently, these muscles are more powerful than the dorsal extensors. From the lateral epicondyle of the humerus, a series of long flexors, the *flexor carpi radialis*, *flexor carpi ulnaris*, and *flexor digitorum communis* (Figure 5-22) fan out to the lower end of the forearm and to the hand.

With the anatomical background gained from the preceding paragraphs, we can now discuss some elementary biomechanical aspects of the forelimb of a salamander, as a basic model of a tetrapod.

Movements of the forelimb occur in two phases (Figure 5-20): propulsive and recovery. Let us consider the *propulsive* phase first. It consists of a backward movement of the humerus accompanied by flexion of the forearm (*antebrachium*) and carpus. During its backward swing, the humerus moves in a horizontal plane, with its distal end at a slightly higher level than the proximal end. At the end of the phase, the humerus has moved backward to such an extent that it is at an angle of about thirty-eight degrees with the long axis of the body. As we have seen, the elliptical articular surface of the head of the humerus permits a wider range of movement in the horizontal plane than in the vertical plane. Backward movement of the humerus is caused by the retractors of the humerus, latissimus dorsi, pectoralis, procoracohumeralis, and the coracobrachialis, acting on the shoulder joint. Except for the coracobrachialis, all muscles insert close to the proximal head of the humerus. Multiple muscular action across the shoulder joint results in better control of movement and velocity, increased power, a wider range of excursions of the humerus, and a greater stability. Because we lack data on the timing of activity of these muscles of the shoulder joint, we are unable to explain the exact mechanisms and functions. We must be aware that muscles mostly function in closely coordinated groups, rarely singly. And the patterns of muscle activity can vary significantly in different situations.

During the propulsive phase, the backward swing of the humerus is accompanied by flexion of the elbow and carpometacarpal joints, while the flexor muscles of the hand are pressing the digits against the ground, thus increasing the friction and preventing backward slipping. Flexion of the elbow joint is accomplished by contraction of the humeroantebrachialis, flexor carpi radialis, and flexor carpi ulnaris. Basically, the flexed forearm acts as a secondary lever, transmitting the propulsive force from the arm to the point of resistance, the substrate. It also supports the weight of the body, which is transmitted along the radius of the forearm through the hand to the ground. During the propulsive phase, the hand and digits are flexed firmly against the ground by the flexor digitorum communis and other smaller flexors. Therefore, the hand acts as an expanded base, serving mainly as a friction pad to prevent backward slipping. It is evident that the arm is the propulsive element of the forelimb. It is the driving lever that, swinging back in a horizontal plane, propels the body forward. The flexed forearm and hand represent a weight-supporting column that transmits the propulsive force to the ground.

The recovery phase starts when the digits are lifted from the ground. The humerus swings forward in the horizontal plane to a position of about a forty-five-degree angle to the long axis of the body. The elbow and wrist are still flexed during the recovery phase (Figure 5-20). However, toward the end of the phase the wrist and digits begin to extend. Placing of the hand on the ground marks the end of the recovery phase, during which there is never a complete extension of the elbow joint (Figure 5-20). The arm is raised in the vertical plane by the dorsal muscle derivative, the dorsal scapulae, while another dorsal muscle, the procoracohumeralis protracts the humerus. Flexion of the elbow joint is decreased and the joint itself stabilized by the forearm extensors, anconeus (triceps), extensor carpi ulnaris, and extensor carpi radialis. Initiating the recovery phase is the extensor digitorum communis, which extends both the wrist and the hand as a whole.

Let us now consider the movements of the hindlimbs. As we can see from Figure 5-24, the movements of the hindlimbs closely resemble those of the forelimbs except that the hindlimbs are raised higher above the ground.

The hindlimb (Figure 5-23) is supported by the *femur* in the thigh; and two bones, a medial *tibia* and a lateral *fibula* are in the shank (*crus*). Six *tarsals* make up the ankle (*tarsus*). *Fibiale* and *fibulare* are the two bones of the proximal row lying opposite the tibia and fibula, respectively. There are three distal tarsals and a centrale, the latter lying between the proximal and distal rows. The foot contains five *metatarsals*. The first digit is the *hallux* (Figure 5-23).

The hindlimb is anchored to the pelvic girdle, which consists of a *puboischium* and an *ilium*. The puboischium is a flat ventral plate, of which the anterior end represents the *pubis* (Figure 5-23) and the posterior end the *ischium*. On either side of the puboischium is a deep depression, the *acetabulum*, that accommodates the head of the femur. Extending dorsally from the lateral side of the puboischium is the ilium, which appears as a bony rod. The ilium is fastened to a strong rib of the sacral vertebra.

Various movements occur in the hip, knee, and ankle joints, all of which are synovial. In the hip joint, the femur is allowed a very wide elliptical movement. The femur moves over a wider range in the horizontal plane than in the vertical plane. Both the knee and ankle joints are restricted to flexion and extension, although some rotation does take place.

As in the forelimb, hindlimb movements are caused by two opposing groups

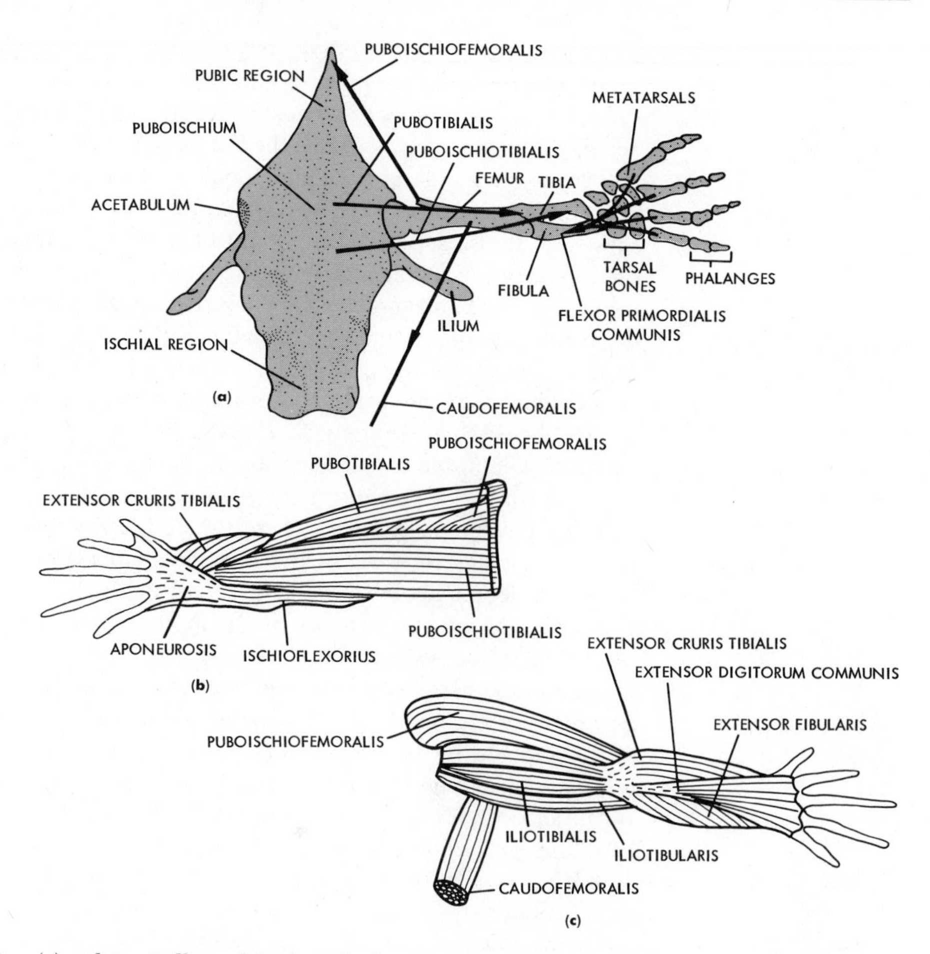

Figure 5-23 (a) pelvic girdle and limb with direction of principal muscles in a representative urodele seen from ventral. (b) ventral view (flexor muscle group) of superficial pelvic limb muscles. (c) dorsal view (extensor muscle group) of superficial pelvic limb muscles (b and c show a five-toed species).

of muscles. Extension is, in general, executed by ventral derivatives, whereas flexion is by dorsal derivatives.

At first, we will discuss the dorsal muscles. Passing from the dorsal side of the pubis, ischium, and ilium to the anterior face of the shaft of the femur is the *puboischiofemoralis internus*, a large, powerful muscle. A deeper dorsal muscle, the *iliofemoralis*, runs from the ilium directly to the posterior aspect of the femur. One must note that these two dorsal muscles have antagonistic functions: the iliofemoralis, because of the insertion on the posterior aspect of the femur, will draw the femur toward the tail; it, hence, is a retractor, whereas the puboischiofemoralis internus is a protractor, moving the femur forward. Three more dorsal muscles originate from the ilium: the *iliotibialis*, passing in front of the knee to attach to the tibia; the *iliofibularis*, running down the posterodorsal border of the thigh to insert on the proximal end of the fibula; and the *ilioextensorius*, running parallel with, and posterior to, the iliotibialis. Distally, the dorsal musculature is continued beyond the knee by muscles of the extensor series, of which the extensor digitorum communis is the chief extensor of the foot. It arises from the lateral epicondyle of the femur and inserts on either side of the bases of the metatarsals. Other major extensors of the leg originating from the lateral epicondyle of the femur are the *extensor tarsi-tibialis* and *extensor cruris tibialis* (Figure 5-23); the latter inserts along the entire lateral border of the tibia.

The ventral muscles of the hip and thigh regions mainly adduct the femur and flex the knee joint. A large muscle, the *puboischiofemoralis externus,* arises from the outer surface of the anterior end of the pubis and inserts on the proximal end of the femur (Figure 5-23). A smaller muscle, the *ischiofemoralis,* emerges from the inner side of the ischium and also runs to the head of the femur. Two ventral muscles, *puboischiotibialis* and *pubotibialis,* run between pelvic girdle and tibia. The latter inserts on the proximal end of the tibia, whereas the former attaches to the shaft of the tibia. To this group belongs a rather odd, straplike muscle, the *ischioflexorius,* that originates from the ischium, passing back of the knee (popliteal region) to insert on the aponeurosis of the flexor primordialis communis. A well-developed *caudofemoralis* runs between the ventral side of the caudal vertebrae and the femur; this important muscle can execute a powerful backward pull of the femur and, hence, contributes significantly in propulsion. Beyond the knee, the most important flexor is the very large *flexor primordialis communis* muscle, which arises from the lateral aspect of the fibula and inserts on the proximal end of the terminal phalanx of each toe. However, the origin of the muscle is subject to considerable intergeneric variation in salamanders. The action of the flexor primordialis communis is not only to flex the tarsus and digits, but also to turn the foot forward (adduct) before it is placed on the ground.

As is the case with the forelimb, hindlimb locomotion can be subdivided into propulsive and recovery phases (Figure 5-24). The propulsive phase begins immediately on the contact of the sole of the foot with the ground. It consists of a backward movement of the femur accompanied by flexion of the leg (crus) and foot (pes). During its backward swing, the femur moves mainly in a horizontal plane. At the end of the phase, the femur has moved to such an extent that it is at an angle of about thirty-three degrees to the long axis of the body (Figure 5-24). Backward movement of the femur is principally caused by the action of the caudofemoralis, which is aided by the iliofemoralis and puboischiofemoralis externus. The backward swing of the femur is accompanied by flexion of the knee and ankle joints, while the flexor primordialis communis is pressing the foot and digits against the ground, thus preventing backward slipping of the foot. Flexion of the knee joint is caused by contraction of the iliofibularis and ischioflexorius. The flexed leg serves as a lever, transmitting the propulsive force from the thigh to the ground. During the propulsive phase, the foot and digits are flexed firmly against the ground by the flexor primordialis communis. The foot and lower leg act as a kind of stationary pivot on which the femur rotates. The femur is the driving lever which, swinging back in a horizontal plane, propels the body forward. Note that the foot is in a forward position parallel to the long axis of the foot and is, hence, in the direction of movement, allowing the muscles to operate under great mechanical advantage.

The recovery phase starts when the foot gradually rolls off the ground (Figure 5-24). During this phase, the limb sweeps forward through almost 170 degrees; the movement depends on the action of the puboischiofemoralis

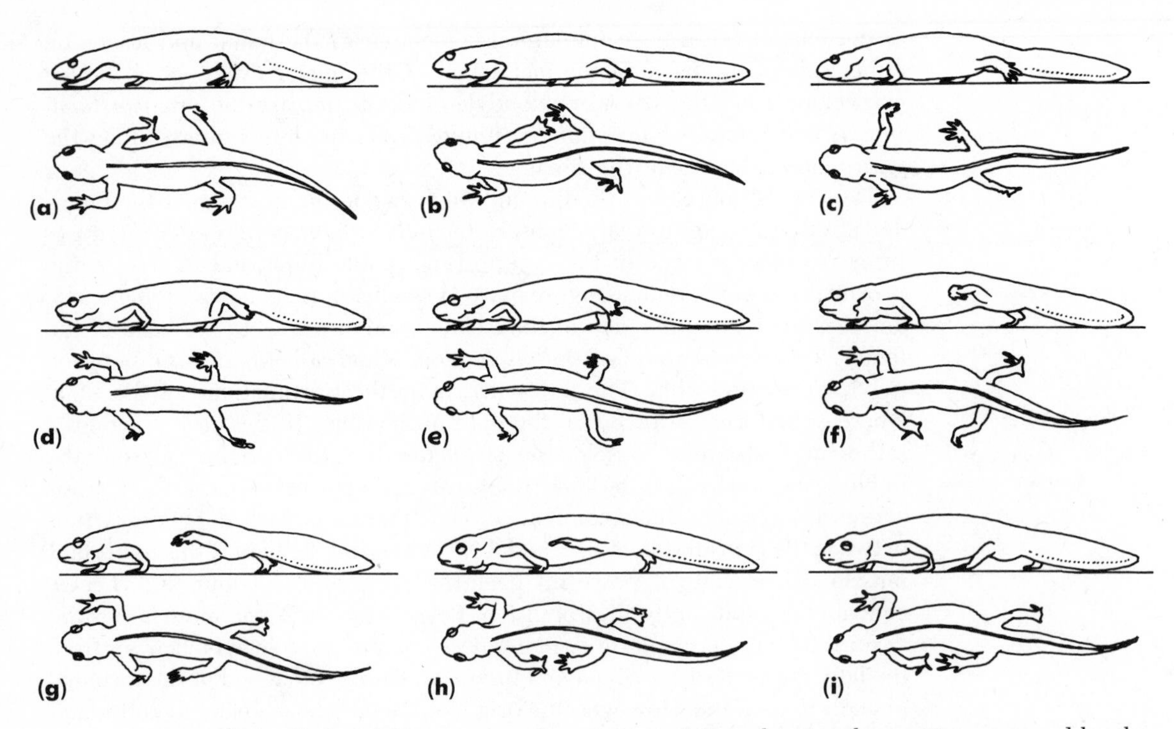

Figure 5-24 Series of diagrams based on motion picture showing the positions assumed by the limbs during the locomotion of a salamander. (After Schaeffer, 1941.)

internus. The upper and lower parts of the hindlimb are brought forward as one unit by the coordinated activity of the puboischiofemoralis internus acting on the thigh and the iliotibialis, ilioextensorius, iliofibularis, extensor tarsi tibialis, and extensor cruris tibialis acting on the crus. When the hindlimb has moved forward to a position at just beyond right angles to the body, the knee is flexed. With the knee flexed, the entire limb is brought forward farther by a sort of crawl-stroke movement, while the extensor digitorum communis extends the foot. As soon as the foot is placed on the ground, the propulsive phase begins.

Finally, it is important for lower tetrapods to increase the length of the stride by lateral undulations of the body. During the forward swing of the recovery phase of one limb, the opposite limb, on the side where axial muscles are contracting, is in contact with the ground and is undergoing propulsive movement. The propulsive leg is on the concave side of the body, while the leg in the recovery phase is on the convex side of the body and, hence, the length of its stride is increased. When the right forelimb is forward, the left hindfoot is also in that position. This diagonal pattern of movement of the limbs is a fundamental feature of most tetrapods.

Evolutionary Origin of the Tetrapod Limb

Although the details of the sequence of stages by which a tetrapod limb evolved from a fish fin are still disputed, it is probable that the ancestral condition is to be found among the crossopterygians, which have a lobed fin.

Lobed fins have a prominent fleshy base, formed by muscles attached to primary fin rays, or pterygiophores, that are organized into a series of three large basals and several series of small radials. The basal pterygiophores are the small *propterygium* anteriorly, *metapterygium* posteriorly, and the large *mesopterygium* centrally (Figure 5-25). Such fins were highly mobile and could

Figure 5-25 Hypothetical role of the basal pterygiophores (shaded) in the evolution of the vertebrate limb according to the axial and basal theories. Po, propterygium; Ms, mesopterygium; Mt, metapterygium. (After Zug.)

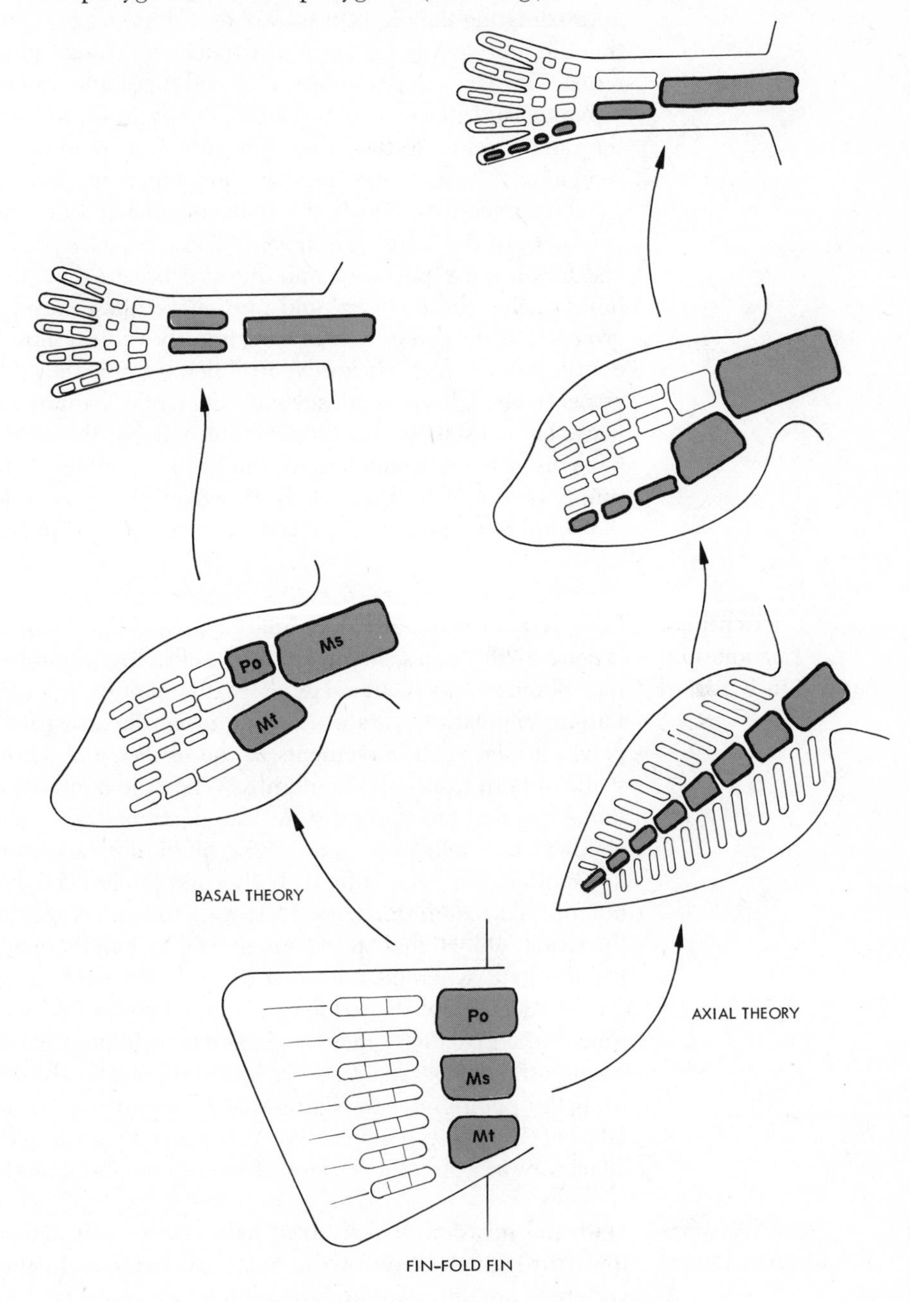

have been used for sculling, rotatory movements, "walking" along the bottom, or in resting with the head raised up. Short, broad, and relatively sturdy pectoral fins evolved as an adaptation to life in shallow waters where the support from the aquatic medium was reduced. The origin of the tetrapod limb has been studied by anatomists, and they have proposed two theories. According to the *basal theory*, in some crossopterygian fishes with short, broad, and sturdy fins (e.g., *Eusthenopteron*), the mesopterygium became recognizable as the humerus or femur, the propterygium as the radius or tibia, the metapterygium as the ulna or fibula, and the radials as a series of distal elements approximating the digits of tetrapods (Figure 5-25). The *axial theory* proposes that the pterygiophores became arranged in a longitudinal series (as in *Sauripterus*). The axial pterygiophores formed the single proximal element (humerus or femur) and all the elements along the posterior margin of the limb, whereas the radials made up the other elements. Concomitantly, the muscles became reorganized around the shoulder and hip joints into groups serving as adjustable braces, by which the body is balanced on its limbs and by whose contractions the latter are moved. Those muscles drawing the leg backward and forward are posterior and anterior braces that protract and retract the limbs during the recovery and propulsive phase, respectively. Those muscle groups that draw the limb toward (mostly ventral muscles) and away (mostly dorsal muscles) from the midventral line become the adductors and abductors, respectively. The musculoskeletal elements of limbs and girdles show a remarkable constancy of a fundamental pattern throughout the tetrapods. We have discussed, at some length, the locomotory apparatus of the urodele as a simply built model of tetrapods. However, there is considerable evidence that the simplicity is more a derived, than a truly primitive, condition.

Reptilian Locomotion: Pattern in a Lizard

Lizards also create lateral undulations of the body when walking and running (Figure 5-26). As in salamanders, areas of weak curvatures can be recognized in the shoulder and pelvic regions that separate regions of maximal bending. Lateral undulations rotate the pectoral girdle and, to a lesser extent, also the pelvic girdle. Such movements of the girdles will increase the length of the stride of both front- and hindlimbs. When the body is stretched, two diagonal limbs (e.g., left hind-, right front-) support the lizard, while the other diagonal couplet (i.e., left front- and right hindlimbs) are swung forward. During locomotion, the lizard lifts its body completely off the ground except for the tail. Initial acceleration from a fast walk to a run is accomplished by increasing the stride rate. Higher speeds are gained by lengthening the stride and changing the gait, whereby only one limb makes contact with the ground. The increased length of the stride can be attributed to the switch from *plantigrade* (sole walking) to *digitigrade* (finger and toe walking). Salamanders are unable to employ the digitigrade mode of locomotion. The ability of a lizard to switch from the plantigrade to digitigrade locomotion is an important innovation, offering the species a mechanism to increase its stride and utilize its limbs more efficiently in a greater variety of environmental situations.

Axial Skeleton and Musculature

With the invasion of terrestrial habitats, we can detect a progressive trend toward regionalization of the vertebral column. In the lizard, two *cervical* vertebrae are differentiated to facilitate movements of the head. A more solid

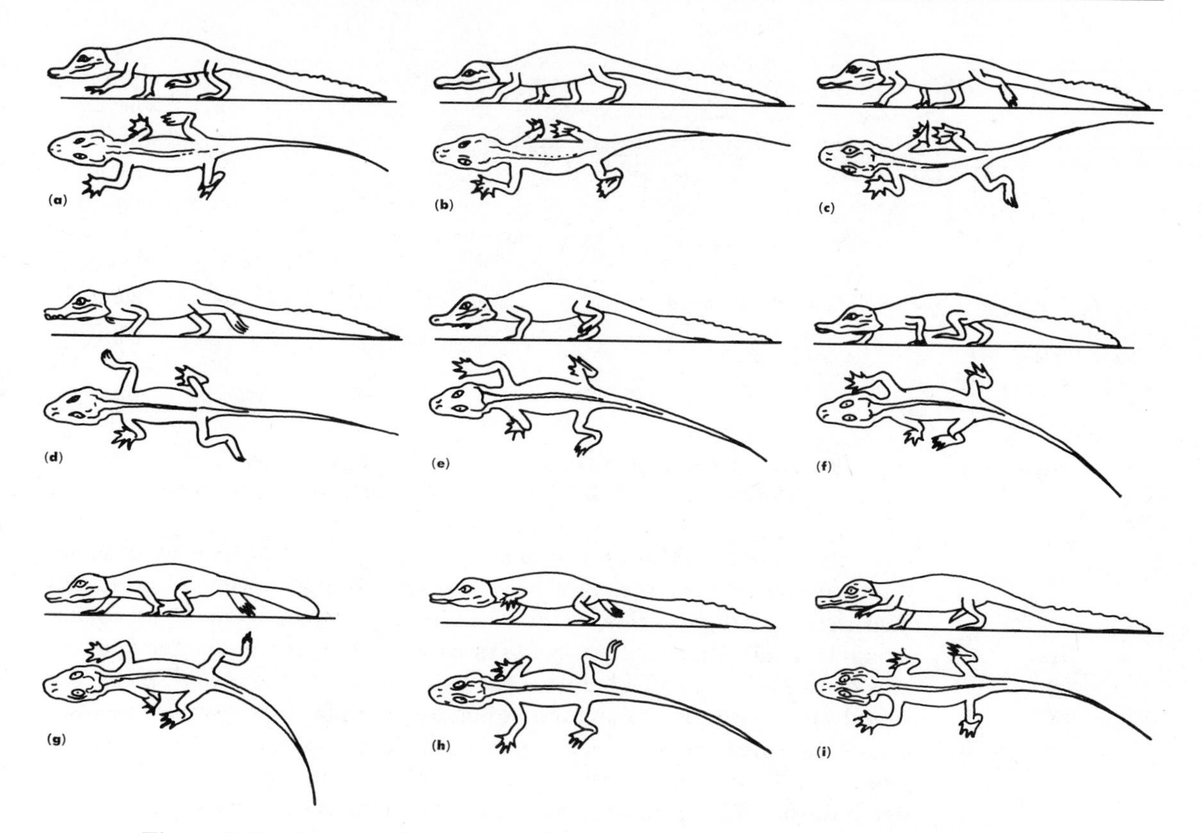

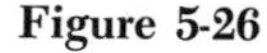
Figure 5-26 Series of diagrams based on a motion picture showing the position assumed by the limbs during locomotion of the alligator. (After Schaeffer, 1941.)

anchorage of the hindlimbs is acquired by direct articulation of the pelvic girdle to two *sacral vertebrae* that are fused with each other and with their ribs. Distinct *thoracic* and *lumbar vertebrae* are differentiated to improve both lung respiration and locomotion. Thoracic vertebrae with articulating ribs provide the basic building blocks for a forced draft mechanism to ventilate the lungs. Aside from their role in respiratory movements, ribs transmit part of the body weight to the front limbs; hence, ribs are connected to the sternum (Figure 5-27). In the lizard, the rib possesses only one head, which is a fusion of the original anteroventral (*capitulum*) and posterodorsal (*tuberculum*) heads. It articulates with a process, the *synapophysis*, that represents a fusion of two processes, the *parapophysis* and *diapophysis* (Figure 5-28). The vertebral column provides a dual function: mobility in the horizontal plane, which allows lateral undulations, and torsional resistance. Mobility in the horizontal plane is made possible by the joints between the vertebral bodies, which are concave anteriorly and convex posteriorly; such vertebrae are named procoelous. The pre- and postzygapophysis with horizontal articular surfaces resist twisting movements between adjacent vertebrae and, yet, allow movements in the horizontal plane.

Concomitant with the emergence of vertebral specializations is the appearance of subdivisions in the epaxial musculature that develop long fibers, some of which attach to neural spines and function as levers that transmit forces to

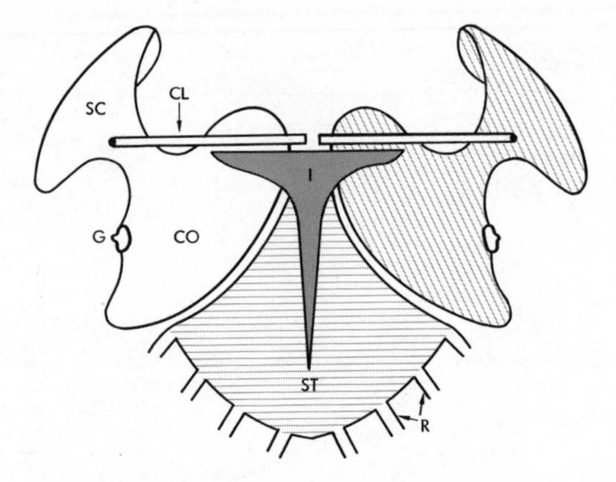

Figure 5-27 Diagram of the shoulder girdle of a lizard. SC, scapula; CO, coracoid; CL, clavicle; I, interclavicle; R, ribs; ST, sternum; G, glenoid. (After Gray, 1968.)

the vertebral bodies. Three major subdivisions (Figure 5-29) can be found in the reptilian epaxial muscle mass. The *iliocostalis* extends downward on to the flank attaching to the ribs; it represents a major component in producing lateral bending. Above the transverse processes of the vertebrae is the *longissimus dorsi.* Most medially, between the longissimus and the neural spines, is a complicated crisscross of short muscle bundles attaching to transverse processes and neural spines; they collectively form the *transversospinalis,* which resists sagittal bending of the vertebral column in the dorsal direction. With the development of ribs, the anterior region, the hypaxial musculature, becomes modified. From the intermediate layer, the obliquus abdominis externus, differentiates the *external intercostals,* while the *internal intercostals* split off from the obliquus internus (Figure 5-29). The external intercostals pull the ribs forward, thereby increasing the volume of the thorax during inspiration.

Hindlimb of the Lizard: Structure and Function

The hindlimbs generate an important proportion of the force for body propulsion. As in the case of the salamander, we can also recognize a propulsive and a recovery phase. The propulsive phase begins when the foot is placed on the ground in a plantigrade position; both the foot and lower leg are extended

Figure 5-28 (a) a generalized tetrapod rib and its vertebral articulations (from Adams and Eddy); (b) lateral aspect of thoracic vertebra of a lizard with a rib possessing one head.

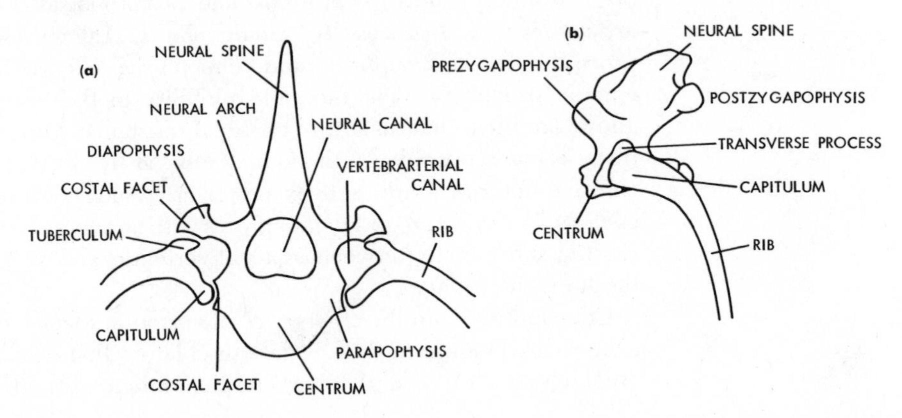

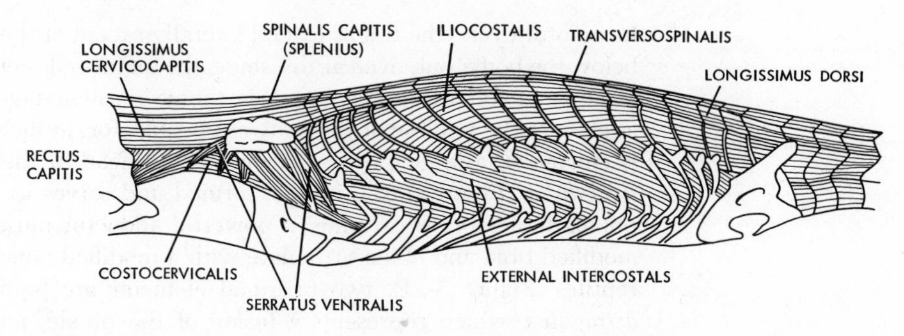

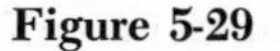
Figure 5-29 Major axial muscles of *Sphenodon* after removal of some superficial muscles. (After Maurer.)

while the femur is flexed (Figure 5-30). As the leg is pulled backward, the foot is progressively flexed on the crus [Figure 5-30(b)]. At the point where maximum propulsion is generated [Figure 5-30(c)], there is considerable flexion at the tarsal joint, accompanied by extension of the foot as the force is transmitted to the ground. During the propulsive stroke, the direction of the foot becomes progressively sideward, being oriented at an angle of about twenty to thirty degrees to the longitudinal body axis. Toward the end of the propulsive phase, all three segments of the limb are extended so that femur and crus form a nearly straight line [Figure 5-30(d)]. The recovery phase starts when the foot is lifted from the ground coincident with the flexing of thigh and crus [Figure 5-30(e)]. Maximal flexion of the thigh, accompanied by its adduction, serves to swing the distal limb segments forward [Figure 5-30(f)]. The legs of lizards describe wide, ellipsoidal arcs during both propulsion and recovery and operate mainly in a horizontal plane.

The pelvic girdle of lizards differs very much from that of the salamander. The ilium is much more expanded for the attachment of more extensive limb muscles on the outer surface and for medial anchorage to the fused sacral vertebrae (Figure 5-31). Both ischium and pubis are well developed, so they can serve as places of origin for the adductor muscles and receive the forces transmitted from the femur. The terminal head of the femur is elliptical, confining thigh movement mostly to the horizontal plane. Throughout the

Figure 5-30 Diagram illustrating the locomotor cycle of the right hindlimb of a lizard. Direction of progression is from left to right. (a) through (d), propulsive stroke; (e) to (a), recovery; A, acetabulum. (After Snyder, 1954.)

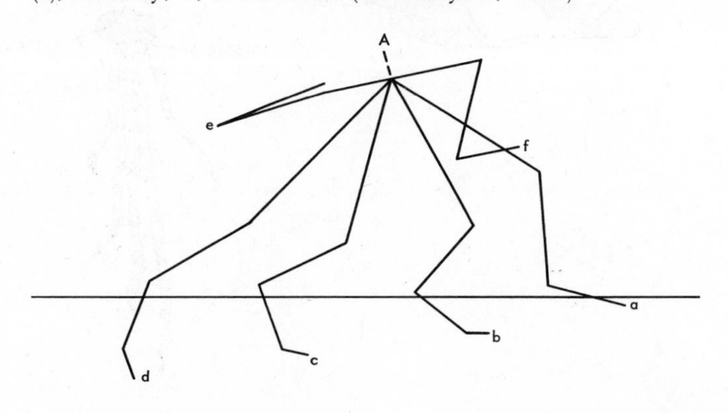

locomotor cycle, the thighs extend laterally except at the time that the foot is below the body; but, even at this stage, the femur is directed lateroventrally at an angle of about forty degrees. It is, therefore, advantageous to place the feet as close to the midline of the body as possible, for, in such a position, leverage is enhanced. a large process medial to the head of the femur, the *internal trochanter* (Figure 5-31), is well formed and serves as the insertion of the puboischiofemoralis externus, a powerful adductor muscle. Distally, the unmodified tibia and fibula articulate with a modified tarsus. In more primitive reptiles (Figure 5-31), two proximal elements are found in the tarsus: the *astragalus*, which represents a fusion of the tibiale, intermedium, and one centrale; and the *calcaneum*, which is simply the fibulare under another name. In lizards, the astragalus and calcaneum have fused into one bone, the *astragalocalcaneum* (Fig. 5-31). The junction between astragalocalcaneum and the tibia and fibula is rendered immobile by strong ligaments. Distally, the astragalocalcaneum forms a true synovial joint only with one element the fourth tarsale (Figure 5-31). It is clear that there is a strong consolidation of the tarsus of the lizard. Consolidation of tarsal elements is accompanied by the development of a synovial joint, which combines stability with precise control of

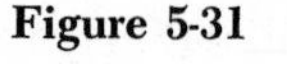

Figure 5-31 (a) Lizard pelvic girdle seen from lateral; (b) anterior aspect of hindlimb skeleton of a lizard.

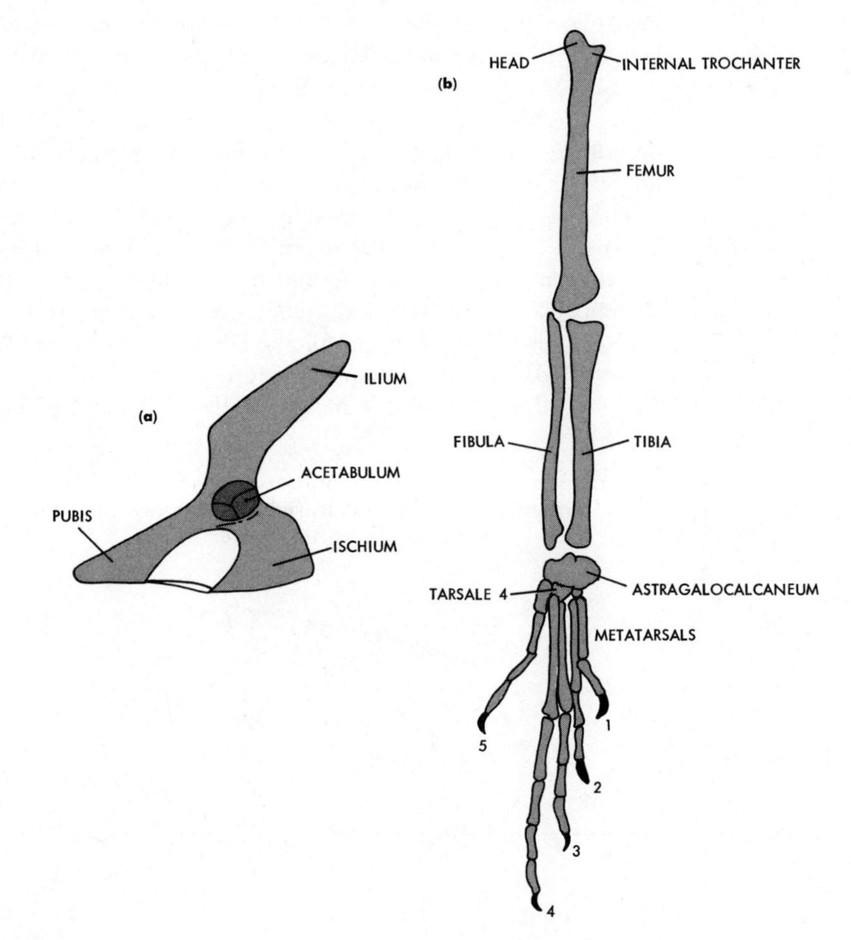

specific movements. If one compares the hindlimb of the salamander with that of the lizard, we can see that the reptilian innovations have occurred mainly in the distal part. Consolidation of tarsal elements has led to the development of a single and stable ankle joint, which contributes to the propulsive capabilities of the reptilian foot and enables the lizard to employ temporary digitigrade locomotion. For the foot to generate a successful thrust, it requires a firmly consolidated lever that, in lizards, is simply the elongated metatarsals. In reptiles, the foot is no longer a flexible base giving support to the propulsive movement of the leg, as described for the salamander.

Major innovations in the hindlimb muscles of the lizard are also mostly confined to the distal parts. The most interesting evolutionary changes of the foot musculature involve the peroneus (which is the extensor fibularis of the salamander) and flexor primordialis communis. The original extensor fibularis muscle, as found in the salamander, undergoes splitting, topographical shifts, and functional reversal in the lizard; the new name for this muscle is the *peroneus.* A posterior subdivision, the *peroneus longus,* arises from the lateral epicondyle of the femur and inserts into a hooked portion of the fifth metatarsal [Figure 5-32(a)]. Being posterior to the axis of rotation of the ankle joint, the peroneus longus becomes a strong flexor (plantar flexor) of the foot. The anterior subdivision, the *peroneus brevis,* arises from the fibula and inserts on

Figure 5-32 Superficial musculature of the left hindlimb of a lizard, dorsal view. (After Snyder, 1954.)

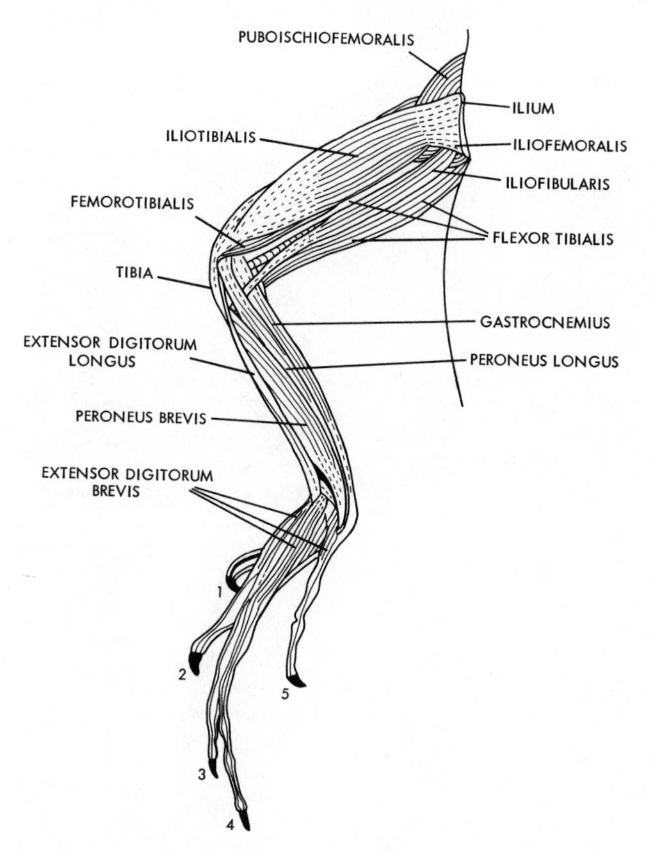

the dorsal surface of the fifth metatarsal, retaining its function as an extensor (dorsiflexor) of the foot. The original flexor primordialis communis (as found in the salamander) splits into two layers in reptiles: the superficial layer, which becomes the *gastrocnemius* (Figure 5-32), and the deeper one, the *flexor digitorum longus*. Because the gastrocnemius inserts into the plantar fascia of the foot (Figure 5-33), its function is plantar flexion of the foot during propulsion. The ability of the foot of the lizard to participate actively in the propulsive effort is correlated with the differentiation of the peroneus longus and gastrocnemius muscles, which, in cooperation with the flexor digitorum longus, can lift the proximal part of the foot off the ground (Figures 5-26 and 5-34). The remaining extensors (extensor tibialis or tibialis anterior, extensor digitorum longus, and peroneus brevis) retain their original extensor, or dorsiflexion, function.

As in the salamander, the muscles of the hindlimb can be arranged into functional groups. We will discuss the muscles in respect to two major functional groupings: propulsive and recovery. Let us first consider the propulsive phase during which the extended leg is retracted. Retraction of the femur during the propulsive phase is brought about by contraction of the *caudofemoralis*, in cooperation with the *ischiotrochantericus* and *puboischiofemoralis externus*, both of which adduct and, to some degree, retract the femur. The

Figure 5-33 Superficial musculature of the left hindlimb of a lizard, ventral aspect. The central portion of the puboischiotibialis has been removed. (After Snyder, 1954.)

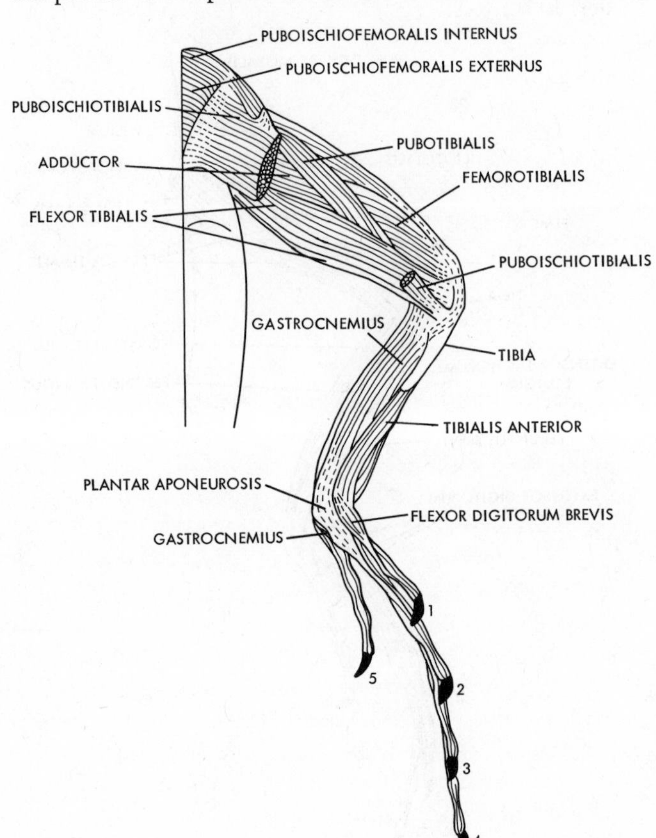

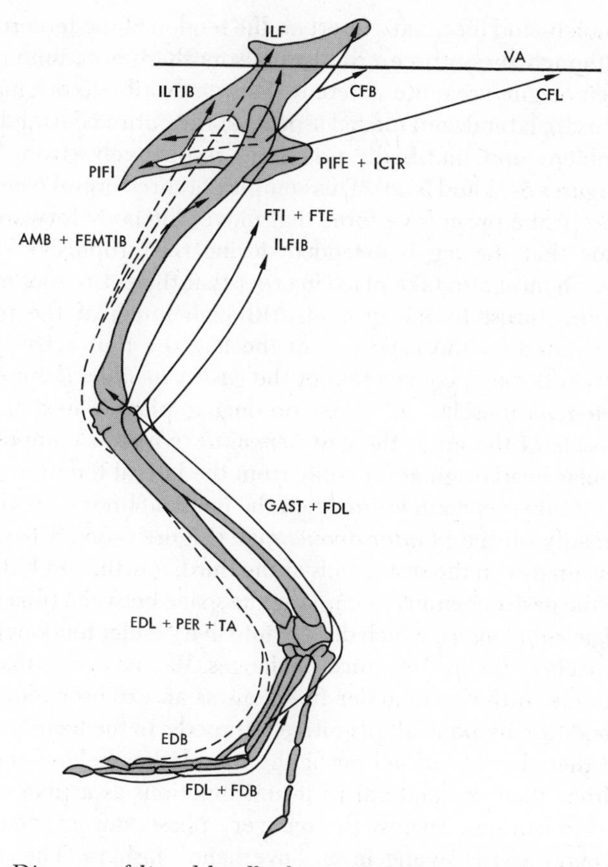

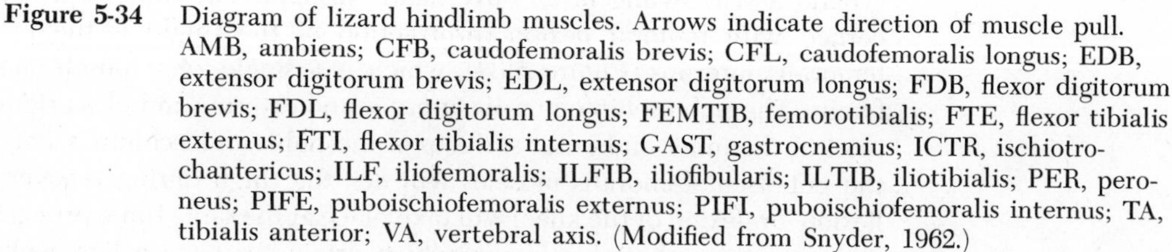

Figure 5-34 Diagram of lizard hindlimb muscles. Arrows indicate direction of muscle pull. AMB, ambiens; CFB, caudofemoralis brevis; CFL, caudofemoralis longus; EDB, extensor digitorum brevis; EDL, extensor digitorum longus; FDB, flexor digitorum brevis; FDL, flexor digitorum longus; FEMTIB, femorotibialis; FTE, flexor tibialis externus; FTI, flexor tibialis internus; GAST, gastrocnemius; ICTR, ischiotrochantericus; ILF, iliofemoralis; ILFIB, iliofibularis; ILTIB, iliotibialis; PER, peroneus; PIFE, puboischiofemoralis externus; PIFI, puboischiofemoralis internus; TA, tibialis anterior; VA, vertebral axis. (Modified from Snyder, 1962.)

anatomical and topographical configuration of these muscles (Figures 5-32, 5-33, and 5-34) resembles that of the homologs in the salamander. The angle of application of the caudofemoralis on the femur is ninety degrees; consequently, the action of this muscle is both rapid and powerful. At the onset of the propulsive phase of the limb, the femur is flexed maximally; in this position, caudofemoralis acts with a very long power arm, resulting in a strong posterior movement of the entire limb. Additional muscles also may retract the thigh, although their primary function is crural flexion; they include the *flexor tibialis* and iliofibularis, which, because of their position and attachment to the crus, assist in femoral retraction at the onset of the propulsive stroke, provided that their flexor action is checked by the opposing group of extensor muscles.

During propulsion, the knee joint is extended by three crural extensors: the ambiens, iliotibialis, and femorotibialis muscles. The femorotibialis, the largest muscle of the thigh, inserts by means of a broad, strong tendon that passes over the anterior surface of the knee to the tibia (Figures 5-33 and 5-34). Both the ambiens and iliotibialis insert on the tendon of the femorotibialis (Figure 5-34). Although these three crural extensors share a common functional insertion, their origins are quite different. The femorotibialis originates from the anterior (dorsal), lateral, and medial aspects of the entire central shaft of the femur. The ambiens and iliotibialis originate, respectively, from the pubis and ilium (Figures 5-32 and 5-34). This complex of three crural extensors provides a large part of the propulsive force that moves the body forward and upward. At the time that the leg is extended during the propulsive stroke, strong plantar flexion must also take place in order that the entire foot may transmit the force of the thrust to the ground. Although much of the force of propulsion is generated by the extensors of the leg, the foot actively participates in this action through contraction of the *gastrocnemius, flexor digitorum longus,* and *peroneus* muscles, all three producing plantar flexion. The most extensive muscle of the leg is the *gastrocnemius,* which is composed of two heads: the fibular head originates mainly from the lateral femoral condyle, and the tibial head takes its main origin from the medial femoral condyle. Both heads insert broadly on the *plantar aponeurosis* (Figure 5-35), a flat, tendinous sheet that terminates on the distal ends of the third, fourth, and fifth metatarsals. Internal to the gastrocnemius, occupying the space between tibia and fibula, is the *flexor digitorum longus,* which divides into five slender tendons inserting on the dorsal aspects of the five terminal phalanges. We have seen that the peroneus longus muscle in the salamander functions as an extensor (dorsiflexor) of the foot, a condition found in all primitive tetrapods. In the lizard, its insertion has shifted so that the line of action of the muscle lies behind the ankle (Figure 5-32) rather than dorsolateral to it and functions as a plantar flexor.

We can now review the recovery phase, during which the foot leaves the ground and is swung in an "overhand" fashion. The muscle primarily concerned with femoral flexion (protraction of the thigh) is the puboischiofemoralis internus (Figure 5-34), which is a single-joint muscle moving the femur. The puboischiofemoralis internus and externus muscles originate from most of the inner and outer surfaces of the pubis and ischium. Contraction of the puboischiofemoralis muscles will flex the thigh during recovery. Crural flexion (or flexion of the knee) will take place at the same time, primarily by the action of the puboischiotibialis, which originates from a ligament between pubis and ischium and inserts on the proximal tibial shaft. The largest and most powerful flexor of the leg is the flexor tibialis internus, a three-headed muscle occupying the posterior aspect of the thigh. All three heads originate from a ligament between ilium and ischium (the ilioischiadic ligament) and insert on the tibial condyles and the proximal tibial shaft. Other crural flexors include the iliofibularis, a long slender muscle [Figures 5-32(a) and 5-34] originating from the ilium and inserting on the proximal shaft of the fibula; the pubotibialis, which runs between the pubic rim and the lateral tibial condyle; and, finally, the flexor tibialis externus, with an origin on the ilioischiadic ligament and an insertion into the gastrocnemius and lateral tibial condyle. Multiple and strong crural flexors are needed by all lizards to bring about rapid flexion during the recovery of the hindlimb and to prevent passive extension of the knee during the propulsive stroke. The hindlimbs of quadrupedal reptiles function as propulsive levers in which strong flexors characterize both knee and ankle joints. Finally, we need to consider dorsiflexion of the foot during the recovery stroke. Three muscles lift up the foot and toes from the ground: the tibialis anterior, peroneus brevis, and extensor digitorum longus and brevis. The tibialis anterior originates from all but the posterior surface of the central tibial shaft and inserts on the dorsal surface of the first metatarsal. It is not necessary to remember the exact topography of the other muscles, but it is important to know that in unison these muscles will dorsiflex the foot.

Forelimb of the Lizard: Structure and Function

Lizards are adapted for moving at relatively great speeds, some attaining speeds of fifteen miles per hour, which, in terms of its own body length, puts them in the same class as the very fastest terrestrial mammals. Their hindlimbs bear a disproportionately large responsibility for drive. Some lizards can run on their hindlimbs and run faster on their hindlimbs than on all fours. When lizards run on their hindlimbs they take strides far longer than the short forelimbs could make. Lizard forelimbs play a dominant role in steering and orientation, although they do contribute to drive also. The gait of the forelimb of the lizard resembles that of the urodele amphibians (Figure 5-26). The propulsive stroke starts when the hand is placed on the ground, with the digits spread and extended cranially; this stage is followed by a retraction of the humerus—at first along a horizontal plane and then dorsally and caudally; the elbow begins to extend; and, finally, the hand lifts off the ground, at which point the recovery stroke begins. During recovery the humerus moves cranially while the wrist and digits are loosely flexed (Figure 5-26).

Nearly all the remarkable locomotory capabilities of a lizard depend on the specialized anchorage of the forelimbs to the axis of the body. As in all tetrapods, the head of the humerus articulates with the glenoid fossa in the pectoral girdle and the main retractor muscles of the limb (latissimus dorsi and pectoralis) originate from the dorsal and ventral surfaces of the body (Figure 5-35). In order to transfer the propulsive forces of the retractor muscles from the ground to the vertebral column, an effective mechanism preventing the backward movement of the pectoral girdle must be well developed. We have seen that in urodele amphibians the fixation of the pectoral girdle is effected by the muscles that attach it to the vertebral column. In contrast to the amphibian

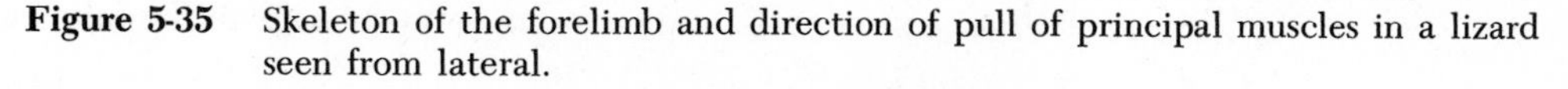

Figure 5-35 Skeleton of the forelimb and direction of pull of principal muscles in a lizard seen from lateral.

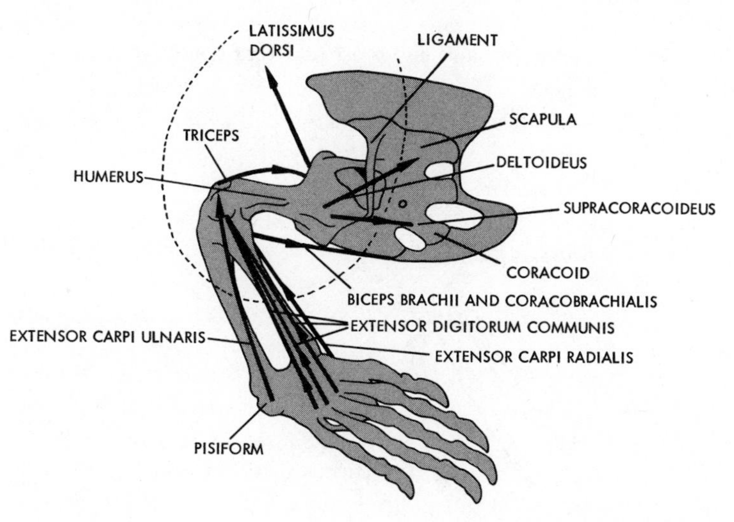

condition, the more robust pectoral girdle of a lizard is firmly fixed to the vertebral column by means of the sternum and ribs (Figure 5-27). More extensive ossification also occurs in the pectoral girdle. The T-shaped *interclavicle*, a median dermal element, resists bending together with the *clavicles*. An ossified *coracoid* is separated from the scapula by a suture on the line of the glenoid and shares the glenoid cavity with it. The glenoid cavity has become an open notch, facing as much backward as outward. This results in a greater freedom of motion of the humerus, enabling the lizard to bring the forelimb closer to the body than a salamander can. The humerus resembles that of the salamander, although it is more elongate and slender. The two expanded ends of the humerus are still seemingly twisted onto one another: the proximal end is essentially horizontal, the distal end tilted so that its lower surface faces the forward slanting forearm (Figure 5-35). The major weight-bearing element in the forearm is the columnar-shaped *radius;* the *ulna* lies lateral to the radius and articulates with the distal edge of the humerus by a notch. Above this notch the ulna projects as the *olecranon,* the "funny bone," serving as an insertion of the triceps muscle, the main extensor of the forearm. In the reptilian wrist, a new element, the *pisiform,* has appeared; the small bone is attached to the outer margin of the carpus (wrist) and forms a point of attachment for the tendons of muscles along this aspect of the limb (Figures 5-35 and 5-36).

No major architectural novelties can be found when one compares the reptilian limb with that of the salamander. However, a true *biceps brachii,* originating from the coracoid and inserting mainly on the radius, makes its first appearance in reptiles (Figure 5-37). As an adaptation to weight bearing, the flexor digitorum communis as found in salamanders has split into two distinct components in the lizard: the *flexor digitorum superficialis* and *profundus.* The latter inserts on the distal phalanges, whereas the former attaches to the proximal phalanges.

Let us first review the major muscles responsible for the propulsive phase. The hand is pressed against the ground by the flexor digitorum superficialis and profundus, and flexion of the elbow joint is accomplished by contraction of the biceps, flexor carpi radialis, and flexor carpi ulnaris. As in the salamander, the

Figure 5-36 Skeleton of hand and wrist of a representative lizard.

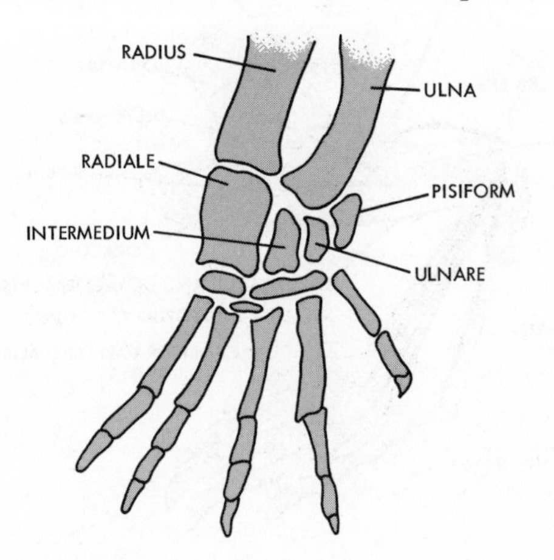

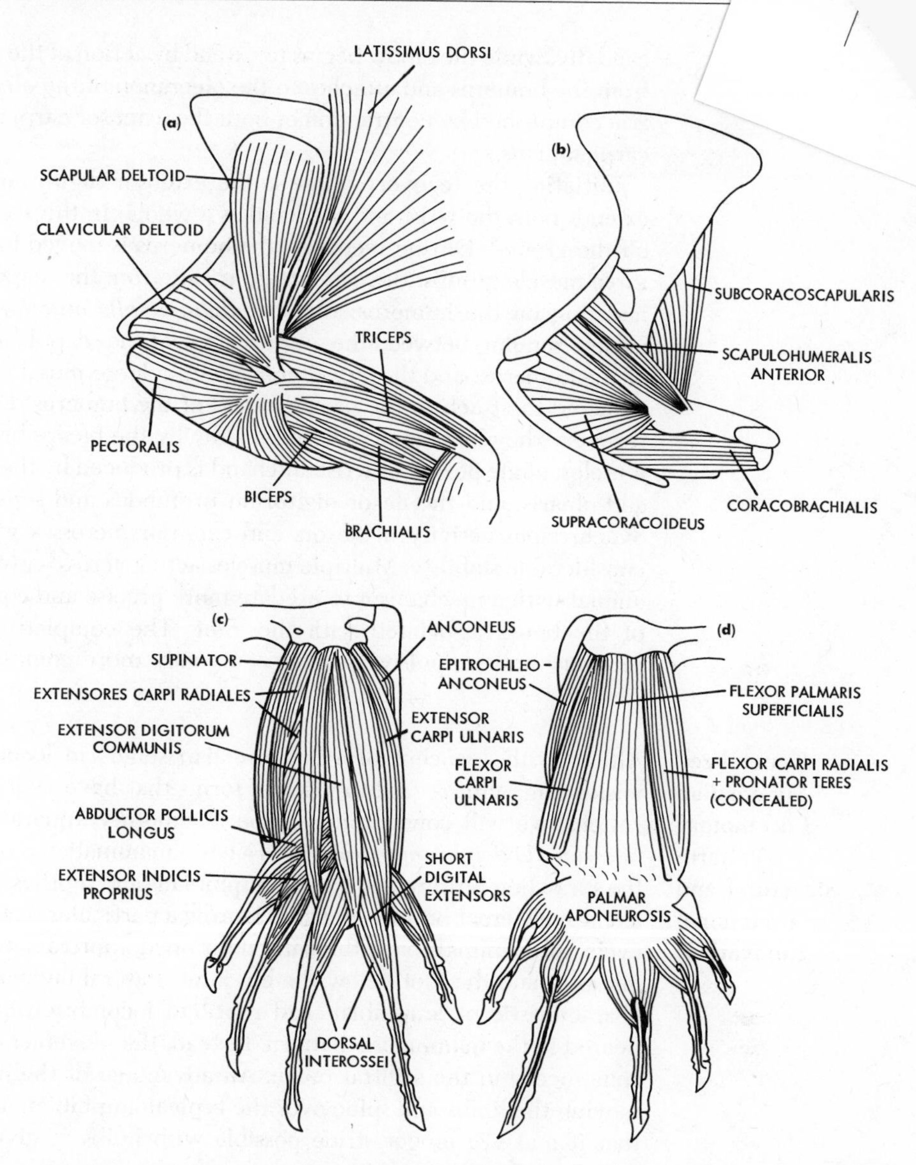

Figure 5-37 (a)(b) shoulder and upper arm muscles in a lizard, seen from lateral. (b) is a deep dissection; (c)(d) muscles of the forearm and hand in a lizard. (c) is a view from the extensor surface. (From Romer, courtesy of Saunders Publishing Co.)

flexed forearm acts as a secondary lever transmitting the propulsive force from the arm to the point of resistance; the expanded hand acts as an expanded base against the substrate. Retraction of the humerus propels the body forward. A powerful retractory force on the humerus is generated by a combination of three muscles: the *latissimus dorsi,* which originates from the lumbodorsal fascia on the back and flank and inserts near the head of the humerus; the *pectoralis,* the origin of which spreads over the sternum and ribs inserting on a prominent process beneath the proximal end of the humerus; and the coracobrachialis, which runs between the back end of the coracoid and the distal end of the humerus. The combined action of the three retractors will move the humerus backward, at first along a horizontal plane and then dorsally and caudally, while the elbow begins to extend by action of the *triceps,* which arises from the humerus and attaches to the olecranon of the ulna; further extension is accomplished by contraction of both the extensor carpi ulnaris and extensor carpi radialis.

Initiating the recovery phase is the extensor digitorum communis, which extends both the wrist and the hand as a whole. In this way, the hand is lifted off the ground. During recovery, the humerus is moved forward primarily by three muscle groups: the *deltoideus,* arising from the scapula and clavicle and inserting on the humerus; the *scapulohumeralis anterior,* a small and deep muscle running between the outer surface of the scapula and posterior aspect of the humerus; and the *supracoracoideus,* a large muscle that arises from the coracoid to attach to the proximal end of the humerus (Figure 5-37). During recovery the elbow joint is flexed mainly by the biceps brachii and *brachialis* muscles, while flexion of wrist and hand is produced by the flexor carpi radialis and ulnaris, and the flexor digitorum profundus and superficialis complexes. Synchronous activity of flexors and extensors across a given joint results in considerable stability. Multiple muscles acting across a given joint provide the animal with a mechanism to execute more precise and controlled movements of the bones associated with the joint. The complete mechanisms of the recovery and propulsive strokes is actually more complex than given here.

Generalized Mammalian Locomotory Pattern: Structural and Functional Innovations

Although the concept of a "mammalian stage" in locomotor adaptation is misleading, with regard to modern forms that have evolved diverse specializations, we will concentrate on the locomotory apparatus of the American opossum, *Didelphis marsupialis,* as a basic mammalian model. Compared with the sprawling posture and gait of amphibians and reptiles, all mammals exhibit a relatively erect posture and gait. During a particular phase in the locomotory cycle, all mammals have the hand positioned approximately below the shoulder joint and the foot below the hip joint. Lateral undulations of the body so characteristic of amphibian and reptilian locomotion have virtually disappeared in the mammalian pattern. Instead, the movements of the mammalian spine occur in the sagittal plane. An advantage of the mammalian mode of moving the limbs and spine over the typical amphibian and reptilian mode is that it makes a longer stride possible with limbs of given length and mass.

Axial Skeleton and Musculature

The reptilian atlas-axis complex is constrained in the range of possible movements of the head because of the kidney-shaped occipital condyle and the presence of proatlas ossicles. Mammals, in contrast, have consolidated the reptilian elements into two vertebrae, the atlas and axis (Figure 5-38). The atlas is a ringlike structure derived from neural arches and intercentrum. The atlantooccipital joint (between the atlas and the skull) permits extensive flexion-extension, whereas the atlantoaxial joint (between the atlas and the modified second cervical vertebra, the axis) is primarily rotatory. At the atlantoaxial joint, loss of the zygapophyses (Figure 5-38) is a prerequisite to the development of rotational movement. Mobility of the head in mammals is far superior to that of lower tetrapods.

With a few exceptions, mammals have seven cervical vertebrae without ribs. The constancy of this number is remarkable, considering that in all other axial regions the number of vertebrae vary. In the trunk two regions differentiate: a

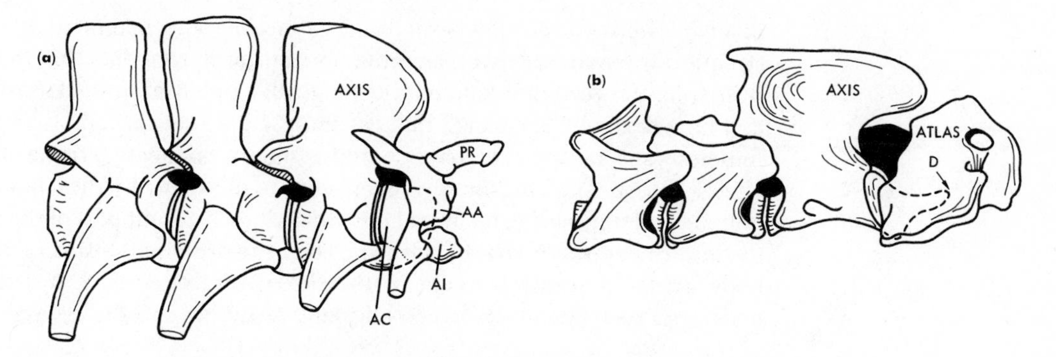

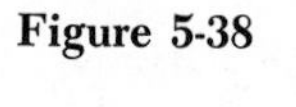

Figure 5-38 Lateral view of first four cervical vertebrae in (a), a pelycosaur and (b), a mammal, illustrating major changes in the atlas-axis complex from pelycosaurs to mammals. The part of the atlas centrum obscured by the atlas arches is represented by a dashed line. AA, atlas arch; AC, atlas centrum; AI, atlas intercentrum; D, dens; PR, proatlas. (From Jenkins, 1970, courtesy of *Evolution.*)

thoracic division with ribs and a ribless lumbar division characterized by long transverse processes. Three to five vertebrae are incorporated in the mammalian *sacrum* to furnish a more solid anchorage of the pelvic girdle with the axial skeleton. The *sternum* has the form of an elongated, jointed, bony rod with which the ribs are articulated at the "nodes." Mammalian ribs are differentiated into two parts: a dorsal bony part, the *vertebral rib,* articulating with the vertebral column; and a cartilaginous ventral part, the *sternal rib,* articulating with the sternum. The ribs are typically bicipital, the posterodorsal head (*tuberculum*) articulating with a process (*diapophysis*) on the neural arch; the anteroventral head (*capitulum*) is housed in depressions of two adjacent vertebrae (demifacets, Figure 5-39). In the lumbar region, the ribs have become reduced in size and fused to the vertebrae as prominent transverse processes.

As stated previously, axial movement associated with locomotion may

Figure 5-39 Costovertebral relationship in a mammal, allowing the necessary mobility of the ribs for respiratory movements.

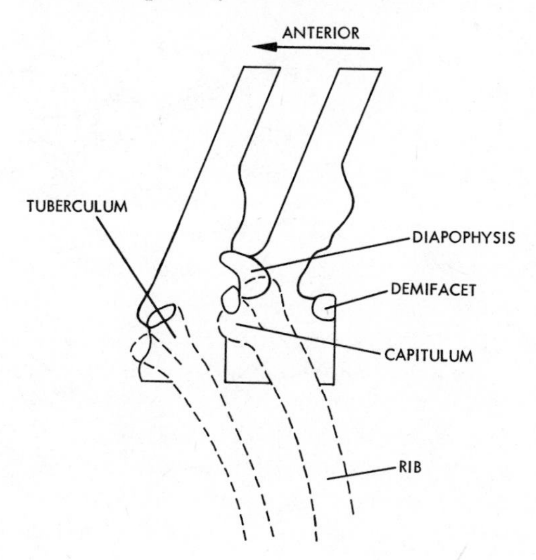

involve extensive sagittal flexion and extension. The mammalian axial construction is remarkably well adapted for sagittal movements (Figure 5-40). In principle, the vertebral column, pelvic girdle, and axial musculature function as a bow flexed by a "string" that is composed of the sternum, the abdominal muscles (especially the rectus abdominis), and the connective tissue of the linea alba (Figure 5-41). The "bow" is connected to the string by the ribs and by the oblique and transverse abdominal muscles, which transmit part of the weight of the intestines directly to the body axis. Thus, the principal static function of the body axis is to resist bending in the dorsal direction; such a resistance is generated by ligaments between adjacent vertebrae and the epaxial muscula-

Figure 5-40 Tracings of X-ray motion picture of a tree shrew running on a branch, showing spinal flexion in the sagittal plane. In this case spinal flexion is used for both lengthening the stride and adjusting to the irregular surfaces of the substrate. When walking slowly, the tree shrew also flexes its spine variably, to make adjustments to the branches of the tree. (From Jenkins, 1974, and courtesy of Academic Press.)

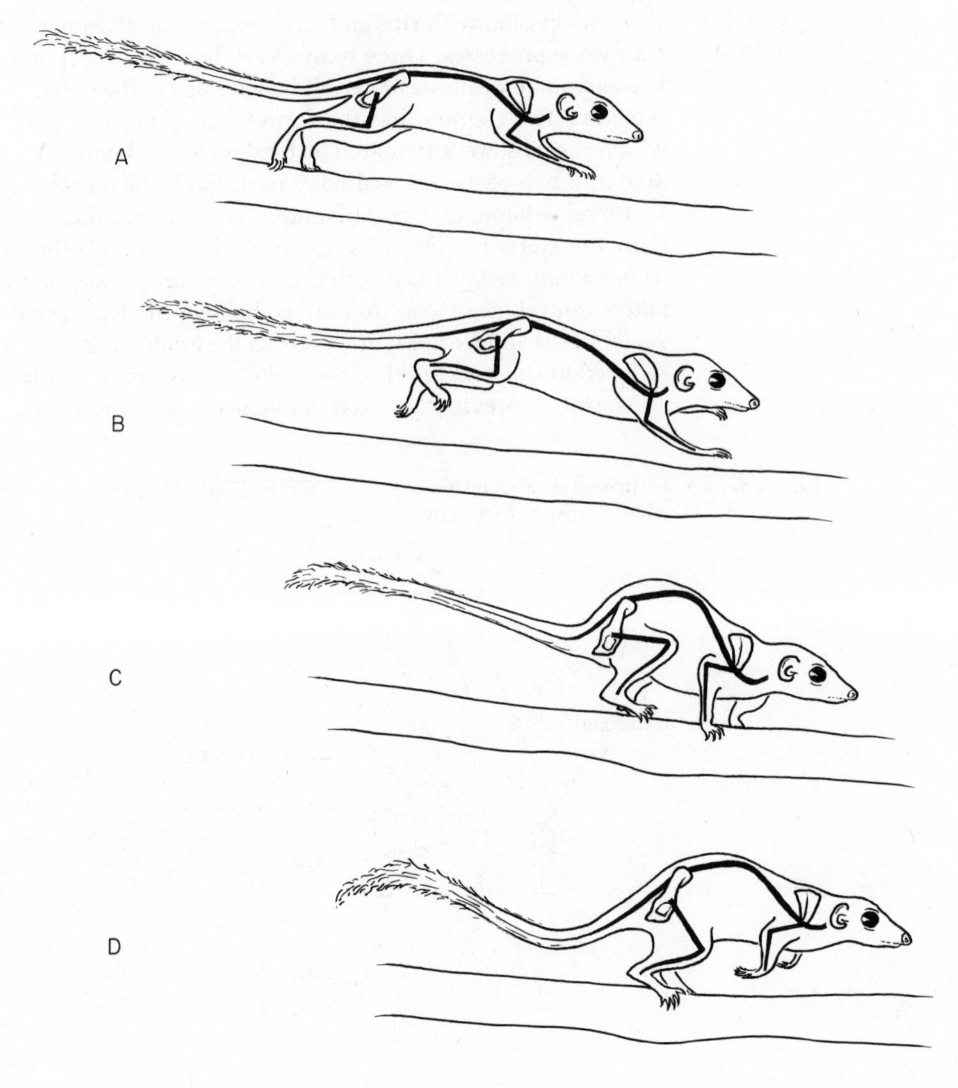

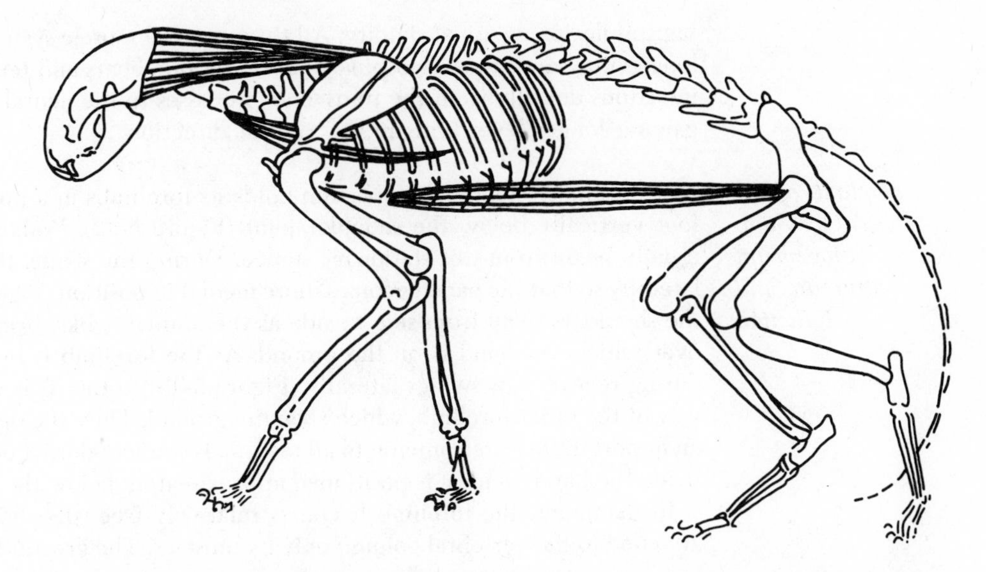

Figure 5-41 Skeleton of a cat with the splenius, and rectus abdominis muscles, to show the bow-and-string construction of the trunk of mammals and the attachment of the head and neck to this construction. (After Slijper, 1946.)

ture. In addition, the body axis is an organ of locomotion, transmitting the thrust from the hindlimbs forward. Such a function is again realized by a resistance to bending in the dorsal direction. During a leaping gallop, flexion and extension in the sagittal plane can be executed efficiently by the bow-and-string type of construction.

As we have seen, the most lateral division of the epaxial musculature, the iliocostalis muscle, is dominant in reptiles. This fact is correlated with lateral undulation. The iliocostalis, acting on ribs that function as levers, effects lateral flexure of the vertebral column with greater mechanical advantage than the more medially situated *longissimus* and *transversospinalis* muscle systems. In mammals, the most medial of the epaxial muscles, the transversospinalis, is commonly better developed (Figure 5-42) than either the more lateral longissimus or iliocostalis. The transversospinalis contributes effectively to

Figure 5-42 Epaxial musculature of an opossum. Transversospinalis is undifferentiated but large. Iliocostalis and longissimus are relatively small if compared with those of reptiles. (Modified after Slijper.)

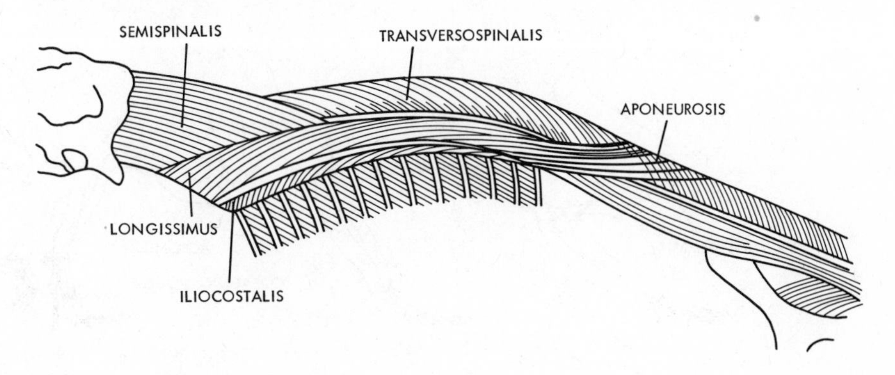

sagittal flexion or spinal rigidity. All three epaxial muscle systems (iliocostalis, longissimus, and transversospinalis) develop long fibers and tend to shift their insertions dorsally from the transverse processes to the neural spines thereby gaining longer lever arms in the sagittal direction.

Forelimb of the Opossum: Movement, Structure and Function

In a stationary stance, the opossum holds its forelimbs in a position with the foot vertically below the shoulder joint (Figure 5-43). Walking movements usually begin from this stationary stance. During the stride, the body moves laterally so that the hand becomes more medial in position (Figure 5-43). Thus, the shoulders sway from side to side as the animal walks, bringing the chest over whichever hand is on the ground. As the forelimb is brought forward during recovery, it swings laterally (Figure 5-43) so that it is well out of the way of the other forelimb, which is on the ground. Thus, the opossum exhibits an important feature common to all mammals studied: during one phase in the stride the hand (manus) is positioned approximately below the shoulder joint.

In mammals, the forelimb becomes relatively free; the shoulder girdle is attached to the vertebral column only by muscles. The greater freedom of the

Figure 5-43 Tracings drawn from X-ray motion pictures of the left forelimb and scapula of the opossum. Four stages of a walking step are shown. The broken lines are parallel to the long axis of the body. A, dorsal view; B, lateral view. (Modified from Jenkins, *Journal of Zoology*, 1971.)

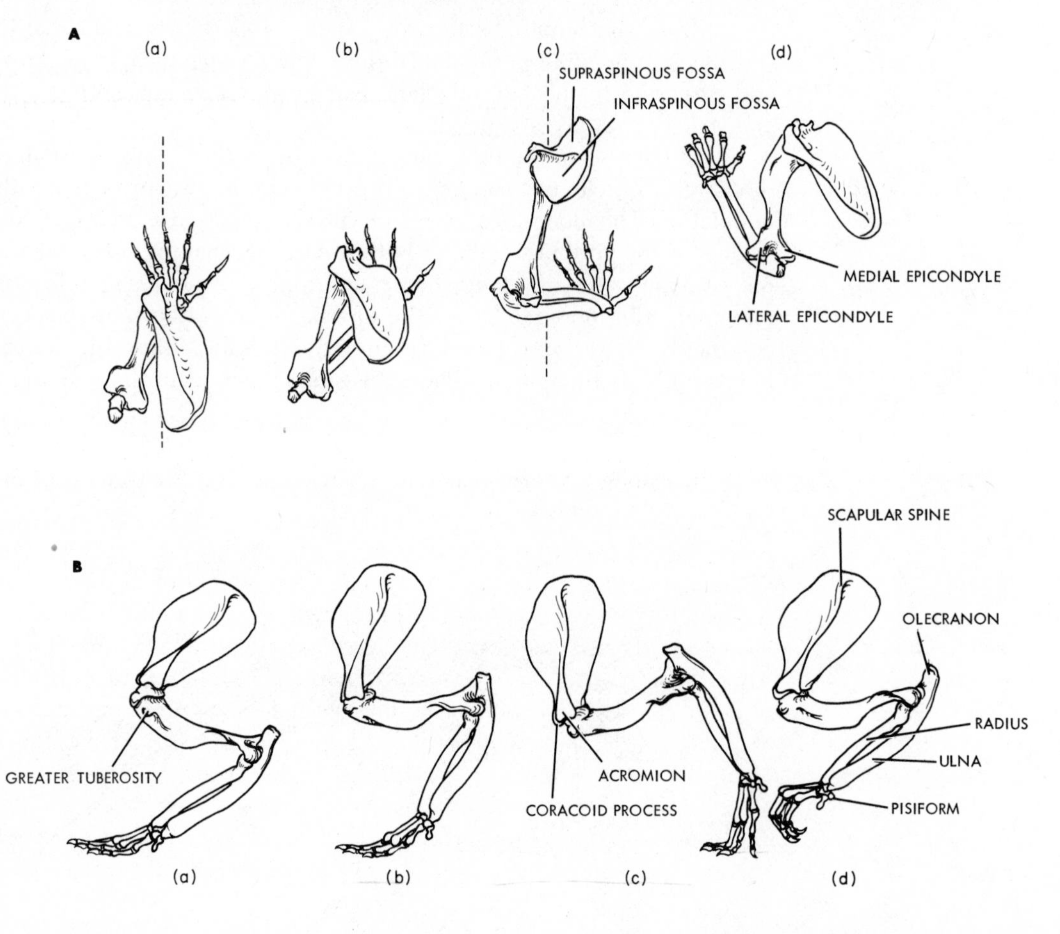

forelimb can be regarded as a versatile design, offering the development of such adaptations as digging, collecting food, or otherwise handling the surrounding world. There is a very major change in the pectoral girdle of mammals if compared with that of the lizard. The girdle is practically reduced to a dorsally situated *scapula* with the glenoid cavity below it. Even the scapula itself has radically changed, for, instead of being a simple plate, it has a prominent ridge, the *scapular spine*, running down its length separating *supraspinous* and *infraspinous fossae*. The entire coracoid plate is reduced to a process like a crow's beak, the coracoid process, attached to the lower margin of the scapula (Figure 5-43). Meanwhile, the point of attachment of the *clavicle* (which is not fully developed in the opossum) to the scapula has differentiated in a distinct process, the *acromion*. This major change in girdle structure is related to the shift in limb posture in mammals, as compared to reptiles, and accompanies major changes in limb musculature. As a mammal walks (Figure 5-43), the scapulae move relative to each other and to the underlying ribs. Actually, there is very little movement at the shoulder joint while the scapula and humerus swing as a unit through about 45 degrees. Rotation about a high pivot (between scapula and ribs), rather than a lower one (the joint between humerus and glenoid cavity), lengthens the stride.

The *humerus* (Figure 5-43) is a typical long bone that is no longer twisted, as in the reptile. At its upper end is the *greater tuberosity* on the lateral side, serving as the point of attachment of muscles fixing the limb to the shoulder and abducting and externally rotating it. The *lesser tuberosity*, on the medial face, carries muscles that serve for adduction and internal rotation. Of course, the main muscles for moving the limb are attached farther down the shaft, where they acquire a greater leverage. Distally, the humerus is expanded into *lateral* and *medial epicondyles*, from which arise, respectively, extensor and flexor muscle masses. The bones of the forearm, wrist, and hand (Figure 5-43) in the opossum are relatively unspecialized. Both *radius* and *ulna* are fully developed, articulating with, respectively, the *radiale* (*scaphoid* in mammalian anatomy, navicular in human anatomy) and the *ulnare* (*cuneiform* in mammalian anatomy, *triquetrum* in human anatomy). This wrist joint allows a considerable range of movement. In general, there is usually little movement at the intercarpal and carpometacarpal joints, except in mammals where the thumb is free. The next important joint is that between the metacarpals and phalanges; here, and at each of the two interphalangeal joints, movement is possible in one plane only.

The mammalian forelimb is not jointed to the main vertebral axis, and the whole transmission of weight is accomplished by five muscles. The *serratus anterior* runs from the ventral part of the ribs to the upper border of the scapula and, thus, makes a sling by which the weight of the body, transferred through the vertebral column to the ribs, is hung on to the limb that, in turn, acts as a pillar (Figure 5-44). The scapula also is attached to the body by two muscles on its dorsal side. The most superficial is the *trapezius*, a broad band of fibers arising from the occiput of the skull and spines of the vertebrae all along the neck, thorax, and cranial part of the lumbar region and converging to insert along the spine of the scapula and the reduced clavicle. The function of the muscle is to hold the scapula against the body. Deep to the trapezius, the *rhomboid* muscles run from the ligamentum nuchae and spines of the anterior thoracic vertebrae to the dorsal border of the scapula; these muscles hold the

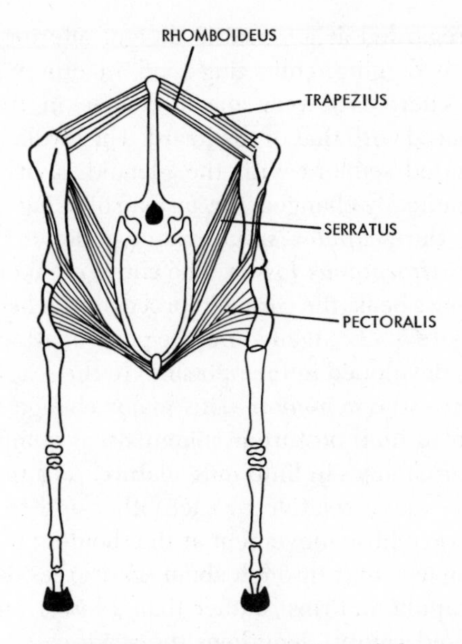

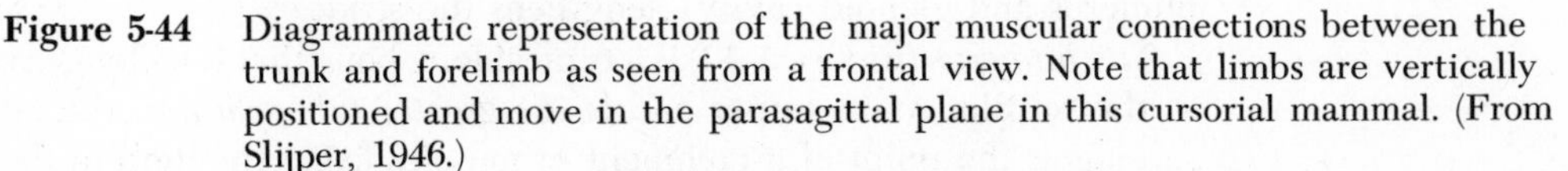

Figure 5-44 Diagrammatic representation of the major muscular connections between the trunk and forelimb as seen from a frontal view. Note that limbs are vertically positioned and move in the parasagittal plane in this cursorial mammal. (From Slijper, 1946.)

scapula against the chest wall. The cranial border of the scapula is held in to the body by the *levator scapulae major* and *minor,* running from the skull to the distal scapular spine and inferior angle of the scapula, respectively.

The muscles that form contractile braces around the shoulder joint (Figure 5-45) include shorter ones that arise from the scapula and longer ones that originate from the axial skeleton. The shorter muscles, whose tendons are inserted close to the proximal end of the humerus, arise from the surfaces of the scapula. On the lateral side, the *supraspinatus* and *infraspinatus* (both derivatives of the reptilian supracoracoideus) are inserted on the greater tuberosity. They are acting as lateral and cranial braces of the humerus on the scapula. Acting as a medial brace is the *subscapularis,* running from the ventral, or costal, surface of the scapula to the lesser tuberosity. The mammalian subscapularis is a homologue of the reptilian subcoracoscapularis. The longer braces include the *pectoralis* and *latissimus dorsi,* respectively, on the medial and posterior aspect.

It is practical to subdivide the movement cycle of the forelimb during forward locomotion into propulsive and recovery phases. During the propulsive phase (Figure 5-43), the hand is on the substrate, and the whole limb is drawn posteriorly and somewhat medially; the elbow and wrist joints are extended, and the scapula is rotated to a vertical position. Once the foot is lifted off the substrate, the recovery phase has begun. The digits and elbow and shoulder joints are flexed and the whole limb is drawn forward and laterally (Figure 5-43); the scapula is rotated in an oblique position, with the glenoid cavity facing anteroventrally.

We will review the major muscle actions that are responsible for the execution of the propulsive phase in the opossum. The latissimus dorsi, arising from the lumbodorsal fascia and ribs to the posterior aspect of the humerus,

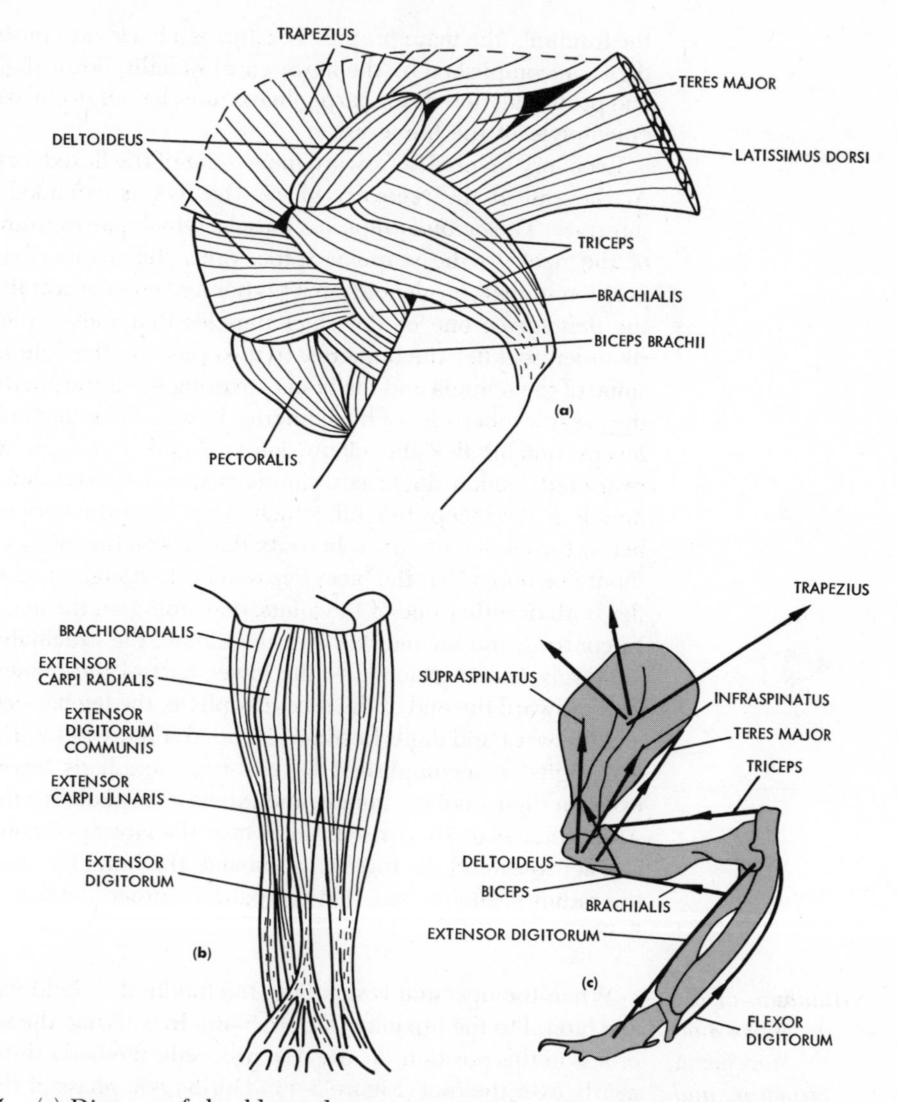

Figure 5-45 (a) Diagram of shoulder and upper arm muscles of an opossum. Left lateral view, superficial layer only; (b) diagram of forearm muscles of the extensor aspect of an opossum. Tendons to the fingers not drawn; (c) diagram of forelimb skeleton with the principal direction of pull of the major muscles (left lateral view).

draws the entire limb back. This action is made possible because the shoulder, elbow, and wrist joints are either extended or stabilized by multiple muscle actions. The scapula is rotated into a vertical position by action of the characteristically mammalian *teres major* muscle, which represents a separate slip of the reptilian latissimus dorsi that has gained contact with the scapula. During the activity of the latissimus dorsi and teres major muscles, passive flexion at the shoulder joint is prevented by action of the subscapularis, supraspinatus, and coracobrachialis, which pass from the coracoid process to the humeral shaft. The elbow joint is strongly extended by the *triceps* muscle, whose three upper origins include a long head from the scapula and lateral and medial heads from the humerus. At the lower end, the tendon of the muscle is inserted into the olecranon process, which projects from the ulna. In the distal part of the forelimb, the main propulsive effort is a backward push of the forearm and digits, accomplished by the flexor carpi radialis, flexor digitorum superficialis, and profundus and flexor carpi ulnaris muscles, all originating from the medial epicondyle of the humerus.

Recovery involves a forward movement of the flexed forelimb (Figure 5-43). At the end of the recovery phase, the foot is extended and placed on the substrate. The *acromiotrapezius* muscle, which passes from the mid-dorsal line of the neck to the scapular spine, pulls the scapula back into an oblique position so that the glenoid cavity is pointed anteroventrally. At the same time, the *deltoideus,* one of the chief muscles that moves the whole arm at the shoulder, will flex the humerus. In the opossum, the deltoideus arises from the spine of the scapula and the fascia covering the infraspinatus and is inserted on the greater tuberosity of the humerus. Lower down the limb, the brachialis and biceps brachii flex the elbow joint (Figure 5-45), in which movement is restricted about a single axis: simple flexion and extension. The anterior flexor muscle is the biceps brachii, which arises from the coracoid process to insert below the elbow into the tuberosity that lies on the medial side of the radius. It should be noted that the biceps crosses both shoulder and elbow joints and can flex both or either one of the joints, depending on the action of other muscles. In contrast, the second flexor of the elbow, the brachialis, crosses the elbow joint only and is attached to the lower part of the humerus and to the ulna. Thus, toward the end of the recovery phase, the leg has been brought forward, and the wrist and digits become extended (Figure 5-43). Extension of the wrist and digits is accomplished by the brachioradialis, extensor carpi radialis, extensor digitorum communis, and extensor carpi ulnaris muscles (Figure 5-45). All of these extensors arise from or near the lateral epicondyle of the humerus and act to extend the forearm and hand; the brachioradialis assists in rotating the radius in such a way that the palm is turned dorsally (supination) (Figure 5-49).

Hindlimb of the Opossum: Movement, Structure, and Function

When the opossum is standing, the hindlimb is held with the foot anterior and lateral to the hip joint (Figure 5-46). In walking, the foot is set down more or less in this position. As the step proceeds, the body shifts to a position more nearly over the foot (Figure 5-46). During one phase of the locomotory cycle, the foot (pes) is positioned approximately below the hip joint, a feature characteristic of all mammals studied. The locomotory thrust is, therefore, transmitted along a parasagittal plane.

The attachment of the pelvic girdle to the vertebral axis by means of modified sacral-transverse processes and ribs allows very little movement. The ilium and ischium are characteristically well developed in quadrupedal mammals (Figure 5-46). The expanded ilia and ischia serve as attachments for the muscles that stabilize the balance of the body on the leg and propel it forward and backward. Between the ilium and ischium is the acetabulum, and the pubis forms a narrow tie number across the midventral line.

The head of the *femur* is carried on a long neck that holds the bone out at a distance from the body and allows a swinging motion of the leg. At the upper end, the femur (Figure 5-46) bears the *great trochanter* on the lateral surface and the *lesser trochanter* on the medial aspect. The lower end is expanded into lateral and medial condyles, each with a facet for articulation with the tibia and fibula. In the lower leg, the tibia is always the better developed one (Figure

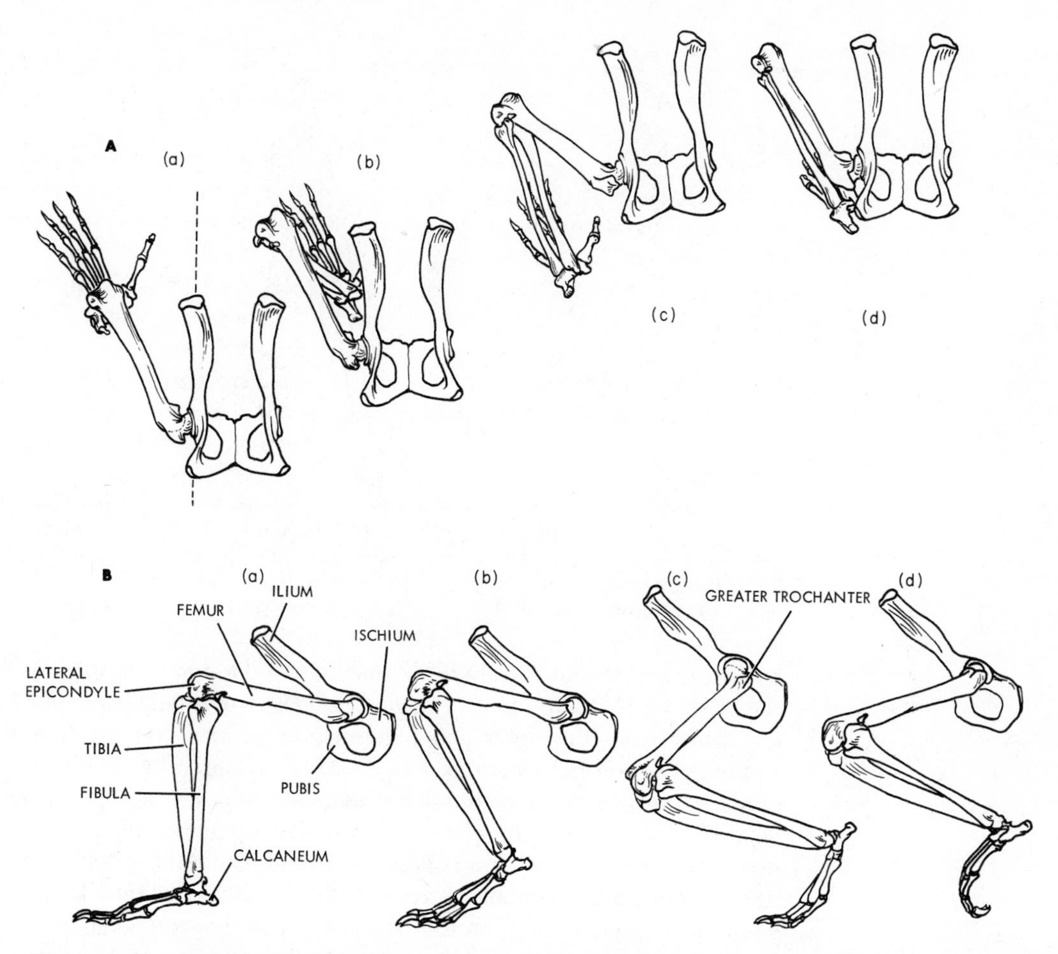

Figure 5-46 Left hindlimb skeleton and pelvic girdle drawn from X-ray pictures of the walking opossum. Four stages of a walking step are shown. The broken line is parallel to the long axis of the body. A, dorsal view; B, lateral view. (From Jenkins, *Journal of Zoology*, 1971.)

5-46) and lies on the medial side. In the opossum, the fibula is well developed (Figure 5-46), although it is reduced in many other mammals. In the ankle joint, we can recognize several characteristically mammalian innovations. The lateral proximal tarsal (fibulare) has become the calcaneum (Figure 5-47), having a prominent and characteristic backward prolongation, the heel, to which are attached the great calf muscles (gastrocnemius, soleus, and plantaris). The medial proximal tarsal (tibiale), fused with the intermedium, has become the astragalus (Figure 5-47), which is typically supported entirely by the calcaneum. In mammals, the astragalus provides the main articulation with the tibia, allowing movement only in one plane. The upward movement of the foot is known as *dorsiflexion,* the downward one as *plantar flexion.*

As in all quadrupeds, the posterior group of muscles around the hip joint is especially large and serves to draw the limb caudally, giving the main locomotor thrust. Included among this group are the *gluteal muscles* (maximus, medius, and minimus) that arise from the sacrum and ilium (Figure 5-48) and insert on the upper femur. These muscles, acting from the femur as a fixed

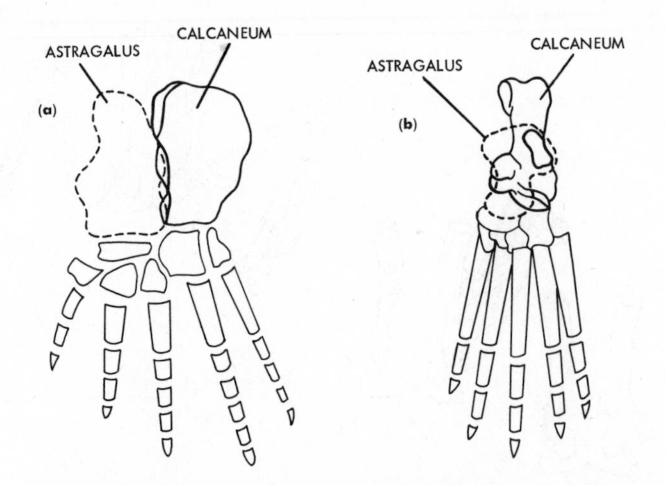

Figure 5-47 Dorsal view of a left foot of (a) pelycosaur and (b) a mammal. The calcaneum is drawn in solid heavy line, the astragalus in dashed heavy line. In pelycosaurs the calcaneum and astragalus were positioned side by side. In mammals the astragalus is typically supported entirely by the calcaneum. (From Jenkins, 1970, courtesy of *Evolution*.)

point, prevent the body from falling medially. Functioning as medial braces of the hip joint are the *adductor* (magnus, longus, brevis) and *pectineus* muscles, all arising from the pubis and ischium and inserting on the femur. Their combined action prevents the body from falling laterally. The more cranial brace of the hip joint includes the *iliopsoas* muscles originating from the ventral surfaces of the ilium and of the lumbar vertebrae and inserting on the lesser trochanter of the femur. They act to stabilize the front of the hip joint. At this point, it is important to focus on the *tensor fascial latae* muscle, which has a long course from ilium to tibia. The *fascia lata* is thickened laterally to form a very strong and broad band of connective tissue on the side of the thigh. This band of fascia, kept tight by the muscle at its upper end, acts as a brace that distributes the weight of the body symmetrically on the femur. As in the forelimb, we can recognize both propulsive and recovery phases during hindlimb movements.

We will first discuss the mechanics of the propulsive phase. During this phase, the leg is pulled backward and extended, thrusting the body forward (Figure 5-46). A caudal group, the "hamstrings," crossing at the backs of both hip and knee joints, changes the angle at which the whole leg is held relative to the body and, thus, produces a main part of the forward thrust. To give these muscles good leverage, the ischium is prolonged backward and the internal muscle architecture essentially is parallel-fibered. All four hamstrings, the *biceps femoris, gracilis, semitendinosus,* and *semimembranosus* (Figure 5-48), arise from the ischium. The former inserts on the lower leg via the fascia lata, and the remaining three insert on the tibia. As mentioned before, during propulsion, these muscles are aided by action of the gluteal muscles. In order for the leg to function as a strut, the knee joint (crossed posteriorly by the hamstrings) must be kept from bending. Such an extending force is generated by the pinnate-fibered *quadriceps femoris,* which is composed of four parts (Figure 5-48): the *rectus femoris, vastus medialis, lateralis,* and *intermedius.* The former arises from the femur. All parts insert into the patellar tendon, a

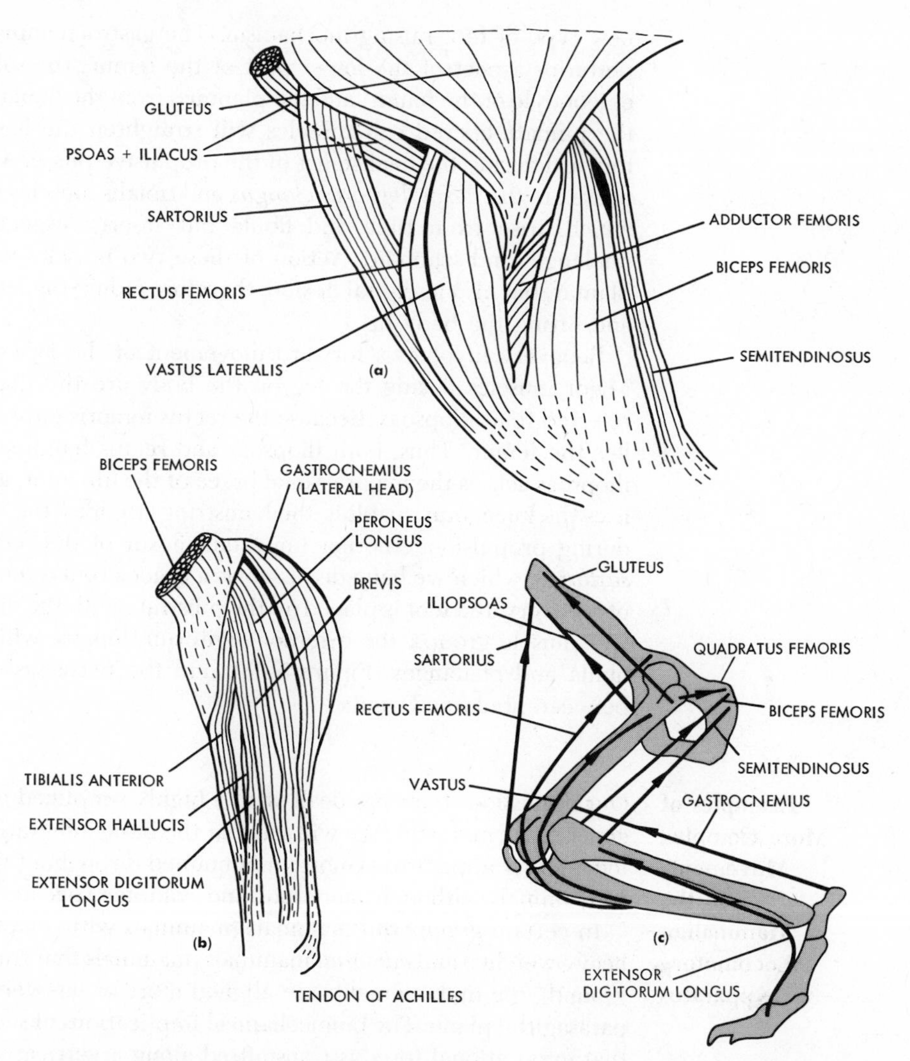

Figure 5-48 (a) limb muscles of the pelvis and thigh of the opossum seen from the left lateral side, superficial layer only; (b) muscles of the lower leg of the opossum seen from an anterolateral aspect; (c) hindlimb skeleton and pelvic girdle with the major direction of pull of the principle muscles.

broad band reaching across the front of the knee and containing a tendon or sesamoid bone, the *patella*, or kneecap. The functional significance of the sesamoid is to increase the working distance of the muscle from the center of rotation of the joint. The patellar ligament proceeds downward to be inserted on the front of the tibia, allowing the quadriceps to produce a large arc of movement of the tibia on the femur. With the changed limb posture and reduction of the tail, in mammals, the originally powerful reptilian leg retractor, the caudofemoralis, has been reduced to a variable and small *pyriformis* (Figure 5-48). Forward thrust is also generated by the calf muscles, gastrocnemius, soleus, and plantaris (Figure 5-48). These muscles no longer extend to the sole of the foot as in reptiles, but instead insert via the tendon of Achilles in the heel tuber (the backward prolongation of the calcaneum), resulting in a new type of foot-raising mechanism. The gastrocnemius originates from the posterior aspect of the lower part of the femur; the soleus from the lateral epicondyle of the femur; and the plantaris from the fibula. Combined action of the pinnate-fibered calf muscles will straighten the leg and push the body forward during the final stages of the propulsive phase. Also active during this phase are the *flexor digitorum longus* and tibialis muscles (Figure 5-48), both of which arise from tibia and fibula but insert, respectively, on the distal phalanges and scaphoid. Action of these two muscles will result not only in plantar, but also in digital flexion, therefore aiding the lengthening of the limb and lifting the body up.

Recovery involves a forward movement of the flexed limb (Figure 5-46). Major muscles flexing the leg on the body are the iliacus and psoas, often referred to as iliopsoas. Because the rectus femoris crosses the hip joint, it can flex the femur. Thus, both iliopsoas and rectus femoris have dual functions: iliopsoas acts as the major cranial brace of the hip joint, and the rectus femoris fixes the knee joint enabling the hamstrings to draw the whole limb backward during propulsion. Another important flexor of the femur is, of course, the *sartorius*, which we have discussed as a brace around the hip. Toward the end of recovery, the foot is placed on the substrate with the digits fully extended by two muscle groups: the extensor digitorum longus, which runs between the fibula and phalanges (Figure 5-48), and the peroneus complex, which runs between fibula and digits.

Examples of More Complex Movements in the Mammalian Locomotory Apparatus

Our discussion so far has dealt with a highly simplified model of walking in a generalized mammal. We will see that the basic integration of the mammalian locomotory apparatus is mostly maintained throughout the adaptive radiation of mammals, although numerous and complex specializations have evolved.

In certain *graviportal* mammals (mammals with adaptations for support of heavy weights) and *cursorial* mammals (mammals that travel far and fast on the ground), the limb elements are aligned more or less vertically and move in a parasagittal plane. The biomechanical implications of such an architecture are that gravitational force is transmitted along a vertical strut, and locomotory thrust is generated along a parasagittal plane (Figure 5-44). In cursorial mammals, the effective lengths of the limbs functioning as levers are increased, and a great deal of energy is saved by tendon elasticity. Kinetic and potential energy lost at one stage during a step may be stored as elastic energy in a stretched tendon and restored to the body in an elastic recoil.

In arboreal primates and man, there is a rotating movement of the radius and ulna, giving the hand an increased mobility. The radius carries the hand and is able to rotate with it around the ulna from the prone position in which the radius lies across in front of the ulna (Figure 5-49). Rotation of the radius to a position parallel to the ulna is termed supination. During supination, the palm of the hand is turned so that it faces dorsally or forward (in man). Movement in the opposite direction is called pronation. Supination and pronation enable the individual to bring the hand into a great variety of positions and to exert a powerful twisting action.

One of the more spectacular modifications of the forelimbs occur in bats (Chiroptera). Flight in bats depends on a wing membrane stretched between

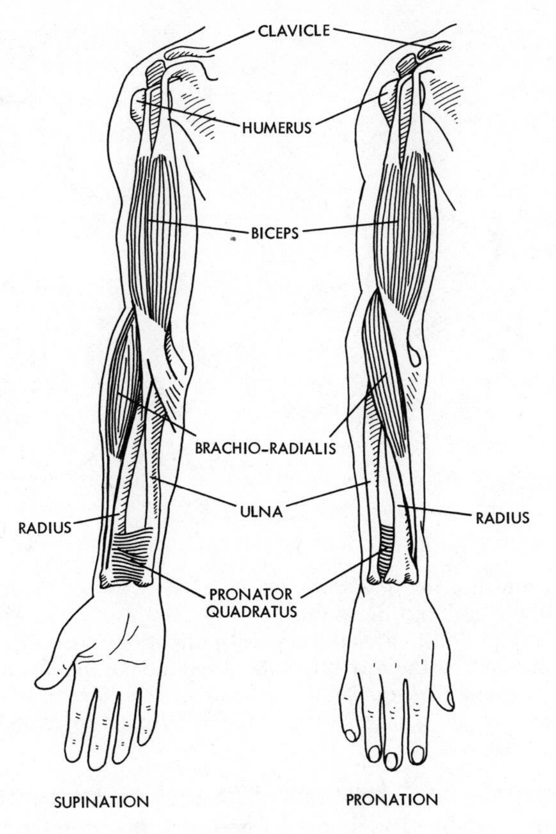

Figure 5-49 Positions of the human ulna and radius during supination and pronation. Note how the distal end of the radius has rotated partially as a result of the action of the pronator quadratus muscle.

limbs and body. In the hand, the second to fifth digits are extremely elongated, with the membrane running between them (Figure 5-50a). The ulna is rudimentary; the proximal end of the radius forms a hinge joint with the humerus. Distally, the radius forms a hinge joint with the carpus. Both elbow and wrist joints are hinges that can bend only in the plane of the wing. Consequently, the forearm cannot be twisted. The extensor carpi radialis has shifted its insertion on the metacarpal of the second digit, swinging it forward. Because of a ligament between the second and third digits, the forward swinging second digit will keep the third (longest) digit extended. At the same time, other muscles pull the hindlimb medially. Opposing forces on the third digit and the hind limb set up tensions in the wing membrane. See Figure 5-50(a) for lines with open arrowheads. Substantial lift is produced by the downstroke, during which the wing membrane balloons upward between the digits. The downward stroke is caused by action of the hypertrophied pectoralis muscle [Figure 5-50(b)], which originates partially from an enlarged sternum with a small keel. The deltoideus, subdivided in several distinct heads, is prominently developed and is mainly responsible for the upstroke. Thus, the structural basis underlying the flight mechanism of bats still conforms to the characteristically mammalian integration.

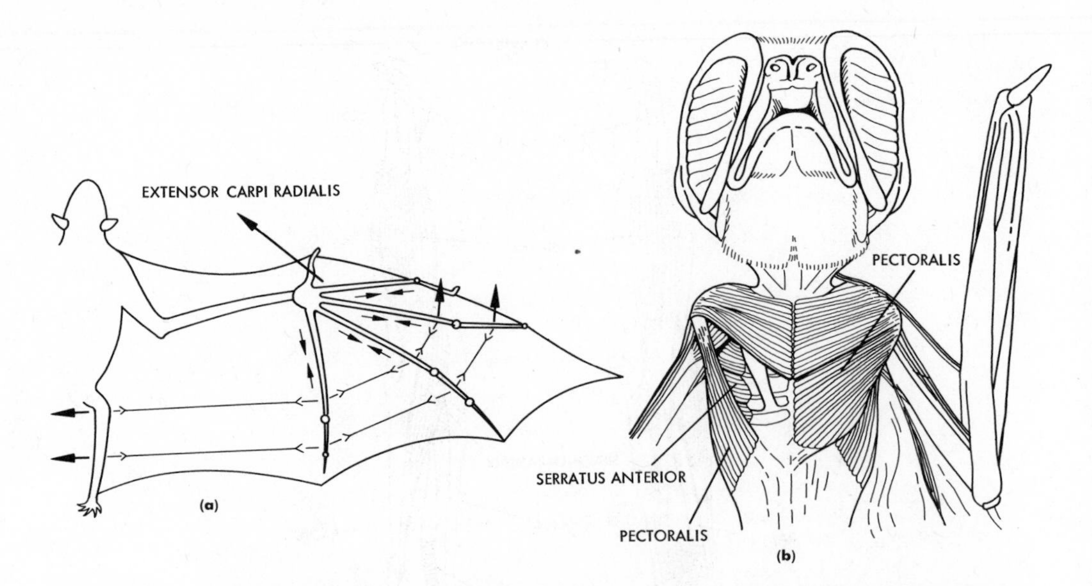

Figure 5-50 (a) Diagram showing how the wing of a bat is spread by action of the extensor carpi radialis pulling the second digit in the direction of the large arrow. The hindlimb is pulled in a nearly opposite direction, creating tension in the directions indicated by the arrows on the wing membrane. (From Pennycuick, *Journal of Experimental Biology,* 1971.) (b) Some of the major muscles active in the movements of the wing of a bat. (Modified from Vaughan, 1970, courtesy of Academic Press.)

The Vertebrate Head Skeleton: Its Basic Components

The vertebrate head skeleton exhibits great structural and functional diversity. Originally, the head skeleton is composed of three basic components (neurocranium, splanchnocranium, and dermatocranium) that are at first rather distinct, both in their phylogeny and ontogeny.

The *skull,* or *cranium,* or *neurocranium proper,* began in the early chordates ancestral to vertebrates as a simple structure protecting and supporting the brain. When three pairs of sensory capsules (nasal, optic, and otic) were added, the neurocranium was formed. In its cartilaginous condition the neurocranium is often referred to as chondrocranium. Vertebrates that do not have bony heads (cyclostomes, chondrichthyes) have well-developed chondrocrania. Most vertebrates that do have bony heads retain a chondrocranium only in larval or fetal life. In the adults of these vertebrates, most of the cartilaginous structure is replaced by bones that ossify within the cartilage. Part of the bony neurocranium is, therefore, preformed in cartilaginous form.

The *splanchnocranium* is generally defined as the skeleton developing in relationship to the pharynx, which is the expanded part of the digestive tube lying between the oral cavity and the esophagus. The splanchnocranium encompasses the skeletal elements supporting the jaws, tongue, gill arches, and gill-arch derivatives. It is composed of seven paired, visceral arches, of which the first one is the *mandibular arch* and the second is the *hyoid arch.* The next five form the gill arches.

The *dermatocranium,* or *dermocranium,* represents all dermal bony elements developing in the dermis of the outer headskin and the dermis underlying the oral and pharyngeal epithelium. It must be emphasized that the dermatocranium makes up an important proportion of the head skeleton, except, of course, in chondrichthyeans and living agnathans.

Bony Elements in the Head Skeleton of a Teleost

Teleosts possess a large number of bones.

BONES CONTRIBUTED TO THE NEUROCRANIUM PROPER BY OSSIFICATION OF CARTILAGE. Figure 5-51 shows an *occipital* group: *supraoccipital,* SOC, forming the roof of the foramen magnum; *exoccipitals* (EOC) bordering the foramen magnum on either side; and *basioccipital* (BOC), forming the floor of the foramen magnum; the *sphenoid* region: a median *basisphenoid* (BS) associated with the pituitary glands. On either side of the basisphenoid are the *pleurosphenoids; orbitosphenoids* present in most vertebrates are absent in this advanced teleost; the *ethmoid* region includes the *mesethmoid* (E) and *lateral ethmoids* (LE), which form the anterior border of the neurocranium.

BONES CONTRIBUTED BY THE SENSORY CAPSULES BY OSSIFICATION OF CARTILAGE. In the *otic* region are two uniformly occurring ossifications, an anterior *prootic* (PO; Figure 5-51) and a posterior *opisthotic.* More dorsally is the *epiotic.* Neither the optic nor olfactory capsules contribute any bones to the neurocranium.

OSSIFICATIONS IN THE CARTILAGINOUS SPLANCHNOCRANIUM. Two bones ossify in the palatoquadrate cartilage of the teleosts. The *quadrate* (Q) serves as an articulation of the skull with the lower jaw, and the *metapterygoid* (MPT) forms one of the elements making up the bony palate. Two ossifications occur in Meckel's cartilage (representing the embryonic lower jaw): the *articular* (A) serves to articulate the lower jaw with the skull at the quadrate of the skull, and a *retroarticular* appears at the back of the lower jaw. The *symplectic* (SY) and *hyomandibular* (HM), both are involved in suspending the jaws, and the *hyoid* supports the tongue. Several separate elements can be recognized in the hyoid (Figure 5-52): the *interhyal* (IH), *epihyal* (EH), *ceratohyal* (CH), *hypohyals* (HHD), and *basihyal* (BH). The next successive five arches are the typical gill arches, of which the first four are usually gill bearing. Each gill arch includes on either side a major dorsal and ventral element (Figure 5-52). The dorsal elements are the *epibranchials* (EB), and the ventral ones are the *ceratobranchials* (CB). There are generally *pharyngobranchials* (PB) turned inward *over* the pharynx. Below the ceratobranchial are short *hypobranchials* (HB) and median *basibranchials* (BB). The fifth gill arch, devoid of gills, is reduced to a ventral element, the *ceratobranchial* (CB5) (Figure 5-52).

Added to the neurocranium and splanchnocranium are numerous dermal bones that collectively are called the dermatocranium. Forming the roof of the skull are from front to rear: the *nasal, frontal,* and *parietal* (PA) (Figure 5-51). Closely associated with the orbit are the *lacrimal* (L), *prefrontal* (PF), and *circumorbitals* (SO). Toward the side of the skull and behind the eye are the *sphenotic* (SPH) and *pterotic* (ST) bones. Fish possess a protective cover over the gill region (Figure 5-51) composed of a *preopercular* (POP), *opercular* (OP), *subopercular* (SOP), and *interopercular* (IOP). The floor of the skull (Figure 5-51) is formed by the *vomer* (PV) and *parasphenoid* (PS). Completing the palate (Figure 5-51) are the *palatine* (P), *ectopterygoid* (ECT) and *entopterygoid* (ENT). Dermal additions to the jaws (Figure 5-51) include the *premaxilla* (PM) and *maxilla* (MX), representing the functional upper jaw; in the lower jaw the *dentary* (D) appears. Tooth plates of dermal origin often are added to various bony elements. For example, tooth plates are added to the ceratobranchials of the fifth arch and the pharyngobranchials of the second, third, and fourth arches. Once equipped with a battery of teeth, these bones function as upper and lower pharyngeal jaws (Figure 5-52 and 5-58).

Basic Components of the Head Musculature

We can recognize four groups (Figure 5-53) of head muscles: (1) intrinsic eye muscles; (2) extrinsic eye muscles; (3) hypobranchial muscles; and (4) branchiomeric muscles.

Intrinsic eye muscles, derived from myotomes, include the iris and ciliary muscles. Other derivatives from myotomes are the extrinsic eye muscles, superior rectus, inferior rectus, internal rectus, external rectus, inferior oblique, and superior oblique. All extrinsic eye muscles occur in all nonblind vertebrates in a rather constant configuration.

HYPOBRANCHIAL MUSCLES (Figures 5-53 and 5-54). These muscles also are derived from embryonic myotomes. They run on the ventral side of the head between the lower jaw and hyoid and pectoral girdle. Running between the hyoid and lower jaw is the *geniohyoideus* muscle (GH) (Figure 5-54). The *sternohyoideus* (SH) (often, and more appropriately, called the rectus cervicis) passes from the pectoral girdle to the hyoid.

BRANCHIOMERIC MUSCLES (Figure 5-54). These muscles develop unlike other skeletal muscles from the ventral subdivision of the anterior visceral mesoderm. Because of this visceral derivation, the gill arch muscles are innervated by special visceral, rather than by somatic, nerves. Most remaining head muscles belong to the branchiomeric category. Before reviewing the muscles associated with the jaws and gill arches in a teleost, we must understand the muscular components of a single gill arch in a shark (Figure 5-54). The major components include (1) *dorsal constrictors* (dorsal to the level of the gill slits); (2) *ventral constrictors* (ventral to the level of the gill slits), which surround the pharynx and serve to compress it; (3) *levators*, which are attached to the dorsal ends of the skeletal arches and elevate the gills; (4) *adductors*, which bend each arch at the middle on each side; (5) *interarcuals*, which similarly bend the upper ends of the skeletal arches backward; and (6) *subarcuals* and *epiarcuals*, which act on elements of adjacent gill arches on, respectively, the ventral and dorsal aspects.

In a representative teleost, we may recognize the following derivatives of the constrictor dorsalis of the mandibular arch, all innervated by the trigeminal nerve: the *adductor mandibulae muscles* (AM) (Figure 5-54), composed of several distinct heads, all associated with the jaws; the *levator arcus palatini*

Figure 5-51 The head skeleton of a representative advanced teleost. (a) lateral aspect of intact skull; (b) lateral aspect of neurocranium; (c) dorsal aspect of neurocranium; (d) lateral aspect of "palate" or suspensory apparatus, with attached opercular series and mandible, upper jaw disarticulated. A, articular; BOC, basioccipital; BS, basisphenoid; D, dentary; E, ethmoid; ECT, ectopterygoid; ENT, entopterygoid; EO, epiotic; EOC, exoccipital; F, frontal; HM, hyomandibula; IOP, interoperculum; L, lachrymal; LE, prefrontal; MPT, metapterygoid; MX, maxilla; N, nasal; OP, opercular; P, palatine; PA, parietal; PM, premaxilla; PO, prootic; POP, preopercular; PS, parasphenoid; PV, vomer; Q, quadrate, SO, circumorbital; SOC, supraoccipital; SOP, suboperculum; SPH, sphenotic; ST, pterotic; SY, symplectic.

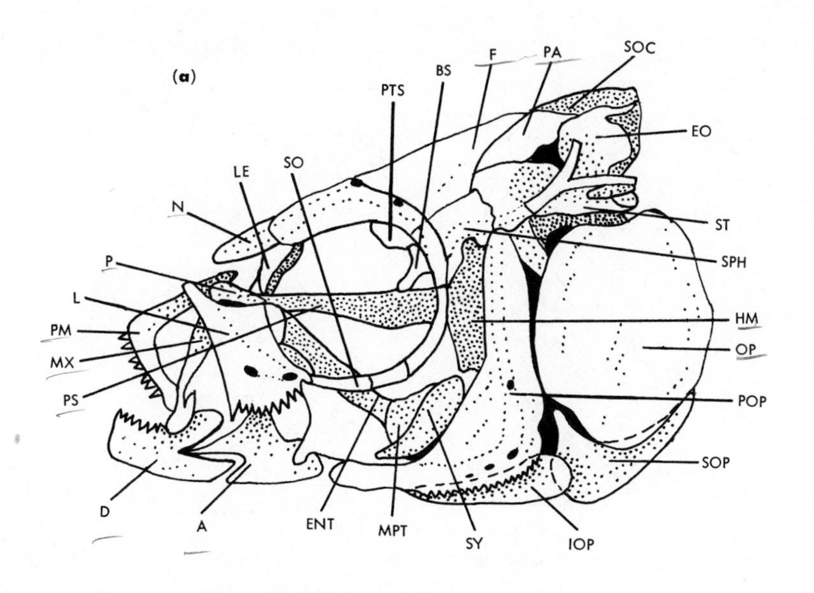
(a)
F
PA
SOC
BS
PTS
EO
LE
SO
N
ST
SPH
P
L
HM
PM
OP
MX
PS
POP
SOP
D
A
ENT
MPT
SY
IOP

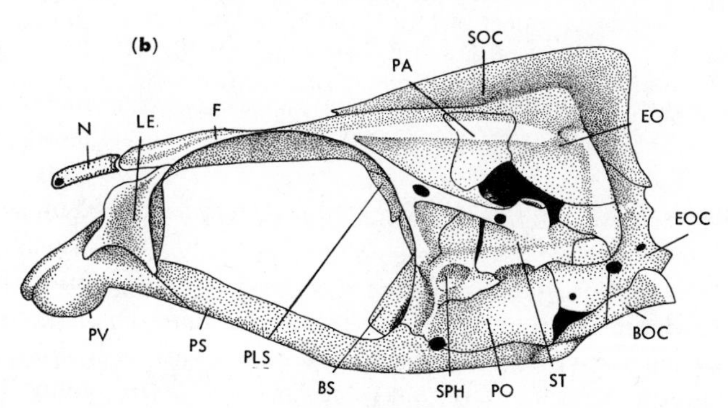
(b)
SOC
PA
N
LE
F
EO
EOC
PV
PS
PLS
BS
SPH
PO
ST
BOC

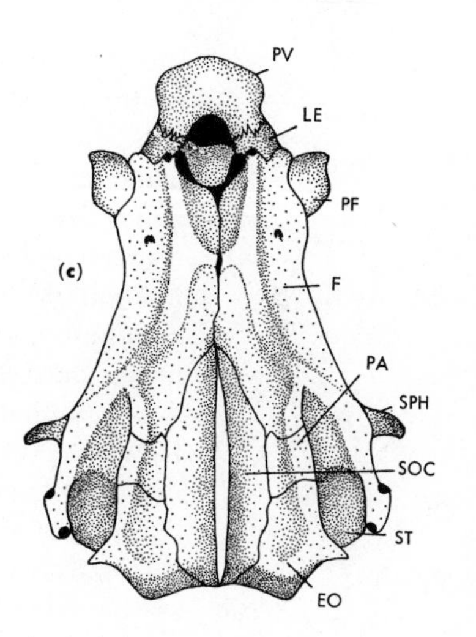
PV
LE
PF
(c)
F
PA
SPH
SOC
ST
EO

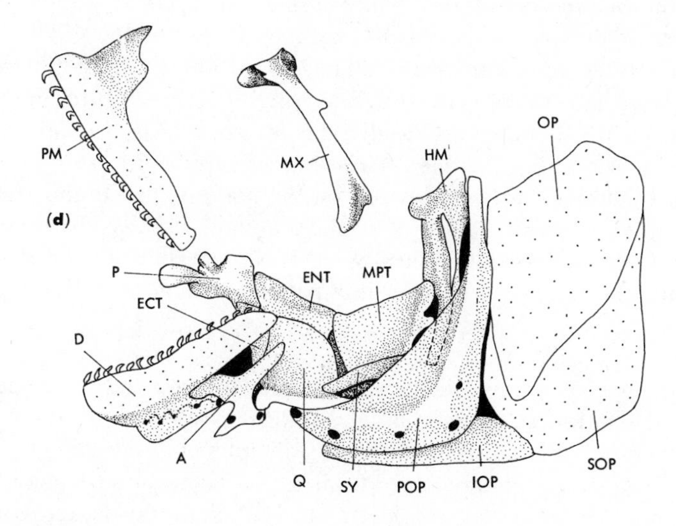
PM
MX
HM
OP
(d)
P
ENT
MPT
ECT
D
A
Q
SY
POP
IOP
SOP

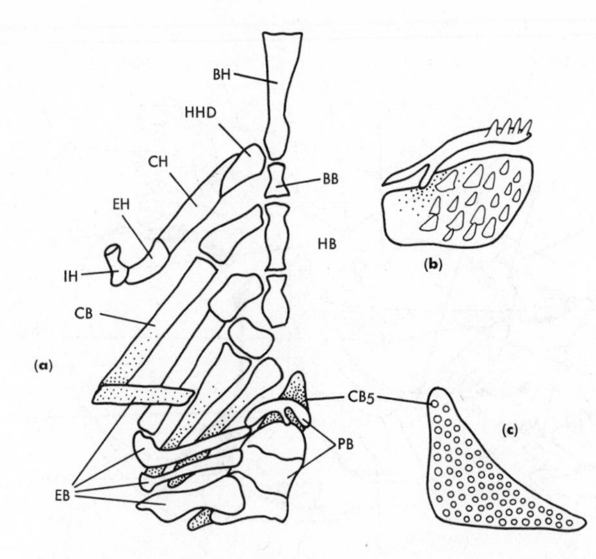

Figure 5-52 (a) Dorsal aspect of left half of hyobranchial skeleton of a representative teleost. (b) ventral aspect of pharyngobranchials with tooth plates, together forming the upper pharyngeal jaw; (c) dorsal aspect of fifth ceratobranchial with toothplate, together forming the lower pharyngeal jaw. BB, basibranchial; BH, basihyal; CB, ceratobranchial; CH, ceratohyal; EB, epibranchial; EH, epihyal; HB, hypobranchial; HHD, dorsal hypohyal; IH, interhyal; PB, pharyngobranchial.

(LAP) (Figure 5-54), which pulls the palate sideways to expand the buccal cavity; and the *dilatator operculi* (DO), which plays a key role in spreading the gill cover to create a suction. Derivatives of the constrictor dorsalis of the hyoid arch, all of which are innervated by the facial nerve, include the *adductor arcus palatini* (AAP) and *adductor operculi* (AO), compressing, respectively, the buccal and gill cavity and the *levator operculi* (LO), which lifts the gill cover. An important derivative of the constrictor ventralis of the hyoid arch is the *hyohyoideus* (HH), which also is innervated by the facial nerve and plays a key role in the respiratory movements of the gill cover.

Most branchiomeric components are well developed in the teleost gill arch. However, the constrictors are reduced to very small muscles controlling the movements of the gill filaments. Dorsally the *levatores* (LE, LI) are well differentiated (Figure 5-55); in some arches they are split into external and internal heads. Epiarcuals are represented by the *obliquus* (OD) and *transversus dorsalis* (TVD) muscles (Figure 5-55). Adductors usually are present between epi and ceratobranchial. Ventrally, the subarcuals become differentiated into obliqui and transversi muscles in a pattern analogous to the dorsal components.

The Determination of Muscle Function As Applied to the Teleost Head

With the emergence of modern electronics, rapid progress has been made recently in the field of functional morphology of the musculoskeletal system. It is, therefore, necessary to understand the basic principles of *electromyography*. When a chemical transmitter is received from the nerve at the surface of the motor fiber, it induces a secondary electrical event that travels across the muscle cell membrane to trigger the shortening responses of bundles of filaments. The over-all summated electrical event in the many muscle fibers is

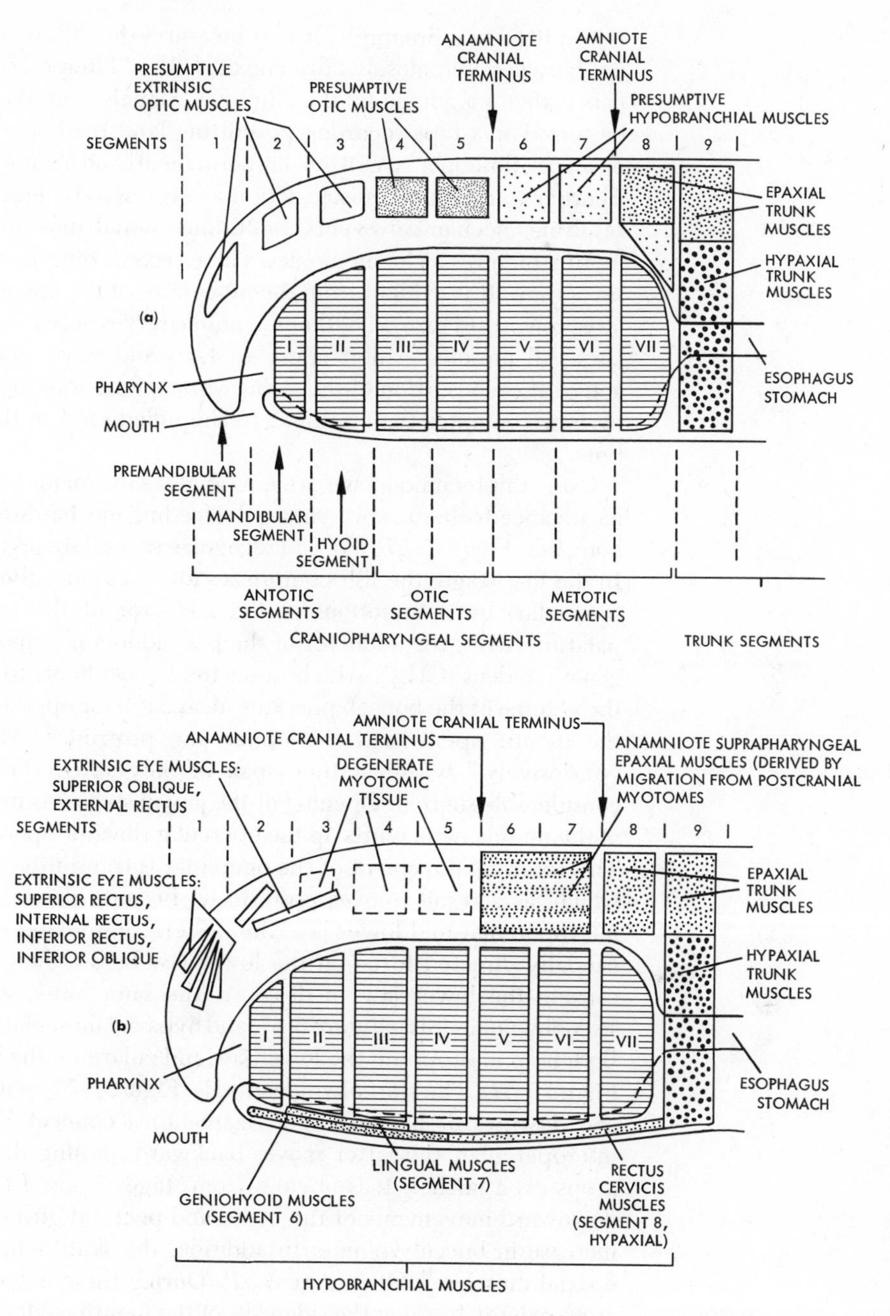

Figure 5-53 Diagrammatic phantom lateral view of craniopharyngeal segments showing (a) embryonic relationships and (b) adult relationships. (From H. M. Smith, 1960, and courtesy of Holt, Rinehart and Winston, Publishers.)

of sufficient magnitude to create localized voltage changes in the muscle. Electromyography (Figure 5-56) is the technique to record these electrical events. A contracting muscle will "fire" an electrical event that spreads through the adjacent tissues and induces voltage change between electrodes placed in or on such tissues. The electrical event picked up by the electrode is transmitted to a preamplifier that measures the difference in electrical potential between two closely adjacent sites. The "filtered" signal of the preamplifiers is then magnified by the amplifier. Signals from the amplifiers can either be stored on a tape recorder, permitting later readout; observed immediately on an oscilloscope; or written out permanently on a paper record (Figure 5-56). In order to correlate muscle action as expressed by electrical events with the resulting mechanical events, multidimensional movements are recorded by motion pictures or X-ray movies. These records must be synchronized with the myograms. It is evident from the myograms of the jaw muscles of a representative teleost (Figure 5-56) that no muscle serves only for opening or closing the jaws. All muscles exhibit complex starts and stops, and several muscles are active at each portion of the cycle. Muscle functions in general are, therefore, far more complex than outlined in simplified form in the treatise on locomotion.

Using this technique, we can now analyze the major biomechanical events in an advanced teleost representing the feeding mechanism of a typically aquatic bony fish (Figure 5-57). We can recognize several stages during a feeding cycle. In the first stage, the fish compresses all its cavities (buccal, pharyngeal, and opercular) by contractions of the adductor of the palate (adductor arcus palatini, AAP), the adductor of the jaw (adductor mandibulae, AM_3), and the geniohyoideus (GHA), which raises the buccal floor. In the next three stages, the volume of the buccal, pharyngeal, and gill (or opercular) cavities increases, the mouth opens, and the upper jaw protrudes. All these events occur "explosively," within a time span of only 110 milliseconds, and create a considerable suction. Opening of the jaw in stage 2 is mainly caused by action of the muscle, which lifts up the opercular (levator operculi, LO, Figure 5-57). This rotating movement of the opercular is transmitted to the interopercular, which, as a result, moves posteriorly. Because there is a ligament between interopercular and lower jaw, the posteriorly moving interopercular exerts a caudally directed force on the lower jaw below the jaw joint (Figure 5-57) causing the lower jaw to drop. At the same time, we see activity in the sternohyoideus (SH) (Figure 5-57) and hypaxial musculature (HY) (Figure 5-57). Both help in dropping the lower jaw and enlarging the buccal cavity (stage 3, Figure 5-57). The sternohyoideus (SH) (Figure 5-57) pulls the hyoid backward and downward. Because of a ligamentous connection between hyoid and interopercular, the latter moves backward, pulling the lower jaw so that it drops even farther. It is obvious from stages 3 and 4 that the backward and downward movements of the hyoid and pectoral girdle result in a significant increase in buccal volume. In addition, the skull is lifted by activity in the epaxial muscles (EM) (Figure 5-57). During these events, the palatoquadrate arch (palate), forming the sidewalls of the mouth cavity, is spread apart by the

Figure 5-54 (a) Diagrammatic representation of the musculature of a single gill arch in a representative shark; (b) lateral view of head musculature of a representative teleost; (c) ventral view of head musculature of a representative teleost. AAP, adductor arcus palatini; $AM_{1\text{-}3}$, adductor mandibulae; BSR, branchiostegal ray; D, dentary; DO, dilatator operculi; EP, epaxial muscles; GHA, GHP, geniohyoideus; HH, HHI, HHS, hyohyoideus; IOP, interoperculum; LAP, levator arcus palatini; LIM, ligament between interoperculum and mandible; LO, levator operculi; MX, maxilla; OP, operculum; PM, premaxilla; SCL, supracleithrum; SH, sternohyoideus; SOP, suboperculum; UH, urohyal of hyoid.

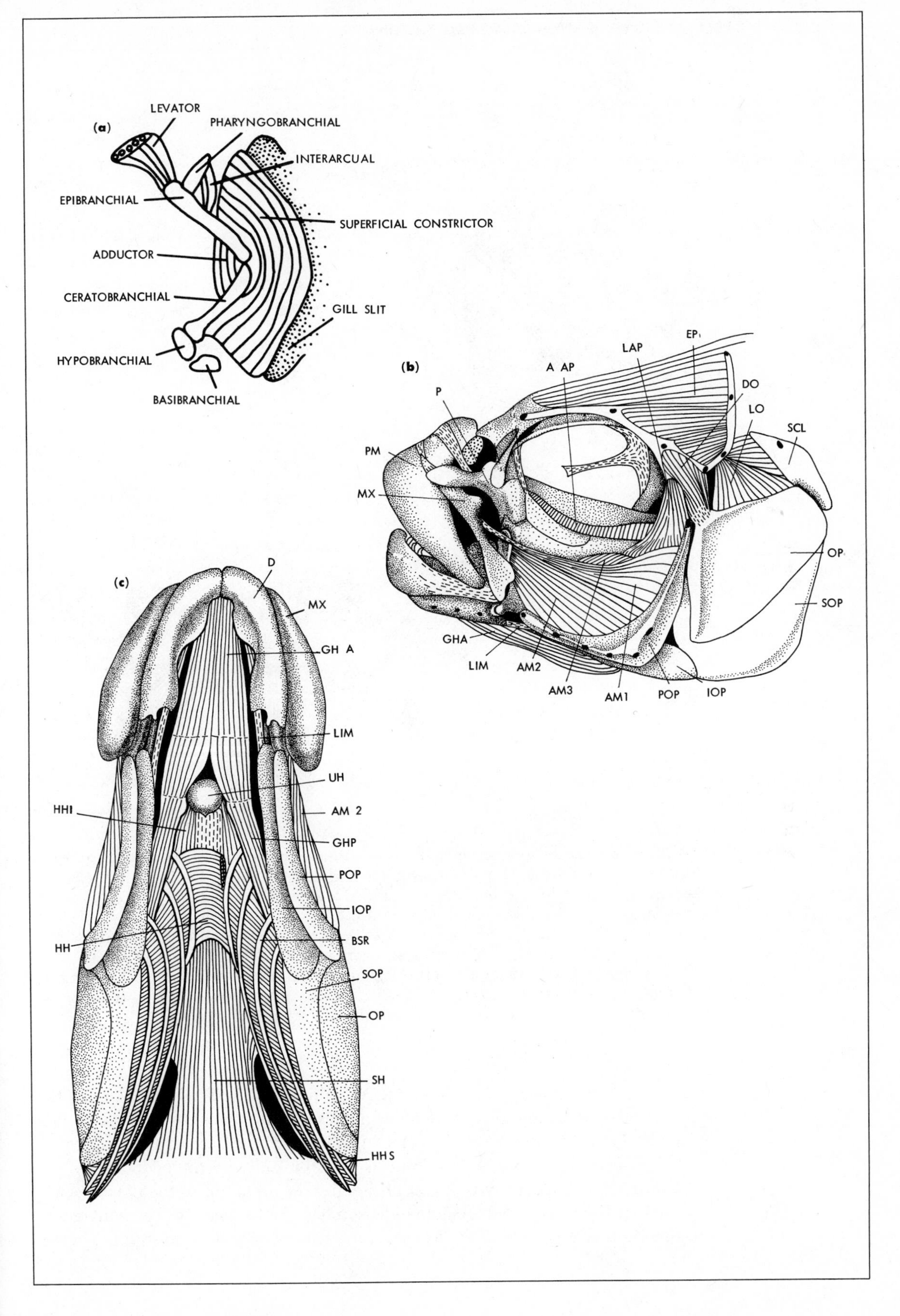
(a)
LEVATOR
PHARYNGOBRANCHIAL
INTERARCUAL
EPIBRANCHIAL
SUPERFICIAL CONSTRICTOR
ADDUCTOR
CERATOBRANCHIAL
GILL SLIT
HYPOBRANCHIAL
BASIBRANCHIAL
(b)
EP
LAP
A AP
DO
P
LO
SCL
PM
MX
OP
SOP
GHA
LIM
AM2
AM3
AM1
POP
IOP
(c)
D
MX
GH A
LIM
UH
HHI
AM 2
GHP
POP
IOP
HH
BSR
SOP
OP
SH
HHS

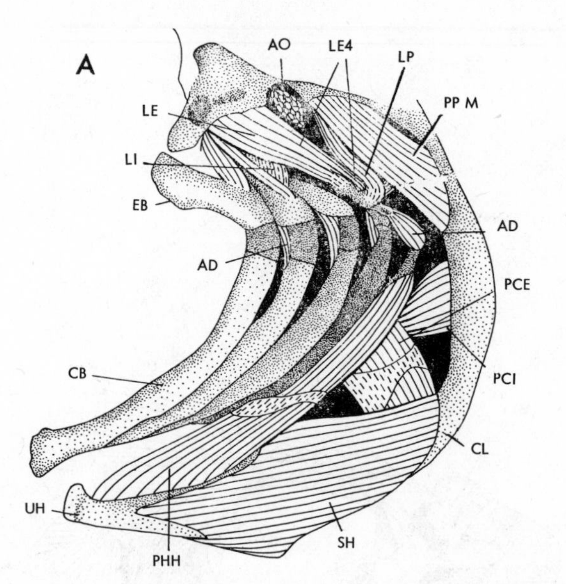

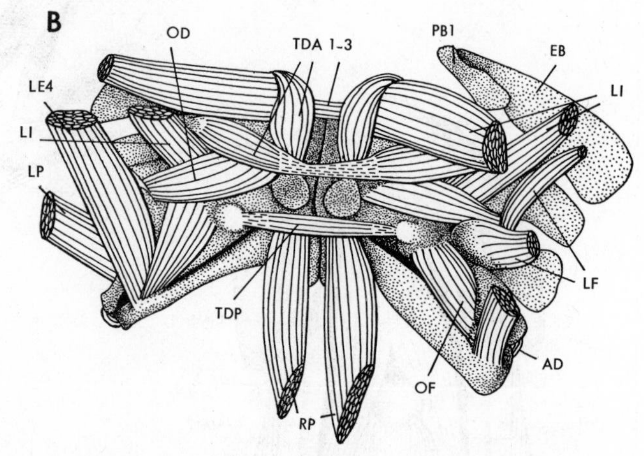

Figure 5-55 Gill arch muscles in a representative teleost. A, lateral view; B, dorsal view. All these muscles are only active during feeding and coughing. During quiet respiration these muscles are not active at all. AD, adductor branchialis; AO, adductor operculi; CB, ceratobranchial; CL, cleithrum; EB, epibranchial; LE and LE_4 levator externus; LI, levator internus; LP, levator posterior; OD, obliquus dorsalis; OF, obliquus posterior; PB, first pharyngobranchial; PCE, pharyngocleithralis externus; PCI, pharyngocleithralis internus; PHH, rectus communis; PPM, protractor pectoralis; RP, retractor dorsalis; SH, sternohyoideus; TDA_{1-3}, transversus dorsalis anterior; TDP, transversus dorsalis posterior; UH, urohyal.

action of the levator arcus palatini (LAP) (Figure 5-57). The opercular cavities are enlarged during stages 2 and 3 by a muscle that moves the opercular cover laterally, the dilatator operculi (DO) (Figure 5-57). After suction has peaked (stage 4 in Figure 5-57), the mouth is closed rapidly by the adductors of the jaw, and the buccal cavity is compressed by the adductor of the palate (AAP) (Figure 5-57) and the geniohyoideus (GH). Note that the geniohyoideus functions as a jaw opener when it contracts together with the sternohyoideus

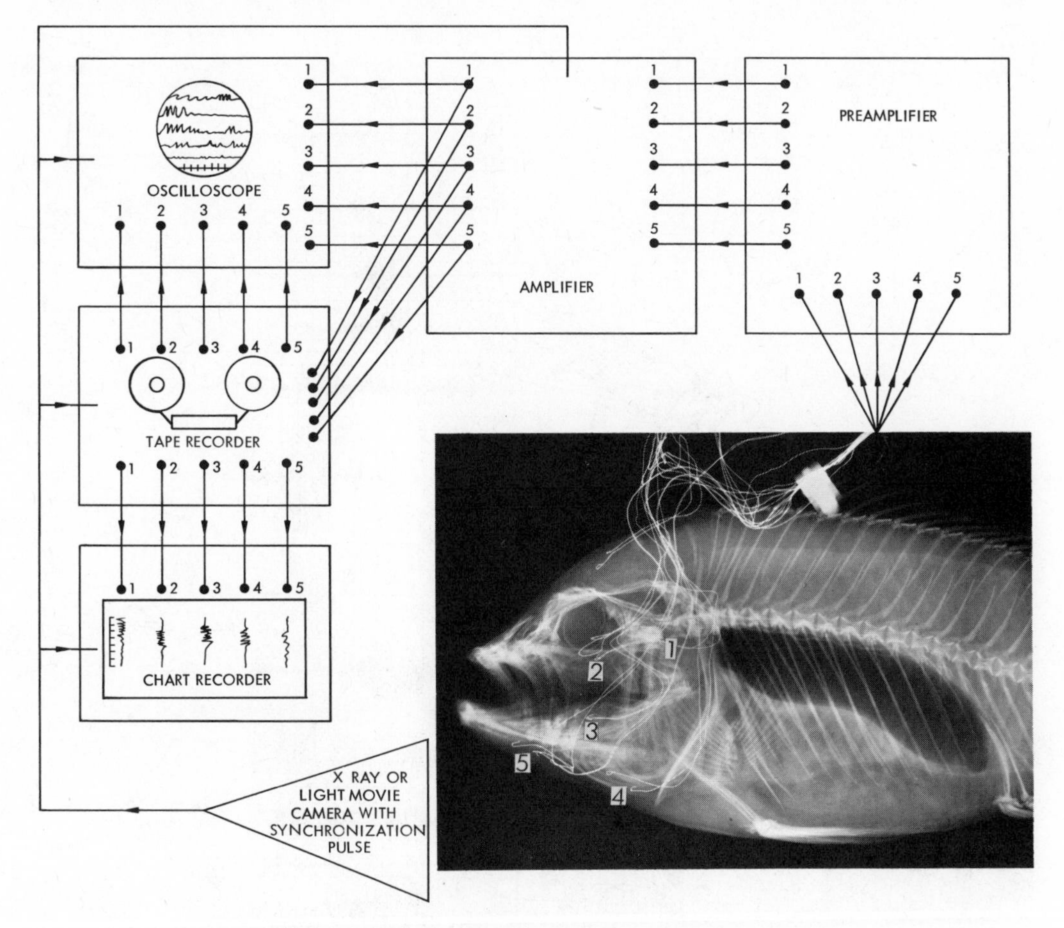

Figure 5-56 Diagram of experimental set-up for electromyographic recording of the head muscles of a teleost. The X-ray picture is that of a living fish with the wire-electrodes in place.

and is unopposed by the jaw adductors (stages 2, 3, and 4); during closing of the jaws, the geniohyoideus pulls the hyoid up and foreward because the mandible becomes a relatively fixed point by action of the jaw adductors (stage 5). By means of this basic mechanism, teleosts can suck in food or gather it with their jaws. Because teleosts lack a mobile and fleshy tongue, food is transported down to the pharynx and esophagus by water currents created by volume changes of the gill cavities and buccal cavity. Back in the pharynx and in front of the esophagus, teleosts possess a "second" set of jaws. Dermal bones carrying teeth fuse with the ceratobranchial of the fifth gill arch and the dorsal gill arch elements of the second, third, and fourth gill arches (Figure 5-52 and 5-58). In most teleosts, these pharyngeal jaws play an important role in transporting the food into the esophagus, thus functioning in a fashion analogous to the mobile fleshy tongue of the higher vertebrates. The upper pharyngeal jaws are operated by the levators (Figure 5-58) functioning mainly as protractors of the upper pharyngeal jaws, and retractory movements are produced by muscles of myotomic origin, the *retractor dorsalis* (Figure 5-58). Coordinated with the

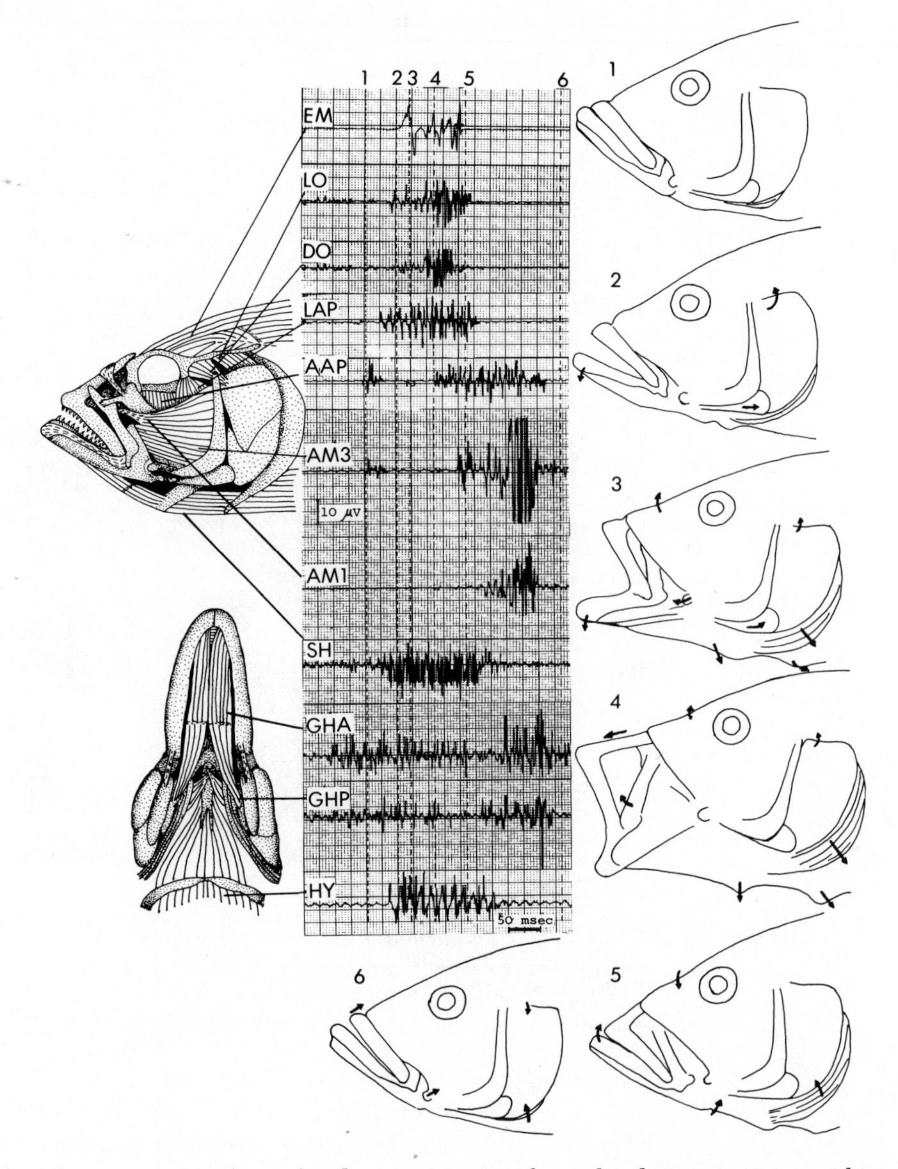

Figure 5-57 Electromyograms (center) indicating activity of muscles during six stages within a feeding cycle in a teleost. At the left side the anatomical locations of the muscles are depicted in lateral (top) and ventral (bottom) views (compare with Figure 5-54). Arranged around the electromyograms are tracings from motion pictures of the six stages of the feeding cycle. Arrows indicate principal movements(s) characterizing the stage. AAP, adductor arcus palatini; AM 1 and 3, adductor mandibulae; DO, dilatator operculi; EM, epaxial muscle; GHA and GHP, geniohyoideus; HY, hypaxial muscle; LAP, levator arcus palatini; LO, levator operculi; SH, sternohyoideus.

movements of the upper pharyngeal jaws are the protractory and retractory motions of the lower pharyngeal jaws, which result in a very effective raking action. Protractory and retractory movements are caused, respectively, by action of a branchiomeric muscle, the *rectus communis*, (Figure 5-58) and a muscle passing from the pectoral girdle to the lower pharyngeal jaw (pharyn-

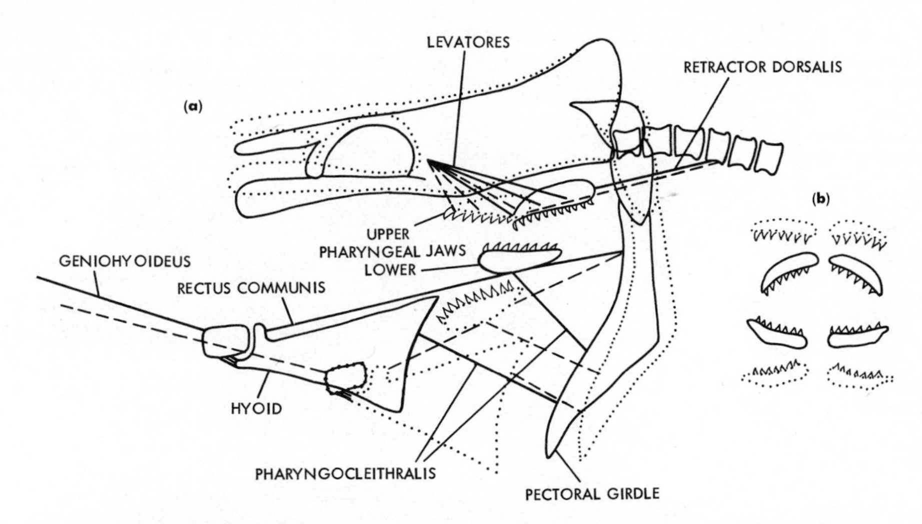

Figure 5-58 (a) Simplified diagram depicting movements of pharyngeal jaws and hyoid from a lateral aspect. Broken lines indicate position and condition of muscles during protraction of the pharyngeal jaws. Solid lines show position and condition of muscles during retraction of the pharyngeal jaws. Bony elements in solid lines represent position during full retraction, while dotted outlines indicate positions during extreme protraction. (b) Front view of upper and lower pharyngeal jaws. Solid lines show position during retraction, dotted lines indicate position during protraction. (From Liem, 1970, courtesy of Field Museum of Natural History, Chicago.)

gocleithralis). It is clear that the raking action of the upper and lower pharyngeal jaws represents a most efficient swallowing mechanism when working in coordination with the directed water currents controlled by changes in volume of the pharyngeal and opercular cavities. The teleostean jaw and swallowing mechanism have proved to be great evolutionary innovations because they have enabled the teleosts to exploit a vast spectrum of foods from ocean depths of nine-thousand meters all the way up to fresh waters at an elevation of five-thousand meters!

The Emergence of the Amphibian Skull

The transition from fish to tetrapod involves the development of evolutionary novelties that lead to the establishment of a new organization. These, in turn, enable the new group to invade and occupy the terrestrial habitats. Here we will focus on the changes in the head skeleton. Because of a very good fossil record, there is little doubt that the ancestral stock from which amphibians have arisen is the lobe-finned fish suborder Rhipidistia. It is, therefore, important to appreciate the basic morphology of the rhipidistian skull. In contrast to teleosts, the rhipidistian dermal skull is much more solid and complete (Figure 5-59); although the eyes are far forward on the head, the postorbital skull and cheek regions are both long and the "face" short. The brain case is built in two quite separate units: (1) the anterior segment is rather firmly bound to the upper jaw and palatal elements; it includes the nasal, ethmoid and sphenoid regions; and (2) the posterior segment includes the otic and occipital regions. The two units articulate in such a way as to allow only dorsoventral rotation of the anterior portion on the posterior as though about a transverse horizontal

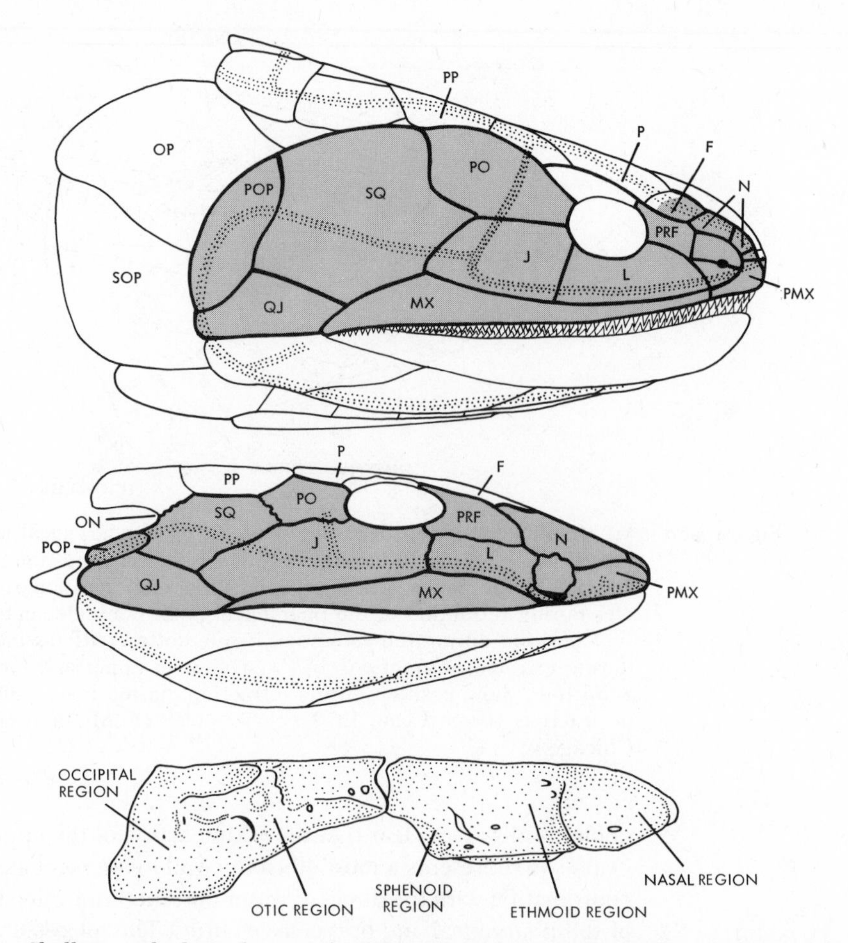

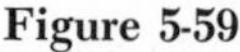
Figure 5-59 Skulls in right lateral view of A, the rhipidistian *Eusthenopteron,* and B, of the primitive amphibian *Ichthyostega* (after Jarvik); C, neurocranium in left lateral view of a rhipidistian showing the intracranial joint allowing kinetic movement (after Thomson). F, frontal; J, jugal; L, lacrimal; MX, maxillary; N, nasal; ON, otic notch; OP, opercular; P, parietal; PO, postorbital; POP, preopercular; PP, postparietal; PRF, prefrontal; QJ, quadratojugal; SQ, squamosal.

hinge. Such a brain case with two units that move in relation to one another is called a *kinetic* skull. Rhipidistians still possess a complete set of gill arches and a morphologically well-differentiated opercular apparatus (Figure 5-59). However, the hyomandibular bone is released from the function of suspending the upper and lower jaws (Figure 5-59).

A principal difference between the rhipidistians and the earliest amphibians lies in the over-all proportions of the skull (Figure 5-59). During the evolution of the Amphibia there has been a considerable increase in the relative length of the anterior portion of the skull (Figure 5-59). In this way, the length of the tooth row and, thus, the size of the dental battery have been increased without an over-all increase in the size of the whole skull. Furthermore, there is a progressive trend toward more extensive fusions of the palate to the cheek and the cheek to the skull roof (Figure 5-60). Concomitant with increased strength of the attachment of the cheek plate to the skull table is the loss of intracranial kinesis in most amphibians. The closer association of the palate with the brain

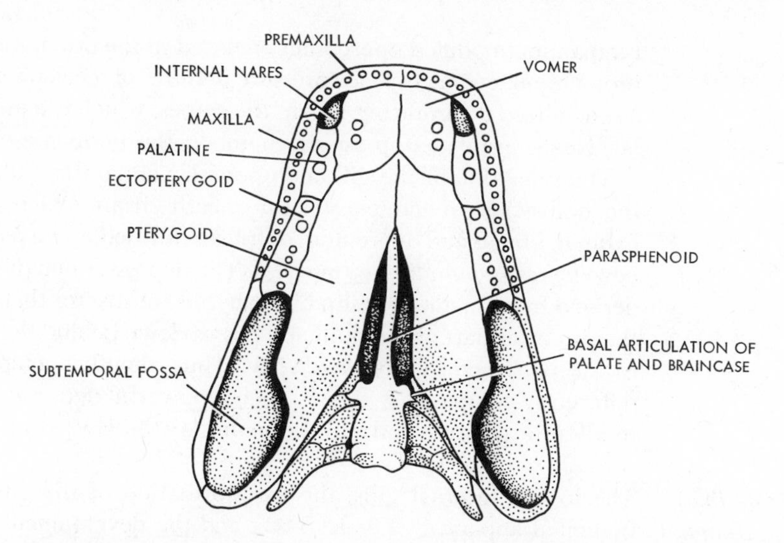

Figure 5-60 Ventral view of the skull of a labyrinthodont amphibian showing the union of palate to neurocranium and upper jaw. (Modified from Watson.)

case released the hyomandibular from its suspensory function to be modified to become a smaller, lighter bone, an ear ossicle (*stapes*). With increasing dependency on the lungs for respiration, the operculum and the visceral arches become superfluous. Their progressive modification or reduction (Figure 5-59) releases space for a major reorganization of the ear (otic) region into an entirely novel, functional system. The posterodorsal margin of the squamosal forms the lower margin of the newly formed *otic notch* (ON), (Figure 5-59b), which encloses the former opercular. Once in this position, the opercular became a new and flexible structure (*tympanum*, Figure 5-61) that was able to vibrate directly in response to air-borne pressure waves and to transfer this motion to the hyomandibular (now the stapes, lying in an air chamber, the tympanic cavity). Thus, during the rhipidistian-tetrapod transition, the basic building blocks for aerial sound perception for vertebrates was formed: the

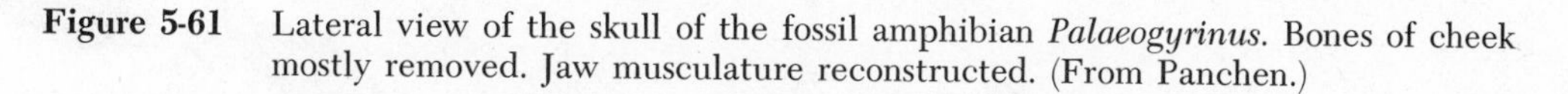

Figure 5-61 Lateral view of the skull of the fossil amphibian *Palaeogyrinus*. Bones of cheek mostly removed. Jaw musculature reconstructed. (From Panchen.)

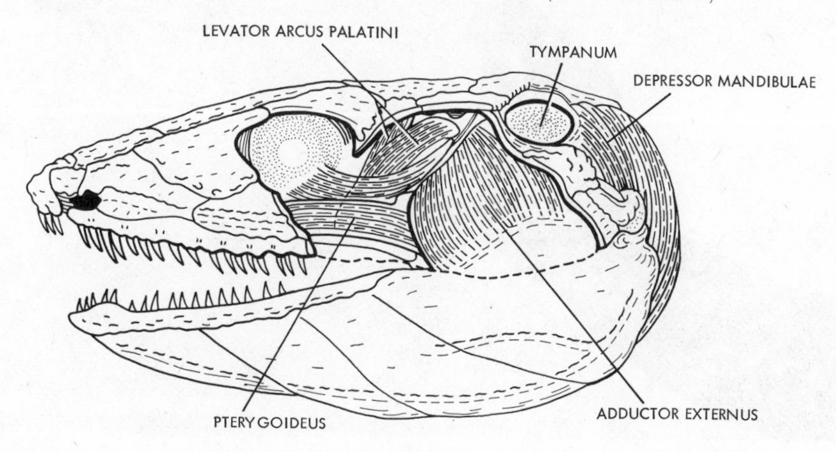

tympanum (modified operculum enclosed in the otic notch), capable of registering minute changes in air pressure in terms of mechanical deformations; and a specialized transmitting organ, the stapes, which is a modified hyomandibular, freely suspended in an air chamber, the tympanic cavity.

The reduction of the gill and opercular apparatus "allows" the muscles of the embryonic hyoid constrictor sheath (from which the muscles of the hyomandibular and operculum of fish are formed) to be rearranged to form the new *depressor mandibulae muscles*. The depressor mandibulae (Figure 5-61) is derived from an anterior slip of the hyoid constrictor that arises laterally from the posterior part of the skull and passes down behind the tympanum to insert to the back end of the lower jaw. Thus, mouth opening proceeds entirely differently compared with fish. Such a powerful depressor would be necessary to affect a positive and rapid opening of the jaws while swimming.

The Amphibian Tongue

The loss of internal gills, the transformation of the gill arches into a hyobranchial apparatus (Figure 5-63), and the development of a tongue (Figure 5-62) in amphibians must have represented a series of interdependent evolutionary changes. A newly evolved tongue in amphibians plays an important role in the acquisition and manipulation of terrestrial food and represents one of the key innovations in vertebrate evolution. Reduced and modified remnants of the anterior gill bars serve, as the hyoid apparatus, to support the tongue (Figure 5-63). The lungless plethodontid salamanders rely mainly on tongue projection to obtain prey (Figure 5-62). They possess an appropriate example of a prehensile tongue.

Figure 5-62 Protruded tongue of a European salamander striking a prey. (Photograph by Dr. Gerhard Roth, University of Munster, Germany.)

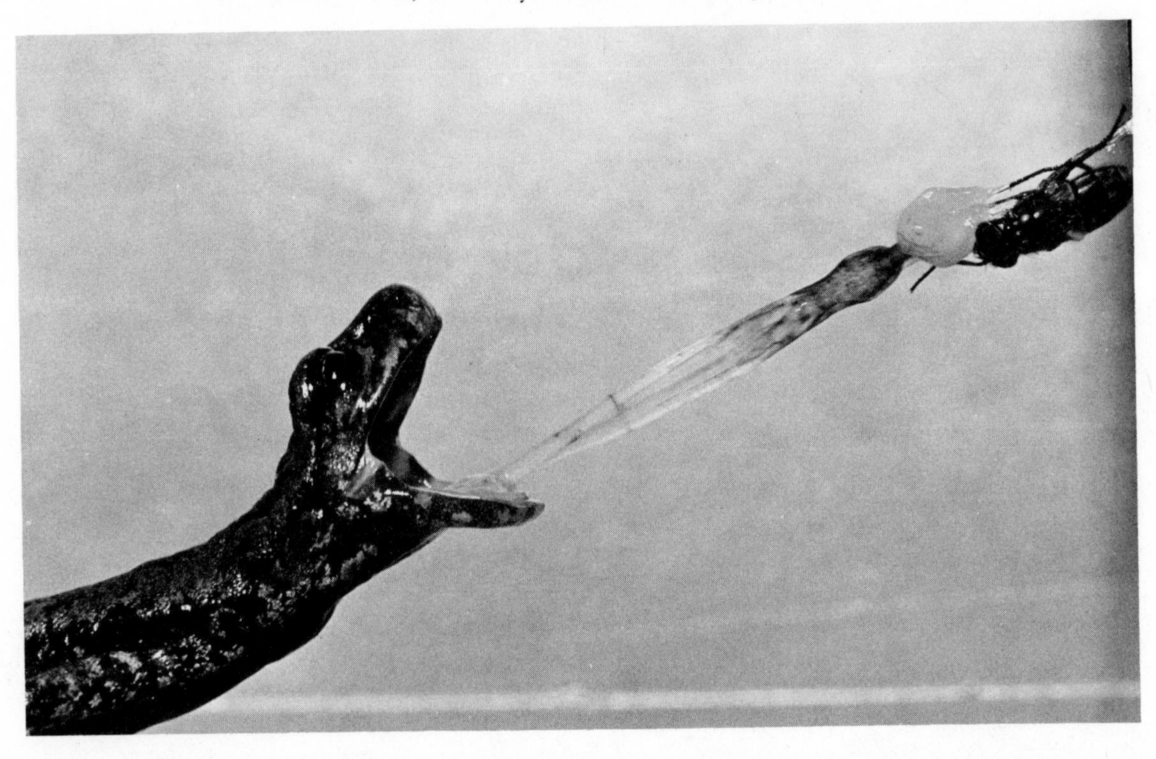

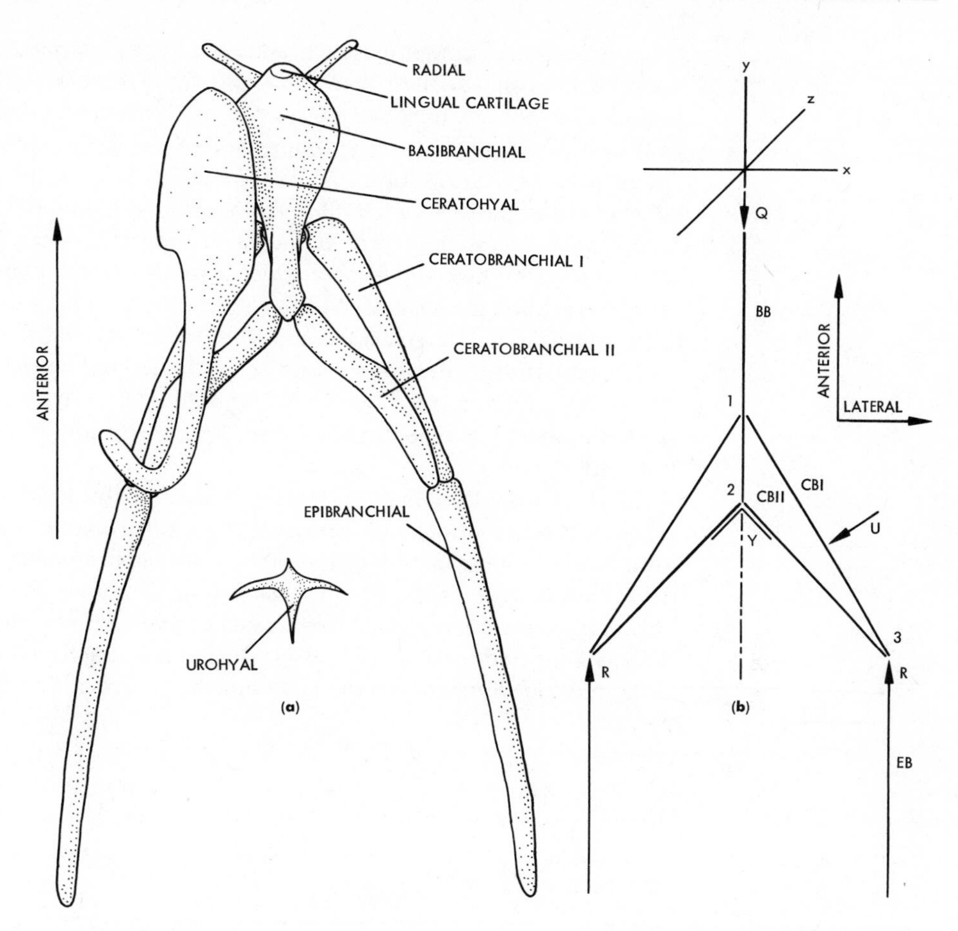

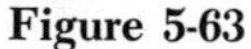

Figure 5-63 (a) Dorsal view of the hyobranchial skeleton of a plethodontid salamander. Only the left ceratohyal is shown. (b) A schematic diagram of the hyobranchial elements. The major articulated elements are represented (BB, CBI, CBII and EB; compare with part a). The fixed sites of articulation are 1, 2 and 3. Arrows Q and R represent forces acting on the skeleton. (After Lombard and Wake.)

In the floor of the mouth lie a pair of *ceratohyals* (Figure 5-63) that do not contact each other nor articulate with other elements. The projectile thrown out of the mouth (Figure 5-62) consists of cartilages as well as muscle, connective tissues, and a rather complex tongue pad. The cartilages that move out of the mouth with the tongue are the *median basibranchial* and two pairs of *ceratobranchials,* which articulate with the posterior portion of the basibranchial (Figure 5-63). Anteriorly, the basibranchial supports the large tongue pad. The first and second ceratobranchial on each side approach each other posteriorly and together articulate with the *epibranchial.* Mechanically, this skeletal framework has the following important characteristics (Figure 5-63): (1) the cartilaginous basibranchials and two ceratobranchials react as if they are rigid; (2) there are three joints, *joint 1* between basibranchial and first ceratobranchial, *joint 2* between basibranchial and second ceratobranchial, and *joint 3* between epibranchial and the ceratobranchials; and (3) the apparatus is bilaterally symmetrical. The principal muscles associated with the hyobranchial skeleton are the *rectus cervicis profundus* and the *subarcualis rectus I*

(Figure 5-64). The former muscle is a direct anterior continuation of the muscle that arises on the ischium. It extends past the sternum to the basibranchial region. The subarcualis rectus I originates in the ventral surface of the ceratohyal and wraps around the epibranchial in a complex spiral (Figure 5-64).

During tongue projection, the hyobranchial skeleton is folded. Folding is achieved by apposition of joint 3 to the midline (Figures 5-63 and 5-65). In the folded state, the second ceratobranchial is superior to the first one. Accordingly, during folding, the basibranchial is rotated about joint 1 such that the distal tip is directed ventrally.

Projection of the tongue is accomplished by the subarcualis rectus I muscles, which squeeze and thrust the epibranchials. Joint 3 on each side will, therefore, ride along a track formed by the lateral wall of the cavity of the subarcualis rectus I muscle. Movement along the track forces the joints toward the midline, and folding occurs.

Contact is made with the prey on the axis of projection and capture is accomplished by a mucous coating of the tongue pad (Figure 5-62). Tongue retraction occurs by action of the rectus cervicis profundus muscle returning the folded skeleton in the mouth. The retractive force is applied along the Y axis on the distal tip of the basibranchial [Figure 5-63(b)]. The direction of the force is indicated by the arrow at Q. During retraction, the epibranchials are separated and directed into the cavity of the subarcualis rectus I muscle. It is

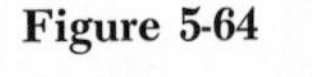

Figure 5-64 Dorsal view of the principal muscles of the hyobranchial skeleton in a plethodontid salamander. Part of right ceratohyal has been removed to show the origin of the subarcualis rectus muscle. (From Lombard and Wake.)

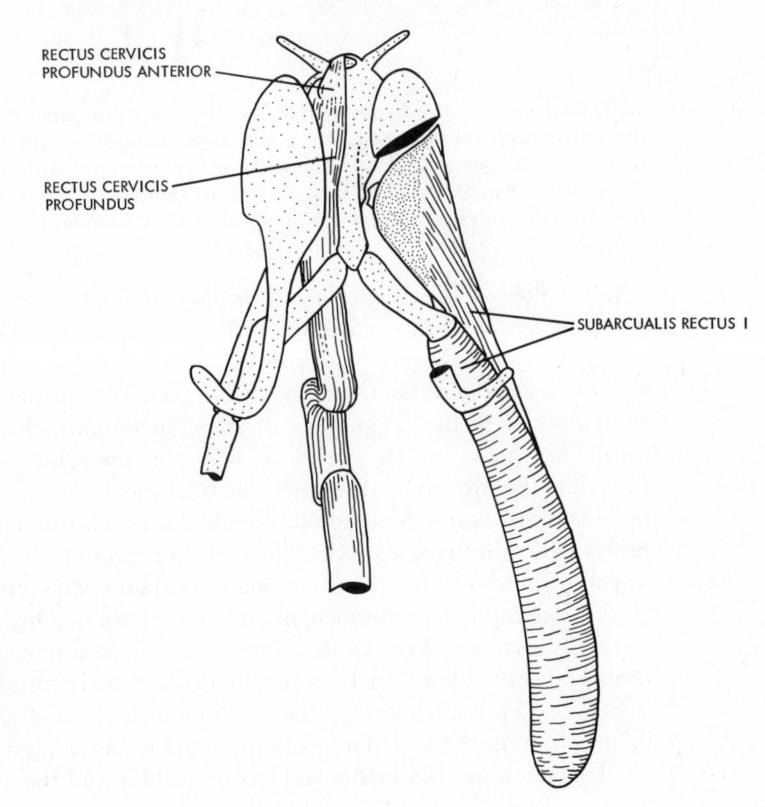

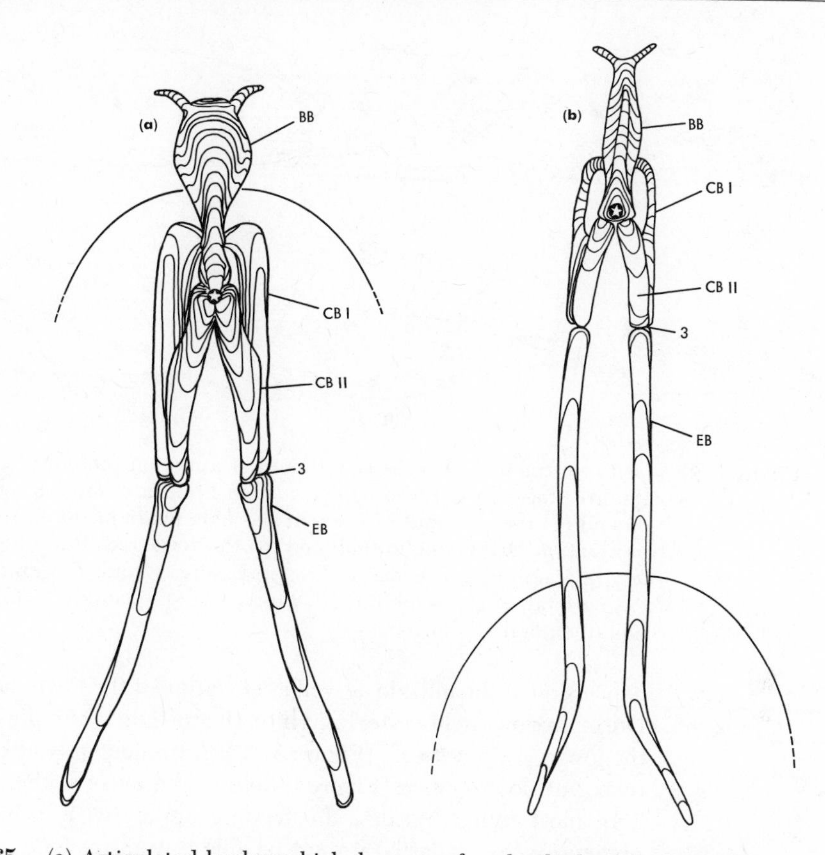

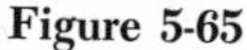
Figure 5-65 (a) Articulated hyobranchial elements of a pleothodontid salamander in a partially projected state. (b) Same structures in a fully projected condition. The curved line represents the margin of the jaw. The star represents the most dorsal point on the reconstruction. Note that joint 3 moves medially during tongue projection. During the folded (projected) state, the second ceratobranchial (CBI) moves superior to the first one (CBII). (From Lombard and Wake.)

important to appreciate the high speed of the feeding sequence, which may last only ten milliseconds. The effectiveness of this system, involving combined use of jaws and tongues, is reflected in the evolutionary success of the terrestrial salamander family Plethodontidae.

The mobile tongue, characteristic of most land vertebrates, is a most essential tool of the feeding apparatus of tetrapods. Aside from its role in the gathering of food, the mobile tongue manipulates food so that it will be positioned properly between the teeth of the upper and lower jaws; in all tetrapods the tongue is an integral part of the swallowing mechanism; in mammals the tongue participates in vocalization; and in man it is an important aid to speech.

Jaw Apparatus in a Representative Lizard

The feeding movements of a lizard proceed in a rather characteristic pattern (Figure 5-66): The prey is grasped in the mouth while the tongue is slid forward beneath it [Figure 5-66(a)]. As the lower jaw is depressed, the muzzle is raised relative to the rest of the skull [Figure 5-66(b)]. At the same time, the tongue pulls the prey upward and backward to disengage it from the mandibular teeth and to bring it further back into the mouth. Closure of the mouth involves

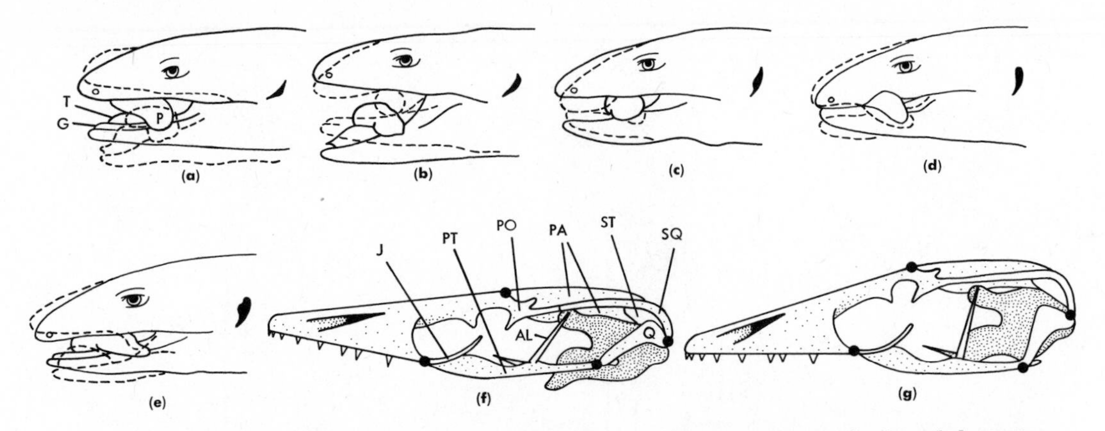

Figure 5-66 (a) to (e) tracings of successive frames of a motion picture of a lizard depicting the five stages in the feeding cycle (from Frazzetta, *Journal of Morphology,* 1962); (f) and (g) the positions of the four movable joints in the lizard skull. (f) depicts protraction of the kinetic skull and (g) the retracted state (from Alexander, courtesy of Cambridge University Press). AL, alisphenoid; G, gums; J, jugal; P, prey; PA, parietal; PO, postorbital; PT, pterygoid; Q, quadrate; SQ, squamosal; ST, supratemporal; T, tongue.

depression of the muzzle as well as elevation of the lower jaw [Figure 5-66(c)]. During a slow, deliberate, hard bite, the muzzle is *sharply depressed,* meeting the lower jaw anteriorly [Figure 5-66(d)]. Immediately after the bite, the jaws relax but do not open [Figure 5-66(e)]. The cycle is then repeated.

In most living reptiles, the feeding apparatus is primarily an organ of prehension. Secondarily, it must be able to transfer food from the jaws to the esophagus by means of the tongue. Food can be torn apart if it is too large to swallow entirely, *but it is not masticated.*

The major skeletal units underlying lacertilian jaw mechanics can be summarized as follows:

1. The *muzzle* is the anterior part of the skull and is composed of the premaxillae [Figure 5-67(a)], nasals, vomers, maxillae, prefrontals, frontals, and lachrymals.
2. The *occipital unit* is made up by the basioccipital, supraoccipital, exoccipitals, basisphenoid, and prootics [Figure 5-67(b) and (c)]. This unit encloses the brain.
3. The *parietal unit* [Figure 5-67(c)] also encloses the brain and consists of parietals, supratemporals, squamosals, and postfrontals.
4. The *basal unit* forming the posterior roof of the mouth [Figure 5-67(c)] is composed of the pterygoid, ectopterygoid, jugal, and palatines.
5. The *quadrate* [Figure 5-67(a), (b), and (c)] is no longer firmly attached to other bones along its length but has movable joints at its ends. Dorsally, there is a joint with the squamosal and supratemporal bones of the parietal unit. This joint is movable, making it possible for the quadrate to move in a pendulumlike fashion. Ventrally, the quadrate articulates with the pterygoid of the basal unit and with the mandible.
6. Each *mandible* articulates with the saddle-shaped ventral tip of the quadrate [Figure 5-67(d)]. The jaw joint is formed by the quadrate superiorly and the articular inferiorly. Behind the jaw joint is a rather prominent retroarticular process that supports the lower edge of the tympanic membrane and

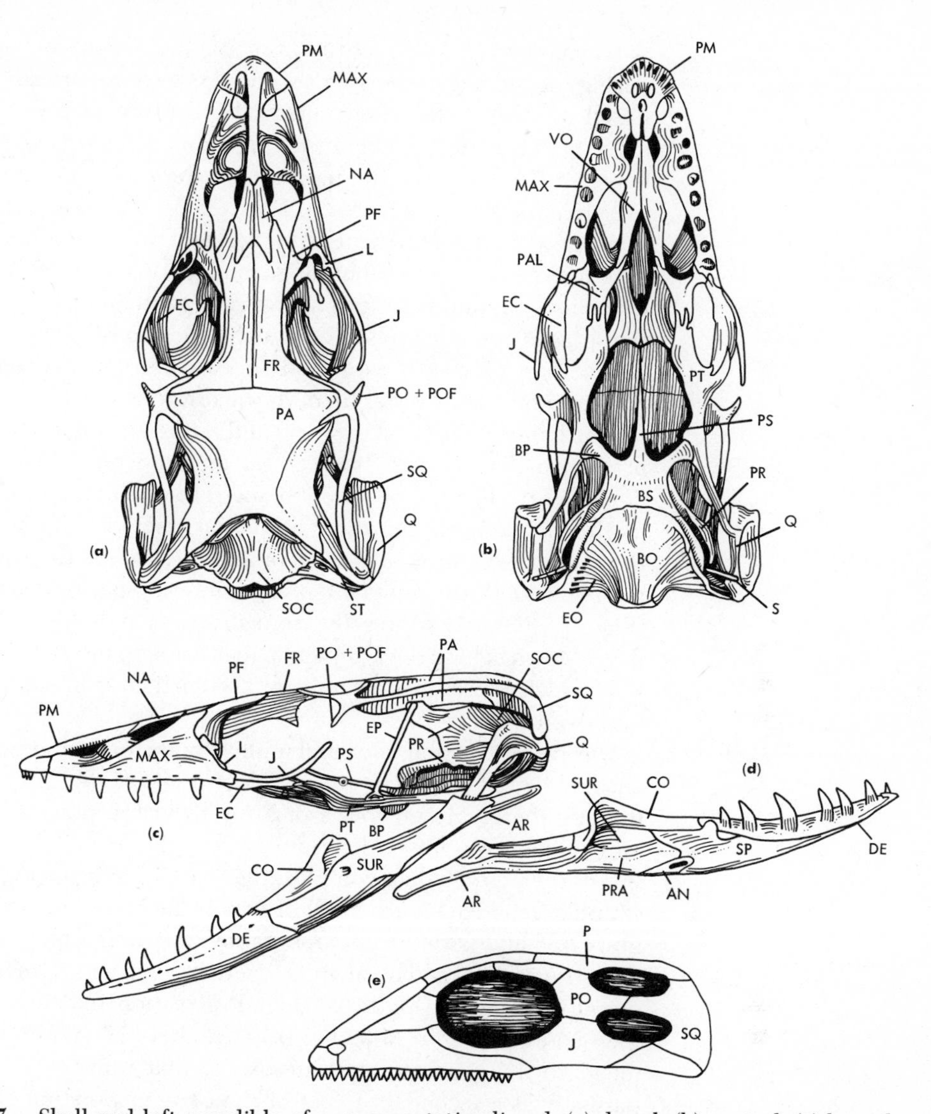

Figure 5-67 Skull and left mandible of a representative lizard. (a) dorsal, (b) ventral, (c) lateral, (d) medial view of mandible, (e) diagram of an original diapsid skull. [(a–d) from Frazzetta, *Journal of Morphology*, 1962]. AR, articular; AN, angular; BP, basipterygoid process; BO, basioccipital; BS, basisphenoid; CO, coronoid; DE, dentary; EC, ectopterygoid; EO, exoccipital; EP, epipterygoid; FR, frontal; J, jugal; L, lachrymal; MAX, maxilla; NA, nasal; P, PA, parietal; PAL, palatine; PF, prefrontal; PM, premaxilla; PO + POF, fused postorbital and postfrontal; PR, prootic; PRA, perarticular; PS, parasphenoid; PT, pterygoid; Q, quadrate; S, stapes; SOC, supraoccipital; SP, splenial; SQ, squamosal; ST, supratemporal; SUR, surangular; VO, vomer.

functions as a lever for the depressor mandibulae muscle acting as the mouth opener. Structurally, the mandible represents a complex assemblage of many elements: the teeth-bearing *dentary* forming the anterior one third of the mandible; the *splenial, angular, articular,* and *prearticular* representing a major part of the medial aspect of the mandible; the *surangular* on the lateral aspect; and finally a dorsal element, the *coronoid.*

We can now reconstruct the principal movements of the skull of the lizard during the feeding cycle (Figure 5-66). As we have already seen, the muzzle is raised relative to the skull during the capture of prey. Such a remarkable movement is made possible by a hinge joint across the skull roof (Figure 5-67) between the frontal of the muzzle and the parietal of the occipital unit. All principal joints are indicated in Figure 5-66, which also shows the movement they allow. Farther forward from the basal unit, there is a flexible region that functions in effect as another hinge joint, allowing the elevation of the muzzle. Of course, the quadrate can swing as a pendulum. In Figure 5-66, the relationships of the units relative to each other and the topography of the interconnecting joints are expressed in two extreme positions: i.e., *protracted* [Figure 5-66(f)], the distal tip of the quadrate swung forward, basal unit moved forward, muzzle rotated upward) and *retracted* [Figure 5-66(g)], the distal tip of the quadrate swung back, basal unit moved backward, muzzle rotated downward). How are protraction and retraction accomplished? We have seen that the mandible is depressed during protraction (Figure 5-66). Depression of the mandible is brought about by contraction of the depressor mandibulae muscle, which represents a modified anterior slip of the hyoid constrictor. It (Figure 5-68) arises laterally from the neck muscles and occipital unit and passes down behind the tympanum (eardrum) to insert on the back of the lower jaw, which possesses a prominent retroarticular process to give the muscle improved leverage.

During protraction, the basal unit is drawn forward [Figure 5-66(f)]. At least one muscle, the *protractor pterygoideus* (Figure 5-68), is in a favorable position to draw the basal unit forward. From the large pterygoid (Figure 5-68), the muscle fibers run upward and forward to the occipital unit. Another subdivision of the pterygoideus complex, the *levator pterygoideus*, is a parallel fibered, straplike muscle (Figure 5-68) of which the fibers run vertically between the pterygoid and parietal. Its topography and architecture are ideally suited to draw the parietal and basal units together. Such a movement does indeed occur during protraction [Figure 5-66(f)]. Protraction, of course, involves rotation of the quadrate [Figure 5-66(f)]. Because there are no direct muscular attachments to the quadrate, we can assume that rotation of the quadrate occurs passively as a secondary effect of the combined action of the protractor and levator pterygoidei.

We have already seen that lifting of the mandible and retraction of the basal unit occur together [Figure 5-66(g)]. Actually, retraction and mouth closure are functionally connected. On the side of the head, the *adductor externus muscle* runs from posteriorly and above, downward, and anteriorly toward the mandible. There is practically no muscular attachment to the dentary! Instead, most adductor musculature converges onto a complex and concentrated tendon (often called the *Bodenaponeurosis*). This tendon lies within the muscle and serves as the insertion of the adductor externus. The tendon in turn attaches to the coronoid bone [Figure 5-67(c)]. Contraction of the adductor externus produces not only a strong elevation of the mandible, but a powerful retraction of the basal unit and the quadrate as well, because of the anteroposterior direction of many of the fibers. Another important specialized subdivision of the adductor complex is the *pseudotemporalis muscle* (Figure 5-68), which, because of its mechanically advantageous vertical position above the lower jaw, elevates the mandible. Finally, at the rear angle of the jaw is the large

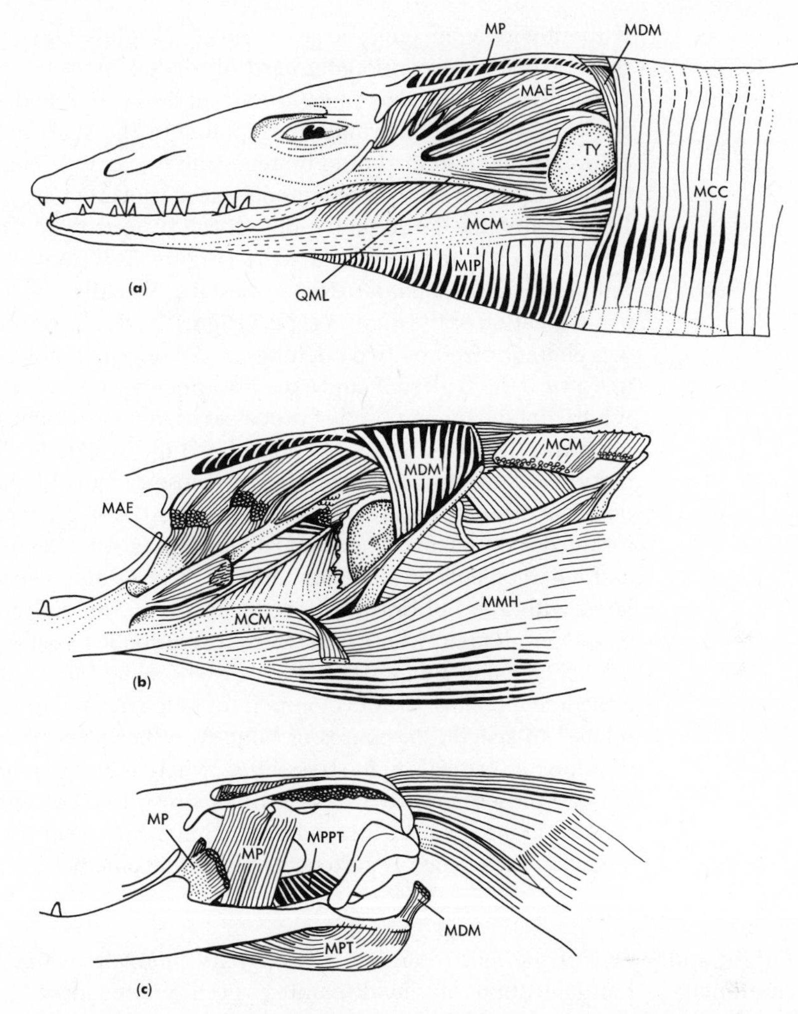

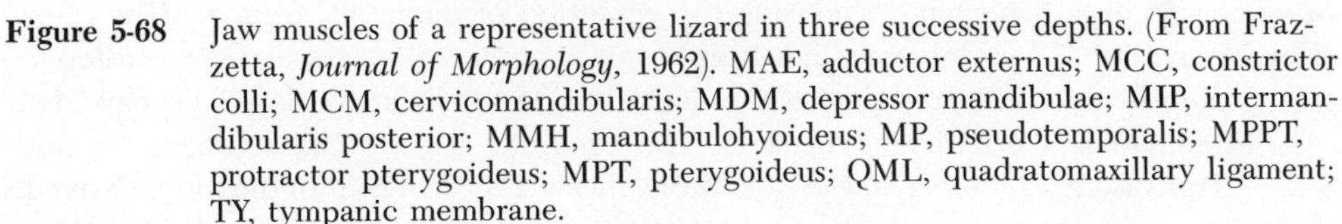

Figure 5-68 Jaw muscles of a representative lizard in three successive depths. (From Frazzetta, *Journal of Morphology,* 1962). MAE, adductor externus; MCC, constrictor colli; MCM, cervicomandibularis; MDM, depressor mandibulae; MIP, intermandibularis posterior; MMH, mandibulohyoideus; MP, pseudotemporalis; MPPT, protractor pterygoideus; MPT, pterygoideus; QML, quadratomaxillary ligament; TY, tympanic membrane.

pterygoideus muscle [Figure 5-68(c)] inserting on the retroarticular process. From this process, the fibers run forward, turn inward beneath the shaft of the mandible, and attach to the pterygoid of the basal unit. It is postulated that action of the pterygoideus will result primarily in elevation of the mandible, although, secondarily, retraction of the basal unit may occur.

Skulls exhibiting the intracranial movement described here for the lizard are said to be kinetic. Various functions have been ascribed to kinesis. Here we list only the three better documented ones: (1) kinesis enables the upper jaws to move down on the prey as the mandible is adducted, resulting in a most efficient grip; (2) in lizards, the independent mobility of the muzzle can work as a release mechanism, allowing the prey to be drawn back from the teeth without widely opening the jaws; and (3) a slightly kinetic skull can absorb the shock when an exceptionally hard bite is taken.

Finally, we will focus on an important innovation in the reptilian skull roof. As we have seen, in primitive amphibians, the skull roof (Figure 5-59) was complete and uninterrupted by fenestrae (windows). The earliest reptiles and turtles still have skulls with unfenestrated roofs that are called *anapsid* skulls. In the lizard, the dorsal skull roof is distinctly fenestrated; the cavity behind the orbit is called the temporal fossa (Figure 5-67) and contains the pseudotemporalis and adductor externus muscles. Actually, the lizard skull represents a modification of the *diapsid* type. Originally, the diapsid skull [Figure 5-67(e)] was characterized by two openings in the temple region. One window is high up toward the skull roof, and a second appears well down the side of the cheek. A horizontal bar composed of processes of the postorbital and squamosal bones separates the two openings [Figure 5-67(a)]. In lizards, the lower bar closing the lower fenestra has disappeared because the jugal bone is reduced leaving but a single arch [Figure 5-67(c)]. Formation of an open temporal fossa allows for the development of mechanically advantageous sites for muscle attachment and enlargements of adductor muscles. The origin of fenestra may be correlated with areas of drastically reduced stress. In "understressed" areas, bone would get thinner, whereas in stressed areas bone hypertrophies. Regions with thickened bone might then serve as more favorable sites for jaw muscle attachments. Thus, the development of fenestrae and fossae very probably are related to specific responses of bone to either stress or the lack of it. Bone develops in critical, high-stress lines, whereas the fenestrae and fossae, in a sense, represent negative areas without stress. In forms with fenestrated skulls, the lateral muscle mass attaches to the inner surface of the temporal arch. The medial muscle mass is fastened to the dorsomedial border of the fenestra.

Origin and Evolutionary Consequences of a Key Innovation: Reptilian-Mammalian Transition

One of the most dramatic evolutionary innovations to emerge is the radical reorganization of the disparate, yet interdependent, feeding and hearing mechanisms during the evolution of mammals from reptiles.

The first major transformation in the evolutionary line leading to mammals occurred during the derivation of cladistically derived synapsids from more primitive Synapsida. The *synapsid* skull is characterized by one temporal fenestra on the side of the cheek (Figure 5-69). In the more derived synapsids, there is a change in the roof and walls of the temporal fossa. The fossa was displaced backward and downward relative to the position of the coronoid region of the lower jaw (Figure 5-69). A major result of this event was the displacement of the jaw articulation downward, relative to the coronoid region, which now forms a coronoid eminence raised far above the jaw articulation (Figure 5-69). The formation of a coronoid eminence increased the relative moment arm of the adductor externus muscle. These structural modifications (posterodorsal enlargement of temporal fossa, lateral temporal fenestra, and development of a coronoid process on the dentary bone) become increasingly more pronounced in the more specialized forms. Thus, a distinct evolutionary trend can be seen (Figure 5-69) culminating in the establishment of a mammal-like morphology for (1) the general shape of the temporal fossa; (2) the area of origin for the adductor externus muscle closely approaching the mammalian temporalis muscle; and (3) the configuration of the lower jaw with

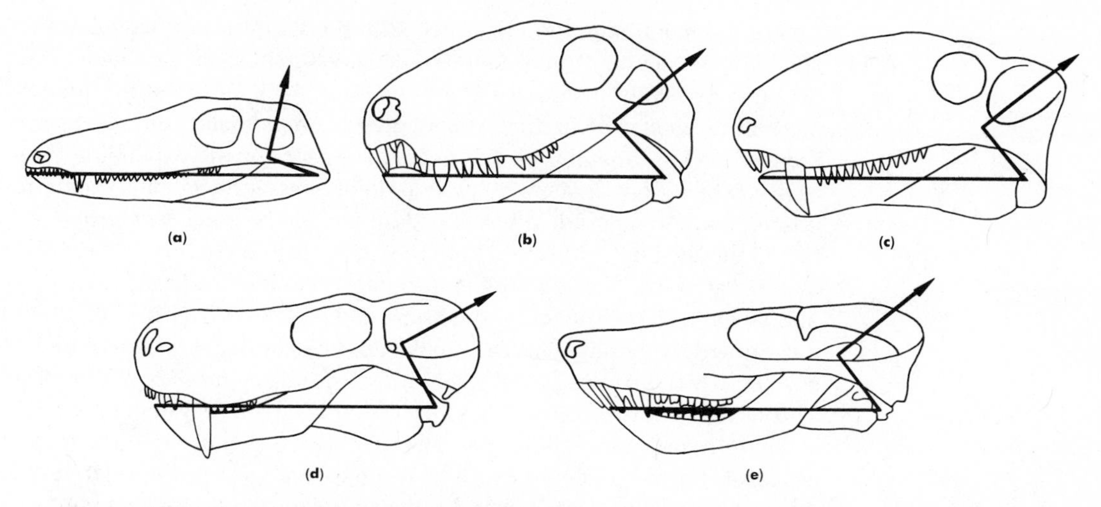

Figure 5-69 A series of synapsid skulls illustrating morphological stages in the modification of the temporal region and its effect on external adductor musculature in lateral view. The line of action and relative moment arm of the external adductor are illustrated. Lengths of the muscle force vectors are not intended to indicate relative force of muscle contraction. (a) A primitive pelycosaur; (b) A sphenacodontid *Dimetrodon;* (c) A primitive therapsid; (d) A gorgonopsian; (e) A bauriamorph. (From Barghusen, 1972.)

a distinct coronoid process on a greatly expanded dentary. In Figure 5-69, we can see the line of action and relative moment arm of the adductor externus muscle in the various forms leading to mammalian morphology. The derived mammal-like synapsids may have preyed on relatively large animals, possibly approaching the size of the predator. If so, the development of a relatively long moment arm for the adductor externus is significant because large forces resisting jaw closure develop during the seizure and killing of large prey. Even more importantly, the posteriorly directed line of action of the adductor externus is appropriately arranged to resist loading applied to the open jaw by the struggling prey.

Among the very advanced Synapsida are the cynodonts, which possess many characteristically mammalian jaw structures. For the first time, the *zygomatic arch* (composed of the jugal and a process of the squamosal) is bowed out away from the lateral surface of the dentary. This created space for a muscle running between the two structures and established the medial surface of the zygomatic arch as a potential area of muscle attachment. The muscle between the zygomatic arch and dentary is the *masseter muscle* (Figure 5-70), which has split off from the original adductor externus, now properly designated the *temporalis muscle.* In cynodonts, the dentary develops a massive coronoid process, representing a replacement of a tendon (the reptilian Bodenaponeurosis) by bone. As a consequence of the changes reflected in the expansion of the dentary (Figure 5-70), and reduction of the other jaw bones (splenial, angular, articular, prearticular, surangular, and coronoid), most of the adductor muscles are inserted on the dentary.

During the changes in jaw structure and function, the squamosal developed a flange that buttressed the quadrate from behind and took up the stresses transmitted to the quadrate from the lower jaw. Thus, the quadrate became reduced in height and loosely attached to the skull. A new articulation of the jaw between dentary and squamosal became established. At the same time, a new and typically mammalian jaw opening mechanism (to be discussed on page 266) emerged, so that the depressor mandibulae muscle was either reduced or lost, opening up the necessary space for the expanding tympanic cavity. Once the dentary-squamosal joint was formed, the quadrate and articular would be gradually reduced in size and become *decreasingly* important in the jaw joint. However, both the quadrate and articular are retained and incorporated by the expanding tympanic cavity. Both elements become middle-ear ossicles, the *incus* (quadrate) and *malleus* (articular). Together with the stapes (hyomandibular), they make up the mammalian chain of middle-ear ossicles, which articulate with each other forming a complex system of levers. This system of levers represents an exceedingly effective system for conducting and amplifying sound vibrations. The new integration of the three middle-ear ossicles is far more effective as a hearing apparatus than the relatively inefficient piston mechanism formed by the reptilian stapes in direct contact with the tympanum.

It is interesting to note that the origin of the mammalian middle-ear mechanism was the outcome of selective pressures acting primarily on the

Figure 5-70 Skeletal and muscular changes in cynodont reptiles, illustrating the evolutionary development of the masseter muscle. Lines of action and relative moment arms of the jaw muscles are indicated in (d), (e), and (f). (g) and (h) respectively lateral and medial aspects of the mandible of a cynodont, showing the dominance of the dentary and reduction of all other elements. [(a–f) from Barghusen, 1972; (g) and (h) from Romer, courtesy of Saunders Publishing Co.]

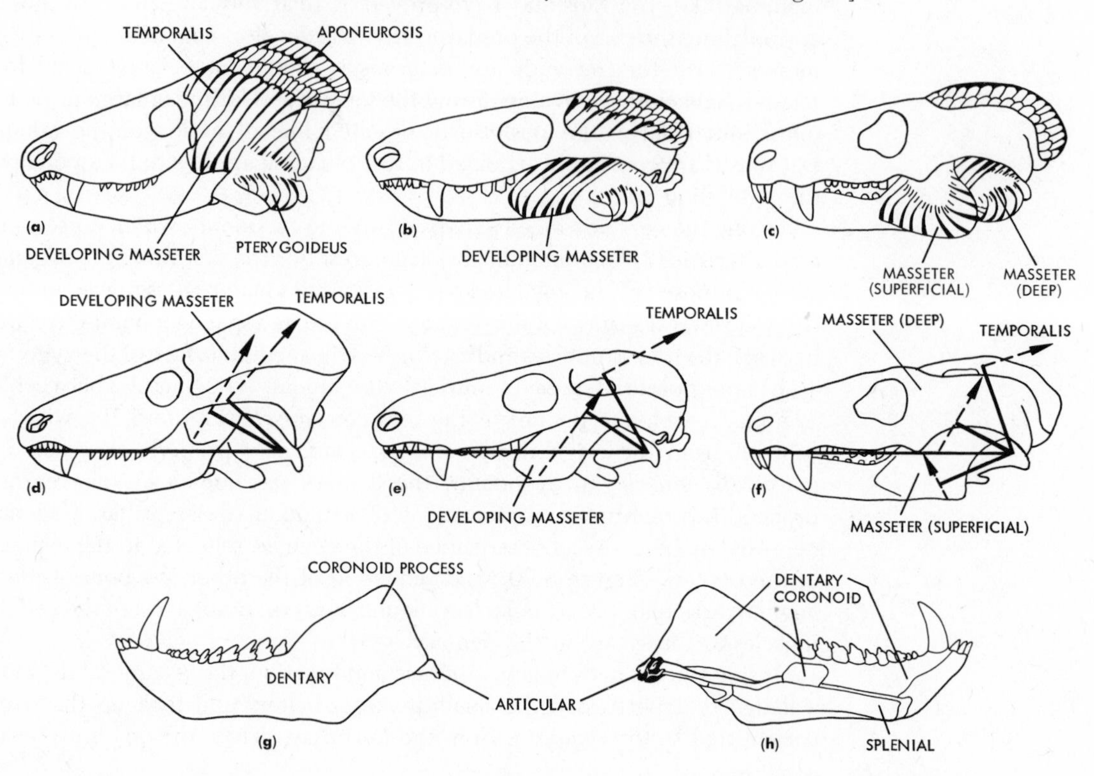

masticatory apparatus of mammal-like synapsid reptiles. Only after the mammalian level of organization was reached in the jaw mechanism did selection act to incorporate what might be called remnants, fortuitously preadapted to conduct and amplify sound, into the hearing mechanism. The selective advantage of a superior hearing apparatus is significant, and it has played an important role in the adaptive radiations of mammals. The emergence of the mammalian level of organization in the feeding apparatus can be regarded as a radical improvement to meet the mechanical requirements of true mastication. Mastication increases the efficiency with which food is made ready for absorption and, therefore, ultimately is related to *endothermy*, which is one of the most far-reaching and dramatic developments in the evolution of the vertebrates.

Generalized Mammalian Feeding Mechanism Expressed in the Opossum

In mammals (except for a few independently modified forms), the feeding apparatus is primarily an organ for the mechanical reduction of food; only secondarily is it an organ of prehension. We can consider the feeding apparatus of an opossum representative of a primitive mammal. This section deals with the structural and functional characteristics of the integrated adaptive complex involved in the masticatory process in the opossum.

An integral part of the jaw apparatus is the *hyoid* complex. It typically includes a main body, the *corpus* (Figure 5-71), formed from median ventral arch elements, remnants of the hyoid arch, and modified branchial arches that extend outward and upward as "horns" or cornua. Situated in the base of the tongue is the body of the hyoid. Dorsally, part of the horn fuses with the ear region of the skull to form the *styloid process* (Figure 5-71). The basal part of the hyoid apparatus lies close to an enlargement of the windpipe, the *larynx*, which is formed by additional cartilages that represent modified visceral arch elements. Included in the larynx are the unpaired *thyroid* (Figure 5-71) and *cricoid* cartilages in line ventrally behind the body of the hyoid and paired *arytenoids* in line dorsally. Together with the laryngeal muscles and vocal cords (attached to the arytenoids), the cartilages form the voice box. The ability to emit a great variety of sounds with their larynx has provided mammals with a remarkable mechanism for social communication, which has played a key role in the evolution of many groups of mammals. Beyond the larynx, the windpipe, or *trachea*, is strengthened by ring-shaped tracheal cartilages.

Although we have discussed the major cranial innovations characterizing the mammals in the previous section, there are several other salient features to be seen in mammals. In the opossum, a zygomatic arch, formed by the jugal and squamosal, is situated below the large temporal fossa (Figure 5-71). Farther forward, the maxilla (bearing the large canine teeth) expands dorsally and reduces the size of the lachrymal. Many elements are lost, including the prefrontal, postfrontal, postorbital, supratemporal, and quadratojugal. A secondary palate is differentiated by the maxillae and palatines, which fold ventrally and medially to produce an elongate shelf lying below the original roof of the mouth. It is clear that the secondary palate is an aid in the maintenance of breathing while the mouth is functioning in eating. Of course, kinetic properties have been completely abandoned and the pterygoids reduced to little more than small wings of bone projecting ventrally from the skull base. The mandible (Figure 5-71) is composed solely by the dentary and

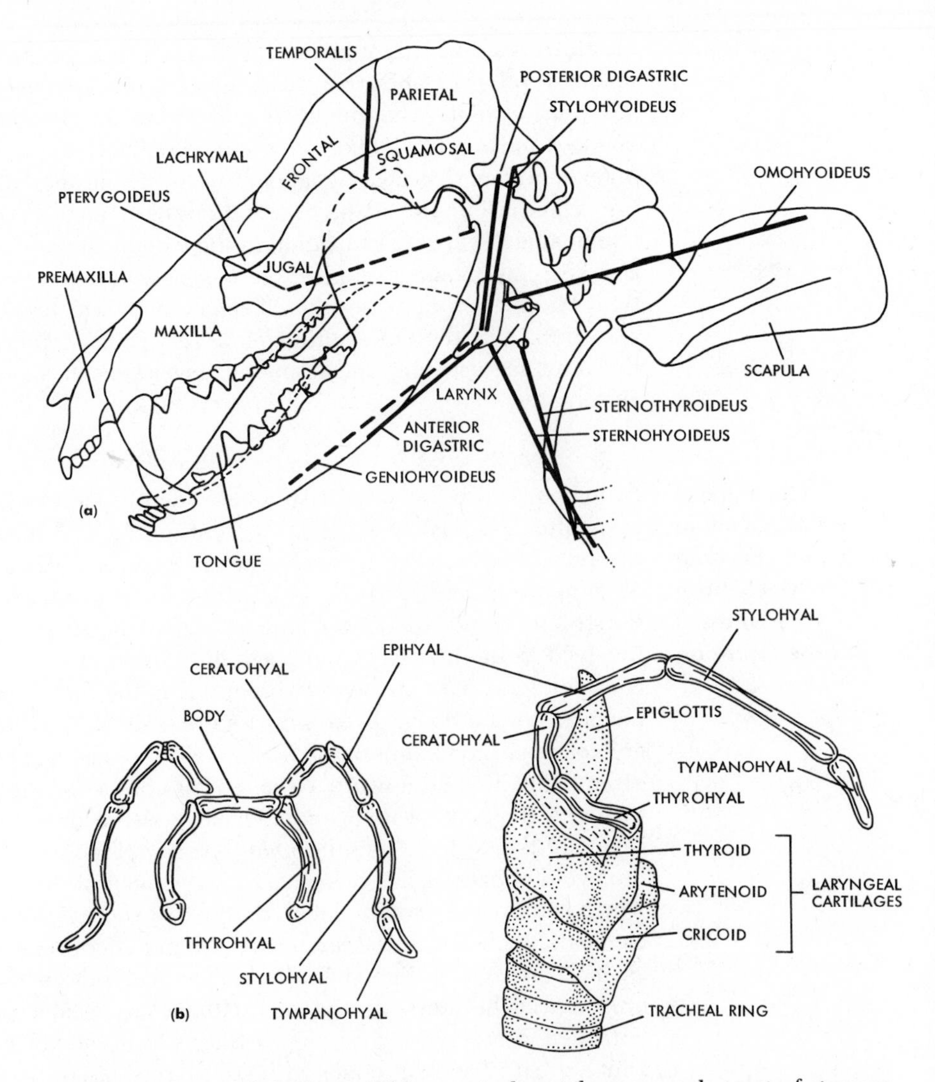

Figure 5-71 (a) Lateral aspect of skull, mandible, cervical vertebrae, scapula, part of sternum and hyoid of the opossum, with general topography of major muscles indicated by heavy lines. (b) Hyoid, larynx and trachea of a representative mammal in dorsal and lateral views. [(a) modified from Crompton et al.; (b) from Walter and Sayles, 1949.]

possesses a well-developed condyle that fits in the glenoid cavity of the squamosal to form the jaw joint.

Mastication involves movements of the skull, mandible, and hyoid relative to each other. Masticatory muscles can be divided into the hyoid and adductor muscle groups. The attachments and general lines of action of the two muscle groups are shown in Figure 5-71, and the general topography is expressed in Figure 5-72. For our purposes, it is not necessary to describe the morphological features in detail.

Movements of the chewing cycle are produced by cranial, lower jaw, and hyoid movement. The cycle can be divided into four distinct stages, based on both the direction and rate at which the upper and lower tooth rows converge or diverge. The stages [Figure 5-73(a)] are (1) fast closing (FC), in which the upper and lower teeth are brought into contact with the food; (2) power stroke

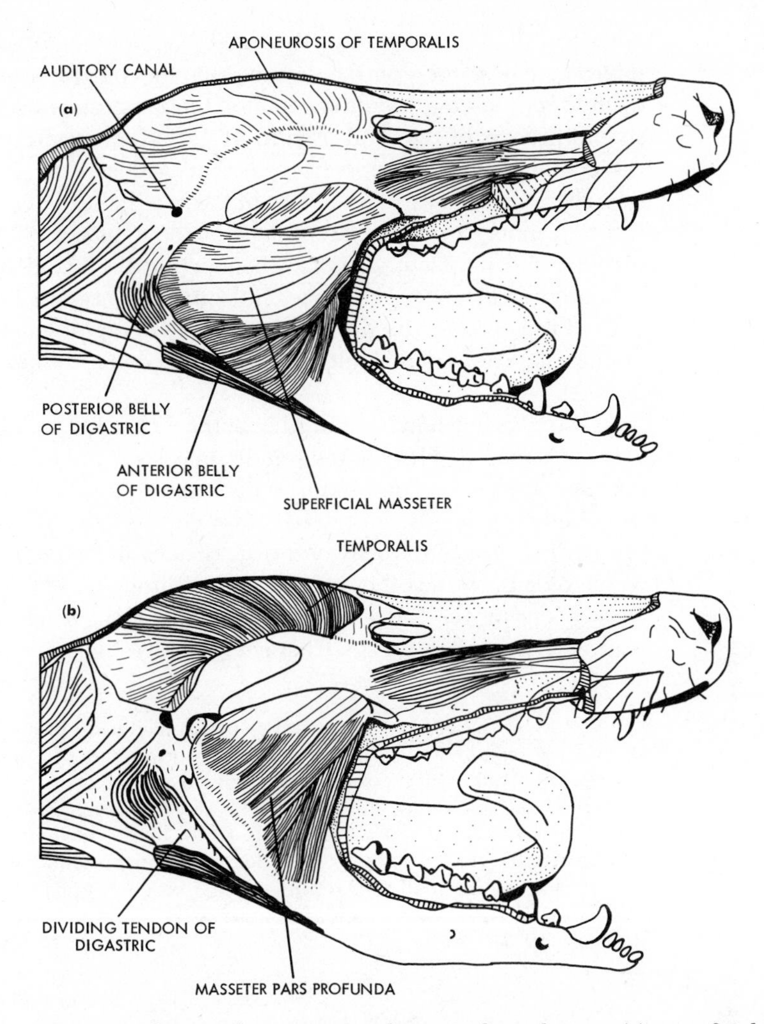

Figure 5-72 Jaw musculature of an opossum shown in lateral view. (a) superficial layer; (b) after removal of aponeurosis of temporalis and superficial head of masseter. (From Turnbull, 1970, courtesy of the Field Museum of Natural History, Chicago.)

(PS), in which the food is punctured, crushed, or sheared (the rate at which the upper and lower teeth converge is much slower than in FC); (3) slow opening (SO); and (4) fast opening (FO). Movements of the hyoid [Figure 5-73(b)] can be summarized as follows: During FC the hyoid moves slightly backward; during the PS it moves downward and backward; in SO it moves upward; and in FO the hyoid tends to move only slightly backward. Thus, the movements of the hyoid are intimately correlated with those of the jaws [compare Figure 5-73(a) and (b)]. The principal muscles (Figures 5-71 and 5-72) controlling the movements of the lower jaw are the *"temporalis,"* a pinnate muscle mass passing from the temporal fossa to either side of the coronoid process of the dentary; the *masseter,* originating from the zygomatic arch and inserting on the ventral edge of the outer face of the lower jaw; and the internal *pterygoideus,* running forward from the inner face of the lower jaw to the small pterygoid bone. The hyoid muscles include the anterior and posterior bellies of the *digastricus,* of which the anterior belly is a flat sheet of muscle attaching posteriorly to a transversely orientated common tendon and anteriorly to the inferomedial border of the lower jaw. In contrast, the posterior belly is cylindrical, running from the common tendon posterodorsally to insert on the styloid process of the skull (Figure 5-71). Other important hyoid muscles are the *geniohyoideus,* passing from the internal face of the mandible near its symphysis to the ventrolateral face of the hyoid; the *omohyoideus,* a broad, thin muscle running from the lateral face of the hyoid to the cranial angle of the scapula; and the *sternohyoideus,* connecting the hyoid with the sternum.

Crompton, Thexton, Hiiemae, and Cook [5] have analyzed the jaw mechanism of the opossum by means of electromyography synchronized with X-ray motion pictures. Their results are summarized in a composite diagram [Figure 5-73(c)].

As the jaws reach maximum gape at the end of FO and then begin to adduct, activity is recorded for the temporalis, masseter, and internal pterygoideus. As the jaws reach tooth-food-tooth contact, the amplitude of the signals increases, reaching peak levels during the slow close of the PS. There is very little activity in the anterior belly of the digastricus and geniohyoideus muscles. The omohyoideus, on the other hand, is active throughout FC but shows increasing activity during the PS. The sternohyoideus, silent for part of FC, reaches its peak activity during the PS. Characteristically, there are two distinct bursts of

Figure 5-73 Jaw and hyoid movement (a, b) in the opossum during part of a typical sequence of chewing on soft food. In (a) the relative movement of upper and lower jaws has been plotted as degrees of gape (vertical axis) against time (horizontal axis). Each chewing cycle has four stages: FC—Fast Closing; PS—Power Stroke, SO—Slow Opening and FO—Fast Opening. In (b) hyoid movements are plotted. (c) Composite diagram of activity of adductor and representative hyoid muscles are plotted diagrammatically. (Data obtained from electromyographic recordings synchronized with X-ray motion pictures, by Crompton, Thexton, Cook and Hiiemae, 1975.)

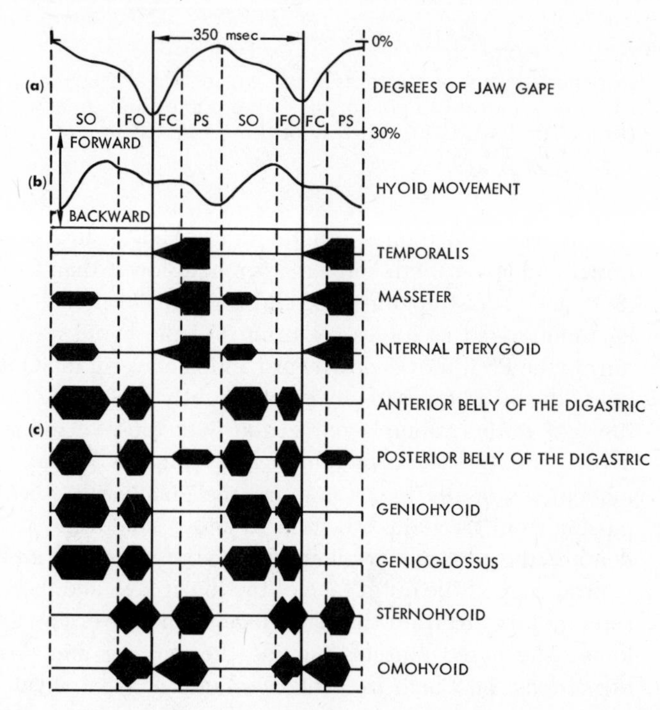

activity in the digastricus and geniohyoideus during the opening cycle [Figure 5-73(c)]: the first during SO and the second during FO. As the jaws reach maximum gape, the digastricus ceases its activity, but the sternohyoideus and omohyoideus are active during FO, FC, and the PS.

Based on these anatomical and experimental data, we can characterize the masticatory cycle of the primitive mammal feeding on soft food as follows:

1. During FC the jaw is elevated by the temporalis, masseter, and internal pterygoideus muscles. All three continue to adduct the jaws against the resistance of the food during the PS [Figure 5-74(a)]. The backward movement of the hyoid during FC is produced primarily by the omohyoideus.

Figure 5-74 Reconstructions from a lateral projection X-ray motion picture of a typical chewing cycle in the opossum, when feeding on soft food, to show the movements of the lower jaw and hyoid during each stage of the cycle. Tracings of the hyoid and lower jaw at the end of each stage (clear outline) have been superimposed on tracings of their initial position (shaded outlines). The solid black arrows show the direction and amplitude of each movement. The lines of action of the muscles active during each stage are indicated by open arrows. The width of the arrow reflects the level of EMG activity, i.e., high levels are represented by wide arrows, lower levels by narrow arrows. (a) Fast Closing; (b) Power Stroke; (c) Slow Opening; (d) Fast Opening. (From Crompton, Thexton, Cook and Hiiemae, 1975).

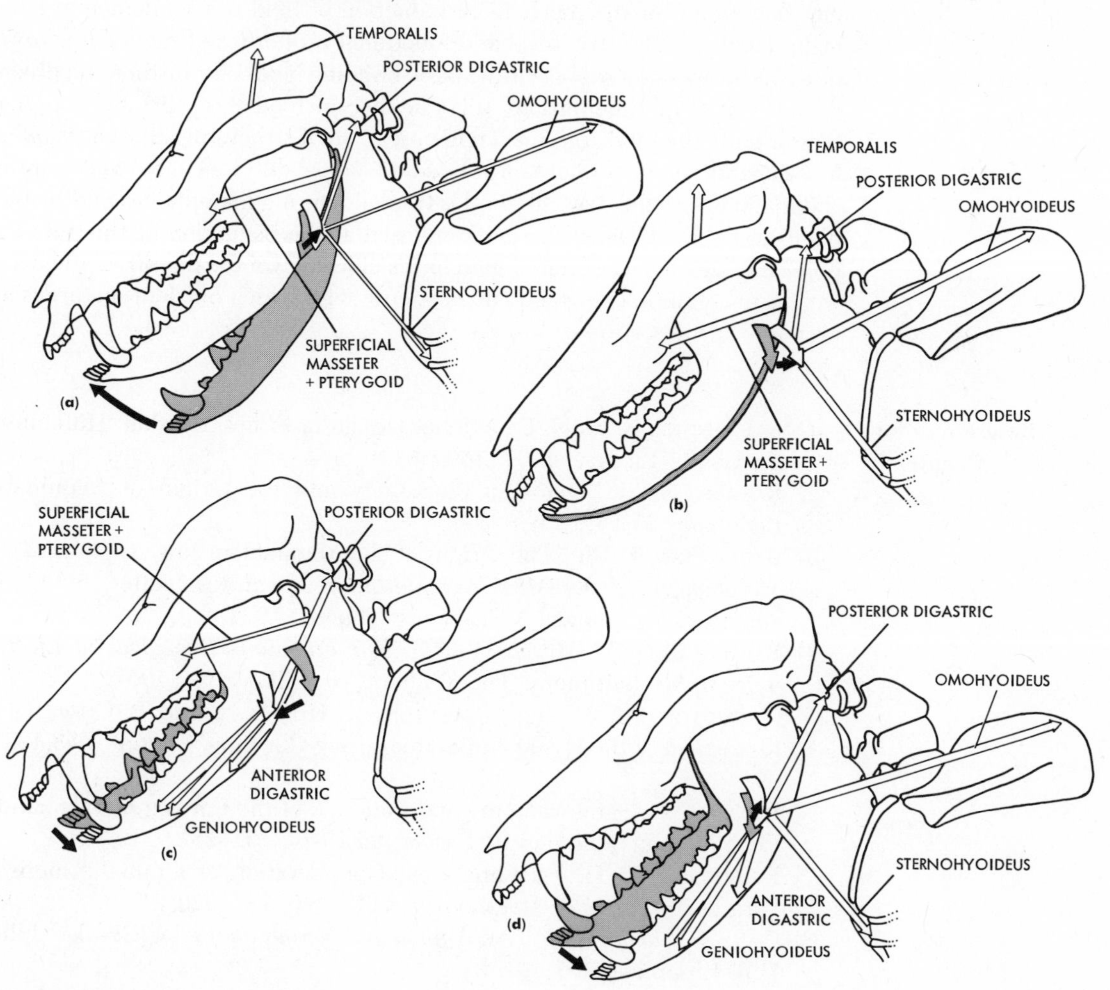

2. During the PS, the sternohyoideus becomes more active, pulling the hyoid bone markedly downward as well as backward [Figure 5-74(b)]. The combination of an extensive upward movement of the lower jaw and a downward movement of the hyoid stretches the digastricus and geniohyoideus muscles.

3. During SO, all the movements of the lower jaw and hyoid are reversed [Figure 5-74(c)]. The hyoid is pulled forward by the anterior belly of the digastricus and geniohyoideus muscles. Activity, even if at a low level, in the masseter and internal pterygoideus in SO, restricts mandibular depression.

4. In FO [Figure 5-74(d)], all the hyoid muscles are contracting. The retraction and elevation of the hyoid by the posterior belly of the digastricus, omohyoideus, and sternohyoideus ensure that the contraction of the anterior belly of the digastricus and geniohyoideus moves their mandibular rather than their hyoid attachment, so causing rapid depression of the lower jaw.

5. It is important to understand that mammalian jaw-opening mechanisms are not effected solely by the digastricus, as generally assumed. Experimental evidence has shown that jaw opening in mammals proceeds by a cooperative effort of both bellies of the digastricus, omohyoideus, sternohyoideus, and geniohyoideus muscles.

The integration of the skull, mandible, hyoid, and associated musculature into one organ for the mechanical reduction of food is a fundamental mammalian feature. The structural and functional properties of the feeding apparatus of the opossum represent the basic building blocks for further and diverse evolutionary experimentations. It is true that the food of reptiles and mammals is essentially the same, often even identical. So, if structural differences reflect functional differences, there must be some radical differences between reptiles and mammals in the way the food is treated. During mammalian mastication, plant and animal tissues are broken down at the beginning of the gut. This process results in an enormous gain in the efficiency of the digestive process, as well as in a gain in the speed and economy with which nutrient materials are prepared for absorption.

References for Chapter 5

[1] Alexander, R. McN. Functional Design in Fishes. London: Hutchinson University Library, 1967. 160 pp.

[2] Bakker, R. T. "Dinosaur Physiology and the Origin of Mammals." *Evolution,* **25**(1971), 636–658.

[3] Barghusen, H. R. "The Origin of the Mammalian Jaw Apparatus." In *Morphology of the Maxillo-Mandibular Apparatus,* edited by G. H. Schumacher. Leipzig: VEB Georg Thieme, 1972.

[4] Basmajian, J. V. *Muscles Alive. their Functions Revealed by Electromyography.* Baltimore: The Williams & Wilkins Co., 1967.

[5] Crompton, A. W., A. J. Thexton, K. Hiiemeae, and P. Cook. "The Movement of the Hyoid Apparatus during Chewing." *Nature,* **258**(1975), 69–70.

[6] Evans, F. G. "The Anatomy and Function of the Foreleg in Salamander Locomotion." *Anatomical Record,* **95**(1946), 257–281.

[7] Frazzetta, T. H. "A Functional Consideration of Cranial Kinesis in Lizards." *Journal of Morphology,* **111**(1962), 287–320.

[8] Gans, C. *Biomechanics: An Approach to Vertebrate Biology.* Philadelphia: J. B. Lippincott Co., 1974.

[9] ———, and W. Bock. "The Functional Significance of Muscle Architecture: A Theoretical Analysis. *Ergebnisse der Anatomie und Entwicklungsgeschichte,* **38**(1965), 115–142.

[10] ———, and T. S. Parsons. *Biology of the Reptilia,* vol. 4. New York: Academic Press (1973).

[11] Gray, J. *Animal Locomotion.* London: Weidenfeld and Nicholson, 1968.

[12] Haines, R. W. "The Shoulder Joint of Lizards and the Primitive Reptilian Shoulder Mechanism." *Journal of Anatomy,* **86**(1952), 412–422.

[13] Hildebrand, M. "How Animals Run." In N. K. Wessells (ed), *Vertebrate Adaptations,* pp. 30–37. San Francisco: W. H. Freeman & Co., Publishers, 1968.

[14] Hopson, J. A. "The Origin of the Mammalian Middle Ear." *American Zoologist,* **6**(1966), 437–450.

[15] Huxley, H. E. "Muscular Contraction." *Endeavor,* **15**(1956), 177–188.

[16] Jenkins, Farish A. "Limb Posture and Locomotion in the Virginia Opossum (*Didelphis marsupialis*) and in Other Noncursorial Mammals." *Journal of Zoology.* **165**(1971), 303–315.

[17] Kummer, B. *Bauprinzipien des saugetierskeletes.* Stuttgart: Georg Thieme, 1959.

[18] Liem, K. F. "Evolutionary Strategies and Morphological Innovations: Cichlid Pharyngeal Jaws." *Systematic Zoology,* **22**(1973), 425–441.

[19] Lombard, E., and D. B. Wake. "Tongue Evolution in the Lungless Salamanders, Family Plethodontidae. I. Introduction, Theory, and a General Model of Dynamics." *Journal of Morphology,* (1976), 265–286.

[20] Nursall, J. R. "Swimming and the Origin of Paired Appendages." *American Zoologist,* **2**(1962), 127–141.

[21] Panchen, A. L. *Batrachosauria.* In *Handbbuch der Palaoherpetologie* (*Encyclopedia of Paleoherpetology*), edited by O. Kuhn, part 5. Portland, Ore.: G. Fischer Verlag, 1970.

[22] Romer, A. S. *The Osteology of the Reptiles.* Chicago. University of Chicago Press, 1956.

[23] Schaeffer, B. The morphological and functional evolution of the tarsus in amphibians and reptiles. *Bulletin of the American Museum of Natural History,* **78**(1941), 395–472.

[24] ———, and D. E. Rosen. "Major Adaptive Levels in the Evolution of the Actinopterygian Feeding Mechanism." *American Zoologist,* **1**(1961), 187–204.

[25] Slijper, E. J. "Comparative Biologic-Anatomical Investigations on the Vertebral Column and Spinal Musculature of Mammals." *Verhandelingen Akademie Wetenschappen Amsterdam,* **42**(1946), 1–128.

[26] Snyder, R. C. "The Anatomy and Function of the Pelvic Girdle and Hindlimb in Lizard Locomotion." *American Journal of Anatomy,* **95**(1954), 1–46.

[27] Thomson, K. S. "The Evolution of the Tetrapod Middle Ear in the Rhipidistian-Amphibian Transition." *American Zoologist,* **6**(1966), 379–397.

6

Digestion and Nutrition

Introduction Like the epidermis, cells that line the gastrointestinal tract have direct encounter with the external world. A major function of the epidermis of most vertebrates (amphibians excepted) is to limit strictly the entry of chemical substances and parasitic organisms from the outside, while permitting egress from the body of specialized secretions (slime, oil, sweat, and so on). In striking contrast, the major function of the epithelium of the digestive tract is to facilitate the entry of a wide range of chemical substances into the body. This is not an easy task. It involves the formation of enzyme-containing secretions needed for the chemical breakdown of food materials. Often it must be supported by the mechanical breakdown of food objects, and transfer processes are needed to bring nutrients into suitable relationship to the surfaces where entry takes place. Various self-protective devices are present that, in general, limit to a certain extent the ingress of harmful substances and invading organisms.

The digestive system also has other, somewhat less conspicuous, roles. It removes undesirable substances from the body through the excretory processes of the liver and the mechanical egestion of food materials taken in but undigested. Relatively large masses of bacteria and of dead cells lost from the walls of the digestive tract itself also are discarded. As will be noted later, the tract contains structures important for a wide range of additional important bodily functions—search for and selection of food, self-defense, respiration, reproduction, and kidney excretion—through activities of some of its specialized regions (*buccal cavity, pharynx, aboral intestine, cloaca*).

The structures of vertebrate gastrointestinal systems reflect the variety and complexity of their functions. In all vertebrates, digestive tracts include spaces (e.g., stomach, intestinal chambers) where ingested food materials are subjected to enzyme-catalyzed reactions that break them down to individual molecules. These processes depend on formation by secretory cells of fluids that contain a range of enzymes and ions dissolved in water. The secretory cells exist individually in the epithelium or are grouped into glandular organs (including the pancreas and the massive liver in all vertebrates, and salivary glands in most vertebrates except fish). In some of the digestive chambers, microorganisms flourish, and there break down nutrients by enzymatic reactions that supplement the effects of the animal's own enzymes. Such microbial processes are exploited to a particularly high degree in ruminant mammals.

There are relatively huge surface absorptive areas across which exchanges occur from the luminal environment within the digestive tract, but still outside the epithelial boundaries of the body, to the neighboring blood stream. These absorptive areas, lying chiefly in the intestines, are characterized by a single epithelial cell layer with rich underlying meshes of blood and lymph capillaries. The transfer of materials across this cellular layer is partly passive—by diffusion from the gastrointestinal lumen—and partly by active, energy-requiring processes within the cells themselves.

The mechanical breakdown of food materials and the transfer of nutrients depend chiefly on the specialized muscles built into, or associated with, the digestive system (Figure 6-1). Ciliated surfaces play some part in transfer, too, especially in lower vertebrates. Characteristically, secretory cells in the epithelium form *mucus*, a slimy, often quite complex, protein-containing fluid that plays an important part in lubricating the nutrient masses in transit and in protecting the delicate epithelia from mechanical and chemical damage.

There is provision for storage in all vertebrate digestive systems. More or less unworked food materials can be held in distensible esophagi or stomachs (most vertebrates), *crops* (expanded chambers in the esophagi of many birds) or skin-lined cheek pouches (some mammals). Molecular-sized nutrients (e.g., glycogen, iron, some fats, and vitamins) are stored in various places—chiefly in the liver. Ingested material that has been thoroughly worked over in upper areas of the tract may be stored in the aboral part of the intestine or in the cloaca, pending completion of water absorption and an appropriate time for its expulsion from the body as feces.

Thus, a complex of basic processes—ingestion, digestion (chemical breakdown), absorption, mechanical transfer, self-protection, excretion, and egestion—are carried out by all vertebrate digestive systems in closely integrated patterns (Figure 6-2). The fundamental origin of the patterning involves the structure of the tract itself, with its supporting vascular systems (blood, lymph). This structural organization is modulated and supplemented by sensitive responses of the individual epithelial and muscle cells of the tract and by adaptive nervous and endocrine control. The over-all patterning is visualized best, perhaps, by reviewing in some detail the form and function of the relatively simple digestive systems of amphibians. Following this, variations and adaptive mechanisms in members of the other vertebrate classes will be noted more briefly.

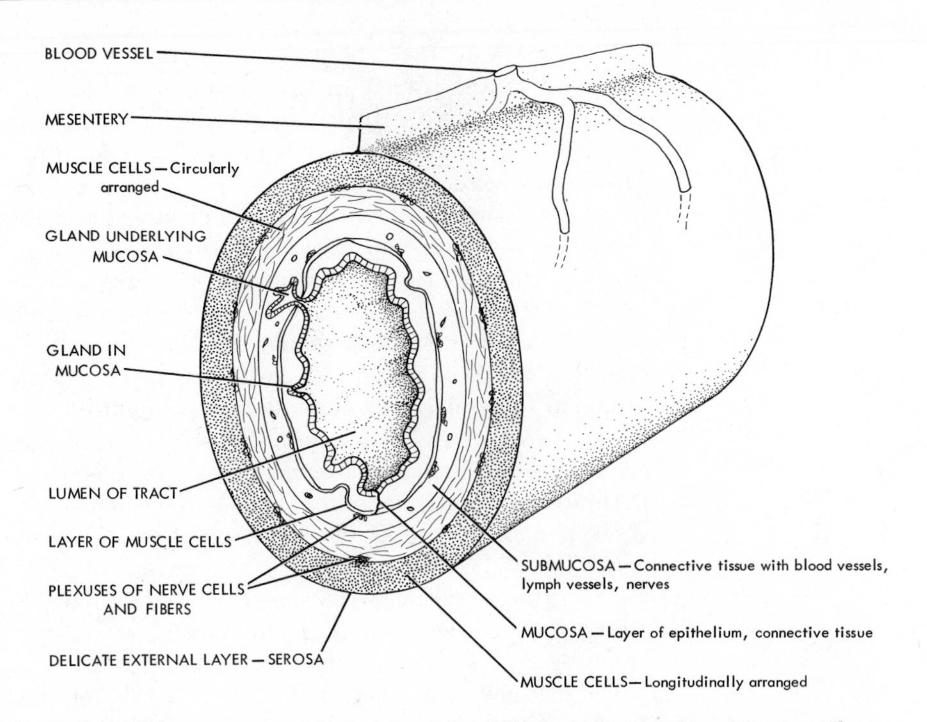

Figure 6-1 Diagram representing some features of the digestive tract of vertebrates. The tubular structure (gut) varies greatly in different animals, and in any one animal from one part to another, in shape (e.g., expanded stomach chamber, narrow intestinal region), function, pattern of lining epithelium and glands, etc. Some features which are quite generally found include: (1) The *tunica mucosa*, an inner lining of epithelial cells underlaid by connective tissue. The mucosa may be deeply folded, or modified by the presence of many special processes, *villi*. The epithelial cells of the mucosa secrete a variety of substances (water, salts, mucus, enzymes), and also carry on the absorptive activities of the intestine. (2) The *tunica submucosa* is a connective tissue layer under the mucosa. It may contain complex glands with ducts leading to the mucosal surface. (3) *Muscle* layers, made up of smooth muscle cells, are usually arrayed in part along the circumference and in part along the major axis of the gut (circular and longitudinal muscles), although other arrangements of muscle cells may occur as well. In general, contraction of the muscle cells in the circular layer results in narrowing and lengthening of the gut segment, while contraction of the longitudinal muscle cells shortens the gut segment. (4) A delicate sheath, the *serosa*, covers the digestive tract when the tract lies free in the body cavity. In those conditions, the gut is attached to the wall of the body cavity by the thin, sheet-like mesentery traversed by large blood and lymph vessels, and by nerves. Rich nets of blood vessels, lymph vessels (except in cyclostomes and elasmobranchs), and nerves lie in and between the connective tissue and muscle cell layers.

Digestive Systems of Amphibians: *Necturus maculosus*

Figure 6-3 shows some characteristic features in the anatomy of the urodele amphibian, *Necturus maculosus*. Typically, in amphibians, the opening of the mouth into the buccal cavity is surrounded by epidermally lined lips. Small, conical teeth are set along the margins of the jaws, and sometimes also in additional rows or groups on the roof of the mouth. These function chiefly in holding the slippery, squirming organisms that are the typical prey of amphibians. Generally, there is a muscular tongue attached to the floor of the

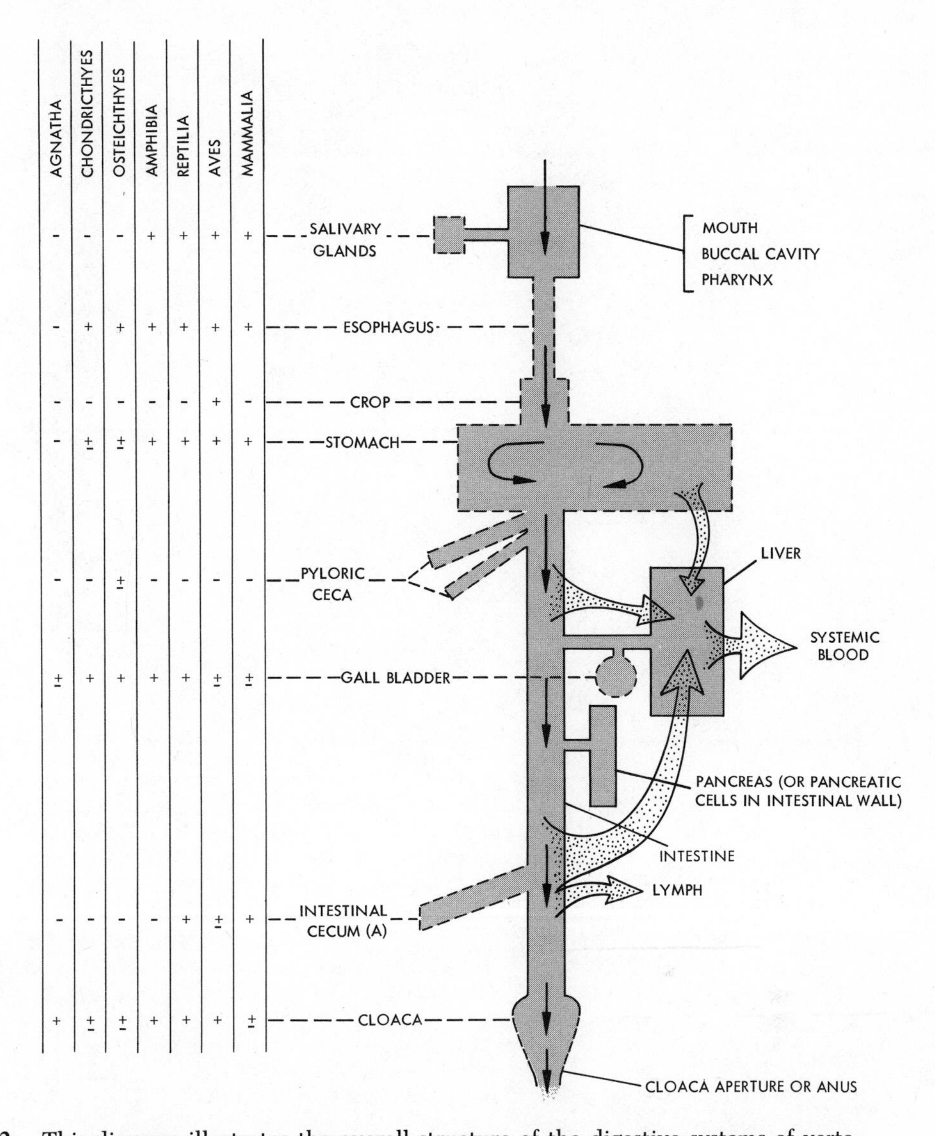

	AGNATHA	CHONDRICHTHYES	OSTEICHTHYES	AMPHIBIA	REPTILIA	AVES	MAMMALIA
SALIVARY GLANDS	-	-	-	+	+	+	+
ESOPHAGUS	-	+	+	+	+	+	+
CROP	-	-	-	-	-	+	-
STOMACH	-	±	±	+	+	+	+
PYLORIC CECA	-	-	±	-	-	-	-
GALL BLADDER	±	+	+	+	+	±	±
INTESTINAL CECUM (A)	-	-	-	-	+	±	+
CLOACA	+	±	±	+	+	+	±

Figure 6-2 This diagram illustrates the overall structure of the digestive systems of vertebrates. Components indicated in solid lines, and named on the right side of the diagram, are present in all vertebrates. Components indicated by dashed lines are absent in some or many vertebrates; in the table on the extreme left of the diagram, their presence (+) or absence (−) is listed for each of the classes. Note that, in some cases, the structure may be either present or absent in different members of a given class, and in that case the indication is (±). The direction of movement of food materials through the tract is indicated by solid arrows, while the flow of absorbed nutrients is shown by stipled arrows. Thus, in a pattern that is consistent throughout the vertebrates, most nutrients are absorbed in the intestine and travel from there to the liver where they may be stored, or modified chemically, or distributed unchanged through the circulatory system to the rest of the body. Some absorption may occur through the walls of the stomach (small organic molecules only); and some nutrients absorbed by the intestine bypass the liver (by entering the lymphatic circulation) and are carried thence to the blood and body as a whole.

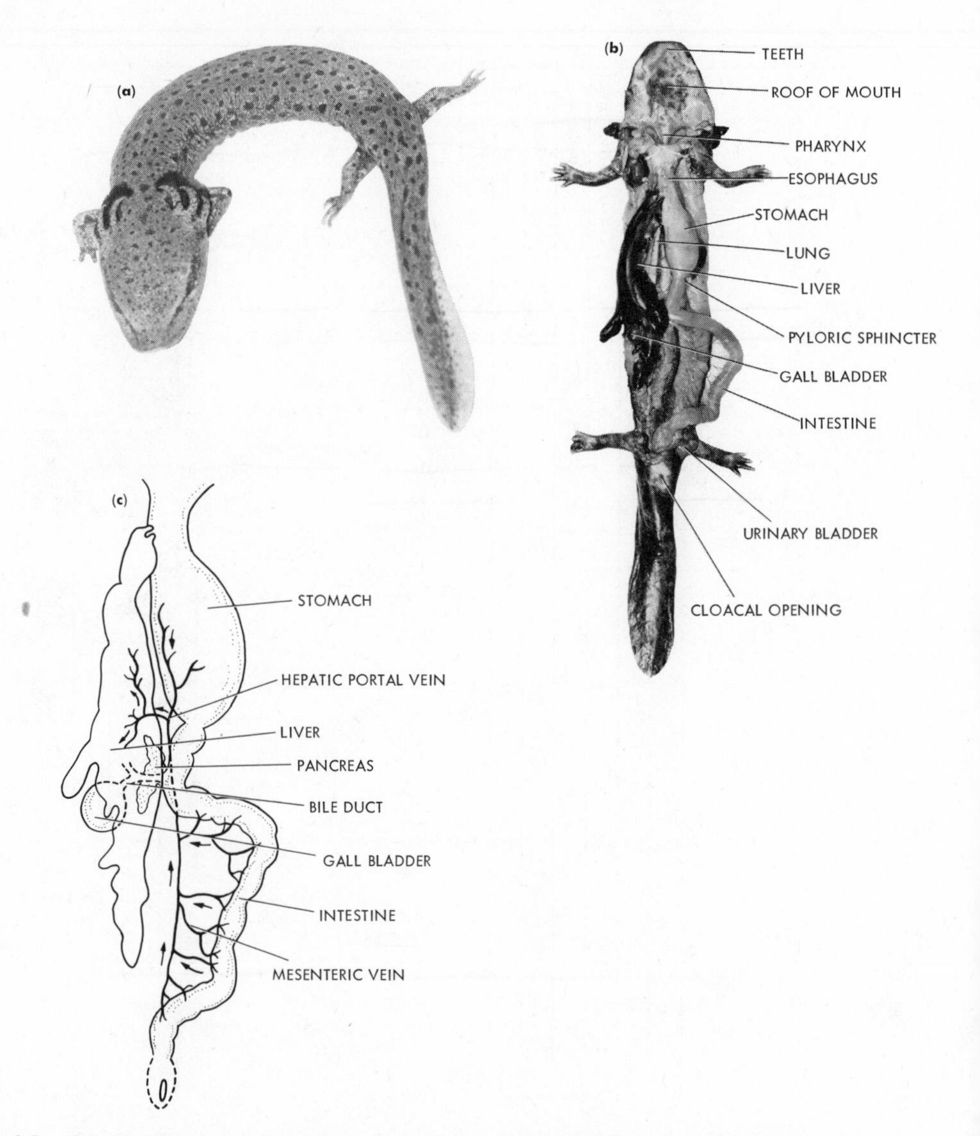

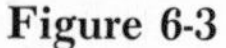

Figure 6-3 (a) *Necturus maculosus* is an aquatic urodele amphibian possessing both lungs and external gills as respiratory organs. The digestive system of *Necturus* is structured in a manner fairly representative of amphibians in general. It is shown in ventral view in (b) after removal of the lower surface of the body, including the lower jaw and ventral wall of the pharynx. In (c) the pattern of veins that carry blood containing absorbed nutrients from the intestine to the liver is indicated in diagrammatic form. The connections between the liver, gall bladder, and intestine are shown as well. (Photographs by Sigurd Olsen.)

mouth that aids in securing nutritive material and guiding it into the pharynx. An especially interesting tongue structure occurs in anuran amphibians (frogs, toads) (Figure 6-4). The tongue is attached only at its anterior region. It can be flicked out of the mouth to trap a flying insect and then quickly retracted by contraction of its muscles to bring the prey back with it into the buccal cavity. In the roof of the mouth are the respiratory openings, or internal nares (*choanae*). Thus, in amphibians, as in some fish and all higher vertebrates, a

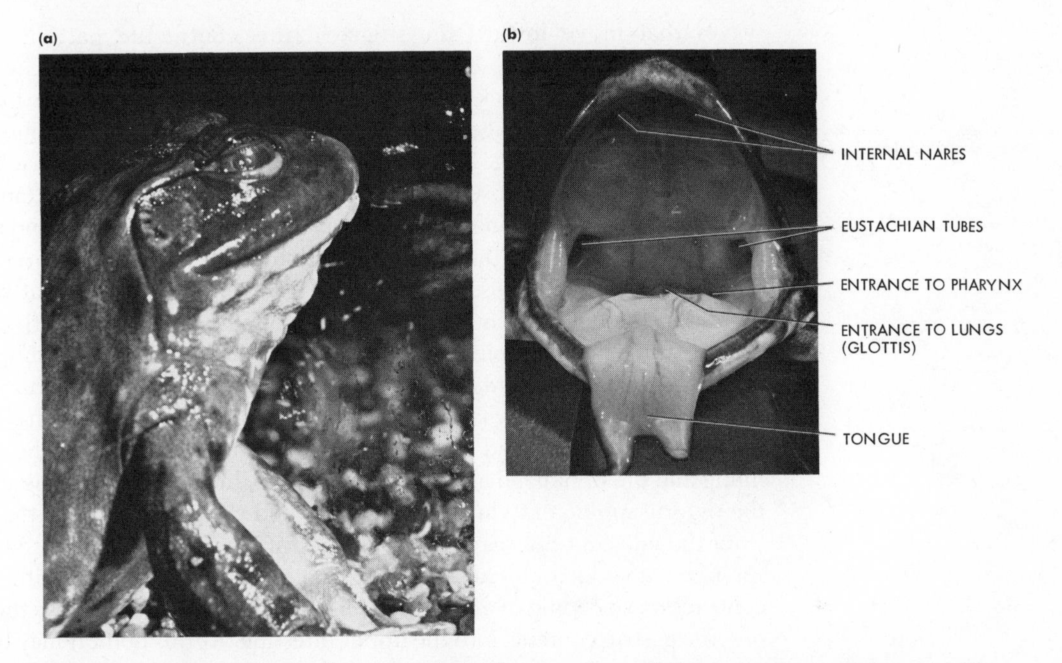

Figure 6-4 The structure of the mouth in the anuran amphibian, *Rana catesbeiana.* (a) The animal is at rest, seen from the side. (b) The mouth is widely open, with the tongue in its fully extended position. When the mouth is closed, the tongue folds back upon itself into the mouth. (Photographs by Sigurd Olsen.)

common channel through the buccal cavity leads to the respiratory and digestive organs.

In most amphibians, much of the buccal cavity is lined with ciliated epithelium. Beating continuously in the direction of the pharynx, the cilia carry particles of food and secretions toward the aboral region of the tract. There are numerous mucus-secreting cells in the epithelium lining the mouth cavity. Larger and more complex salivary glands lie adjacent to the buccal cavity, and their ducts open into it. Richly supplied with blood vessels, like the buccal mucosa itself, these glands are capable of secreting large amounts of fluid. They are absent in some aquatic amphibians, and it is inferred that their chief role in terrestrial forms is to provide the fluid that is so helpful in swallowing food. Amphibian salivary glands also may produce starch-splitting enzymes, the importance of which is doubtful, as most digestion of carbohydrates and other food components occurs in the intestine. By complicated muscular contractions involving the tongue and the floor of the mouth, ingested food is thrust into the pharynx. This is an expanded chamber, lined with ciliated, multilayered and mucus-secreting epithelium, that leads into the next part of the digestive system (the esophagus) and also carries air to the lungs. During swallowing, the transit of food into the lungs is prevented by muscular contractions that close off the glottis, the entryway into the lungs.

From the pharynx, food passes into the *esophagus*, a relatively short, thin-walled tube lined with ciliated, stratified epithelium. Scattered gland cells in the mucosa form mucus and proteins that can change to proteolytic enzymes (pepsinogens). Layers of muscle cells in the wall contract rhythmically in waves that move toward the stomach (this contractile pattern is termed peristalsis). Food in the esophagus travels with these peristaltic waves and generally cannot go back into the mouth because the esophagus has a thick band of circularly arranged muscle cells (a *sphincter*) at its junction with the pharynx. By contracting, this can effectively close off the route back into the upper part of the tract. Once the food mass has reached the stomach, it is restrained from reentering the esophagus by contraction of a second sphincter at the junction between the esophagus and stomach. In passing, it may be noted that such sphincters, muscular structures guarding entry or exit at particular regions of the tract, are important features of all vertebrate digestive systems.

The stomach of amphibians is an expanded chamber of a simple, long, spindle shape in *Necturus* and other urodeles, and of similar form, although often wider and somewhat curved, in the shorter-bodied anurans. Here the food mass is retained and flooded with gastric juice secreted by the epithelial lining and the organized glands of the wall. At the aboral end of the stomach is the pyloric sphincter, which can contract and effectively seal off the stomach from the adjacent intestine. Much of the time, while digestion proceeds, this sphincter is indeed closed. From time to time, in correlation with waves of contraction sweeping over the stomach walls, it relaxes and allows the passage of some gastric contents into the upper intestine. At this point it may be well to consider more specifically the role of the stomach. In amphibians and other vertebrates with stomachs (later it will be noted that in some groups of fish the stomach is absent), this organ appears to have a major common function: the maintenance of collected food materials and their regulated supply to the dominant digestive-absorptive organ, the intestine. This storage function is correlated, in almost all normal vertebrates, with the secretion of gastric juice rich in hydrochloric acid. It has been inferred that the original role of this strongly acid gastric secretion may have been to prevent bacterial action on nutrients in the stomach. This is especially important in vertebrates that store food in the stomach for relatively long periods of time (from one to several days in fish, amphibians, and reptiles). At the same time, the low pH is important in protein digestion. This is another function of the stomach, inferred by some zoologists to be a more recent innovation in evolution. For, in addition to hydrochloric acid and the ubiquitous mucus, gastric secretory cells form *pepsinogens*, which quickly turn into proteolytic enzymes (pepsins) on egress from cells and exposure to acid. Optimally active at the low pH of the stomach contents, pepsins catalyze the breakdown of proteins in the ingested food materials. Generally, considerable protein digestion has occurred by the time the gastric contents are ejected from the aboral (pyloric) region of the stomach into the oral region of the small intestine.

In *Necturus* and other urodele amphibians, the small intestine is a short, somewhat coiled tube with a relatively narrow diameter (Figure 6-3). It tends to be longer and more coiled in anurans. The mucosa lining its walls is folded or ridged in patterns that vary from one species to another but always functions to increase its active surface area. Sometimes the surface shows multiple tiny, fingerlike extensions, each termed a *villus*. Within the villi, ridges, and folds, and everywhere else under the surface, are rich networks of blood and lymph capillaries. Lining the surface are cells of characteristic fine structure, with membranes facing the intestinal lumen, that are modified to increase surface area by the formation of myriad tiny projections, or *microvilli* (Figure 6-5).

These give the impression of striation when seen at the magnification of the light microscope. Cells with microvilli were described at an early stage of study as having a *brush border*. Mucus-secreting cells and simple glands are present as well. Into the small intestine lead ducts from the *gall bladder* and pancreas, bringing bile from the liver and pancreatic juice from the pancreas. With the intermittent contractions of the muscles of the intestinal walls, the material within the intestinal lumen is thoroughly mixed with the variety of enzymes, bile salts, ions, and secretions of the intestine, liver, and pancreas, and digestion is carried nearly to completion. At the same time, nutrients and components of the mixture traverse the epithelial lining of the intestine into the tissue spaces and from there go into the blood and lymph vessels underlying the epithelium.

Beyond the small intestine, the contents are moved by contractions of the muscular walls into the short, wide-bored large intestine. Here digestion continues and may be supplemented by bacterial decomposition. Absorption of water and salts condenses the gut contents, conserving these valuable inorganic constituents. It is not known whether absorption of organic materials also occurs in the large intestine of amphibians, but in general such absorption is not a function of this region of the digestive tract of vertebrates.

Finally, moving from the large intestine, the now useless remnants of the ingesta—fragmented chitinous skeletons of insects, undigested bones and skin of small vertebrates, cellulose fibers from plants, bacteria, bits of dead cells—enter the cloaca. The cloaca is a short, broad chamber with connections from the ureters, reproductive system, and the urinary bladder, as well as from the major inlet from the large intestine. Through its single orifice, the egesta of the digestive tract (feces) as well as sperm, ova, and urine gain exit from the body.

The liver, as shown in Figure 6-3, is a large gland of somewhat irregular form, lying along the stomach and upper part of the intestine. Into it, through the hepatic portal vein, flows the venous blood draining the absorptive surfaces of the intestine. This blood traverses a vast series of irregular channels lying among and in intimate proximity to the liver cells (Figure 6-6). Newly absorbed nutrients pass from the blood into the liver cells, there to be reprocessed (as in the synthesis of blood proteins from absorbed amino acids by liver cells) or stored (as in the deposition of liver glycogen, formed in part from glucose absorbed from the intestine). Some of the nutrients bypass the liver cells, of course, and travel through the great hepatic veins toward the heart, from there to be distributed throughout the body. The liver cells also form the special secretion, *bile*. This contains components of importance in the over-all digestive process: bile salts that aid in the attack of enzymes on fats, various ions, and water. It should be mentioned, too, that bile formation participates in the process of excretion (this major regulatory function will be discussed more fully in Chapter 8). Thus, liver cells extract from the blood and secrete into the bile certain waste organic compounds: for instance, breakdown products of hemoglobin metabolism (bile pigments). Traveling from the liver to the intestine, these compounds then move down the digestive tract and eventually are discarded from the body.

As bile is formed, it accumulates in special channels, bile capillaries, that lie among the liver cells in patterns roughly parallel to, but separate from, the blood capillaries. The secretion then passes from the bile capillaries into larger and larger ducts and finally into the hepatic duct leading out of the liver.

Shortly after emerging from the liver, the hepatic duct joins a second duct (cystic duct) from the gall bladder, a spherical storage chamber for bile. The duct emerging from the confluence of the hepatic and cystic ducts (common bile duct) leads to the intestine (Figure 6-2). The mechanisms of the vertebrate gall bladder and bile ducts have been studied most fully in mammals. It seems, however, that the system functions similarly in other classes because the general structure and the functions of bile appear to be closely parallel among vertebrates. Even the lack of a gall bladder in certain birds and mammals is not a major exception to the generality of bile processing because the absence can be interpreted as an adaptation to the special dietary patterns of these animals. In any case, for lack of adequate studies on amphibians, most of the rest of this account of bile formation and transport must be based on observations of mammals, but with the implication that it is relevant to amphibians as well. At the junction between the common bile duct and the intestine, a specialized array of smooth muscle cells acts as a sphincter. This is of functional importance because contraction of these muscles (closure of the sphincter) blocks the flow of bile into the intestine and, thus, forces bile back into the gall bladder where it is stored. While in the bladder, the bile may be concentrated and changed in composition by activities of the cells of the bladder wall. When partially digested food (especially if rich in fatty materials) enters the duodenum, hormonal and nervous stimuli act on the gall bladder and sphincter. Bile is driven out, by contraction of bladder muscles, through the common bile duct and the now opened sphincter into the intestine. As a whole, these structures—bile ducts, gall bladder, and sphincter—provide a means of supplying bile to the intestine in close correlation with digestive demands.

The pancreas is an irregularly shaped gland, far smaller than the liver, lying close to the small intestine (Figure 6-3). In amphibians, as in other vertebrates, it is made up of cellular structures, the *acini*, of a tubular or roughly spherical shape (Figure 6-7). The pancreatic cells that form each acinus lie in close association with a rich capillary network and produce a large volume of secretion that is poured into an associated duct system. The secretion contains

Figure 6-5 (*opposite*). (a) In cross-section, the small intestine of the amphibian *Amphiuma* (photomicrograph, enlargement 225 ×) shows deep folds of the inner surface lining the lumen. This inner surface is a layer of lightly staining epithelial cells, where much of the production of enzymes takes place. Mucus is formed here too, and absorption of water, salts, and nutrients occurs across the luminal boundary of the cells.

(b) A small section of the intestinal wall of *Amphiuma* is shown in a micrograph, magnified 442×. The luminal surface shows a characteristic *brush border.* Several specialized cells may be seen in the process of discharging large masses of mucus into the intestinal lumen.

(c) Further details of the amphibian intestinal cell are indicated in this diagram derived from electron micrographs of amphibian intestinal tissue. The brush borders, when seen at high magnification, prove to consist of great numbers of fine processes, *microvilli.* At very high magnification these are seen to be embedded in and covered by a surface coat which is thought to have a protective function. Within the cell lie characteristic structures important in absorption and in synthesis of digestive enzymes, e.g., smooth and rough endoplasmic reticulum, Golgi apparatus, and numerous vacuoles.

The histological material shown in (b) was made available through the kindness of Mrs. Deborah Christensen.

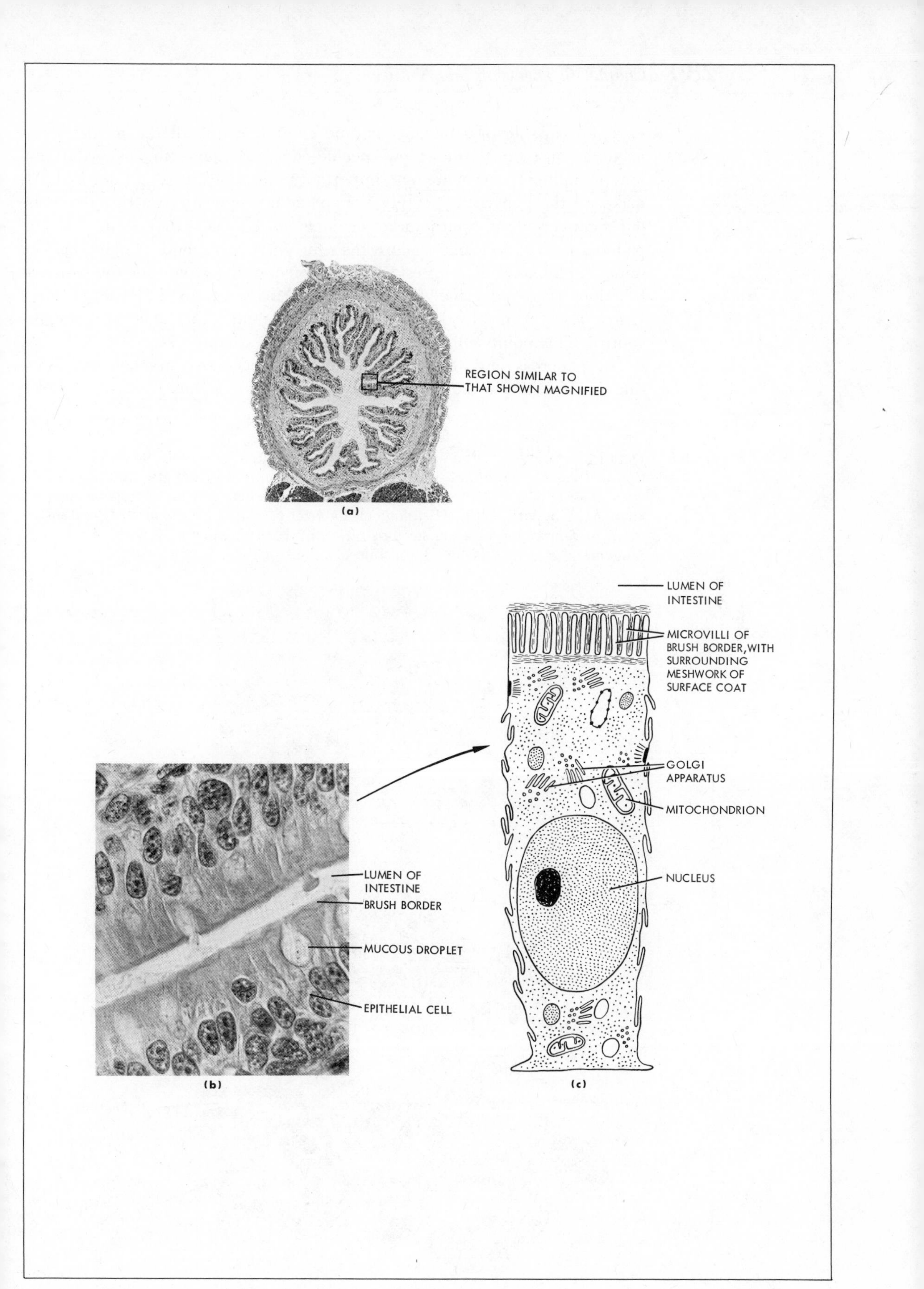
REGION SIMILAR TO
THAT SHOWN MAGNIFIED
(a)
LUMEN OF
INTESTINE
BRUSH BORDER
MUCOUS DROPLET
EPITHELIAL CELL
(b)
LUMEN OF
INTESTINE
MICROVILLI OF
BRUSH BORDER, WITH
SURROUNDING
MESHWORK OF
SURFACE COAT
GOLGI
APPARATUS
MITOCHONDRION
NUCLEUS
(c)

enzymes capable of attacking organic compounds of all the major classes occurring in food. Besides these specific chemical agents, mucus, water, and ions (including bicarbonate ions, important in the adjustment of intestinal pH) make up the pancreatic secretion. Each acinus opens into a tiny duct; the ducts converge and then join larger ducts until, at last, they form the main pancreatic duct (or ducts—more than forty-five are found in individuals of some amphibian species). Besides its conspicuous digestive role, the pancreas functions as an endocrine gland. Special cell masses, termed islets (or *islets of Langerhans*), form blood-borne hormones of great importance in metabolic control. This point will be discussed further in Chapter 11.

The pancreas—like the liver and, in fact, the entire stomach and intestine—is attached to the body wall by a *mesentery*, a thin, transparent tissue

Figure 6-6 Detail of structure of the amphibian liver. (a) Photomicrograph (96×) of a section of liver of *Amphiuma*, showing dense arrangement of hepatic and pigment cells (macrophages). Numerous blood and bile channels can be distinguished among the hepatic cells. (Histological material prepared by Deborah Christensen.) (b) Redrawing of a classic illustration by Eberth, showing details of the arrangement of cells and blood and bile channels of the liver of a frog.

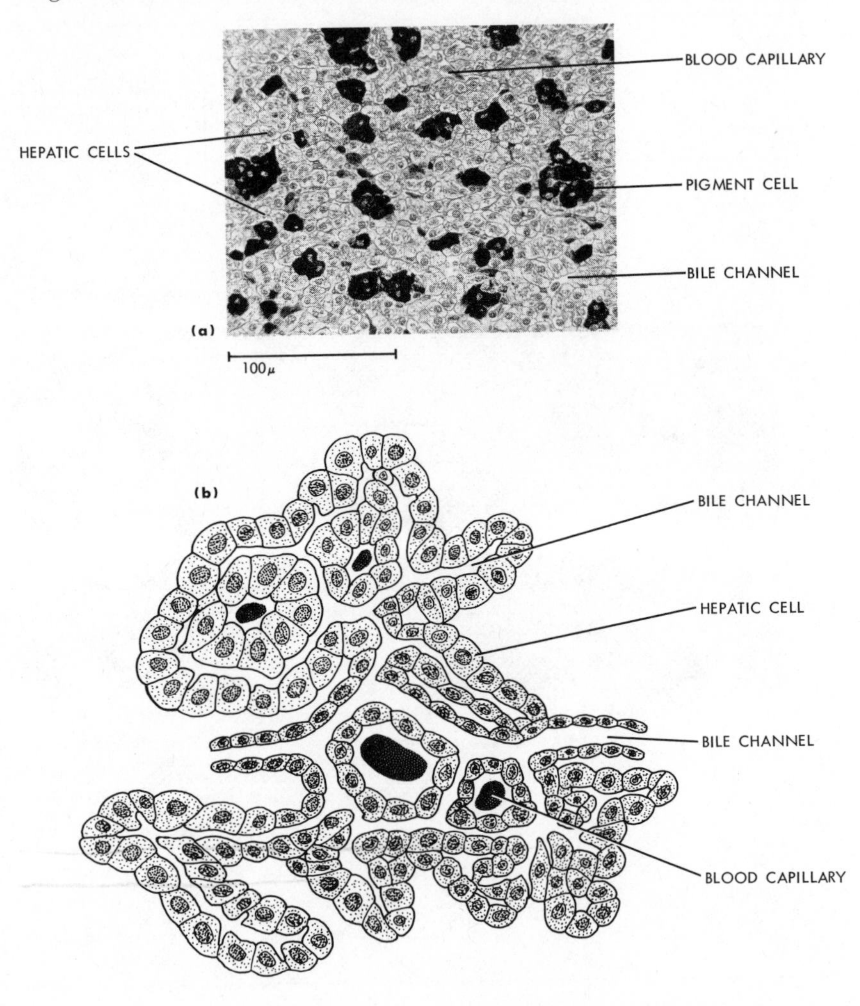

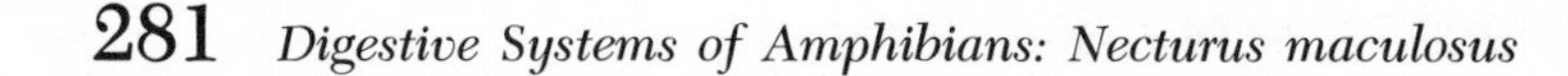

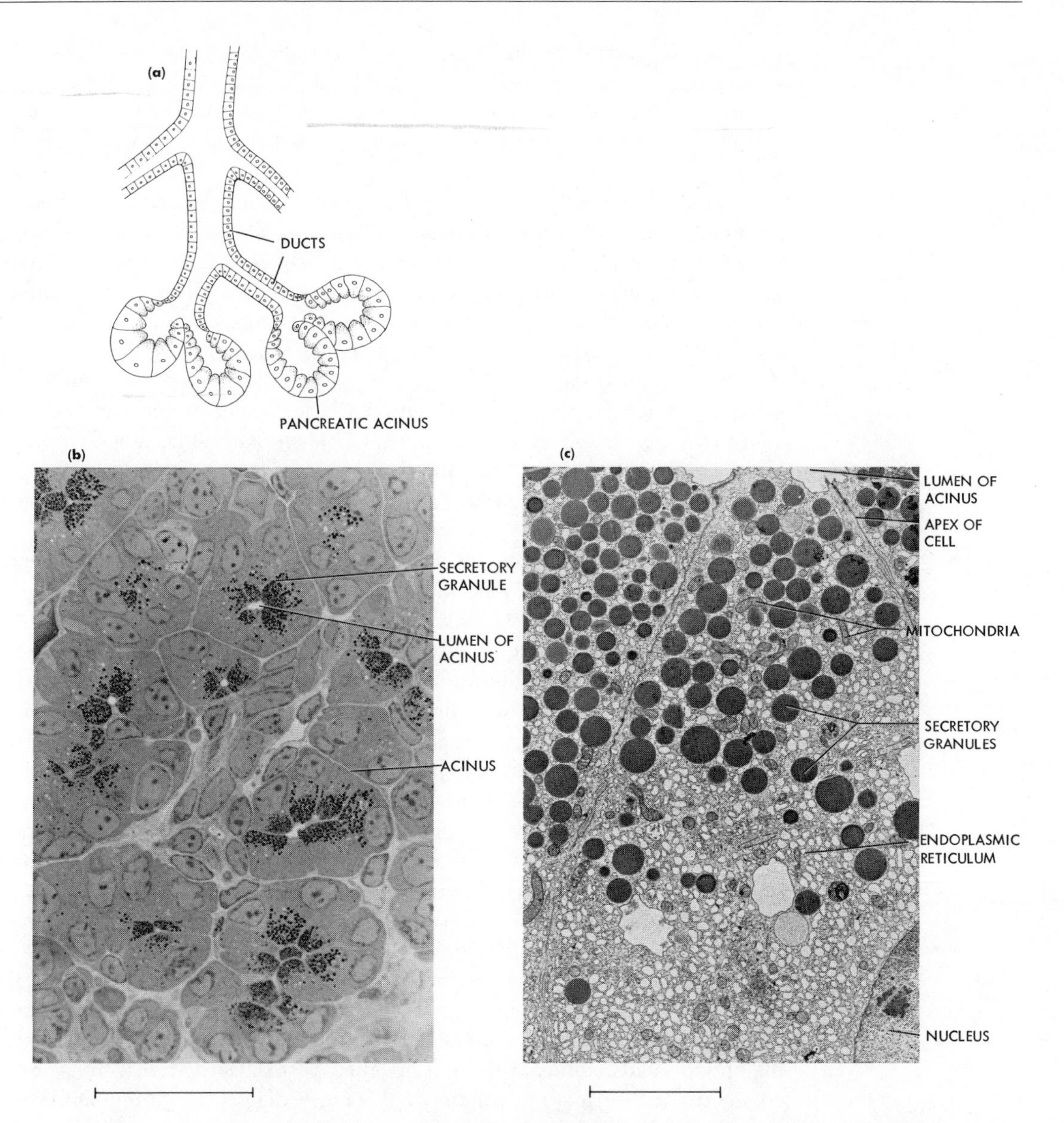

Figure 6-7 The pancreas of amphibians is built up of a complex system of branching cellular ducts leading into tubular acini where the secretion is formed. This is diagrammed in (a). (b) is a photomicrograph of pancreatic tissue of the urodele amphibian *Amphiuma means.* Proteins that are converted on secretion into active digestive enzymes, are packed into special granules, secretory (or zymogen) granules. (c) This electron micrograph of a small part of the *Amphiuma* pancreas shown in (b) illustrates the dense packing of the apical region of the cell with secretory granules. (Histological material of (b) prepared by Deborah Christensen, photographed by Sigurd Olsen; (c) electron micrograph by Deborah Christensen.)

sheet. Within this mesentery travel the nerves and the blood and lymphatic vessels. Embryologically, the mesenteries arise from the lining of the coelom, but they are expanded and modified in shape and attachment to the body wall with the growth and final assumption of the adult form of the tract and its associated glands.

The entire gastrointestinal tract is richly supplied with arteries. The vascular supply is important because of the high energy demands of the secretory, absorptive, and muscular activities of the tract. With the exception of its oral and aboral ends, the digestive system is drained through a special system of veins into the great *hepatic portal vein*, which leads to the liver (Figure 6-3). Thus, it brings directly to the liver cells the venous blood heavily enriched with the many substances absorbed through the intestinal mucosa. In amphibians, as in some fish and all higher vertebrates, the intestinal surface is also drained by a second transport system, the complex of lymphatic vessels. Lymph capillaries underlying the mucosal epithelium are the chief route of entry into the body of fatty materials. Lymph, often termed *chyle* when laden with fats during digestion, flows from the mucosa through lymphatics (*lacteals*, a name suggesting the milky appearance of the fat-rich chyle) of increasing size. These eventually reach the great lymph vessels (*cisternae chyli*, single or paired *thoracic ducts*) that, bypassing the liver, deliver lymph directly to the great veins draining into the heart.

Changing Structure

It is of interest to consider the changing pattern of digestive-tract structure through the course of the individual animal's life. The lining (epithelium and glands) of the digestive tract of *Necturus* and other amphibians originates embryologically from endoderm lining the *archenteron*. Only the most anterior and posterior sections of the tract (mouth, salivary glands, part of the cloaca) have epithelia formed by inturning of the surface ectoderm. During development the archenteron lengthens and progressively takes on rudiments of form, with an expanded forechamber (pharynx), a narrower posterior section of the foregut, and a narrow hindgut region. From the body surface anteriorly a pit (*stomodeum*) deepens toward the pharynx. Subsequently, this stomodeum becomes the buccal cavity. Where it fuses with the pharyngeal wall a membrane forms and then breaks down so that the buccal cavity opens directly into the pharynx, at which point there is a junction between ectoderm and endoderm. Similarly, at the posterior end of the body, a surface pit (*proctodeum*) forms and deepens. Its ectoderm fuses with the hindgut and the membrane between them breaks through so that the gut opens directly into the proctodeum (differentiating as the cloaca) and through it to the outside. Surrounding the endodermal regions of the gut are mesodermal derivatives—muscles and connective tissue making up the bulk of the digestive system. Differential growth molds the tract into its characteristic form; accessory glands, which originally budded from the endoderm, grow to adult mass and function as liver and pancreas; and the definitive patterns of blood supply, gland secretion, and neural control are established.

In other vertebrates, too, the glandular and absortive surface is of endodermal origin, with much of the rest of the tract arising from mesoderm. The details of gastrointestinal development vary greatly among vertebrates, depending on the amount of yolk in the egg and the pattern of development characteristic of the particular vertebrate class (refer to Chapter 3). Often, young vertebrates have patterns of digestive form and function quite different from those of adults of the same species. The most dramatic change is seen in fish and amphibians that have discrete larval and adult stages, separated by clear-cut processes of *metamorphosis*. For instance, the anuran amphibians, frogs and toads, with aquatic, filter-feeding larvae (tadpoles), and terrestrial, carnivorous adults show a dramatic change in gastrointestinal activity and structure at metamorphosis. The narrow, simple foregut of the larva changes into the esophagus and expanded stomach; the long, coiled midgut shortens and widens as it becomes the adult intestine; and the surface and gland cells break down to be replaced by adult structures (Figure 6-8). In urodele larvae, although the general anatomy of the digestive tract is altered less drastically at metamorphosis, the surfaces and the glands undergo as thorough a cellular reorganization as they do in anuran amphibians. In other vertebrates, too, adjustments occur. The lamprey's larva (Ammocoete), feeding by a filtration process on algae, bacteria, and other tiny particles, changes into an active, predatory adult that attaches by a sucker to another organism (e.g., a teleost fish) and travels with it, rasping away at its flesh and sucking its blood. The transition involves complete reorganization of the mouth and pharyngeal region (Figure 6-9). The intestine lengthens and develops the characteristic folding of the internal surface. Both liver and pancreas undergo structural change. The process as a whole occurs rapidly—in the course of a few weeks in the case of the lamprey *Lampetra*, for instance. It is a straining and dangerous time of structural revolution, involving not only the digestive tract but the entire body as well. In some nonparasitic lampreys, metamorphosis results in the formation of a nonfunctional intestine, so the adults cannot feed, and die soon after reproduction.

In other classes of vertebrates, the alterations of the digestive tract with development are generally less striking and far reaching than those described for amphibians and cyclostomes. Two specific cases may be cited. In some reptilian species, the dentition changes with age (immature *versus* adult animal), apparently in correlation with changing patterns of nutrition (e.g., insectivorous *versus* herbivorous habits). In mammals, the type of cellular enzymes formed by the stomach changes with maturation (predominance of infantile *rennin*, an enzyme specific for attacking milk protein, *versus* adult type *pepsin*).

Once growth has reached an advanced stage or is complete, the basic structure of the gastrointestinal tract generally shows little further change. In all vertebrates, however, the moment-by-moment form of the gut varies with the degree of distension of various chambers and the intermittent, often rhythmic, contractions of the muscular walls. Internally, the delicate mucosa is subject to continuous replacement of its cells. Unfortunately, very little is known about this process in nonmammalian vertebrates. In mammals, it has been studied extensively with estimates of the average lifetime of individual cells obtained from observations of average frequencies of mitosis or by the use of radioactive substances incorporated into the cellular structure. The cells of the buccal epithelium are replaced in the course of about 8.5 days (rabbit), and the stomach surface and gland cells in periods ranging from 6.5 to 1.8 days. The lining of the small intestine turns over relatively rapidly, with replacement times measured as approximately 1.4 to 2.8 days in various mammals. In view of these rapid cell replacement cycles, and the great energy demands placed on the intestinal mucosa in absorption of nutrients and formation of secretions, it is understandable that this structure is one of the most metabolically active of all mammalian tissues (details in other vertebrates are still few and far between). It is also one of the most sensitive to disease and damage, e.g., from starvation, ionizing radiations, and so on.

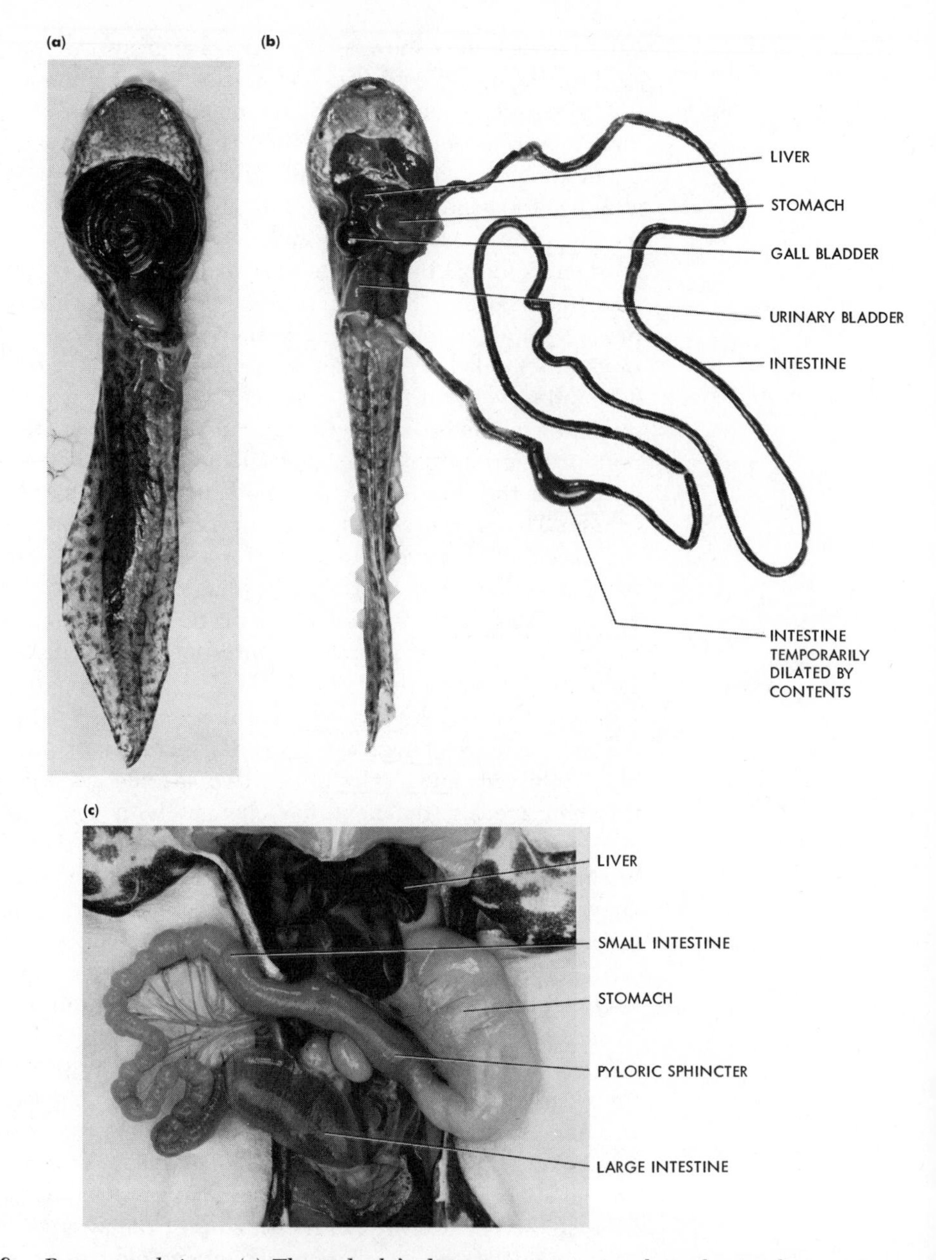

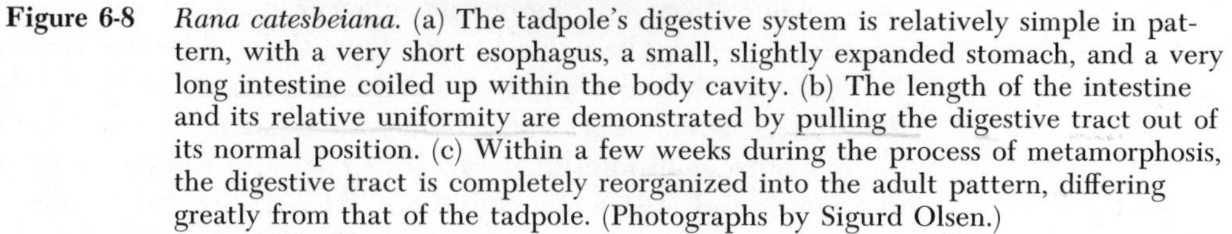

Figure 6-8 *Rana catesbeiana.* (a) The tadpole's digestive system is relatively simple in pattern, with a very short esophagus, a small, slightly expanded stomach, and a very long intestine coiled up within the body cavity. (b) The length of the intestine and its relative uniformity are demonstrated by pulling the digestive tract out of its normal position. (c) Within a few weeks during the process of metamorphosis, the digestive tract is completely reorganized into the adult pattern, differing greatly from that of the tadpole. (Photographs by Sigurd Olsen.)

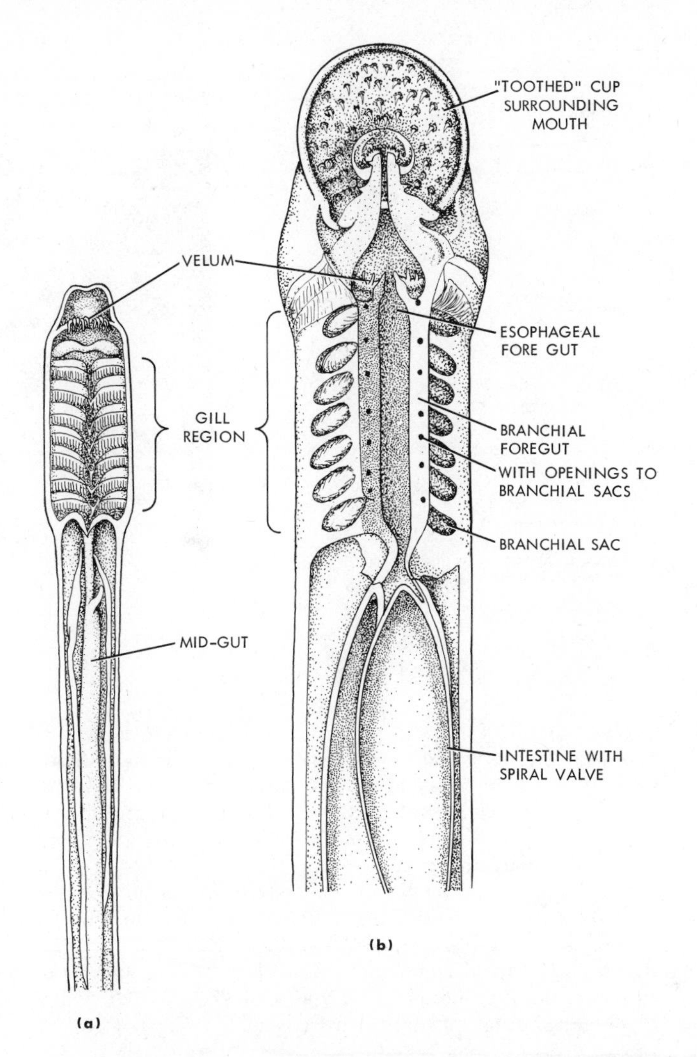

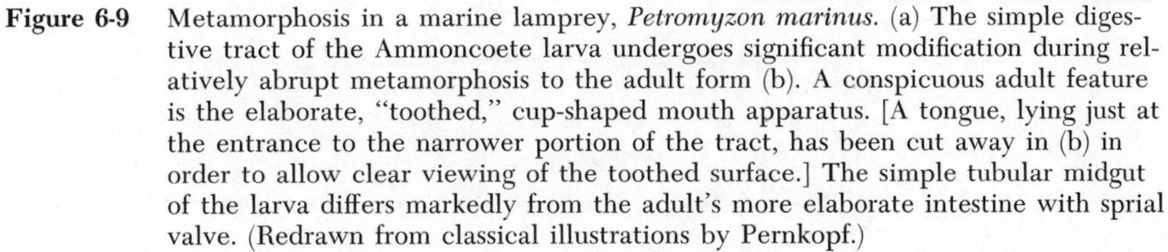

Figure 6-9 Metamorphosis in a marine lamprey, *Petromyzon marinus.* (a) The simple digestive tract of the Ammoncoete larva undergoes significant modification during relatively abrupt metamorphosis to the adult form (b). A conspicuous adult feature is the elaborate, "toothed," cup-shaped mouth apparatus. [A tongue, lying just at the entrance to the narrower portion of the tract, has been cut away in (b) in order to allow clear viewing of the toothed surface.] The simple tubular midgut of the larva differs markedly from the adult's more elaborate intestine with sprial valve. (Redrawn from classical illustrations by Pernkopf.)

Turning now to a comparison of gastrointestinal tracts among different vertebrate classes, the most striking generalization to note, perhaps, is the basic similarity of plan among all vertebrates. Tunicates and cephalochordates, too, show some important features of their digestive apparatus in common with vertebrates, as may be seen by comparing Figures 2-7 and 2-8 with vertebrate digestive systems illustrated in this chapter.

Digestive Systems of Cyclostomes

The cyclostomes have relatively simple gastrointestinal systems (Figures 6-10 and 6-11). The mouth is, of course, jawless, and is little more than a round opening into the pharynx. It may be surrounded externally by tentacles (as in

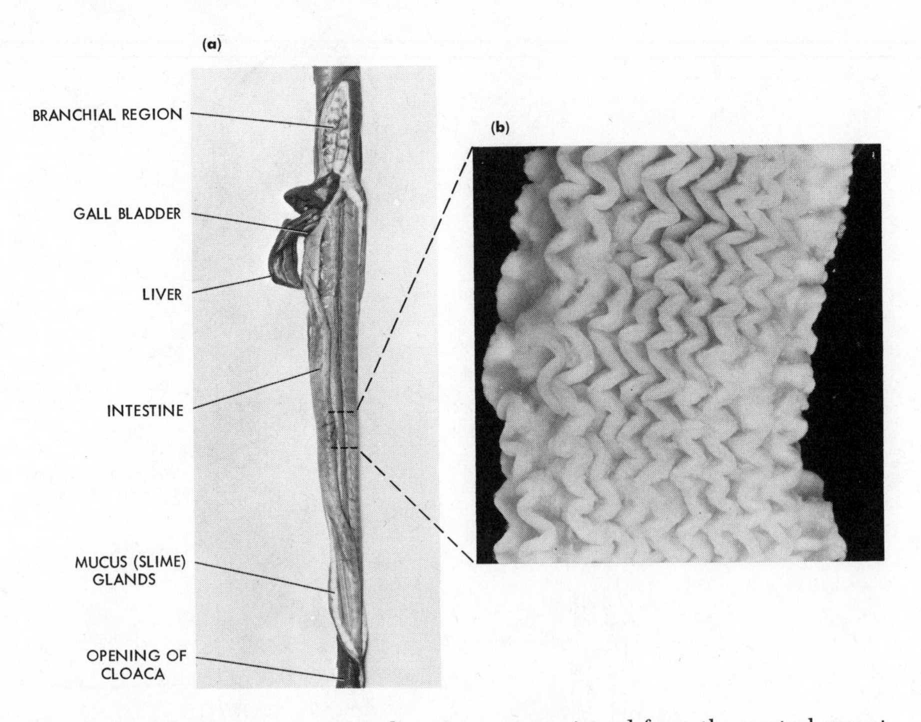

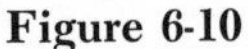

Figure 6-10 Pacific hagfish, *Eptatretus stouti;* digestive system viewed from the ventral aspect. (a) From the buccal cavity, a narrow esophagus (not shown) runs parallel with and dorsal to the array of gill pouches seen at the top of the photograph (see also Figure 7-16b). The esophagus leads directly into the intestine, for there is no stomach. There is a well-formed liver and gall bladder, but no pancreas; rather, pancreatic cells are located within the wall of the intestine. The intestine continues throughout the length of the body to a cloacal opening just anterior to the short tail. (b) Internally, the intestinal wall shows a patterning of intricate folds. These folds greatly increase the surface area. The epithelial lining of the intestine is made up of columnar cells with a dense stand of microvilli on the luminal surface, and of glandular cells that form digestive enzymes. (Material supplied by S. Vigna; photographs by Sigurd Olsen).

hagfish) or by a more elaborate, funnel-shaped "hood" (as in parasitic lampreys). The latter may be studded with small, horny toothlike projections containing keratin. It serves as a sucker, an organ of attachment to the lamprey's prey. At the entrance to the mouth is a muscular structure, often called a tongue, carrying keratin "teeth." This is used to scrape or tear away at the dead or dying tissues of organisms (hagfish) or fish and other living prey (lampreys). Associated with the toothed tongue of lampreys are paired glands that secrete an anticoagulant substance into the host.

The pharynx of the hagfish has numerous (six to fourteen) gill openings on each side and, as is typically the case in vertebrates, it is shared by the respiratory and digestive systems. This sharing may have originated within the group of chordate animals by adaptation of the pharynx from its primary nutritive role to allow exploitation of its structure in respiration. Evidence for such a primary role of the pharynx and its mucus-secreting endostyle exists in the lamprey larva and in tunicates and cephalochordates. The manner in which the pharynx operates in respiration and the structure of the gills will be considered in more detail in the next chapter. In lampreys, although the

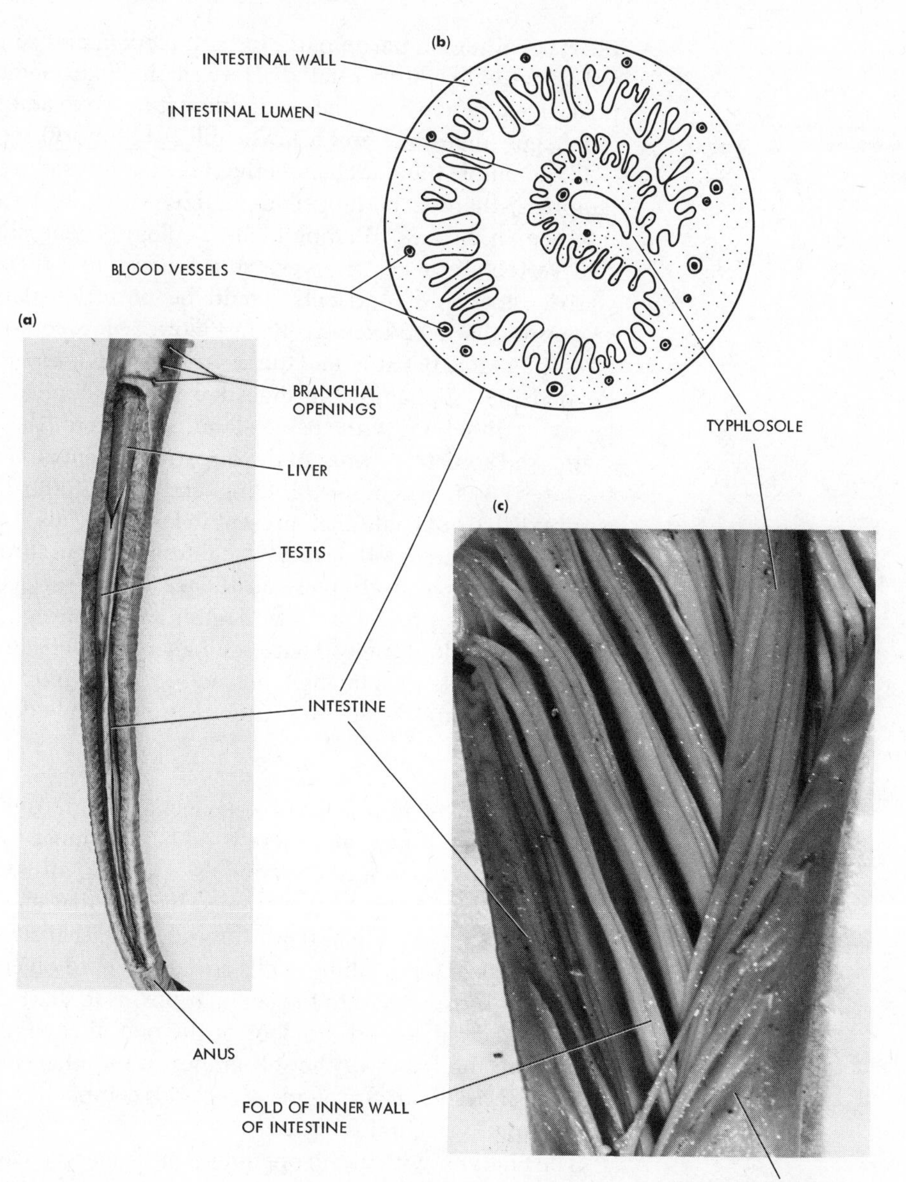

Figure 6-11 Digestive tract of an adult Pacific lamprey, *Entosphenus tridentata.* (a) A narrow esophagus leads through the branchial region to the beginning of the intestine; this is at the level of the last branchial opening, shown at the top of the photograph. There is no stomach. A liver, without a gall bladder, discharges bile into the intestine. The surface of the simple, straight tube of the intestine is greatly expanded internally by many longitudinal folds, and by a major folding of the inner wall, the typhlosole. (b) The structure of the intestinal wall and typhlosole is indicated in diagrammatic form in cross-section. The typhlosole runs the length of the intestine, gradually twisting, in a spiral pattern, along its course. (c) A short stretch of the inner wall of the intestine is shown in detail. Much of this inner surface is lined with ciliated cells and gland cells; together, these function in transport, digestion, and absorption of nutrients. Underlying the mucosal layer is muscle. Blood vessels in the intestinal wall and typhlosole transport absorbed nutrients to the rest of the body. The intestine is shown enlarged about 60×, as compared with its size in (a). (Photographs by Sigurd Olsen.)

pharyngeal region participates in both feeding and respiration, the processes are structurally differentiated. Here, a shelflike membrane develops in the pharyngeal region of the larva, dividing it into two adult structures—an upper esophagus (digestive) and a lower gill (respiratory) section.

In the simple tubular channel that follows the esophagus, in both hagfish and lampreys, there is nothing comparable with the stomach as described, for instance, in the case of amphibians. Zoologists generally infer that the primitive vertebrate ancestors were stomachless, like the cyclostomes of today. Gastric storage of nutrients would be of little adaptive value in jawless organisms that feed exclusively and more or less continuously on small particles—plankton, detritus, and the tissue and blood rasped and sucked from prey. Beyond the esophagus, the digestive tract is typically lined with a single-layered epithelium with many gland cells. Here, the main secretory, enzymatic, and absorptive work of the digestive system is done. The inner surface of the intestine shows complex folding, and a longitudinal fold (sometimes called a spiral valve or typhlosole), present in lampreys, also functions to increase the active surface area of the tract. The intestine opens through a small orifice, the anus, into a cloacal groove, and thence to the surface of the body. A well-formed liver is present in both hagfish and lampreys. Hagfish also have gall bladders for bile storage. There are pancreatic cells in cyclostomes that form digestive secretions, but these are not collected into a single gland structure. Rather, they are found in scattered groups embedded in the walls of the intestine.

Digestive Systems of Jawed Fish

Among the jawed fish, sharp divergences are seen from the tract structure as it occurs in cyclostomes, and there is perhaps as much variation in the forms of digestive tracts among these animals as among all the rest of the vertebrate classes. The mouth opening is framed by jaws that may carry teeth ranging in form from simple conical teeth through long, sharply pointed fangs to blunt, platelike tools for crushing shells and other hard objects (Figure 6-12). These teeth may be restricted to the jaw margins, as in sharks, or occur grouped or in rows on the palate and the floor of the buccal cavity. In some teleosts, teeth arise from the bones of the gill arches in the pharynx. The mouth cavity of jawed fish is devoid of a tongue and glands comparable with the salivary glands of tetrapod vertebrates.

The pharynx, with its gill openings, shows marked variation in structure. The esophagus is a relatively short, simple tube lined with mucus-secreting cells, and it is often ciliated. Beyond it, in most fish groups, lies a stomach (Figure 6-13). This is reasonably interpreted as an evolutionary adaptation to the

Figure 6-12 (*opposite*). Great variety exists in the dentition of fish. (a) and (b) show views of the teeth of the carnivorous elasmobranch, the mako shark, *Isurus glaucus*. The sharply pointed teeth with wide bases are attached by connective tissue to the cartilaginous jaws. Many of the teeth shown would not be seen in the open mouth of the living animal, since they would be covered with the oral mucous membrane. They are replacement teeth, which progressively move into position along the jaw margins as older functional teeth are shed. (c) and (d) show dentition of a teleost, the cod *Gadus*. The teeth are fine and numerous, serving in the retention of captured prey (invertebrates, small fish) in the mouth. (Shark jaws prepared by Arthur D. Welander; cod skull obtained from Richard C. Snyder; photographs by Sigurd Olsen.)

(a)

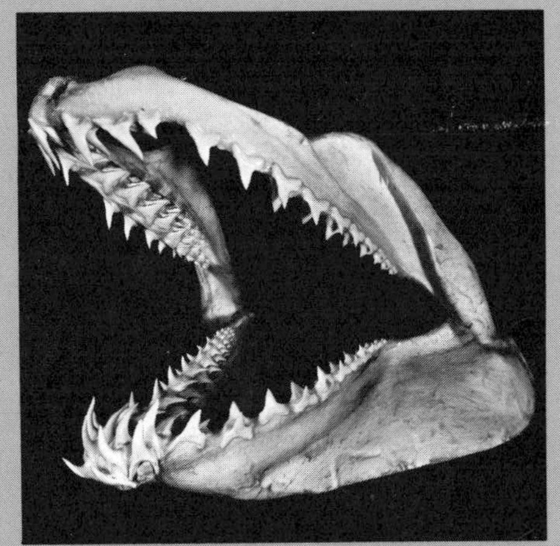

(b)

28 cm

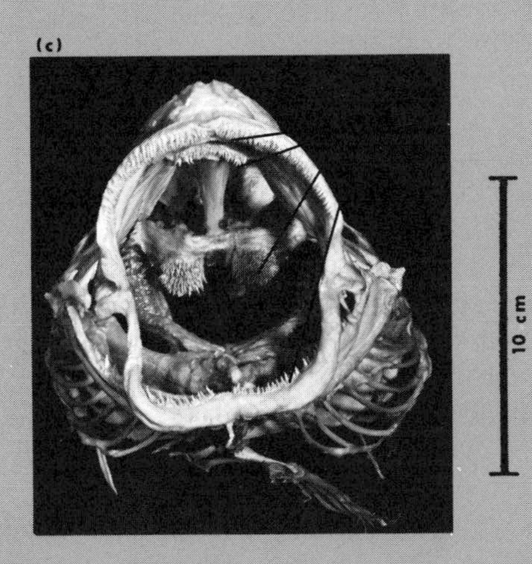

(c)

10 cm

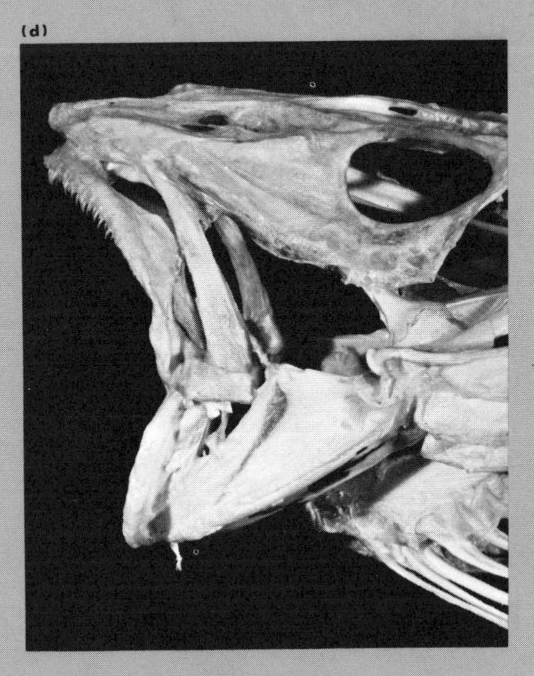

(d)

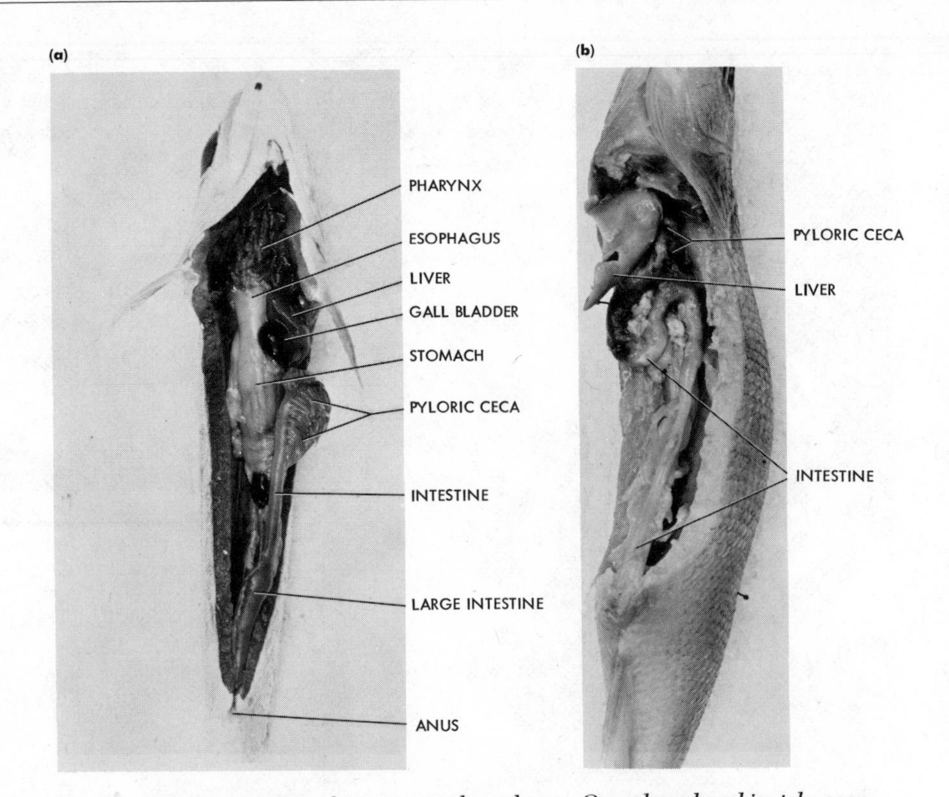

Figure 6-13 (a) The digestive system of a young coho salmon, *Oncorhynchus kisutch,* seen from the ventral view after removal of the ventral body wall. The numerous pyloric ceca greatly increase the functional surface of the tract. (b) In the perch, *Perca* sp., the surface area of the intestine is increased by lengthening; since the intestine is longer than the body cavity, it bends back upon itself. Pyloric ceca are few in number (three in this animal), and are relatively short and broad. (Photographs by Sigurd Olsen.)

ability of these jawed vertebrates to capture and swallow, intact or in large pieces, other organisms of significant size. In a few groups (including lungfish and a few teleosts), however, there is no stomach. Because of the inferred close evolutionary relationships of these forms to animals with well-developed stomachs, most zoologists agree that stomachlessness in these exceptional fish is a secondary adaptation. The walls of the stomachs of fish are lined with secretory cells forming hydrochloric acid and pepsinogens.

Great variety exists in the external form and internal structure of the intestine among fish. Throughout there is a trend toward increasing surface area, either by elaboration of the internal surface, by increase in the length of the intestine, or by use of both of these structural devices. Where the intestine lengthens, it tends to coil back upon itself within the confined space of the body cavity. In many bony fish, elongate blind sacs—pyloric ceca—open from the intestine near its junction with the stomach. They vary in number from a few up to a thousand. In these ceca, as within the intestine itself, digestion and absorption take place. Of the various arrangements for increasing active intestinal surface, the internal elaboration and folding of the mucosa is thought to be most primitive because it occurs in cyclostomes and elasmobranches and rarely in fish considered more advanced from an evolutionary standpoint. Thus, in a manner somewhat reminiscent of cyclostome lampreys, elasmobranchs often have elaborate outfoldings of the intestinal mucosa, curving in a spiral valve about the inside of the tract. Sometimes such a sprial valve is coiled around upon itself so that it nearly fills the lumen of the intestine, offering a greatly expanded surface for secretion and absorption (Figure 6-14). By contrast, the spiral valve is absent in teleosts, but as in amphibians the intestine shows a trend toward increased length. Teleost intestines may be more than twice as long as the entire body, and lengthening is particularly notable in the case of herbivorous fish. At least in part, this is believed to represent a functional adaptation to the special problems of digestion of refractory plant materials. Comparable trends, originating independently, are seen in other vertebrate classes in which some forms adopted a diet of plant material (see, for instance, the discussion of mammalian digestive tracts later in this chapter).

The intestine opens into a wider large intestine (or a short terminal chamber, the *rectum*), which is similar to comparable structures in amphibians. The large intestine or rectum then leads into a cloaca (sharks) or directly to the outside through a small anal opening (bony fish). The liver, with its gall bladder, is a large, consolidated organ (Figure 6-15), and the pancreas is usually recognizable as a single glandular structure, although its tissue is scattered along the intestinal walls and in the mesenteries of a number of fish, including lungfish and some teleosts. There is also great variation in detail of form, duct structure, and blood supply of these digestive glands.

Figure 6-14 Cross-section through the intestine of an elasmobranch (*Scyllium*) showing the elaborate structure of the spiral valve, which folds away from the gut wall at S, and is wound upon itself to fill much of the space within the intestine. (Redrawn from the classic illustration by Jacobshagen.)

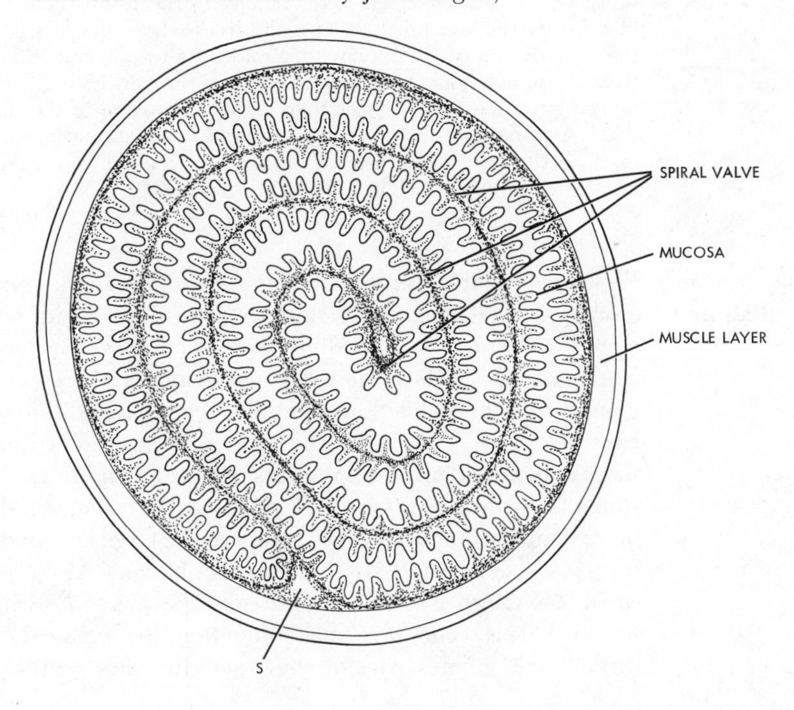

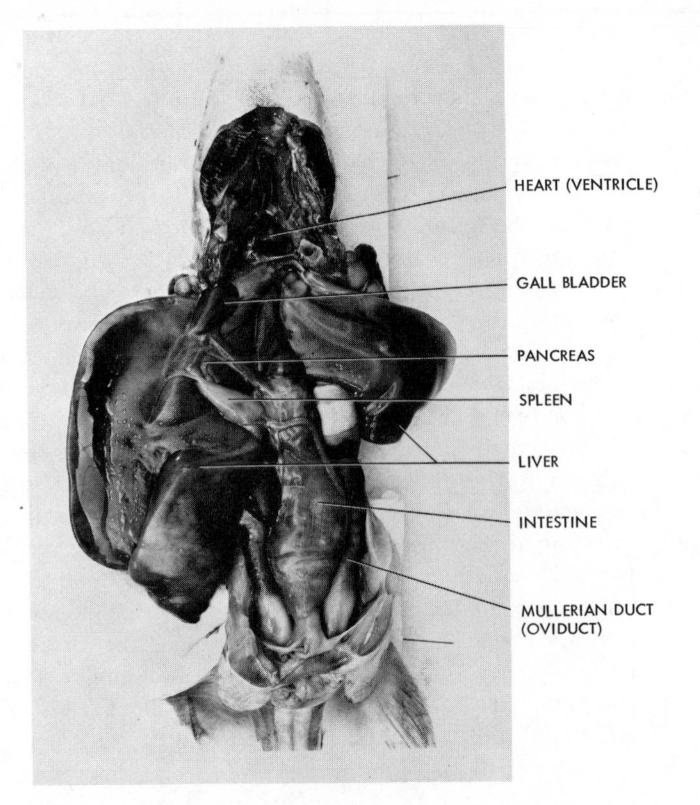

Figure 6-15 An example illustrating the extremely large size that can be attained by the liver in correlation with its storage role. The viscera of this adult female specimen of the holocephalan ratfish, *Hydrolagus colliei*, are shown in ventral view after removal of the ventral body wall. Ratfish are carnivores, and the liver is an important site for storage of nutrients, chiefly in the form of fats. Thus, oil droplets can be seen on the ventral surface of the liver where it was slightly bruised during the preparation of the specimen. Besides its role in nutrition, fat stored in the livers of some chondrichthyan fish is important in increasing the animals' buoyancy, since fat is lighter than water. The form of the digestive tract as a whole is simple, without clear division of stomach and intestine. A spiral valve expands the internal surface of the tract. (Specimen provided by A. O. D. Willows; photograph by Sigurd Olsen.)

Digestive Systems of Reptiles and Birds

Reptiles resemble terrestrial anuran amphibians in many aspects of their gastrointestinal structure. The jaws are rimmed with teeth that usually are conical in form (teeth have disappeared in turtles). The mouth is equipped with a tongue, and well-formed salivary glands are present. In snakes, salivary glands may be modified into highly specialized poison glands that aid in the capture of prey. In some reptile species the stomach is larger and considerably more elaborate than in amphibians (Figure 6-16). An example is the alligator's stomach, with a heavy-walled, muscular forechamber differentiated from a more delicate, aboral pyloric chamber. The large intestine, too, is often greater in size and complexity than that of amphibians. At the junction between the small and large intestines, there may be a saclike extension of the large intestine. This *cecum* presumably functions to increase the over-all volume and surface area of this part of the tract. In some reptiles, the cloaca, at the

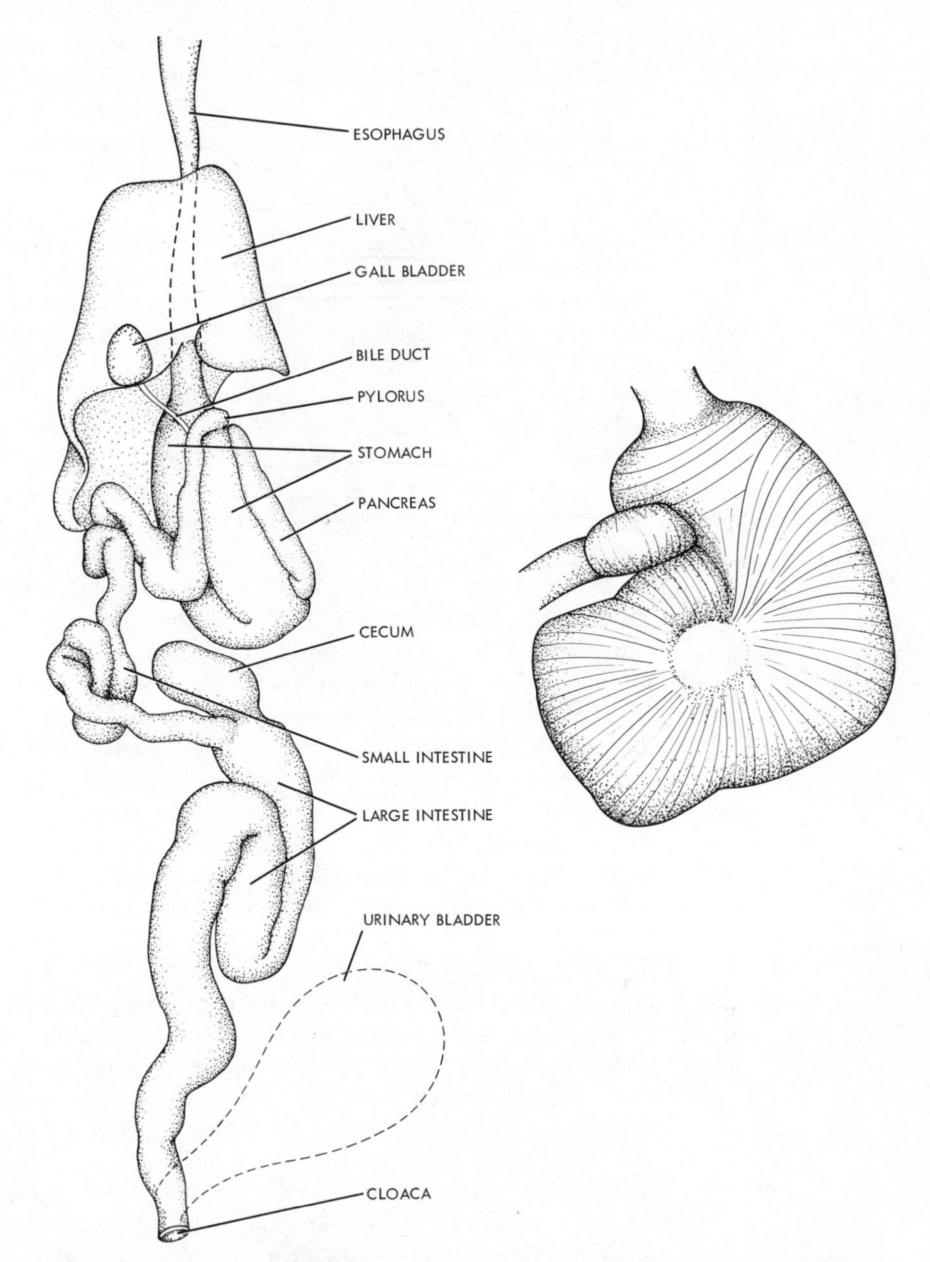

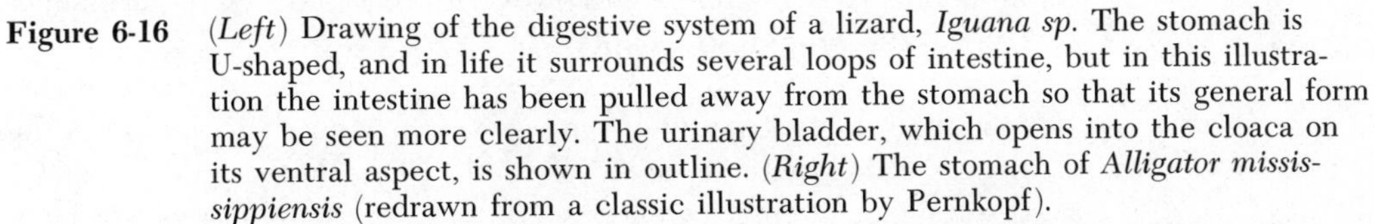

Figure 6-16 (*Left*) Drawing of the digestive system of a lizard, *Iguana sp.* The stomach is U-shaped, and in life it surrounds several loops of intestine, but in this illustration the intestine has been pulled away from the stomach so that its general form may be seen more clearly. The urinary bladder, which opens into the cloaca on its ventral aspect, is shown in outline. (*Right*) The stomach of *Alligator mississippiensis* (redrawn from a classic illustration by Pernkopf).

termination of the digestive system, is found to be quite specialized. Paritally separated cloacal chambers receive the egesta (*coprodeum*), urine, and reproductive cells (*urodeum*). In common, these chambers lead into a terminal cloacal region and from there to the outside.

In birds, as might be expected from their unique exploitation of the possibilities of flight, the mass of machinery surrounding the mouth and in the buccal cavity is greatly reduced. Teeth, which were present in primitive birds (*Archeopteryx, Hesperornis*), are replaced functionally in modern birds by horny beaks, as well as by other adaptations of the aboral gut (to be discussed later). As compared with other vertebrates and their toothed ancestors, modern birds show a general lightening and reduction of the jaws. Salivary glands discharge into the mouth lubricating and digestive fluids (saliva). The esophagus is relatively long and narrow, in correlation with the functional importance of the neck in birds (Figure 6-17), and often there is an aboral region of great expansion to form a crop. Food can be stored here, a feature of adaptive value in birds with their exceptionally high energy demands for flight and the physiological control of body temperature.

The stomach of birds tends to be elaborate. It usually has an orally located section with delicate walls lined with secretory cells (*proventriculus*) and a more posterior section with heavy walls differentiated as a muscular grinding organ. This thick-walled *gizzard* is of great importance in aiding mechanical breakdown of food prior to digestion. Needless to say the toothless bird is unable to use mouth chewing in the preparation of food for digestion. Instead, strong contractions of the gizzard wall, often accompanied by the grinding action of small stones swallowed by the bird and lodged in this organ, provide the mechanical trituration necessary for subsequent extraction of nutrients from resistant food sources such as seeds and insects. The presence of the gizzard, like the loss of teeth, appears to be an important adaptation that serves to decrease the weight of the head. At the same time the trend toward aggregation of food-storage functions in the thoracic and upper abdominal regions (in crop, proventriculus, and gizzard) may aid in adjusting the bird's center of gravity for special flight requirements.

Comparing the rest of the gastrointestinal tract of birds with the tract of reptiles, the observer is apt to be impressed most by the greater lengthening of the small intestine. In the bird, too, the tract is highly active, with rapid muscular movements and fast digestive processes correlated with high body temperature. At the junction between the small and large intestine elaborate ceca may be found, although these are not present in some species, including hummingbirds and some pigeons. The pancreas and liver are generally similar to these organs in other vertebrates, although the gall bladder is sometimes absent. As in reptiles, the cloaca is differentiated into a coprodeum and urodeum. Especially in the coprodeum and adjacent regions of the aboral large intestine, water is reabsorbed from urine entering from the ureters and urodeum. This is significant in birds with physiological requirements for water conservation (see also Chapter 8).

Patterns of Dentition

Mammals differ strikingly from birds in the complexity and importance of the buccal cavity with its array of teeth. These latter may be elaborated into defensive or offensive weapons—*incisors*, fangs, and tusks—or be specialized for such functions as grinding leafy plant material (herbivores) or gnawing hard objects, i.e., chiefly wood (rodents, see Figure 6-18). At this point, it may be well to review some major features of vertebrate dentition before proceeding in a brief survey of the teeth of mammals, for teeth are of great interest to comparative anatomists. They are durable structures that have contributed much to the study of paleontology. In a certain sense, teeth are conservative in

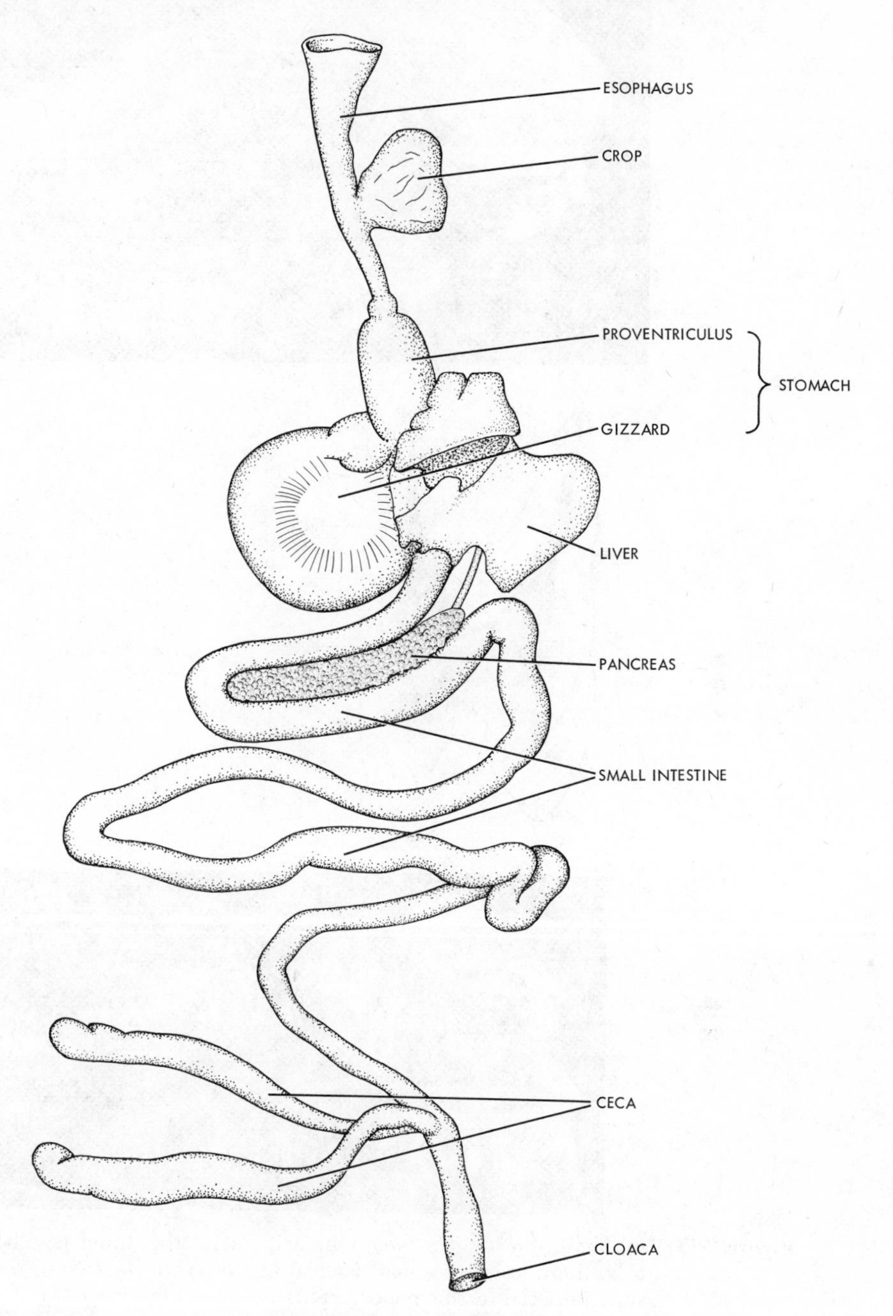

Figure 6-17 Digestive tract of a bird, the Japanese quail *Coturnix coturnix japonica,* seen from the ventral view. The liver has been pulled back to show details of the stomach structure, and the intestine and the ceca have been displaced from their normally compact arrangement in the abdominal cavity to show the length and continuity of these structures.

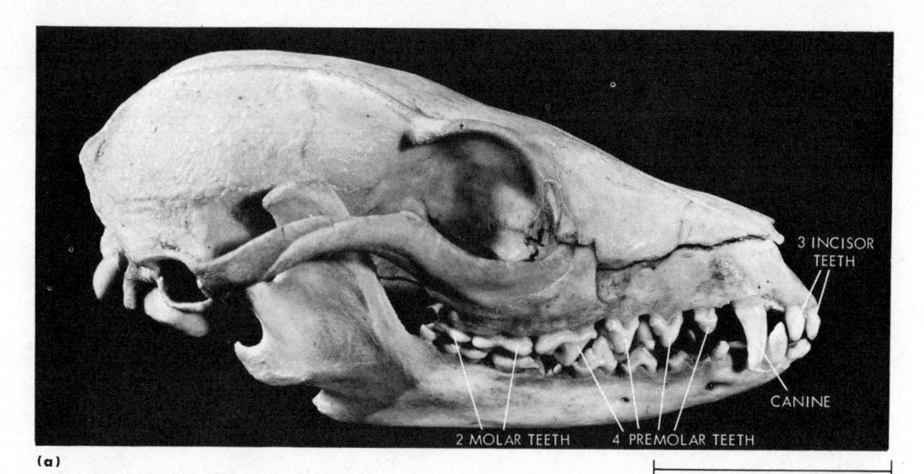

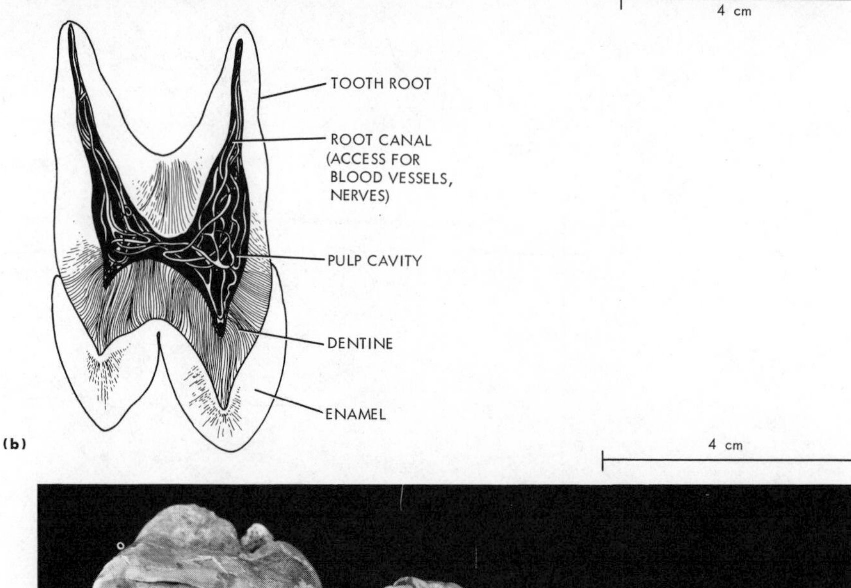

Figure 6-18 (a) Skull of a fox, *Urocyon.* On each side of the upper jaw, beginning most anteriorly, there are three small conical *incisor* teeth (I—two of these appear to be superimposed in this photograph); a long *canine* tooth (C); four *premolar* teeth (P); and two *molar* teeth (M) of more complex form. On each side of the lower jaw there are three incisor, one canine, four premolar, and three molar teeth. Anatomists describe dentition patterns of mammals by a series of numbers representing first the upper jaw (I-C-P-M) and then the lower jaw in a *dental formula* given in the form I-C-P-M/I-C-P-M. Thus, for this fox, the dental formula is 3-1-4-2/3-1-4-3.

(b) Drawing of a ground section (oriented in the plane tongue-cheek) of a human maxillary molar tooth. The roots are normally attached to the socket in the jaw bone by a special bony substance, *cementum.* (Redrawn from D. Permar, *Manual of Oral Embryology and Microscopic Anatomy*, 3rd edition, Philadelphia: Lea and Febiger, 1963, with the kind permission of Professor Permar and the publisher.)

(c) *Left:* A molar tooth of a horse, *Equus caballus*, showing elaborate pattern-

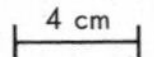

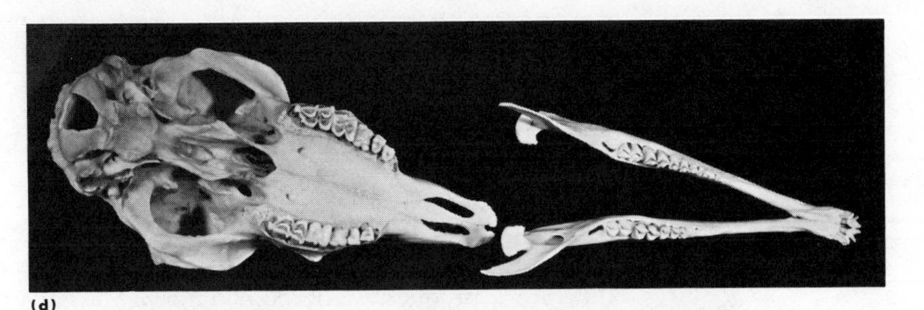

(d)

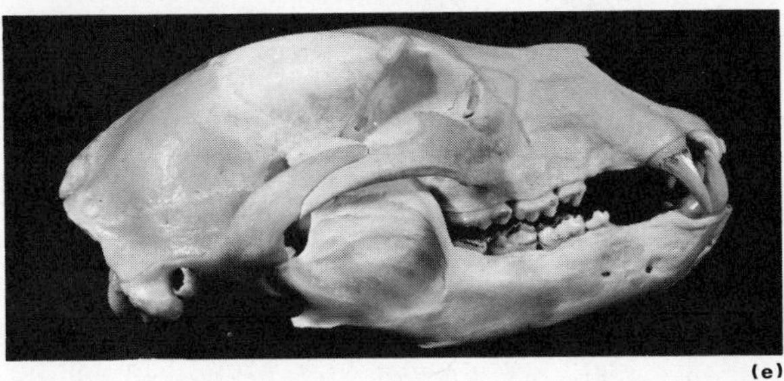

4 cm

(e)

4 cm

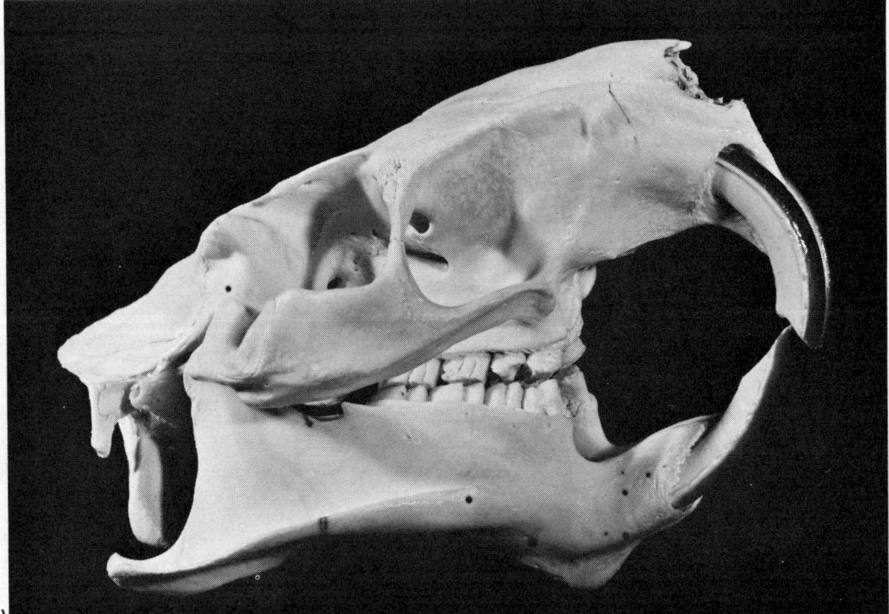

(f)

ing of enamel ridges. These function in the grinding of the plant material that makes up the horse's diet. *Right:* The skull and lower jaw of an insectivore, the Townsend mole *Scapanus townsendii*. The dental formula is 3-1-4-3/3-1-4-3. The teeth are relatively unspecialized in form. To appreciate this, compare the incisors with those of the nutria (f), the canines with the canine teeth of the bear (e), and the premolars and molars with those of the sheep (d). Note that the horse tooth and mole skull were photographed simultaneously, so their comparative sizes are represented accurately.

(d) Lower jaw and skull (seen from the palatal aspect) of an artiodactyl, the domestic sheep *Ovis aries*.

(e) Skull of the black bear *Euarctos americanus*.

(f) Skull of a rodent, the nutria *Myocastor coypu*.

(Skeletal material obtained from Richard C. Snyder. Photographs by Sigurd Olsen.)

form, so the comparison and contrast between teeth in individuals and in members of differing species, genera, and higher taxonomic groupings provide significant evidence about relationships and divergencies. At the same time, teeth show beautiful functional adaptiveness in relation to their capture of prey, crushing and maceration of food materials, and other processes. Thus, they are a productive source of information about the feeding and related activities of animals. Indeed, there are few structures that demonstrate more clearly than teeth the intimate interplay and interdependence of form and function. In the brief treatment necessitated by the limitations of space and objectives of this book, facets of tooth anatomy can only be touched on. The interested reader will find references to more comprehensive discussions of dentition at the end of this chapter. Further study of this subject is sure to increase the comprehension of broad principles as well as narrower details of comparative anatomy.

In a review of the dentition of vertebrate animals, it may be recalled that teeth are or once were present in gnathostome fish and in some orders of all higher vertebrate classes. They have been lost secondarily, however, in some amphibians (toads), some reptiles (turtles), and in all modern birds. In fish and amphibians, teeth are entirely mesodermal in origin. They develop in the dermis of the jaws, palate, and/or pharynx with a hard, mineralized *dentine* forming the main body of the tooth. Dentine varies in composition from one animal to another, but in general it has a low content (about 30 per cent) of water and organic materials (proteins, organic acids, and so on). Its main constituents are inorganic salts, which are in crystalline form; a complex inorganic calcium phosphate salt, hydroxyapatite, resembles dentine in structure. Often, dentine is highly organized in the form of parallel arrays of tubules. It surrounds a pulp cavity lined with special dentine-building cells (*odontoblasts*) where blood vessels and nerves come into close proximity with the inner layers of dentine. In reptiles and mammals, dentine is covered with a layer of very hard mineralized material, *enamel*, secreted by cells of ectodermal origin. These cells, *ameloblasts*, are lost after the growing tooth erupts into position in the mouth, so that enamel, once fully formed, cannot grow or be replaced. Enamel consists of salts to the extent of about 96 per cent of its mass, chiefly in crystalline form, and is seen in mammals in clearly prismatic arrays. Calcium, magnesium, sodium, phosphate, and carbonate have been identified as constituents of enamel. It provides an exceedingly resistant covering for the underlying dentine and contributes to the long life of definitive mammalian teeth. In all vertebrates except mammals, formation of new teeth is more or less continuous throughout the animal's life. In contrast, mammals generally develop two sets of teeth, the early "milk," or deciduous, teeth are formed first and then are replaced by a not quite full complement of permanent teeth. These latter generally function throughout the major part of the animal's life. In mammals the teeth may, but rarely do, continue to grow from their roots throughout life. This is the case, for instance, with the incisors (front gnawing teeth) of rodents.

Vertebrates, in general, show quite a wide range of placement of teeth and in the means of their attachment to the supporting structures. Often, the teeth adhere firmly to the underlying bone through fibrous and cement attachments, but in certain fish and reptiles the attachment may be ligamentous and, thus, less rigid. When dentition is "hinged" in this manner, it may be quite movable.

In some species, teeth are set in sockets in bone (thecodont pattern), whereas in others they are attached on top or along the inner (tongue) side of the jaw (patterns termed, respectively, acrodont and pleurodont). In fish, teeth may occur in rows along the jaw margins, over the entire palate, and on gill arches in the pharynx. They may be very numerous, numbering in some cases hundreds or thousands in a single individual. Typically, in amphibians and reptiles, the teeth are fewer, acrodont or pleurodont in arrangement, and more restricted than in fish in placement along the jaw margins; palatal teeth occur in some reptiles. In mammals, the number of teeth is relatively small (generally forty-four or less), and their position is restricted to the jaw margins (Figure 6-18). The one or several roots of each tooth are set in bony sockets. If it is possible to speak of a typical form of tooth in fish, amphibians, and reptiles, this could be described as conical. Mammals have some conical teeth as well (the complement of front incisors and the *canine* teeth), but the cheek teeth (*premolars* and *molars* along the sides of the jaws) are far more complex in structure. Their opposing surfaces have multiple cusps and ridges with patterns that are constant within a given species, although varying from species to species. Superimposed on the basic tooth forms and patterns are features correlated with function. Among fish and reptiles, some species feeding on hard, resistant material (e.g., plants, shelled invertebrates) show flattened, platelike, grinding teeth. Herbivorous mammals have specialized premolars and molars with ridged and very resistant grinding surfaces. In carnivorous mammals, the canine teeth may be elongated and sharply pointed, serving as weapons for attack on prey. Some other tooth adaptations will be touched on later in this chapter.

Digestive Systems of Mammals

The tongue of mammals is often highly specialized, functioning in the selection of food (tasting), capture of prey, grasping of food, and in vocalization. Chemical receptors are arranged in taste buds on the surface of the tongue to assist in recognizing and choosing the food. Information from these receptors also is used in the reflex activation of muscle movement and secretion of other parts of the digestive tract. The secretion of saliva moistens and lubricates the food and begins digestion of its carbohydrates. Thus, the mouth of the mammals is a complex structure with diverse and important digestive functions (Figure 6-19).

The esophagus tends to be long and slender, but no specialized crop is found in mammals. The stomach differs from the bird's stomach, showing no division into proventriculus and gizzard. Rather, the stomach of mammals is often a relatively compact organ, even though it may be differentiated by contour and by patterning of the tissues of the wall into regions located orally (cardiac region), centrally (fundic), and aborally (pyloric), this latter part lying adjacent to the pyloric sphincter (Figure 6-20). As in other vertebrates, the mammalian pyloric sphincter, which opens and closes depending on the degree of filling and the digestive condition of the stomach, has an important part in controlling the supply of nutrients to the small intestine. Perhaps the most striking adaptations of mammalian stomachs are seen in the case of herbivores with their enlarged and specialized stomach chambers where symbiotic microorganisms are maintained (Figure 6-21). Bacteria and protozoans are of great importance to the mammal because of their unique biochemical capabilities. Supplementing the limited enzyme syntheses of their mammalian hosts, these

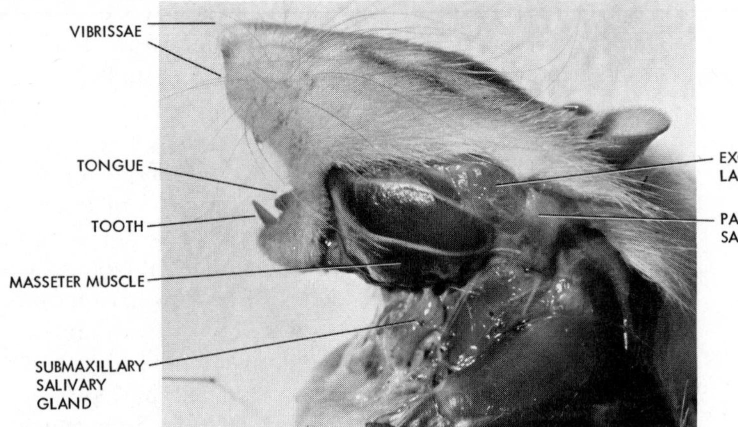

Figure 6-19 Some of the structures involved in the diverse functions of the mouths of mammals are shown in this lateral view of the partially dissected head of a rat, *Rattus norvegicus*. Selection of food is carried out in large part by chemical receptors on the tongue and in the nasal passages. Chemical information is supplemented with input from the eyes and from mechanical receptors signalling movement of the vibrissae. The jaws, with sharp teeth, are moved by muscles including the powerful masseter muscle, which plays a major part in jaw closure. Saliva, flowing into the mouth through ducts from three major pairs of salivary glands (submaxillary, parotid, sublingual—the latter not shown) provides mucus-containing saliva aiding in swallowing the food, and also containing enzymes for breakdown of carbohydrates. (Photograph by Sigurd Olsen.)

organisms form cellulases and other enzymes that break down the resistant chemical compounds of plant cell walls and fibers and, thus, increase greatly the efficiency of use of plant materials by herbivorous mammals.

As in birds, the small intestine of the mammal tends to be extremely long. The internal surface is further increased by the presence of innumerable villi, tiny fingerlike projections of the mucosa into the intestinal lumen. Often considerable specialization in structure and function is seen from oral to aboral end of the small intestine. The *duodenum*, the region nearest the pyloric sphincter, is relatively short and thick-walled and has the highest rate of muscle activity, absorption, and secretion of any region of the tract. In some mammals, a middle section (*jejunum*) can be easily differentiated, and beyond this the terminal section (*ileum*) is also distinctive in contour, structure, and function (for instance, bile salts are strongly absorbed in this section of the intestine alone). Both the small and large intestines are far more elaborate in herbivores than in carnivores. All these structural features are correlated with the high energy demands and specialized dietary requirements of these homeothermic vertebrates.

The general structure of the liver and pancreas is comparable with that seen in most other vertebrates, although in some birds and in certain mammals (deer, horses, some rodents) the gall bladder is absent. It has been suggested that this peculiarity can be correlated with the lack of a need to store and concentrate bile during times of gastrointestinal inactivity.

A large semiblind chamber, the cecum, is generally attached to the large intestine at its junction with the small intestine, and at its opposite end may terminate in a further blind extension, the *appendix*. These structures appear to

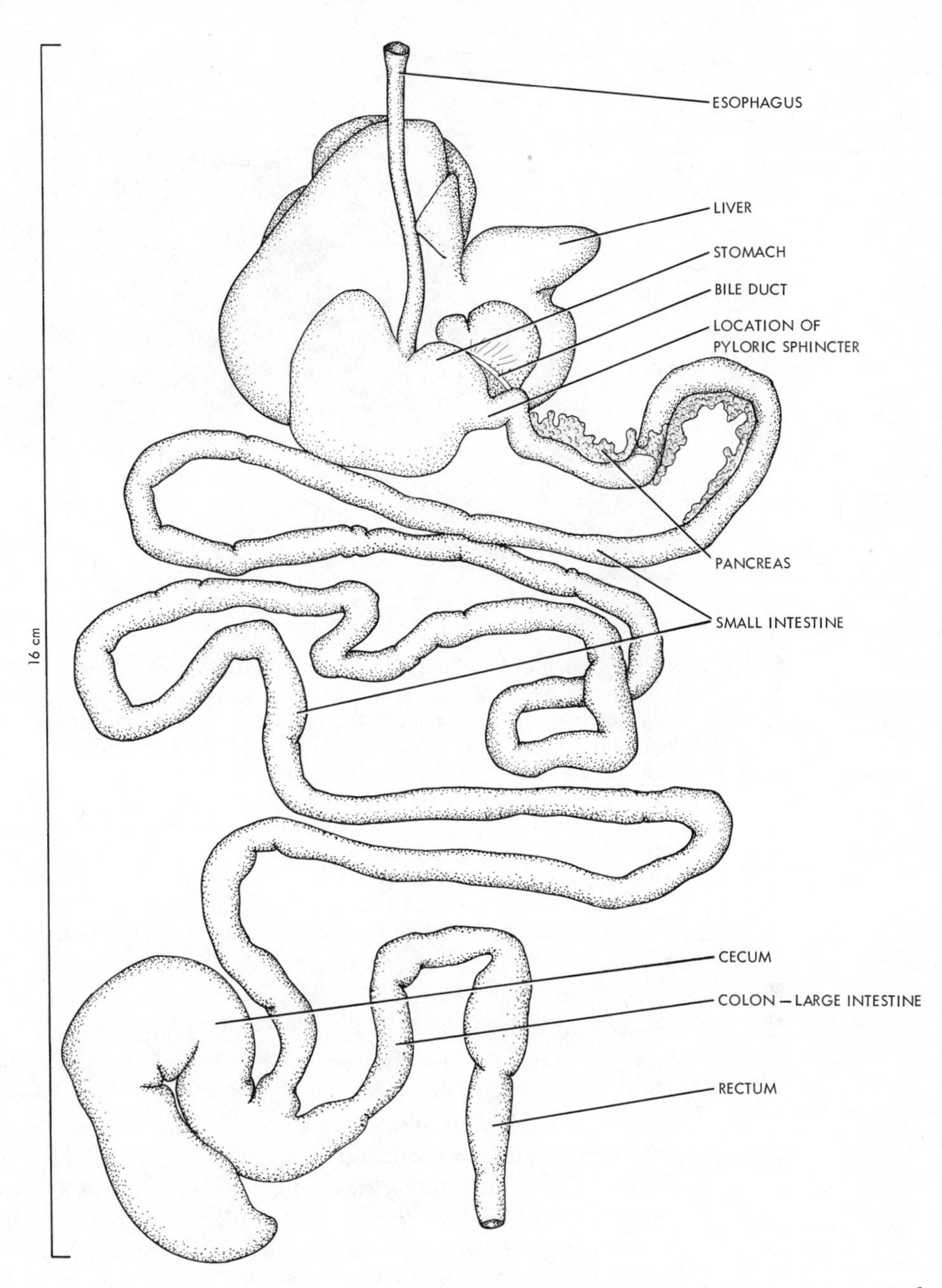

Figure 6-20 Digestive system of a mammal, the Norway rat, *Rattus norvegicus,* seen in ventral view, and freed of mesenteries. The liver has been reflected back and orally, to reveal the stomach and intestine. There is no gall bladder in the rat; rather, bile flows through the delicate bile duct into the small intestine in the region of the pancreas.

function as digestive ceca generally do, to provide additional volume for processing of the nutrient mass by digestive enzymes and microorganisms and to give additional absorptive area. The structures are particularly well developed in herbivores, and are smaller (the appendix may even be greatly reduced and essentially nonfunctional) in carnivores and primates, including man.

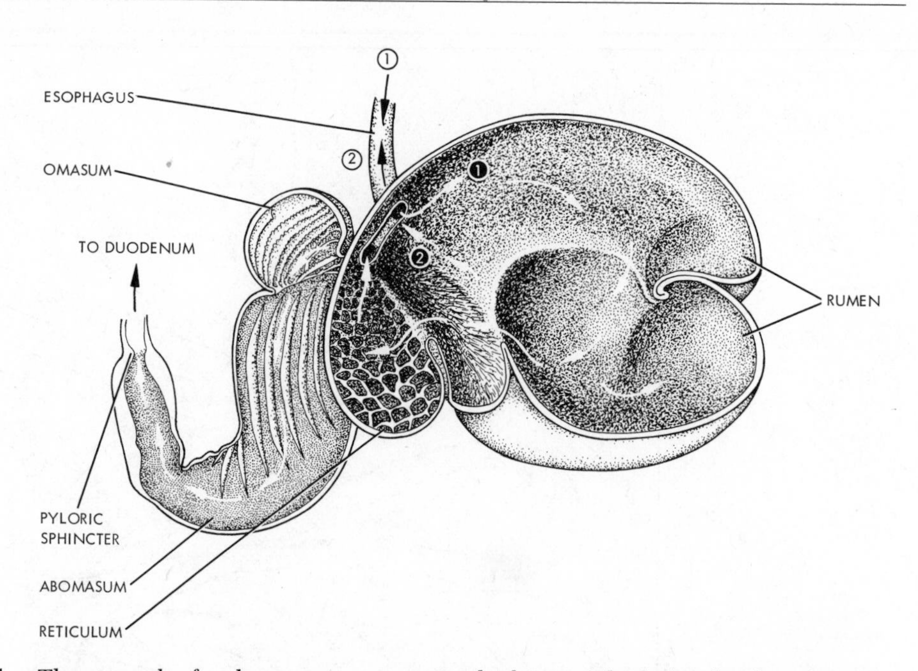

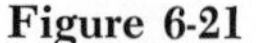
Figure 6-21 The stomach of a sheep, a representative herbivore. The large chambers lying between the esophagus and the abomasum-pylorus region (parts which are closest in function to simpler mammalian stomachs) are specialized for culture of hugh colonies of microorganisms. The routes of the ingested material is suggested by the white arrows. Food is swallowed, passing from the esophagus into the rumen where mechanical breakdown and attack by microorganisms is initiated (arrow path marked ①). Repeatedly, the partially digested material is regurgitated from the rumen, up through the esophagus, to the mouth for further chewing (arrow path marked ②), swallowing, and further digestion. These elaborate chemical and mechanical processes result in exceptionally efficient use of plant material by the sheep. (This figure is slightly modified from the classical illustration by Pernkopf.)

As in all other vertebrates, the mammal's large intestine functions chiefly as a site of water and ion absorption. Only in monotremes and marsupials does it lead into a cloaca. In more advanced mammals, its contents pass into the *rectum*, a smaller, simpler chamber lined with stratified epithelium, and thence through a sphincter-bounded opening, the anus, to the outside.

In this account, similarities among vertebrate digestive systems have been emphasized. Needless to say, the detailed functioning of the gastrointestinal tract is highly varied and sensitive to the environmental conditions under which the individual lives. The tract is richly supplied with nerves, some operating independently, some controlled by the central nervous system. Most of the extrisic control is from the autonomic division of the central nervous system. Often, however, the extreme oral and extreme aboral ends of the tract are powered by striated (voluntary) muscle, and are under the animal's voluntary control. Although the smooth muscle cells of the digestive tract show a great deal of intrinsic contractility, shortening more or less rhythmically and so mixing, churning, and moving the contents along the tract, the over-all patterns of movement are integrated by nerve plexuses within the gut wall or by nerves connecting with the central nervous system. Thus, the waves of contraction that sweep in a consistent direction along the tract wall (peristalsis) depend on both intrinsic and extrinsic nerves. Furthermore, autonomic nerves (chiefly the paired vagus nerves) take part in integrating the motility of different regions of the tract: examples are the relaxation of the stomach walls as food approaches it from the esophagus, and stomach contractions that eject the contents into the small intestine when the oral regions of the intestine have emptied. In a wide range of vertebrates the stomach makes strong, repeated contractions when it has been empty for some time. In man, these contractions are often recognized consciously as "hunger pangs," and they seem to play a part in signaling the need for food (onset of hunger). Although we cannot, of course, know whether vertebrates other than ourselves feel hunger as we do, some experimental results suggest that similar gastric contractions precede the onset of feeding behavior in a variety of vertebrate animals. Thus, nerves have a function in signaling the need for food, as well as in coordinating feeding behavior and digestive processes.

Neural control of the gastrointestinal tract is supplemented by chemical control. The food materials themselves, especially after partial digestion, may stimulate gland secretion. Cells of the walls of the small intestine, stomach, and other gastrointestinal regions form endocrine secretions (*secretin,* identified in all vertebrate classes, and *gastrin, enterogastrone, pancreozymin-cholecystokinin,* and so on, identified in various vertebrate classes, but studied thus far chiefly among mammals). These endocrine hormones travel through the blood to the gland and muscle cells of the stomach, small intestine, liver, gall bladder, and pancreas and modulate the responses of these structures with respect to current digestive demands. Hormonal control is specific and may be quite complex. For instance, the hormone secretin is absorbed into the blood from the duodenal mucosa when acidic materials enter the duodenum from the stomach. Secretin then travels through the circulation and stimulates increased secretion by the pancreas and liver. In contrast, when fatty materials enter the duodenum from the stomach, the inhibiting hormone *enterogastrone* is released from the intestinal mucosa and acts on the stomach to inhibit gastric secretion; this mechanism has been studied fairly adequately only in mammals. The person who wishes to appreciate fully the structure of the gastrointestinal tract must take into account its dynamic organization. The anatomy of the tract fluctuates with its changing pattern of activity. Thus, for instance, the color, length, contour, and position of the intestine, and the pattern of flow through its blood and lymph vessels, change greatly from time to time, depending on the animal's current state of gastrointestinal function and neural and endocrine control.

Diversity of Functions

Even though much more could be said of the basic structural characteristics of the digestive system, the preceding account may give an adequate introduction to gastrointestinal tracts from the point of view of their primary role in animal nutrition. Before completing this chapter, however, it may be useful to refer briefly to some of the accessory roles that the system has taken on from time to time in the course of vertebrate evolution. In some cases the variety, adaptiveness, and uniqueness of these roles are quite surprising. They are a reminder of a very important characteristic of evolutionary adaptations—the tendency of these adaptations to be "opportunistic," that is, to take advantage of existing structures and functions to arrive at solutions to new biological problems.

Table 6-1 Functions of Components of the Digestive System.

Main Digestive Functions	*Regions Where These Functions Are Carried Out*
Capture of food	Mouth, buccal cavity
Transport of food materials	Mouth, buccal cavity, pharynx, esophagus, crop, stomach, intestine, cloaca, rectum
Mechanical breakdown of food	Buccal cavity, stomach (gizzard)
Secretion, enzymatic breakdown of food materials (digestion)	Buccal cavity (if salivary glands are present), stomach, intestine, pancreas, liver (bile salts aid fat digestion), pyloric ceca, intestinal ceca
Absorption of nutrients, salts, water	Stomach (small organic molecules only), intestine, pyloric, and intestinal ceca; cloaca
Storage of food materials; of absorbed nutrients	Esophagus, crop, stomach, cloaca, rectum, gall bladder (bile); liver (absorbed nutrients)
Control of microorganisms	Stomach (acid; and culture of organisms in ruminant mammals); intestine; lymphatic tissue in walls of tract
Control of activity of digestive tract	Buccal cavity (neural signals to aboral tract); stomach (neural signals for feeding, satiety; neural signals and hormones controlling digestive activity); intestine (hormones controlling secretion of liver, pancreas; contraction of gall bladder)
Other Functions	*Regions Where These Functions Are Carried Out*
Respiration	Mouth, buccal cavity, pharynx
Defense, offense	Mouth, buccal cavity, poison glands
Excretion	Liver; rectal gland
Reproduction	Buccal cavity, crop, stomach (feeding and care of young); cloaca (transport of sperm, ova)
Formation of blood cells and proteins	Liver (intestinal wall in some fish)
Vocal signaling	Mouth, buccal cavity, pharynx

Thus, the arming of the mouth cavity with teeth or beaks, of adaptive value in capturing prey, is completely compatible with the use of the mouth cavity and teeth as defensive weapons as well as tools for nest building (fish, birds), and for burrowing or gnawing trees (rodents). The salivary glands, evolved chiefly in connection with the formation of lubricating fluids in terrestrial vertebrates, are adapted in certain reptiles as poison glands. In some birds salivary glands function in producing cementing substances for nest building. This recalls a very widespread association of the digestive tract with reproduction and care of the young. In some fish species the fertilized eggs are retained in the mouth of one of the parents until they have developed far enough to be capable of independent existence. Many birds and some mammals feed their offspring with food regurgitated from the crop or stomach, and in pigeons and some other bird species the esophagus or crop form special secretions (e.g., pigeon crop milk) used for feeding the young. Widely throughout the vertebrate series the reproductive and digestive systems are also intimately associated where they share the common exit chamber, the cloaca. Universally, the digestive and respiratory systems are interwoven in the buccal-pharyngeal region. Correlated with this is the widespread use of the buccal cavity, with its teeth and tongue, in vocal signaling. The buccal cavity plays an important part as a site of sensory receptors (Chapter 9). Although these function chiefly in food selection and in the integration of the activities of the entire digestive tract, they also have wider roles in the economy of many vertebrates. In animals of most vertebrate classes, the walls of the tract contain aggregations of lymph cells that are important in protecting the organism against invading microorganisms. Sometimes these are found at the junction of the buccal cavity and pharynx, as in the case of the mammal's *tonsils*. The walls of the small intestine, too, often show lymphocyte aggregates or lymph nodes (mammals). Finally, red blood cells are formed in the intestinal walls (cyclostomes, lungfish) and in the liver (fish, amphibians, vertebrate embryos). All in all, the digestive system is a remarkably versatile participant in the complex interplay of vertebrate organ systems.

References for Chapter 6

[1] Andrew, W., and C. P. Hickman. *Histology of the Vertebrates.* St. Louis: The C. V. Mosby Co., 1974.

[2] Bonneville, M. A. "Fine Structural Changes in the Intestinal Epithelium of the Bullfrog during Metamorphosis." *Journal of Cellular Biology,* **18**(1963), 579.

[3] Brauer, R. W. "Liver Circulation and Function. *Physiological Reviews,* **43**(1963), 115.

[4] Ito, S. "Structure and Function of the Glycocalyx." *Federation Proceedings,* **28**(1969), 12.

[5] Pernkopf, E. "Beiträge zur vergleichende Anatomie des vertebraten Magens." *Zeitschrift für Anatomie und Entwicklungs-geschichte,* **91**(1930), 329–390.

[6] Peyer, B. *Comparative Odontology.* Chicago: University of Chicago Press, 1968.

[7] Porter, K. R. "Independence of Fat Absorption and Pinocytosis." *Federation Proceedings,* **28**(1969), 35.

[8] Sellers, A. F., and C. E. Stevens "Motor Functions of the Ruminant Forestomach." *Physiological Reviews,* **46**(1966), 634–661.

7

Respiration and Circulation

Introduction: How Success Depends on Joint Efforts

Just as division of labor and specialization characterize successful modern society or a smaller unit of a successful enterprise, so the success of the vertebrate animal is dependent on a division of labor among specialized cells, tissues, and organs that serve specific functional tasks. Structure and function become inseparable, and the student of morphology, or structural patterns, must never lose sight of the functional objectives of the various structural components.

Respiration, unfortunately, does not have a single well-defined meaning in biological contexts. In its widest interpretation it incorporates all operations that effect the exchange of respiratory gases between the external environment and the metabolizing cells. Sometimes it is used in reference to the biochemical degradation of substrate in cellular metabolism. That meaning of the term will not be treated in this text.

External respiration is that part of the exchange of gases between the organism and the environment that is responsible for loading oxygen and unloading carbon dioxide. This process takes place in the specialized structures we call respiratory organs, and the design of these organs is our present concern. The circulating blood is the medium that connects the interior of the organism with the environment. Thus, oxygen is loaded from the environment into the blood, while, conversely, carbon dioxide is unloaded in the opposite direction. The equally important processes, the unloading of oxygen and loading of carbon dioxide at the transport terminals serving the metabolizing cells, occur in the fine-bored, thin-walled capillaries of the vascular system. How the capillaries are designed to promote efficient exchange between the metabolizing cells and the circulating blood will be discussed in the section on circulation. It will become increasingly apparent, however, that the efficiency of these processes depends inseparably on the joint functions of the organs of respiration and circulation. Both organ systems are designed to promote effective transport between the external environment and the living cell.

It is important to realize that the ultimate physical process that completes this transport chain is simple, passive diffusion. Diffusion is the random movement of matter (molecules) from regions of high concentration to regions of low concentration. The rate of diffusion depends on several factors formally expressed in Figure 7-1. You will see throughout these chapters that the objective of both the respiratory and circulatory organs is to aid the process of diffusion. Figures 7-1 and 7-2 indicate the principal ways in which the structure of these organs can promote effective diffusion exchange. A comparison is made of diffusion through an unspecialized body surface and a surface spe-

Figure 7-1 Diffusion through a homogenous barrier is expressed by Ficks law in which S is the amount of substance diffused, D is the diffusion coefficient, A is the surface area available for diffusion, C is the concentration at distance 1 from the barrier and t is time (duration) of diffusion. The ratio $^{dc}/_{dl}$ is called the concentration gradient.

Ficks equation and Figure 7-1 show that:

(1) A large surface area allows more substance to diffuse through, i.e., $S \alpha A$.

(2) If diffusion length is shortened while the concentration difference across the barrier remains unchanged the concentration gradient will increase; so will the amount of S, i.e., $S \, \alpha \, \frac{dc}{dl}$.

The same result occurs if the length remains unchanged and the concentration difference increases.

(3) More time will allow more substance to diffuse through, i.e., $S \, \alpha \, t$.

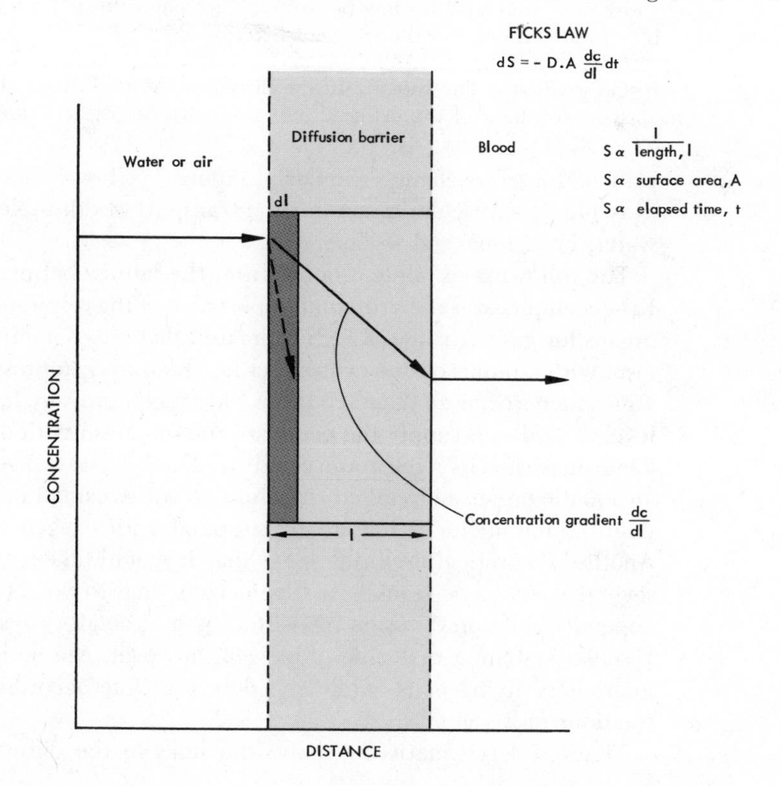

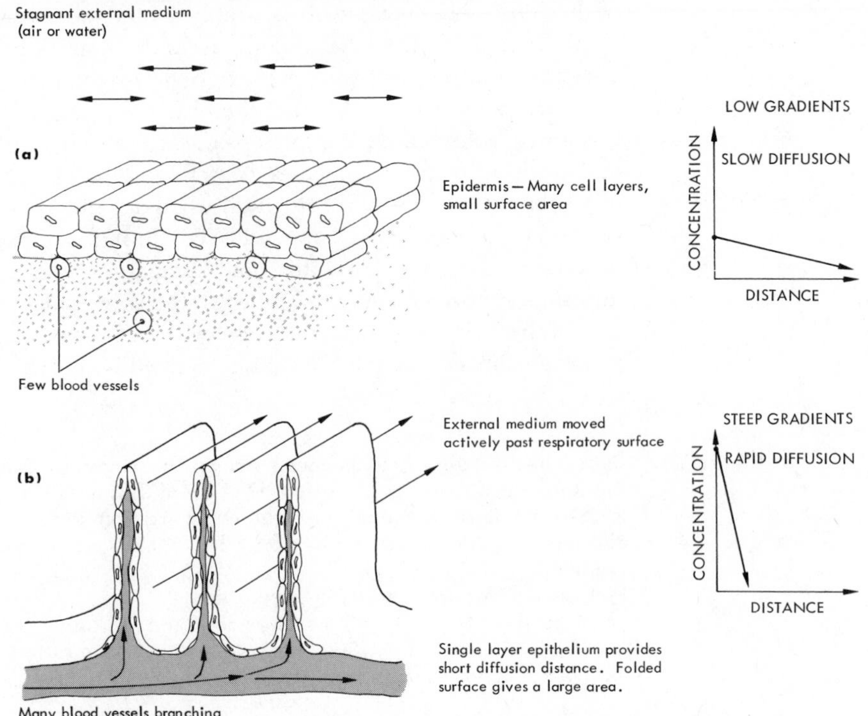

Figure 7-2 (a) An unspecialized body surface is inefficient in gas exchange because it provides a small surface area and low concentration gradients of the respiratory gases. The gradients are low because of the long diffusion distance, the sparse blood circulation, and the stagnant external medium. (b) A body surface specialized for gas exchange is efficient because of a large surface area and steep diffusion gradients. The short diffusion distance, the rich blood circulation, and the active movement of the external medium cause the steep diffusion gradients.

cialized for gas exchange. Similarly, Figure 7-3 shows the difference between avascular and vascular tissues for the transport of diffusible matter like gases, water, heat, ions, and so on.

The following example, quoted from the late Nobel prize winner, August Krogh, emphasizes the profound importance of the respiratory and circulatory organs for gas exchange. Krogh estimated that a hypothetical spherical organism with a radius of one centimeter needs an oxygen pressure of twenty-five atmospheres or more than 125 times the oxygen pressure in ambient air at sea level, in order to supply the center of the organism with oxygen by diffusion alone unassisted by a respiratory and circulatory system. Krogh calculated that an aquatic organism living in air-saturated water cannot have a radius greater than 0.5-millimeter if it is to be supplied with oxygen by diffusion alone. Another theoretical example shows that it would take a molecule of oxygen several years to be transported from your lung to your toe if the transport depended only on diffusion unassisted by a circulatory system. Aided by the vascular system, a molecule of oxygen, however, can be brought within ten microns of living cells, while the diffusion time accordingly is reduced to fractions of a second.

Figure 7-4 schematically shows the links in the diffusion path from the

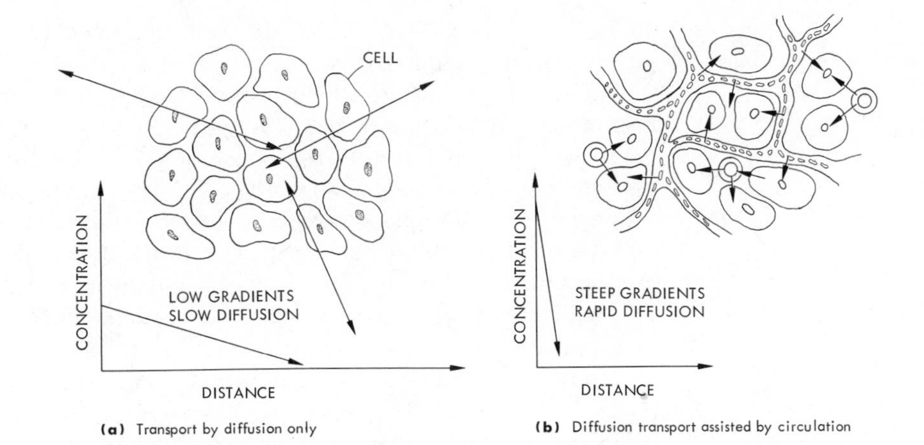

Figure 7-3 (a) Tissue without blood circulation. If masses of cells (tissues) receive nutrients and exchange gases and metabolites by diffusion only, the transport is slow and inefficient and cannot support the metabolic activity of an active tissue.
(b) Tissue with blood circulation. If blood capillaries penetrate tissues to the vicinity of individual cells, transport is rapid and efficient. Interruption of the blood supply to metabolically active tissue causes the cells to die or be damaged irreversibly.

Figure 7-4 The transport path for respiratory gases in a vertebrate. Diffusion transport exchanges gases with the environment at diffusion sites I and II. The rate of diffusion is increased by bulk, convective transport of air or water and blood effected by a ventilatory pump (pump I) and a perfusion pump (pump II). Blood constitutes the internal transport medium. The presence of hemoglobin in blood greatly increases its gas transporting capacity and also facilitates the rate of diffusion across diffusion sites I and II.

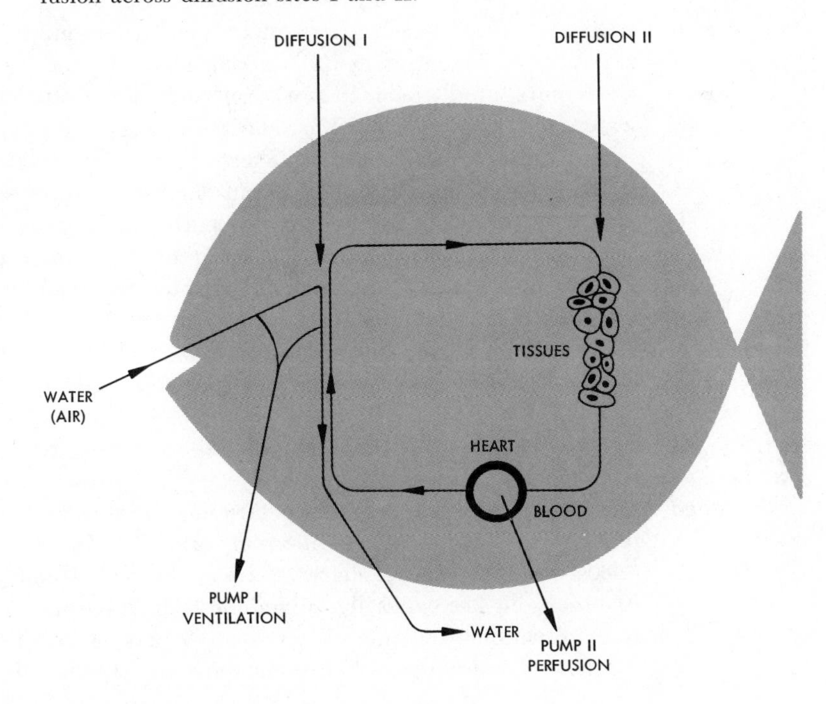

ambient medium to the metabolizing cells in a hypothetical vertebrate—fish looking, but representative for all vertebrates.

The circulating blood connects the cells with the environment at the respiratory exchange surfaces. Two principal diffusion sites occur in the pathway: diffusion 1 in the respiratory organs and diffusion 2 between the vascular system and the metabolizing cells. The rate of diffusion across these sites is aided by the phenomenally great arborization of the vascular system into the small and thin-walled exchange vessels, the capillaries. The resulting closeness between blood in the capillaries and the air or water at diffusion site 1 as well as between blood and the extracellular fluid bathing the cells at diffusion site 2 promotes rapid diffusion.

Two important pumps in the pathway—pump 1, the ventilatory pump, and pump 2, the circulatory pump, or heart—cause convection and the bulk transport of water or air and blood to the diffusion sites where the concentration gradients are steep, resulting in effective gas transport.

The need to assist diffusion transport by convective bulk transport by specialized organs is essential in the vertebrate animal. The architecture and bioengineering of these organs at representative levels of vertebrate organization form the basis for the following two sections.

The Respiratory System

External respiration or gas exchange in chordates as a rule takes place in specialized organs called respiratory organs. These include various forms of gills and lungs designed for aquatic and aerial gas exchange, respectively. Gas exchange also can take place through nonspecialized parts of the general body surface, the skin. In some amphibians, such nonspecialized body surfaces constitute the primary or only site for gas exchange.

The Physics of Respiratory Media

Air and water differ in essential respects as media for gas exchange. In particular, the concentrations of oxygen in air and air-saturated water at twenty degrees centigrade differ about thirtyfold, and oxygen diffuses 300,000 times more rapidly in air than water. These facts vastly increase the volume of water aquatic animals need to ventilate to extract amounts of oxygen comparable to those of air breathers. There are other disadvantages to the aquatic animal; the density of water is about one thousand times that of air, and the viscosity of water at twenty degrees centigrade exceeds that of air fifty times. For these reasons, the energy costs of moving water past the respiratory exchange surfaces, i.e., the work of breathing, is markedly higher for the water breather than the air breather.

We will see, subsequently, how these physical differences between water and air have entered into the design of gills and lungs as respiratory organs.

The Organs for Aquatic Gas Exchange

Functional gills are found in all fish, although they are much reduced in some of the air-breathing forms such as the lungfish and several tropical fresh-water teleosts. Gills can be external, protruding directly from the branchial region into the ambient water, or internal, protected by specialized skin folds in teleosts or the general pharyngeal wall in cyclostomes and elasmobranchs. External gills are generally vulnerable to injury from mechanical contact or from predation by enemies. External gills can be moved in a wavelike action by specialized muscles to effect convection or stirring of the external water.

However, much more effective stirring occurs in internal fish gills by active ventilatory movements that renew the water passing the respiratory exchange surfaces. Functional external gills are also present in larval amphibians and are retained in some neotenic adult urodeles.

The Structural Design of Gills

How can the structure of gills promote exchange diffusion between the external environment and blood? We would anticipate the following structural factors to be important.

1. A large surface area for diffusion.
2. Short diffusion distances and low diffusion barriers.
3. Structural support for an effective ventilatory pump—i.e., the continuous renewal of water facing the external exchange surface. Efficient gas exchange depends on the effective matching of the ventilatory current with the blood current on adjacent sides of the gill surfaces. We refer to this as ventilation-perfusion matching.
4. Numerous small-calibered and thin-walled vascular channels, or blood spaces, with a large surface area in gills and a large ratio of surface to volume present to match water and blood.
5. Moreover, there must exist effective structural support for a continued and adjustable passage of blood through the exchange surfaces. We refer to this as a controllable blood perfusion. It means that the volume flow of blood pumped by the heart must be adjusted to the over-all gas-transport requirements of the organism. At the same time, the flow of water, the ventilation volume, must be controlled to meet the same end. We will see later how these two pumps also must be adjusted to each other to achieve maximum efficiency in gas exchange.
6. Special to fish gills and distinct from the lungs of air-breathing vertebrates, the spatial organization of the water current and the blood current is arranged in a counterstreaming fashion. We refer to the typical fish gill as having countercurrent gas exchange.

Let us examine typical elasmobranch and teleost gills in the light of the foregoing considerations. In fish the gill structures are supported by, and attached to, a series of symmetrically placed branchial arches. In the more primitive elasmobranch fish, the two rows of primary filaments of a typical branchial arch are attached to opposite sides of a supportive interbranchial septum (Figure 7-5). In teleosts this interbranchial septum is reduced or absent, and each primary filament is supported by an independent gill ray.

The tips of the primary filaments in all types of gills are usually in contact with the tips of the primary filaments of the neighboring branchial arches [Figure 7-5(a)]. The individual filaments, which look something like the leaves of a book, are stacked on top of each other. Both sides of each primary filament bear numerous secondary lamellae [Figure 7-5(b)]. The secondary lamellae are the actual site of gas exchange. In active species of fish they are extremely numerous and provide a surface for exchange that may exceed the entire surface area of the fish itself by more than ten times (Table 7-1). In some fish the distance from water to blood is less than one micron. Note that the water will pass on both sides of the secondary lamellae and, thus, the diffusion distance between water and blood is further reduced. It is of equal importance that the large number of secondary lamellae reduces the distance across which the gases

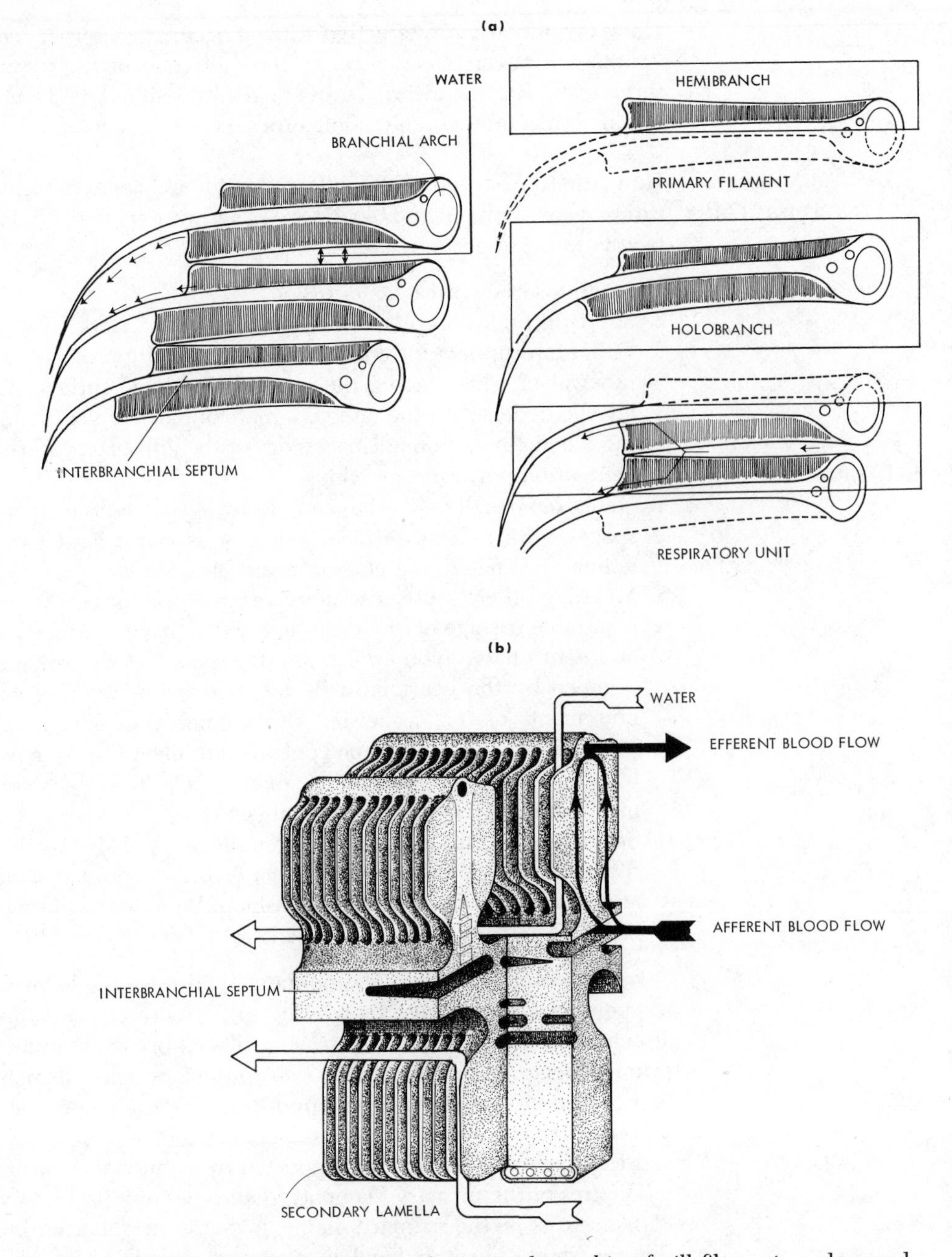

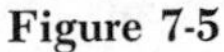

Figure 7-5 (a) Successive branchial arches showing relationship of gill filaments and secondary lamellae to the interbranchial septum in an elasmobranch fish. Note that two successive hemibranchs between interbranchial septa make up one respiratory unit. The arrows indicate the pattern of water flow. (b) Relationship of blood flow and water flow in a respiratory gill unit. The solid arrows indicate blood flow, open arrows the water flow.

Table 7-1 Dimensions of Fish Gills in Relation to Normal Activity Habits. [From G. Hughes (1966)]

	Toadfish	*Fluke*	*Sea Robin*	*Tautog*
Behavior	sluggish	sluggish	moderately active	moderately active
Weight (g)	305	404	365	466
Total number of filaments	656	1706	1996	1580
Total length of all filaments (mm)	4752	9747	14,022	12,537
Secondary lamellae (per mm)	11	19.3	20	19
Gill area (mm²/g body weight)	151	247	432	450
	Butterfish	*Mullet*	*Mackerel*	*Menhaden*
Behavior	moderately active	very active	very active	very active
Weight (g)	261	250	226	525
Total number of filaments	1548	2190	2814	2298
Total length of all filaments (mm)	9855	16,399	19,271	25,138
Secondary lamellae (per mm)	32	26.5	29.4	27
Gill area (mm²/g body weight)	461	1010	1040	1241

must diffuse in the water. This is important because water may otherwise become too distant from the exchange surfaces to participate in actual gas exchange with the blood. When exchange is limited by diffusion distance, this is referred to as functional or physiological dead space, as distinct from bypass of water referred to as anatomical dead space (Figure 7-6). The water taken into the mouth during breathing is guided through progressively smaller channels. When passing between the secondary lamellae, the width of the water channel can be as short as one fiftieth of a millimeter in some fish.

The large number of secondary lamellae results in a large total cross-sectional area irrigated by the water flow and a low velocity of the water past the exchange surfaces of the gills. Both of these factors will promote an efficient exchange diffusion.

Table 7-1 illustrates how the dimensions of fish gills, which we have discussed to be important determinants of the efficacy of the gills in gas exchange, bear a relationship to both the habitat and behavior pattern of the fish. Species with sluggish habits and low oxygen requirements have coarse gills with few and thick filaments and lamellae. Conversely, active fish have numerous filaments, which result in a large surface area. The distance from the water to the blood side of the secondary lamellae is in some species less than 1.0 micron (e.g., tuna).

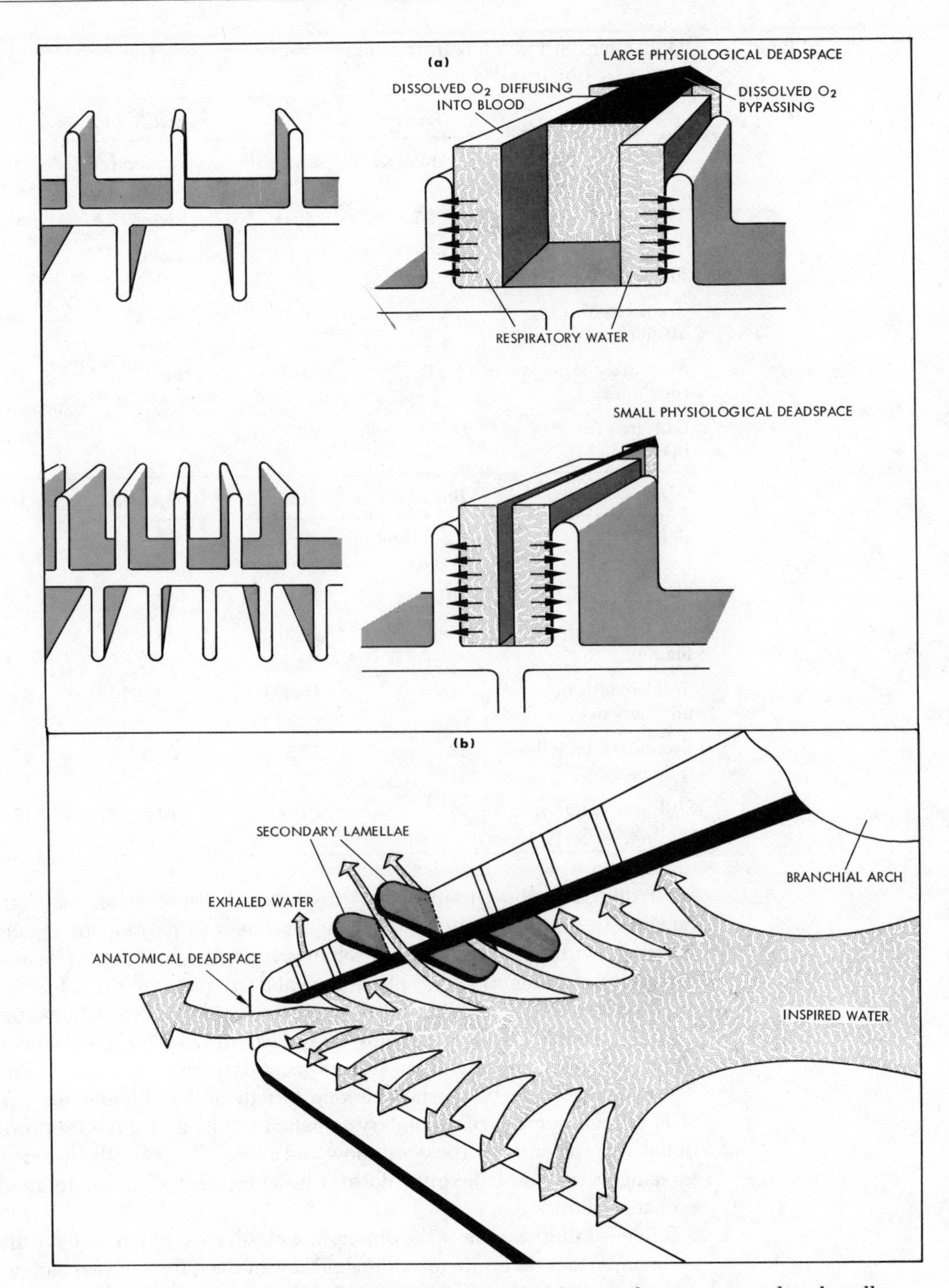

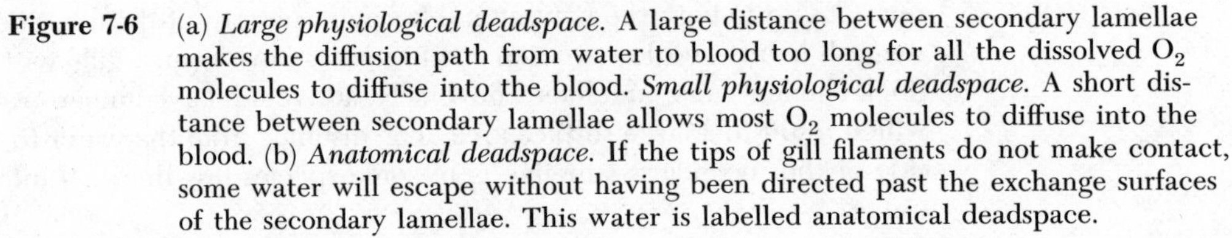
Figure 7-6 (a) *Large physiological deadspace.* A large distance between secondary lamellae makes the diffusion path from water to blood too long for all the dissolved O_2 molecules to diffuse into the blood. *Small physiological deadspace.* A short distance between secondary lamellae allows most O_2 molecules to diffuse into the blood. (b) *Anatomical deadspace.* If the tips of gill filaments do not make contact, some water will escape without having been directed past the exchange surfaces of the secondary lamellae. This water is labelled anatomical deadspace.

In an active fish such as a mackerel, the number of secondary lamellae can be extremely high and exceed forty per millimeter. Their shape is asymmetrical, with their greatest height facing the water current [Figure 7-7(b)]. This will help to assure that the water flow is broken up to eliminate possible stratification that would reduce the diffusion rate. The resistance to water flow in each of these narrow channels can be considerable, but the remarkably large number, more than a quarter of a million in a small fish, will assure an adequate flow of water at acceptable energy cost to the fish. However, it is important to realize that the low concentration of oxygen in water, compared to that in air, will necessitate a ventilation volume past the gills many times the volume of air that an air breather needs to ventilate to extract a comparable amount of oxygen. Also remember that the one thousand times higher density of water compared to air will additionally increase the energy cost of breathing in water. In accordance with this high energy cost, the ventilatory reserve allowing increased ventilation during periods of increased need for gas exchange is much smaller in water breathers than in air breathers.

Many of the most active pelagic fishes swim with partly opened mouths and, thus, reduce the requirement for pumping water over the gills simply by making use of the swimming action for renewal of the water surrounding the gills. The fastest swimmers, like the tuna, have added structural support to prevent deformation of the filaments and lamellae by the force of the great swimming speed.

Some structural features of gills reduce the apparent disadvantage of breathing in water compared with air. The counterstreaming arrangement of the water and blood flow is one very important factor. Figure 7-7 compares schematically a countercurrent and a concurrent exchange system with diagrams of their exchange efficiency.

The unidirectional path of the respiratory water current is indispensible to a countercurrent exchange. Water enters the mouth or spiracles and continues past the gill surfaces before it is expelled from the gill slits or the opercular openings. Unlike conditions in air-breathing vertebrates, the unidirectional water passage results in a continued renewal of water at the exchange surfaces. In the bidirectional airway of lung breathers, freshly inspired air is prevented from reaching all the way to the exchange surfaces of the lung before it becomes admixed with the gas trapped in the airways. The latter is designated as dead space.

The continued passage of the unidirectional water current past fish gills is achieved by the rhythmic activity of skeletal musculature that causes a double pumping action. A buccal or orobranchial pressure pump and an opercular parabranchial suction pump, aided by valves in the mouth or spiracle and at the opercular or gill slit openings, assure a fairly continuous and steady passage of water. Water is prevented from entering the esophagus by a sphincter valve demarcating the posterior end of the pharyngeal cavity. The passage of the water from the buccal cavity across the gill surfaces is illustrated in Figure 7-7.

Note that the direction of the water flow between the secondary lamellae depends on the close apposition of the distal tips of succeeding primary gill filaments. This apposition will prevent the discharge of water before it has made contact with the gas-exchanging surfaces of the secondary lamellae. Such closure becomes crucial for the efficiency of the gas exchange. It is structurally supported by specialized abductor muscles of each branchial unit (Figure 7-8).

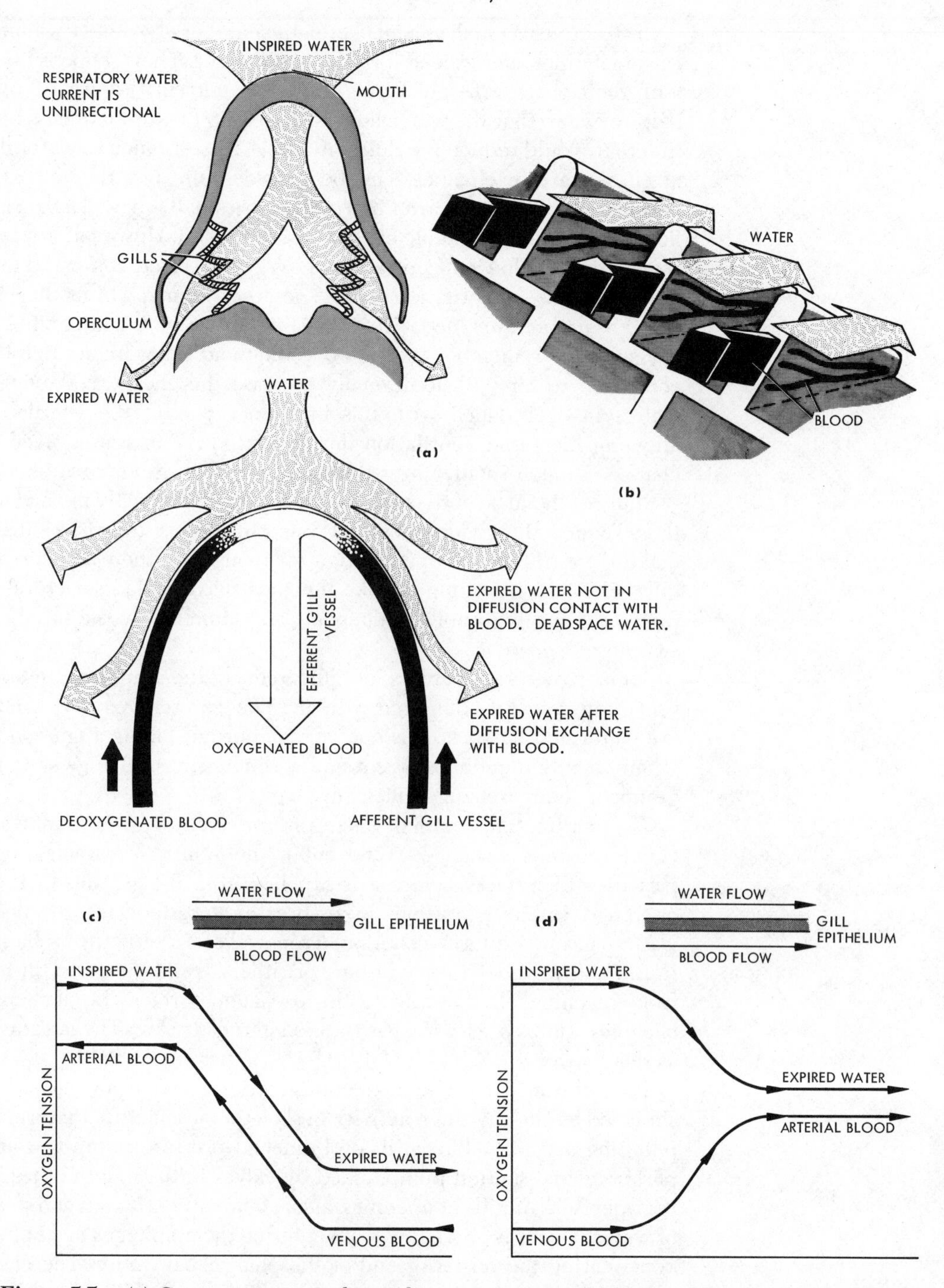

Figure 7-7 (a) Countercurrent exchange between water and blood in fish gills. (b) Blood and water meet in a counterstreaming arrangement in the secondary lamellae. (c) In a countercurrent arrangement the arterial blood can reach O_2 tensions higher than those in expired water. On its way through the gill, blood continues to make diffusion contact with inhaled water. (d) When water and blood meet in a concurrent exchange, the arterial blood can only reach O_2 tensions approaching those in expired water. Oxygen gain is hence much less than in a countercurrent system.

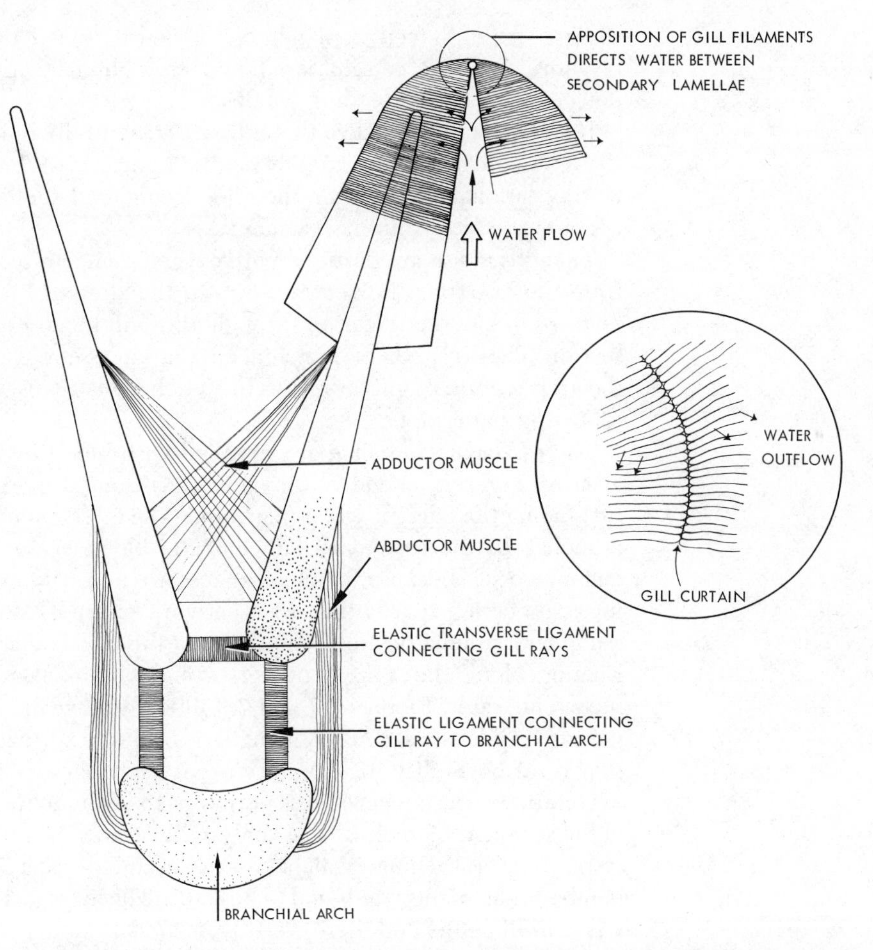

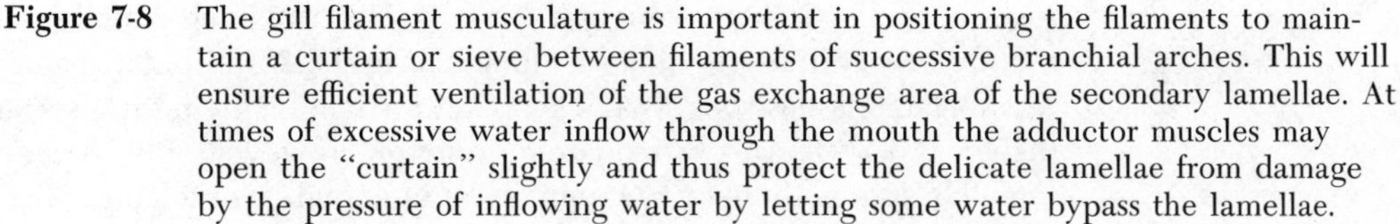

Figure 7-8 The gill filament musculature is important in positioning the filaments to maintain a curtain or sieve between filaments of successive branchial arches. This will ensure efficient ventilation of the gas exchange area of the secondary lamellae. At times of excessive water inflow through the mouth the adductor muscles may open the "curtain" slightly and thus protect the delicate lamellae from damage by the pressure of inflowing water by letting some water bypass the lamellae.

After having traversed the secondary lamellae where the gas exchange takes place between the water and the blood, the respiratory current continues behind the operculum for final discharge or, in the elasmobranchs, is guided in channels between interbranchial septa before being expelled from the gill slits.

The gills make up a delicate organ vulnerable to mechanical injury from external objects or predators. Consequently, in adult fish they are confined to the protected pharyngeal cavity. In this location, however, the gill surfaces are in danger of being smothered by ingested food. Among the structural measures to circumvent this are the lateral position of the gills and the presence on the branchial arches of gill rakers, which act as filtering devices. The gill rakers are developed to various degrees in different fish in apparent coordination with their habitat. Bottom-living species from muddy habitats have well-developed gill rakers. They can also be secondarily well developed in forms that actually depend on them as filters for feeding purposes, such as in the large basking sharks. These fish feed on plankton and their flattened, tapering gill rakers may be more than ten centimeters long, resembling in appearance the filtering devices in plankton-feeding whales.

It is important to realize that gills are structurally built to function in water that gives the fish approximately neutral buoyancy. When fish gills are exposed to gravitational effects in air, their fine secondary lamellae rapidly collapse and become ineffective in gas exchange.

The gills in fish are perfused with oxygen-deficient blood channeled directly from the heart via the ventral aorta to the afferent branchial arteries. Many features of blood circulation through gills will be described in the subsequent section. Those aspects of branchial circulation essential to the understanding of the architecture of gills as respiratory exchange organs will be discussed in the following paragraphs.

The secondary lamellae are supplied with blood by the afferent branchial arteries (Figures 7-5 and 7-10), which pass along the primary filaments on the side facing the effluent water (e.g., along the interbranchial septum in elasmobranchs). In the secondary lamellae the blood ceases to circulate in well-defined vessels because the blood spaces are more akin to lacunae or irregular interconnecting spaces than to a network of true vascular capillaries. Each secondary lamella, in turn, empties into the efferent primary filament artery running along the margin of the primary filament on the side facing the incurrent water. Figures 7-7 and 7-9 illustrate that the direction of the blood passage inside the secondary lamella is opposite to that of the water; thus, a structural possibility for countercurrent exchange of gases is provided.

Detailed experiments on eels permit us to evaluate the total external surface of the secondary lamellae. The analysis illustrates the remarkably large area of respiratory epithelium available for gas exchange. In a 500-gram eel, the total number of filaments was found to be 2,100. The average length of each filament was 7 millimeters and there were 18 secondary lamellae per millimeter. This gives a total of 264,600 lamellae ($2100 \times 7 \times 18$). Each lamella was approximately 0.60-millimeter along its long axis and 0.45-millimeter along its short axis. A lamella has roughly the shape of a triangle, giving it a surface area of $0.45 \times 0.6 \times 0.5 = 0.135$ square millimeter. However, only 75 per cent of the lamellar area is thought to be involved in actual gas exchange. Because each lamella effectively offers two sides for gas exchange, we get a total available surface equal to $264{,}000 \times 0.101 \times 2 = 52{,}920$ square millimeters, or approximately 530 square centimeters. This area represents many times the general body surface area of the fish itself. Because of the internal pillar cells supporting the frame of the internal lamellar space the blood inside the lamellae will be dispersed over an area less than that available for irrigation by water from the outside. Consequently, the surface area of the internal blood space is estimated to be about 50 per cent of the external surface.

We contended early in this chapter that an important design requirement for a respiratory organ was to provide a close matching of water and blood over a large exchange surface offering minimal diffusion distance and resistance. Figure 7-8 shows schematically how the water and blood when matched in the lamellae of the gills have been distributed across remarkably large areas. This not only serves to provide a large exchange surface area and a very short diffusion path between water and blood, but the matching process also is favored by an expansion of the cross-sectional areas of the fluid columns and,

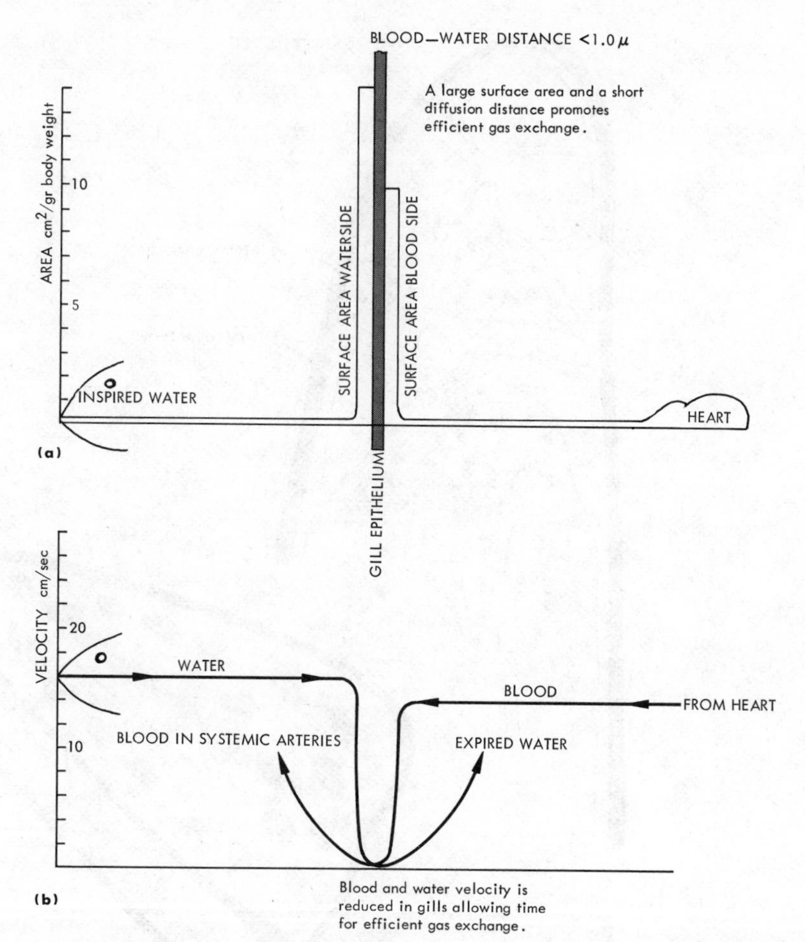

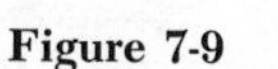

Figure 7-9 Structural factors important for the matching of water and blood in fish gills. (a) The surface expansion of the blood and water columns in the gills promotes efficient gas exchange by providing a large surface area and (b) a low velocity of the flowing fluids.

thus, by a marked reduction in the velocity of the flowing water and blood. This velocity relationship follows from the simple principle of continuity in flowing incompressible fluids. We have all observed how the water slows down when a creek widens. The decreased velocities of water and blood are of obvious advantage by allowing more time for the gas exchange diffusion.

Recent work on circulation through the primary filaments of teleost gills has brought evidence suggesting that blood circulation through the gill filaments may follow alternate routes either around the tip of the margin of the filament or through a central nonrespiratory compartment and, thus, avoid contact with water at the respiratory exchange area [Figure 7-10(a) and (b)]. The important inference to be made from such an arrangement is that the fish may be able to execute some kind of regulation as to the portion of blood directed to the respiratory exchange spaces or directed past these to the efferent side of the branchial circulation. One important advantage of such branchial vascular shunts in fish may be related to osmotic regulation. The blood and other body fluids in both fresh and salt-water fish are not in osmotic

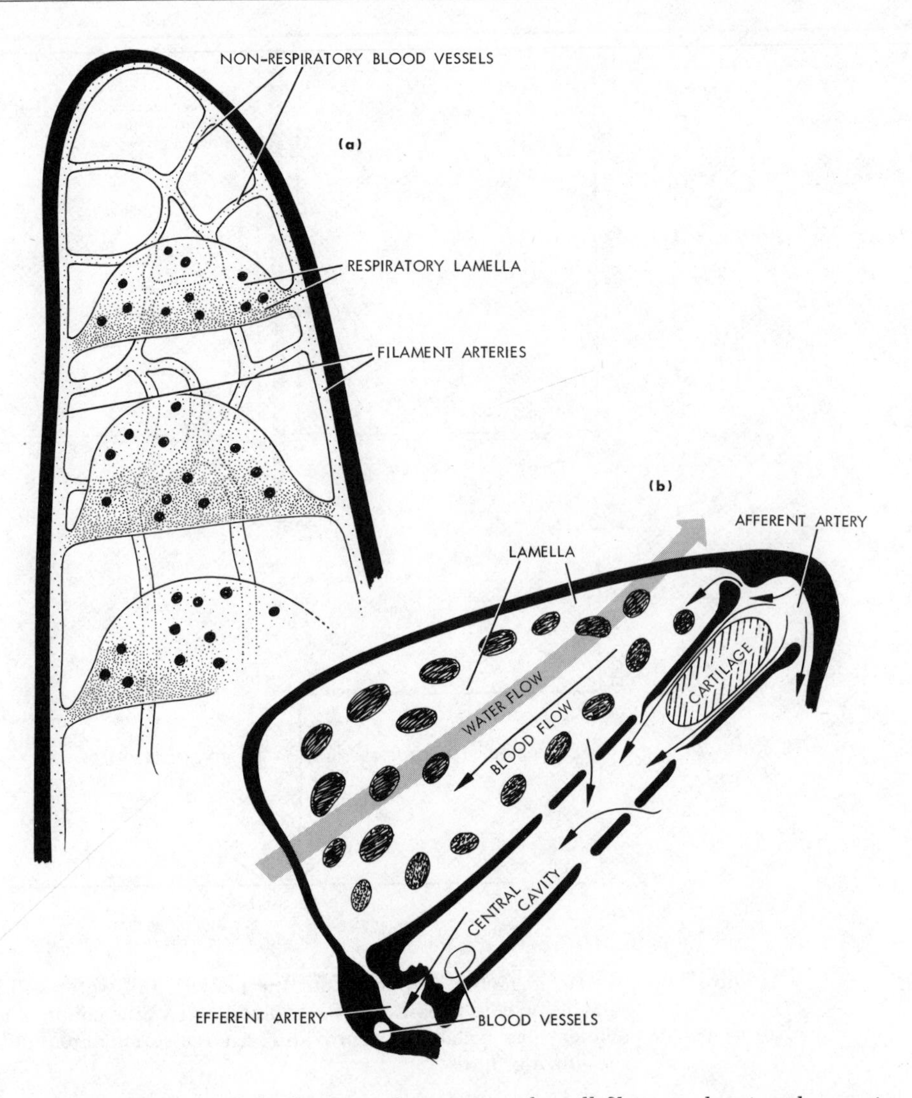

Figure 7-10 (a) Simplified schematic drawing of a portion of a gill filament showing the respiratory and nonrespiratory (shunt) blood paths. (b) Drawing of a gill filament showing a secondary lamella in cross-section. The vascular channels are marked with arrows and the large gray arrow indicates direction of water flow between the lamellae.

equilibrium with the surrounding medium but have an osmotic concentration intermediate between fresh and salt water. During passage through the respiratory exchange spaces in the gills, the blood will, thus, be subjected to osmotic gradients and the result will be an undesirable flux of water into the blood of fresh-water fish and out of the blood of salt-water fish. Hence, the salt-water fish will constantly lose water that has to be compensated for by intake of external sea water, which in turn will pose problems of ionic imbalance because of the high salt content in sea water. Paradoxically, then, we can say that a fish in sea water lives in a "dry" environment. Conversely, the fresh-water fish will be constantly diluted and will have the problem of retaining enough salts. It is interesting that cells in the respiratory epithelium of teleost fish gills have specialized for active excretion or absorption of salts in marine and fresh-water fish, respectively (see Chapter 8). The vascular shunts of the secondary lamellae just described may be important in allowing blood to be shunted past the exchange spaces and thus minimize the stress on osmoregulatory mechanisms. Such shunting must, however, be compromised with the actual needs of the fish for respiratory gas exchange.

The presence of discrete anatomical shunts allowing the bypass of respiratory portions of the secondary lamellae has been questioned by other recent investigators who claim that changes in blood flow through fish gills can be accommodated simply by recruitment of more or less secondary lamellae. This implies that during conditions of low blood flow through gills, the secondary lamellae toward the distal segment of the filaments are not perfused.

Still another possibility exists for blood flow through fish gills to make variable contact with the respiratory water current and, thus, represent a shunt in a functional sense. This may occur if the size of the blood space within the secondary lamellae varies. An increased width of the blood space might allow a portion of the increased blood volume passing through the secondary lamellae to do so with reduced exchange efficiency with water, due to the slowness of diffusion in blood as well as in the water outside the exchange surfaces. This type of shunting caused by a diffusion limitation is commonly referred to as a physiological shunt or dead space to make it distinct from the anatomical shunt or dead space described earlier (Figure 7-6)

Fish gills show structural adaptations to environmental conditions and to general activity and behavior. Thus, many fish in oxygen-deficient waters show especially well-developed lamellae that provide a large surface area for gas exchange. However, many teleost fish in tropical oxygen-deficient waters have developed accessory means for gas exchange—utilizing the atmospheric air. In these forms, the gills show a tendency to degenerate and become less useful in aquatic gas exchange. Pelagic fish, which often are very active swimmers, show structural adaptations resulting in large gill areas and short diffusion paths. Mackerel and herring are good examples of this category. In some fast-swimming fish like the swordfish, a peculiar gill structure has developed to withstand the deforming forces of a high swimming speed.

The Ultrastructure of Gill Lamellae

The gas-exchanging surfaces of the secondary lamellae are made up of two epithelia that become continuous along the distal margin of the lamellae. The two epithelia are held apart by a special type of cell, pillar cells. Pillar cells are spaced randomly and extend from one epithelium to the other. The resulting space between the two epithelia takes the form of a lacunar compartment. This is the compartment that constitutes the blood space. Below each epithelium lies the basement membrane. In Figure 7-11(a) and (b) electron micrographs of a secondary lamella from the teleost *Gadus pollachius* are shown. One can see that the blood space is uniformly delineated by a thin sheath growing out from the main body of the pillar cells. This outgrowing lining is called a pillar flange and is less than 0.5 micron thick. The flanges of adjacent pillar cells interconnect and, thus, come to represent the continuous boundary of the blood space. The average cell-body to cell-body distance of the pillar cells in this particular fish is about ten microns. The width of the lacunar blood space is, in general, just large enough for the red blood cells to pass through (Figure 7-11). Between the pillar flanges and the epithelia is a basement membrane that

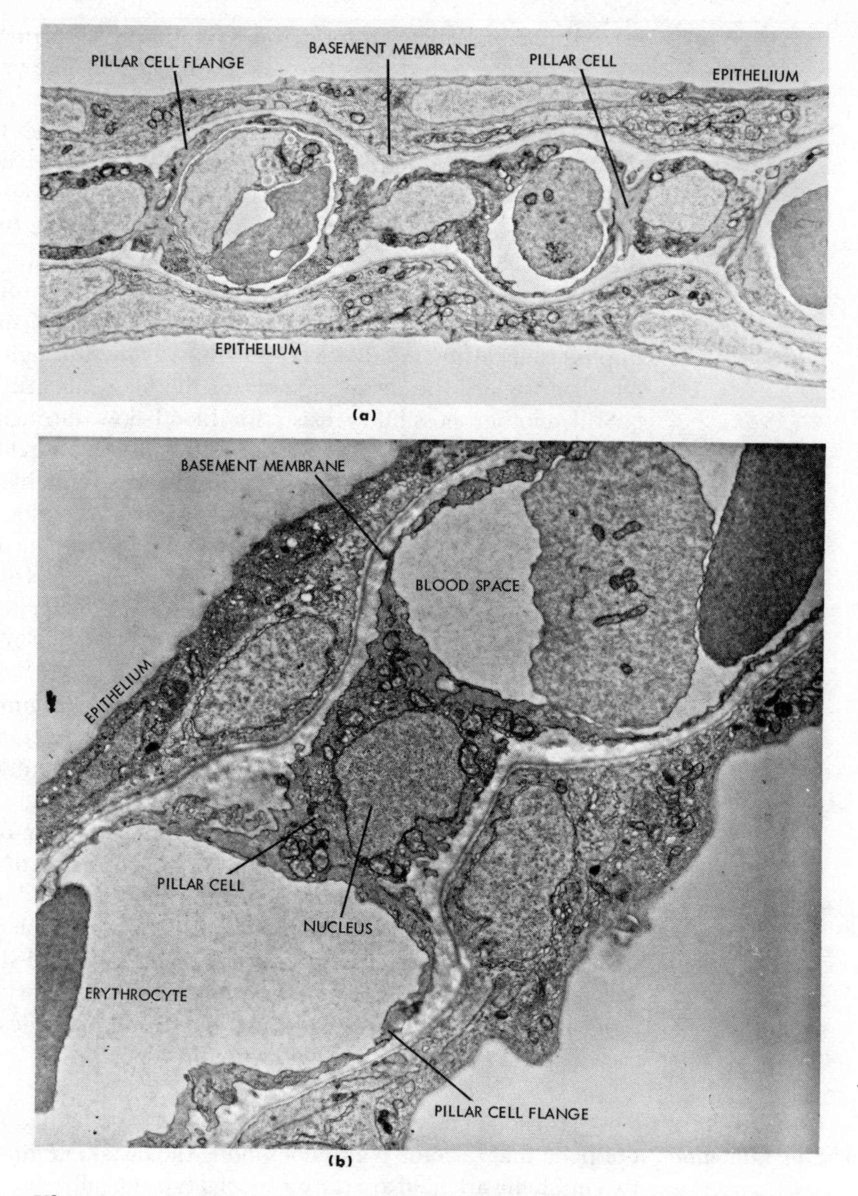

Figure 7-11 Ultrastructure of blood-to-water diffusion path in a teleost fish gill (*Gadus pollachius*). (a) Low power electron micrograph showing the general structure of a secondary lamella. Three pillar cells can be seen, with their flanges meeting and joining to delimit the blood space. Erythrocytes and a white blood cell are present. Note also the basement membranes and epithelia (4,000×). (Courtesy Professor G. M. Hughes.)

(b) An approximately median section of a pillar cell body and its flanges. Note the large nucleus, and the abundant mitochondria and tubules set in a dense cytoplasmic matrix. A column is present, enclosed in an infolding of the plasma membrane. Note the junction of one of the pillar cell flanges with that of an adjacent pillar cell. In the epithelia, note the location of the nuclei opposite the pillar cell body.

typically contains collagen fibers. The basement membrane shows a tendency to become thickened where it is juxtaposed to the pillar cell body. The epithelia facing the waterside of the lamellae are often made up of single cell layers but also may constitute several cell layers in complex interdigitation. The flat nuclei of the epithelial cells characteristically tend to be located on a common axis with pillar cells and the nuclei of epithelial cells on the other face of the lamella. This arrangement is of obvious advantage because the denser cellular elements then come to exert a minimal resistance to diffusion between the blood space and the water. The electron micrographs reveal the remarkably short and direct diffusion path between the water and the blood. Several recent authors report an average water to blood diffusion distance of less than one micron. It also has been suggested that the pillar cells may have an additional importance in regulating local blood flow through the lamellae by intermittent shortening and, thus, restricting the blood passage. The presence of distinctly fibrillar material in these cells offers a structural basis for this suggestion.

The Phylogeny of Vertebrate Gills

In embryological life all vertebrates develop a series of lateral visceral clefts and separating arches in the pharyngeal region. There is a tendency for the number of clefts to be larger in the more primitive forms. Some cyclostomes may have as many as fourteen, whereas teleost fish show only five or six pairs. Among air-breathing tetrapods, the visceral clefts disappear in adult life, with the exception of the first cleft, which is homologous with the Eustachian tube and parts of the middle ear. The clefts are formed when the endodermal pharyngeal pouches push through the mesenchyme and meet ectodermal invaginations. The gills formed in the clefts are, in general, considered to be of ectodermal origin (Figure 7-12).

The cyclostomes have gills that in many respects differ from the basic design apparent in vertebrates. We will begin a brief survey of gills in vertebrates, therefore, with the elasmobranch fish, which are considered to represent archaic conditions. Vertebrate gills are formed in successive bilateral branchial clefts, or perforations, between the pharynx and the exterior. The structural supports for the gills are the branchial arches or gill arches. These arches make up the part of the tissue space between the successive branchial clefts. The first pair of clefts in elasmobranchs has become highly specialized and is called the spiracles. Functionally, the spiracles are a site of entrance for the respiratory water current in addition to the mouth opening. Fast-moving sharks commonly swim with their mouths partly open; their habits have largely eliminated the need for spiracles, which have become vestigial or lost. On the other hand, the mouth of bottom-dwelling forms among sharks and skates could easily be smothered with mud and debris, and in such forms the spiracles have become very large and are the primary entrance for the respiratory water current.

The first branchial arch, the mandibular arch, occupies a region between the mouth and the specialized spiracular opening. Typically, in sharks there are five pairs of branchial clefts behind the spiracles. The number may be larger in some primitive sharks; for example, members of the primitive genus *Heptanchus* show seven clefts. The branchial arch between the spiracle and the first regular branchial cleft is referred to as the hyoid arch (Figure 7-13). In sharks each of the branchial clefts, except the spiracles, has a wide internal connection with the pharynx, whereas the external opening is commonly

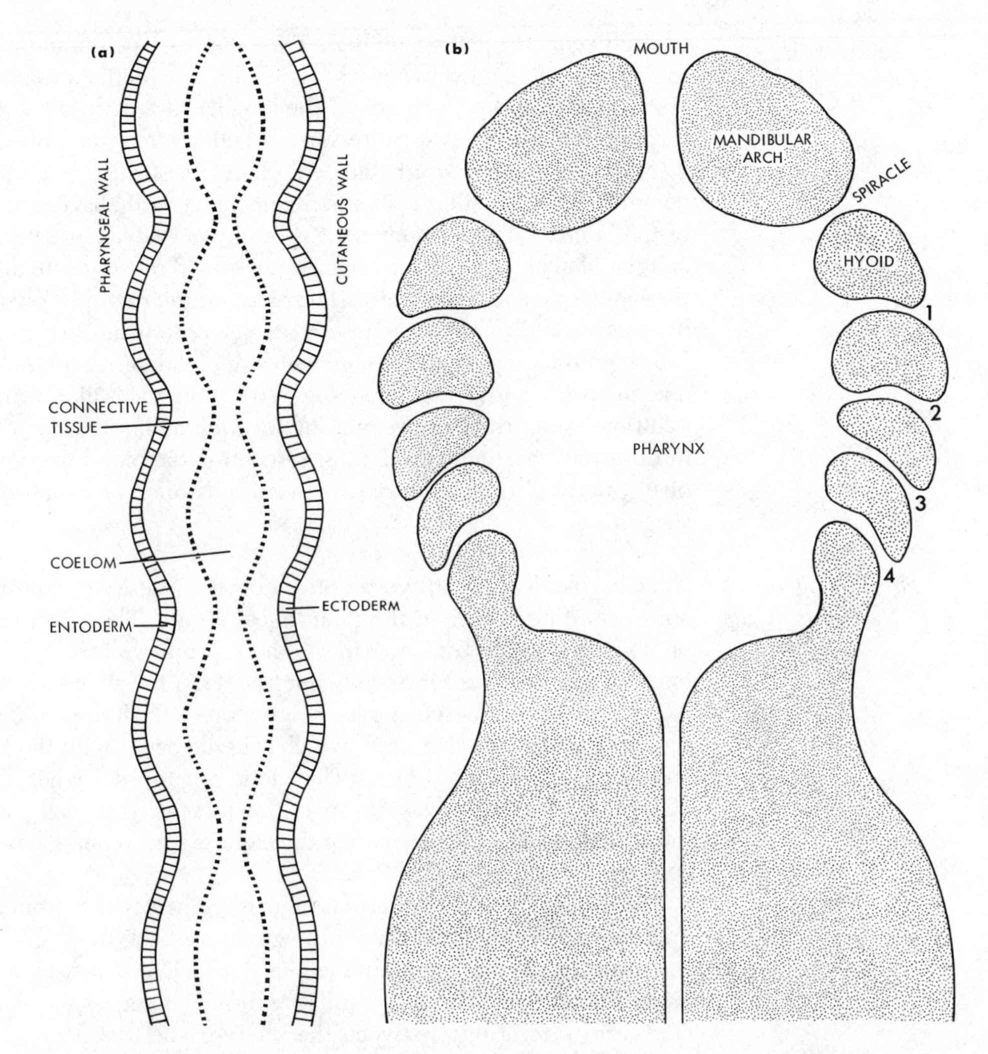

Figure 7-12 Horizontal sections through the branchial region of developing fish embryos: (a) beginning evaginations of the pharyngeal wall; (b) formation of gill slits completed. (In part, from Romer.)

narrowed to a slitlike aperture placed laterally. In some sharks, like the large basking sharks, the slits may be larger and traverse most of the dorsoventral aspect of the fish. In the dorsoventrally flattened skates and rays, the external gill slits are located on the ventral side. Characteristically, in sharks the gill slits are separated by membranous outgrowths from the arches. These structures, called interbranchial septa, represent the boundary between each gill cleft and the next [Figures 7-13(a) and 7-14(a)]. The distal ends of the septa, which are usually thin, make up an epithelial cover over the succeeding gill cleft. The gills are represented by a folded stack of plates or filaments radiating from the branchial arch. These filaments, in turn, project secondary lamellae from their upper and lower sides. The blood vessels and nerves supplying the respiratory portion of the gills run along the branchial arches. The arches also provide the basis for various placed supporting structures referred to as gill rods or rays. In elasmobranchs each typical branchial arch projects a double row of gill filament piles located one on each side of the interbranchial septum. Such a gill is called a holobranch. On some of the branchial arches only one side has developed filaments; this is a hemibranch. The common classification of a structural gill unit as one complete holobranch does not correspond to a functional gill unit that consists of two hemibranchs from succeeding branchial arches [Figures 7-5(a) and 7-13(a)].

The markedly modified spiracular opening in the elasmobranchs sometimes carries a rudimentary hemibranch on its anterior wall; this structure is known variously as a mandibular or spiracular pseudobranch. Its function is not known, but it can hardly be respiratory because the blood supplying it, with few exceptions, comes from the efferent side of the branchial circulation and, thus, has been arterialized already by passage through gills. The hyoid arch usually bears only a posterior hemibranch. The last gill slit has only one hemibranch projecting from the preceding branchial arch.

Members of the Holocephali, which are related to the elasmobranchs and with them make up the chondrichthyian order, differ from the elasmobranchs by lacking the spiracle altogether, and also by having one gill slit less than the

Figure 7-13 (a) Horizontal section through the head of an elasmobranch fish. (b) Horizontal section through the head of a teleost fish.

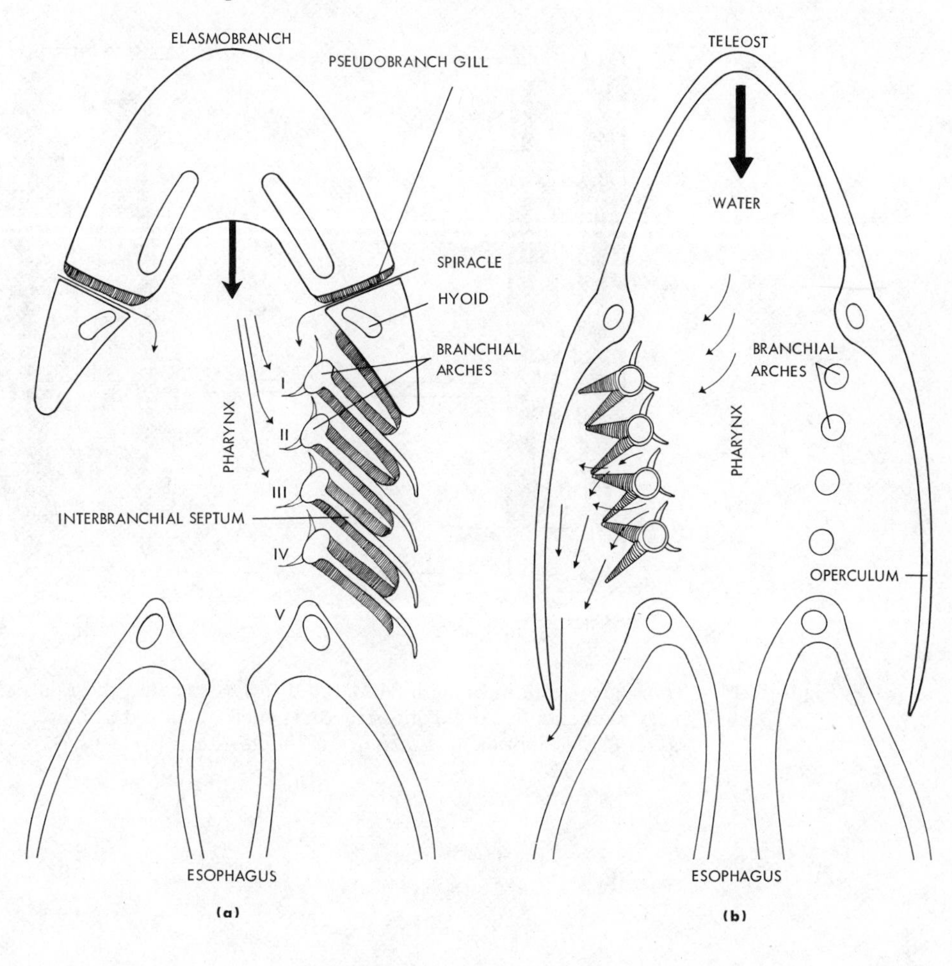

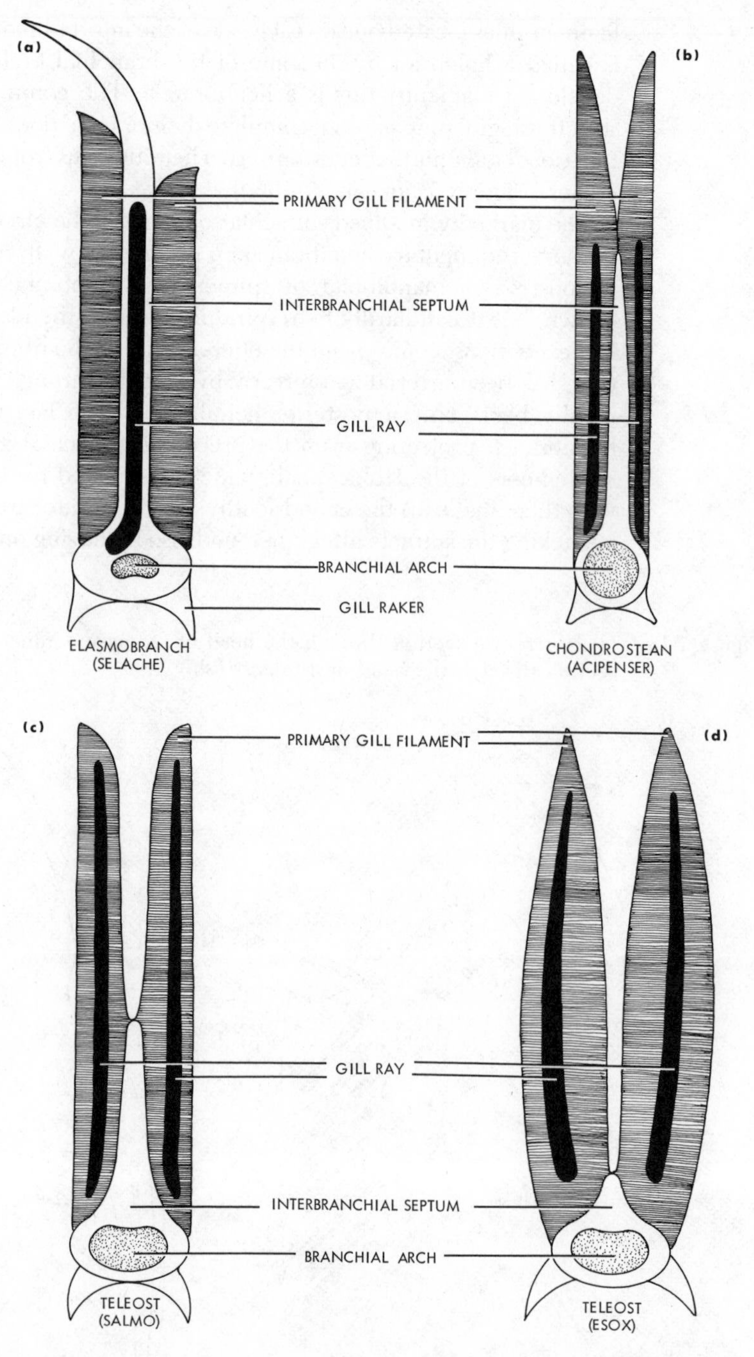

Figure 7-14 Transverse sections of branchial arches in fish showing the relationship of the primary filaments to supporting structures such as the interbranchial septum and gill rays. (a) Elasmobranch (*Selache*); (b) chondrostean (*Acipenser*); (c) teleost (*Salmo*); (d) teleost (*Esox*).

elasmobranchs. Notably, the Holocephali also have a large skin fold protecting the branchial region. This skin fold, or operculum, has its origin in the hyoid arch and grows posteriorly over the gill clefts. The operculum unites dorsolaterally and ventrally to the body wall, but the lateral posterior margin is free and is the site for discharge of the expired water.

diff in Telosts

In the bony fish, a number of differences stand out from the basic gill structure of the elasmobranchs. The more primitive Chondrostei have retained the spiracles, but the interbranchial septum has been further reduced so that the gill filaments project freely beyond the outer margins of the septa. This development progresses further in teleosts, and in some the interbranchial septum is reduced to almost nothing (Figure 7-14). With this development, the gill rays have an increased structural importance in support and orientation of protruding hemibranchs. The fifth branchial arch usually remains gill-less. There are, however, several variations in the distribution of hemibranchs on the various branchial arches among different fish. Many attempts have been made to relate these differences to phylogenetic ranking, but the discussions are mostly speculative.

An important structural change of major functional consequence in the bony fish is encountered in the operculum, which becomes well developed, movable, and structurally supported by specialized opercular bones and muscles for movement. The opercula in bony fish are more than the protective covering in Holocephali. They are essential in the breathing mechanism, providing a branchial chamber suction pump that, linked with the pharyngeal pressure pump, is responsible for propulsion of the respiratory water current (Figure 7-15). The gills of *Neoceratodus,* the only lungfish that is primarily an aquatic breather, conform to the basic plan of elasmobranchs. There are five branchial clefts and the hemibranchs are largely attached to well developed interbranchial septa. A specialty of the gills in *Neoceratodus* is the extension of the branchial filaments to the dorsal and ventral walls of the branchial clefts; thus, the hemibranchs are continuous between interbranchial septa. In the other genera of lungfish, *Protopterus* and *Lepidosiren,* a marked gill reduction accompanies an increased development of lungs in accord with the dominance of aerial breathing. In both genera, the gill reduction results in two branchial arteries passing to the efferent branchial side without the intervention of gills (see the next section, on circulation).

Figure 7-16 shows the respiratory water current in elasmobranchs and the mechanical events associated with the breathing cycle.

Gills of Cyclostome Fish

The gills in cyclostomes are specialized and have little apparent structural homology to gills in other fish. Their unique features may result from the semiparasitic feeding habit of attaching and partly imbedding their heads into their prey. In addition, the entire muscular apparatus involved in moving the respiratory water current of the jawless cyclostomes is different from that in other fish. The cyclostomes, comprising the lampreys and the myxinoids (hagfish) are often referred to as marsipobranchs because of their peculiar pouchlike gill structure. The gills are contained in spherical pouches that connect either internally with the pharynx through individual openings (myxinoids) or by union with a special diverticulum of the pharynx that ends blindly and is located ventrally (lampreys). This diverticulum, referred to as the branchial canal, lies below the regular pharynx and communicates anteriorly

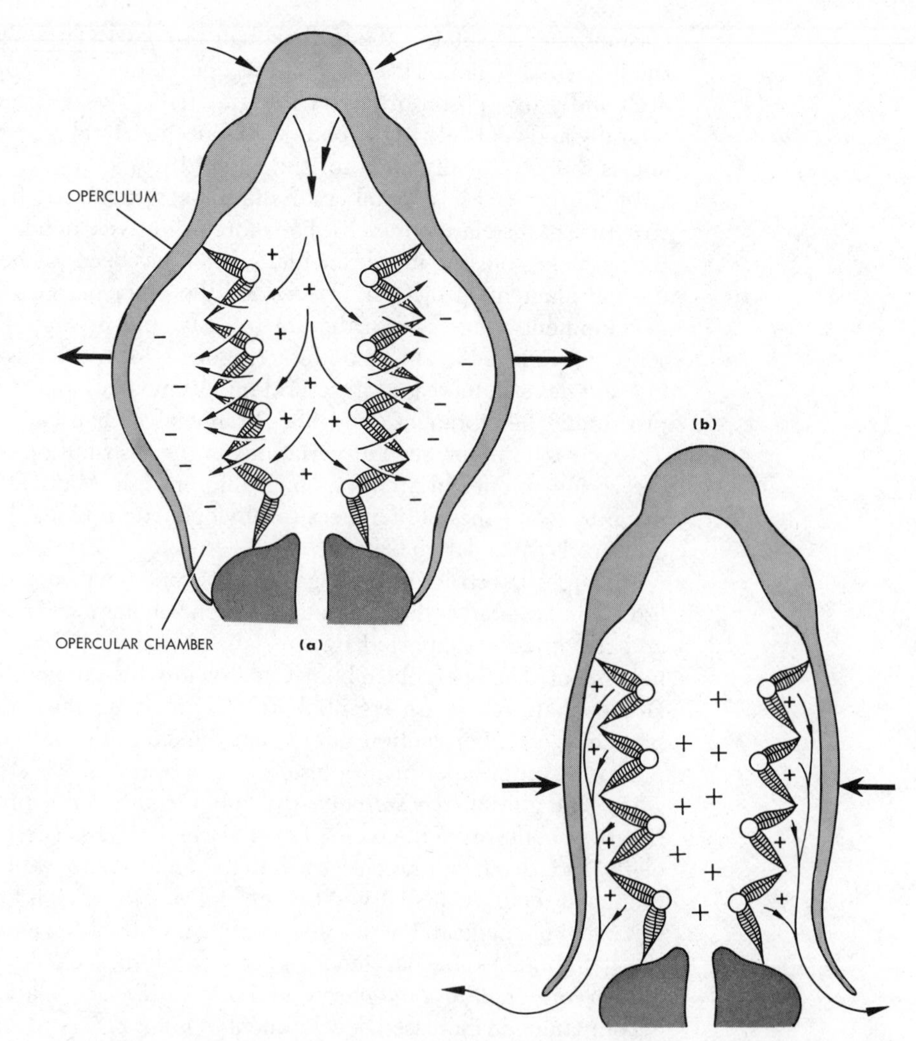

Figure 7-15 Horizontal section through head of teleost fish showing how the pharyngeal and opercular pumps propel the respiratory water current. Pressures indicated by + and − relative to the ambient pressure surrounding the fish. (a) Dilatation of opercular chamber will cause a suctional attraction for water across the gills. (b) Contraction of opercular chamber will cause water to escape along a pressure gradient from the buccal chamber to the outside.

with the esophagus. In adult lampreys, all the gill pouches open separately to the exterior through small orifices covered by fine skin folds that act as valves. In the myxinoids, the expired water is discharged through one external aperture made up by a common duct from each side. These ducts in turn are formed from the fused canals leading from the gill pouches on each side [Figure 7-17(a)]. The branchial filaments in cyclostomes are represented by a continuous system of tightly packed epithelial ridges investing the entire internal surface of the gill sacs. This epithelium is reported to be partially ciliated in lampreys, a feature that is typical of their very specialized larval form, the ammocoete. The interbranchial septa are thick and well developed and constitute the external boundaries of the gill units. Each of these units includes a specialized branchial cartilaginous skeleton and branchial muscles in addition to a complicated system of vascular channels that include some large peribranchial sinuses. When well-developed muscles of each sac rhythmically contract, water is discharged through the external orifices. During inhalation, the water current enters the pouches through the same orifice from which it is expelled. The forces involved in inhalation are thought to result from a passive recoil of the branchial pouches after the very active constriction during expiration. In this way the lampreys employ a bidirectional respiratory passageway that is unique practice among water-breathing vertebrates.

In myxinoids, a fairly wide nasal aperture provides entrance for the water

Figure 7-16 Diagram of a dogfish in side view (left) and horizontal view (right) to show the path of the respiratory current. The horizontal sections pass through the external gill slits and illustrate the changes in volume of the parabranchial and oro-branchial cavities. Pressures in these cavities are indicated as + or − with respect to zero pressure outside the fish. Full line arrows show the movements of the mouth and branchial regions, and their thickness indicates the relative strength of contraction. Valve in spiracular channel (a) open and (b) closed. (After G. M. Hughes, *Comparative Physiology of Vertebrate Respiration.* Cambridge: Harvard University Press, 1963, p. 145).

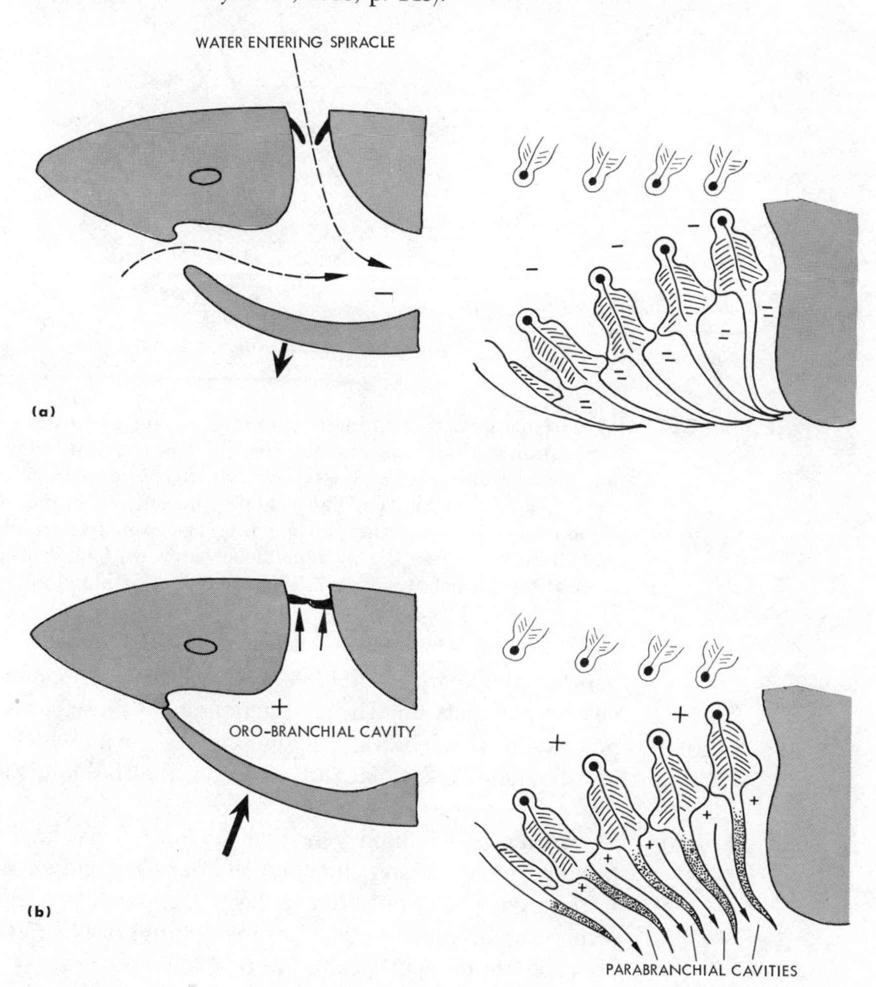

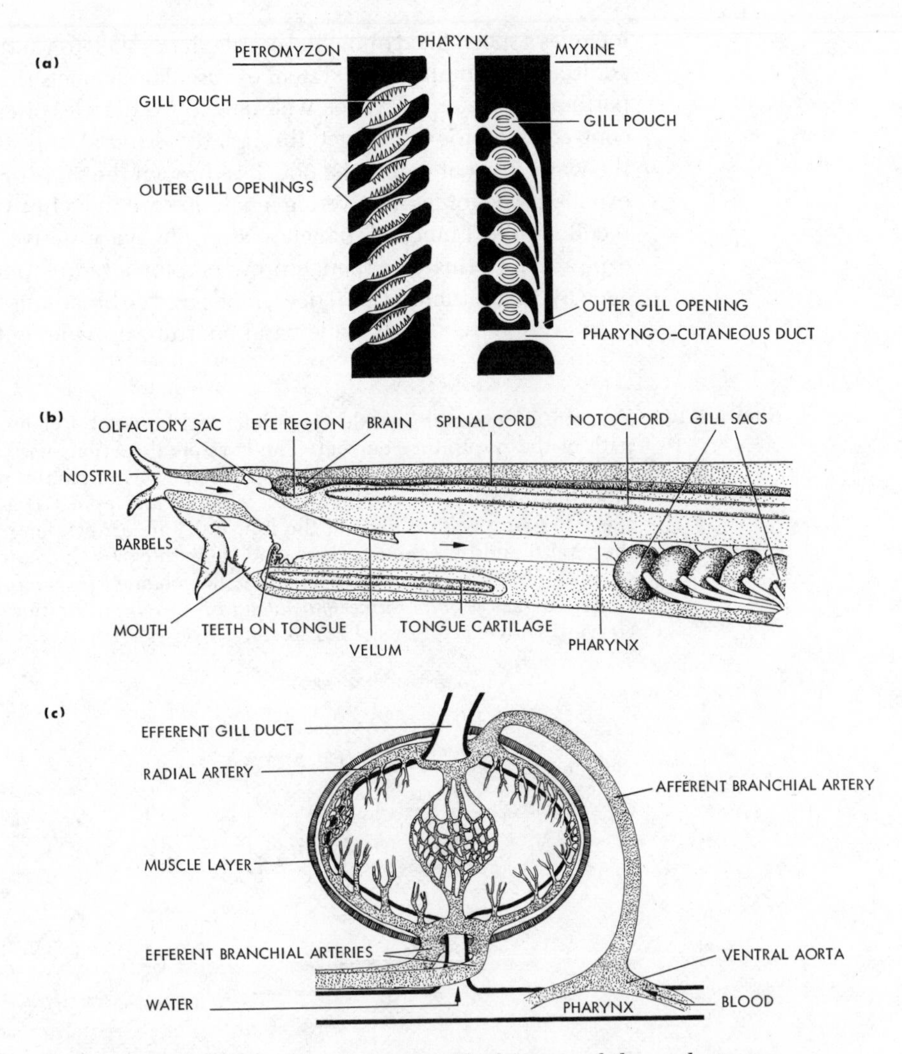

Figure 7-17 (a) Horizontal section through the branchial region of the cyclostomes *Petromyzon* and *Myxine* showing the gill pouches and inflow and outflow channels of the respiratory water current. (b) Respiratory tract of Myxine showing the respiratory water passage. The velum represents a unique pumping structure propelling the water into the gill pouches. These in turn are also contractile and contribute further to the movement of water. (c) Detail of a gill pouch in Myxine illustrating the counterstreaming arrangement of blood and water.

current, even when the mouth parts are imbedded in prey during its semiparasitic feeding. An effective water current is maintained by the rhythmic velar movements aided by contractions in the muscular walls encircling the gill pouches. As in teleosts and elasmobranchs, the matching of blood and water in myxinoid gills takes place in countercurrent fashion [Figure 7-17(b) and (c)].

The Rise of Air Breathing

More than 350 million years ago in the Silurian and Devonian geological periods, the prevailing climatic conditions brought about a marked reduction in the oxygen concentration of large fresh-water basins. As a result, aquatic respiration became marginal for many forms of life and a direct utilization of atmospheric oxygen became necessary for survival.

Fossil evidence, as well as a variety of extant bony fish, show several types of structural adaptations for the direct utilization of atmospheric oxygen. These adaptations include modifications of the buccal cavity or arborizations from pharyngeal and opercular cavities for direct gas exchange between blood and rhythmically ingested air. In other modifications, diverticula from the upper gastrointestinal tract developed as receptacles and gas exchangers for air. The stomach or cloaca of other fish similarly developed specialized sections that could function in aerial gas exchange; still others developed a marked vascularity of the skin that was effective in direct aerial gas exchange.

Today, many fish show these and other structural adaptations for air breathing, but only one of the early types evolved further toward the tetrapod condition in vertebrate evolution. This is the type seen in extant lungfish, some holostean fish, and a few teleosts in which a diverticulum of the foregut—later referred to as a lung, air bladder, or swim bladder—developed as a rhythmically ventilated respiratory organ. Why this solution was selected in the further evolution of vertebrates remains speculative. However, with the development of a separate vascular supply and a separate vascular return from the air-breathing organ to the heart, such as seen in the lungfish, the efficiency of air breathing made a great advancement in terms of oxygen delivery to the tissues, which was needed to support the increasing metabolic levels and the higher operational ability of tetrapod vertebrates.

A structural homology between the air bladder of fish and the lungs of lungfish and tetrapods seems well established. The swim bladder in fish has been credited with at least four different functions. First, it serves as a hydrostatic organ in buoyancy control; this is its primary function in modern teleosts. Secondly, the swim bladder may function as a respiratory organ or as an oxygen storage organ. Thirdly, it is important as a sense organ for pressure perception and as an amplifying structure in sound perception. Lastly, the bladder itself can be an organ of sound production.

Cyclostome and chondrichthyian fish show no trace of a swim bladder. The swim bladder of fish as well as the lungs of tetrapods are of endodermal origin and develop as an outgrowth from the gut. In some fish, it may retain a connection with the gut, the pneumatic duct, and is called a physostomous air bladder. In others, the pneumatic duct degenerates and the bladder is called physoclistous. In still others, the entire swim bladder atrophies and disappears. Figure 7-18 shows the various types of air bladders and their connection with the gut.

In this chapter we are primarily concerned with the structural design of respiratory organs. We will concentrate on the air bladder as a respiratory organ and as a forerunner of the higher vertebrate lung. It seems well established that the air bladder was an accessory respiratory organ when it first appeared with certainty in the early crossopterygian fish. This view is supported by the early evolution of these fish in swampy, oxygen-deficient tropical waters. Their accessory respiratory organs were probably essential and permitted survival while other fish became extinct. The presence of lunglike air bladders also is suggested in fossil evidence, particularly from the Devonian period. Although a respiratory function of the air bladder has been lost in the great majority of modern fish, it has been retained in some archaic forms like *Polypterus, Amia, Lepidosteus,* and the lungfish. In the latter it has become the principal organ of gas exchange. The diversity in adaptive adjustment and the

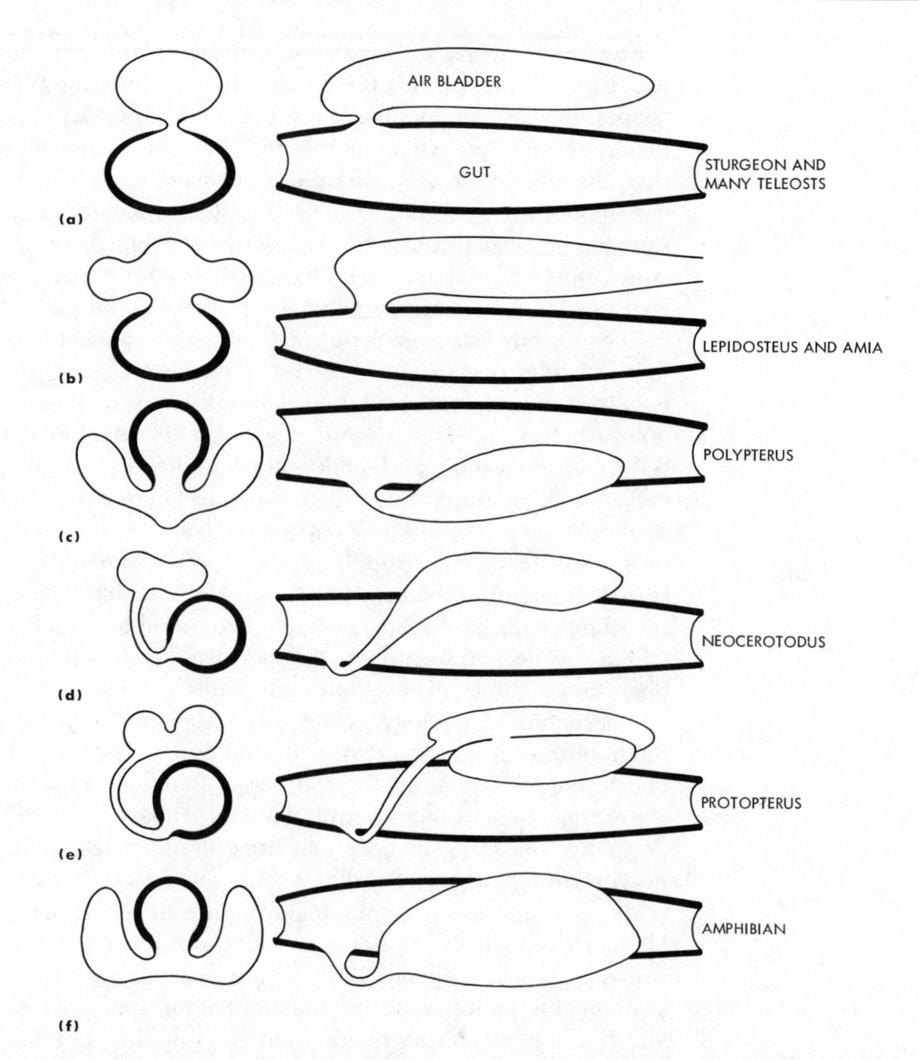

Figure 7-18 Phylogeny of the swim bladder. Schematic cross-sections and longitudinal sections of various air bladders: (a) typical physostome teleosts, (b) holostean type, *Amia* and *Lepidosteus.* (c) *Polypterus;* (d) *Neoceratodus,* the Australian lungfish; (e) *Protopterus,* the African lungfish; (f) amphibian.

many modes of accessory air breathing displayed by modern tropical freshwater fish all indicate that the transition to aerial breathing was gradual and that the original means of aquatic gas exchange with gills regressed slowly and assisted the new mode of aerial gas exchange in its development. Because of the differences in the physical properties of oxygen and carbon dioxide in their relation to water and air as respiratory media, the developing air-breathing organs initially were primarily oxygen-absorbing organs, and the retained aquatic organs assisted effectively in the elimination of carbon dioxide. The long and gradual development of air breathing was probably crucial for selection of the final structural design of lungs and for the drastic change in the piscine pattern of blood circulation that was necessary for efficient gas transport.

The Structural Design of Lungs

It was noted earlier that the oxygen concentration in the atmosphere is about thirty times greater than that of air-saturated water at about twenty degrees centigrade. This reduces the requirement for ventilation, or renewal of the external medium to the respiratory surfaces in air breathers. On the other hand, the truly terrestrial vertebrates faced a new problem of water balance caused by the dessicating effect of terrestrial life. This was different from the problems of maintaining osmotic and fluid balance in both fresh- and salt-water aquatic animals. Finally, exposure to air creates entirely new conditions in terms of gravitational effects and differences in density of the media and the tissues involved in gas exchange. The aquatic fish is under no significant gravitational stress because of the nearly uniform densities of water, blood, and the tissue intervening between them. Thus, only minimal forces are acting to deform the extremely delicate respiratory surfaces suspended in water. Conversely, in the air breather, blood, air, and the tissues separating them are all of different densities, and the resulting vertical pressure differences and shearing forces act to deform the respiratory surfaces.

The hazards of desiccation and the need for mechanical protection and structural support to minimize these gravitational effects required that the air-breathing organ be internal. This, in turn, called for an elaborate system of passageways connecting the internal gas-exchanging membranes with the external air. Because the gas-exchanging surfaces must be moist, the geometry and other characteristics of the airways had to meet this need. The problem of water conservation is probably the reason why no air-breathing vertebrate has developed a unidirectional airway system in which the inhaled air passes across the respiratory surfaces and escapes via tubes different from those by which it entered the organ. One important consequence of the bidirectional airway is the inevitable presence of a gas volume trapped in the airways after an expiration and before the next inspiration. This nonfunctional, or dead air, space is referred to as the anatomical dead space. The presence of such a dead space will cause oxygen tension inside the lung to be lower than in outside air; conversely, the carbon dioxide tension in the lung will exceed that of ambient air.

The anatomical dead space in a human lung is represented by a volume of about 150 milliliters, which is a considerable portion of the 500 milliliters inhaled with each breath. We refer to the latter as the tidal volume.

In the following description of the mammalian lung, we will see how these general requirements have influenced its design. In a later section, we will discuss the phylogenetic development of the lung and its relation to special environmental conditions. It was emphasized earlier that the respiratory system shares responsibilities with the circulatory system in the transfer of gases between the external environment and the immediate cellular environment. This is done by matching large volumes of gas with large volumes of blood in the gas-exchanging organ. This organ, in turn, is designed to effect the rapid transfer of gases by diffusion between the two media.

The respiratory system in air-breathing vertebrates is made up of two main parts; a system of conducting and distributing tubes leading into a transitional zone made up of respiratory bronchioles and alveolar ducts that, in turn, terminate in the small, thin-walled air sacs, the alveoli, where the actual gas exchange takes place. The upper airway, or respiratory tract, includes the nasal and mouth passages. Air taken in through these openings converges into the trachea. The trachea, in turn, divides into a right and a left bronchus. Each of these divides again into two, and so on, until in man there are more than twenty subdivisions. You can calculate that from the single primary tracheal tube there arise about one million terminal tubes or bronchioles connecting to the alveolar system. For each of these terminal airways, there are many blind, thin-walled alveolar sacs (Figure 7-19). Man has about 300 million alveoli in two lungs. Their diameter ranges between seventy-five to three hundred microns, and their total surface area can be as much as eighty square meters, or more than forty times the general body surface area. In mammals as well as in reptiles and amphibians, the lung mass is proportional to body mass, and the lung volume particularly in mammals is also proportional to body mass. The respiratory surface area of the lung, on the other hand, bears a relation to the metabolic rate of the animal. An increased lung surface area results when the size of individual alveoli diminishes. A small shrew has alveoli of twenty-five to thirty microns in diameter, whereas a sluggish mammal like the manatee has alveolar diameters up to two thousand microns. A comparison between mammalian species of equal size but different activity pattern shows that the active species may have up to 70 per cent greater respiratory surface area, and that the thickness of the barrier separating blood from air in the alveoli is also lesser in the active form.

In both gills and lungs the initial column of water of inspired air is distributed to cover remarkably large areas for diffusion exchange. In gills the inspired water ventilates the exchange surfaces directly, whereas in lung breathers the inspired air reaches the exchange surfaces of the lung by diffusion. There are important differences in how the surface expansion in respiratory organs is structurally supported. Because the density of the respiratory medium and tissues is uniform in the water breather, there are no deforming forces and it is, thus, possible to obtain a large surface expansion simply by guiding the water through finer and finer gates. Note that this is not a system of tubes or channels, but of gates provided by the spaces between the secondary lamellae. Note also that in the tubular airway system there are more than twenty subdivisions of the airways from the trachea to the terminal bronchioles [Figure 7-19(c)]. Another important difference is related to the bidirectional passage of the external medium in the air-breathing animal. The air has to leave the same way it enters [Figure 7-19(b)]. No actual gas exchange occurs in the conducting airways in lungs, and it is important to keep down their volume so that as much of the air inhaled as possible reaches the terminal air spaces. However, if the tubes are made very narrow in order to reduce the dead space, the breathing muscles will be required to work too hard to move the air in and out. The moved air must reach as many alveoli as possible; but as long as the lungs occupy a sizeable space, the distance from the trachea to the

Figure 7-19 (*opposite*). (a) Gas exchange in the lung depends on matching blood and air over large surface areas. (b) Air flow to the lung and alveoli is bidirectional. Air enters by the same route it leaves. The blood flow through the lung is unidirectional. The blood volume in the lung is effectively replaced with incoming blood at each heart beat. (c) In the human lung the trachea divides into two bronchi which by twenty or more subdivisions make up a million bronchioles terminating in about 300 million dead-end air sacs, the alveoli. (d) Gas exchange takes place in the alveoli where the diffusion path between blood and gas is about 0.5 micron. Human alveoli vary in size from 75 to 300 microns.

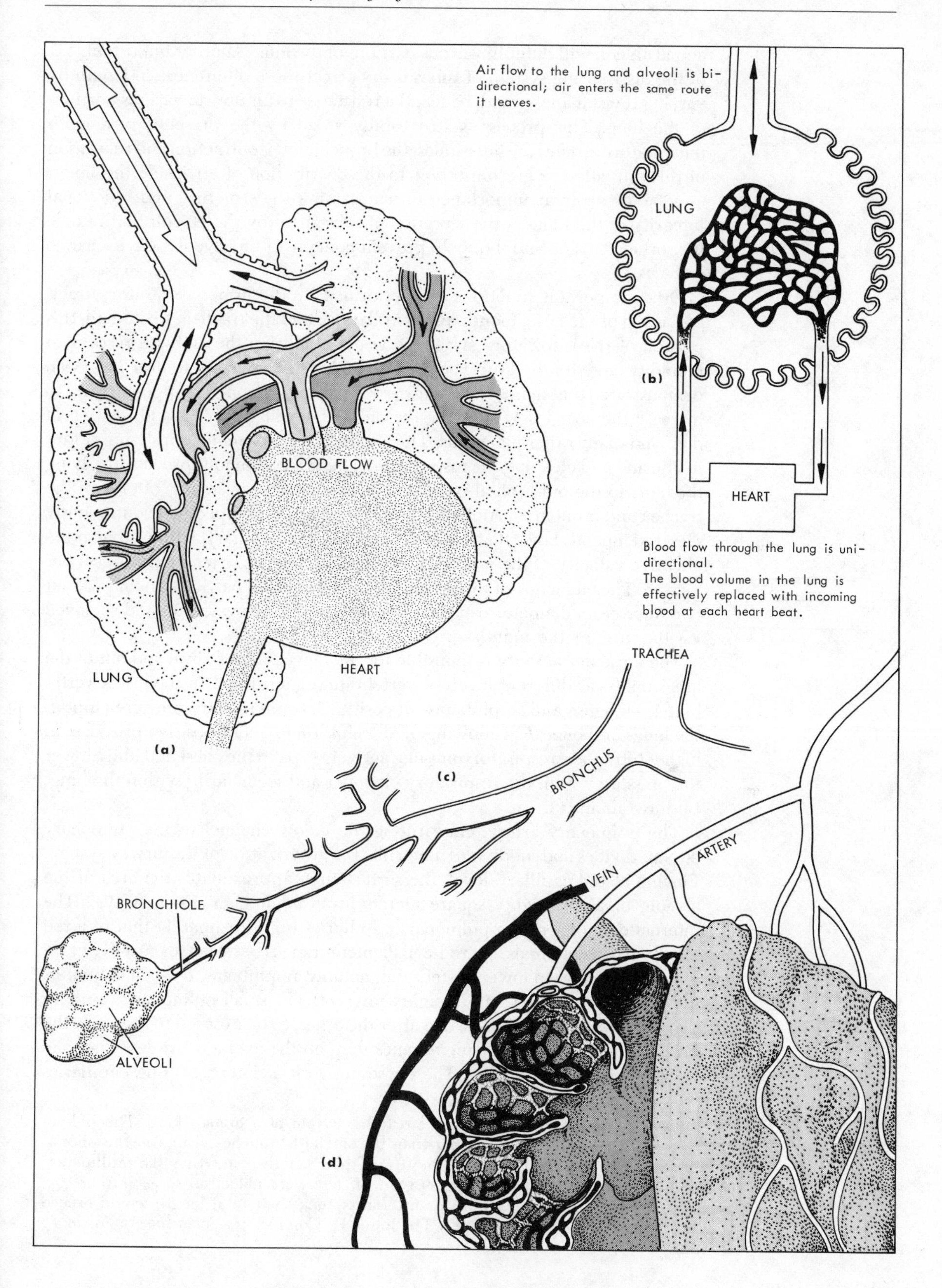

Air flow to the lung and alveoli is bi-directional; air enters the same route it leaves.
LUNG
(b)
HEART
Blood flow through the lung is uni-directional.
The blood volume in the lung is effectively replaced with incoming blood at each heart beat.
BLOOD FLOW
LUNG
HEART
(a)
TRACHEA
BRONCHUS
(c)
ARTERY
VEIN
BRONCHIOLE
ALVEOLI
(d)

actual alveoli will differ in various portions of the lung. Another important task of the distributing system of tubes is to correct these differences in length by varying their diameter and, hence, the resistance to air flow in various portions of the lung. This process is structurally aided by the presence of smooth musculature around the finer tubes, the bronchioles. Contraction and relaxation of this musculature are important to the distribution of air inside the lungs.

A factor of great importance in maintaining the structure and functional integrity of the lungs is the secretion of substances by the alveoli that reduce the surface tension and thus help prevent collapse of the alveolar gas exchange surfaces.

Other important functions are performed by the upper respiratory tract, consisting of the nose, mouth, pharynx, larynx, and the trachea. First of all, this portion of the respiratory system serves to condition the air with respect to humidity and temperature before it reaches the alveoli. It also has been demonstrated that many vertebrates use evaporation from the upper airways to increase the heat loss; this is the reason for fluttering in birds and panting in dogs and many other mammals. The airways are also tremendously important in filtering and cleaning the inspired air. This is a continuous process involving the hairs in the outer nasal passages and the active movements of cilia in the trachea and bronchi. Particles in the inspired air are trapped in a continuously secreted mucous layer that is moved upward by this ciliary action. In man the effective velocity of the mucous film can be as high as sixteen millimeters per minute. The efficiency of the process is remarkable; all particles down to about two microns in diameter can be transported up the airway tract and removed via the nose or the mouth.

The muscular activity responsible for the movement of air in and out of the lungs differs at different levels of vertebrate organization. In the lower vertebrates—lungfish and amphibians—a positive intrapulmonary pressure empties the lungs and buccal swallowing of air refills them under positive pressure. In higher tetrapods, respiratory muscles actively expand the chest and, thus, lower the pressure within the respiratory chamber and secondarily within the lungs to draw air into them.

The pulmonary artery, constituting the inflow channel to the pulmonary circuit, divides and subdivides much like the arborization of the airway system. Finally, at the capillary level, the surface area approximates the area of the alveoli, or about eighty square meters in an adult man (Figure 7-19). The internal diameter of the pulmonary capillaries is approximately that of a red blood cell. In mammals, the red-cell diameter ranges between seven and twelve microns, whereas in lower vertebrates, notably amphibians, red-cell diameters may be as large as seventy to eighty microns. The small pulmonary capillary diameter forces the red cells, one after the other, to be exposed to the air in the alveoli, separated only by membranes that, on the average, are less than one micron thick (Figure 7-20). The dense network of interdigitating capillaries

Figure 7-20 (*opposite*). X-ray photograph of the arterial system of a human lung. The pulmonary capillaries are actually finer than the smallest branches visible in the photograph. Their diameter is from 5 to 10 μm, their length is less than 0.5 millimeter, and their wall thickness is about 0.1 μm. If they were placed in series, rather than in parallel as they are in the arterial system, their total length would exceed several hundred miles. (Courtesy Dr. Julius H. Comroe, Jr., *Scientific American*, February, 1966.)

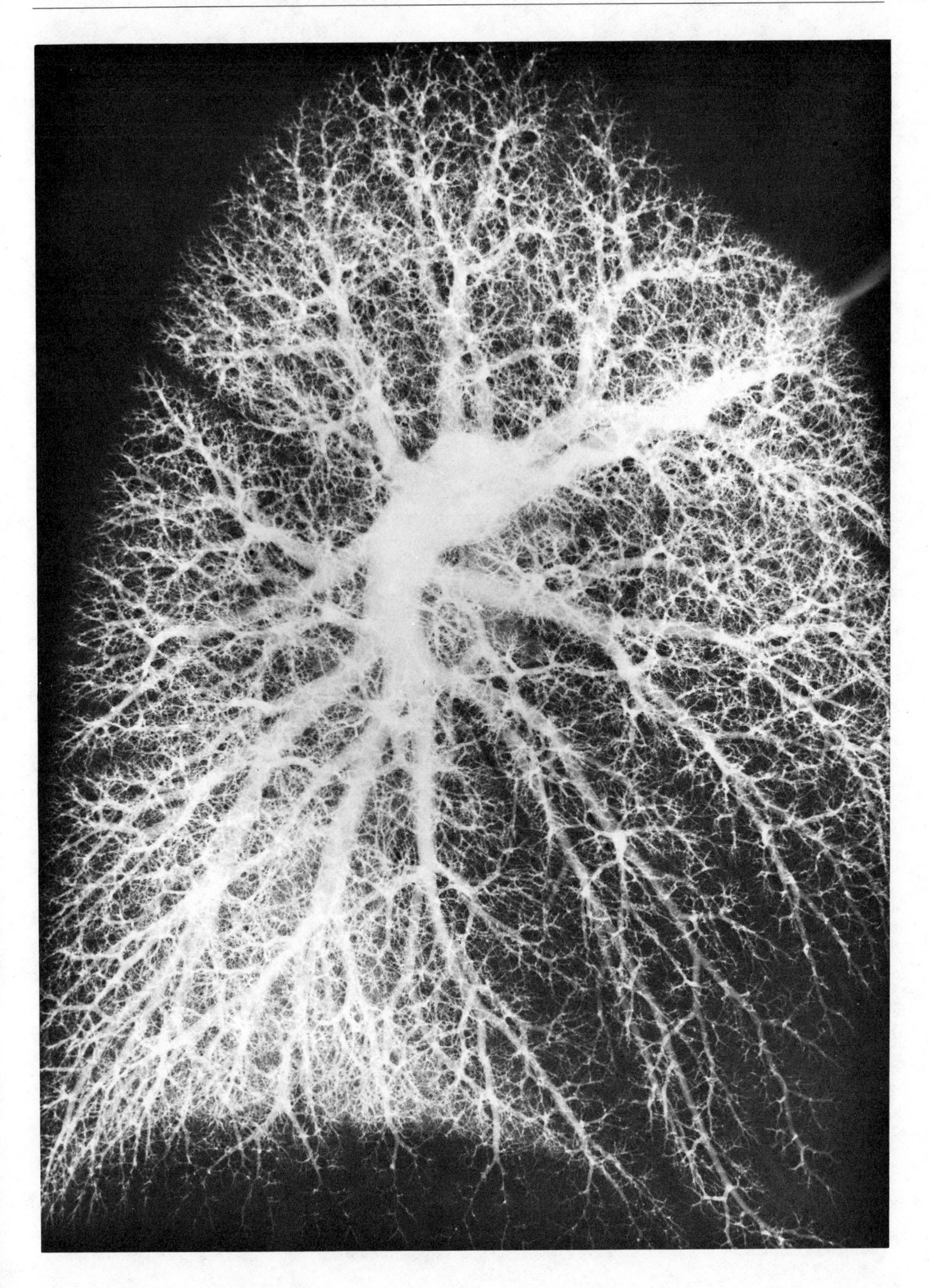

literally covers the alveolar walls. In mammals, the average capillary diameter is about eight microns. It can be estimated that each alveolus is surrounded by 4.7×10^{-7} milliliters of blood, or about 2,300 red blood cells at any one time. When recalling that one milliliter of blood contains five billion red cells and that in resting man four thousand to five thousand milliliters of blood pass through the lung every minute, the number of capillaries and the surface area for gas exchange become staggering. The volume flow of blood through the mammalian lung is matched by a nearly equal volume of ventilated air.

We have all experienced the varying demands for oxygen according to our physical activity. During hard exercise we need more oxygen in order to metabolize more foodstuffs to provide for the increased energy needed. It is surprising that the organ systems are capable of changing their performance to allow for such sudden increases in our activities. Man, for example, can increase the blood flow through his lungs ten times during hard exercise. A corresponding increase in the volume of air ventilated must accompany this response. These facts emphasize the remarkable efficiency in the structural design of the lungs. Similarly, when mammals are exposed to reduced oxygen concentration in inspired air—e.g., at a high altitude—the ventilation volume goes up.

The submicroscopic structure of the tissues separating blood from alveolar gas has been clarified in recent years with the aid of electron microscopy. An important parallel to conditions in gills is represented by the continuous epithelial lining of the pulmonary alveoli. Figure 7-21 shows how the cytoplasm that envelops the epithelial cell nucleus forms flattened extensions that reach out to cover the pulmonary alveoli. These nuclei bulge into the alveolar recesses with the important result that the thin cytoplasmic extensions will cover the contact surface of the pulmonary capillaries where the actual gas exchange takes place. It has been estimated that approximately two thirds of the total surface of one alveolus will represent a direct interface with blood capillaries. Thus, this area constitutes the maximum actual respiratory surface.

Figure 7-21 The distance separating blood and air in an alveolus is extremely short. Only three layers intervene between air and blood in the most direct diffusion path: the alveolar epithelium, a basement membrane, and the endothelium of the blood capillary. The total diffusion distance in a human lung may be as short as 0.2 μm. [After Frank N. Low, *Anatomical Record,* Vol. 139 (Jan.-Apr., 1961), p. 106.]

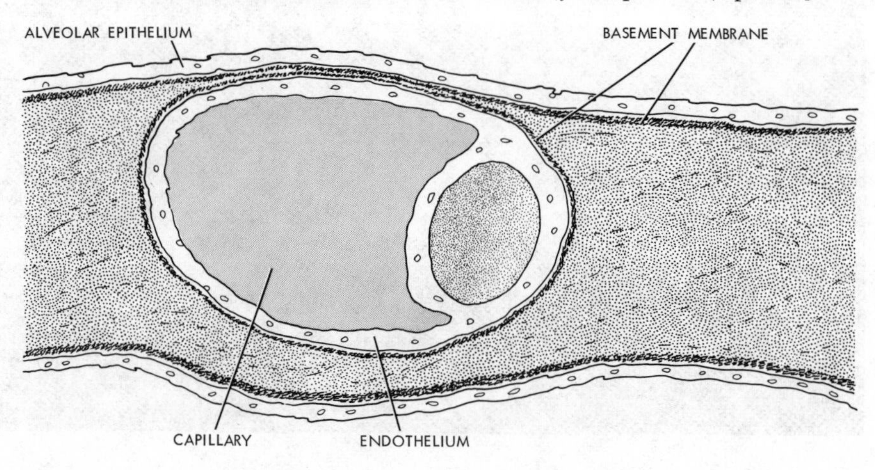

Table 7-2 Approximate Length of the Normal Diffusion Pathway in Respiratory Organs of Representative Vertebrates. [From H. Schulz (1959)]

	Fish (airbreather, haplochromia)	*Amphibian (toad)*	*Bird (pigeon)*	*Mammals (rat)*	*(human)*
Capillary endothelium (angstroms)	1000–4000	2000	385–750	180–1000	200–4000
Basement membrane (angstroms)	600	7000–23,000	295–470	650	1100–1600
Alveolar epithelium (angstroms)	1500–15,000	4000–5000	145–175	500–1000	400–650
Total diffusion pathway (microns)	0.31–2.0	1.3–3.0	0.1–0.14	0.13–0.26	0.36–2.5

However, relative modifications in this exchange area are considered to be one of the adjustments that air-breathing vertebrates make when their demands for gas exchange are increased drastically. The blood-to-air pathway is extremely thin, and the boundary is made up of three different layers. Facing the blood side is the endothelium of the pulmonary capillary. Between this endothelium and the cytoplasmic extensions of the alveolar epithelium is a homogeneous basement membrane (Figure 7-21). Table 7-2 provides a comparison of the thickness of these three layers in four representative vertebrate classes. The pigeon shows by far the shortest diffusion distance, less than 0.14-microns. Among the species studied, the toad has the longest blood-to-air pathway, up to three microns. Data in Table 7-2 suggest that the length of the blood-to-air distance shows adaptive adjustment to the respiratory demands of the different species more than to their phylogenetic ranking.

There is one barrier to the exchange diffusion of the respiratory gases that we have not yet discussed: the red cell itself. Recent studies have suggested that the diffusion barriers in the blood-to-air pathway are so small that the resistance to gas movement into the red cell itself may be the greater, and actual limiting, factor in the whole diffusion process. The remarkable chemical compound, hemoglobin, which is located inside the red cells and gives blood its red color, increases the capacity of man's blood to carry oxygen by almost forty times. The reason for the relatively slow exchange of oxygen and carbon dioxide with the hemoglobin in red cells is that the gases react chemically with the hemoglobin in addition to being transferred by simple physical diffusion. This process requires considerably more time than diffusion alone and becomes an important consideration when we compare gas transport in vertebrates, inasmuch as red cells are reduced more than tenfold in density and one thousandfold in volume when comparing some amphibians to mammals. Gas transport in mammals is, hence, favored by the smaller size of their red cells, which will provide a much larger surface area per volume unit of red cell than in the lower vertebrates.

The Phylogeny of Vertebrate Lungs

Paleontological evidence suggests that lungs were present in certain freshwater placoderms and in most crossopterygian fish. Lungs are also present in extant dipnoans and some primitive actinopterygian fish. They appear in all vertebrates as a ventral outgrowth from the foregut or lower pharynx. In some forms, the developing lungs resemble the gill pouches in their early outgrowths from the pharynx. On the basis you will read in some textbooks the incorrect statement that primitive lungs may have developed as a modified pair of posterior gill pouches.

The Lungfish

In *Neoceratodus,* the Australian lungfish, the lung has not assumed an important role as a respiratory organ but functions merely as an accessory O_2 absorber to the gills when the surrounding water becomes severely hypoxic or rich in carbon dioxide, or when the fish swims vigorously. It is an unpaired, richly vascularized, thin-walled sac located dorsally to the gut and extending the entire length of the body cavity. If the lung is cut open along a midventral, nonvascular strip of connective tissue, transverse septa are seen that divide the lung bilaterally into a series of compartments. Medial reticulated septa further increase the internal surface (Figure 7-22). The pulmonary arteries follow the medial stripe and give off lateral branches that supply fine vessels to the internal surface of the lung, including all the septa. The lung narrows anteriorly and connects, on the right ventrolateral side, with the pharynx through the pneumatic duct. A slitlike muscular glottis marks the entrance from the pharynx to the pneumatic duct.

Histologically, the lung contains an epithelium, connective tissue, and smooth muscle, in addition to the blood capillaries. The smooth muscle is present in the lung wall but is more prominently developed in the septa. Its function in the mechanics of breathing is to maintain the prevailing positive pressure, essential for exhalation, inside the lung. It may have an additional role in buoyancy control by changing the volume of the lung.

The lung structure of the African (*Protopterus*) and South American

Figure 7-22 The lung in *Neoceratodus* showing how the septa and trabeculae compartmentalize the lung and increase the internal surface area.

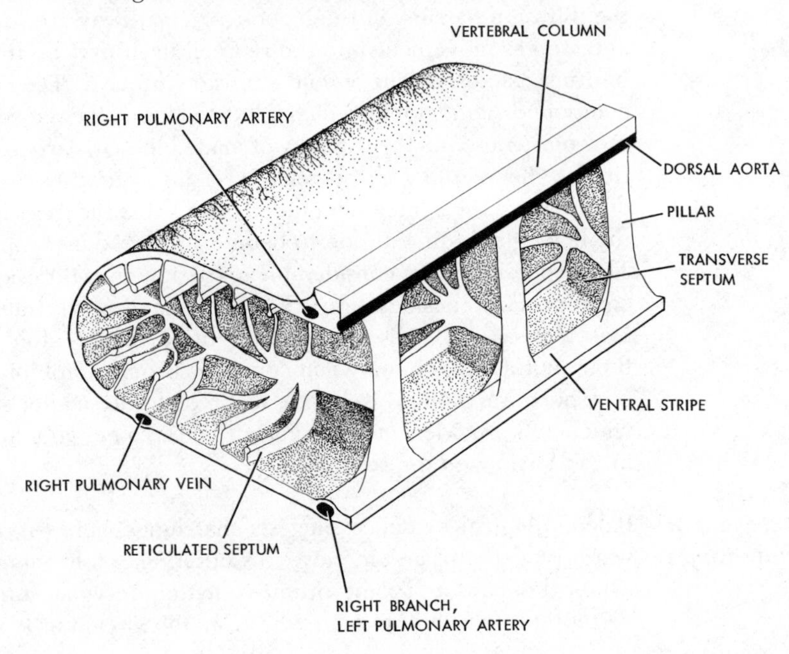

(*Lepidosiren*) lungfish shows a considerable advance over *Neoceratodus*. This development correlates with the greater emphasis on pulmonary breathing that has become the main method of oxygen absorption in these forms. In contrast to *Neoceratodus*, these lungfish can subsist on pulmonary breathing alone and will in fact "drown" if they are prevented from reaching the surface to breathe air. The lung is a paired structure, except for a fused anterior section, and the external lung surface is correspondingly much larger than that of *Neoceratodus*. Moreover, the internal septation and compartmentalization are far more developed and the gas exchange sites are more akin to alveoli than are the compartments between the septal ridges of the *Neoceratodus* lung (Figure 7-23). Note that in the actual gas exchange areas of the lung the blood is separated from the alveolar air only by the capillary wall and by thin extensions of the alveolar epithelial cells. The nuclei of the epithelial cells show a

Figure 7-23 (a) Ventral view of lung in *Protopterus aethiopicus*, (b) Schematic drawing of the anterior region of the lung in the African lungfish, showing how the trabeculation creates alveolar-like pockets. (c) Close-up photo of the dense trabeculation inside the lung of the African lungfish. [Parts (a) and (b) redrawn and part (c) reproduced from Monique Poll, "Étude sur la structure adulte et la formation des sacs pulmonaires des Prototypères" (*Ann. in 8vo Zool.* 1962, **108**) by kind permission of Dr. H. Poll, Musée Royal de l'Afrique Central, Tervuren, Belgium]

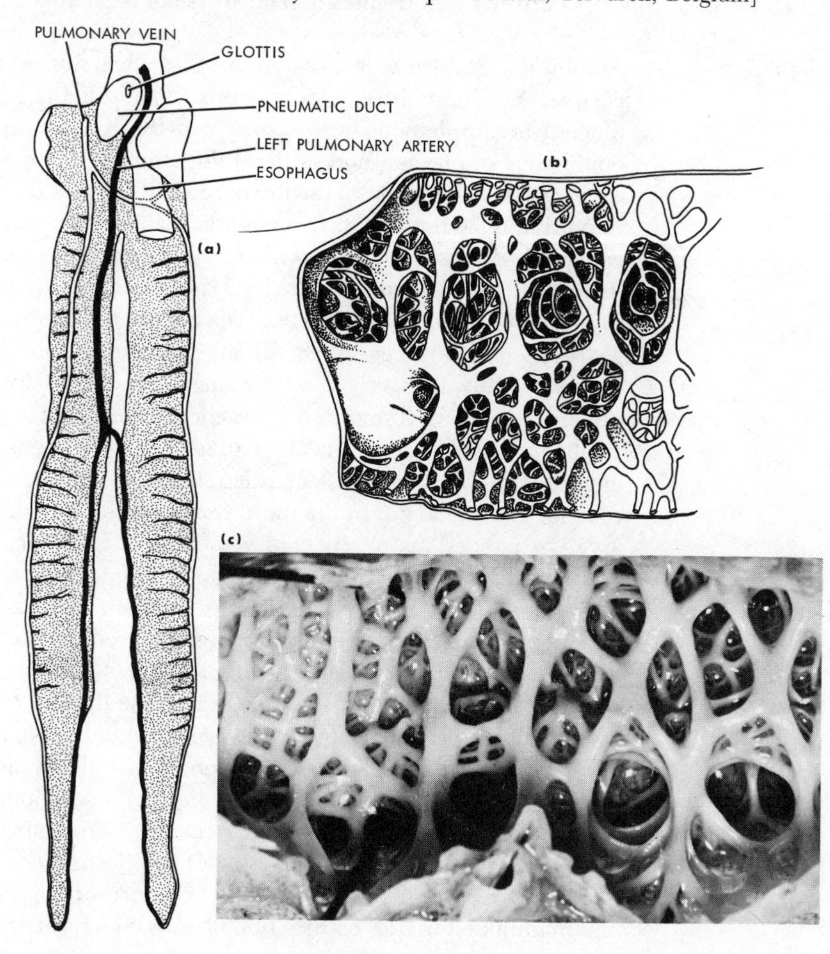

clear tendency to be removed from the most direct diffusion path between blood and gas. This tendency, which is also apparent in fish gills, is typical of tetrapod lungs as well.

The mechanics of breathing in the lungfish—e.g., in *Protopterus*—is based mainly on the same muscular and skeletal elements as in teleost and elasmobranch fishes. This is particularly true for the branchial irrigation, but also the aerial breathing cycle depends on movements only slightly modified from an aquatic breathing cycle. Inspiration of air occurs by a buccal force-pump mechanism, a method also passed on to amphibians, whereas expiration from the lung occurs by releasing the gas behind the closed glottis because following an inspiration this gas is under a positive pressure. The compressing forces are set up in part by the surrounding water but also from contraction of the smooth musculature embedded in the lung itself. The ribs take no part in the breathing of lungfish, and the rise of air breathing was, hence, not associated with an aspiratory method of ventilation.

A comparison of the lungfish thus, shows that increased dependence on aerial breathing is correlated with an increased surface area available for gas exchange in the lung. Secondly, the blood vascular system shows an increase in the vascularization of the lung, matching the increase in the blood-air interface for gas exchange. Thirdly, the diffusion barriers and the length of the diffusion path between blood and air tend to become smaller.

Lungs in Tetrapods

Amphibian lungs are marked by a great diversity in structure in apparent correlation with multiple modes of breathing utilizing gills, skin and the buccal mucosa to supplement gas exchange in the lung. Some forms show a clear dominance of one method of breathing and have entirely discarded another. Thus, we find lungless salamanders relying on gills or skin for gas exchange. Accordingly, lung structure in amphibians varies markedly and shows, in general, little advance over conditions in lungfish. Internal septal arrangements increase the surface area of the lung, but the internal divisions of the lung fail to provide direct, narrow passageways from the external environment to the gas exchange surfaces. Such direct communication seems essential for an efficient renewal of fresh air to the lungs. Yet, in amphibians we see the earliest phylogenetic manifestation of a trachea, although not differentiated into the tracheobronchial tree typical of the higher tetrapods. The surface area of amphibian lungs bears a clear relationship to the dependence of the lung in over-all gas exchange. In the more terrestrial forms, such as toads, the surface area is expanded mainly by enlargement of the primary alveolar septa and by subdivisions of these into finer secondary and tertiary septa (Figure 7-24).

Although the effective length of pulmonary capillaries in the frog lung is about equal to that in mammals (i.e., six hundred to eight hundred microns, the thickness of the blood-to-air barrier (1.3 to 3.0 microns) is on the average four to six times that of the mammalian lung. The density of capillaries is also less than in higher vertebrate lungs and is distributed mainly on the alveolar septa. Figure 7-25 shows a photomicrograph of a capillary network in a frog lung.

The mechanics of breathing are similar to those in lungfish; a buccal positive pressure pumping or swallowing mechanism forces air into the lungs. Expiration occurs in response to an opening of the glottis, and the air escapes because of a positive pressure inside the lung. This pressure is, at least in part, created and maintained by the contraction of smooth muscle in the lung itself and by

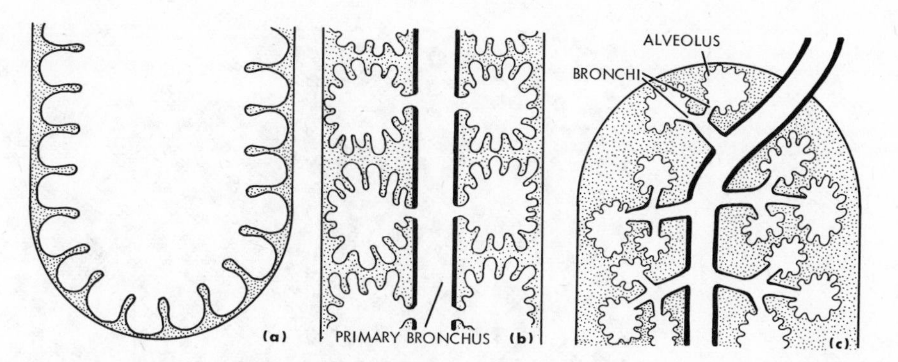

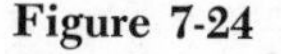
Figure 7-24 The evolutionary development of the internal lung surface and connecting airways in tetrapods. (a) In the primitive amphibian stage, ridges and septa provide surface expansion, but there are no direct passageways to bring the inhaled air into rapid communication with the gas exchange surfaces. (b) The stage represented by some reptiles shows further compartmentalization of the lung, and more direct communication from the primary bronchi to the gas exchange surfaces. (c) In the mammalian stage, further bronchial subdivisions provide large surface areas for gas exchange, and direct airways (leading through many orders of bronchi) route the inhaled air to the alveolar surfaces.

the recoil from the previous inhalation, as well as by extrapulmonary compression forces from the body wall or, if the animal is partly immersed in water, by external hydrostatic pressure. Until recently, it was alleged that amphibians—e.g., frogs—would inhale gas reduced in oxygen because some of the expired gas would mix with the gas held in the buccal cavity and, thus, contaminate the gas to be inhaled with the subsequent inspiration. Recent work (Figure 7-26) has modified this view. Significant mixing is avoided because the exhaled gas escapes rapidly in a forward direction partly directed by the vertically placed glottis. The resultant jet-streamlike expiration will, hence, be confined to the upper half of the buccopharyngeal airway and effectively bypass the posterior portion of the buccal cavity where the gas ready for the next inhalation is temporarily stored. Following exhalation, the buccal floor is raised and the lung filled under pressure.

After inhalation, a period follows when the glottis remains closed, but oscillatory pumping movements will drive air in and out of the open nostrils. These oscillations have been thought of as part of an olfaction, or "sniffing," process, but in the light of recent work this phase must also serve in flushing out residues of expired air in the buccal cavity, thus preparing for the next ventilation cycle of the lung.

Reptiles, like amphibians, constitute a diverse group; however, in reptiles the lung has become the principal and, with few exceptions, the only organ for gas exchange. Consequently, we see both structural and functional changes that increase the efficiency of the lung. The structural subdivision of the internal lung space has advanced considerably from the amphibian stage, although primitive reptiles like Sphenodon show a far smaller ratio of surface area to lung volume than do advanced forms of lizards and crocodilians. Asymmetry in the size of the paired lungs appears generally in vertebrates, but it is most explicit in reptiles where elongated forms like snakes and some lizards show one lung to be completely atrophied or much smaller than the other. Snakes and lizards also tend to have the gas-exchange surfaces concentrated in the

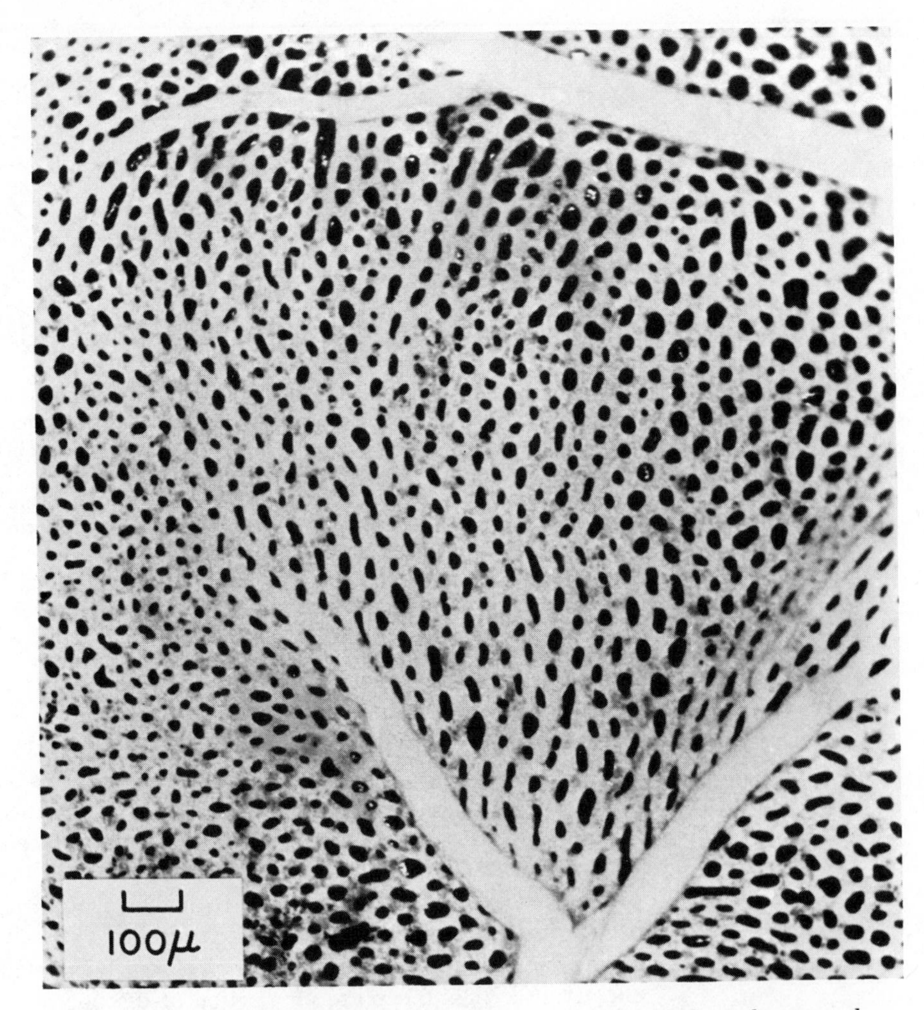

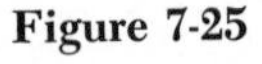
Figure 7-25 A photomicrograph of a capillary network in the frog lung. Along the upper border a major arteriole supplies the capillary network. Two collecting venules are shown at the bottom of the photograph. (Reproduced from Maloney and Castle, *Respiration Physiology,* **7:** 137, 1969.)

anterior end of the lung. Typically, the trachea may extend halfway down the length of the lung. In such cases, the anterior portion is referred to as a tracheal lung. The posterior portion of these lungs is conspicuously free from vascularization and appears to function as a storage chamber for air (Figure 7-27).

In Chelonia (turtles) and crocodilians the lung shows a more uniformly distributed parenchyma. In line with this, the airways seem to be more structurally complex, with more distinct secondary bronchi branching off from the primary bronchi that extend from the trachea (Figure 7-27).

Also, in reptiles the size of the alveoli increases with body size, but only to a body weight of about one kilo. Maximum alveolar diameters are about two thousand microns. Agile active species, particularly among lizards, show smaller alveoli and larger lung-surface area. A comparison of lungs in all nonflying vertebrates shows that lung mass remains in direct proportion to body mass. The size and number and, thereby, the surface area of the alveoli

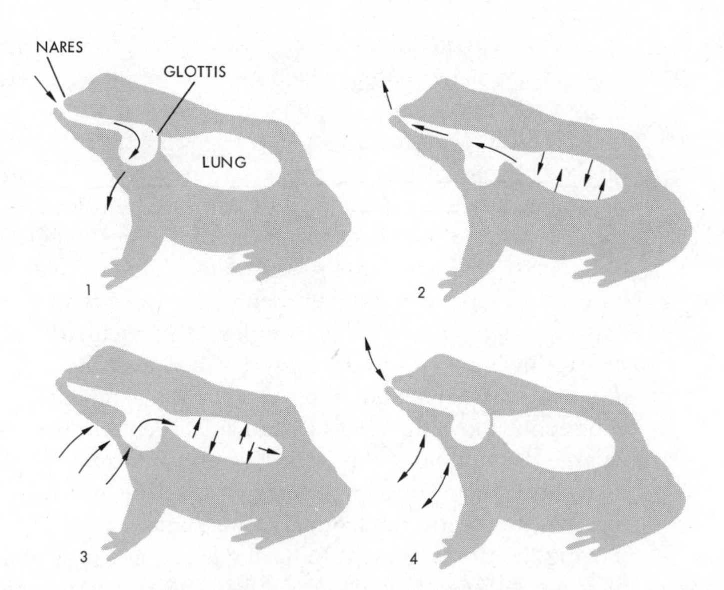

Figure 7-26 Stages in the breathing cycle of the frog *Rana catesbeiana.* (1) By lowering of the buccal floor air is drawn into the buccal cavity filling in particular its posterior portion. (2) The glottis opens and air escapes from the lungs and exits by the nostrils. The rapid flow of expired air passes largely dorsal to the main chamber of the buccal cavity. (3) The nostrils are closed, the buccal floor is raised, and air pressed (pumped) into the lung passes the open glottis. (4) While the glottis is closed the buccal floor is lowered and air re-enters the buccal cavity. Stage 4 is repeated in an oscillatory manner until a new ventilation cycle of the lung is started. (Redrawn from Gans *et al., Science,* **163:** 1223–1225, 1969.)

show various relationships to body size, metabolic rate, behavior patterns, and environmental factors, all with the apparent goal of achieving a stable ratio of metabolic rate to the surface area available for gas exchange.

Of perhaps greater functional advantage and advancement from the amphibian stage than the structural features of the reptilian lung are the supporting structural changes that allow reptiles to inhale air by suctional attrac-

Figure 7-27 A longitudinal section through a snake or lizard lung shows an anterior portion where alveoli are dense and in direct communication with primary bronchi. In its posterior part this type of lung becomes sac-like with sparse vascularization.

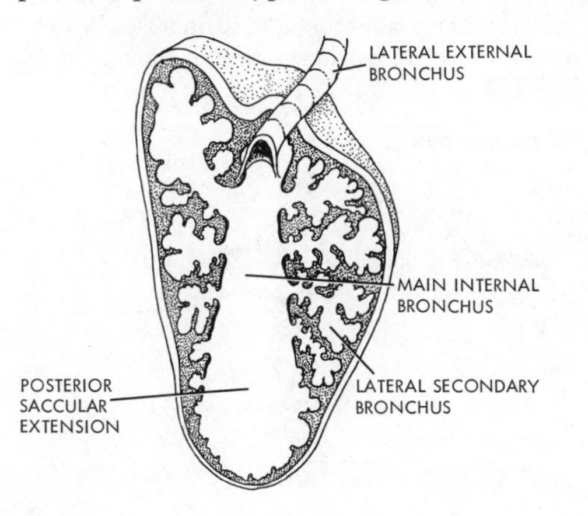

tion. This is accomplished by an expansion of the general body wall that is brought about by muscular actions. The resulting internal pressure changes draw air into the lungs. Expiration is effected either by passive recoil of the body wall or, as in snakes, by active contraction of specialized muscles. This method of ventilating the lungs is far superior to the positive pressure breathing in lungfish and amphibians. In squamate reptiles, the musculature moving the ribs, assisted by other specialized muscles, is responsible for the expansion of the body wall and the ensuing suctional inhalation of air into the lung. The absence of a true muscular diaphragm in vertebrates below mammals is variously compensated for in reptiles. Most peculiarly in crocodilians where a massive liver adheres to the posterior margin of the lungs, a special muscle *M. diaphragmaticus* (not homologous with the mammalian diaphragm) will by its contraction move the liver in a caudad direction. This movement, in turn, will enlarge the volume of the pleural cavity and cause a suctional inflow of air to the lung. The reverse process is effected by contraction of the transverse musculature of the abdominal wall (Figure 7-28). The costal musculature moving the ribs is thought to have a lesser, and only a supportive, role in the mechanics of breathing in crocodilians. Breathing in turtles, where the ribs are fused to the compact shell, must, of necessity, be different from that in other reptiles. The shell makes a compact casing for all internal organs of the turtle. Only posterolateral and anterior apertures where the limbs protrude contain soft, movable tissue. Posteriorly, two sets of muscles by their contraction cause volume changes inside the shell. Anteriorly, muscles effecting an outward and inward movement of the pectoral girdle will similarly contribute to the volume changes and, thus, cause ventilatory movements of the lung based on an aspiratory type of inhalation. Unlike mammals in which the lungs are completely separated from the abdominal cavity by the diaphragm, in reptiles the breathing movements will transmit pressure changes on all the visceral organs [Fig. 7-28(b)].

Bird lungs are the most specialized of all vertebrate lungs. Their structural design meets the demands of flight and allows the rapid adjustments that are required by the changes in the metabolic requirements for gas exchange.

Differing from the usual organization of vertebrate lungs into airways and

Figure 7-28(a) Lung inflation in crocodilians is largely the result of contraction of the muscle, *M. diaphragmaticus*, which by moving the liver backwards will expand the pleural cavity and suck air into the lungs. The reverse movement is effected by contraction of the transverse abdominal wall muscles. Movements of the ribs by the intercostal musculature play a lesser role in breathing. (Redrawn after Gans, *Evolution,* **24:** 723–734, 1970.)

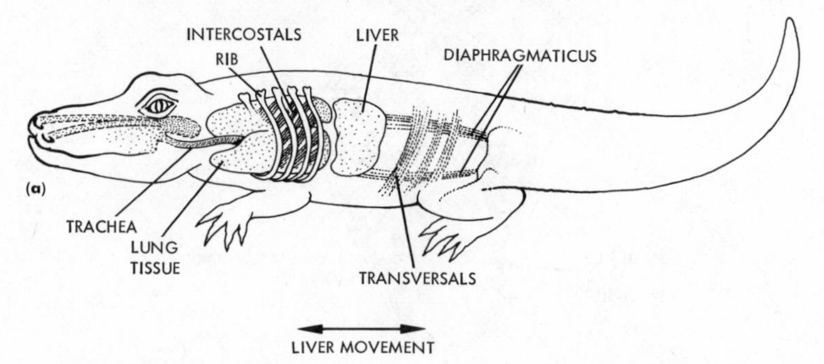

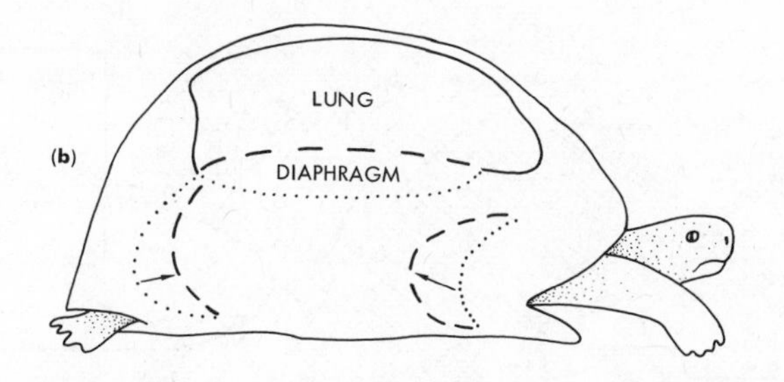

Figure 7-28(b) The compact shell and fused ribs of the tortoise allow volume changes inside the shell only by movement of the tissues associated with the limbs. The dotted lines show the position of these tissues at inspiration and dashed lines at expiration. These movements are transmitted into volume changes of the lung by a displacement of the lung's ventral borders through attachment to a nonmuscular "passive diaphragm." (Redrawn after Gans & Hughes, *J. Exp. Biol.*, **47:** 1–20, 1967.)

gas-exchange surfaces, the bird lung also shows an additional structural component: its unique system of air sacs. These make up the terminal air spaces that communicate, on the one hand, with the exterior through the lungs and tracheobronchial system (Figure 7-29) and, on the other, with spaces in some of the larger bones. The air sacs are thin-walled with smooth nonvascular surfaces. They surround most of the viscera, and their total volume capacity exceeds by far that of the lungs themselves. The matching of blood and gas for actual gas exchange takes place in the lungs, which are relatively small, situated dorsally in the thoracic region and closely adhered to the overlaying ribs and thoracic vertebrae. The two primary bronchi branching from the trachea pass through the two lungs where they lose their cartilagenous support and give rise to a

Figure 7-29(a) Structure of the lung and airways in birds. The extensive system of thin-walled air sacs fills a large portion of the body cavity of birds. The air sacs are connected to the main primary bronchus as well as directly to the lung. The air sacs are ventilated to keep the residual air space and dead space of the lung itself very low.

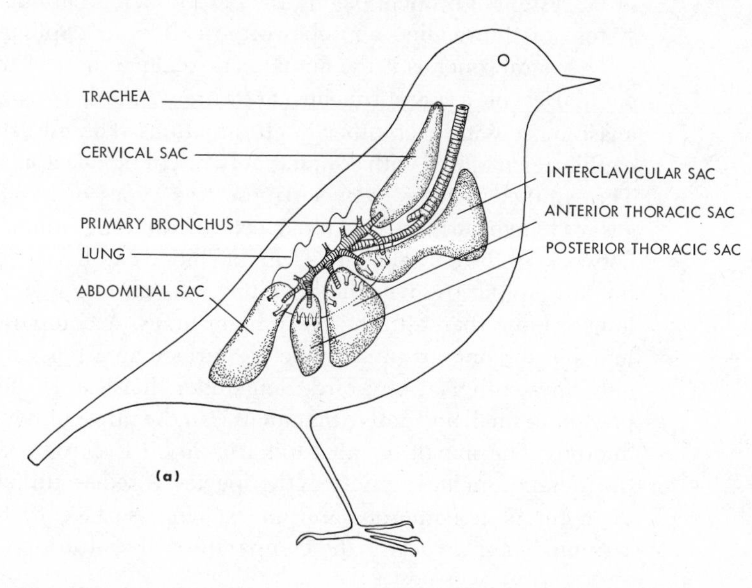

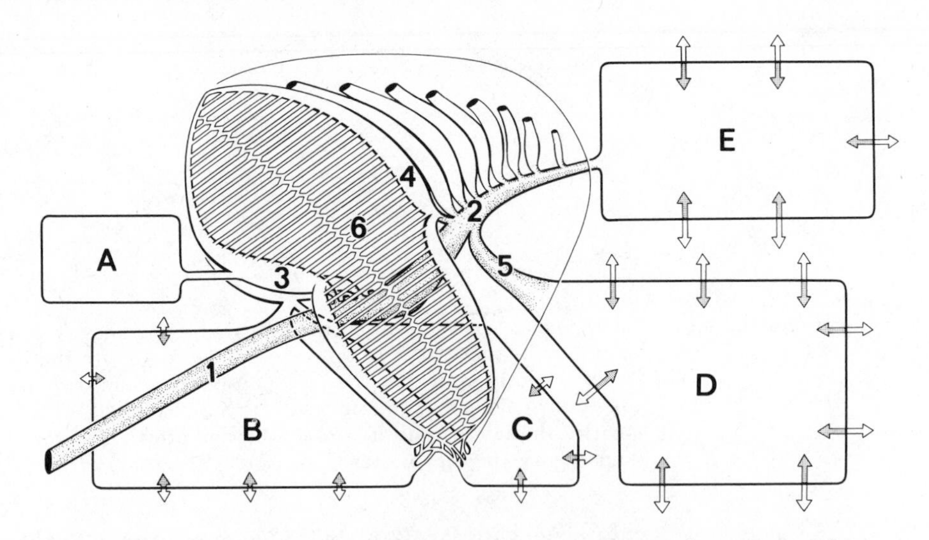

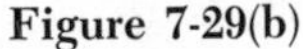

Figure 7-29(b) Schematical drawing of the bird lung and its connections to the air sacs. The trachea (1) continues after bifurcation into the primary bronchus (2) which gives off the ventrobronchi (3) at the lung hilus. From the posterior primary bronchus originate the dorsobronchi (4) and the large laterobronchus (5) which enters the posterior thoracic air sace (D). The primary bronchus enters the abdominal air sac (E). The anterior air sacs are connected to the ventrobronchi (3), the cervical sac (A) to the first, the interclavicular (B) and anterior thoracic (C) sacs to the third ventrobronchus. These latter sacs (B, C) are in addition connected to the parabronchial net at the lung margin. The ventrobronchi and the dorsobronchi are connected by numerous parabronchi in parallel arrangement (6). The whole bronchial system is connected in parallel with the air passage through the primary bronchus into the posterior air sacs. The length of the white arrows indicates the estimated dilatation of the air sacs in inspiration, and the length of the dotted arrows the estimated compression in expiration. (After H. R. Duncker, *Respiration Physiology*, **14:** 44–63, 1972.)

branching system of tubes, the secondary bronchi. From these tubes a third-order system of bronchi begins, the parabronchi, that finally gives rise to a fine system of thin-walled air tubes often called air capillaries [Figure 7-29(b)].

The parabronchus is the actual gas-exchange unit of the bird lung. The air capillaries of a parabronchus are three to ten microns in diameter and anastomose with each other in all directions. The interstices between the air capillaries are filled with a similar network of blood capillaries [Figure 7-29(c)]. The contact surfaces between these two types of capillaries constitute the gas-exchange surfaces. Fifty per cent of the lung volume is occupied by the networks of the densely packed air and blood capillaries. The small diameter of the air capillaries gives the bird lung a much larger surface area per unit of lung volume than all other vertebrate lungs. A comparable size mammalian lung has in comparison an effective surface area per volume unit of the lung only one-tenth that of a bird. Remember that the smallest alveolar diameter present in small and active mammals is in the range of twenty-five to thirty-five microns. The size of the air capillaries in birds shows adaptive variation with the metabolism and activity of the species. A sedate turkey has an air-capillary diameter of ten microns and an exchange-surface to lung-volume ratio of 188 mm^2/mm^3, whereas the comparable values for a pigeon are three to five microns and 302 mm^2/mm^3. A similar comparison, relative to body weight, gives a hen 18,3 cm^2 gas exchange surface per gram body weight, whereas the values for the pigeon are 40,3 cm^2 and for the diving coot 86,2 cm^2. The size of an air capillary in the bird lung results in an exceedingly small radius of curvature, which causes the surface tension of the capillary to be so high that a re-expansion of a collapsed air capillary would be impossible. This is probably the reason why the air capillaries exist as rigid tubes and the bird lung has a nearly fixed volume throughout the breathing cycle. Air flow through the parabronchi can be undirectional because they are open-ended, in distinction to the alveoli of all other lungs, which are closed and, thus, requires a bidirectional airflow. This fact increases the gas exchange efficiency of the bird lung.

The large-caliber airways, nonfunctional in actual gas exchange, continue through the lung parenchyma into the two abdominal air sacs; other air sacs have smaller connections with second- or third-order bronchi of the lungs. Secondary connections with the system of parabronchi are sometimes referred to as recurrent bronchi.

Inspiration in birds is effected by a lowering of the breastbone. This will

Figure 7-29(c) Semischematic drawing of the lung and air sacs of a large, stork-like bird. The anterior portion of the dorsobronchi (4) and their parabronchial net (6) connecting to the ventrobronchi (3) have been removed. The origin of the ventrobronchi after the entrance of the primary bronchus (2) into the lung hilus and the origin of the dorsobronchi (4) and laterobronchi (5) from the posterior primary bronchus are demonstrated. The first ventrobronchus opens far cranially via the ostium into the cervical air sac (A), the third ventrobronchus medial to the lung hilus by ostia into the interclavicular (B) and anterior thoracic (C) air sacs. The additional ostia of both these air sacs are connected to lateral branches or parabronchi of the first and second ventrobronchi. The posterior thoracic air sac (D) is always connected by the great laterobronchus (5), and the abdominal air sac (E) to the primary bronchus. (After H. R. Duncker, *Respiration Physiology*, **14:** 44–63, 1972.)

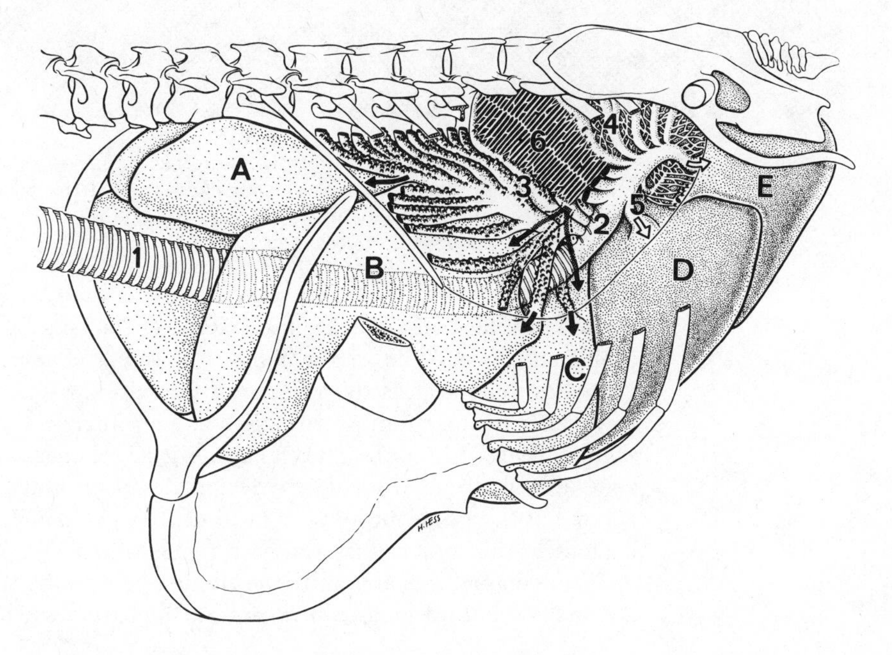

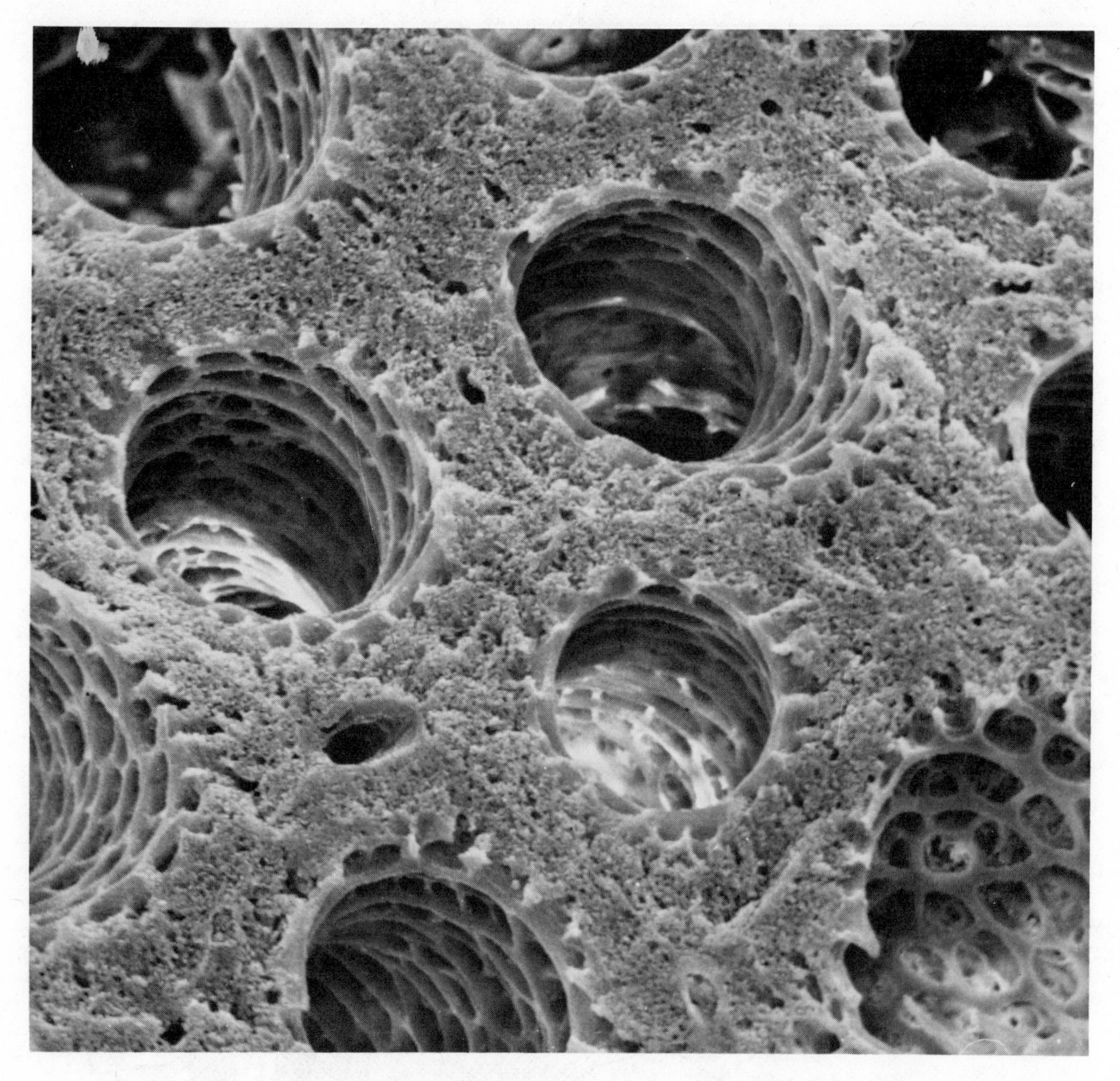

Figure 7-29(d) The parabronchi of the bird lung is where gas exchange takes place. Air can flow unidirectionally through them. The air capillaries of the parabronchi are invested with blood capillaries. This micrograph was made by Professor H. R. Duncker using scanning electron microscopy.

expand the chest and increase the volume of the air sacs. For reasons already explained, the volume of the lungs themselves changes little and actually decreases somewhat as the volume of the entire system of airways and sacs increases. Conversely, during exhalation, the volume of the lung itself increases slightly. This is important because the gas-exchange surfaces then will be ventilated with gas from the posterior air sacs during exhalation also. The volume of the posterior air sacs is, in fact, large enough so that the lung air content in this way can be renewed twice within one breathing cycle.

Through a bellows action by the rib musculature on the extensive air sacs, the air flow in bird lungs can reach very high values. The system appears to

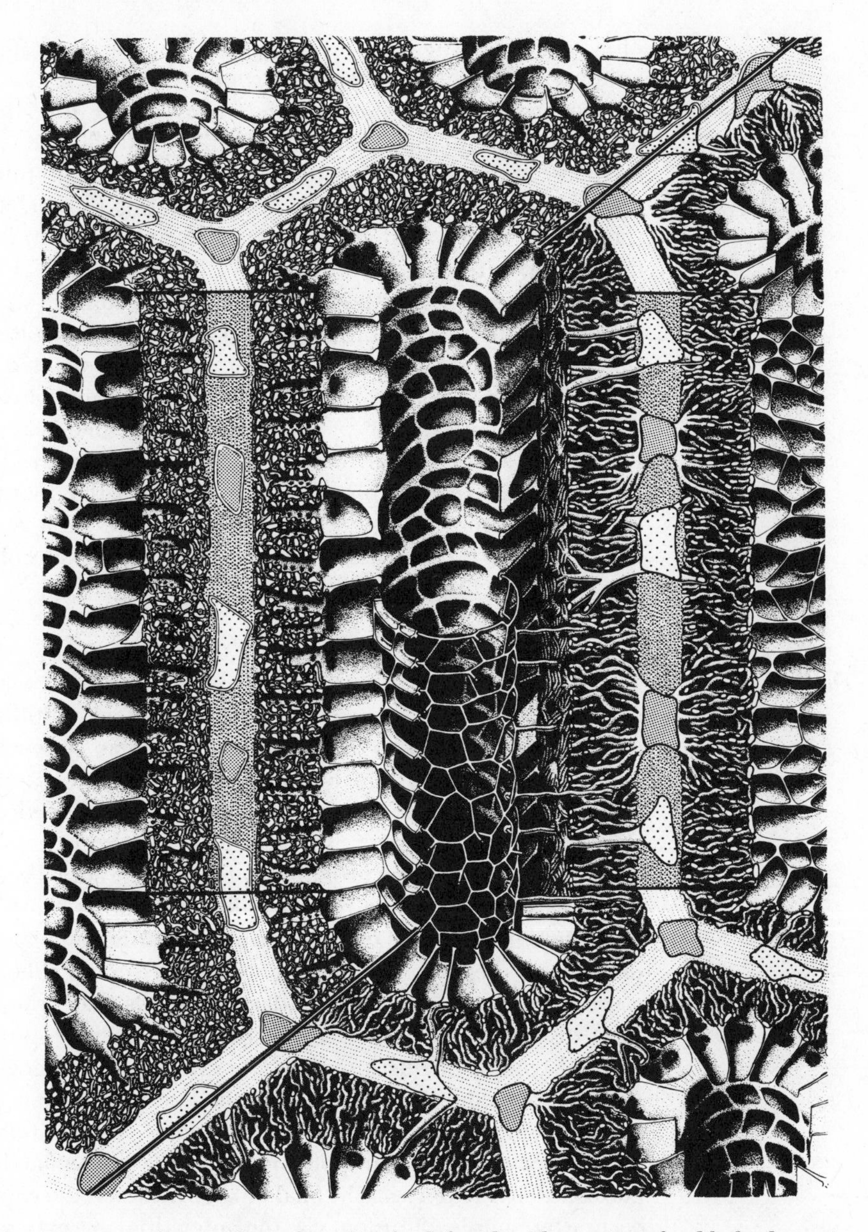

Figure 7-29(e) Drawing of a parabronchus. On the left side: The atria, infundibula departing from them, and the three-dimensional air capillary meshwork arising from the infundibula. On the right side: in the interparabronchial septa, the arterioles (dense stippling), from which the capillaries originate and run radially to the lumen. The infundibula lie between the capillaries, which are surrounded by a well-developed three-dimensional air capillary network. The long straight capillaries, seldom interconnected, are collected at the lumen by the venules, which penetrate the parabronchial wall and flow into small veins (light stippling) located in the interparabronchial septa. The septa of the atria and their smooth muscle bundles are supplied by a loose capillary network. (After H. R. Duncker, *Verh. Anat. Ges.*, **68:** 517–523, 1974.)

offer a unique structural advantage in that the dead space and residual air volumes of the gas-exchange portion of the lung can be kept very low. Moreover, the interdigitating system of exchange tubes that allows air to flow constantly in both directions is an excellent and efficient structure for diffusion exchange of the respiratory gases. In addition to the usefulness of the air sacs in increasing the total air capacity and facilitating the exchange in flow through tubes, they help to reduce the specific weight of birds and to regulate their body temperature.

The mammalian lung appears to have evolved from structures most closely represented by those in crocodilian-like reptiles. In addition to internal septation by repeated subdivisions of the airways connected to the trachea and terminating in the blind alveoli, the lungs of mammals are distinctly lobulated. The human lung, which typifies the mammalian lung, was described in detail in the early part of this chapter. It should be emphasized, however, that the powerful muscular diaphragm that separates the thoracic and abdominal cavities in mammals provides the structural basis for a very efficient suctional attraction of air. The ventilatory capacity of mammals probably exceeds that of all other vertebrates.

The Circulatory System

The vertebrate cardiovascular system is responsible for the transport of a variety of substances and energy, such as heat, within the organism and between the external environment and the organ systems. In a mammal, like man, the vascular system provides about sixty thousand miles of transport channels for the multitude of transport functions needed within a complex organism. A closed system of tubes, the blood vessels, constitutes the transport channels. Energy for the propulsion of blood, the actual carrier in the transport operations, comes from a rhythmically contracting heart. It is common to all the different transports carried out by the cardiovascular system that sites for pickup and delivery are represented by a remarkable arborization of the vascular channels into fine capillary networks that bring the blood within a short distance of the external environment in the respiratory organs and to internal compartments in immediate closeness to the cells.

In the preceding section we described how the respiratory and circulatory systems jointly exchange the respiratory gases, oxygen and carbon dioxide. One terminal of this vital transport operation lies in the fine-bore capillaries that perfuse all living tissues. The final linkage between the capillaries and the cells occurs by simple diffusion via the interstitial or extracellular fluid. These tissue-capillary networks, the systemic capillaries, also function as terminals for the delivery of nutrients, the metabolic fuel, which are absorbed from ingested food in the gastrointestinal system. The transport proceeds through important processing plants such as the liver and other glands before the absorbed nutrients are carried onward to their destination in the metabolizing cells. The cardiovascular system also transports enzymes, hormones, and other biogenic regulators between their production sites and the cellular environment.

Carbon dioxide is not the only metabolic end product carried by the cardiovascular system. Heat is constantly being produced and must be distributed and eliminated for animals to maintain their internal temperature balance. Excess water and electrolytes are transported to the kidneys, the main plant for adjusting the fluid balance of the body. Kidney function requires that the cardiovascular system maintain the blood pressure needed in the filtering units, the glomeruli, to free the blood of unwanted waste. This filtering process does not, however, discriminate adequately between substances that need to be eliminated and many that need to be retained. Consequently, an important function of the cardiovascular system in the tubular portions of the kidneys is to take back, or reabsorb, into the blood those essential constituents lost in filtration. Details of the structural basis for these processes will be described in Chapter 8.

Finally, the cardiovascular system performs a vital service to the health of the organism by providing transport in immunological reactions. The blood itself serves as substrate for many such reactions and also as a reservoir for leucocytes and other bacteriophage elements used in defense against disease. The morphology of the blood, the transport carrier, are in many respects equal in importance to the structure of the transport network, the blood vessels and the heart. The cardiovascular system must be safeguarded against damage. Consequently, all vertebrates store and circulate substances in the blood that will stimulate clotting reactions to prevent blood loss if a cut or injury occurs.

The operational complexity and varied purposes of the cardiovascular transport functions require careful coordination and control of the specific transports. Such control is executed by mechanisms built into the structural components of the cardiovascular system itself and by the autonomic nervous system and several hormones.

We first need to understand the structural design of the various anatomical segments of the cardiovascular system in the light of their functional importance. The vascular channels make up two separate but interrelated systems, the blood vascular system and the lymphatic system. Blood vessels are differentiated into arteries and arterioles carrying blood to the capillaries, and venules and veins bringing blood back to the heart. Finally, the blood itself constitutes a tissue vital to the transport operations.

The Heart: Source of Energy for Blood Movement

The vertebrate circulatory system is closed—that is, the blood that leaves the heart will return again to the heart by a continuous system of tubes. Vertebrate hearts are, in essence, two-stroke pumps with a filling phase and an emptying, or ejection, phase. Valves are located at appropriate places at the inflow and outflow ends of cardiac chambers to assure the unidirectional passage of blood. Valve actions are purely passive and depend only on the pressure changes at the two sides of a valve (Figure 7-30).

Vertebrate cardiac muscle, or myocardium, has an inherent ability to contract spontaneously in a rhythmic fashion. This autogenic excitation depends structurally on specialized cardiac tissue, referred to as a pacemaker or nodal tissue, discretely located at the venous inflow end of the heart (Figure 7-31). This tissue generates the waves of electrical excitation that result in subsequent contraction. The even spread of the excitation wave, which assures a coordinated contraction, is aided by a conducting system also derived from modified cardiac tissue, often referred to as Purkinje tissue or conducting tissue (Figure 7-31). Vertebrates without well-developed conducting tissue depend on a remarkable property of cardiac cells to follow the contraction of neighboring cells in an orderly fashion. This sequence depends on particularly close contact between the cell membranes of neighboring cells and not, as was believed

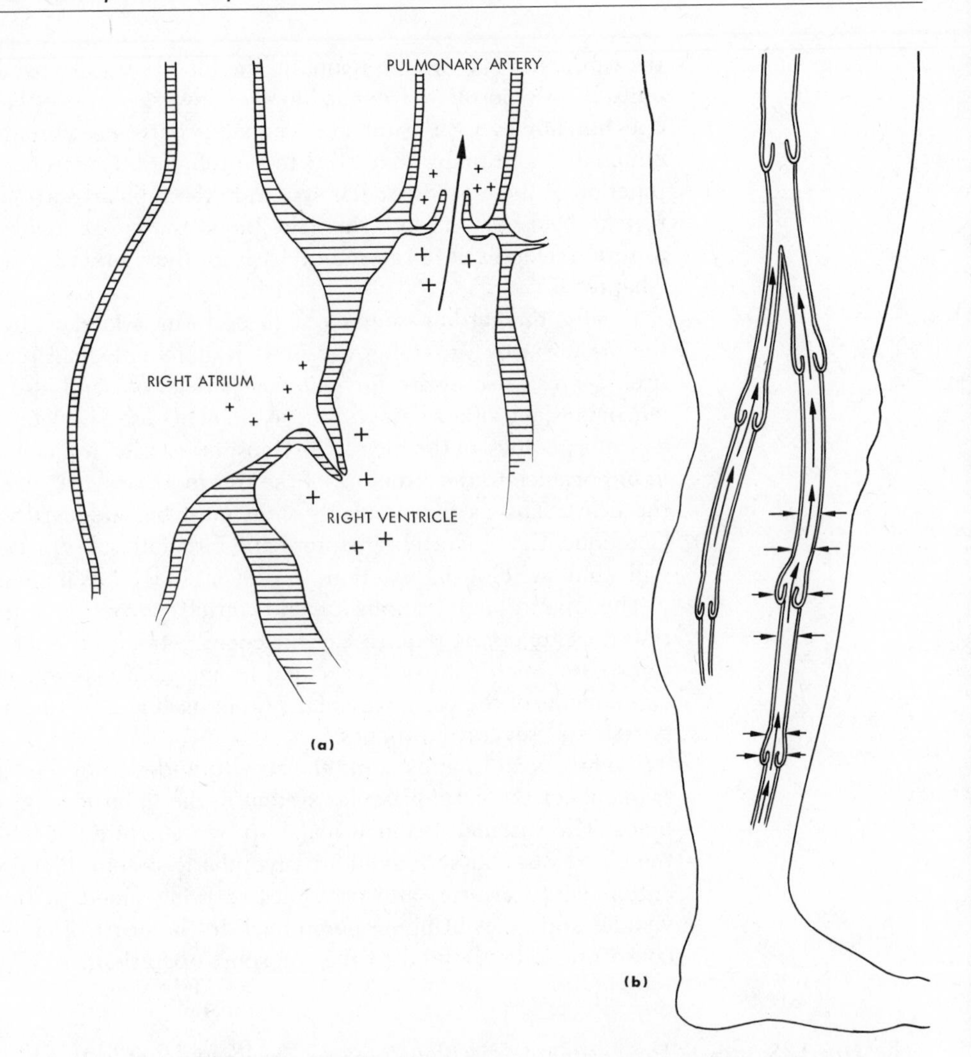

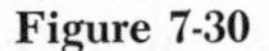

Figure 7-30 Valve action in the cardiovascular system. (a) Cardiac valves are passive structures. They are forced open or closed by pressure differences across their two sides. For example, when the right ventricle of a mammalian heart contracts, the intraventricular pressure exceeds both the right atrial and the pulmonary arterial pressure. Thus the inflow (atrioventricular) valve closes, and the outflow valve to the pulmonary artery opens. (b) Valves are located at strategic points in the veins of the extremities of mammals. Any activity involving the use of the skeletal muscles surrounding the veins will massage the blood in the direction of the heart. Venous valves are also present in the segmental veins of the fish. Use of the swimming muscle exerts a pumping action on the veins and promotes the return of venous blood to the heart.

before cardiac tissue was studied with the aid of electron microscopy, on cytoplasmic continuity into a syncytial arrangement of the cardiac cells.

Myocardium has characteristics in common with both skeletal muscle and visceral smooth muscle. In functional properties and control, it resembles smooth muscle, but it is more akin to skeletal muscle in its red color, cross striations, and the speed and vigor of contraction. The heart, unlike most other organs, must be on continuous duty and its high metabolic activity requires an effective supply of nourishment and an exchange of respiratory gases and

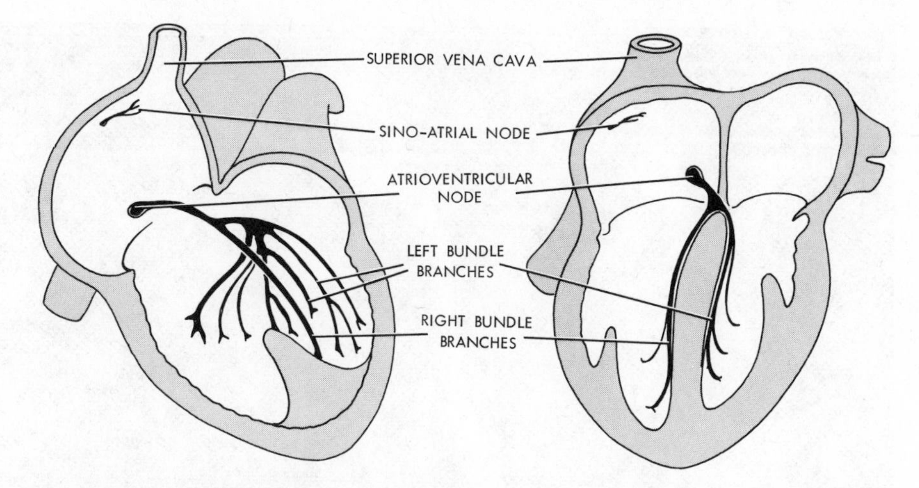

Figure 7-31 Pacemaker and electrical conduction system of the heart. The pacemaker and conduction system of the vertebrate heart generates and transmits the electrical impulses which cause contraction of the cardiac muscle. In lower vertebrates the pacemaker is in the sinus venosus, in higher vertebrates it is located at the superior inflow end to the right atrium and is called the sino-atrial node. The atria have no conduction system, but when the impulse reaches the atrioventricular node, the continued wave of excitation spreads along specialized conducting tissue (Purkinje fibers) in the ventricle. The electrical events of cardiac contraction can be recorded, and is called an electrocardiogram.

metabolic waste products. The coronary vessels serve as transport channels for this important process. Figure 7-32 shows a coronary angiogram of a shark heart.

The Large Arteries: Pressure Chambers for Blood

The arteries leaving the heart are more than passive conduits for the blood on its way to the capillaries. Figure 7-33 summarizes the structural characteristics and the changing function of the arterial tree as we move from the heart to the capillaries. The large component of elastic tissue in the central arteries enables these vessels to serve a crucial function as pressure chambers, temporarily storing the kinetic energy released by the contracting heart as potential energy in their elastic walls. This, in effect, will assure that when the outflow of blood from the heart ceases between beats, the potential energy stored in the arterial walls and imparted by the preceding heart beat will continue to drive the blood through the peripheral vessels; thus, the intermittent flow of blood in the central arteries is converted to a more continuous flow in the peripheral arterioles and capillaries. As the larger elastic arteries stiffen from calcium deposits with age (arteriosclerosis), this important function becomes less efficient and the strain on the peripheral blood vessels greater. If the strain becomes excessive a blood vessel may break. This may be serious; if it occurs in the brain we refer to this as a stroke.

The Smaller Arteries, and Arterioles: Resistance Vessels That Control Blood Flow

You will note from Figure 7-33 that the structure of the arterial wall changes with increasing arborization and reduction in the caliber of the arteries. The elastic connective tissues are far less prominent, whereas the smooth muscle component becomes more conspicuous as arteries become smaller. We will discuss how this reflects on alteration in function with size of the arteries.

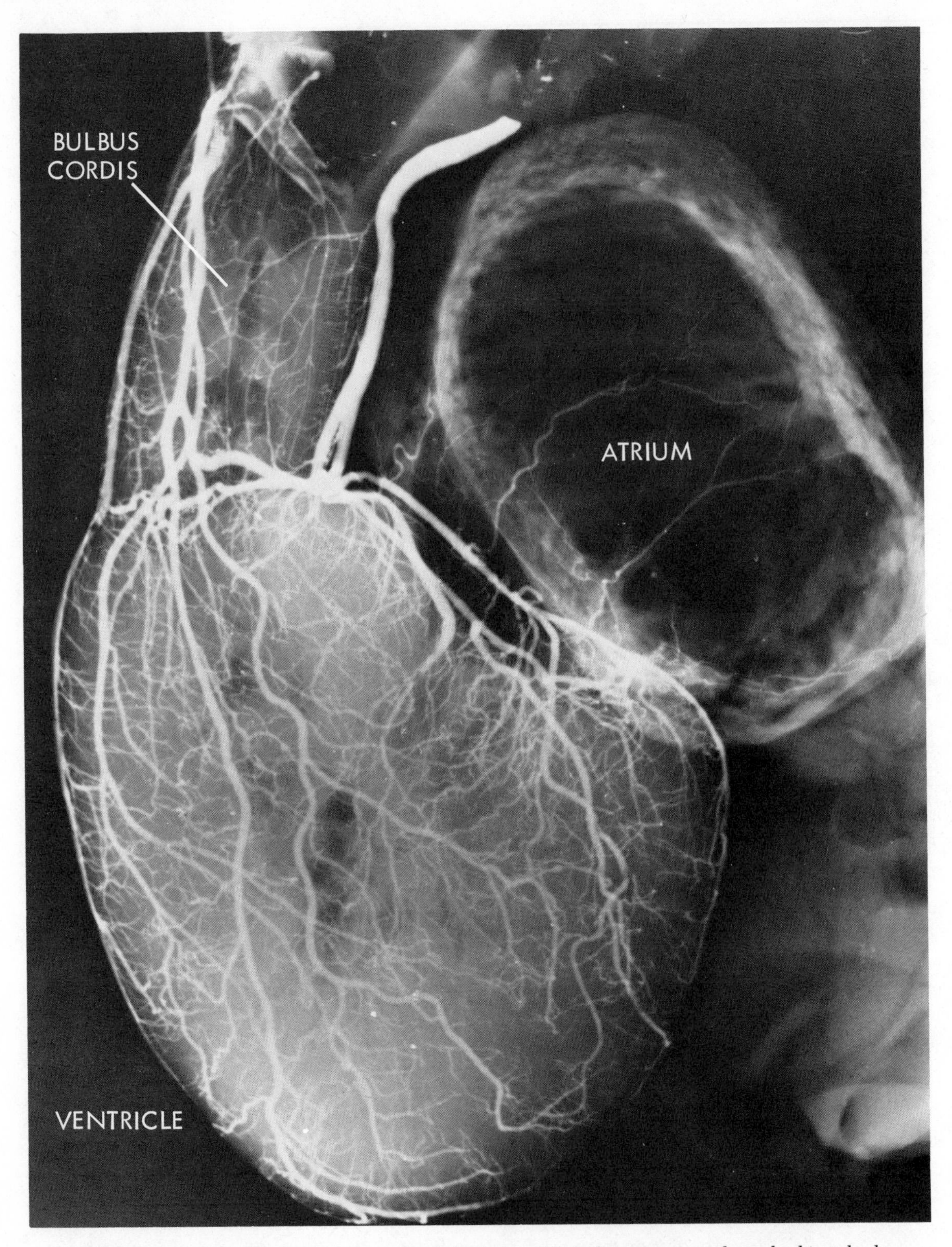

Figure 7-32 The coronary arteries of the heart of *Selache maximus*, a large basking shark, have been injected with a radio-opaque medium for X-ray analysis. The coronary vessels bring a continuous supply of nourishment and oxygen to the heart muscle.

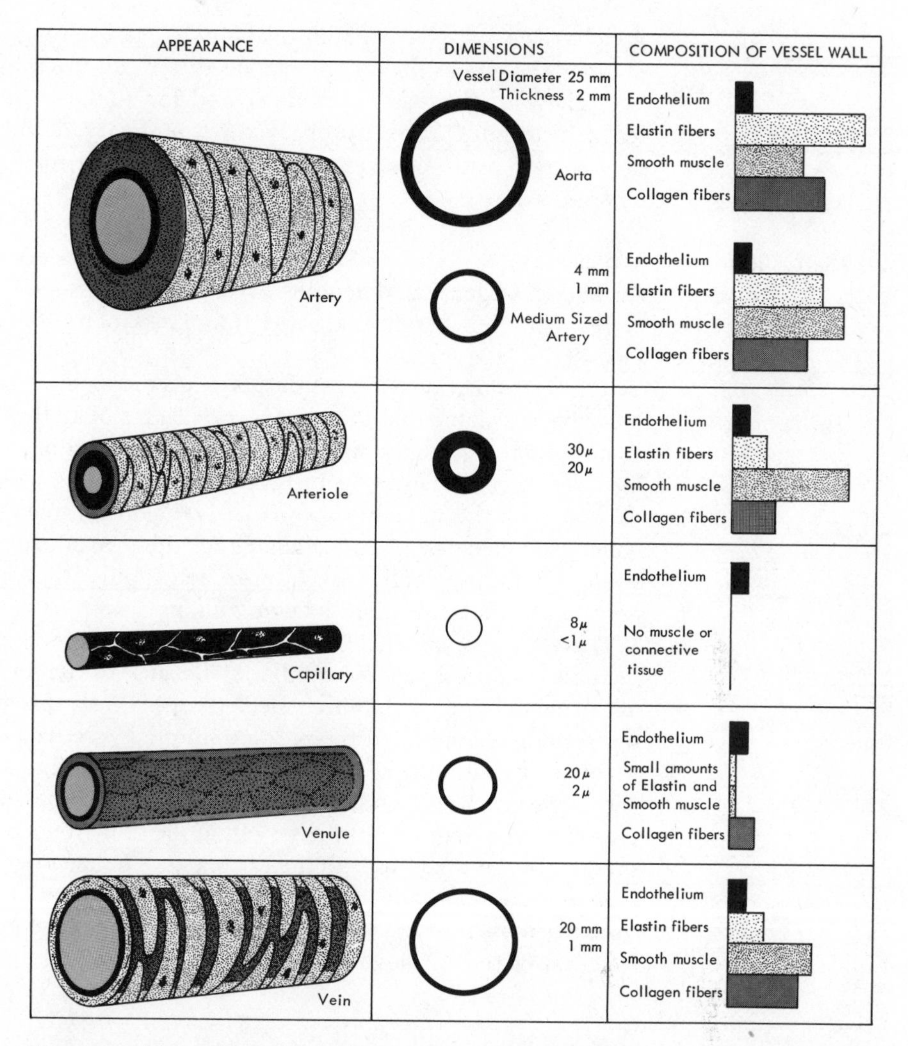

Figure 7-33 Structure of blood vessels in various parts of the vascular system. Diagram showing how the appearance, dimension, and composition of blood vessels change in accordance with their various functions. The entire vascular system is lined with a sheath of endothelial cells whereas the amount of smooth muscle and connective tissue varies.

The demands that various organs and tissues make on the transport service of the cardiovascular system constantly change. Obviously, the skeletal musculature needs more blood when it is active and the gastrointestinal system needs more blood right after a meal. The coordination of these varying requirements is handled largely by the nervous system, but the change in actual blood-flow distribution is executed by the smaller muscular arteries and arterioles. By actively changing their caliber, these vessels alter the resistance to blood flow in accordance with shifting requirements. An important safety and survival factor is built into this system. If simultaneous calls are raised for increased blood flow to many organs, the cardiovascular system may not be able to fill the requests. The autonomic nervous system responds by instructing the arterial smooth muscle to give priority to the more vital organs. The arterioles of various tissues also have a built-in ability to exert self-control over the amount of blood the tissues they serve may need. We call this self-control autoregulation. These are the reasons why the blood flow to the central nervous system or to the heart muscle never drops below a minimal requirement, whereas blood flow to skin and skeletal muscle can be reduced to very low levels in an emergency.

The Capillaries: The Transport Terminals

When a blood vessel divides, the combined cross-sectional area of the two branches is larger than that of the parent vessel. Repeated subdivisions of blood vessels will, thus, result in a greatly increased total cross-sectional area of the blood path (Figure 7-34).

If the cross-sectional area of the aorta in a mammal is one square centimeter, it can be calculated that the cross-sectional area of all the capillaries combined is more than eight hundred times larger. It has also been meticulously calculated that the aorta of a dog, by repeated subdivisions, makes up 1,200,000,000 capillaries in the mesenteric vascular bed alone (Table 7-3). Where accurate counts and calculations are available for other mammals, it has been demonstrated that, in skeletal musculature, each capillary serves a volume of tissue only ten times larger than its own volume.

These simple geometrical relationships indicate some extremely important features in the design of the capillaries. Because the larger blood vessels, as well as the many millions of interconnected capillaries, represent a closed continuous system of fluid-filled tubes, some simple hydrodynamic principles apply. The law of continuity of flow in closed tubes states that a volume passing one end of the system per unit time must appear at the other end and must pass any cross-sectional segment of the system in the same time. Stated differently, the product of the total cross-sectional area and the mean velocity of the moving fluid must remain the same throughout the system. Figure 7-34 helps to visualize these important features of fundamental consequence to the transport function of the cardiovascular system.

Table 7-3 Geometry of the Mesenteric Vascular Bed of a Dog. [Reprinted from A. C. Burton, *Physiology and Biophysics of the Circulation* (Chicago: Yearbook Medical Publications, Inc., 1966), p. 217.]

Type of Vessel	*Diameter (mm)*	*Number*	*Total Cross-Sectional Area (cm^2)*	*Length (cm)*	*Total Volume (cm^3)*	
Aorta	10	1	0.8	40	30	
Large arteries	3	40	3.0	20	60	
Main artery branches	1	600	5.0	10	50	
Terminal branches	0.6	1,800	5.0	1	25	
Arterioles	0.02	40,000,000	125	0.2	25	
Capillaries	0.008	1,200,000,000	600	0.1	60	740 (capillaries through vena cava)
Venules	0.03	80,000,000	570	0.2	110	
Terminal veins	1.5	1,800	30	1	30	
Main venous branches	2.4	600	27	10	270	
Large veins	6.0	40	11	20	220	
Vena cava	12.5	1	1.2	40	50	
					Total 930	

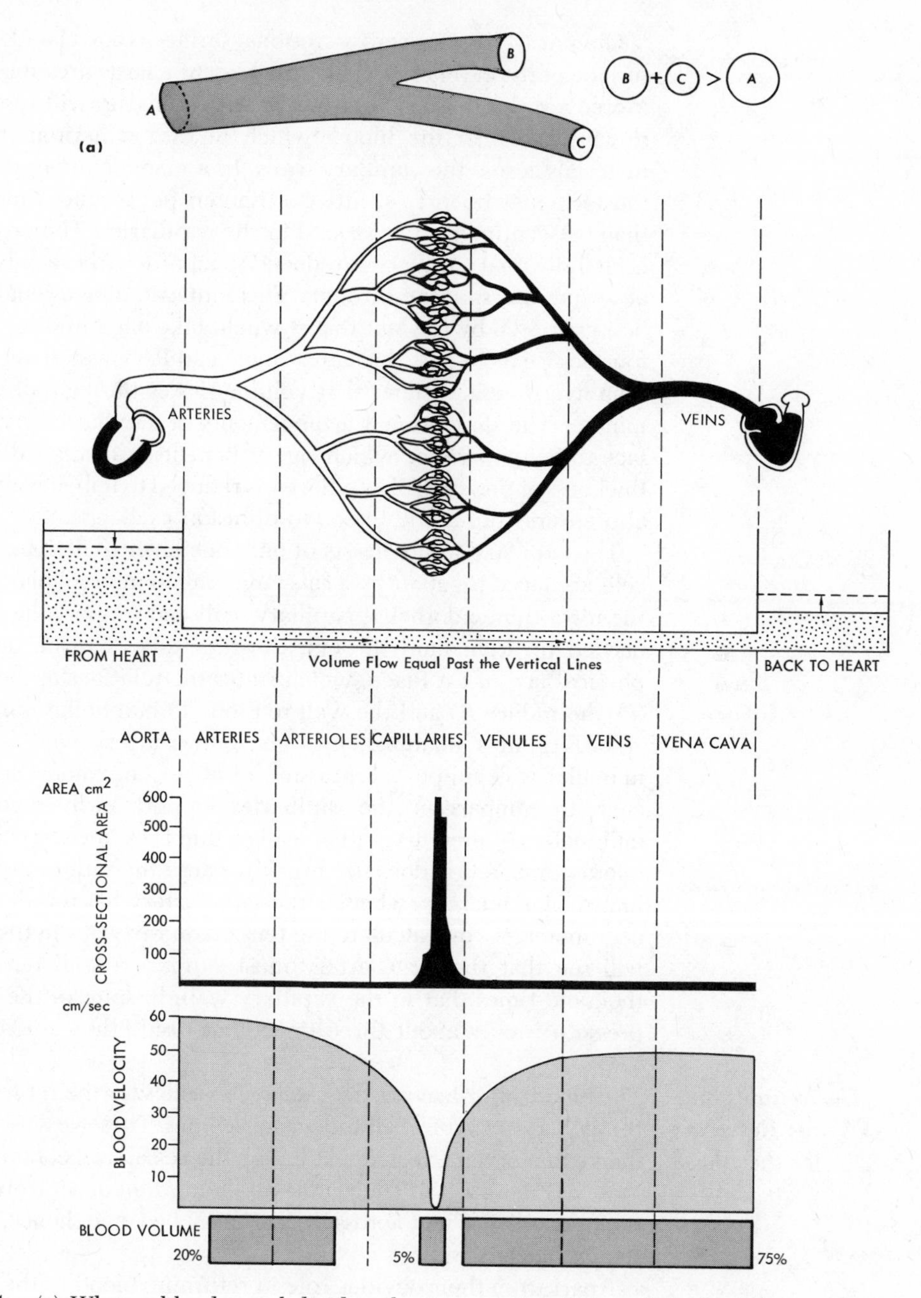

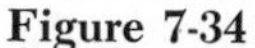

Figure 7-34 (a) When a blood vessel divides, the cross-sectional area of the branches is always greater than that of the parent vessel. $b + c > a$. (b) A schematic representation of the systemic vascular bed in a dog. During any given period of time, the blood flow past any vertical section through the parallel arrangement of vessels must be equal to the volume of blood pumped from the heart or returning to the heart. For any given pressure, the velocity of a fluid is inversely proportional to the cross-sectional area of the tubes through which it flows. Therefore the enormous increase in cross-sectional area produced by the arborization of the arterial tree will cause the blood to slow down in the capillaries. The volume of blood held by the arteries is relatively fixed at about 20% of the total. The volume of blood held by the veins is variable, but under normal conditions they usually contain about 75% of the total.

Obviously, the large cross-sectional surface is correlated with a huge over-all surface of the capillaries. This will present a large area for exchange diffusion. Moreover, the small size of individual capillaries will result in a remarkable slowing down of the blood, which in turn will promote the exchange of materials across the capillary walls. In a mammal (dog) the blood velocity in the aorta may be forty or fifty centimeters per second, which is reduced to less than 0.1-centimeter per second in the capillaries. The cross-sectional area of each individual capillary is reduced to an order where only one red blood cell at a time can squeeze through. This fantastic dimensional change is perhaps best expressed by the fact that it would take one milliliter of blood more than fourteen years to pass through a single capillary; yet, in an adult human, more than five thousand times that volume passes through all the capillaries each minute. The dimensional arrangements of capillaries give them a high surface-to-volume ratio, which again benefits exchange diffusion. Lastly, the thickness of the walls of capillaries is reduced to unbelievable thinness, and this also ensures rapid and effective diffusion exchange.

The capillary wall consists of only one layer of flat endothelial cells. These cells are laced together by a substance called intercellular cement. The reason the ultra-thin endothelial capillary wall can support the substantial pressure needed to drive fluid through the thin tubes can be understood from the physical law of La Place, which states the relationship between the pressure (P), the radius (R) and the wall tension (T) in a hollow tube or compartment ($T = PR$). In a mammal, a large central artery with a thick, elastic, and muscular wall supports a pressure of about one hundred millimeters of mercury. In comparison, the capillaries support an average pressure of thirty millimeters of mercury with a wall so thin it can be seen only with a powerful microscope. Yet, it does not break because the radius of the capillary is only four or five microns, whereas the large artery has a radius of more than one centimeter. If you calculate the tension on the walls in the two examples, you will see that the large artery must support a wall tension more than ten thousand times that in the capillary wall, in spite of the fact that the inside pressure is only about three times that inside the capillary.

The Venules and Veins: Reservoir for the Blood

The blood, after leaving the exchange vessels for the return to the heart, flows through a system of small tubes, the venules. These vessels show a structure not unlike that of the arterioles, although the vessel wall contains much less smooth muscle (Figure 7-33). The venules are important in controlling the lower end of the pressure gradient across the capillary bed, and, hence, the fluid balance in the exchange vessels.

Apart from their obvious role in returning blood to the heart, the veins are an important reservoir for blood. Their vessel wall is structured to give them a large and variable capacity (Figure 7-34). The veins in a higher vertebrate contain 75 per cent or more of the total blood volume. Because their capacity is so large, only a small reduction in the caliber of veins will shift a large volume of blood toward the heart. When situations call for an increased outflow from the heart, the first important step in the regulatory mechanisms is an adjustment in the caliber of veins. The resistance to blood flow is low in veins because of their large caliber. Only a small pressure gradient normally suffices to drive the blood back to the heart.

Veins in the extremities of most vertebrates have valves to prevent an engorgement of venous blood. Such valves allow the surrounding skeletal musculature to exert rhythmic pressure on the veins, with a squeezing effect on the blood in the direction of the heart. You will have noticed that long, passive standing in an upright position will cause your feet to swell, because the fluid balance has been disturbed by an inadequate return to the heart of the venous blood columns in the legs and feet. It was long thought that venous valves were present only in terrestrial animals, because they experience net gravitational forces. Recent studies have revealed, however, that fish have venous valves, and some species even have valves in segmental arteries. In aquatic animals, which experience no net gravitational forces on their hydrostatic columns of blood, valves assist the unidirectional flow of blood. Swimming movements causing changing extravascular compressive forces are very important to venous return in fish.

The Lymphatic Vessels: Fluid Balance Vessels

The lymphatic system is both anatomically and functionally closely associated with the veins. Lymph vessels share in the circulatory transport function by returning constituents of the capillary filtrate to the blood and removing foreign particles and abnormal fluid aggregations from interstitial tissue spaces. Figure 7-35 illustrates the close anatomical association between lymphatics, arteries, and veins. Some of the lower vertebrates, notably amphibians but also fish, have lymph hearts that aid in the circulation of lymph.

The Blood: Transport Carrier

Blood is a highly functional tissue in regard both to its liquid part, the plasma, and to its cellular constituents. The percentage of the total volume of blood occupied by the cells is called the hematocrit. Among the vertebrates it ranges from 10 per cent in some fish to more than 60 per cent in diving mammals. The specific gravity of cells is higher than that of plasma, a fact that makes it easy to separate the two fractions by centrifugation. The cell fraction is dominated by the red cells, the erythrocytes. In man there are normally about five million red cells per cubic millimeter. The normal number of white cells or leucocytes is only one six-hundredth of that, ranging from five thousand to eight thousand per cubic millimeter (Table 7-4). The chief function of the various types of leucocytes is to defend the body against disease. They do not directly assist in the transport functions of the blood, but their mobility in the blood path makes their defense against pathogenic bacteria more efficient. The third cellular constituent, the thrombocyte (platelet in mammals), is more numerous than white cells but far fewer than red cells. Their role is in hemostasis: the formation of blood clots to prevent bleeding when blood vessels become severed or injured. The red cells, or erythrocytes, in contrast, are involved largely with the transport functions of the circulatory system. In vertebrates, the red cells gain their color from hemoglobin, a protein (globin) combined with four iron-containing groups (pyrrole groups). Hemoglobin possesses the ability to associate reversibly with oxygen and, thus, greatly increases the capacity of the blood to carry oxygen. Hemoglobin, also of utmost importance in carbon dioxide transport, combines directly with carbon dioxide and acts as an important buffer to neutralize acids. The oxygen-carrying capacity of vertebrate bloods ranges from three to four volumes per cent (ml O_2 per 100 ml blood) in some sluggish fish to more than 35 per cent in very active or diving

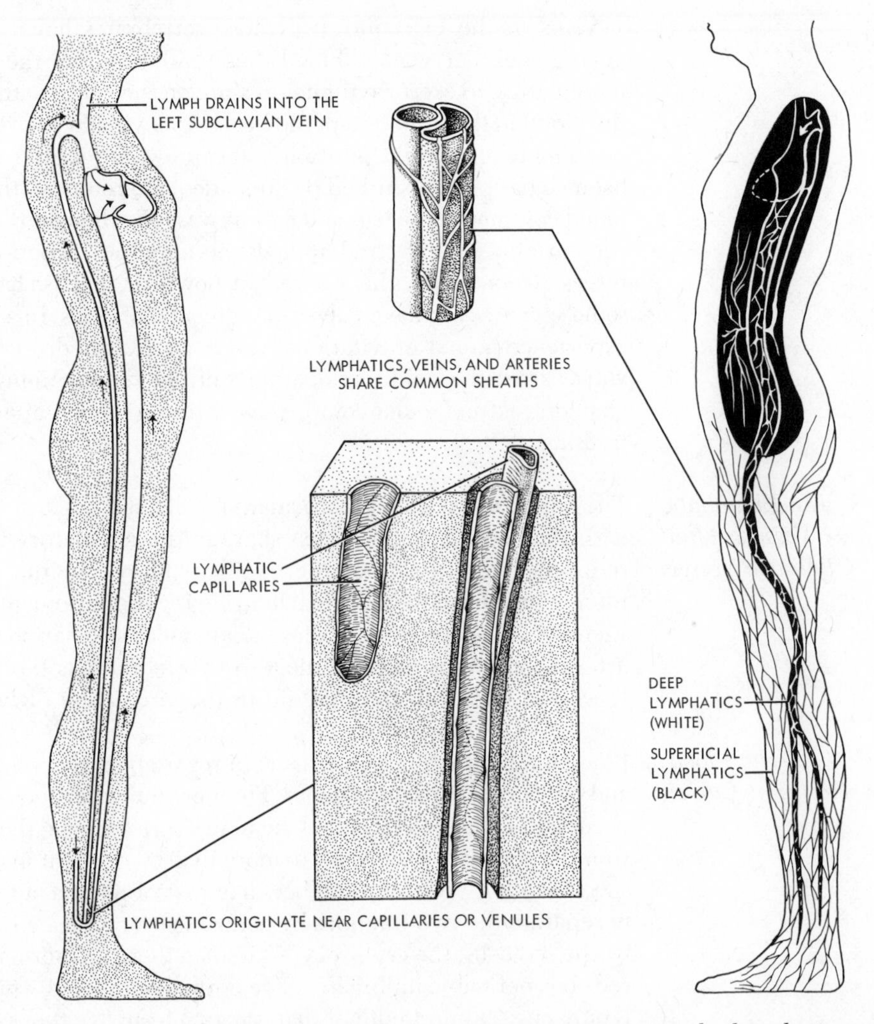

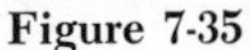

Figure 7-35 The lymphatic system is essentially a paravenous system since the lymphatic capillaries lie in close association with the capillaries and veins of the blood vascular system; the collecting lymphatics tend to accompany veins and arteries and drain into the central veins. Like the veins, the lymphatics consist of both deep and superficial vessels. (Redrawn from R. F. Rushmer, *Cardiovascular Dynamics*, 2nd edition. Philadelphia: W. B. Saunders Co., 1961.)

mammals and birds. In contrast to conditions in vertebrates, most invertebrates carry their respiratory pigments, like hemoglobin, in free solution in plasma, rather than enclosed in cells.

The size and shape of vertebrate red blood cells vary considerably (Figure 7-36). In vertebrates below mammals, the red cells are nucleated and usually oval and flattened in shape. The mammalian red cell is generally a circular biconcave disc lacking a nucleus in its mature form. It shows an almost unbelievable efficiency in packing the important hemoglobin molecules. As much as 25 per cent of the available volume is taken up by hemoglobin molecules, whereas water fills 70 per cent, and only 5 per cent is taken up by other cellular constituents. In lower vertebrates, these proportions are different. The nonnucleated type of red cell is one important reason why the

Table 7-4a The Average Number of Erythrocytes per Lymphocyte in Various Vertebrate Classes.

Class	*Erythrocytes per Lymphocyte*
Fish	5–12
Amphibians	20–70
Reptiles	100
Birds	70–200
Mammals	350–2000

Table 7-4b Total Erythrocyte Count and Size in Various Vertebrates.

	Number (per mm^3 of blood)	*Long Diameter × Short Diameter (microns)*
Fish		
Lamprey	133,000	15 × 15
Ray	230,000	25 × 14
Flounder	2,000,000	12 × 9
Amphibians		
Salamander (Amphiuma)	95,000	78 × 46
Frog	400,000	22 × 15.5
Reptiles		
Turtle	630,000	21.2 × 12.4
Birds		
Hummingbird	6,500,000	
Ostrich	1,600,000	18 × 9
Mammals		
Man	4,500,000 5,000,000	7–8

oxygen-carrying capacity in mammals can attain such high values. The size of most mammalian red cells is about eight to ten microns in diameter, but in some species such as goats it may be as small as four microns. In contrast, the nucleated cells of fish and amphibians can attain diameters more than ten times, and volumes more than one thousand times, greater than those common in mammals.

Blood cells in vertebrates have a limited life span, and new cells continue to develop and differentiate throughout the life of the organism. Like the rest of the tissues making up the circulatory system, the blood cells originate from mesoderm. The tissues in which blood cells are formed later are called hemopoietic tissues. In vertebrate embryos, hemopoietic tissues are very widespread, whereas they become more restricted in the adults. In teleost fish, the kidneys show hemopoietic activity through the whole life cycle. In higher vertebrates, the bone marrow is a most important site for blood-cell production. Blood-forming tissues also have aggregated to form discrete hemopoietic organs like the spleen, for the formation, storage, and destruction of blood cells. In mammals, specialized lymph nodes produce lymphocytes.

Plasma, the other fraction of blood, is about 90 per cent water by weight. Solid constituents (7 per cent in mammals) are largely the plasma proteins that

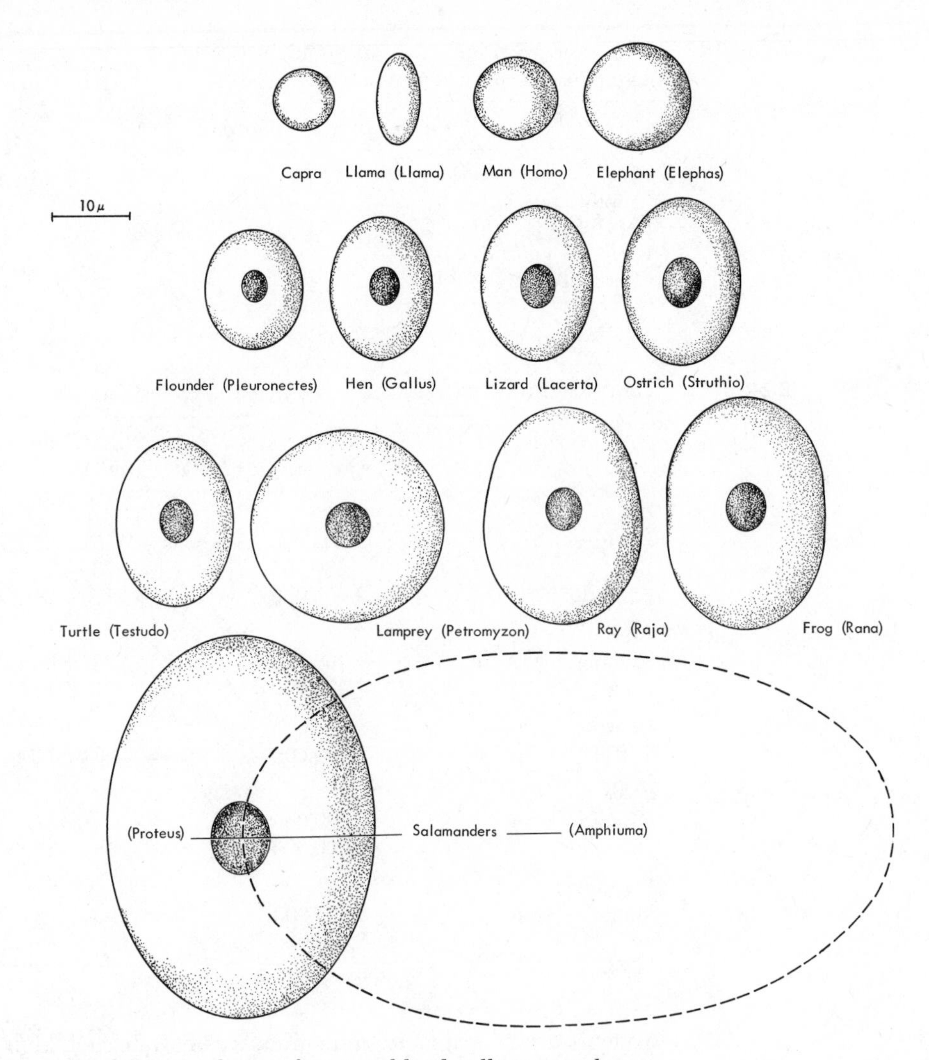

Figure 7-36 Morphology and size of various blood cells in vertebrates.

serve an important function in the fluid balance of the circulatory system. Because most of the protein molecules are too large to be filtered through the capillary walls, they remain inside the vessels where they create a colloid osmotic pressure greater than the osmotic pressure in the extravascular fluid. Because of this osmotic force, fluid filtered from the capillaries by the higher hydrostatic pressure inside them than in the interstitial space will be reabsorbed and kept in balance with fluid filtration. The most important plasma proteins in vertebrates are the albumins and globulins. Besides maintaining the osmotic balance, the globulins, at least in higher vertebrates, function in antibody reactions to protect against disease. Fibrinogen is another protein that is crucial to the blood-clotting mechanism. Lipoproteins are important carriers of lipids to the various sites of synthesis in cells. All the plasma proteins are important agents that assist hemoglobin, carbonates, and other buffers to maintain the blood pH within narrow limits.

Phylogeny of the Vertebrate Circulatory System

The intimate functional connection between the respiratory and circulatory systems also is expressed in their phylogenetic development. In fact, changes in the mode of breathing have been the primary and most decisive factor in the evolutionary transformation of the cardiovascular system.

The branchial arches provide the structural basis for the respiratory organs in fish. Because the process of gas exchange with the environment is inseparably related to the transport function of the cardiovascular system, it is not surprising that the vascular channels, the blood vessels, bringing blood to the branchial region, show a close spatial relationship to the branchial arches so that the paired vessels of the system, the aortic arches, match the number of branchial arches. After traversing the branchial region, the vessels converge and connect to the dorsal aorta where the distributing arterial system originates. This fundamental plan is similar in the embryological development of all vertebrates. Subsequent changes in this vascular arrangement have occurred in response to a reduction or redistribution of the gas-exchanging surfaces of the gills or to the gradual development of aerial gas exchange and lungs in lungfish and tetrapods.

The vascular system in amphioxus is said often to resemble that in ancestral vertebrates. Amphioxus lacks a discrete heart; blood is propelled forward in the ventral aorta by a peristaltic contraction of the vessel itself. The fifty or so pairs of gill arches receive paired afferent vessels (aortic arches). Where these branch off from the ventral aorta, contractile enlargements contribute to the further propulsion of blood (Figure 7-37). The branchial vessels converge to form the dorsal aorta, which is paired in the branchial region. The forward extensions of the dorsal aortae correspond to the vertebrate internal carotids that supply much of the circulation to the head. Paired segmental arteries and unpaired arteries to the viscera branch off from the median dorsal aorta in its posterior

Figure 7-37 Schematic representation of the central circulation in *Amphioxus*.

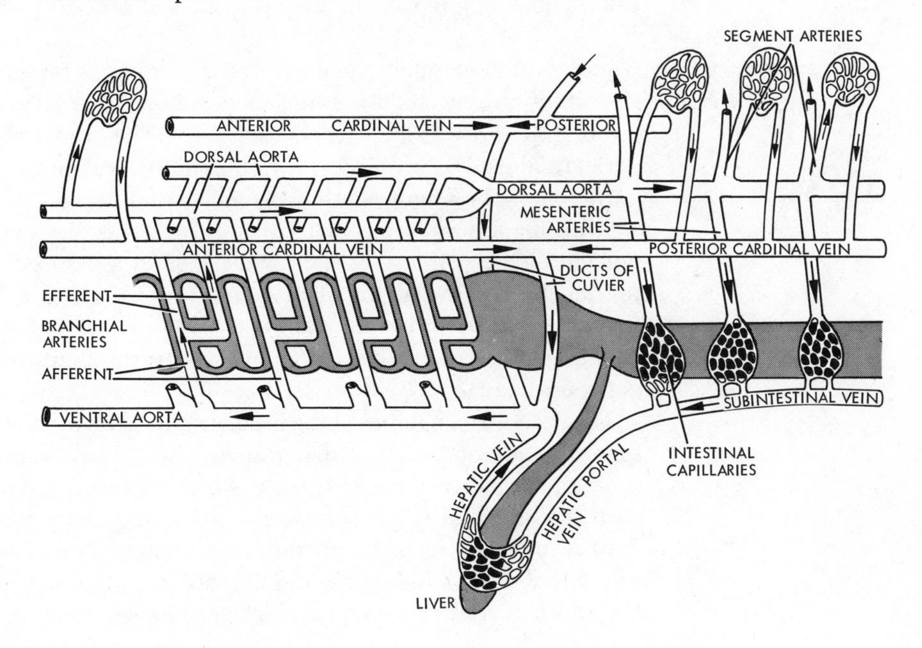

course. The return of venous blood to the ventral aorta goes through posterior cardinals and subintestinal veins. The subintestinal veins resolve in the liver; they recollect there as hepatic veins. These, together with the postcardinals and the precardinals, unite to form the ducts of Cuvier, which continue into the ventral aorta via the sinus venosus (Figure 7-34). Amphioxus does not show a renal portal system.

The fish heart, like that of all vertebrate embryos, develops from the ventral median blood vessel in the branchial region. During development, the heart becomes oriented in a posterior direction in apparent correlation with the adult type of breathing. In gill-breathing aquatic species it remains in its anterior location, whereas in lung-breathing forms it has a more posterior position. The progressive development and differentiation of a distinct neck region in tetrapods contributes further to the posterior location of the heart.

The typical fish heart often is referred to as a venous heart because blood returning to it is deoxygenated blood coming from the tissues. Similarly, it is called a single heart because it pumps blood of only one type and because the entire output from the heart is routed through the aortic arches perfusing the gills where gas exchange takes place. The postbranchial circulation of blood leaving the gas exchange surfaces is called the systemic circulation. In the circulation of fish the gas exchange vessels are characteristically in direct series with the systemic vessels in one continuous circuit (Figure 7-38). After lung breathing became established in higher tetrapods, the two types of vascular bed were still connected in series but in a double circuit in which the heart connected the two (Figure 7-38). In vertebrates that depend on bimodal gas exchange, such as lungfish, amphibians, and reptiles exclusive of crocodiles, the respiratory vascular bed and the systemic vascular bed are coupled not in series but in parallel because the heart is incompletely divided in two halves. During the evolution of air breathing, different vascular arrangements developed, some of which can be seen in air-breathing fish such as the bowfin and garpike, the bichir (*Polypterus*), the electric eel, and others.

Circulation in a Typical Fish

Circulation in elasmobranchs (sharks and skates) is typical of fish with a heart consisting of four chambers: the sinus venosus, the atrium, the ventricle, and the bulbus cordis (Figures 7-39 and 7-40). The whole organ is enclosed in a pericardial cavity surrounded by a lining, the pericardium, that, in many fish, particularly the chondrichthyians and chondrosteans, is tough and makes a rigid casing around the heart. This pericardial cavity develops as an anterior portion of the primary coelom and becomes separated from the abdominal portion by the development of a septum transversum. In chondrichthyians (sharks) and chondrosteans (sturgeons), the pericardial cavity retains a small communication with the abdominal cavity through a duct, the pericardio-peritoneal canal.

The receiving chamber of the fish heart, the sinus venosus, is thin-walled and sparsely equipped with cardiac muscle fibers. Blood enters the sinus venosus through the paired ducts of Cuvier, which, in turn, connect with anterior and posterior cardinal veins, the lateral veins, and the subclavian veins [Figure 7-40(a) and (b)]. In addition, the sinus venosus of elasmobranchs receives a substantial venous inflow through the hepatic veins entering directly through the septum transversum. This entrance is guarded by peculiar muscular

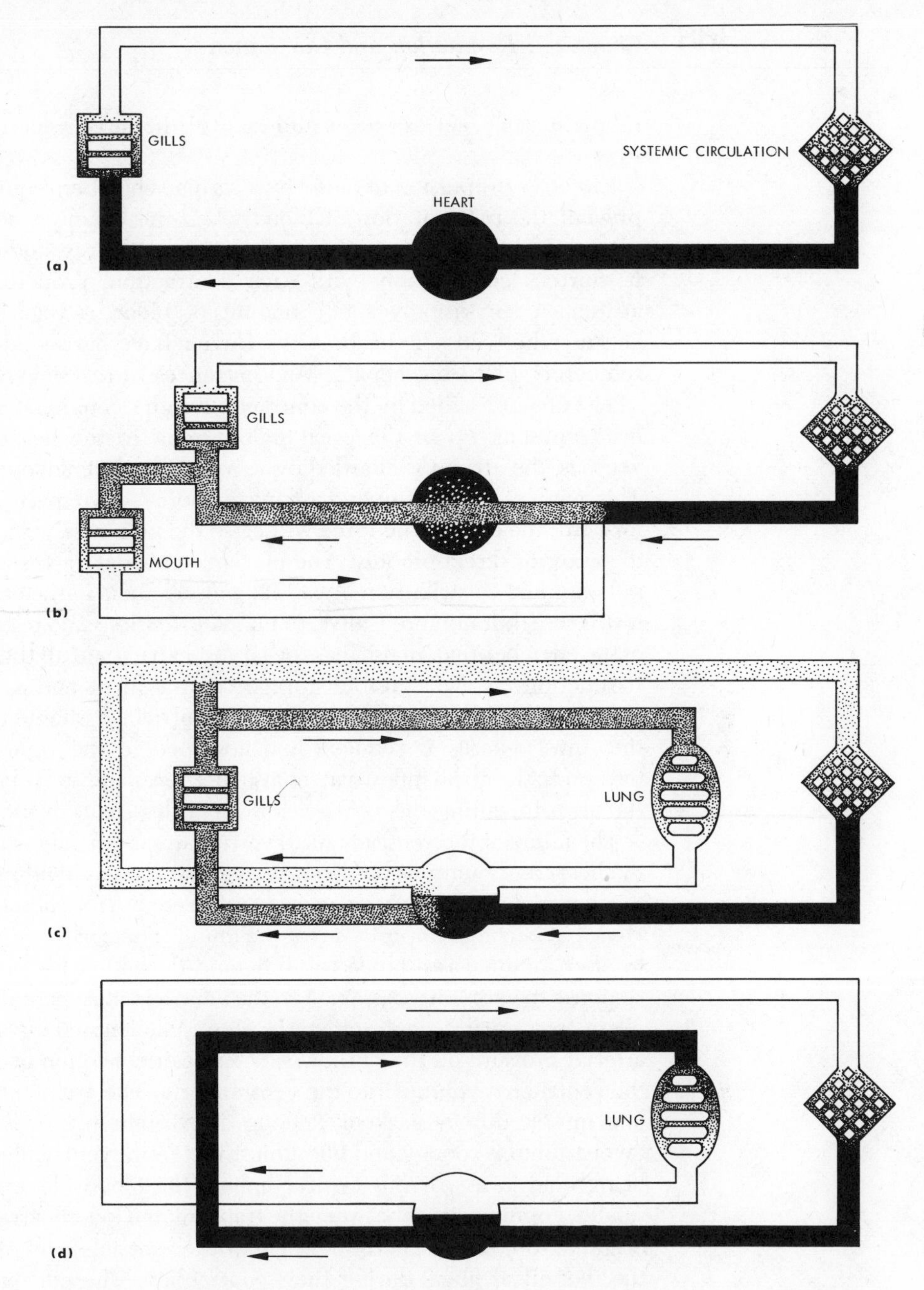

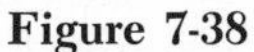

Figure 7-38 Evolution of vascular circuits in vertebrates. (a) Typical in series arrangement of the branchial (gills) and systemic circulation in a fish. During the evolution of air-breathing mechanisms many fish were bimodal breathers, retaining functional gills while developing air-breathing organs. New vascular circuits developed to these organs. The arrangement in (b) is typical of the teleost *Symbranchus marmoratus*, which can use its gills and a vascular oral mucosa for air breathing. (c) Only in the lungfish did the new air-breathing organ have a vascular circuit with direct connection back to the heart (a pulmonary vein). This condition was conducive to development of an incomplete separation of the two sides of the heart resulting in a partial separation of the well-oxygenated (pulmonary venous) and poorly oxygenated (systemic venous) blood on their passage through the heart. (d) In the homeotherm vertebrates, birds and mammals, the pulmonary and systemic vascular beds have attained complete separation in the heart. No admixture of the two types of blood occurs.

sphincters that act as valves and control the large venous flow through the liver.

Effective propulsion of blood by a cardiac chamber requires valves that will prevent the regurgitation of blood when the chamber contracts. The sinus venosus lacks such valves at the inflow side, but X-ray studies have shown that it shortens considerably with each contraction, propelling blood into the atrium. A surprisingly small amount of blood is regurgitated posteriorly because the walls of the ducts of Cuvier have narrowed and the muscular sphincters guard the hepatic vein entrances to the sinus venosus.

The atrium is filled by the contraction of the sinus venosus and the pressure head building up in the great inflow veins to the heart. Unlike the sinus venosus, the atrium is guarded by valves on both its inflow and outflow sides. This enables the chamber to have a more active part in pumping blood forward; the contractile force will close the inflow valves and propel blood in the anterior direction only. The presence of atrial valves is correlated with a more uniform distribution of cardiac muscle fibers in the atrial wall. This wall is still thin, but numerous diverticulated extensions add to the over-all volume of the chamber that, in fish, has the largest capacity of all the cardiac chambers. It functions as a distensible storage compartment and as a most important chamber for filling the main pump, the ventricle. It should be noted that when the sinus venosus is reduced and absorbed in the right atrium of higher tetrapods, the atrial inflow valves also disappear and the relative importance of the atria for filling the ventricles is much less than in fish.

The filling of the ventricle in all vertebrates is crucial for the performance of the heart as a pump. It has been shown that increased filling and distention of the ventricle increase the energy released by its subsequent contraction. Hence, it becomes important that the muscle fibers of the ventricle be arranged to offer minimal resistance to filling and distention because filling occurs in response to very low pressure. Yet, the ventricle must contain powerful muscle fibers that during contraction develop sufficient energy to overcome the arterial pressure on the outflow side and cause ejection of a major portion of the ventricular volume into the ventral aorta. The architecture of the ventricular muscle differs markedly among the vertebrates. In fish, the ventricular myocardium is spongy and fills almost the entire ventricular space. Instead of being held in a spacious central lumen, the blood fills up the thousands of smaller compartments between the trabeculated loosely arranged muscle fibers (Figure 7-39). This structural design of the ventricle may offer less tension on the individual fibers during their contraction. The efficient atrioventricular valves guard the entrance from the atrium to the ventricle, and these close

Figure 7-39 (*opposite*) (a) The fish heart typified with the large basking shark, *Selache maximus*. The four chambers of the heart, the sinus venosus, atrium, ventricle, and bulbus cordis, are invested in a tough semirigid pericardium. (b) The sinus venosus is a spacious chamber sparsely equipped with muscle fibers. Blood enters the sinus venosus from the ducts of Cuvier and through the sphincters marking the entrance from the hepatic veins. (c) The atrium has the largest capacity of all chambers in the fish heart. It is thin-walled but well equipped with musculature. Valves prevent regurgitation to the sinus venosus during atrial contraction. Atrioventricular valves prevent backflow to the atrium during ventricular contraction (d) The ventricle has a dense outer layer of compact myocardium. Most of the ventricle consists of a trabeculate sponge-like myocardium. A small central lumen shows the opening of the atrioventricular valve.

during the powerful ventricular contraction with its rapid development of high pressure.

Ventricular outflow occurs through the fourth chamber of the fish heart, the bulbus cordis. In elasmobranchs, this chamber represents a direct forward continuation of the cardiac muscle. In higher fish, such as the teleosts, the bulbus cordis is reduced and replaced by a bulbous swelling of the smooth

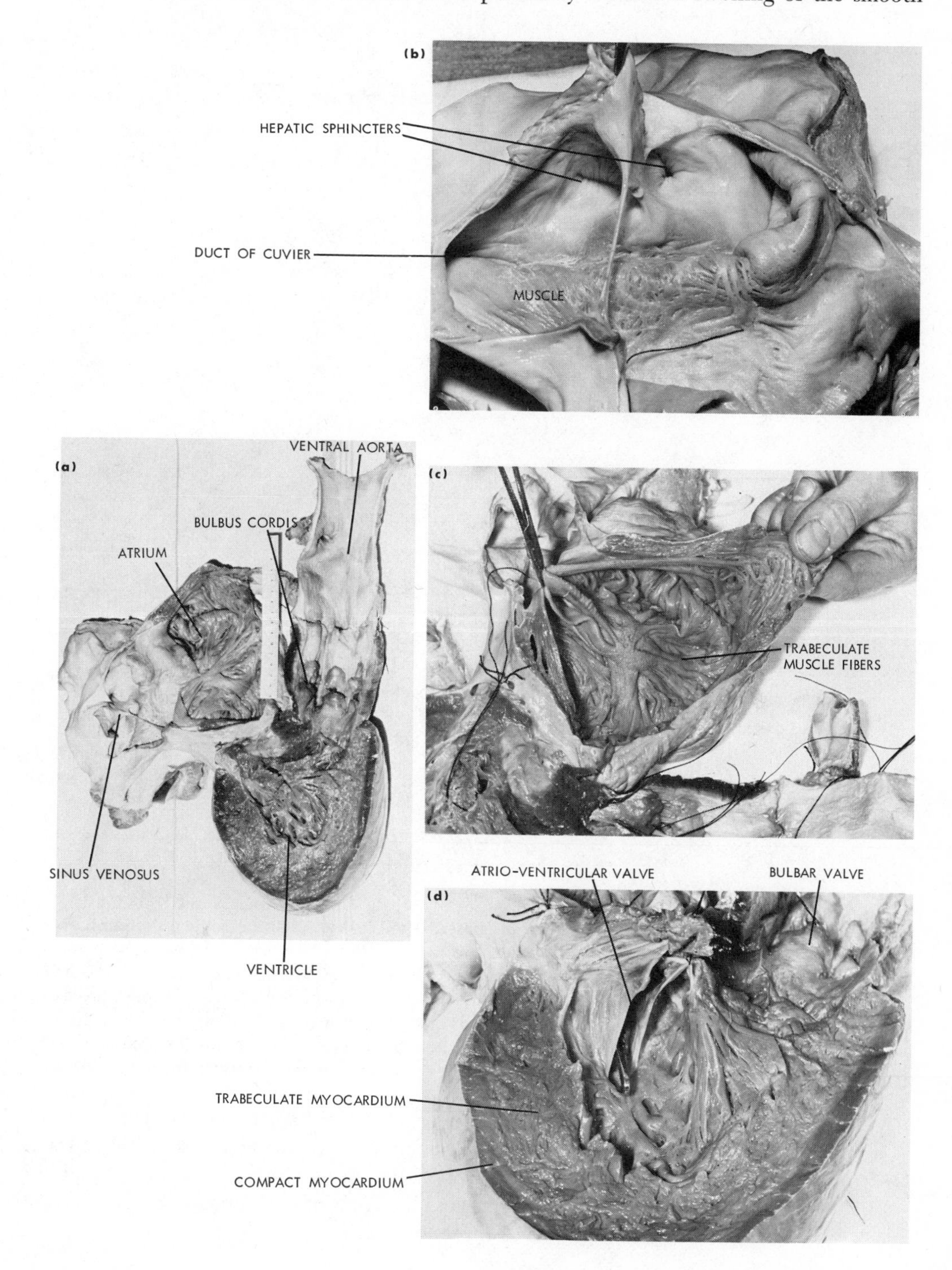

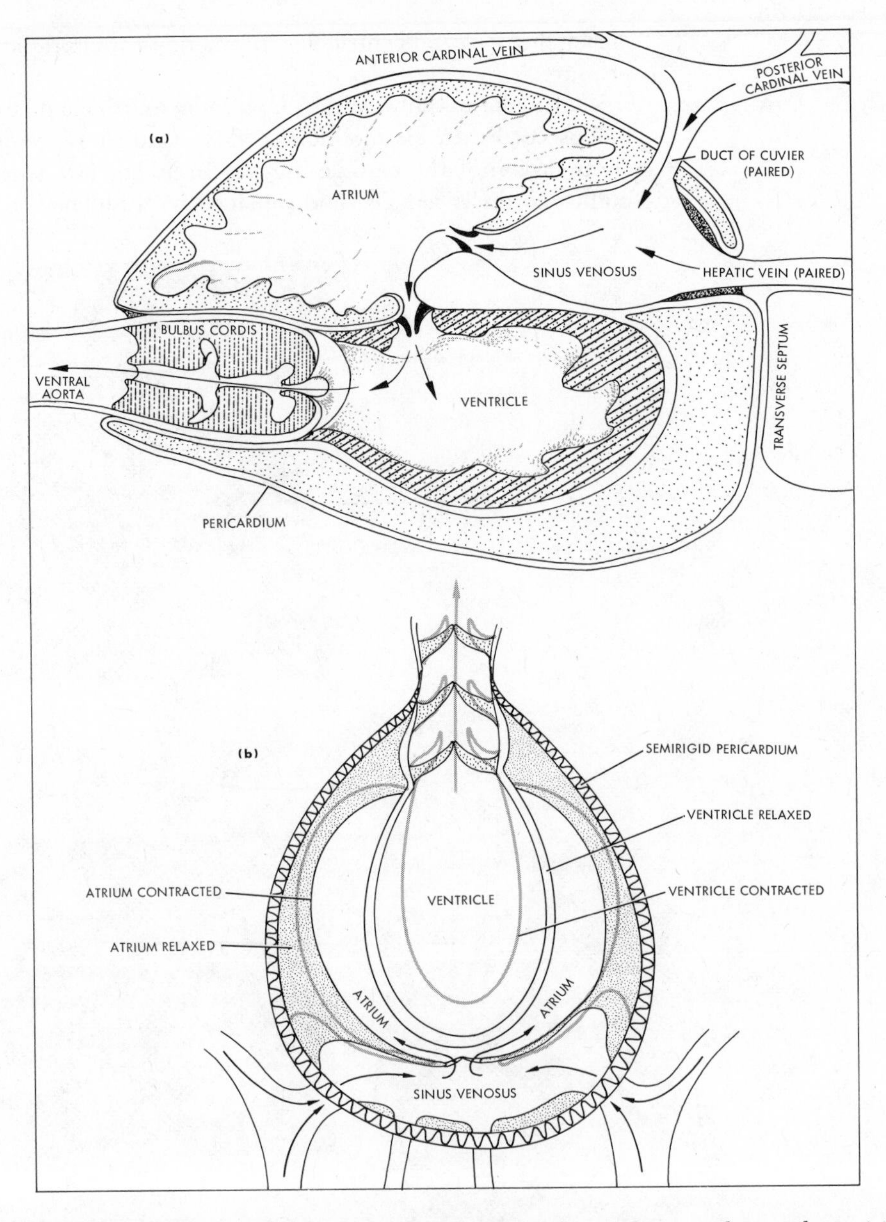

Figure 7-40 (a) Schematic representation of the elasmobranch heart showing the tough semirigid pericardium surrounding the cardiac chambers. (b) A subambient pressure always prevails inside the pericardium. This will steepen the pressure gradient toward the heart along the veins and promote venous return. When the ventricle contracts blood is ejected into the ventral aorta and intrapericardial pressure is reduced still further. This pressure transmits through the thin walls of the sinus venosus and atrium and causes a suctional attraction for venous blood. In this manner the heart functions both as a pressure pump and a suction pump.

muscle and elastic tissues of the ventral aorta. This latter structure, referred to as the bulbus arteriosus, belongs with the arterial system and not with the heart itself. The bulbus cordis of the elasmobranchs houses a variable number of outflow valves that prevent backflow of the blood ejected by the ventricle. This heart chamber has three main functions.

1. Its musculature supplies the necessary support for the potency of the valves. The distention of the bulbous tube during ventricular contraction might otherwise interfere with the valve closure.
2. The bulbus cordis is an effective pressure chamber; it converts the abrupt ventricular pressure rise to a more extended depulsated pressure that will prolong the over-all ejection time from the heart.
3. The bulbus cordis adds a small volume of blood to the volume ejected by the ventricle with each contraction and, thus, functions as an auxiliary pump.

The single ventricular pump of the fish heart functions as a pressure pump because it must develop a pressure in excess of the high arterial pressure to eject its stroke volume. The pumping action of the heart and the expulsion of blood into the ventral aorta has an additional indirect effect that is particularly important in the chondrichthyians. In these forms, which have a rigid, tough pericardium, a subambient pressure prevails in the pericardial space. The reduction in the size of the ventricle during its contraction will cause a transient further reduction of pressure in the pericardial fluid surrounding the heart. In turn, this low pressure surrounding the heart will be transmitted into the thin-walled receiving chambers, as well as to the larger central veins. The resulting decrease in central venous pressure and the consequent increase of the pressure gradient toward the heart will promote the venous return and the filling of the heart (Figure 7-40). In this way, the heart becomes a suction pump for venous blood returning to the heart while it performs the opposite role of a pressure pump for ejecting blood from the heart. This double effect, possible only because of the rigid pericardium surrounding the heart, is also present to a lesser degree in teleost and dipnoan fish, as well as in some amphibians. The pericardium also has an important cushioning effect on the heart, suspended as it is in a fluid-filled compartment.

In all the gnathostome fish, six pairs of aortic arches develop embryologically, but among the adults there is great variation and modification in the number retained. No extant adult gnathostome has a first aortic arch. In some forms, the anterior margin of the spiracle, which should be perfused by the first aortic arch, shows a small gill, which is supplied by blood from the efferent section of the second aortic arch or by a vessel arising from the dorsal aorta. This small gill is, hence, not perfused with deoxygenated blood from the ventral aorta like other gills and is, therefore, called a pseudobranch. The efferent part of the first aortic arch when it exists in some fish makes up part of the supply to the brain. The second aortic arch is typically present in sharks and skates (chondrichthyians) where it perfuses a hemibranch between the spiracle and the first gill slit. In other fish, the second aortic arch has a more inconsistent pattern. However, all fish as a rule show fully developed aortic arches 3 through 6.

In all fish, the dorsal aorta continues forward and makes up the internal carotid arteries (Figure 7-41). In its posterior course, the dorsal aorta is the distributing vessel for the remainder of the entire arterial (systemic) circulation. Associated with the appearance of paired fins in chondrichthyians are the subclavian and iliac arteries branching from this vessel. Other branches include three arteries supplying the gastrointestinal tract, the coeliac and the anterior and posterior mesenteric arteries. In addition, there are renal arteries to the kidneys, intersegmental arteries to the body wall, and arteries to the gonads.

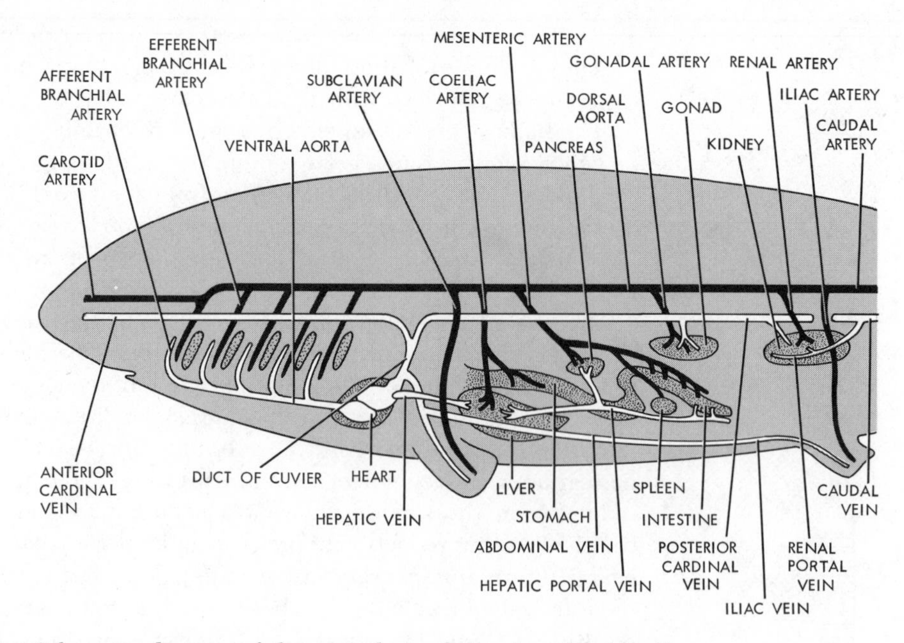

Figure 7-41 Schematic diagram of the central circulation in a fish (shark).

Arteries break down into exchange vessels or capillaries in gills and other respiratory organs, as well as in the general metabolizing tissues. Other capillary beds when they occur are known as portal vascular beds.

The hepatic portal system, which is present in all vertebrates, receives its main blood supply from the veins draining the intestines and continues into a second capillary network in the liver. Leaving the liver, the blood collects into larger veins that finally make up the hepatic veins draining into the larger systemic veins or directly into the sinus venosus (elasmobranchs). The functional significance of the hepatic portal system lies in its direct dispatch of the nutrients absorbed by the intestines to the liver for further processing and storage [Figure 7-42(a)].

Another type of portal circulation, the renal portal system, exists in fish and lower tetrapods. This system is present when the entire caudal venous backflow, or a part of it, is routed through capillaries in the kidneys [Figure 7-42(a)]. When leaving the kidneys, these veins become confluent with the posterior cardinals (fish) or the posterior vena cava (tetrapods). The tissues perfused by a portal system do not receive the needed oxygen from its portal afferent supply because this blood is already deoxygenated. However, in addition to the inflowing venous blood, these organs receive a supply of systemic arterial blood via vessels branching off from the dorsal aorta (hepatic and renal arteries).

Figure 7-42 (*opposite*) Diagrammatic outline of the transformation of the cardinal veins into the vena cava in representative vertebrates. The hepatic and renal portal systems and their transformation are also depicted. (a) Fish stage; (b) lungfish and lower tetrapod stage; (c) mammalian stage.

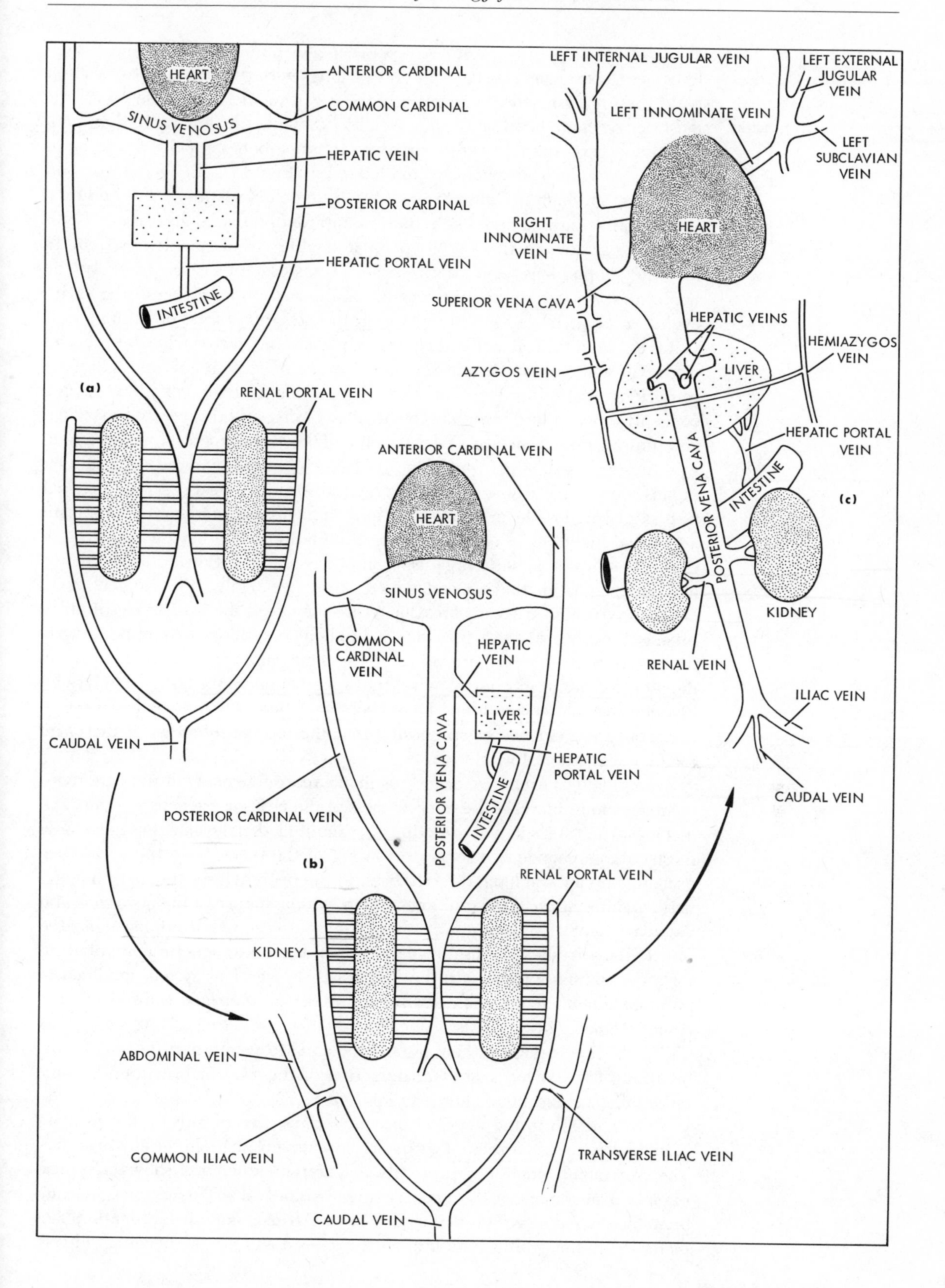

HEART
ANTERIOR CARDINAL
COMMON CARDINAL
SINUS VENOSUS
HEPATIC VEIN
POSTERIOR CARDINAL
HEPATIC PORTAL VEIN
INTESTINE
(a)
RENAL PORTAL VEIN
CAUDAL VEIN
LEFT INTERNAL JUGULAR VEIN
LEFT EXTERNAL JUGULAR VEIN
LEFT INNOMINATE VEIN
LEFT SUBCLAVIAN VEIN
HEART
RIGHT INNOMINATE VEIN
SUPERIOR VENA CAVA
HEPATIC VEINS
HEMIAZYGOS VEIN
LIVER
AZYGOS VEIN
HEPATIC PORTAL VEIN
POSTERIOR VENA CAVA
INTESTINE
(c)
KIDNEY
RENAL VEIN
ILIAC VEIN
CAUDAL VEIN
ANTERIOR CARDINAL VEIN
HEART
SINUS VENOSUS
COMMON CARDINAL VEIN
HEPATIC VEIN
LIVER
HEPATIC PORTAL VEIN
POSTERIOR VENA CAVA
INTESTINE
POSTERIOR CARDINAL VEIN
(b)
RENAL PORTAL VEIN
KIDNEY
ABDOMINAL VEIN
COMMON ILIAC VEIN
TRANSVERSE ILIAC VEIN
CAUDAL VEIN

Circulation in Lungfish: The Rise of a Double Circulation

The usefulness of all respiratory organs depends on how the vascular system links the gas exchange surfaces in gills or lungs with the cells and tissues that build the organism. Ideally, a respiratory organ should be supplied with the most deoxygenated carbon dioxide-rich blood, while the flow path of oxygen-rich blood from the respiratory organ to the metabolizing cells should be as direct as possible. This scheme is fulfilled in fish (Figures 7-38, 7-41), birds, and mammals, although in different ways: by a single circulation in fish and by a completely separated double circulation in birds and mammals (Figure 7-38). The gradual change from a single to a double circulation is apparent in lungfish, amphibians, and reptiles. This transformation was probably the single most important adaptive change of internal organization in the entire evolution of vertebrates; it started with the adoption of air breathing.

It is essential to realize that the rise of air breathing started while vertebrates were still aquatic and before they could move effectively on land. Climatic conditions existing 350 million years ago brought about very low oxygen concentrations in large tropical fresh-water basins, and this created marginal and limiting conditions for vertebrate life. The development of air breathing was, thus, originally the result of lack of oxygen in water and the air-breathing organs were primarily organs for oxygen absorption. However, the original respiratory organs, the gills, had been long in the making and served important additional functions in carbon dioxide elimination and other excretory processes. In addition, aquatic organisms had arrived at solutions to fluid balance and ionic regulation that depended on the presence of gills. It was logical, then, and also typical of early air breathers, that they keep the primary respiratory organs, the gills; in fact, they had a bimodal breathing, extracting oxygen directly from the atmosphere but retaining their gills for elimination of carbon dioxide and osmotic control with the aquatic medium. The piscine swim bladder and the tetrapod lung have been described as homologous structures and available evidence has suggested that the original function of the swim bladder was respiratory.

A bimodal gas exchange called for major rearrangements of the cardiovascular system to increase the effectiveness of the new gas exchange organ. The first important steps in that direction are exemplified in the extant lungfish. The most notable change is a separate routing of the blood from the aerial gas-exchange organ, the lung, directly back to the heart. This is the first phylogenetic manifestation of the pulmonary vein, which, in turn, is the first structural vascular adaptation to minimize the mixing of oxygenated and deoxygenated blood. Many fresh-water fish have developed elaborate structural adaptations for air breathing, but only in the lungfish is the vascular system modified to promote the efficiency of the air-breathing organ in oxygen transport to the tissues (Figures 7-38, 7-43).

While they retained their organs for aquatic respiration (gills), the air-breathing fish showed a gradual reduction of these as air breathing became more efficient and important (Figure 7-43).

The reduction in the distribution and surface area of gills on the primary branchial arches is correlated with corresponding changes in the aortic arches. The Australian lungfish is primarily a water breather and shows a typical piscine arrangement of the primary aortic arches, all of them diverging into branchial exchange vessels in the gills. In the African lungfish, air breathing has increased in importance and branchial arches 3 and 4 have become gill-less, with the result that aortic arches 3 and 4 represent direct shunts to the dorsal aorta.

In lungfish and all tetrapods, as well as in some primitive fish showing a respiratory function of the swim bladder, the air-breathing organs are supplied with blood from arteries branching from the most posterior epibranchial arteries, derived from the sixth aortic arch (Figure 7-43). (These arteries are called pulmonary arteries in lungfish and tetrapods.)

With the development of separate specialized vessels conveying blood to the lung and back from the lung to the heart, we see structural changes in the heart itself for the accommodation of two blood streams of different quality. Thus, the atrium of the typical fish has, in the lungfish, become partially divided by a septum. The sinus venosus connects the large systemic veins to the right portion of the partially divided atrium, and the pulmonary vein carrying oxygenated blood empties to the left part of the atrium. The partial atrial septum and the direction and dynamics of inflow prevent extensive mixing of the two types of blood in the atrium. Following atrial contraction, this separation is maintained in the ventricle, which also shows a partial septum extending from the apex of the heart (Figure 7-44). Ventricular contraction expels the blood in a laminar, streamlined flow pattern so that extensive mixing is avoided also in the anterior, undivided part of the ventricle. In the bulbus

Figure 7-43 Central circulation in the African lungfish, *Protopterus*. Note that gills are absent on two branchial arches. A pulmonary artery branching off from the most posterior epibranchial artery and a pulmonary vein returning directly to the heart are features among extant vertebrates first seen in the lungfish.

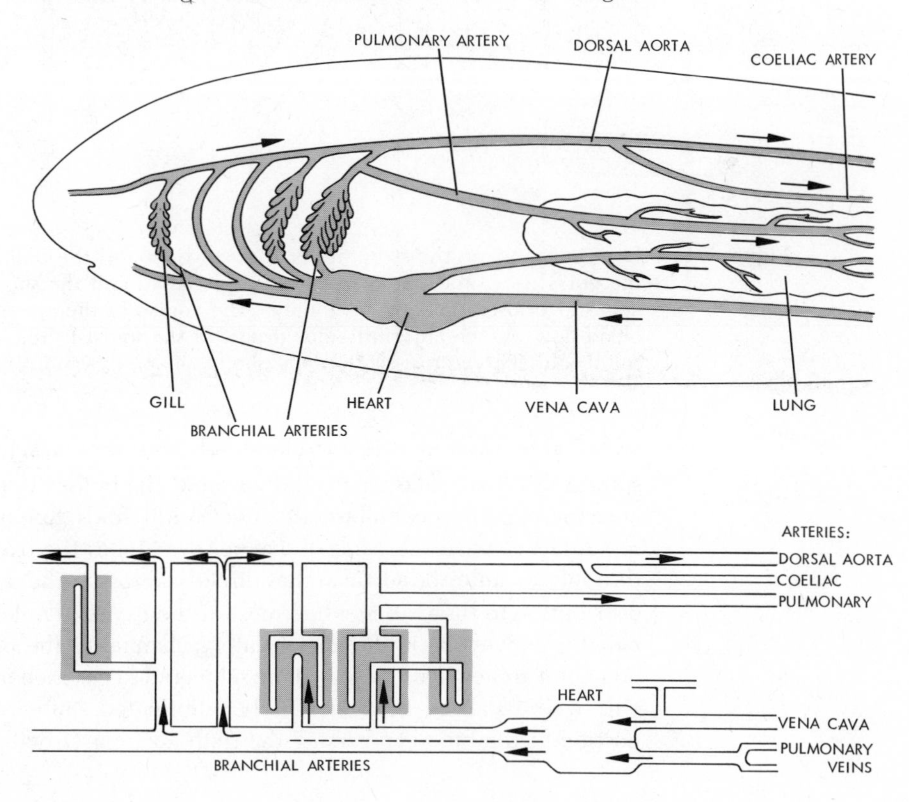

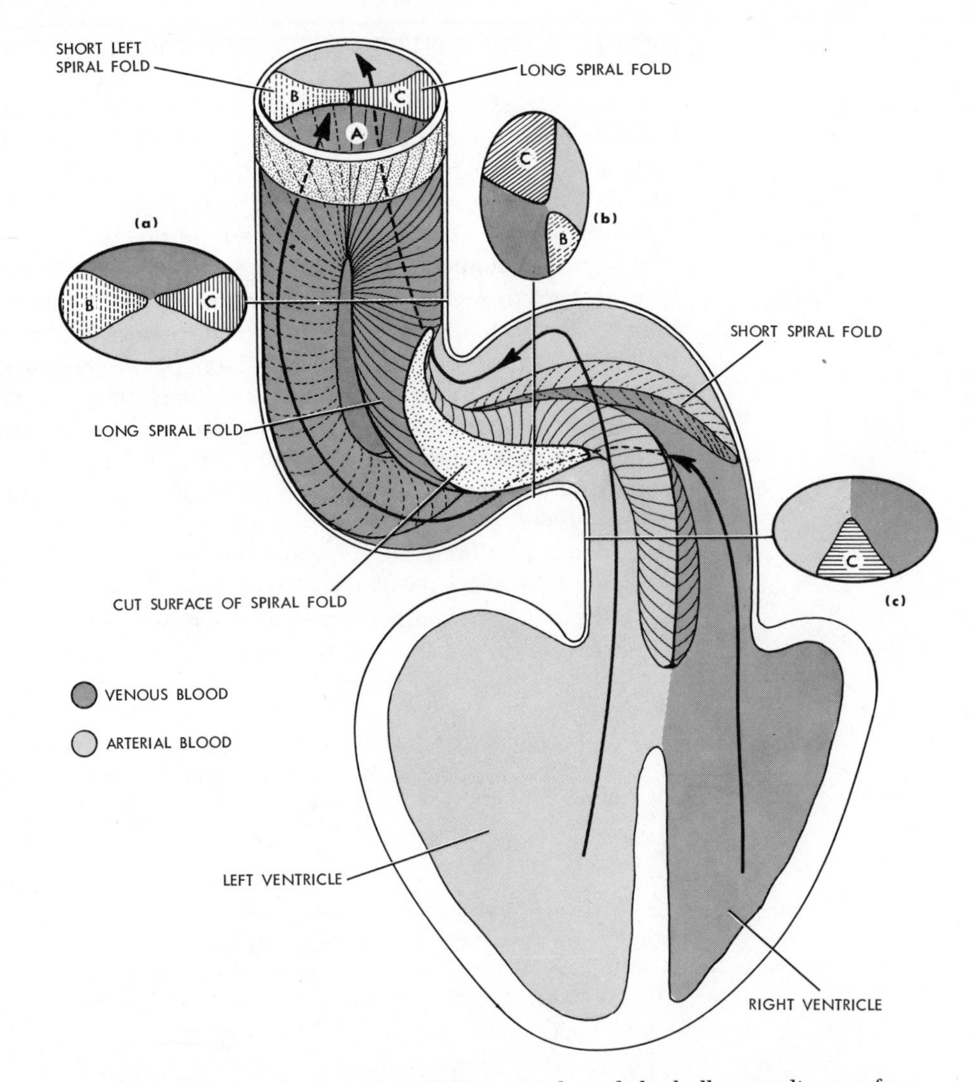

Figure 7-44 Diagram showing the structure of the ventricle and the bulbus cordis seen from the dorsal side, and the course of the spiral fold and the shorter left fold in the conus arteriosus. The arrows indicate the course of the arterial and the venous blood flow. A, the anteriorly joined part of the folds; B, the short left fold; C, the spiral fold. Three cross-sections are shown at (a), (b), and (c).

cordis, the separation of the two blood streams is largely maintained by the spiral folds that make a partial division of the bulbus in two channels. In the anterior end of the bulbus, the two spiral folds fuse to form completely separated channels. A ventral channel dispatches the oxygen-enriched blood through the anterior aortic arches directly across to the dorsal aorta for direct distribution to the tissues, while most of the oxygen-depleted systemic venous blood is conveyed via the dorsal bulbus channel to the posterior gill-bearing branchial arches. This blood will be subjected to gas exchange in the remaining gills where carbon dioxide will be eliminated. Subsequently, the greater portion of the blood traversing the sixth aortic arch will be diverted via the pulmonary arteries to the lungs where the major portion of the oxygen absorption occurs.

Circulation in the lungfish *Protopterus* and *Lepidosiren* shows a high degree of selective passage through the heart. Blood rich in oxygen and low in carbon dioxide is conveyed predominantly to the anterior systemic aortic arches with little recirculation of blood to the lung. In turn, the most oxygen-depleted blood is channeled to the posterior arches. It is, however, important to realize that this arrangement can change, depending on conditions in the external environment. These fish may experience aerated water in which a larger fraction, but not all, of their respiratory needs can be supplied by aquatic respiration. Conversely, they may experience very adverse conditions with oxygen-poor and carbon dioxide-rich water that compels them to rely more on aerial respiration. Finally, they may become air exposed during drought and, thus, be entirely at the mercy of their lungs for gas exchange.

Accordingly, the lungfish need the flexibility their vascular system offers in dispatching blood to the alternative organs for gas exchange. These animals are also first to show modifications in the typical piscine pattern of veins; the most notable change is the appearance of the large caval vein in replacement for the right cardinal vein in the more primitive fish (Figure 7-42).

Amphibian Circulation

The circulatory system of amphibians shows little advance over that in lungfish in regard to a complete separation of the systemic and respiratory vascular circuits in the heart. The differences reflect the diversity in modes of respiration seen in amphibians, such as external gills in neotenic salamanders, buccopharyngeal breathing, pulmonary breathing, and cutaneous breathing. This diversity also is reflected in the pattern of the aortic arches and their branches. In larval and neotenic urodeles (Figure 7-45), aortic arches 3, 4, and 5 usually branch into exchange vessels in the gills, whereas the sixth aortic arch functions as an afferent vessel to the lungs. However, a connection remains between the sixth aortic arch and the dorsal aorta (ductus arteriosus, ductus botalli) (Figure 7-45). In adult, gill-less urodeles, all aortic arches become direct, uninterrupted vessels. Arch 3 serves as parent vessel for the internal carotid system (Figure 7-45), and only in some species does this arch retain a connection with the dorsal aortic system through the ductus caroticus. Aortic arches 4 and 5 pass directly to the dorsal aorta. Arch 6 also retains a connection (ductus arteriosus), but this duct is functionally not very significant. The principal role of arch 6 is to serve as parent vessel for the pulmonary arteries. In adult anuran amphibians (Figure 7-45) further modifications include a loss of aortic arch 5 and the ductus aorticus. In addition to perfusing the lung, arch 6 gives rise to vessels perfusing the skin, which in many amphibians has a respiratory function, particularly in the elimination of carbon dioxide.

Turning to the venous backflow to the amphibian heart, the pulmonary vein of amphibians drains separately into the left atrial space. The systemic veins empty via the sinus venosus in the right atrial space. Blood returning from the skin is not routed to the left atrium but is mixed with systemic venous blood before entering the heart. The usefulness of the amphibian skin as a respiratory organ, thus, is limited to elevating the oxygen content of the systemic venous blood and has no direct part in providing oxygen-rich blood selectively to tissues in need of it. It seems as though skin is too nonspecialized and too

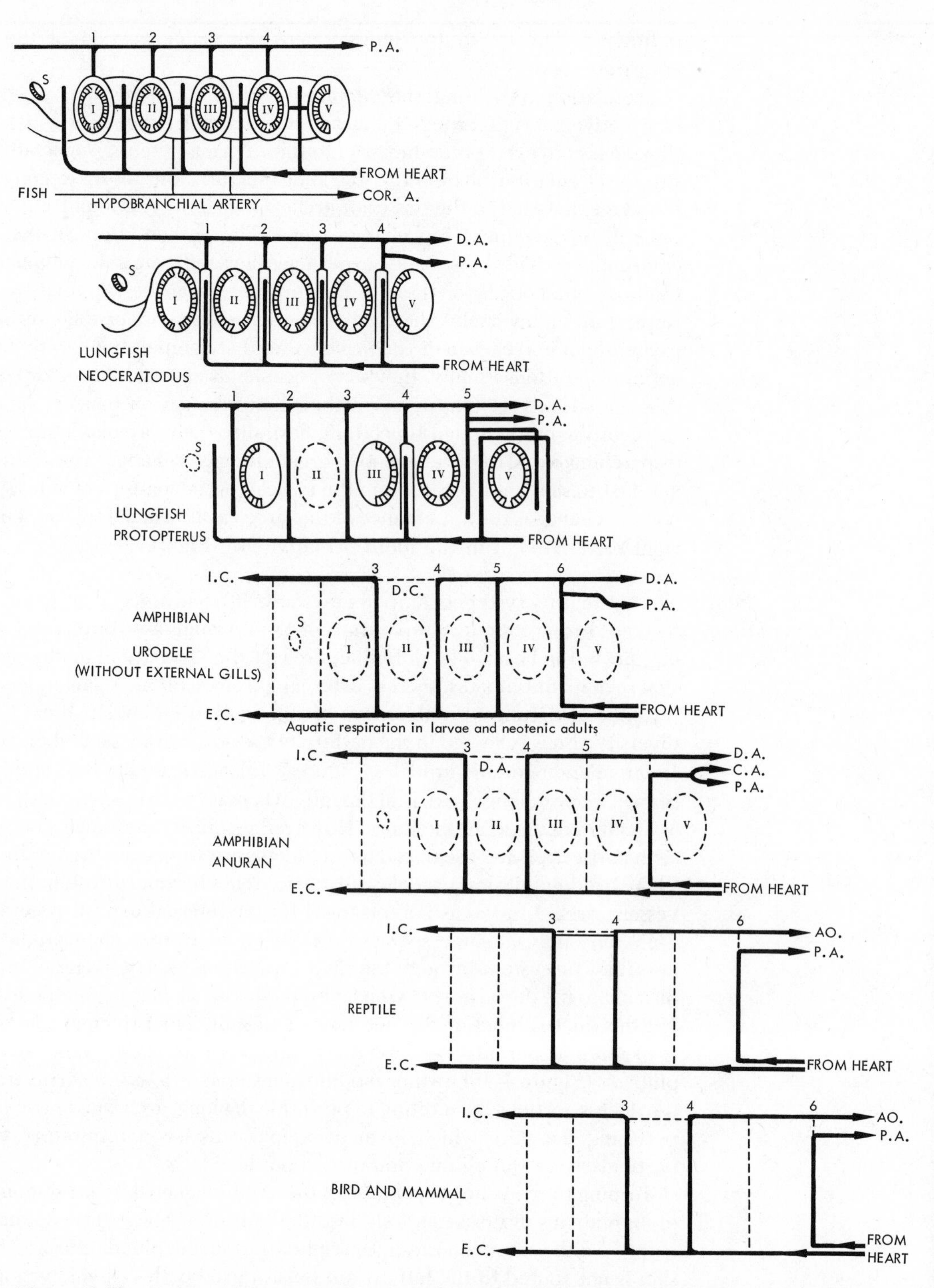

Figure 7-45 Diagrammatic representation of the transformation of the aortic arches in vertebrates. Aortic arches are numbered with arabic numerals, gill slits with roman numerals. The dashed lines indicate extinct conditions. A, artery; AO, aorta; CA, cutaneous artery; COR A, coronary artery; DA, dorsal aorta; DC, ductus caroticus; PA, pulmonary artery; S, spiracle.

involved in other functions to have been connected into a vascular circuit that joins the rest of the circulation only in the heart.

The heart itself shows a functional division of the atrial space into two compartments. The atrial septum in some salamanders is reported to be fenestrated, but little or no mixing of blood inside the atrium appears to result from this. Notably, the amphibian ventricle shows no ventricular septum. However, the densely trabeculated, spongelike myocardium tends to arrest temporarily the blood inside the numerous small crypts and spaces of the ventricle and, thus, minimize the mixing of the two types of atrial blood as they enter the ventricle. During ventricular contraction, the blood is ejected in laminar stream lines, which also helps in preventing mixing. The bulbus cordis in amphibians is shorter than in the lungfish, and the aortic arches have a separate origin closer to the base of the heart. As in the lungfish, the spiral fold guides the separate outflows of the two types of blood confined to the left (oxygen-rich) and right (oxygen-depleted) sides of the ventricle.

Various studies based on indicator techniques such as radiological (X ray) and dye-dilution experiments as well as analysis of blood gases in the various inflow and outflow vessels of the heart have documented a high degree of selective perfusion through the amphibian heart in spite of its only partial anatomical separation. The separation is, however, labile and depends on physiological factors such as blood pressure levels and vascular resistance in the various arterial trunks originating from the bulbus cordis. The selective outflow pattern tends to minimize recirculation to the lungs by channeling the oxygen-rich left atrial blood to the carotid and aortic arches, whereas the oxygen-poor blood reaching the heart from the systemic veins is selectively channeled to the sixth aortic arch where the pulmonary and skin circulations originate.

Often you will see the argument that amphibians really do not gain efficiency in oxygen transport from having a partially separated perfusion through the heart because the oxygen gained from gas exchange in the skin becomes mixed with venous blood in the larger systemic veins before reaching the heart. This statement is somewhat presumptuous, inasmuch as the importance of skin respiration for oxygen absorption, particularly in anurans, often has been overestimated. Also, the general elevation of venous oxygen content afforded by skin breathing will benefit the over-all oxygen transport. Figure 7-46 delineates the general course of the larger blood vessels in amphibians.

Reptilian Circulation

In reptiles, several major modifications appear both in the arrangement of the larger blood vessels and in the heart itself. The changes clearly indicate that reptiles have nearly completed the transition from water breathing to air breathing. The lungs are now the principal, if not the exclusive, organs for respiratory exchange with the environment.

It is convenient to discuss the Chelonia (turtles, tortoises) and Squamata (snakes and lizards) separately from the crocodilians because the latter have attained a complete ventricular septum and, hence, distinct right and left ventricles. In all reptiles the large outflow arteries take origin directly from the heart and not from the bulbus cordis, which has regressed completely in the adult. Embryological development, however, shows the bulbus to be absorbed and to have an important part in the formation of the large arterial trunks and the outflow tracts from the heart. With the elimination of the bulbus segment, a major advance was made toward a complete double circulation. Blood, thus, leaves the reptilian heart in discrete vessels, and the aortic arches are no longer connected directly in parallel with each other.

However, as is apparent from Figure 7-47, the ventricular chambers of reptiles, except the crocodilians, are in anatomical continuity. This fact may perhaps explain why it, up until recently, was thought that a high degree of mixing of oxygen-rich and oxygen-poor blood must take place inside the ventricles of reptiles. Recent experimental studies, however, including blood pressure measurements and oxygen analysis of the blood in the various heart

Figure 7-46 General outline of the major arteries (a) and veins (b) in anuran amphibians.

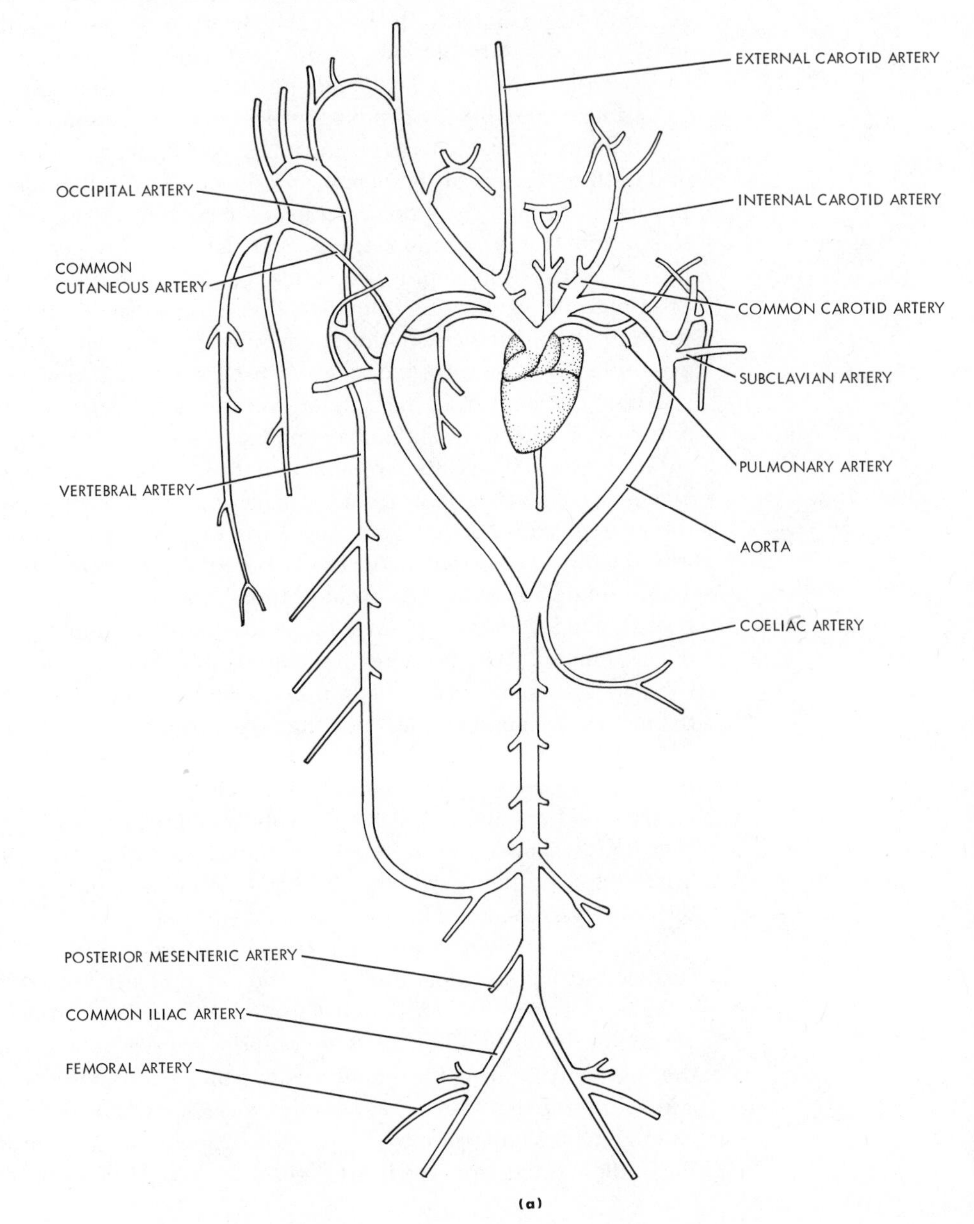

chambers and larger vessels, have revealed that such mixing is surprisingly small. The factors preventing mixing can be explained best by following the course of blood through the heart of a squamate reptile (Figures 7-47, 7-48). For details, see White [20]. Figure 7-47 shows schematically the ventricular compartments that are not drawn to size, in order to emphasize the pattern of blood flow. The left and right atrial spaces are completely separated and the ventricle is subdivided into three compartments. A prominent muscular ridge MR divides the ventral ventricular compartment in two spaces, the cavum pulmonale to the left and the cavum venosum to the right. The muscular ridge

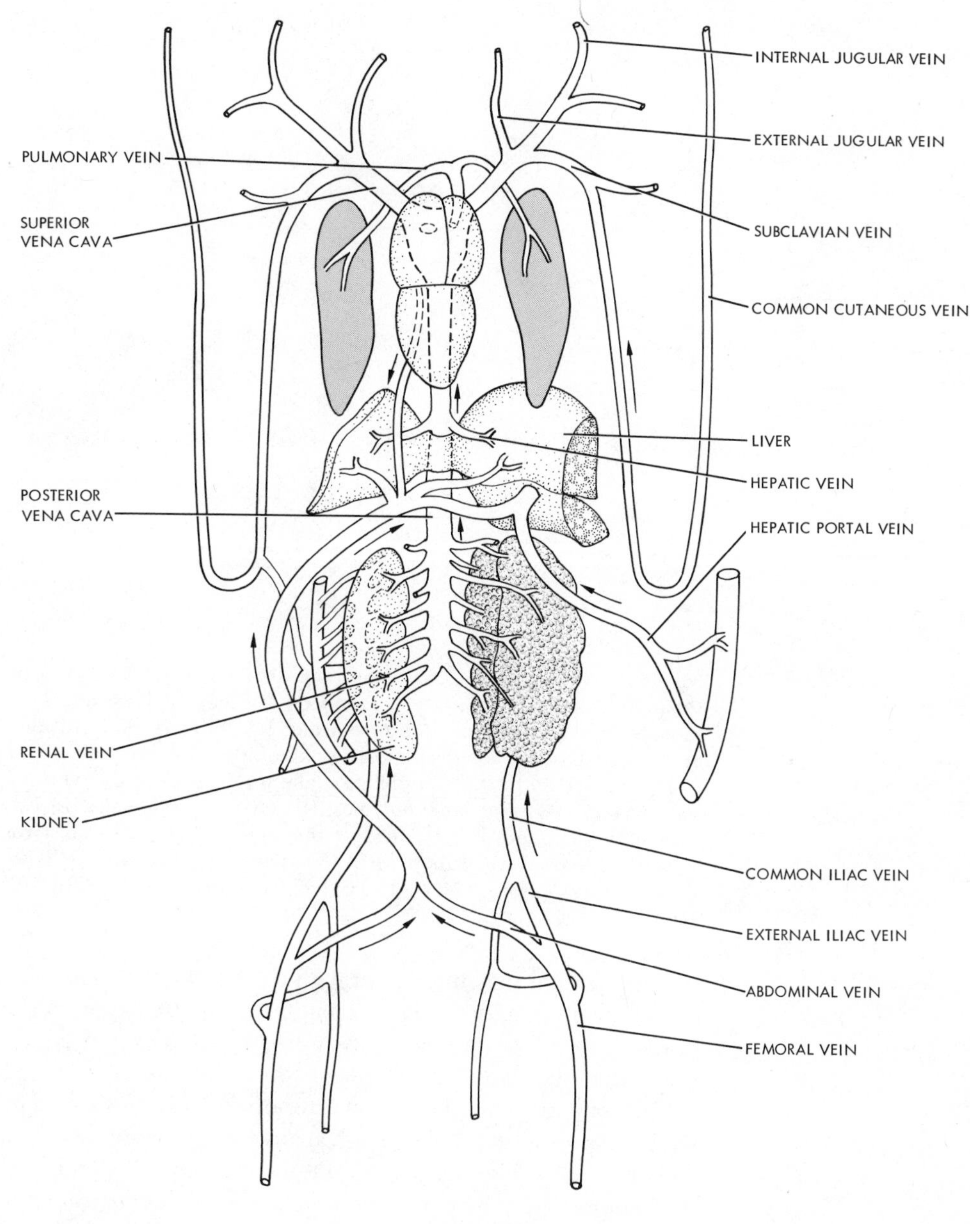

(b)

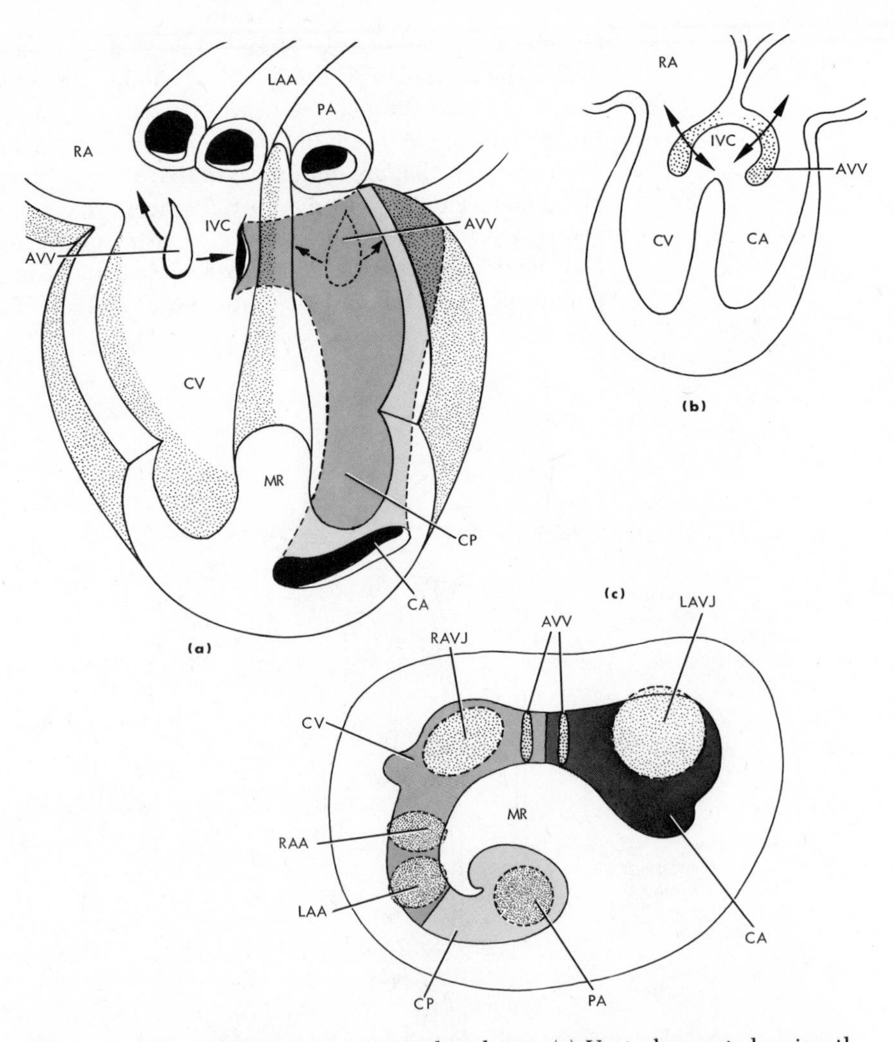

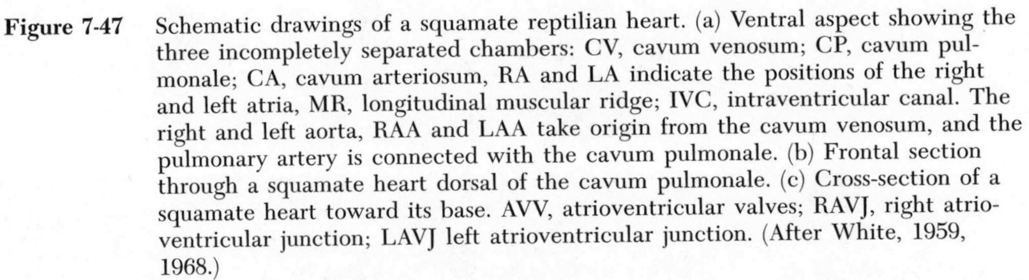

Figure 7-47 Schematic drawings of a squamate reptilian heart. (a) Ventral aspect showing the three incompletely separated chambers: CV, cavum venosum; CP, cavum pulmonale; CA, cavum arteriosum, RA and LA indicate the positions of the right and left atria, MR, longitudinal muscular ridge; IVC, intraventricular canal. The right and left aorta, RAA and LAA take origin from the cavum venosum, and the pulmonary artery is connected with the cavum pulmonale. (b) Frontal section through a squamate heart dorsal of the cavum pulmonale. (c) Cross-section of a squamate heart toward its base. AVV, atrioventricular valves; RAVJ, right atrioventricular junction; LAVJ left atrioventricular junction. (After White, 1959, 1968.)

provides no permanent contact with the ventral ventricular wall and the two spaces are, thus, only partially separated. In particular, in the apical part of the heart, there is free communication around the posterior edge of the muscular ridge when the ventricle is distended during filling. Dorsally to the left, we find a third compartment, the cavum arteriosum. This compartment communicates with the cavum venosum through an opening, the interventricular canal (IVC), and also has communication with the left atrium (LA), while the right atrium (RA) empties into the cavum venosum. The valves guarding the atrioventricular orifices take on special importance because when their large median flaps

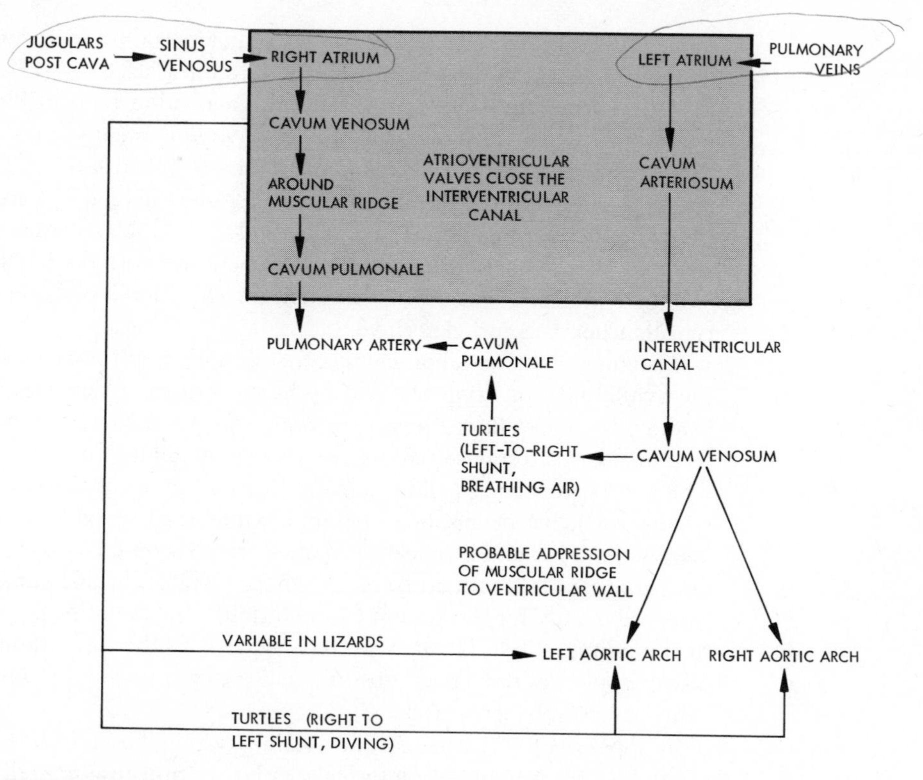

Figure 7-48 Schematic representation of the course of blood through the hearts of non-crocodilian reptiles. The events of ventricular diastole are within the solid rectangle, the systolic events are outside. (After White, 1968.)

are open during atrial contraction, they become adpressed against the apertures of the interventricular canal and, thus, temporarily close the canal. During this phase of cardiac contraction, left atrial blood will, hence, fill the cavum arteriosum and be confined there. Right atrial blood first will fill the cavum venosum and continue past the free edge of the muscular ridge into the cavum pulmonale. When the ventricle contracts, ejection of blood into the pulmonary circuit will precede that into the systemic arterial trunks because the resistance to outflow is much less in the pulmonary arteries. This, in turn, will favor the ongoing displacement of the venous blood from the cavum venosum to the cavum pulmonale from where it exits to the pulmonary arteries (Figure 7-49). It is also of great importance that when the ventricle shortens during contraction, the muscular ridge will make contact with the ventral ventricular wall and prevent a backflow from the cavum pulmonale to the cavum venosum. Meanwhile, as the ventricle continues contraction, the atrioventricular valves have swung back and closed the atrioventricular openings, creating a free passage through the interventricular canal. Blood previously confined to the cavum arteriosum will, hence, flow through the canal and leave the heart through the aortic orifices, filling both the right and the left aorta. This pattern of flow through the heart assures that oxygen-rich blood returning from the lungs to the left atrium is selectively dispatched to the systemic right and left aortae, whereas deoxygenated blood is predominantly channeled from the right atrium to the pulmonary arteries. However, some investigators have reported that under certain circumstances, such as diving in aquatic forms, blood from the left aorta has a lower oxygen content than that from the right aorta, but a higher value than in blood from the pulmonary artery. Physiological studies have documented that these changes occur in response to an increased resistance to blood flow in the pulmonary circuit during diving. This, in effect, will reduce the tendency for blood in the cavum pulmonale to be ejected selectively into the pulmonary circuit, and some admixture back through the cavum venosum into the left systemic aorta will take place. Figure 7-48 summarizes the blood pathway through the reptilian heart as just described.

It seems to be a common characteristic of all respiratory organs that when the availability of oxygen at the exchange surfaces decreases, the arterioles leading to the capillary exchange vessels will constrict and reduce blood flow. In essence, this response will insure that the matching process between blood flow and respiratory gas flow is maintained at an optimal level. For instance, when a turtle has been submerged for a while, the oxygen in the lung has been largely depleted and little can be gained from sending blood through the lung until another air breath is taken. The energy cost of cardiac pumping is, hence, reserved mainly for the systemic circulation, which continues to be important to the animal while blood is shifted away from the lungs. Such shunting of blood would not have been possible had the systemic and pulmonary circuits been completely separated.

In the crocodiles, the cardiovascular system has advanced still further (Figure 7-50). A complete interventricular septum results in distinct left and right ventricles. However, conditions still differ from those in birds and mammals because of some peculiar relationships of the great vessels to the ventricles. The right aortic arch, which gives rise to the entire circulation in the anterior direction, originates from the left ventricle, and the left aortic arch arises from the right ventricle together with the pulmonary artery (Figure 7-50). Another unique crocodilian feature is a communication, the foramen Panizzae, between the right and left aortae at the base of the heart where they originate from the ventricles. This condition may at first sight seem to make it inevitable that deoxygenated blood from the right ventricle will be ejected

Figure 7-49 (*opposite*) Radio-opaque contrast medium has been injected into the right jugular vein of the large lizard, *Varanus niloticus*. (a)-(d) show successive stages of the passage of the contrast medium through the heart.

(a) Shows contrast medium in the sinus venosus with some reflux into the posterior vena cava. The right atrium is contracting and filling the cavum venosum of the ventricle. The pulmonary artery is visible from the previous ventricular contraction.

(b) The ventricle is now contracting and the cavum pulmonale is clearly delineated. Contrast medium is expelled into the pulmonary arteries with no admixture to the systemic aortic arches. Note the distinct demarcation between the right atrium and the ventricle indicating closure of the atrioventricular valves during ventricular contraction.

(c) Ventricular contraction almost completed and there is still no admixture to the systemic arteries. The sino-atrial valve is now open and the right atrium is being filled anew.

(d) In this frame the pulmonary veins are visible alongside the pulmonary arteries. The contrast medium has completed circulation through the lungs and has returned to the left atrium from where it has been expelled via the cavum arteriosum into the systemic arches which now have become visible.

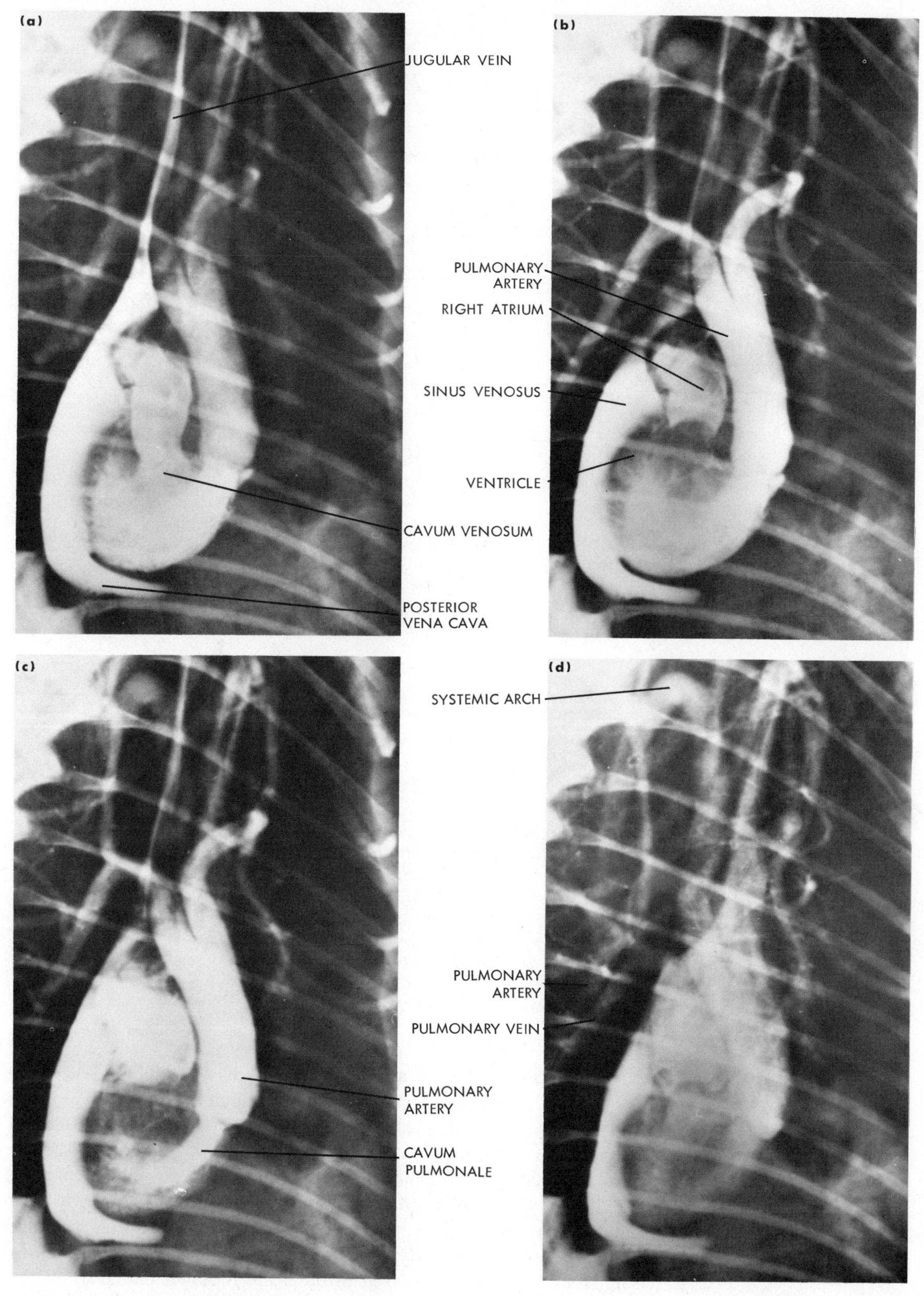
(a)
JUGULAR VEIN
CAVUM VENOSUM
POSTERIOR VENA CAVA
(b)
PULMONARY ARTERY
RIGHT ATRIUM
SINUS VENOSUS
VENTRICLE
(c)
PULMONARY ARTERY
CAVUM PULMONALE
(d)
SYSTEMIC ARCH
PULMONARY ARTERY
PULMONARY VEIN

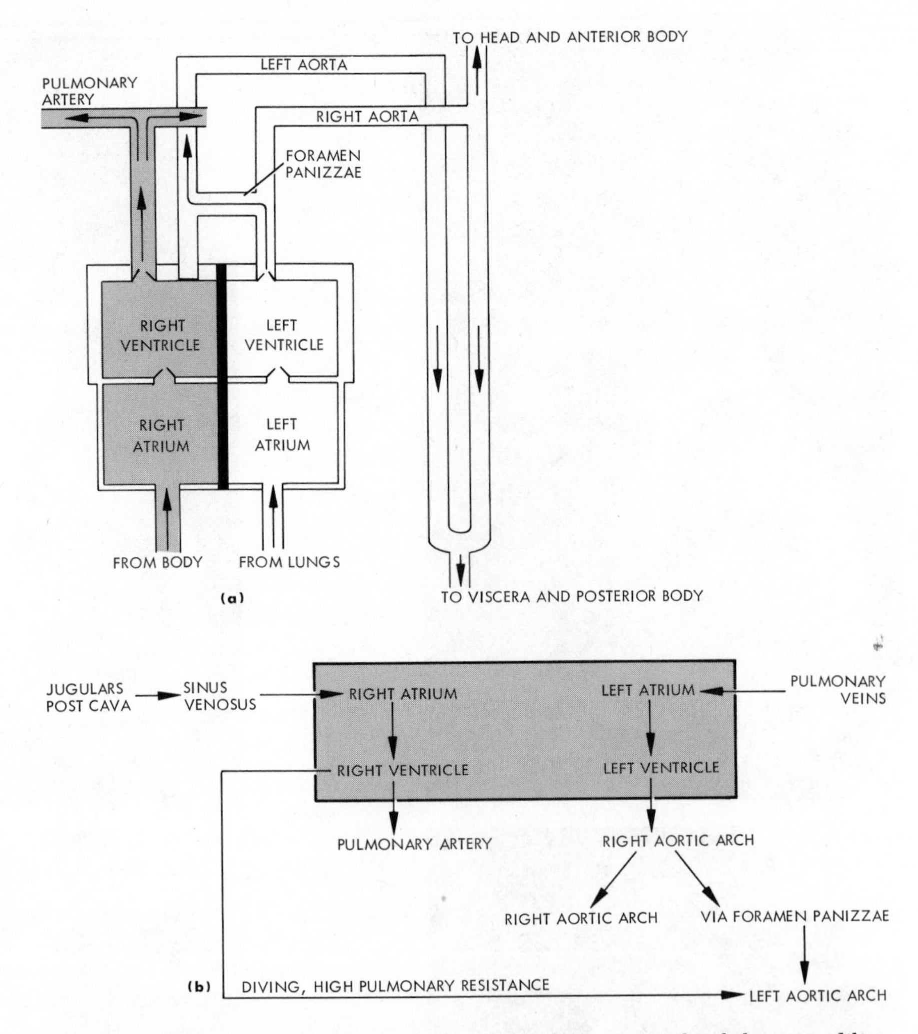

Figure 7-50 (a) Sketch depicting the cardiac chambers and greater vessels of the crocodilian heart. During normal frequent breathing the pressures developed by the left ventricle and prevailing in vessels emanating from the left ventricle exceed those on the right side. This will prevent the left aortic valves from opening and deoxygenated blood from being shunted from the right to the left side. (b) Schematic representation of the course of blood through the crocodilian heart. The events of ventricular diastole are within the solid rectangle, the systolic events are outside. (After White, 1968.)

through the left aorta and cause admixture with right aortic blood where the two aortas join in the descending dorsal aorta or through the foramen Panizzae. Again, physiological experiments have revealed a different story. First of all, it has been shown that the oxygen content of blood in the right and left aortae under normal conditions corresponds with that in the left atrium. In contrast, the much lower oxygen content in the pulmonary artery corresponds with that in the right atrium. This evidence leaves little doubt that both aortae are filled from the same ventricle, obviously the left. The left aortic arch must, therefore, receive its blood from the right arch through the foramen Panizzae.

But why does not blood from the right ventricle reach the left aorta? The answer can be found in the pressure relations inside the heart chambers and the great vessels. Already, with the appearance of a functional lung in the lungfish, the vascular resistance of the pulmonary vascular bed is lower than that in the systemic circuit. This tendency grows stronger as the degree of anatomical separation between the two circuits increases. With complete separation, as in mammals and birds, the two circuits convey the same flow rate of blood, but the driving pressure in the pulmonary circuit is about one tenth of that in the systemic circulation. The crocodilians represent a crucial case in this progressive development. When crocodiles are at the surface or on land and breathing is rhythmic and uninterrupted, the pressures prevailing in the right ventricle and pulmonary artery are lower than those in the left ventricle and right aorta. The higher pressure dveloped by the left ventricle and passed on to the right aorta will transmit through the foramen Panizzae to the left aorta. Because cardiac valves open and close passively in reponse to the pressures acting on each side of them, this can only mean that the valves between the right ventricle and the left aorta will not open (Figure 7-50). Functionally, the left aorta becomes connected to the left ventricle through the foramen Panizzae when crocodiles breath rhythmically.

However, the cardiovascular system in crocodiles, as in other vertebrates, responds to a regulatory mechanism seeking to adjust the flow of blood through the gas exchange surfaces so that deoxygenated blood is matched with gas in the lung to achieve optimal efficiency in gas exchange.

Crocodiles are habitual divers and experiments have demonstrated that during diving or when the crocodile is made to breath a gas low in oxygen concentration the blood oxygen content in the two aortas changes from similar values to a much lower value in the left than the right aorta. This finding is suggestive that blood in the left aorta represents a mixture of deoxygenated blood originating in the right ventricle and oxygenated blood ejected by the left ventricle.

Experiments also have shown that when the oxygen concentration in the lung air is reduced for example during diving, the resistance to blood flow of the pulmonary vascular circuit increases. This, in turn, will cause the right ventricle to contract more forcefully. As the pressure developed by the contracting right ventricle increases and becomes higher than the pressure in the left aorta, the valves separating the left aorta from the right ventricle will open and some of the deoxygenated blood will escape that way. Consequently, left ventricular blood may become a mixture of blood from the left and right ventricles. This may not be as detrimental as it seems because in this way less blood will be conveyed via the pulmonary arteries to the lungs at times when the oxygen concentration in the lung air is low and little oxygen can be absorbed.

We see in this response pattern another example of how a seemingly imperfect vascular arrangement that can cause mixing of blood between the systemic and pulmonary circuits may allow the animal a flexible perfusion pattern that makes possible a saving of oxygen and energy during the variable conditions of oxygen availability that often prevail during normal behavior of lower vertebrates.

Circulation in Birds and Mammals

Birds and mammals are the only vertebrates with completely separated pulmonary and systemic vascular circuits. Their four-chambered heart is a double pump connecting the two circuits in series. The two circuits differ markedly in their functional and structural characteristics and these differences are reflected also in the two pumps of the heart. The systemic circuit is a high-pressure circuit with arteriovenous pressure gradients many times steeper than in the pulmonary circuit. Yet, because the circuits are in functional continuity they must transport the same volume of blood per unit time. Figure 7-51 indicates how the structural design of the right and left ventricles reflects their functions as a volume pump and a pressure pump, respectively. The right ventricular compartment is enclosed as a narrow space between two large surfaces. The chamber, thus, has a large surface-to-volume ratio. The thickness of the wall, the myocardium, is moderate, and when the two curved sides of the chamber are pulled toward each other a large volume is displaced against the small resistance prevailing in the pulmonary arteries. The action of the ventricle resembles that of a bellows used to kindle fires. It is interesting that in mammals with a pathologically high resistance in their pulmonary circuit, the right ventricular wall gradually thickens and becomes similar to the left ventricle. The left ventricular cavity resembles a cylinder with a conical end invested by a heavy layer of muscle (Figure 7-51). This shape gives the chamber a much smaller ratio between its surface area and volume than that of the right ventricle. The left ventricle develops a high pressure during its contraction. This is consistent with the higher energy needed to overcome the high resistance of the systemic circulation. The normal left ventricle adapts

Figure 7-51 Components of ventricular contraction. (a) Blood is ejected from the right ventricle by shortening of the free wall. Compression of the right ventricular cavity may be supplemented by traction exerted on the free wall by left ventricular contraction (bellows action). (b) Left ventricular ejection is accomplished primarily by a reduction in the diameter of the chamber with some additional shortening of the longitudinal axis. (From R. F. Rushmer, *Cardiovascular Dynamics*, 2nd edition. Philadelphia: W. B. Saunders Co., 1961.)

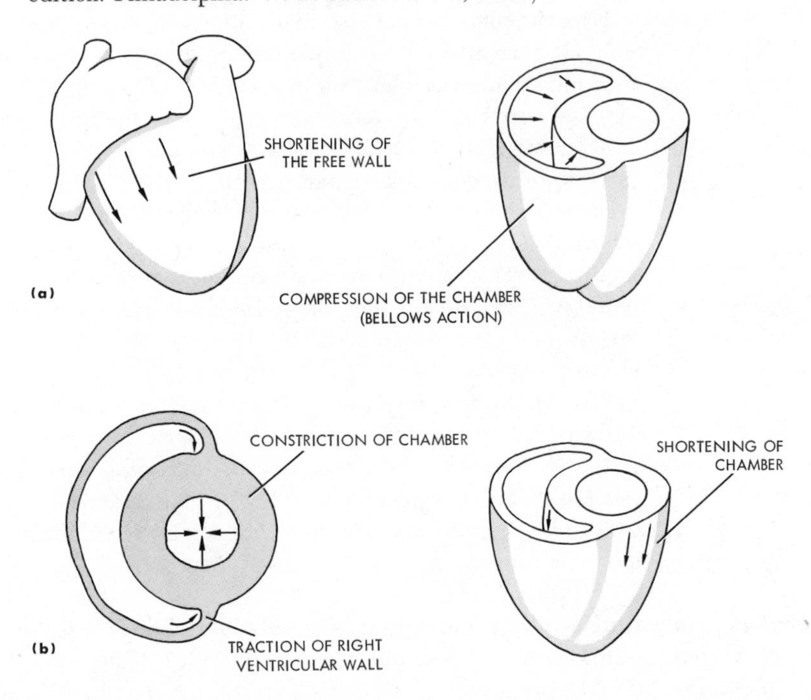

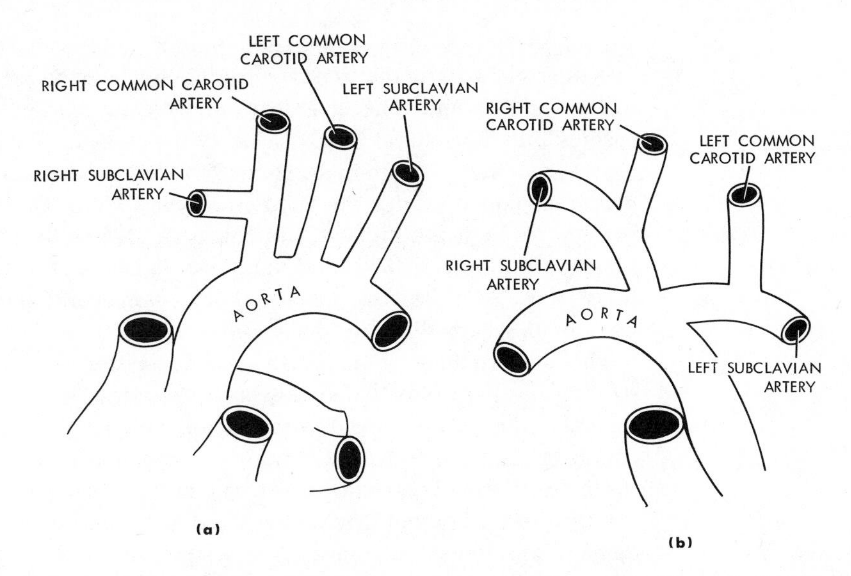

Figure 7-52 (a) Branching of major arteries from the mammalian aorta (human). (b) Branching of major arteries from the avian aorta (duck).

poorly to ejecting, changing large volumes, and if such loads are imposed on it under abnormal conditions of disease, the ventricle dilates and takes on a shape that gives it a larger ratio of surface area to volume.

In lower vertebrate hearts, which have only one ventricular chamber, we see a better potential for adapting to a volume load than in the left ventricle of mammals. This may be the reason why lower vertebrates can adjust to a higher cardiac output by increasing their stroke volume, whereas higher vertebrate hearts predominantly increase their output by increasing the heart rate.

The arrangement of the arterial outflow trunks in birds and mammals has clearly developed from conditions in lower tetrapods. As in reptiles, the ventral aorta and the cardiac bulbus segment are replaced by arterial trunks arising directly from the heart. The external and internal carotids have the same associations with aortic arches 2 and 3 as outlined for lower vertebrates (Figure 7-45). Aortic arch 4 persists on the right side in birds as the main systemic aorta originating from the left ventricle. The fourth aortic arch on the left participates in the formation of the left subclavian artery.

You will read often that the theory of the origin of birds from an extinct group of sauropsid reptiles is supported by the loss of the left aorta, which assumes that this vessel carried oxygen-poor blood from the right ventricle and, hence, was detrimental to efficient oxygen transport. In light of the recent findings about circulation in reptiles, this view is no longer tenable. The arterial system in mammals differs from that in birds by a persisting left aorta, rather than a right one (Figure 7-52). The sixth aortic arch constitutes the pulmonary arteries in both birds and mammals, which conveys blood from the right ventricle to the lungs. In embryological life, the left sixth arch retains the primary connection with the dorsal aorta, the ductus arteriosus; this allows bypassing the lungs. Soon after birth, the ductus arteriosus becomes progressively narrower and later occluded, persisting only as a nonfunctional ligament.

In the venous system in birds and mammals, the caval veins empty directly into the right atrium, the sinus venosus having been absorbed and incorporated in the right atrium. The venous channels show marked differences between species but, in general, birds show two precaval veins entering the heart against one in most mammals. Hence, the left subclavian and jugular veins of mammals connect with the right precaval system. Birds show some venous passage through the kidneys, but this is in all likelihood not functional as a renal portal system. Birds and mammals have functional hepatic portal systems; these systems, the hepatic portal veins and the hepatic veins, are homologues with those in lower vertebrates.

The performance of the cardiovascular system in birds and mammals is strikingly superior to that of lower vertebrates. More than anything is this related to their acute homeostatic control, which maintains very stable internal conditions in temperature, osmolarity, blood composition, and other homeostatic functions. Their superior freedom from environmental stress is also manifest in a high mobility on land and in air and water. All these functions demand an efficient cardiovascular system.

A well-trained man during hard exercise has, for instance, a cardiac output in excess of twenty-five liters per minute, which is more than five times the resting output from the heart. Similarly, in fast-flying birds, the heart rates can promptly change several hundred per cent in response to flight. Such sudden and profound adjustments in cardiovascular performance are entirely lacking in the lower vertebrates, whose activities are more at the mercy of environmental changes. The crucial role of a continued high blood supply to vital organs of mammals is perhaps best exemplified by the loss of consciousness and ensuing death if the blood flow to the central nervous system or to the heart muscle itself is interrupted for more than a few seconds.

References for Chapter 7

[1] COMROE, J. H., Jr. "The Lung." *Scientific American,* **217**(1966),

[2] DUNCKER, H. R. "Structure of Avian Lungs." *Respiratory Physiology,* **14**(1972), 44–63.

[3] GANS, C. "Respiration in Early Tetrapods. The Frog Is a Red Herring." *Evolution,* **24**(1970), 723–734.

[4] GOODRICH, E. S. "Vascular System and Heart." In *Studies on the Structure and Development of Vertebrates.* Vol. II. New York: Dover Publications, Inc., 1958.

[5] HUGHES, G. M. *Comparative Physiology of Vertebrate Respiration.* Cambridge: Harvard University Press, 1963.

[6] ———. *The Vertebrate Lung. Oxford Biology Readers.* Edited by J. J. Head, New York: Oxford University Press, Inc., 1973.

[7] ———, and M. MORGAN. "The Structure of Fish Gills in Relation to Their Respiratory Function." *Biological Review,* **48**(1973), 419–475.

[8] JOHANSEN, K. "Air Breathing Fishes." *Scientific American,* **219**(1968), 102–111.

[9] ———, and D. HANSON. "Functional Anatomy of the Hearts of Lungfishes and Amphibians." *American Zoologist* **8**(1968), 191–210.

[10] ———, and R. STRAHAN. "The Respiratory System of *Myxine glutinosa* L." In *The Biology of Myxine,* edited by A. Brodal, and R. Fänge. Oslo: Universitets-forlaget, 1963.

[11] RANDALL, J. D. "Functional Anatomy of the Fish Heart." *American Zoologist,* **8**(1968), 179–189.
[12] ROBB, J. S. *Comparative Basic Cardiology.* New York: Grune & Stratton, Inc., 1965.
[13] RUSHMER, R. F. *Cardiovascular Dynamics.* Philadelphia: W. B. Saunders Company, 1961.
[14] SATCHELL, G. H. *Circulation in Fishes.* Cambridge Monographs in Experimental Biology, no. 18, New York: Cambridge University Press, Inc., 1971.
[15] SCHMIDT-NIELSEN, K. "How Birds Breathe." *Scientific American,* **225**(1971), 72–79.
[16] ———. *Animal Physiology: Adaptation and Environment.* New York: Cambridge University Press, 1975.
[17] STEEN, J. B. *The Comparative Physiology of Respiratory Mechanics.* New York: Academic Press, Inc., 1971.
[18] TENNEY, S. M., and J. B. TENNEY. "Quantitative Morphology of Cold-Blooded Lungs: Amphibia and Reptilia." *Respiration Physiology,* **9**(1970), 197–215.
[19] WEIBEL, E. R. "Morphometric Estimation of Pulmonary Diffusion Capacity. V. Comparative Morphometry of Alveolar Lungs." *Respiration Physiology,* **14**(1972), 26–43.
[20] WHITE, F. N. "Functional Anatomy of the Heart of Reptiles." *American Zoologist* **8**(1968), 211–219.
[21] WIGGERS, C. J. "The Heart." *Scientific American,* **62**(1957), 3–12.
[22] ZWEIFACH, B. W. "The Micro-Circulation of the Blood." *Scientific American,* **64**(1959), 2–8.

8

Excretion and Osmoregulation

Introduction Vertebrate cells carry out a vast range of activities in precisely regulated patterns. In view of the complexity of their functions, perhaps it is not surprising that these cells are often exceedingly sensitive to slight variations in their chemical organization and, reflecting this, to the composition of their immediate surroundings. Isolated from the vertebrate body, nerve, muscle, or gland tissues can carry out at least part of their normal activity with varying degrees of success. Thus, even under these circumstances nerve cells can conduct nerve impulses, muscles can contract, and glands can secrete characteristic fluids; but these lingering functions remain only while the tissues are maintained in solutions whose composition is controlled with respect to ions, nutritive supplies, oxygen, carbon dioxide, and other metabolic products, pH, and osmotic pressure. There is evidence for the existence of intrinsic, self-regulating mechanisms operating to stabilize the chemical organization of individual cells because slight variations in their immediate environments can occur while the tissues maintain nearly constant composition. On the other hand, this intrinsic regulatory ability is quite limited even in the case of the sturdiest of vertebrate cells. Their capacity to survive depends on support by extracellular fluids that are, in turn, subject to continuing regulation (Figure 8-1). Thus, in the normal active vertebrate, the body fluids often show relatively little variation even under the drastically varying conditions of the external world. This is the result of the effective operation of the homeostatic mechanisms of the organ systems of the intact animal. The immediate environment of the cells—tissue (or interstitial) fluid—is maintained in rapid exchange with the mass of body fluids as a whole by the circulation of blood and lymph through the vascular systems of the body. The composition of blood

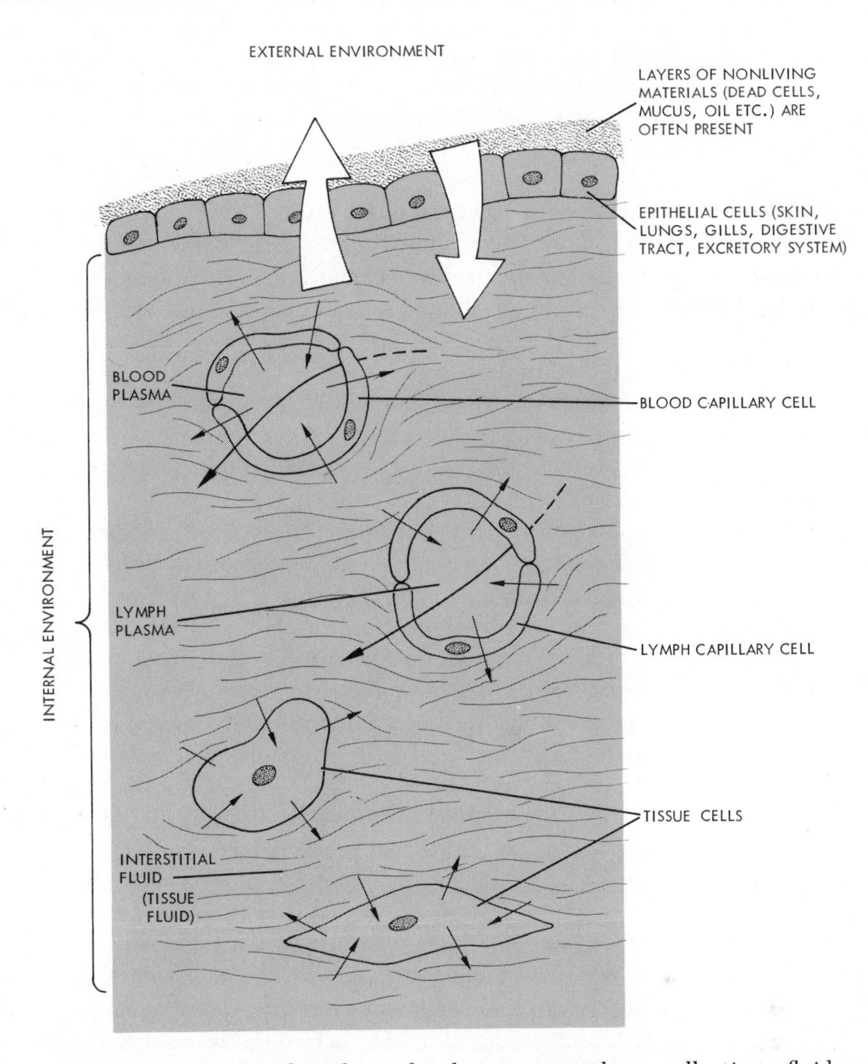

Figure 8-1 Diagram representing the relationship between vertebrate cells, tissue fluids, and the external environment. Exchange of chemical components between the organism and the external environment takes place at specialized epithelial layers of the skin, and in the respiratory, digestive, and excretory systems. Tissue fluids in every part of the body are maintained in physiological exchange with fluids in the body as a whole by circulation of the blood (all vertebrates) and lymph (bony fish and all tetrapods). All of the fluids of the body outside the boundaries of cells are grouped under the common name *extracellular fluids,* and represent the immediate *internal environment* of the cells. The extracellular fluids include not only the three types shown in the diagram (blood plasma, lymph plasma, interstitial fluid), but also intraocular fluids (aqueous and vitreous humors), joint fluids, and fluids in the pericardial cavity and major body cavities (peritoneal, pleural fluids).

and lymph, in turn, is maintained by controlled exchange with the external environment. Into the body come oxygen and other gases, water, nutrients, and salts, chiefly through the respiratory and gastrointestinal organs (also, in amphibians in particular, through the skin). Outward from the body go carbon dioxide (chiefly through the respiratory system and, in amphibians, through the skin), substances that have entered the body in excessive amounts, and waste products of metabolism. The chief route for ridding the body of these excessive, or waste, materials is through organs of excretion and osmoregulation.

Vertebrate Mechanisms for Excretion and Osmoregulation

Among the structures of the excretory and osmoregulatory systems, the *kidneys* are named first because they are so versatile and effective and occur throughout the range of vertebrate animals. Typically, the kidneys form urine containing a relatively high concentration of nitrogenous wastes (chiefly ammonia, urea, and uric acid) and various other excessive, or waste, solutes (organic acids, phosphate and sulfate ions, and so on). Their function in ridding the body of such substances is supplemented in fish by the gills, for here ammonia can diffuse directly into the surrounding water of the environment. To a rather slight extent, too, the liver plays a part in excretion. This point has been touched on already in Chapter 6. Another major regulatory function of the kidneys is their important part in controlling the osmotic pressure of the body fluids. In most vertebrates the urine, as it leaves the kidney, is hypotonic to the blood plasma. Thus, it can operate effectively to remove from the body an excess of water as compared with solutes. Such regulation toward increasing the osmotic pressure of the body fluids is of great value to animals living in fresh water or in any situation in which the surroundings are hypotonic to body fluids. Indeed, many biologists believe that this function of vertebrate kidneys gives strong support to the idea that vertebrates first evolved in fresh water. As yet we do not know enough to decide between this hypothesis and the concept that the sea was the first environment of primitive vertebrates. In any case, in the course of their evolution, vertebrates have radiated into the full range of habitats available on earth, from fresh water through marine and terrestrial regions, including terrestrial deserts. Typically, their kidneys alone are not able to cope with the full impact of osmotic stress imposed by environmental conditions (deficiencies of water or solute, or excess of water or solute in the environment), and a variety of other osmoregulatory processes have a part in their survival. These will be considered briefly, before the more detailed account of the structure and function of vertebrate kidneys.

Fish show many structural-functional adaptations to the environmental osmotic stress that supplement kidney function. These include the following: (1) tolerance by the tissues of high osmotic pressure of body fluids, maintained slightly above the level of sea water by retention of salts (marine cyclostomes—hagfish) or urea (cyclostomes—lampreys, elasmobranchs, holocephalan and coelacanth fish); (2) transfer of ions by special cells in the buccal and pharyngeal gill regions *out* of the body (marine bony fish) or *into* the body (fresh-water fish); (3) formation of a strongly hypertonic secretion by gland cells, such as those of the *rectal gland* associated with the posterior gastrointestinal tract (elasmobranchs; see Figure 8-2); and (4) structural adaptations of the skin, including the presence of mucus-secreting gland cells. These adaptations render the skin moderately impermeable (cyclostomes) or highly impermeable (gnathostome fish) to both water and solutes. In the cyclostome hagfish, mucus secretion is especially copious and tenacious and so rich in salts that some biologists believe that it may be a route for salt excretion (and, thus, of osmoregulation).

The great majority of amphibians live in environments where the fluids with which they come in contact are hypotonic to their own body fluids. Sodium ions (with chloride ions) are actively taken up by skin cells, and water moves readily in either direction through the relatively permeable skin. Adjustment of plasma composition is carried out effectively by kidneys because the hypotonic urine they form meets the major need for regulation (excretion of the net excess of water). Amphibians of a few species (*Rana cancrivora, Bufo viridis*) have adapted to life in salt water. These animals retain salt or urea and, thus, appear to have adjusted to the high osmotic pressure of the environment in a manner reminiscent of the cyclostomes and elasmobranchs. Amphibians living in dry terrestrial conditions (some frogs and toads) survive by systematic collection of water from external sources and by rigorous restriction of water loss from the body. They have no special ways of excreting solutes in excess of water.

In their highly successful adaptations to life on dry land and in the oceans, reptiles, birds, and mammals show a wide variety of mechanisms for osmotic regulation.

1. Reptiles and birds conserve water by excreting nitrogen waste in a nearly solid (nonosmotic) form, such as uric acid, and by absorbing water from the urinary bladder, lower digestive tract, and/or cloaca. The structure of these organs will be discussed later in this chapter.
2. Reptiles and birds have special salt-secreting glands, located in the head, that form fluids of osmotic pressure well above that of blood plama and, thus, remove salt from the body in excess of water (Figure 8-3). These glands have been studied most in marine reptiles and birds, where they are most conspicuous.
3. In all these forms the skin structure is relatively impermeable so that, in general, neither water nor salts diffuse in significant amounts through the skin. In mammals, however, another type of skin adaptation mediates temperature regulation but also affects osmotic regulation. This adaptation is the development of sweat glands that can actively secrete water and salts (chiefly sodium and chloride ions) when the skin or body temperature rises. In some cases, sweat secretion represents a major threat to salt and water balance and must be compensated by other osmoregulatory mechanisms if the animal is to survive.
4. Birds and mammals have a specialized kidney structure that makes possible the formation of urine hypertonic to body fluids. This can be explained best in connection with the following more detailed description of vertebrate kidneys.

An understanding of the vertebrate kidney is furthered by comparisons among different forms, taking into account the variety of operations carried out by these organs in the range of vertebrate animals. Looking backward, as it were, in an evolutionary sense, the excretory structures of tunicates and cephalochordates seem to have little in common with vertebrate kidneys (Figures 2-5, 2-9, and 2-10) and so will not be discussed further here. On the other hand, kidneys throughout the entire range of vertebrate classes present striking common features: patterns of development and basic structures and functions. At the same time, of course, differences exist among the vertebrates, and often these are correlated with adaptive processes that enable the animal to exist in its particular environment.

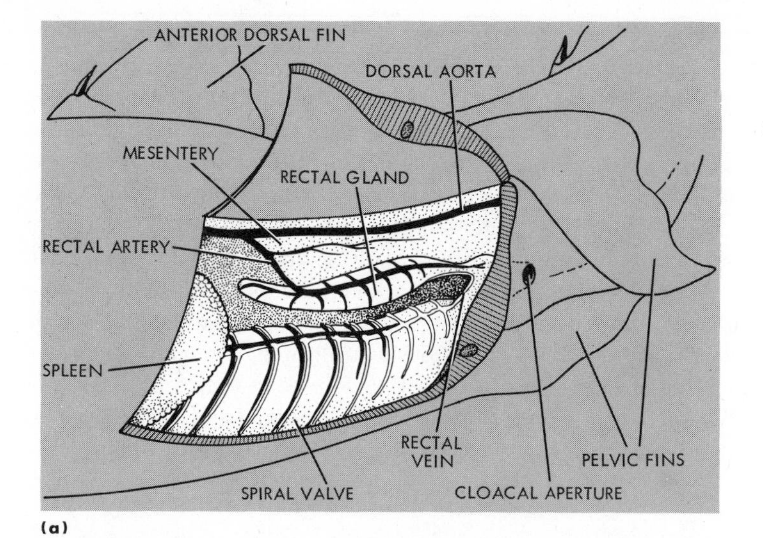
ANTERIOR DORSAL FIN
DORSAL AORTA
MESENTERY
RECTAL GLAND
RECTAL ARTERY
SPLEEN
RECTAL VEIN
SPIRAL VALVE
CLOACAL APERTURE
PELVIC FINS
(a)

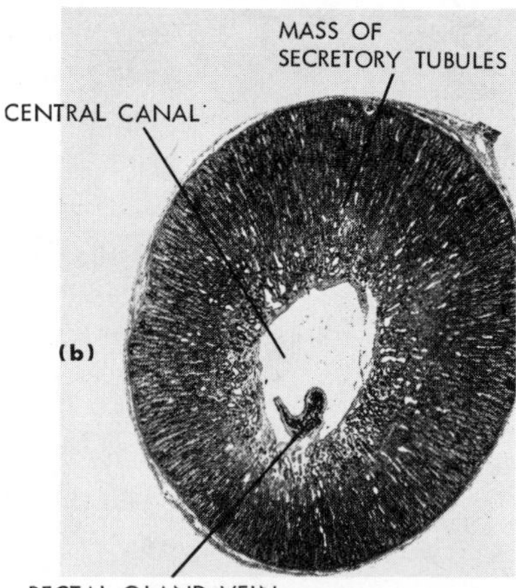
MASS OF SECRETORY TUBULES
CENTRAL CANAL
(b)
RECTAL GLAND VEIN

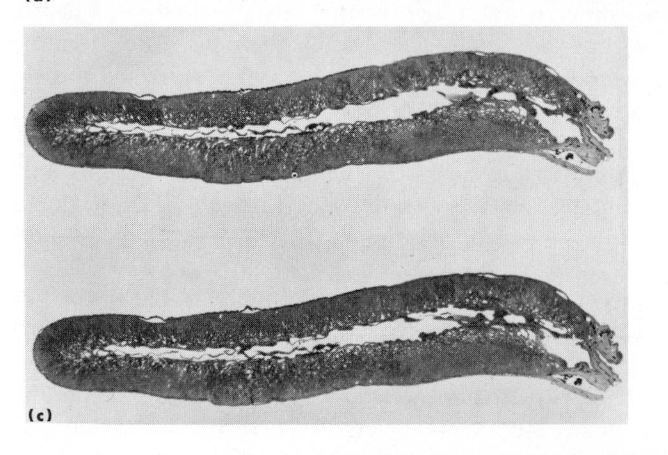
(c)

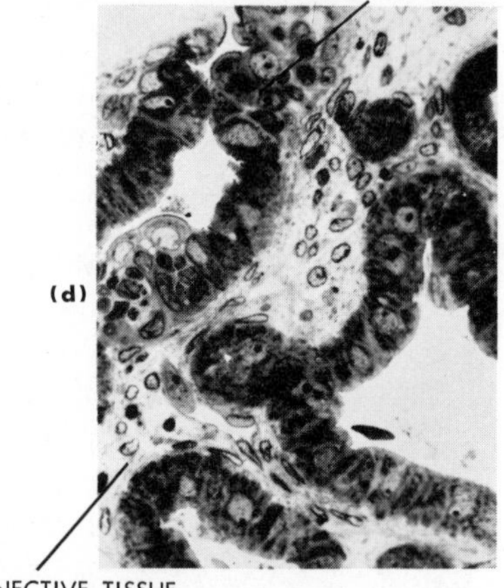
SECRETORY CELL
(d)
CONNECTIVE TISSUE

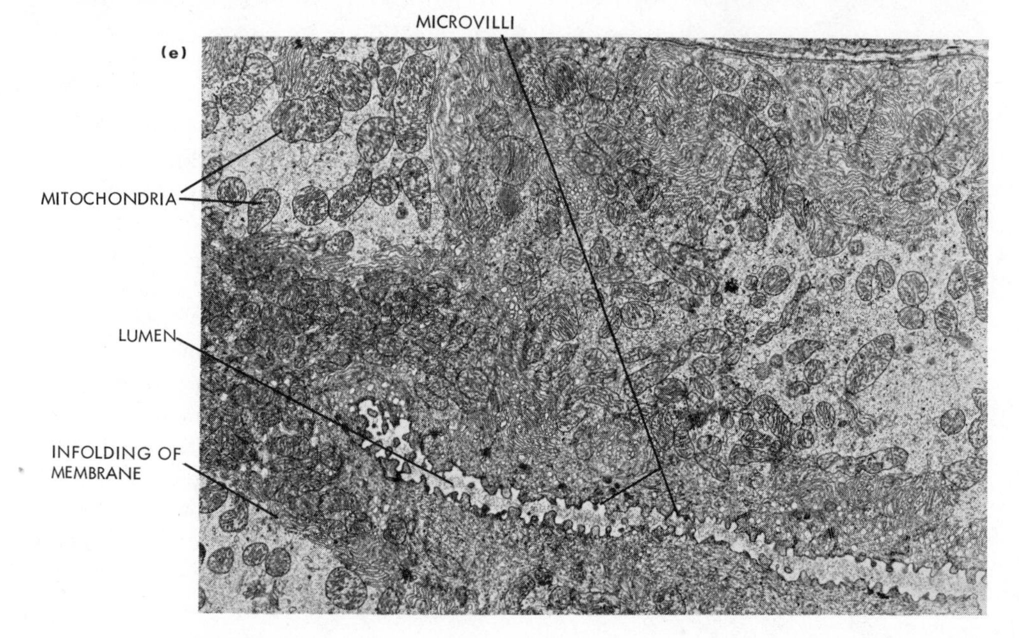
MICROVILLI
(e)
MITOCHONDRIA
LUMEN
INFOLDING OF MEMBRANE

Figure 8-2 (*opposite*). In elasmobranchs, specialized rectal gland tissue secretes a fluid of high sodium chloride concentration into the posterior digestive tract. This function was first discovered by J. W. Burger and W. N. Hess, in 1959. Subsequently, the structure of the gland has been revealed by careful studies of R. E. Bulger and others. The location of the rectal gland of the spiny dogfish, *Squalus acanthias,* is shown in (a). The gland drains into the intestine through a duct, which carries secretion from the gland's central canal. The secretion is formed in a dense mass of branching secretory tubules emptying into the central canal. There is a rich blood supply, with arteries bringing blood in from the external surface of the gland, and drainage through veins leading into the rectal vein lying in the central canal. (b) and (c) are low-power micrographs of the gland in longitudinal- and cross-section. In (d) the secretory tubules are seen at high-power magnification (light microscope). Each tubule has a wall consisting of a single layer of cells. Connective tissue, with blood vessels, surrounds the tubules. At yet higher magnification, in the electron micrograph in (e) the cells of the rectal gland may be seen to be filled with mitochondria and numerous membranes. The basal cell membranes are folded deeply into the cytoplasm to form a complex structure with extensive surface area. There are microvilli at the apical border where the cells open into the lumen of each tubule. These anatomical features of the cells, and of the gland as a whole, are correlated with the energy-demanding process of forming a secretion of high salt concentration. (All figures were kindly supplied by Dr. Ruth E. Bulger.)

Patterns of Kidney Development

In the development of vertebrate animals, mesodermal tissue located on each side of the body gives rise to renal organs. Typically this tissue (variously called nephrogenic tissue, nephrotome) lies between the epimere (more dorsally and centrally located) and the lateral plate (more ventrally located) mesoderm. In embryos, the most anterior region of the nephrogenic tissue differentiates into one or a few tubules, the group of tubules on each side making up a pronephros. From each pronephros a cellular duct (variously called archinephric duct, pronephric duct) arises and grows backward to reach and open at the posterior end of the body. During the animal's further development, the functions of these primitive kidneys are gradually supplemented (some cyclostomes, and a few bony fish) or replaced altogether (all other vertebrates) by kidneys developed from more posterior regions of the nephrogenic tissue. Thus, in fish and amphibians there are two major stages of renal organization: the pronephros stage and the stage of functioning of the definitive kidney that operates throughout adult life (*holonephros* or *opisthonephros*) (see Figure 8-4 and Table 8-1). Reptiles, birds, and mammals show a somewhat more complex pattern of kidney development, for here three stages of renal organization are recognizable: (1) anterior, nonfunctional pronephroi arise first, and then (2) functional *mesonephroi* form more posteriorly in the nephrotome. Each mesonephros is finally superseded in function by a (3) *metanephros,* which is the last to form and forms most posteriorly from nephrogenic tissue. Even in the lower vertebrates in which they are transiently functional, the pronephroi generally degenerate sooner or later after the adult kidneys have become established. Similarly, in reptiles, birds, and mammals, excretory functions of the second-stage kidneys (mesonephroi) are taken over by the later developed metanephric kidneys. Typically, vertebrates have effective kidney function from a very early stage in development, and the "backward" shift of functional tissue allows the uninterrupted operation of the urgently important regulatory functions of the kidneys throughout all phases of the animal's active life. One might guess that the changing structure of the renal organs is correlated with changing demands for regulation, paralleling a progressively altered kidney function. So far, however, we know too little about the operation of embryonic kidneys either to confirm or disprove this hypothesis.

Just as consistent patterns of kidney structure in relation to development are observed, so also is there a striking consistency of structure correlated with function in vertebrate renal organs as a whole. They carry out their unique activities chiefly by processing blood plasma, removing from it excess or injurious substances (water and solutes), and starting these on the way to discard from the body as the final excretory product, urine. It should be noted,

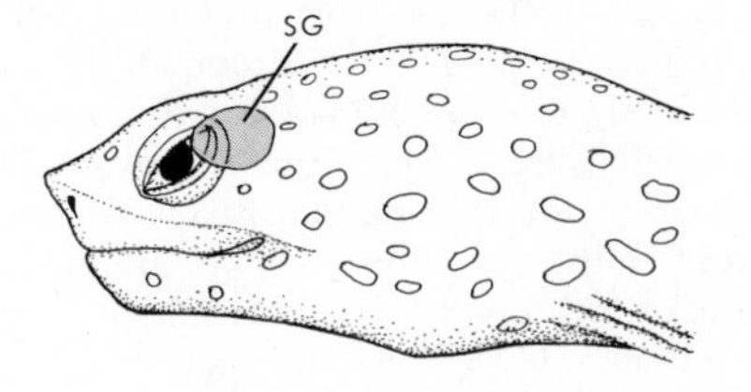

(a)

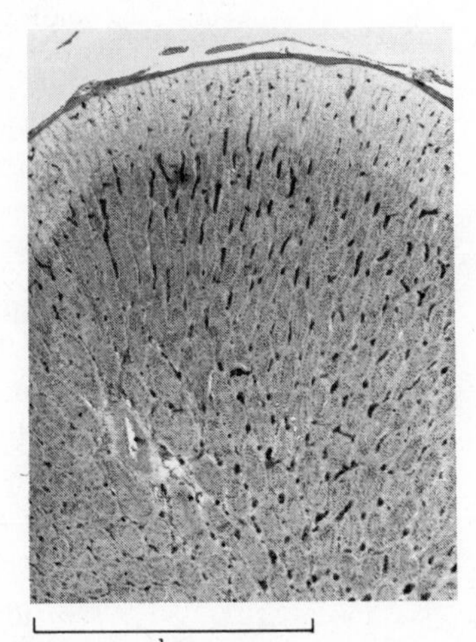

(c)

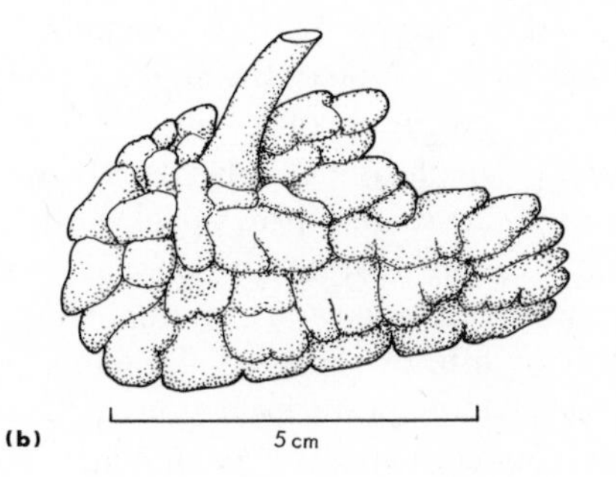

(b)

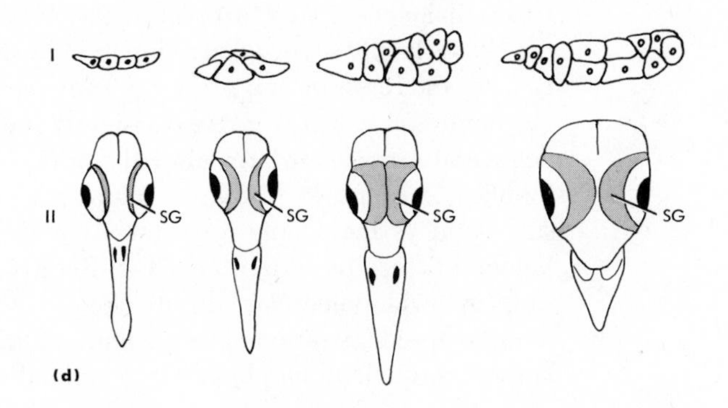

(d)

Table 8-1 A Summary of Vertebrate Kidney Types. Vertebrate kidneys differ not only in structure and function, but also in their mode of origin from mesodermal nephrogenic tissue. Types that are considered to differ significantly in ontogeny and in evolutionary history are listed in this table, which can be reviewed in connection with Figure 8-6. To simplify the description, each kidney type is listed as an individual, although, of course, almost all vertebrates have paired kidneys.

Kidney Type	*Brief Description*
Pronephros	Embryonic kidney. It developed from the most anterior region of the nephrogenic tissue, and often consists of only a few tubules segmentally arranged. This is the functional kidney of embryos of many fish and amphibians. A pronephros arises and remains transiently in early development of amniotes but apparently never functions as an excretory organ.
Holonephros	Kidney of the larvae of cyclostomes, some amphibians. The entire mass of nephrogenic tissue gives rise to this kidney, which is usually of simple form with a single tubule in each segment.
Opisthonephros	Kidney of many adult fish and amphibians. It developed from the main mass of nephrogenic tissues exclusive of the part that gave rise to the pronephros. Except in cyclostome hagfish, in which the tubules of the opisthonephroi are paired and segmentally arranged, there are generally many tubules developed from each nephrogenic segment.
Mesonephros	Kidney of embryonic reptiles, birds, mammals. It developed from a median region of nephrogenic tissue. There are great numbers of tubules without evidence of segmental arrangement.
Metanephros	Kidney of adult reptiles, birds, mammals. It developed from the most posterior region of the nephrogenic tissue with great numbers of tubules and no evidence of segmental arrangement.

Figure 8-3 (*opposite*). Special salt-secreting glands (labelled SG in the drawings) are found in the region of the orbits of many reptiles [(a), the head of a turtle] and birds (d) adapted to life in or near salt water. The structure of these glands is richly tubular. (c) A photomicrograph of a salt gland of the turtle, *Caretta caretta*, shows many tubules cut in a variety of planes. Great numbers of these tubules may be built into rather conspicuous lobulated glands [(b), *Careta caretta*]. Ducts lead from the glands to orbital or nasal openings, and thence the secretion flows away from the body. In the late 1950's K. Schmidt-Nielsen and R. Fänge and their associates discovered the nature and important regulatory role of this secretion. It contains sodium and chloride ions at concentrations that are high relative to the plasma levels, and is formed in response to rising osmotic pressure of the body fluids. In a study of salt glands from twenty-one species of shore birds of the order Charadriiformes, H. Staaland found that the relative sizes and rates of secretion of the glands are correlated with the ecology of the birds. (d) The birds' skulls are represented in row II at half normal size, while above each bird's skull, in row, I, is shown a cross-section of one of its salt glands, magnified five times. The birds are arranged from left to right in order of increasing adaptation to marine life; they are (left to right): the wood sandpiper, *Tringa glareola;* the common sandpiper, *Actitis hypoleucos;* the knot, *Calidris canutus;* and the little auk, *Plautus alle.* [Part (a) is from K. Schmidt-Nielsen, *Scientific American*, January, 1959, p. 114; part (b) is from K. Schmidt-Nielsen and R. Fänge, *Nature,* Vol. 182, p. 783 (1958); part (d) is from H. Staaland, *Comparative Biochemistry and Physiology,* Vol. 23, p. 936 (1967); all three parts are reproduced by permission of the authors and the journals. The photomicrograph (c) was prepared from histological material kindly supplied by Dr. K. Schmidt-Nielsen.]

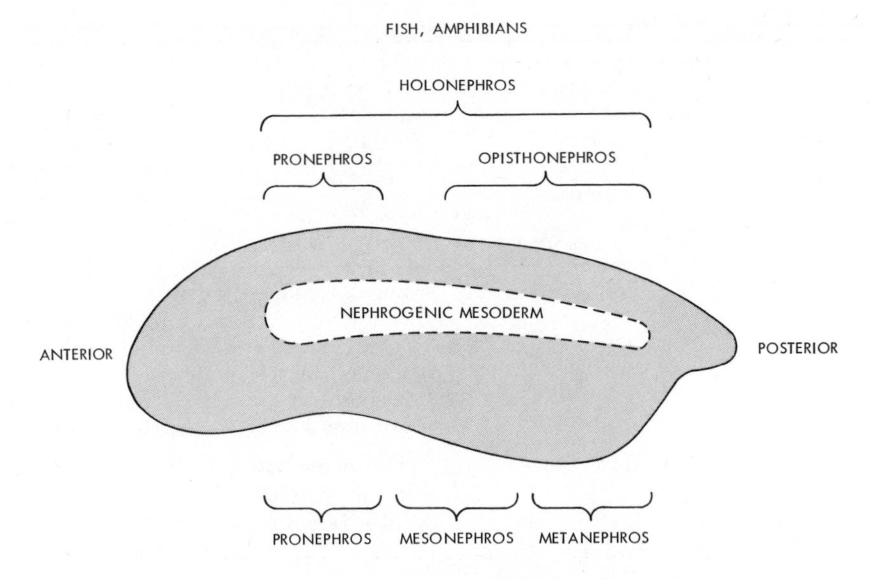

Figure 8-4 The nephrogenic tissue of vertebrate embryos (roughly indicated in the diagram) gives rise to definitive kidneys in a variety of ways. In all vertebrate classes, the pronephric tubules appear in the most anterior segments. When nearly the entire mass of nephrogenic tissue gives rise to the adult kidney, this is termed a *holonephros* (as in cyclostomes. "Holo-" signifies "whole," whereas "nephros" is derived from the Greek word for kidney). In many fish and amphibians, the more posterior region of the nephrotome develops into the definitive kidney, or *opisthonephros* ("opistho-" signifies "behind"). In reptiles, birds, and mammals, there is a progressive developmental sequence from the nonfunctional pronephros, through the embryonic mesonephros, to the adult metanephros. The structures also form in the direction anterior-to-posterior. ("Pro-" signifies "before"; "meso-" is "middle"; "meta-" is "beyond.") The extent of the nephrogenic tissue giving rise to the various kinds of kidneys is indicated only roughly, for in fact there is both overlap and omission of segments where different kidneys of different vertebrates originate. Note, too, that the form of the "generalized vertebrate embryo" shown, as well as its nephrogenic tissue, are symbolic, rather than accurate.

however, that in most vertebrates (mammals excepted) urine, before it is finally voided, can be further modified by reprocessing in special organs such as the *urinary bladder*, the terminal end of the digestive tract, and the cloaca.

Kidney Structure and Function

Throughout the entire range of vertebrate classes and in all stages of development, the kidneys have three major types of basic functional structures: (1) *glomeruli*, (2) tubules, and (3) collecting ducts. The glomerulus is a tuft of capillaries fed by a small arterial branch. Typically, it lies in intimate contact with the end of a kidney tubule, and the details of this relationship between glomerulus and tubule are so important, from a functional standpoint, that they will be considered more fully later in this chapter. Within the glomerular capillaries, blood pressure is relatively high, and some of the blood plasma is filtered off through the walls of the capillaries and collected in the adjacent tubule. In the most primitive vertebrates (cyclostome hagfish), the tubules are so rudimentary that the filtrate may be expelled from the body more or less unchanged. In most vertebrates, however, the composition of the filtrate is greatly modified as it flows along the tubules, which are channels formed of a single layer of cells surrounding a central space, or lumen. The tubules are richly supplied with capillaries. Beyond the tubules and connecting with them are collecting ducts, channels of increasing diameter and wall thickness. They lead into the major urinary ducts (e.g., mesonephric ducts and *ureters*), and these, in turn, lead directly or through temporary storage chambers (bladder, cloaca) to the outside of the body. There is immense variation among vertebrates in the details of structure, assembly, and operation of these elements, in the number of each built into the functioning kidney organ, their relationships with the blood vessels that supply them, and the manner and degree to which they are controlled by the nervous and endocrine systems.

In the primitive kidneys (pronephroi) of cyclostome larvae and embryos of some other fish and amphibians, the glomeruli and tubules may not be directly connected (Figure 8-5). The glomeruli, arising from branches of the dorsal

Figure 8-5 The glomerulus, or tuft of capillaries, of a pronephric duct may not (a) or may (b) be enveloped in the expanded end of the tubule. Fluid is filtered from the glomerulus into the coelom (passing from there into the tubule through the nephrostome) or directly into the tubule. In embryos with functional pronephroi, the individual pronephrons on each side drain into an archinephric duct, which connects all of them with the posterior of the body, usually through the cloaca; urine emerges from here to the outside.

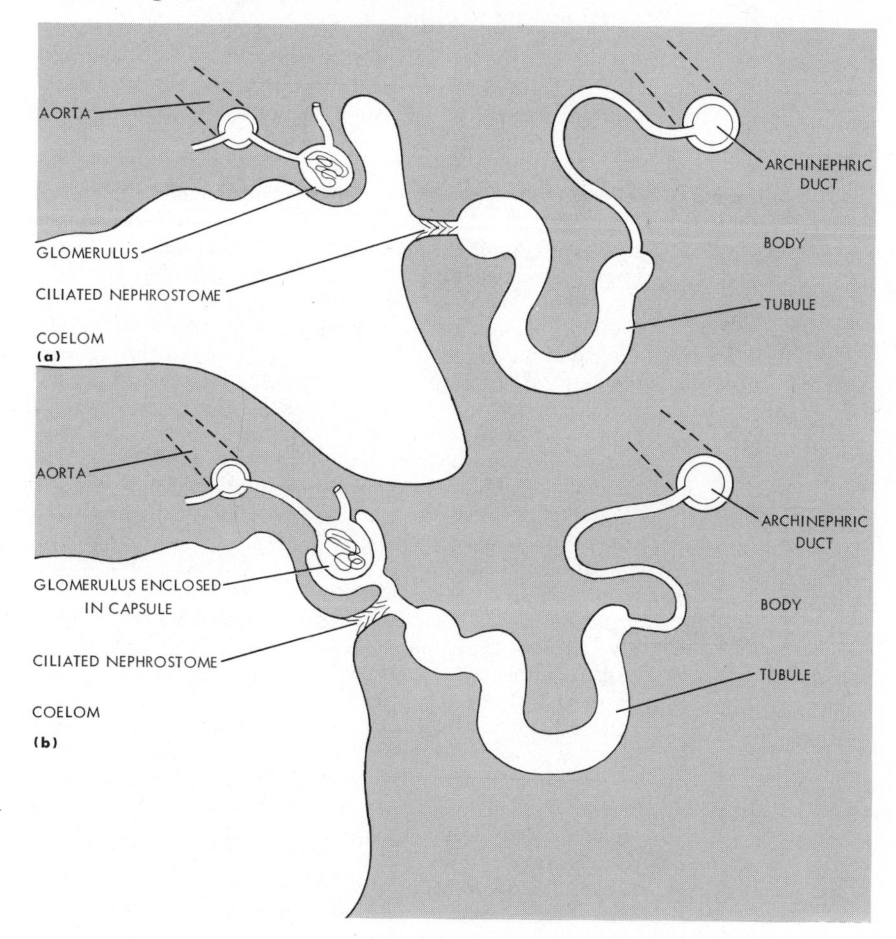

aorta, lie in the roof of the coelum, and fluid from plasma filters from them directly into the body cavity. Nearby pronephric tubules take up coelomic fluid continuously through ciliated openings, *nephrostomes*, communicating freely with the coelom. Once it has entered the tubules, the coelomic fluid becomes the raw material that is converted into urine. A different and more integrated relationship between the glomeruli and tubules appears in the case of other pronephric tubules and in the adult-type kidneys of almost all forms. Here, typically, each glomerulus is enveloped in a delicate sheath formed by the closed end (*Bowman's capsule*) of the nephric duct (the end farthest from attachment to the collecting duct). Some of the components of the fluid flowing through the capillaries filter out between or across the capillary endothelial cells, and then through a differentially permeable *basement membrane*. This structure permits passage of low-molecular-weight substances (e.g., water, inorganic ions, simple sugars) but prevents the passage of large molecules (e.g., most of the blood proteins). Water and solutes then pass directly into the kidney tubule lumen between and across the delicate cells of Bowman's capsule, lying adjacent to the basement membrane. The process as a whole is termed *glomerular filtration*, and is the start of the kidney excretory process in the majority of vertebrates. The filtering unit, the glomerulus together with its intimately associated Bowman's capsule, is termed the *renal corpuscle*. The renal corpuscle and its attached tubule constitute the *nephron*, the basic functional unit of the kidney.

Ideas about the mechanism of operation of the renal corpuscle, as summarized in the preceding paragraph and in Figure 8-6, have been built up gradually over the course of a century or so through the use of many observa-

Figure 8-6 (*opposite*) (a) The thin-walled, expanded end of the renal tubule (Bowman's capsule) completely covers the capillaries of the glomerulus with a delicate sheath, so that a complex structure, the renal corpuscle (a), is formed. Generally, the afferent arteriole, leading into the glomerulus, has a larger cross-sectional area than the efferent arteriole which drains the capillary tuft. This situation contributes to the maintenance of high blood pressure within the capillaries. At this high pressure, plasma components selectively filter out through the capillary wall (arrow marked ①). The movement is opposed by osmotic pressure of plasma proteins which are restrained from passing the glomerular membranes, and by fluid pressure built up with the capsular space (pressures opposing filtration pressure are suggested by the arrow marked ②). In the net, about one-fifth of the plasma traversing the mammal's glomeruli is filtered off into the capsules to flow away through the proximal convoluted tubules. (b) Some details of the filtration membrane (magnified about 30,000×) are shown, summarizing data obtained with the electron microscope and by study of chemical and filtration characteristics of renal corpuscles. The capillary endothelial cells lie closely adjacent to the *basement membrane*. This is a thin, continuous layer consisting of protein (collagen), glycoprotein, and lipid, probably secreted by both the capillary and capsular cells. It appears to be a major filtration barrier, so its chemical organization must participate in limiting the size and form of molecules which can traverse the filtration barrier as a whole. On the side opposite the capillary cells, the basement membrane is in close contact with many fine projections sometimes called "feet," of the capsular epithelial cells (podocytes). In the diagram, the capsular cells are shown in a single plane of section, so the projections appear, in some cases, as isolated masses although, of course, in the real three-dimensional structure they are continuous with the rest of the cells' cytoplasm. (c) Photograph of a renal corpuscle and adjacent cut segments of tubules from the kidney of a rat (magnified 442×). (Histological material prepared by Mrs. Deborah Christensen.)

tional techniques. Even the general form of the renal corpuscle offered certain clues to its function. By the middle of the nineteenth century, the great German physiologist Carl Ludwig had suggested, on the basis of this form, that urine must originate by filtration of fluid from the capillaries through the flat cells of Bowman's capsule into the nephric tubule. Later, this hypothesis was supported by a growing mass of experimental evidence. It was shown, for instance, that if other factors are kept constant urine formation is dependent on the arterial pressure (necessary for filtration), and it ceases altogether when this

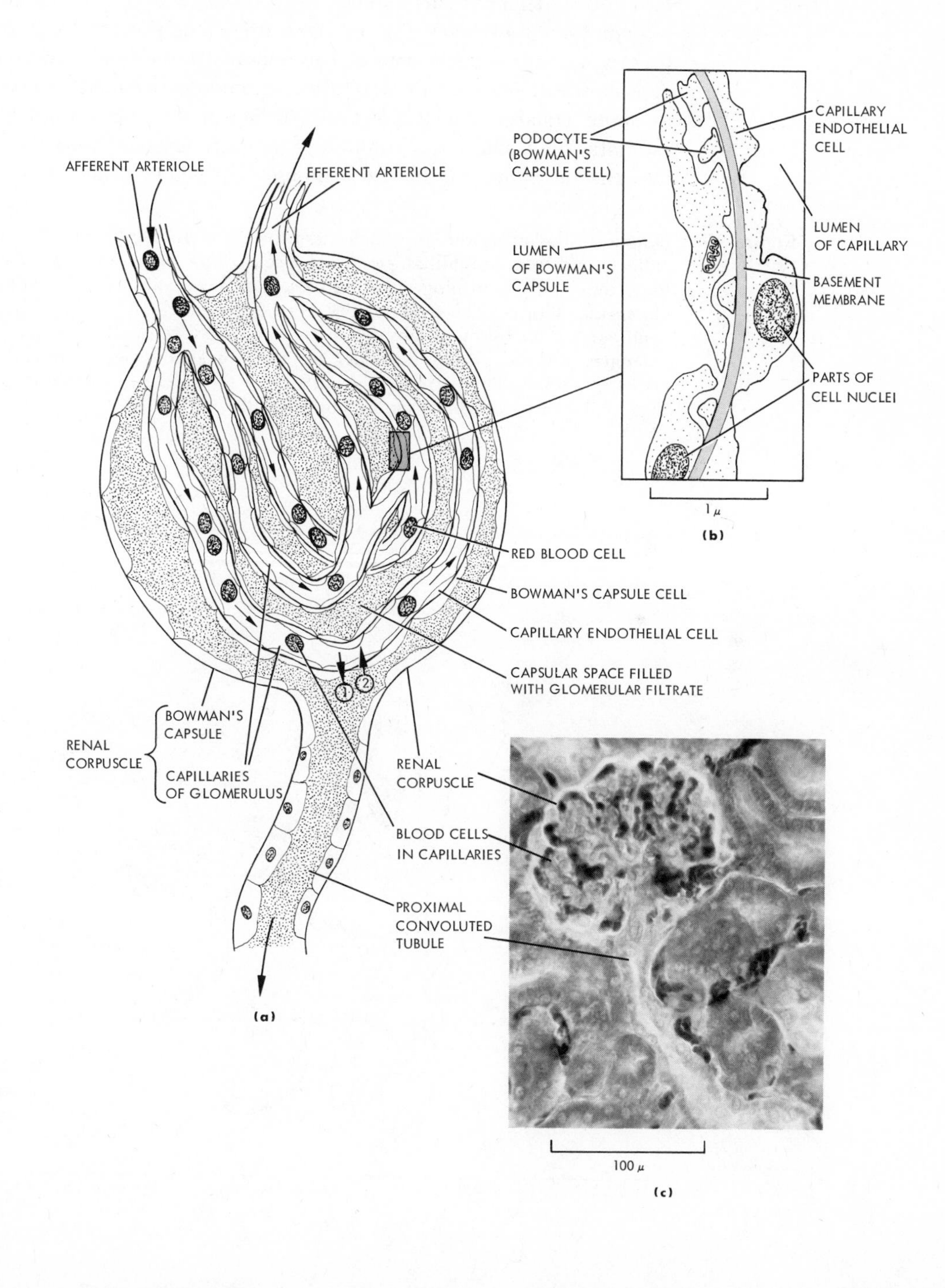

pressure falls too low. Fluid has been collected directly from Bowman's capsule by aspiration through fine capillary pipettes inserted through the capsular wall (Figure 8-7). Microchemical analyses then confirmed the hypothesis that the fluid arises by filtration through the capillary and capsule walls, for the composition of capsular fluid is essentially the same as the composition of plasma, although plasma proteins and other large molecules are generally absent from (or present at only low concentration in) the capsular fluid. Many physiological studies and intensive examinations with electron microscopy have rounded out our current knowledge of the renal corpuscles of vertebrates.

Even though each renal corpuscle is exceedingly small (for instance, about 0.1- to 0.3-millimeter in diameter in mammals), the total filtration surface of capsule-sheathed capillaries is relatively immense. In man it has been estimated as about equal to the entire body surface area. Correspondingly, indirect estimates show that large volumes of fluid are filtered across the membranes into the beginning of the renal tubule. The values listed in Table 8-2 are

Figure 8-7 Diagram illustrating the approach taken by A. N. Richards and his associates to collect fluid from amphibian renal corpuscles. This pioneering work was of great importance in establishing filtration as a main mechanism for the formation of glomerular filtrate. The experiments were begun in the 1920's, and by now have been carried far by many physiologists. Fluid has been collected also from different parts of the nephron and collecting ducts in many species, including a variety of mammals. The extremely small size of the nephron and the experimental tools make this work most taxing and difficult.

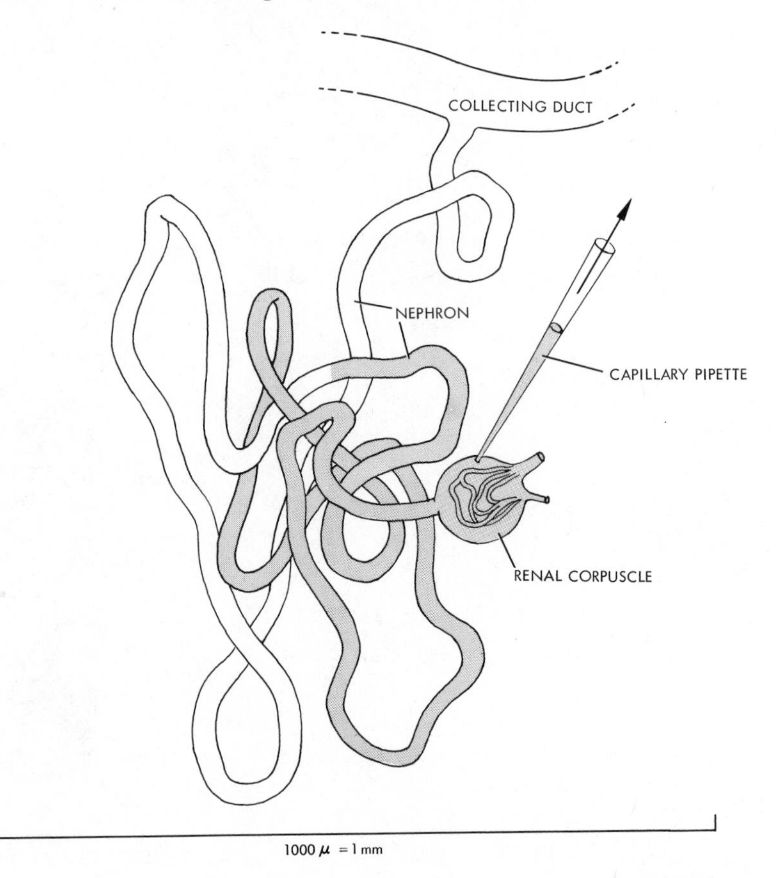

Table 8-2 A comparison of Urine Flow and Glomerular Filtration Rate as Measured Under Experimental Conditions in a Variety of Vertebrates. These figures may be thought of as representative, but they do not give an adequate picture of the variation among members of a class or between individuals within a species, or of changes in a single individual over a period of time. Yet an animal's renal function changes with the character of the environment, with health, activity, age, and many other variables. Note that in birds and mammals the glomerular filtration rate is so high that a volume equivalent to several times the entire mass of fluid in the body is filtered into the tubules during each twenty-four-hour period.

	Urine Flow (ml/kg in 24 hr)	*Glomerular Filtration Rate (ml/kg in 24 hr)*	*Per cent of Body Fluid Volume, Formed in 24 Hour*
Some fresh-water and terrestrial vertebrates:			
Elasmobranch (*Pristis microdon*)	150–450	450	56
Teleost (*Ameiurus nebulosus*)	154	225	28
Amphibian (*Rana clamitans*)	317	822	100
Bird (*Gallus domesticus*)	15–32	3170	490
Mammal (*Canis familiaris*)	15–100	6190	890
Some marine vertebrates:			
Cyclostome (*Myxine glutinosa*)	5.4		
Elasmobranch (*Squalus acanthias*)	1.5	50	7
Teleost (*Lophius piscatorius*)	27	0	0
Mammal (Seal)	12	3600	525

especially striking in the case of birds and mammals, with their relatively high arterial pressures and, consequently, high filtration pressures in the glomerular capillaries.

Fluid entering the capsule flows away from the glomerulus through the tubular lumen. In its path it is surrounded by tubular cells, and these have most significant and varied functions in modifying the composition of the fluid; gradually that process converts it into urine. In almost all vertebrates (cyclostome hagfish appear to be the exception), the tubular cells carry out energy-requiring processes that transfer substances filtered into the luminal fluid back through the tubular cells toward the blood stream. Such reabsorptive processes "save" useful substance, including glucose, amino acids, and ions such as potassium and sodium from the possibility of discard from the body. Urea, phosphate, sulfate, ascorbic acid, and other solutes also can be reabsorbed actively in some vertebrates. Furthermore, the tubule walls act as selectively permeable barriers between tubular fluid and body fluids (blood plasma), so that water, urea, and a variety of other substances may diffuse passively across the cells back into the blood stream. In addition, certain tubule cells are capable of transferring solutes actively in the direction plasma-to-tubule lumen. This process often is termed *renal secretion*. The capabilities of tubules to secrete vary widely from one region to another and from one organism to another. Generally, the proximal tubule, the region lying nearest to the renal corpuscle, shows the most cellular activity. In some cases, metabolic wastes such as urea and phenolic compounds may be secreted into the tubules, as may

Figure 8-8 (*opposite*). Nephron structure in a mammal. (a) Simplified diagram of a complete nephron with its blood supply. The richness of the capillary network around the tubule is suggested. These capillaries intermesh, receiving blood from several different arterioles besides the particular arteriole which drains the tubule's own renal corpuscle. (b) Diagram suggesting some important features of a proximal convoluted tubule cell (shown as if magnified ca 30,000×), as indicated by electron micrography. The luminal surface of the cell is covered with fine projections, mirovilli, each about 1 mμ long. Together, the microvilli greatly increase the surface area bounding the tubular lumen (by as much as forty-fold, according to calculations of J. A. C. Rhodin, who has estimated that a single cell may have as many as 6500 microvilli). The microvilli form a mass visible in the light microscope and termed the *brush border* of the proximal tubular cells. The nucleus lies toward the base of the cell next to the basal layer where deep infoldings of the cell membrane surround numerous mitochondria aligned with the long axis of the cell. Surrounding the entire tubule is a relatively thick, continuous *basal lamina* (basement membrane). The microvilli of the brush border are important features of proximal tubules of almost all vertebrates. They are generally believed to participate in the major transport roles of these cells, which secrete some substances into the urine, and reabsorb back into the blood many components of the glomerular filtrate. Likewise, functionally important characteristics of the cells are the basal arrays of numerous mitochondria arranged in folds of the basal membranes (seen in the light microscope, these arrays of mitochondria and membranes are termed *basal striations*). These appear to be correlated with the major transport activities of the tubular cells in mammals (and in the few teleosts and amphibians so far studied). According to recent work of B. Schmidt-Nielsen and others, in reptiles and birds the lateral boundaries, rather than the basal region, show specialized folding, and mitochondria are more uniformly distributed rather than being concentrated in the basal region of the cells. Somewhat similar basal arrays of infolded membranes and mitochondria (or lateral infoldings) occur in distal tubules of vertebrate kidneys, as well as in salt glands of reptiles and birds. (c) Micrograph showing a region of the outer zone of the kidney (cortex) of a rat (magnification 96×). Many renal corpuscles are visible, packed among segments of complexly coiled tubules. The structure of the renal tubule, as diagrammed in (a) has been established by a study of serial sections such as this, as well as by microdissection, delicate maceration of the tissue, and other techniques to reveal the continuity of individual tubules. (Histological material prepared by Mrs. Deborah Christensen.)

certain ions (phosphate, sulfate in some marine fish). Abnormal compounds, such as some organic dyes and the antibiotic penicillin, may be secreted as well. Many vertebrate nephrons show another region of marked secretory activity. This is the distal tubule, the portion that joins directly with the collecting duct.

In some instances, tubular cells can form new compounds by virtue of their metabolic activities. These compounds generally are produced in relation to special regulatory functions of the kidney. For instance, many vertebrates excrete large amounts of ammonium ions in the urine as part of the over-all mechanism for regulating body acid-base balance. The highly toxic ammonium ions are not present in great enough concentration in plasma to account for the amounts excreted in the urine, and this observation led to the discovery that ammonium ions actually are synthesized in kidney tubule cells. Kidney cells also form enzymes and hormones that aid in the control of the circulation as a whole (renin, erythropoietin). Thus, the cells that make up the walls of the tubules assist in a wide variety of ways in regulating body-fluid composition. They modify the glomerular filtrate in fine detail, extracting some components

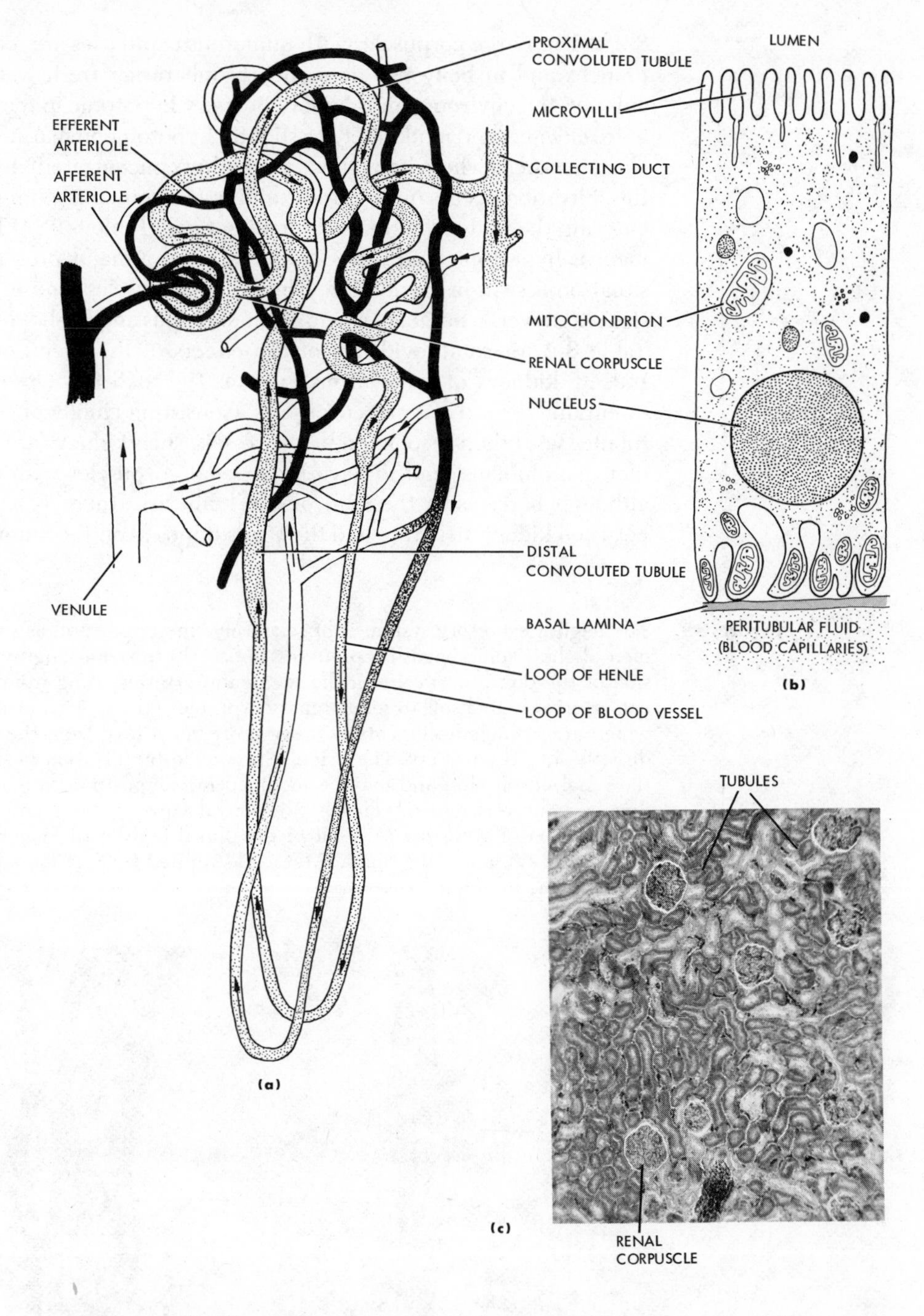

from it and adding others to it. These functions depend on special characteristics of renal cells, which vary distinctively from one region of the tubule to another (Figure 8-8).

Renal Corpuscles The account so far has summarized typical common features of nephrons among a wide range of vertebrate animals. When we consider in greater detail the kidneys of different vertebrates, important variations in nephron structure are noticeable at once. In cyclostomes, the renal corpuscles are large (Figures

8-9, 8-10). Large corpuscles with numerous capillaries are seen also in elasmobranchs and in bony fish permanently inhabiting fresh water. For all these animals, the environment is hypotonic (very hypotonic in the case of fish living in fresh water and moderately to slightly hypotonic in marine cyclostomes and elasmobranchs whose body fluids are highly concentrated). In correlation with this, filtration occurs quite briskly, aiding the animal to eliminate excess water entering the body from the surrounding fluid (Table 8-2). Where the environment is hypertonic and water conservation is a major problem, a contrasting situation is seen in marine fish. The renal corpuscles tend to be far smaller in size and fewer in number than in elasmobranchs and fresh-water bony fish. (See Table 8-2 for some evidence of the effects of this trend on filtration rate.) Indeed, kidneys of some forms, such as the toad fish *Opsanus tau,* may be essentially devoid of renal corpuscles. Consisting chiefly of compact masses of tubules with their associated blood vessels, such kidneys are termed aglomerular. Amphibians tend to have large renal corpuscles with many capillaries, although here, as in the case of the bony fish, there is a clear correlation between kidney structure and the stress imposed on the animal by its environ-

Figure 8-9 The hagfish excretory system is of relatively simple pattern, as shown in this specimen of the Pacific hagfish, *Eptatretus stouti.* Ureters run longitudinally on either side of the dorsal aorta. From the segmental arteries rising from the aorta, afferent arteries branch off to giant renal corpuscles (about 1.5 mm in diameter) which are arranged segmentally. Very short ducts lead from the renal corpuscles directly into the ureters. The urine is produced by filtration of blood plasma into the renal corpuscles, and may be modified in composition as it flows through the ducts and ureters into the cloaca. (a) Ventral view of the hagfish. (b) Dissected, fixed section of a narrow segment of the dorsal body wall; magnified, as compared with (a), about 18 times. (Material supplied by S. Vigna and S. Ergezen; photographs by Sigurd Olsen.)

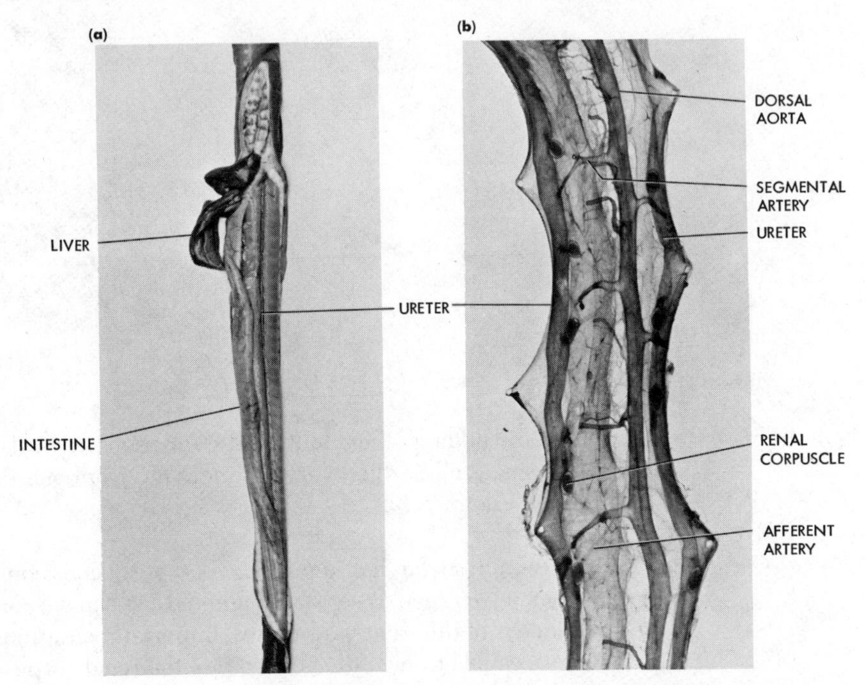

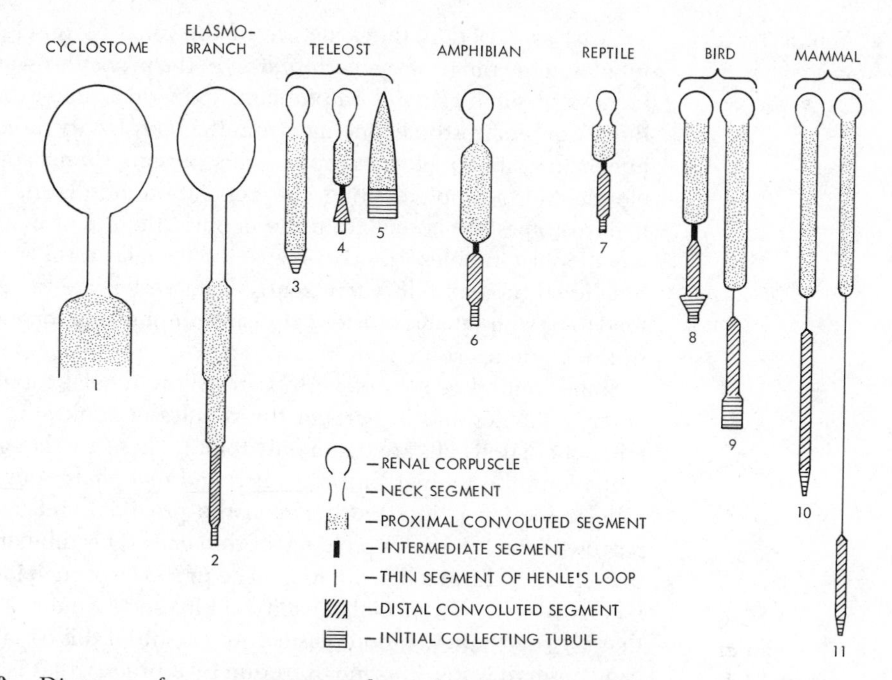

Figure 8-10 Diagrams of representative nephrons from members of the various vertebrae classes. Note the variation in relative sizes of glomeruli, and the trend towards increased complexity of tubular structure.

1. Cyclostome *Bdellostoma stouti*
2. Elasmobranch *Raja stabulaforis* (*R. laevis*)
3. Teleost *Myxocephalus octodecimspinosus*
4. Teleost *Ameirus nebulosus*
5. Teleost *Opsanus tau*
6. Amphibian *Rana catesbiana*
7. Reptile *Chrysemys marginata*
8. Bird *Gallus domesticus*
9. *Gallus domesticus*
10. Mammal *Lepus cuniculus*
11. *Lepus cuniculus*

[From C. L. Prosser and F. A. Brown, *Comparative Animal Physiology.* (Philadelphia: W. B. Saunders Co., 1961.)]

ment. Thus, salamanders and frogs living in fresh water have large renal corpuscles and quite high rates of filtration as compared with amphibians adapted to arid conditions. Frogs living in desert areas (two *Cyclorana* species of Australia) show such a reduction of their renal corpuscles that they are sometimes described as aglomerular. In reptiles and birds, as compared with other vertebrates, there is a consistent trend toward reduction of renal corpuscles, paralleling the successful adaptation of these forms to life on land. Reptiles tend to have few renal corpuscles, and in birds the renal corpuscles, although very numerous, are of small size. In both these classes, the glomeruli have few capillaries, and the capsule is partly filled with connective tissue. Mammals are, of course, as successful at land dwelling as are the other amniotes, but mammalian renal corpuscles do not show a trend toward reduction. They are generally large, well vascularized, and numerous. Marked changes in the patterns of tubular function have enabled the kidneys of mammals to adjust to terrestrial life. These adaptations will be described in more detail in the next paragraph.

Renal Tubules

At least as varied as the structure of the renal corpuscle is the structure of tubules in the range of the vertebrates. In the pronephros, and even in the adult kidneys of some fish and amphibians, individual tubules may have a ciliated funnel, or nephrostome, opening from the body cavity directly into the tubular lumen (or tubular blood supply). This permits direct adjustment of the coelomic fluid, supplementing the regulation of plasma composition. Such nephrostomes are absent from the nephric tubules of many fish, and from all metanephric tubules. In other ways, too, tubular form as well as the shape of individual tubular cells vary greatly from vertebrate to vertebrate. Nephrons tend to be simpler and shorter in fish and amphibians, longer and more complex in the amniotes (Figure 8-10).

Some limited groups of vertebrates show special tubular regions. In pronephric tubules and in parts of the tubules of adult amphibians and reptiles areas of ciliated cells are commonly found. These are thought to aid the flow of fluid along the tubular path. In birds and mammals, a unique and important tubular section is inserted between the proximal and distal tubules. It is of narrow diameter, built up chiefly of thin cells. This tubular region is bent back upon itself to form a hairpin loop. The presence of such loops (loops of Henle) is correlated with a special capacity of bird and mammal kidneys to form urine that is hypertonic to blood plasma. As a result of this capability, these animals can conserve water in urine excretion by a process that is not available to any other vertebrate. At this point it may be well to refer again to Table 8-2 for comparison of the rates of glomerular filtration and urine flow in various vertebrates. The difference between these figures represents tubular conservation of fluid.

Because the structure and function are most elaborate and have been studied most fully in mammals, the tubular loops will be discussed here in relation to the mammalian kidney. In this organ, most of the renal corpuscles and proximal and distal tubules are packed into the surface layer or *cortex*. The loops of Henle, collecting ducts, and associated blood vessels occupy the more central volume (or *medulla*) of the kidney (Figure 8-11). Thus, after passing through the proximal tubule, the glomerular filtrate formed in the cortex plunges down into the medulla along the tubular loop, turns back along this same loop to reenter the cortex in the distal tubule, and finally flows back into the medulla as it traverses the collecting duct on its way to the ureter and the urinary bladder. The hairpin loops of Henle, which are paralleled by blood vessels of comparable hair-pin form (vasa recta), have long been known to exist. Only in the last decades, however, has it become clear that they have special functions that allow the formation of urine that differs from blood plasma in osmotic pressure so that, in the net, water may either be conserved (formation of hypertonic urine) or excreted in excess (formation of hypotonic urine). This functional importance of kidney tubular architecture is so interesting that further discussion is profitable (Figure 8-12). Along the tubule, cellular transport processes remove sodium ions from the fluid in the lumen and transfer these ions into the interstitial fluids. Generally, sodium ions are accompanied by chloride ions (or other anions). It has been established experimentally that the result of this ion transfer is a build-up of a gradient of osmotic pressure in the fluid in the tubules and in the interstitial spaces. Thus, although these fluids are essentially isosmotic with circulating systemic blood in the cortex, they become progressively more hypertonic to blood plasma in the medulla where the loops of

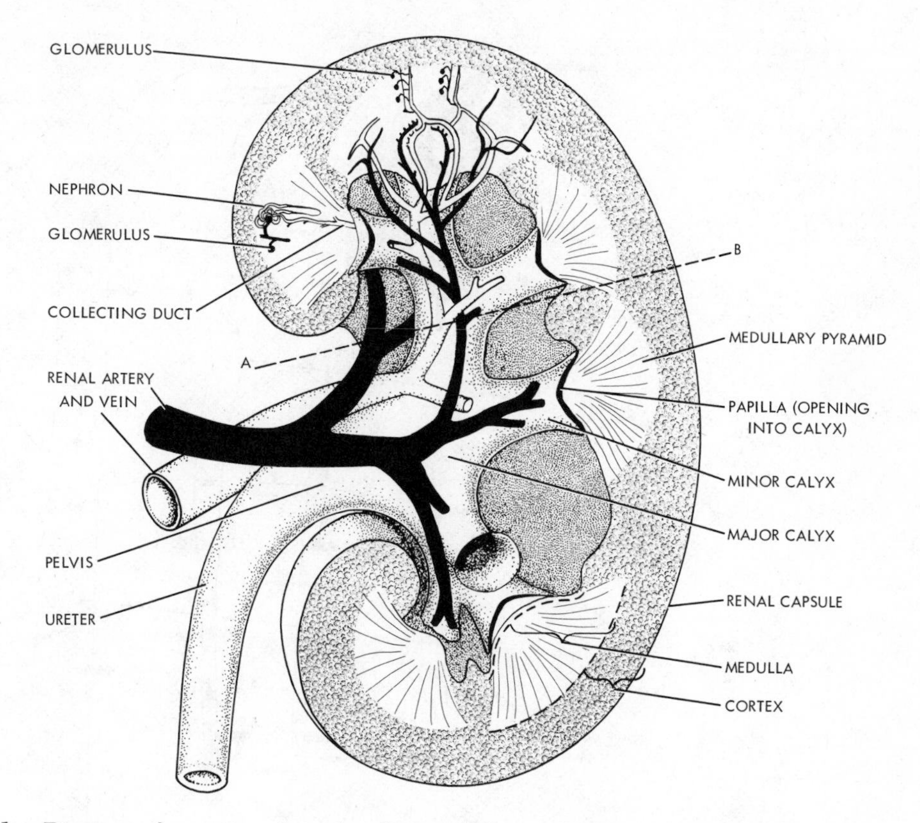

Figure 8-11 Diagram of a section through a human kidney, suggesting the compact organization of nephrons, draining-collecting ducts, and blood vessels in the mammalian kidney. The renal corpuscles, proximal convoluted tubules, and distal convoluted tubules of about one million nephrons are packed into the *cortex*, or outer layer of the kidney. Many of the thin loops and collecting ducts pack into parallel arrays in the *medula*, giving the medullary *pyramids* a readily observed appearance of striation. The collecting ducts of the pyramids empty into a few large channels, *minor calyces*, each minor calyx then merging with others to form *major calyces*. These, in turn, open into the great common chamber, or *pelvis*, leading to the ureter. The pelvis is the site of entry into the kidney of the renal artery and vein, which feed and drain the vast branching system of blood vessels culminating in the glomerular and tubular capillaries.

In the diagram, the part of the kidney shown above the line AB is highly schematized to indicate the relative positions of glomeruli, thin loops, collecting ducts, and other structures within the mass of the kidney. Many of the structures are magnified and simplified to assist the comprehension of their relationship to one another. Below the line AB, the structures shown are represented more realistically, appearing much as they would if you were to view an actual human kidney in section. [Modified from H. W. Smith, *Principles of Renal Physiology* (New York: Oxford University Press, 1956).]

Henle lie. This phenomenon is symbolized in Figure 8-12 by the large stipled arrow, representing the gradient of tonicity with increasing values occurring in regions more distal to the cortex (approaching the entry to the ureter).

The exact mechanism whereby this build-up of tonicity occurs is not fully established. One important theory is that it depends on a countercurrent multiplier effect directly attributable to the shape, cellular activities, and permeability characteristics of the tubular loops. According to this theory, the

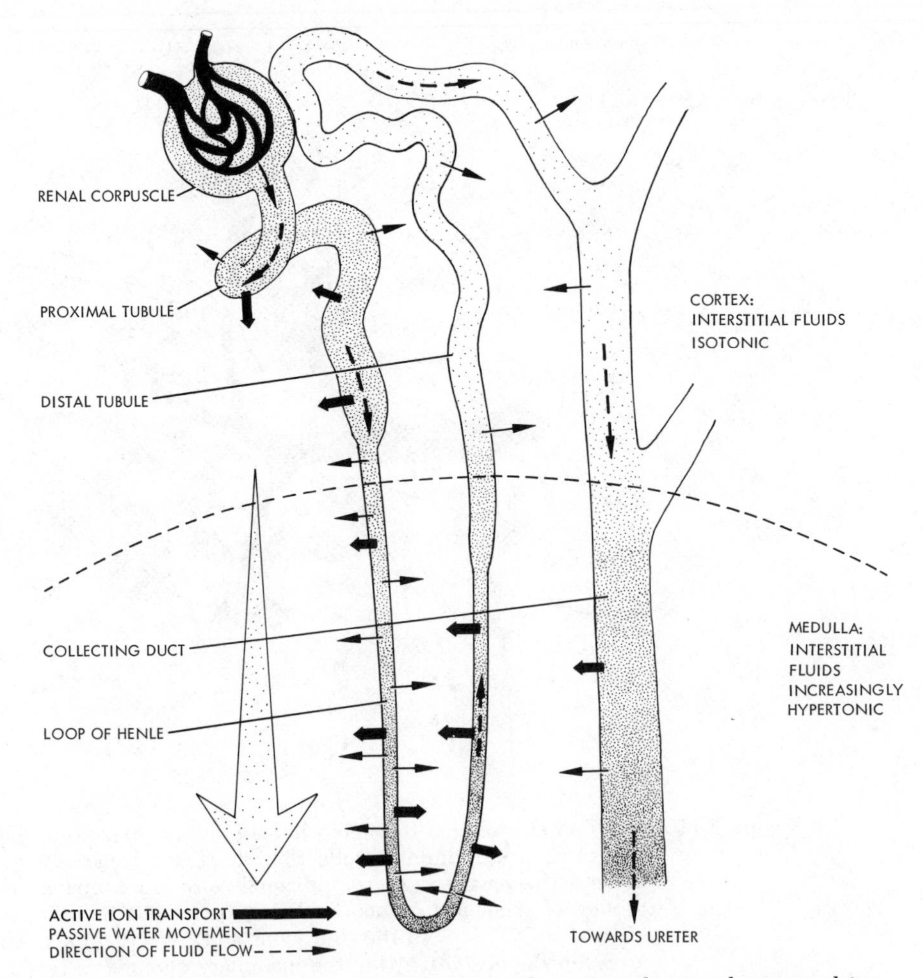

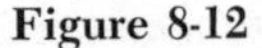

Figure 8-12 Diagram representing processes occurring in the mammalian nephron, resulting in formation of urine of variable tonicity (varying degrees of water conservation or excretion). Sodium ions and accompanying anions are transferred out of the tubule along its entire length (dark arrows). As fluid flows along the tubule (dashed arrows) in the medullary region, it encounters a progressively solute-enriched, hypertonic environment of extracellular fluids (direction of osmotic gradient indicated by a large stippled arrow). Water traverses the walls of the tubule by passive diffusion (thin arrows) in the proximal tubule and the first part of the loop, but in the ascending limb the tubular walls must become less permeable because the tonicity of the fluid in the tubule begins to drop while ionic transport continues. Fluid entering the collecting duct may be quite hypotonic, and if it reaches the ureters unchanged, the urine excreted from the body is hypotonic (i.e., excess body water is excreted). If, however, as suggested in this diagram, the walls of the collecting ducts are in a state of high water permeability (resulting from high level of antidiuretic hormone in the blood), water flows from the collecting ducts back into the medullary interstitial fluids. Thus, the final urine is *hypertonic* (i.e., body water is conserved).

cells in the descending limb of each loop reabsorb sodium ions actively, and anions and water move with these into the interstitial fluids. In at least parts of the ascending limbs of the loops, the cells continue to transfer ions out of the lumina, but here water and ions do not move by passive diffusion easily through the relatively impermeable walls of the tubules, so the fluid outside the tubule becomes enriched in ions, whereas the fluid inside the tubule becomes more dilute. One major net effect of the cell activity of the ascending limb is to increase the tonicity of the fluids surrounding the descending limb. This, in turn, may increase the tonicity of the tubular fluid delivered by the descending limb to the ascending limb. The reader may recognize that the over-all picture represents a classical situation for countercurrent build-up of a concentration gradient—in this case, of sodium ions and anions. According to the general hypothesis presented here, the progressive buildup of tissue-fluid tonicity in the medulla is favored by the fact that the blood vessels also form hairpin loops parallel to the tubules. If, as current experimental evidence suggests, the walls of the blood vessels are quite freely permeable to water, blood can flow through them with continuous adjustment of the osmotic pressure of the plasma to conform with the pattern of interstitial tonicity. Thus, the retrieval of water, lost from the blood as it flows through the descending part of the loop, occurs as the blood traverses the ascending part of the vascular loop. In net effect, the loop shape of the blood vessels protects the osmotic gradient dependent on activity of the tubular loops of Henle, rather than allow the gradient to be dissipated by the flow of blood through the tissues.

To complete this interpretation of the function of the tubular loops in enabling the kidney to form hypertonic urine, it is necessary to take into account the special role of the collecting ducts. These lie parallel to the tubular loops and, thus, traverse an area where the interstitial fluid is hypertonic to blood plasma. They have the important physiological property that the cells making up their walls have variable permeability to water. This variability is controlled by the supply of a posterior pituitary hormone (antidiuretic hormone) in the blood reaching the collecting-duct cells (See also Chapter 12.) When the collecting ducts are relatively *permeable* to water (as in dehydrated animals, with a high rate of release of antidiuretic hormone from the posterior pituitary), water is reabsorbed from the fluid traversing the collecting ducts and a *hypertonic* urine is formed. If, on the other hand, the collecting ducts are in a phase of relative *impermeability* to water (as in hydrated animals, with suppression of the secretion of antidiuretic hormone by the posterior pituitary), the dilute, sodium- and chloride-depleted urine flows through the collecting ducts and on toward the outside of the body.

Finally, looking at the entire mechanism, we can see that the varying capability of the mammalian kidney to form urine more concentrated, equally concentrated, or less concentrated than the blood plasma results from an interplay of the structural organization of the tubules, collecting ducts, and their blood supply on the one hand, and on the other hand of the functional capacity of tubular cells to transport ions actively and the controlled and varying permeability of the collecting ducts. Other factors, such as the transport of urea, may play a part as well. Even though the exact details of the concentrating mechanism of the kidney are not as yet agreed on by physiologists, the general situation should prove, in the long run, to be similar to this account. In any case, it accords with interesting findings about structural variations among mammalian kidneys. Thus, mammals that can form highly concentrated urines (such as some species adapted to life in the desert) have extra long tubular loops organized together in an elongated *renal papilla* (Figure 8-13). Similarly, mammals that must form large volumes of hypertonic urine have great numbers of tubular loops and a thick renal papilla. On the other hand, some mammals have adapted to life in a water-rich environment (*Hydromys*, the Australian water rat, is an example), and in these the tubular loops and renal papillae may be poorly developed. Such differences in this aspect of kidney structure are indicated in Figure 8-13.

Without going into detail, it should be noted that the kidneys of birds also have tubular loops, although the lobulated form of the avian kidney is corre-

Figure 8-13 (a) As shown in this figure, based on work of I. Sperber, the form of the medulla of the mammalian kidney may be modified in animals inhabiting dry environments, as well as in those that live on diets rich in salts and other solutes. The medullary countercurrent system is expanded and protrudes into the ureter as an asymmetrical, though more or less conical structure, the *papilla* (some animals have several papillae; see Figure 8-11). The length and mass of the papillae are approximately correlated with the power of the kidney to form hypotonic urine. Thus, the more effective water conservation in mammals adapted to relatively water deficient environments is based on characteristic structural modifications of the metanephroi. (Courtesy of Dr. B. Schmidt-Nielsen; slightly modified from *Physiological Review*, **38:** 139, 1958.)

(b) Shows typical anatomical divisions of the mammalian kidney; it is in fact an enlargement of the kidney diagram given for Macroscelides under the column heading "Dry habitat" in part (a).

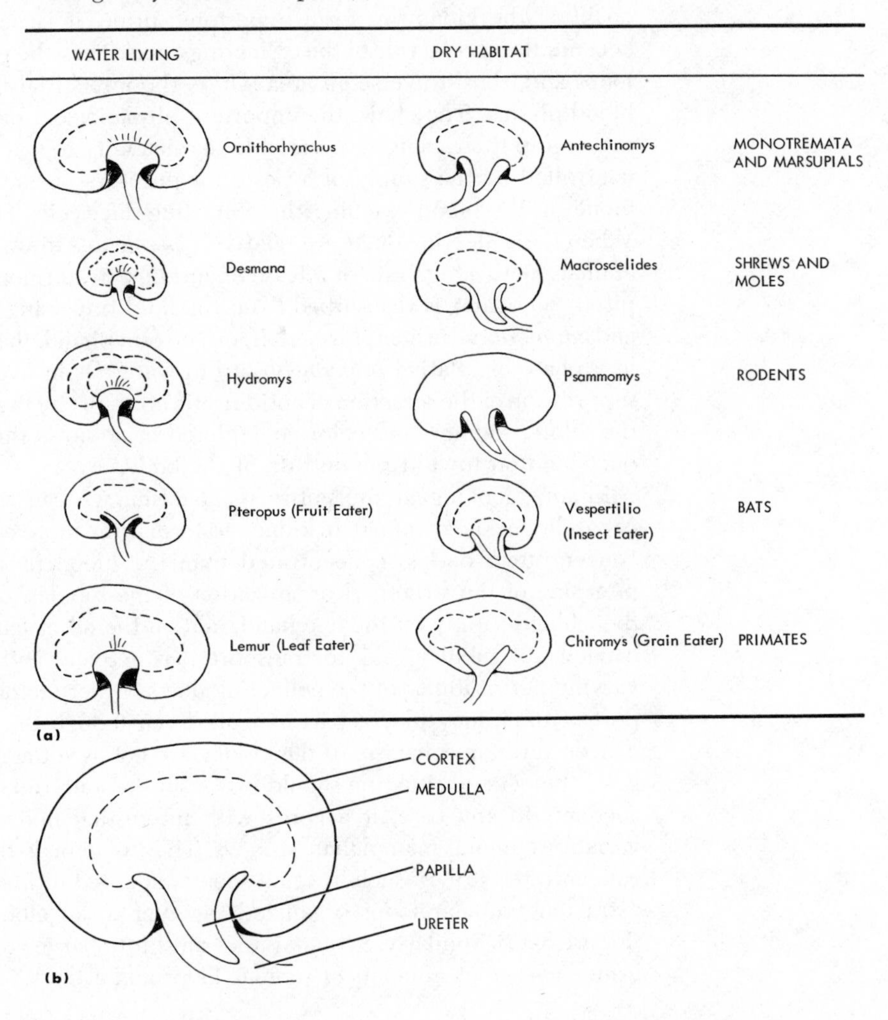

lated with a somewhat different patterning of tubule form and tonicity gradient development. In general, bird kidneys also show correlation of structure with adaptation to environmental osmotic stress. Tubular conservation of water tends to be less effective in birds than in mammals. As will be noted later, however, birds have some interesting accessory devices to facilitate the conservation of water.

Kidney Circulation and Innervation

Intimately bound up with the structure of the nephrons are the blood vessels that supply them. The process of filtration of blood plasma in the renal corpuscle to form glomerular filtrate is dependent on the exceptionally high pressure of the blood in the glomerular capillaries, as well as on the thinness and permeability of the capillary-capsular membranes. Blood flow to the tubules is of prime importance in nourishing the cells, in supplying solutes for excretion, and as a route for return to the body as a whole of valuable substances reabsorbed from the tubular lumina. The general architecture of the blood supply is fairly consistent throughout the vertebrate series. Each kidney is supplied by one (or more) great arterial branch from the dorsal aorta, which breaks up into finer and finer divisions, at last giving rise to the tufts of glomerular capillaries. As noted earlier in this chapter, these may be absent or much reduced in kidneys with suppressed glomerular function such as those of marine fish and desert amphibians. The glomerular capillaries generally drain into still smaller arterioles (efferent arterioles) that lead, in turn, into capillaries meshing about the nephric tubules. In cyclostomes and mammals, these arteriole-fed capillaries furnish the only blood supply to the tubules. However, in members of all the other vertebrate classes, blood is supplied to the tubules entirely or in significant part by renal portal veins, branches of the great veins draining the posterior part of the body. Of prime importance in amphibians, the renal portal system also participates, to a varying extent as compared with arteries, in supplying blood to the renal tubules of reptiles and birds. In birds, venous valves may control the flow of blood from the posterior part of the body to, or bypassing, the kidneys. There is no renal portal system in mammals. Blood that has traversed the capillaries around the tubules drains into venules. These join larger vessels that, in turn, give rise to the renal arteries and nerves.

The nerves to the kidney appear to function chiefly in the regulation of blood supply. In amphibians, for instance, the pressure and flow of blood in the glomeruli and, therefore, the rate of glomerular filtration, are under nervous as well as hormonal control. Thus, blood supply is acutely sensitive, varying from moment to moment, and it also responds to changing external conditions. In contrast to the knowledge of nerve regulation of kidney function in amphibians, little is known about this particular role of nerves in most vertebrates.

Overview of Kidney Structure

Among vertebrates, variation exists in the degree of compactness of the organization of the kidneys. An over-all impression of these structures and their relationships with reproductive systems are given in Figures 12-9 and 12-11, pp. 570 and 574. In fish and some amphibians, the renal tissue is elongate and organized rather loosely with connective tissue (Figures 8-14 and 8-15), whereas the kidneys are quite compact in anuran amphibians (frogs, toads) and in reptiles, birds, and mammals (Figures 8-16 and 8-17). Generally, the more

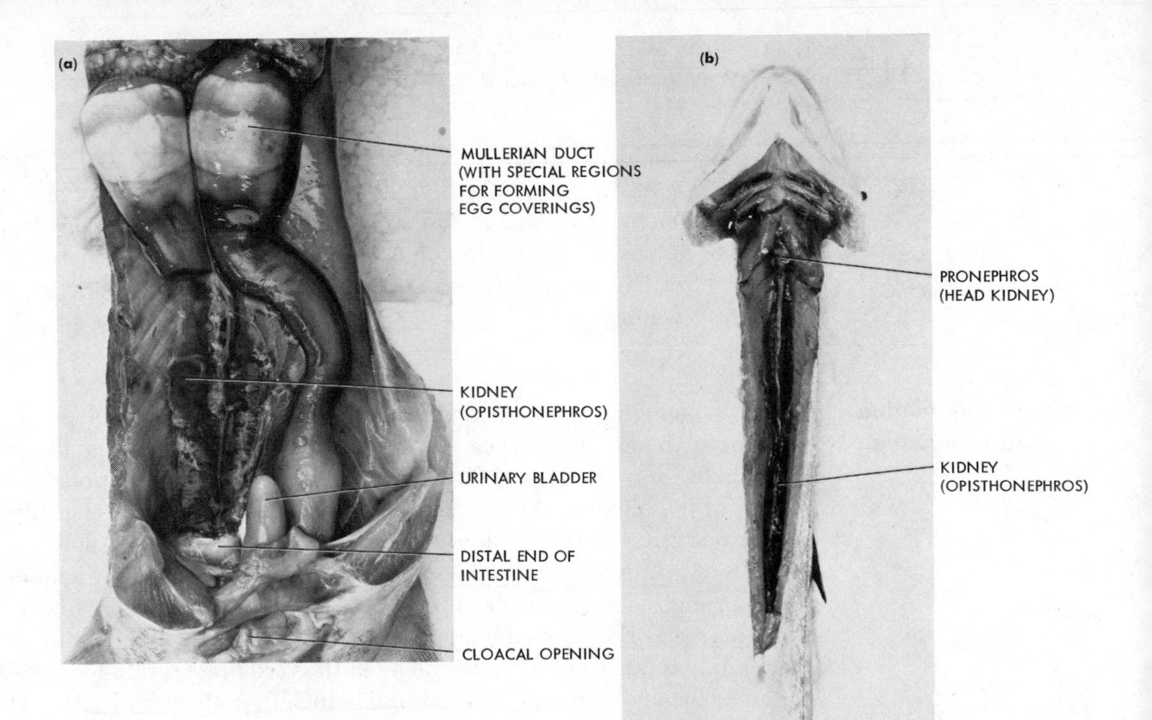

Figure 8-14 Kidneys and associated structures of fish. (a) Holocephalan ratfish, *Hydrolagus colliei*, seen in ventral view; the digestive tract and part of the reproductive system on the right side of this mature female have been removed to show the kidneys (opisthonephroi). Urine flows through the nephric ducts to the urinary bladder, thence through the urinary pore (separate from the openings of the oviducts) to the outside. (b) Teleost coho salmon, *Oncorhynchus kisutch.* The kidneys run throughout nearly the full length of the body cavity. Urine drains from each opisthonephros through nephric ducts to the outside through the urogenital pore. The pronephros (head kidney) does not function in excretion, but rather as a lymphoid organ and endocrine gland. Among fish in general, much variation exists in the structure of the renal system. Usually, as in the two species illustrated here, paired kidneys are located dorsally in the body cavity, but even here variety exists among different fish species. In some, the two kidneys fuse into a single organ. One of the most unusual situations is seen in the coelocanth *Latimeria chalumnae*, where there is a single kidney located posteriorly and ventrally in the body cavity. (Photographs by Sigurd Olsen.)

Figure 8-15 (*opposite*) Structure of amphibian excretory system. (a) A female *Necturus maculosus;* ventral view of the posterior region of the body cavity after removal of the gastrointestinal tract. The nephrons are relatively large in amphibians, and a row of renal corpuscles can be seen even with the unaided eye, arranged along the medial aspect of each kidney. These, and deeper lying corpuscles, lead into tubules; some of the tubules open by ciliated funnels, nephrostomes, directly into the body cavity. The tubules converge to form many larger collecting ducts, and these in turn lead into Wolffian ducts, or ureters, running along the lateral margins of the kidneys. The paired ureters lead into the cloaca (not shown in this figure), opening into it on its dorsal surface. On its ventral surface, an opening from the cloaca leads into the urinary bladder. Urine formed originally in the renal corpuscles and nephrostomes may be stored in this bladder. The narrow anterior segment of each kidney consists of a single row of renal corpuscles with associated tubules and ducts leading to the ureters; in the male, though not in the female, this region is modified to assume a role in reproduction, since it participates in transport of the sperm from the testes to the outside (see Chapter 13). (b) A section of the kidney of the urodele *Amphiuma* sp., magnified 112 X, showing two renal corpuscles, many tubules cut at various angles, and a nephrostome with its funnel opening to the surface of the kidney. (c) The nephrostome shown in (b), enlarged to reveal the cilia in the mouth of the funnel; their beat draws fluid from the body cavity into the tubule leading from the funnel. (Photographs by Sigurd Olsen; histological material for (b) and (c) supplied by Deborah Christensen.)

compact kidneys show a high degree of orientation of tubule, collecting duct, and blood vessel systems, as in reptiles, birds, and some mammals with their lobulated kidneys, and in the clear distinction of cortex and medulla in birds and mammals. In vertebrates with compact kidneys, a connective tissue capsule invests each organ. The capsule may protect the kidneys from mechanical stress and may limit the variations of blood and tissue fluid pressures within the organ. Connective tissue, lymphoid and blood-forming tissue, and reproductive tissue may be incorporated into the over-all structure of the kidneys in various animals belonging to different vertebrate classes.

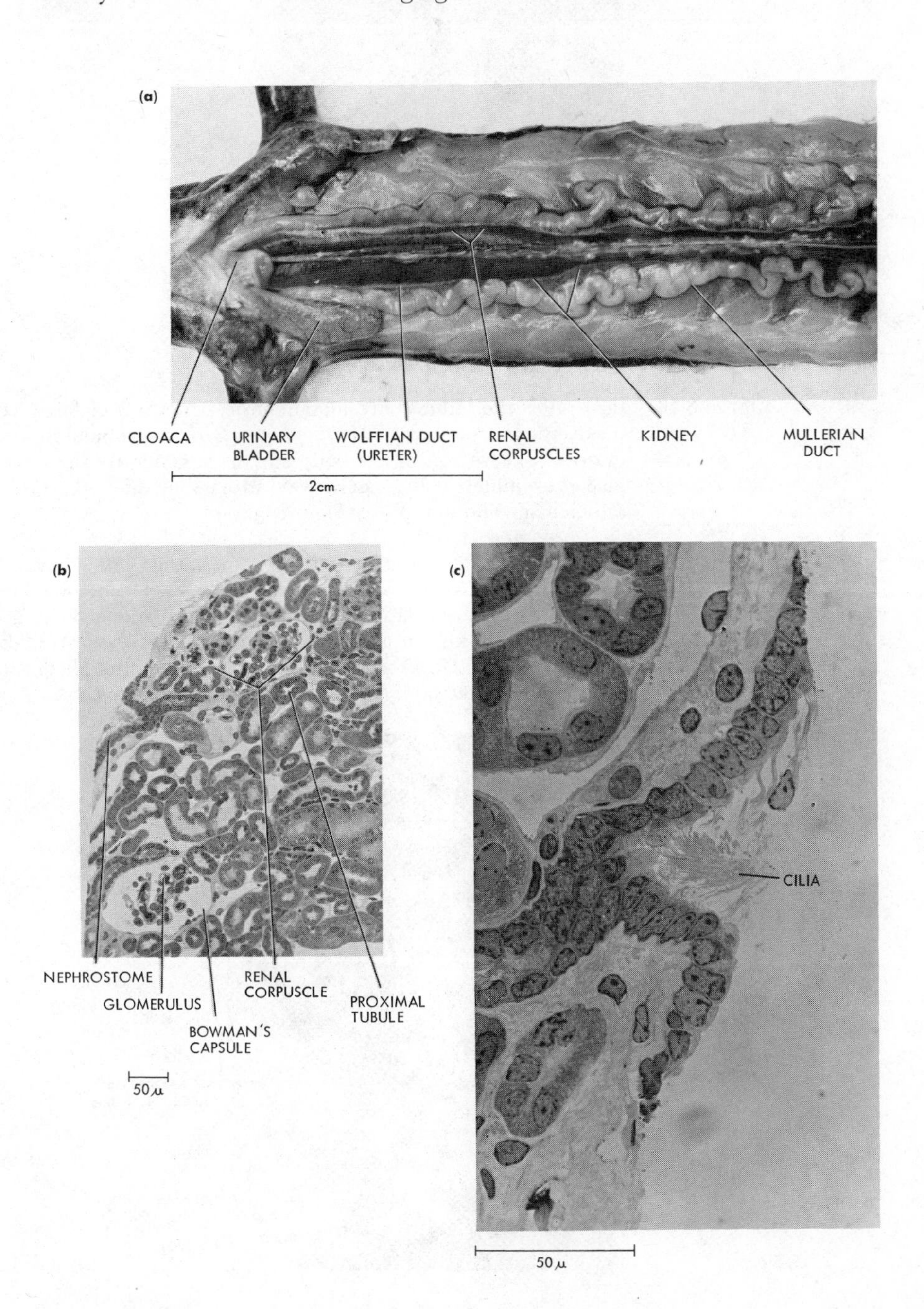

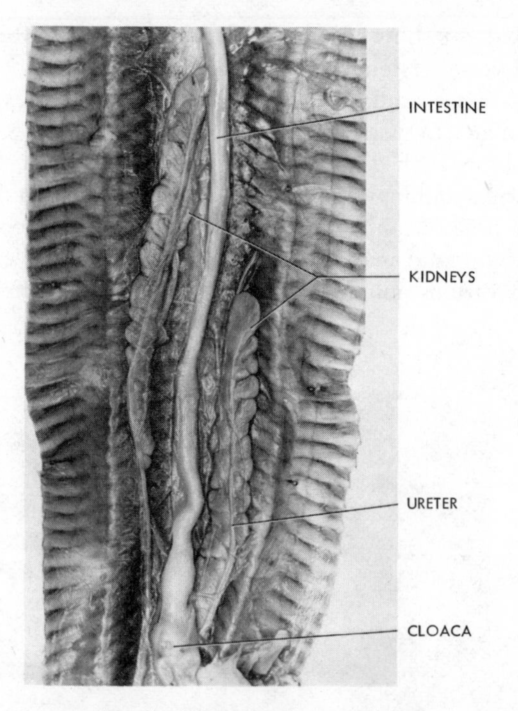

Figure 8-16 In reptiles the kidneys are metanephric structures of fairly compact form. In this snake, the garter snake *Thamnophis sirtalis*, the lobulated kidneys lie dorsally in the posterior region of the body and, as is generally the case in snakes, they are placed asymmetrically. There is no urinary bladder; the ureters open directly into the cloaca. (Photograph by Sigurd Olsen.)

Figure 8-17 The urinary system of a mammal, *Rattus norvegicus*, viewed from the ventral aspect (the digestive system has been removed). From the compact kidneys, ureters lead to the urinary bladder. The size of the bladder varies greatly depending on its degree of filling; in this specimen the bladder was strongly contracted and thus very small. Urine passes from the bladder through the urethra, thence discharges to the outside. (Photograph by Sigurd Olsen.)

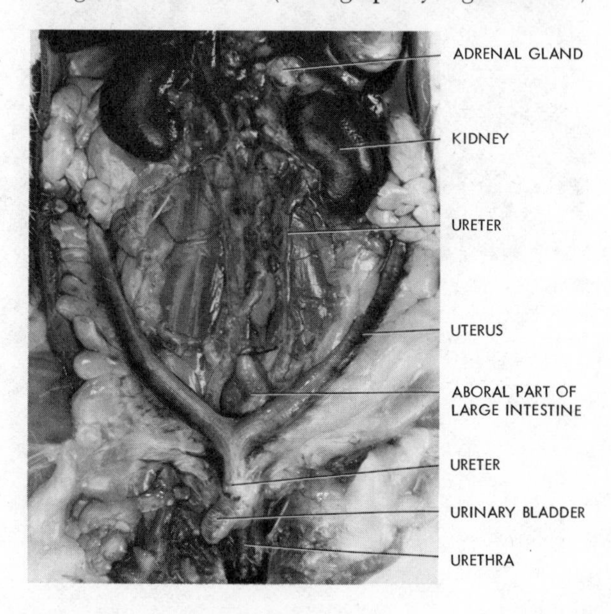

The Participation of Ducts, Urinary Bladders, and Cloacae in Excretion

When urine leaves the nephrons and collecting ducts, it traverses a longer or shorter path before being voided from the body. It passes first through urinary ducts (archinephric ducts in vertebrate embryos, many adult fish, and some amphibians; opisthonephric ducts in other fish and amphibians; and metanephric ducts or ureters in amniotes). Then it may be voided directly from the body through a special opening behind the anus (bony fish), or it may enter a storage chamber. This may be a urinary bladder or urogenital sinus, an expanded region of the ureters, and, thus, as in the case of some fish, of mesodermal origin. In tetrapods, the urinary bladder is of endodermal origin. In amphibians, it arises as an outpocketing of the ventral wall of the cloaca, and urine must traverse the cloaca on its way to storage there. Reptiles and mammals have urinary bladders that are derived embryologically from the allantois and so also are of endodermal origin. Most birds as well as some reptiles (snakes, crocodiles) are devoid of such a bladder, and their urine flows directly from the ureters into the cloaca. From its storage site, the urine finally emerges from the body through the cloacal orifice (amphibians, reptiles, birds, monotremes) or through a *urethra* leading to the outside (higher mammals). The function of urine-storage structures appears to vary from one vertebrate form to another, but probably they generally allow a final reprocessing of the urine formed by the kidneys (Figure 8-18). Ions such as sodium and chloride may be actively reabsorbed into the body fluids through the walls of the large intestine, cloaca, or bladder (amphibians, fresh-water dwelling reptiles, birds), and water may also be reabsorbed. In some cases, the bladder wall is known to be sensitive to hormone (posterior pituitary) control, with increasing water retention when blood levels of the hormone rise. This adjustment parallels the effects of the hormone on kidney tubules (Chapter 12). Bladder- and cloacal-water conservation is of urgent importance in desert-dwelling amphibians and in reptiles and birds. Indeed, in reptiles and birds so much water may be reabsorbed in the cloaca that the final urinary product emerges as a semisolid paste consisting of uric acid, salts, and other excretory materials, with only a minimal amount of water. Such reabsorptive functions are not found in the urinary bladders of higher mammals, but even here the bladder may have important functions (Figure 8-19). It retains the urine until a time suitable for its ejection from the body (by a complicated process, mediated by nerves and muscles of the bladder wall, urinary tract, and body as a whole, termed *micturition*). Urine voided at appropriate times serves as a behavioral cue in many mammals, and the intermittency of urine flow may protect tissues lying near the urinary orifice from constant dampness and excessive threat of bacterial invasion.

A variety of vertebrate animals use the urinary passages as channels for releasing reproductive cells from the ovaries and testes. Thus, in some fish and amphibians the oviduct arises by longitudinal division of the archinephric duct. The male reproductive tract in most vertebrates includes ducts originating from the archinephric ducts. In some fish and amphibians, the sperm and urine traverse common pathways, but in most other vertebrates special tubes for the transfer of urine arise after incorporation of the archinephric ducts into the reproductive system. Some nephric tubules develop into sperm transit routes in many vertebrates. Where this structure is present, sperm generally emerge into the cloaca, but in higher mammals sperm travel through the urethra on their way to the outside. The final form of the excretory apparatus is more or less

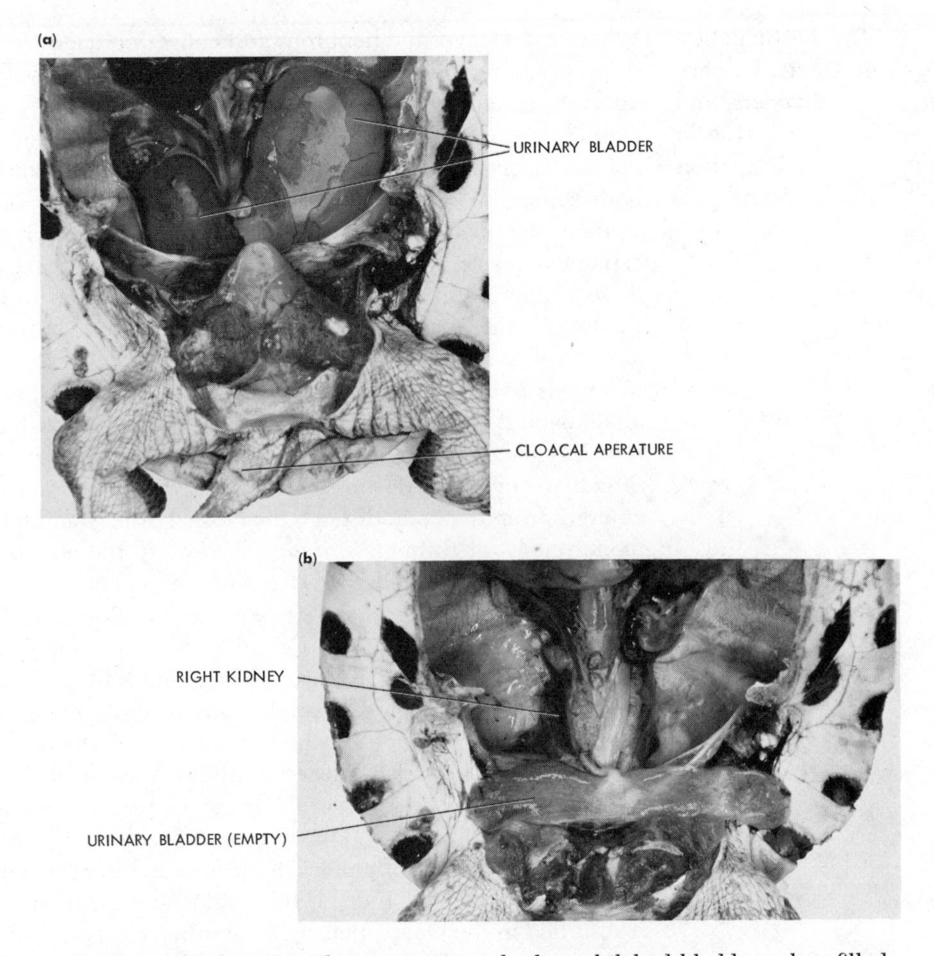

Figure 8-18 (a) In the painted turtle, *Chrysemys picta,* the large bilobed bladder, when filled, occupies much of the body cavity. Only a portion of the posterior part of the body is shown in this illustration, seen from the ventral view after removal of the plastron. (b) The bladder, when emptied, is seen to be a delicate, thin-walled structure. In this turtle, as in many other vertebrates, the urinary bladder serves as a storage and reprocessing site from which salts can be recovered before the urine is voided from the body. In some cases, water, rather than solutes, is specifically conserved by the bladder. (Photographs by Sigurd Olsen.)

dependent on its interaction with the reproductive system. These points will be discussed more fully in Chapter 12.

Perspective on Excretion and Osmoregulation

As was suggested at the beginning of this chapter, the kidneys, despite their highly effective organization for homeostatic regulation, never operate alone. In all vertebrates they are supported by extrarenal excretory mechanisms ranging, for instance, from the liver's excretion of organic compounds to the secretion of sodium and chloride ions by fish gills and rectal glands, and by reptile and bird nasal glands. To understand excretory and osmoregulatory functions completely, the interplay between these various regulatory structures must be taken into account. Perhaps a specific illustration will clarify this

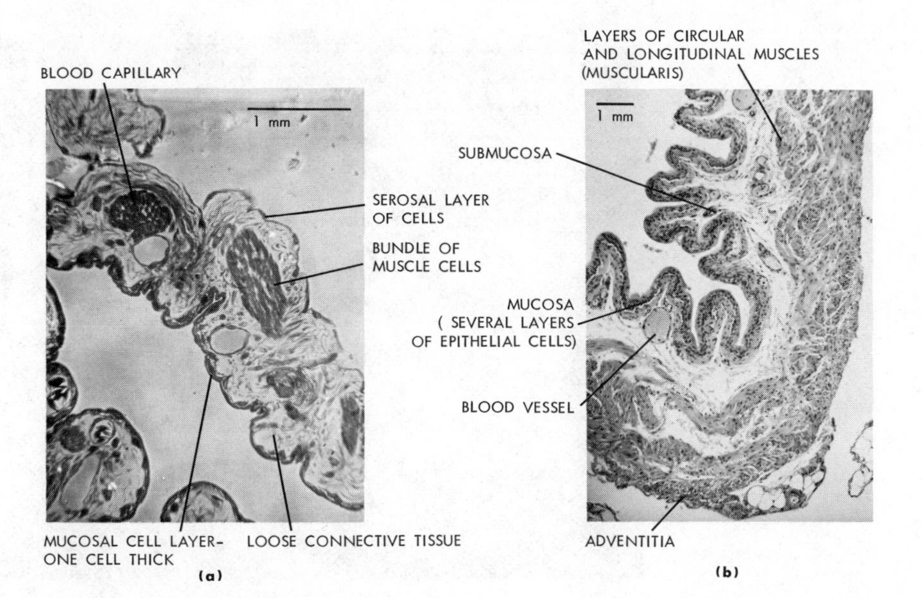

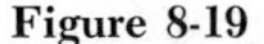

Figure 8-19 The structure of the urinary bladder ranges widely from vertebrate class to class. The amphibian bladder wall is thin, with a single layer of absorptive (mucosal) cells lining the bladder cavity. Loose connective tissue, a few muscle cell bundles, and a thin layer of serosal cells make up the rest of the wall. Fluid is readily absorbed through the delicate wall, especially when the bladder is filled and the wall has become thin as a result of stretching. (a) Is a photomicrograph (270×) of a small part of the unstretched bladder wall of the large toad, *Bufo marinus.* The micrograph was taken from an unstained epoxy section with phase contrast microscopy. (Courtesy of Dr. Alexander Leaf.)

In the mammal, the bladder wall is relatively thick and muscular. The many-layered epithelium lining the bladder cavity changes in thickness as the bladder fills and empties (transitional epithelium), and the inner layers of the wall can be thrown into deep folds, or stretched, depending on the degree of bladder filling. The heavy muscular layer plays a major part in controlling bladder volume. (b) is a photomicrograph (100×) of a small part of the unstretched bladder of the laboratory mouse, *Mus musculus,* stained and shown at about one-third the magnification used for the toad bladder micrograph (Courtesy of Dr. Katharine P. Hummel.)

point. As was noted earlier in this chapter, various marine fish have essentially aglomerular kidneys. This condition is reasonable because of the highly concentrated (i.e., relatively salt-rich and water-deficient) environment in which they live. Rather surprisingly, however, fresh-water fish of a few species, such as the pipefish, *Microphis boaja,* also have aglomerular kidneys. Furthermore, some fish, including the eel, *Anguilla rostrata,* can adapt to life both at sea and in fresh water without corresponding changes in the structure of their renal corpuscles. In all these exceptional cases, compensatory adjustments must be sought in the tubular regions of the kidneys and in extrarenal mechanisms. In most instances, the details of these adjusting mechanisms are unknown.

The functions of the extrarenal salt-secreting structures and of the kidneys are at least partly controlled by endocrine secretions (chiefly from the pituitary and adrenal glands). As might be expected, the regulatory mechanisms, in general, are highly sensitive to conditions of the external and internal environments. For instance, the tubules of the nasal salt glands of marine reptiles

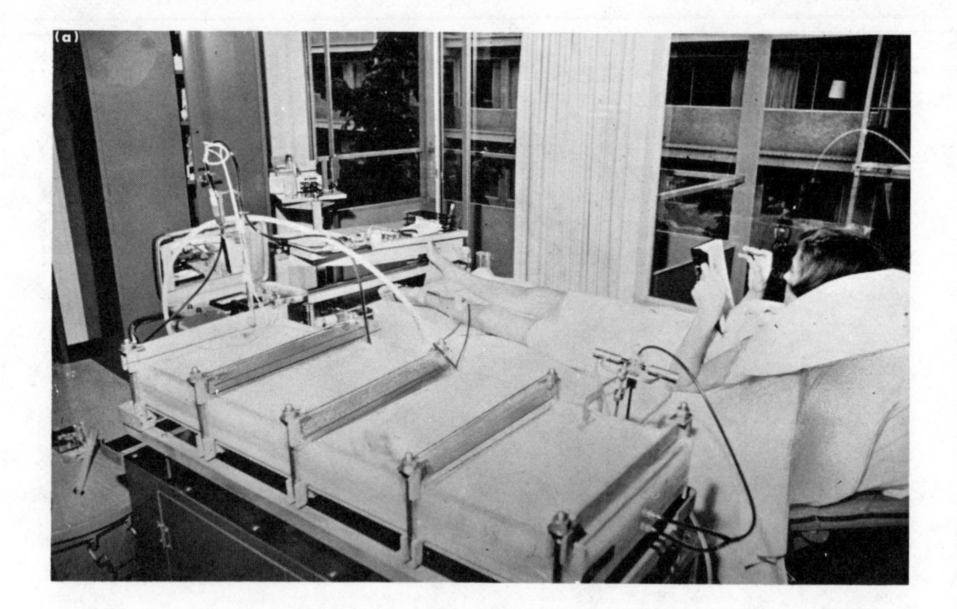

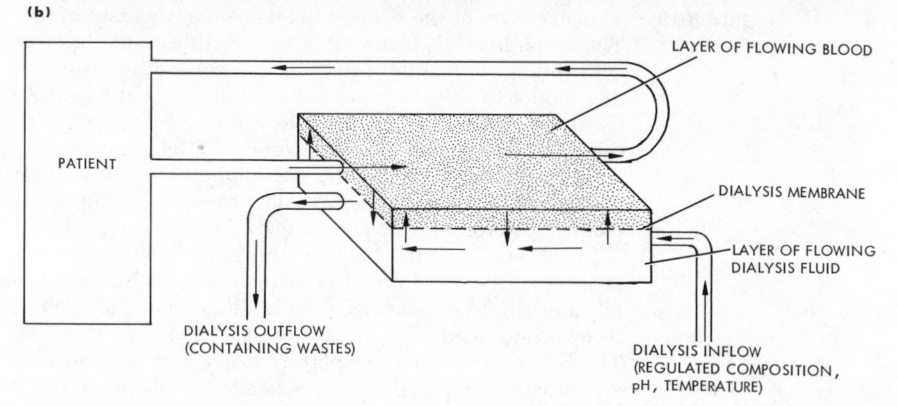

and birds enlarge as the animals adapt to increasingly hypertonic environments. In mammals, healthy kidney tubules enlarge when the diet is excessively rich in protein (which creates an increased demand for excretion of nitrogenous waste) or when part of the kidney tissue is damaged or destroyed. As in every other phase of the life of vertebrates, the intimate interplay between structure and function is clearly reflected in the details of the intricate, varied, and labile organization of the excretory and osmoregulatory systems.

Figure 8-20 is a simplified illustration of one type of man-made "kidney" used by physicians to substitute for the urgently important excretory functions of patients with impaired renal operation. Comparing and contrasting real and artificial kidneys may serve to bring into clear focus some basic aspects of vertebrate renal structure and function.

References for Chapter 8

[1] Bentley, P. J. "Adaptations of Amphibia to Arid Environments." *Science*, **152**(1966), 619–623.

[2] ———. *Endocrines and Osmoregulation*. New York: Springer Verlag, 1971.

Figure 8-20 (*opposite*) Artificial "kidneys" have been designed to substitute for the impaired kidneys of people suffering from certain kinds of kidney disease, poisoning, and circulatory shock. (a) Is a photograph of a patient with an artificial kidney in operation. The blood (rendered incoagulable) is led from an artery, such as one in the lower arm or leg, into a dialyzing system. Here the blood comes in contact with a porous membrane (e.g., cellophane) of large surface area (e.g., about 400 cm^2). Waste materials, as well as other components of plasma, filter through the membrane into a mass of balanced salt solution containing glucose, and with pH and osmotic pressure at levels characteristic of the normal person. Plasma proteins and other very large molecules cannot penetrate through the membrane, and so are retained in the circulating blood plasma. Normal plasma constituents, such as sodium and chloride ions and glucose, even though they may be lost through the dialysis membrane, are restored to the blood by diffusion from the balanced solution of the bathing medium which contains these substances at normal blood concentrations. Waste materials of low molecular weight, such as urea and uric acid, pass through the dialysis membrane and are distributed through the bathing medium. Thus, they are effectively removed from the blood before it returns to the body through the vein. Note the countercurrent arrangement of flow of blood and dialysis fluid through the vein. The rates of flow of dialysis fluid are high (e.g., 400–500 ml/min).

(b) Schema indicating components of an artificial kidney. The artificial kidney is quite selective in removing harmful wastes from the body, while retaining valuable components of the blood plasma. In its functioning, it can be compared with a vertebrate kidney operating with effective glomerular filtration, and with blood flow to the tubules intact, but without any active cellular processes (reabsorption, secretion) of the tubular cells. In this analogy, the dialysis fluid may be compared with the urine of intact vertebrates. The analogy is not exact, however, and the reader is invited to analyze it further, considering for instance, the relative efficiency, demands for water supply, and other aspects of the operation of a normal kidney as contrasted with the artificial kidney.

[3] Brodal, A., and R. Fänge. *The Biology of Myxine.* Oslo: Universitetsforlaget, 1963.

[4] Bulger, R. E. "Fine Structure of the Rectal (Salt-Secreting) Gland of the Spiny Dogfish, *Squalus acanthias.*" *Anatomical Record,* **147**(1963), 95–107.

[5] Chan, D. K. O., and J. G. Phillips. "The Anatomy, Histology and Histochemistry of the Rectal Gland in the Lip-Shark *Hemiscyllium plagiosum.*" (Bennett), *Journal of Anatomy,* **101**(1967), 137–159.

[6] Chase, S. W. "The Mesonephros and Urogenital Ducts of *Necturus maculosus Rafinesque.*" *Journal of Morphology,* **37**(1923), 457–532.

[7] Doyle, W. L. "The Principal Cells of the Salt-Gland of Marine Birds." *Experimental Cell Research,* **21**(1960), 386.

[8] Dunson, W. A., and A. M. Taub. "Extrarenal Salt Excretion in Sea Snakes (Laticauda)." *American Journal of Physiology,* **213**(1967), 975–982.

[9] Fox, H., "The Amphibian Pronephros." *Quarterly Review of Biology,* **38**(1963), 1–25.

[10] Fraser, E. A. "The Development of the Vertebrate Excretory System." *Biological Review,* **25**(1950), 159–187.

[11] Philpott, C. W. "Halide Localization in the Teleost Chloride Cell and Its Identification by Selected Area Electron Diffraction." *Protoplasma,* **60**(1965), 7–23.

[12] ———, and D. E. Copeland. "Fine Structure of Chloride Cells from Three Species of Fundulus." *Journal of Cell Biology,* **18**(1963), 389–404.

[13] Schmidt-Nielsen, K. "Organ Systems in Adaptation: The Excretory System," in *Adaptation to the Environment* by D. B. Dill, E. F. Adolph, and C. G. Wilber, *Handbook of Physiology,* sect. IV. Washington, D.C.: American Physiological Society, 1964.

[14] ———, C. Barker Jorgensen, and H. Osaki. Extrarenal Salt Excretion in Birds." *American Journal of Physiology,* **193**(1958), 101–107.

[15] ———, A. Borut, P. Lee, and E. Crawford. "Nasal Salt Excretion and the Possible Function of the Cloaca in Water Conservation." *Science,* **142**(1963), 1300–1301.

[15] Staaland, H., "Anatomical and Physiological Adaptation of the Nasal Glands in Charadriiformes Birds." *Comparative Physiology and Biochemistry,* **23**(1967), 933–944.

9

The Sensory Receptors

The Role of the Nervous System

The *nervous system* and the *endocrine system* are the integrating systems of the vertebrates. In its role as a coordinator, the nervous system evokes actions and reactions to the myriad of environmental stimuli that are continually bombarding the organism. These responses are directed primarily (1) to the survival of the organism (e.g., movements, protection against enemies, drive for the acquisition of food, maintenance of body temperature in warm-blooded animals) and (2) to the survival of the species (e.g., reproduction: mating, nesting, and care of offspring). The nervous system is organized (1) to sense and to test the environment and to project the resulting influences as *input* to the central nervous system; (2) to process and to store this input within the *central nervous system* for the immediate and future use by the organism: (3) to transmit *output,* which regulates the organism's activities by stimulating or inhibiting the actions of the muscles and glands of the body, and (4) to interact with the endocrine system.

Input (Sensory or Afferent Information) to the Central Nervous System

The nervous system is continually monitoring the environment, both external and internal to the organism, through many sensors (sensory endings). Through these sensors pertinent environmental energies are selected and coded into nerve impulses that are transmitted as *input* by the peripheral nerves to the central nervous system. This input may be perceived at conscious levels in the brain or utilized at unconscious levels. Conscious sensations are the organism's unique interpretation of the environmental stimuli; they are not precise copies of the environment. For example, pain is felt, yet environmental pain as such is nonexistent; a sound is heard, yet a sound is actually not a sound, but rather a wave-length frequency (vibration).

Processing and Storage of Information within the Central Nervous System

The input from the periphery is projected to central nervous system processing stations, which are organized groups of neurons called *nuclei, ganglia,* centers, *neuronal pools,* and cortical areas. These stations act as complex analytic computers and intricate control systems. The new input, acting on the reactive matrix of the nervous system, may influence and activate the expression of the organism's actions. These influences may have an *immediate effect* (e.g., from a simple knee jerk to the reactions exhibited by a hungry predator at sensing prey) or a *delayed effect* (e.g., reactions after the recall of an event that occurred in the distant past). The input may be "held in storage" and later utilized in behavioral patterns, emotional responses, learned skills, memories, and abstractions.

Output (Motor or Efferent Actions)

The actions and reactions of the nervous system are overtly expressed exclusively through muscular activity, glandular secretion, *neurosecretion,* electric discharges in certain fish, and luminescent discharge of a photogenic organ in some deep-sea fish. The output from the central nervous system to the effectors is mediated largely through the peripheral nervous system and, in part, through the endocrine system.

The nervous system and the endocrine system are interacting feedback systems. In fact, the endocrine system is primarily controlled by and integrated with the nervous system. Because of this interaction, the endocrine system is often called the neuroendocrine system.

Classification of Sensory (Afferent) Input

The varieties of information received by the body as input from environmental stimuli are sensed, appreciated, and interpreted by the organism in many ways. To codify the product of these afferent stimuli, several classifications have been proposed. Although none is wholly satisfactory, each makes significant points concerning sensory receptors (nerve endings), sensors, or sensations.

Viewed subjectively, afferent stimuli have roles in the conscious or in the unconscious. Because sensations are basically the organism's subjective interpretations, it is difficult for man to conceive of the manner in which each vertebrate perceives conscious sensations. The unconscious sensations are integrated into activities such as muscular coordination, respiration, cardiovascular tone, alimentary canal motility, and secretory activities. The senses and the sensors may be named according to the organ system with which they are associated; somatic and visceral. *Somatic* refers to the extremities, body wall, and portions of the head; *visceral* refers to the vital organs of the digestive, circulatory, excretory, respiratory, and endocrine systems. Senses with a wide distribution in the body are called *general senses,* those limited to a restricted region are called *special senses.* The general senses (and their receptors) include pain, temperature appreciation, touch, visceral sense, body sense, position sense, and unconscious senses associated with reflex activities. Those of special sense include hearing, balance sense by vestibular receptors, appreciation of water currents (monitored by the inner ear and lateral line), sight (monitored by eye), smell (monitored by olfactory and vomeronasal mucosa), and taste (monitored by *taste buds*). General senses (and their receptors) are classified further as general somatic senses and general visceral senses; for example, pain in the skin and joints is a general somatic sensation, whereas pain in the stomach is a general visceral sensation. The special senses are either somatic senses (appreciation of water currents, hearing, balance, and sight) or visceral senses (smell and taste).

All sensory receptors are biologic transducers by which the stimulus of one form of environmental energy initiates a reaction that may lead to the generation of a nerve impulse. Each receptor is especially sensitive to a specific form of energy—e.g., the sensors in the eye to the electromagnetic waves in the visual spectrum. The lowest-intensity stimulus to which a specific receptor will respond to this most sensitive form of energy is called the adequate stimulus.

The General Senses

The general senses may be classified as *exteroceptive senses, proprioceptive senses,* and *interoceptive senses.* Exteroceptive cutaneous senses are the product of many external environmental energies that stimulate receptors located in the skin (Figure 9-1) and include the general sensations of pain, warmth, cold, and light touch. The movement of hairs and vibrissi and the contact on the skin without deforming it produces light touch. The proprioceptive senses (kinesthetic and deep sensibility) result from the stimulation of receptors by energies generated within the somatic structures (Figure 9-2); these include both conscious and unconscious sensations derived from sensory receptors in muscles, tendons, joints, skin, and connective tissues. Conscious senses include those of position and movements of body parts (joint sense), vibratory sense

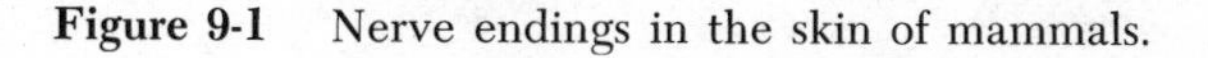

Figure 9-1 Nerve endings in the skin of mammals.

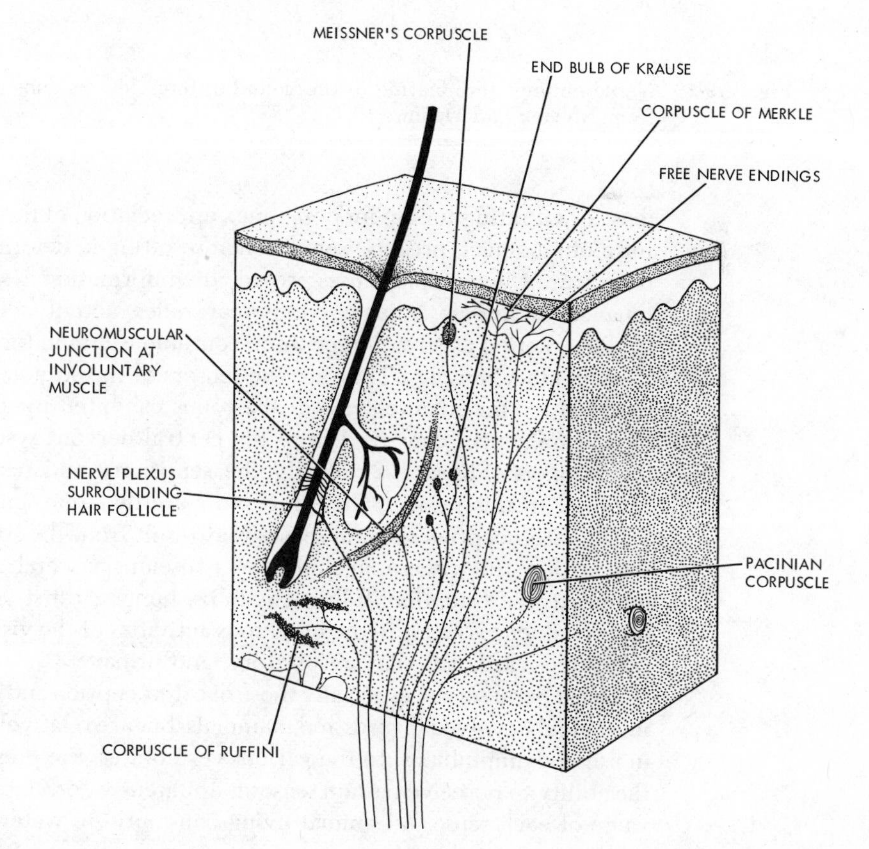

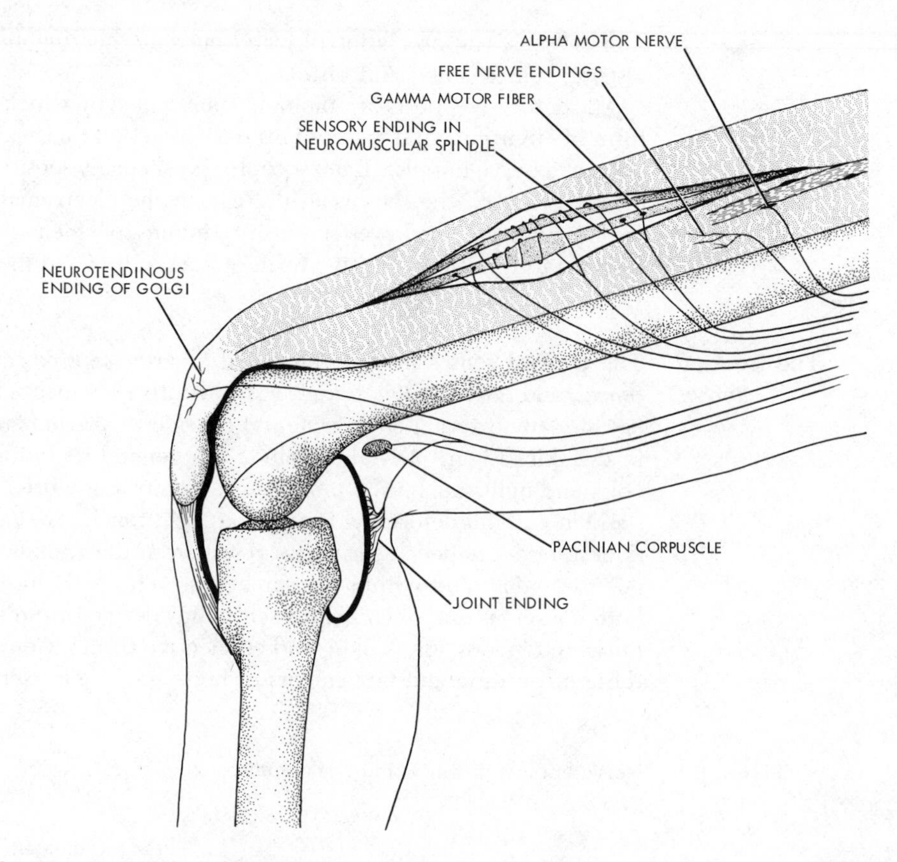

Figure 9-2 Nerve endings terminating in the voluntary muscles, tendons and joints. (Adapted from Noback and Demarest.)

(base of vibrating tuning fork on bone), appreciation of the weight, shape, and form of an object, pressure sense (touch resulting in deformation of skin), and deep pain. Unconscious senses provide the information essential for the maintenance of muscle tone and muscular reflex activity. The neuromuscular spindles, acting as stretch receptors, continuously monitor the tension within the voluntary muscles. They are the sensors of the stretch reflexes, which are essential for the maintenance of the tonus exhibited by all skeletal muscles. Neural influences conveyed from the central nervous system by the gamma motor fibers (Figure 9-2) modify the sensitivity and responsiveness of the spindles. In this respect, these receptors are dynamic sense organs.

The interoceptive visceral senses that result from the stimulation of receptors within the viscera, include the conscious visceral sensations of pain, temperature, distention, fullness, cramps, hunger, thirst, and satiety. Unconscious influences result in the numerous activities of the visceral organ systems (digestive, cardiovascular, respiratory, and urinary).

The general senses, especially those of exteroception and proprioception, are most highly evolved in birds and mammals, but are relatively poorly developed in fish and amphibians. In the various vertebrates, the presence or absence of the ability to perceive certain senses is apparently correlated with the survival value of each sense. An animal living constantly in water is subject to equal

pressure over its body and to only gradual changes in environmental temperature; hence, fish have probably not evolved a variety of elaborate general senses (e.g., two-point discrimination and temperature). In contrast, terrestrial vertebrates receive a variety of continuously changing stimuli from both exteroceptors and proprioceptors about external and internal environments. The exteroceptors sense temperature changes and touch from the differential contact of objects on the body surface (for example, contact of feet on ground); the proprioceptors monitor for data from the muscles and joints as they constantly readjust to the alteration in the tension of muscles and the movement of joints. In general, fish and amphibians have many free nerve-ending receptors. Relatively simple proprioceptors (neuromuscular spindles) are found in the muscles of amphibians. In addition to simple, free nerve endings, birds and mammals have a wide variety of specialized sensory receptors (Figures 9-1 and 9-2). These include Meissner's corpuscle (touch), Paccinian corpuscle (pressure), Merkel's corpuscle (touch), corpuscle of Ruffini (warmth), end-bulb of Krause (cold), neuromuscular spindle (tension within muscles), neurotendinous endings of Golgi (tension within tendons), and others. Transitional forms exist as well.

Free nerve endings are the primary pain receptors; they are located in many regions including the skin, pulp of a tooth, cornea of an eye, and the intestinal tract. Each of these receptors may be excited by thermal, mechanical, and chemical stimuli. Tissue damage may evoke pain sensations; the traumatized tissue releases chemicals, such as histamine, that stimulate the pain receptors. The thermal modalities are sensed by heat and cold receptors located mainly in the skin. Besides the capsulated endings, the corpuscle of Ruffini and end-bulb of Krause noted before, free nerve endings also are thermal receptors. Heat may elicit sensations of both pain and temperature. Within the skin of man, cold receptors are said to be more abundant and more superficially located than warmth receptors. Following stimulation of the cold receptors, reflexes can be evoked that tend to conserve heat; hair and feathers are erected to create a dead air space around the animal; cutaneous blood vessels constrict to reduce the radiation of heat into the air; and the muscle contractions of shivering produce heat. Following the stimulation of heat receptors, reflexes are evoked that tend to dissipate heat, cutaneous blood vessels dilate to increase radiation of heat from the body, and the evaporation of water from the body surface is increased following panting and sweating.

The concept that each nerve ending is exclusively the receptor associated with a specific sense (modality) is known as the law of specific nerve energies. This has been implied before in associating a modality with a receptor (e.g., touch with Meissner's corpuscle). A modern concept is the pattern theory of sensation, which relates a group of nerve endings with each modality. In this concept, a small complex of endings is associated with a *spot,* called warm spot, cold spot, or touch spot. The differential stimulation of the receptors within each spot in different combinations results in various nuances associated with a sensation (e.g., sharp, burning, and dull pain).

A specialized heat receptor, called the pit organ or facial pit, is present in a depression in front of the eye in certain snakes called pit vipers (rattlesnakes) and some boid snakes (pythons). With this receptor, these snakes can detect the presence, even in the dark, of warm-blooded mammalian or avian prey, such as a mouse. These ophidian pit organs are sensitive to wave lengths longer than those of visible light. In essence, they are sensitive radiant heat receptors that can detect a minute change of but 0.003 degree centigrade of heat energy. In boid snakes, these sensors are a series of pits in scales located within the mouth.

Special Senses The receptors of the senses are innervated by branches of the cranial nerves (Figure 9-3): smell by the *olfactory nerve* from the olfactory mucosa and by the terminal nerve from the vomeronasal organ; vision by the *optic nerve* from the eye; sound, head position, and appreciation of water currents by the *vestibulocochlear nerve* from the inner ear and the lateral line organ; and taste by some fibers in the *facial nerve, glossopharyngeal nerve,* and *vagus nerve* from the taste buds. The receptors of smell and taste are the chemists; they are chemoreceptors or special visceral afferent receptors reacting to chemical stimuli. (They are visceral because of their association with sensing for food.) The receptors of the optic system and the vestibulocochlear systems are the

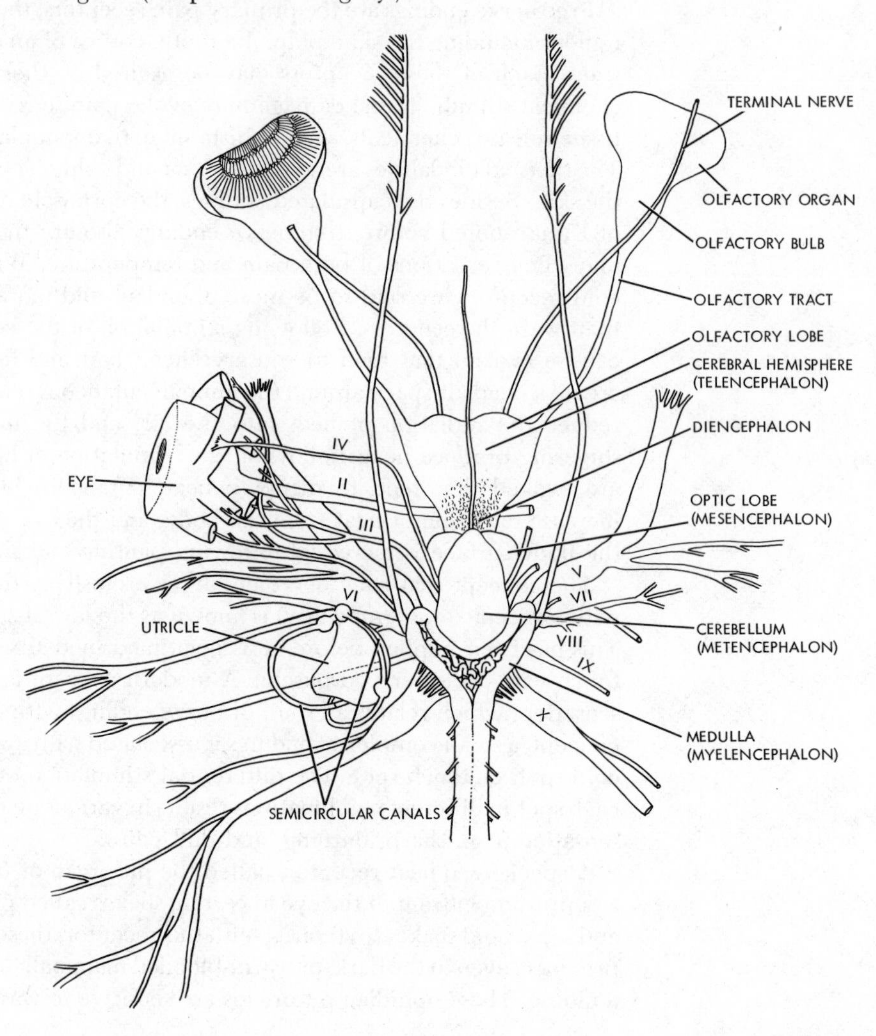

Figure 9-3 Brain and cranial nerves of the shark *Carcharius littoralis*. Cranial nerves, I through X. (Adapted from Kingsley.)

physicists, or special somatic afferent receptors, reacting to physical stimuli. The eye is a photoreceptor and the inner ear and lateral line organ are mechanoreceptors; the former react to light, the latter to vibration.

The Chemical Senses

Vertebrates sense a wide variety of chemical constituents from simple inorganic compounds (e.g., salt and acids) to complex organic compounds. The chemical senses are commonly divided into three categories: common (general) chemical sense, olfaction (smell), and *gustation* (taste). These three chemical senses are arbitrary and overlap. The chemoreceptor sensors have moist surfaces. This moisture, a product of supporting cells and glands, acts as a solvent in which the stimulating chemiclas are dissolved (Figure 9-5). In this medium, the active radicals of the odorant chemicals can adequately stimulate receptor sites on the cells of the sensors. The role of these sensory systems is more than just to identify chemical entities; they are the activators and the sensitizers of the nervous system, evoking many complex behavioral patterns associated with such activities as the search for food, the sexual responses of animals during the mating season, the *agonistic* actions in the presence of an enemy, the preservation of an animal's status within a social group (pecking order), and the identification of offspring, parents, or friendly species.

The olfactory sensors are "distance chemical receptors" stimulated by ultraminute concentrations of odors conveyed through the water to the nasal mucosa of aquatic animals and through the air to the nasal mucosa of terrestrial animals. The gustatory sensors are contact receptors stimulated by the direct contact of the chemical substances, which are dissolved in the secretions surrounding such sensor cells as the taste and olfactory neurons (Figures 9-4 and 9-5). In general, the olfactory system is more sensitive than the gustatory

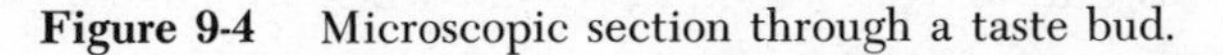

Figure 9-4 Microscopic section through a taste bud.

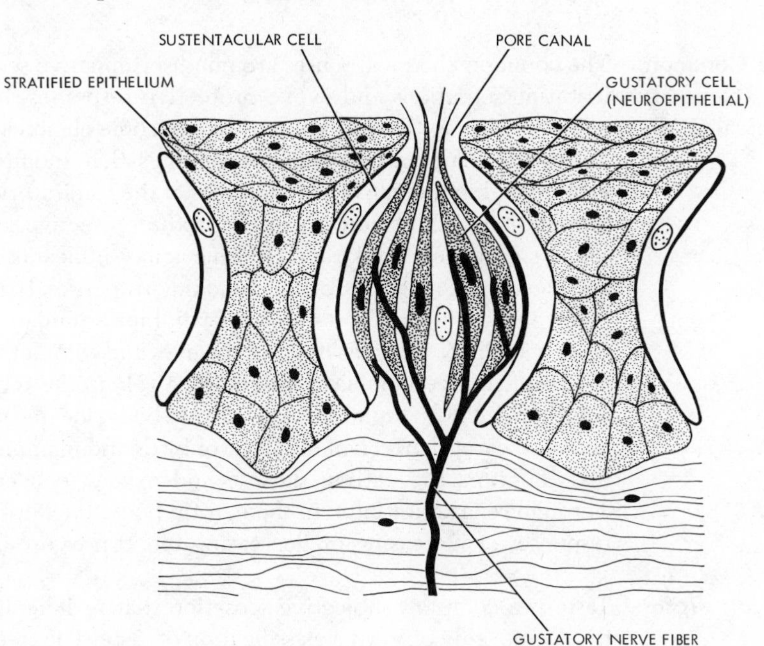

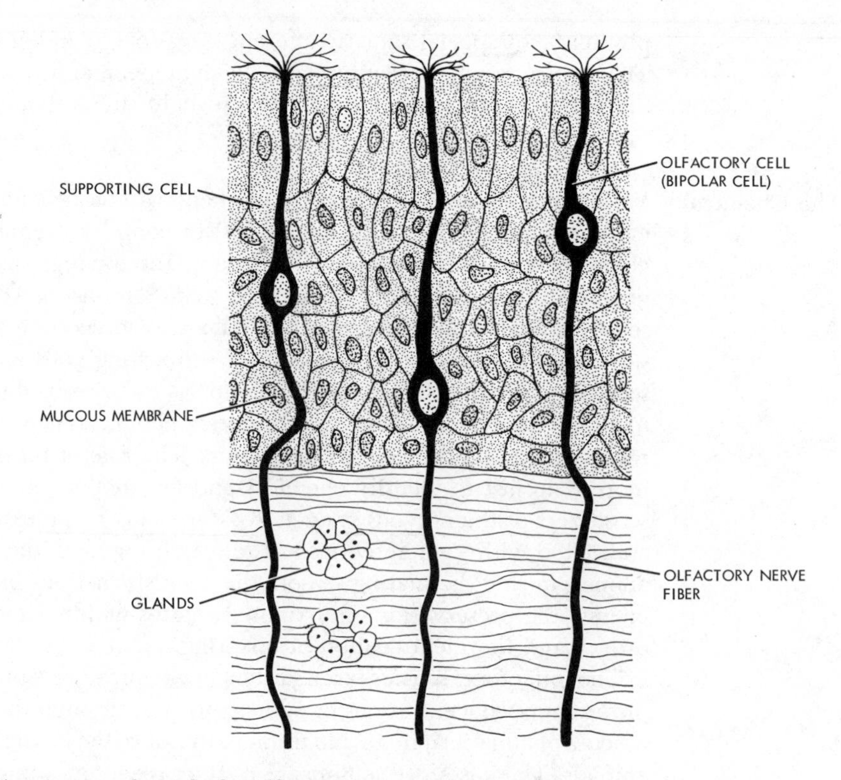

Figure 9-5 Section through the olfactory mucosa.

system; quantitative estimates indicate that it is from ten to twenty thousand times more sensitive (ten thousand in man). This explains why we smell food more than we taste it, and why food tastes flat when we have a cold.

The Common (General) Chemical Sense

The common chemical senses are nondiscriminative senses that often produce irritating sensations and evoke protective responses. In fish and amphibians, nerve endings in the skin are sensitive to some chemical changes in the water. Terrestrial mammals have nerve endings that monitor the chemical sense located within the mucous membranes of the respiratory and digestive systems. Stimulation of these endings results in the protective secretion of mucous and serous substances. Smoke and chlorine act as suffocants on the respiratory tract and evoke mucous secretions and choking responses. Irritants such as ammonia, acids, and astringent food such as persimmon stimulate the common chemical sensory endings. Tear gas irritates the eyes and stimulates the secretion of tears.

Chemical receptors have a significant role in the regulation of the internal environment of an organism. The carotid body, located at the bifurcation of the common carotid artery in the neck of birds and mammals, is a chemoreceptor that monitors the carbon dioxide and oxygen concentration of the blood. Through reflexes initiated in the carotid body, the appropriate cardiovascular responses of increasing or decreasing blood pressure and blood flow follow.

Taste

Taste is a complex, subjective sensation that is difficult to define objectively; there is no truly objective classification of taste. The so-called elementary taste sensations of sweet, sour, salt, and bitter are incorporated into a subjective classification that is used for want of a better substitute. In all probability there are no primary taste qualities comparable to sounds defined in frequencies of cycles per second.

The peripheral receptor for taste is the brandy-snifter-shaped taste bud (Figure 9-4). Each bud is a "contact chemical receptor" composed of up to twenty-five neuroepithelial taste cells and a number of so-called *sustentacular cells;* the latter are actually immature taste cells that replace the aged taste cells. In mammals, each mature taste cell is replaced in about ten to twelve days. The distal tip of each taste cell has a surface specialization of microvilli that interact with sapid substances, either as whole molecules or as ions. Structural differences among the various taste buds do not seem to exist. Current theory suggests that each taste cell reacts to several taste qualities and that the nuances of taste are transmitted by a variety of patterns of sensory impulses from each taste bud.

In most fish and amphibians the taste buds are found not only in membranes lining the mouth, pharynx, and gill apparatus, but may be widely distributed in the body skin. In all vertebrates these taste buds are innervated by the three cranial nerves, facial, glossopharyngeal, and vagus. Taste buds on the rostrum of the sturgeon, a bottom feeder, enable it to sense food by tasting the water before reaching the food. A catfish, which has taste buds scattered over its body surface, tail, and barbels, can immediately be attracted to swim to that side on which a barbel has been stimulated with a drop of meat juice. By tasting the water, many fish maintain contact with their school. After detecting by taste the presence of a predator fish, certain fish are alarmed into an escape reaction. Mammals, whose taste buds are primarily concentrated in the surface epithelium of the tongue, also have taste buds in the palate, mouth, and pharynx. Birds have a few taste buds on the back of the tongue and the palate.

Smell

The olfactory sensors, including the olfactory mucosa and the vomeronasal mucosa, are fundamentally similar in living vertebrates (Figure 9-5). The main difference among the various animals is not in this relatively stable sensory apparatus, but rather in the degree of complexity of the chambers housing these receptor membranes. Vertebrates with a well-developed sense of smell are called *macrosmatic animals;* most mammals, including carnivores, ungulates, and rodents, are macrosmatic. Those vertebrates with a poorly developed sense of smell are called *microsmatic animals;* these include birds, primates, and the baleen whales. The toothed whales are anosmatic mammals with no sence of smell.

In fish, the olfactory mucosa, which comes into direct contact with water, is restricted to a single area. In elasmobranch and teleost fish, the sensors are located in the blind olfactory pit (nasal sac); water flows in and out of this chamber without passing into the mouth. In lungfish, the mucosa lines a nasal cavity that extends from the exterior to the mouth; as in many higher vertebrates, the medium containing the odors passes over the olfactory sensors before entering the throat. The olfactory nerve projects to the olfactory bulb (Figure 9-3). Unlike the conditon in higher vertebrates, there is no separation into a main olfactory bulb and an accessory olfactory bulb. The nerve fibers from the olfactory bulb project via the olfactory tract to the olfactory cortex of the brain.

In most tetrapods, two sensory areas are associated with the olfactory apparatus: an olfactory mucosa with nerve fibers projecting to the main olfactory bulb via the olfactory nerve, and a vomeronasal (Jacobson's) mucosa with nerve fibers projecting to the accessory olfactory bulb via the terminal nerve (Figure 9-3). Whereas the olfactory mucosa is always located within the nasal cavity, the vomeronasal mucosa, which is always separated from the olfactory epithelium, is located either in the ventral medial aspect of the nasal cavity or in an outpocket of the oral cavity. The precise function of the vomeronasal organ of Jacobson is still vague. The organ is well developed in amphibians, lizards, snakes, monotremes, marsupials, insectivores, bats, prosimians, and New World monkeys. It differentiates and then becomes rudimentary in crocodiles, birds, most aquatic mammals, Old World monkeys, and man. The vomeronasal organ is located dorsal to the oral cavity. In reptiles, the cavity of this organ is continuous with the oral cavity. The well-developed vomeronasal organs in lizards and snakes are used to sense chemicals picked up from the air by the protruded tongue. After testing the air the snake's forked tongue is retracted and inserted into the two vomeronasal pockets; the dissolved chemical particles on the tongue stimulate the sensors of the organ. In such an animal, these sensors facilitate smelling and, hence, identification of the food in the mouth.

The olfactory mucosa is composed of bipolar neurons and supporting cells (Figure 9-5). Its bipolar neurons are unique; they are the only nerve cells that perform the dual roles as sensors in direct contact with the external environment and as transmitters of the codes directly to the olfactory bulb of the central nervous system (Figure 9-3). A peripheral process of each bipolar cell terminates in a swelling called an olfactory vesicle (rod) composed of a tuft of one to fifteen cilia (olfactory "hairs" or streamers), depending upon the species. These hairs form a feltwork embedded in the odor-absorbing secretion coating the olfactory mucosa. The ultrasensitive odorant-receptor interaction is presumed to occur at the olfactory hairs. Associated with the olfactory mucosa of many tetrapods are special glands of Bowman, which have an obscure function. These glands are not found in the olfactory mucosa of fish or in the vomeronasal mucosa.

Olfaction has a significant role in many behavioral patterns. Many fish utilize olfactory cues (1) to maintain contact with their territory or their school, (2) to sense enemies and to trigger appropriate escape reactions, (3) to detect food and prey, and (4) to recognize the dominant and submissive members (pecking order) of the social community to which they belong. This last use is vital to the maintenance of the social structure of the fish community. Olfactory cues may be used by migrating fish such as salmon to identify their home and spawning grounds. Apparently the young salmon is imprinted (a permanent change in an animal's behavior as the result of the impact of a sensory cue at a critical time in its life), probably during the first weeks of life, by the odors of the tributary stream of its youth. Later, as a migratory adult, the innate drive to return to the spawning grounds is guided primarily by swimming against a current (*rheotaxis*). The selection of the correct tributary at each fork in the stream is thought to be a function, in part, of the olfactory system; the fish makes a positive response to the key odor cues and a negative response to their absence. Upon entering a tributary lacking the key odor cues, the fish retreats and tests other tributaries until the key cues are sensed. Key odors need not be consciously recognized.

Although most birds are microsmatic, some have a keen sense of smell; these include the nocturnal New Zealand kiwi, vulture, condor, albatross, petrel, and others. The macrosmatic "nose mammals" probably have the most elaborate, acute, and evolved olfactory discriminatory sense of all vertebrates. Elephants, deer, antelopes, sheep, and dogs are in this category. The stalker, to avoid detection, must approach the "nose species" from a downwind position. By smelling her newborn infant immediately after birth, the mother seal registers a scent imprint; with this imprint she is able, by the sense of smell, to recognize and to identify her pup from the others in the rookery. Each of these species has extensive and elaborate scrolls of bone (turbinals, conchae) to increase the surface area of the nasal cavity in order to accommodate the enlarged olfactory epithelial mucosa.

The Eye: The Photoreceptor

Paired lateral eyes are present in the vertebrates. They have secondarily degenerated into vestigial and near-functionless organs in species such as the hagfish, cave-dwelling and abyssal fish, and underground-living amphibians, reptiles, and mammals (e.g., moles). Median eyes, known as pineal and parietal eyes, are found in lampreys and some lizards.

The eyes are the sensitive detectors of light waves. The photoreceptor cells in the eyes of all vertebrates are stimulated by approximately the same range of the radiomagnetic spectrum, from 760 to 390 millimicrons. In a general way, this spectral sensitivity corresponds to the transmission spectrum of water; this suggests that the vertebrate eye was initially evolved to function in the milieu of the primordial vertebrates, shallow water.

The eyes of vertebrates are constructed with similar structural features. Within this basic pattern, a vast array of adaptive variations have evolved to enable the eye to function in many ecological niches. The complete structure of the eye comprises the eye proper (eyeball) and accessory parts such as the eyelids, several types of glands and their ducts, and the extraocular muscles. The optic nerve is made up of fibers projecting from the sensory retina of the eye to optic centers in the brain.

Size and Shape of the Eye

There is no correlation between eye size and body size; in some birds, each eye may occupy one third the volume of the head. The size of the eyeball is related to the image size and has adapted to the light-gathering demands of the animal. A large eye, with its proportionately large image-producing retina, can register a large copy of the environmental field. *Nocturnal* animals have large eyes with which to capture more light out of the dark; their eyes are efficient visual receptors for night vision.

The eyes of vertebrates have three basic shapes: spherical, ovoid, and tubular. Many vertebrates and all mammals have roughly spherical eyes (Figure 9-6), varying from a slightly elongated eye (cat) to a slightly oblate eye (horse). Many *diurnal* birds have flattened, ovoid eyes. Some birds (owls, eagles, and hawks) and some abyssal fish have tubular eyes [Figure 9-6(e)]. Eyes have evolved to gather and to focus light (called *accommodation*) upon the photoreceptive membrane, the *retina.* To accomplish this the eyeball has two segments: the accommodation segment formed by the cornea and lens and the photoreceptive segment formed by the *retina, choroid,* and *sclera* (Figures 9-6 and 9-7). In each eye, the proximity of these two segments is such that the accommodation segment focuses the environmental field of vision upon the retina of the photoreceptive segment.

The shape of the eye is largely maintained by the forces exerted by the internal intraocular pressure from within the eye on the soft tissues of the coats of the eyeball. The natural shape of the eye is generally a sphere. The tubular eye is an adaptation to meet the functional demands of a large eye in a relatively small skull by constricting (hourglass effect) the eye at the junction of the accommodation segment (cornea) and the receptor segment (sclera, choroid, and retina). In effect, the advantages of the large spherical eye are retained in the tubular eye, with only a slight, if any, change in the size and curvature of the cornea and of the retina because of this constriction. By such adaptation the volume of the eye is definitely reduced. This is accomplished, in part, with the aid of a constricting continuous ring of bones or cartilages, called scleral ossicles [Figure 9-6(e)], located adjacent to the sclerocorneal junction; by this means, the intraocular pressure maintains a tubular eye instead of a spherical one. This is analogous to placing a constricting circular band around an inflated balloon.

Basic Structure of the Eye

The eye is composed of three tunics: (1) an outer fibrous tunic, the cornea and sclera; (2) intermediate tunic, the *uvea* (choroid) and (3) an internal tunic, the retina (Figures 9-6 and 9-7). The outer tunic, a thick lamina of dense connective tissue, consists of the opaque sclera (the white of the human eye) and the transparent *cornea.* This tunic acts as the skeleton of the eye and resists the intraocular pressure. The uvea consists of the vascular and pigmented choroid layer, the ciliary body (with muscles essential in accommodation), and the *iris diaphragm* surrounding the pupil. The internal tunic is the retina, which is made up of the light-sensitive receptors (rods and cones) and many nerve cells arranged in layers. The *lens,* located behind the cornea and the iris, is suspended by fine, taut, zonula fibrils that extend from the circumference of the equator of the lens to the ciliary body. The region between the lens, ciliary body, and iris is the posterior chamber and that between the iris and the cornea is the anterior chamber. These two chambers are filled with a fluid called the aqueous humor, which is largely formed by the *ciliary body.* From the ciliary body, the aqueous humor passes successively through the posterior chamber, pupil, anterior chamber, and the meshwork of the iridocorneal (filtration) angle; from here the humor diffuses into Schlemm's canal and the venous drainage (Figure 9-8). The large region between the lens and the retina is filled with a gelatinous matrix called the vitreous humor, which consists of mucoproteins.

To perform its functional role efficiently the eyeball must fulfill several criteria.

1. Light must reach the photoreceptors in the retina.
2. Stray and unwanted light must be prevented from exciting the receptors.
3. The size and shape of the eye must permit light from the visual field to focus an image on the retina. The curvatures and location of the cornea, lens, and retina must be precise within narrow limits.

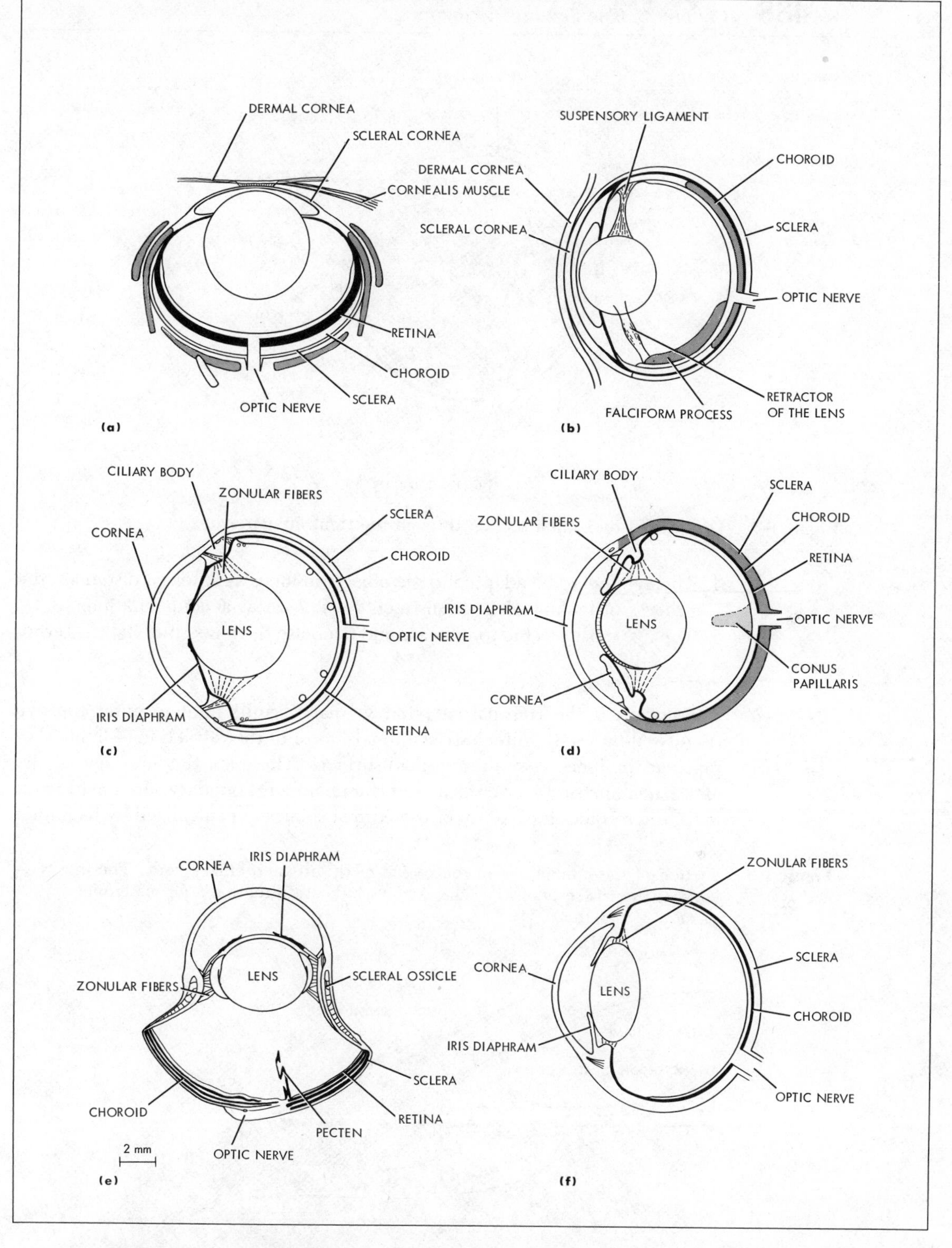

Figure 9-6 Diagrams of sections through the eyes of (a) lamprey, (b) teleost, (c) frog, (d) snake, (e) eagle-owl, and (f) placental mammal. The cornea may be derived from dermal connective tissue (dermal cornea) and the scleral connective tissue (scleral cornea). (After Duke-Elder and after Marshall.)

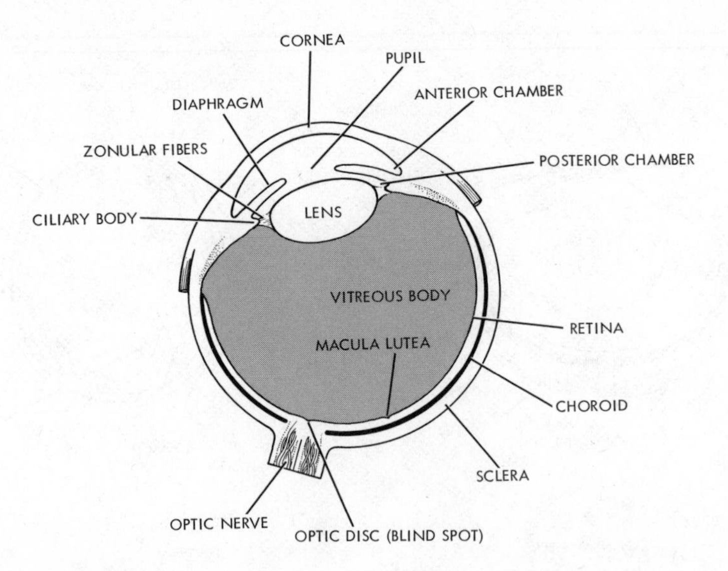

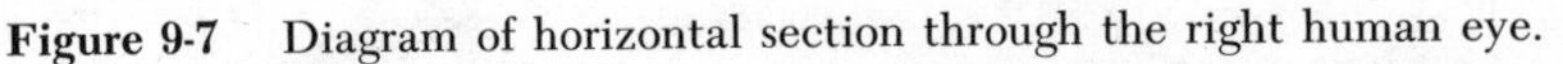

Figure 9-7 Diagram of horizontal section through the right human eye.

4. The eye should be adapted to see objects in focus at different distances; the process of making these adjustments is known as accommodation.
5. The eye must be able to increase and decrease the amount of light entering it.

Cornea The cornea is the transparent window that permits light to enter the eye (Figures 9-6 to 9-8). In terrestrial animals, most of the refraction of light takes place at the inner and outer corneal surfaces. These are the interfaces at the air-cornea junction and cornea- anterior chamber boundary. In aquatic animals, the refracting capacity of the cornea is largely neutralized by its imme-

Figure 9-8 Section of the human eye in the region of the iris and ciliary body. The free edge of the iris rests on the lens. Arrows indicate direction aqueous humor passes.

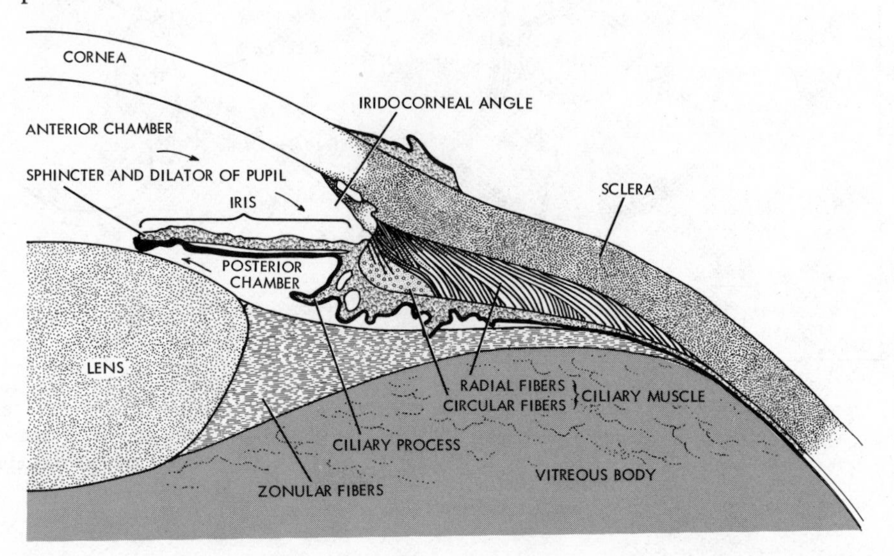

diate contact with water. In land animals, the refraction of the light at the cornea and the lens is collectively integrated to focus and to produce a small environmental image on the retina. In true nocturnal vertebrates and abyssal fish, the cornea is large, permitting the eye to trap as much available light as possible. The transparency of the cornea is due to the laminar organization of its collagen matrix, its lack of any blood vessels, and the state of hydration of its matrix. The avascular cornea is nourished by the aqueous humor of the anterior chamber and by the lacrimal secretion (tears) in which it is bathed. Carotenoid pigments are responsible for the yellow tint of the corneas of some teleost fishes.

Pupil and Iris The pupil is the aperture surrounded by the iris diaphragm, through which the light enters the eye before reaching the lens (Figure 9-9). Muscles within the iris can alter the size of the pupil; muscles that constrict the pupil are sphincters; and those that enlarge the pupil are dilators. In many vertebrates, iridic sphincter muscles are present, whereas iridic dilators are absent. Except for the reptiles and birds in which the iridic muscles are striated and voluntary, all other vertebrates have smooth, involuntary iridic muscles.

In any group of vertebrates, the pupil assumes one of several shapes (Figure 9-9): a vertical slit as in the cat and some snakes, a rectangle as in many ungulates and some whales, or a circle as in man and other primates. The slit pupil is the more efficient because both maximum dilation and constriction can be attained; a narrow slit can be constricted almost completely. The circular pupil cannot contract to a pinpoint because, in the fully contracted iris, the tissues at the edge of the pupil bunch up to prevent complete constriction. The "color of the eye" is largely determined by pigments within the iris. An albino animal has pink eyes because the color of the blood is not obscured in the pigmentless iris.

Accommodation Accommodation is the mechanism by which the light entering the eye, after being refracted by the cornea, is focused on the retina by the lens with such fidelity that the visual field is "in focus." Lenses vary in size and shape in each

Figure 9-9 Diagrams of musculature of mammalian iris. (a) Round pupil of diurnal and nocturnal mammals. Note the sphincter muscle (solid lines) and symmetrical dilator muscle (broken lines). (b) Vertical slit pupil (cat) of nocturnal mammals that live in daylight. the scissor-like action of the two bundles of sphincter muscle (solid lines) compresses the pupil. Note the symmetrical dilater muscle (broken lines). (c) Horizontal pupil of ungulates, some whales and other mammals. The sphincter muscle (solid lines) is anchored to connective tissue (dotted lines). Note the dilator muscle (broken lines). The dilated pupil outlines a circle; the contracted pupil outlines a horizontal slit. (After Walls.)

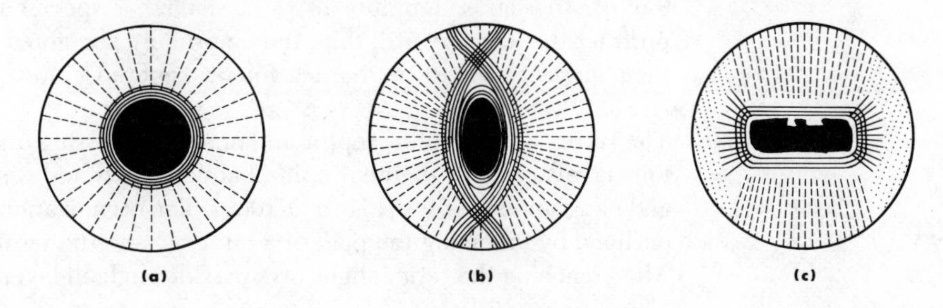

of the classes of vertebrates. In general, aquatic vertebrates are shortsighted. This condition is probably associated with the fact that distant vision is not possible or useful in water; these animals have spherical lenses in order to obtain a great degree of refraction from the increased curvature and, thereby, compensate for the lack of refraction at the cornea (Figures 9-6 and 9-7). Nocturnal terrestrial animals have large, spherical lenses that are usually close to the retina. Such lenses serve two functions: they permit a maximum amount of light to enter the eye and also a maximum refraction to focus the image on the retina. Other terrestrial animals have lenses ranging from ovoid (carnivores and herbivores) to flattened shapes (diurnal primates including man). In the four-eyed fish *Anableps*, the upper half of each eye (the air half) has an ovoid lens with a short axis from the cornea to the lower half of the retina, and the lower half of each eye (the water half) has a spherical lens with a long axis from the lens to the upper half of the retina. In this fish, the water line cuts through the middle of the cornea, and it can see both above and below the water at the same time. In general, the lens has a greater role in refracting light during focusing in aquatic vertebrates than in terrestrial tetrapods.

Fine adjustments in accommodation are effected by changing the curvatures of the cornea or the lens or by adjusting the relative position of the lens to the cornea. The focusing of objects close to the eye (nearsighted vision) may be accomplished by one or more of the following ways: increasing the curvature of the cornea, increasing the curvature of the lens, or by decreasing the distance between the cornea and the lens. By each of these means the image is focused on the retina instead of behind the retina. In the lamprey, the curvature of the cornea can be flattened by the contraction of a muscle attached to it; thus, the nearsighted lamprey accommodates for distant vision [Figure 9-6(a)]. In birds, reptiles (except snakes), and mammals, the lens is an elastic resilient structure; its curvature can be altered by muscles located in the ciliary body [Figure 9-6(c),(d),(e),(f)]. In these animals, the ciliary muscles are relaxed during distant vision and are contracted to accommodate during near vision. In birds and reptiles, the effective ciliary muscle compresses the lens in its equatorial plane; the result is a rounder lens [Figure 9-6,(d) and (e)]. After the muscle relaxes, the lens returns to its former shape. In mammals, in the normal relaxed eye adjusted for distant vision, the ciliary muscles are relaxed; in this state, the normal tension exerted through the taut zonula fibers on the lens maintains a flattened lens. In nearsighted vision in mammals, the ciliary muscles contract to bring the ciliary body closer to the lens. This reduces the tension exerted by the zonula fibers on the lens (the muscle acts as a sphincter) and because of its intrinsic elasticity the lens rounds up. In lampreys and teleost fish, a special muscle contracts to retract the lens backward; thus, these normally nearsighted animals retract their lenses for distant vision [Figure 9-6(b)]. In sharks, amphibians, and snakes, a special muscle contracts to protract the lens forward; thus, these normally farsighted vertebrates protract their lenses closer to the cornea for nearsighted vision.

Retina The retina is the photoreceptor and neural processing organelle. It comprises four layers of cells: pigment epithelial cells, *rods* and *cones*, bipolar neurons, and ganglionic neurons [Figure 9-10(a)]. The organization of the retina can be outlined by following the path of a light ray and the resulting neural activity. After reaching the retina, light rays pass through all layers of the retina to the

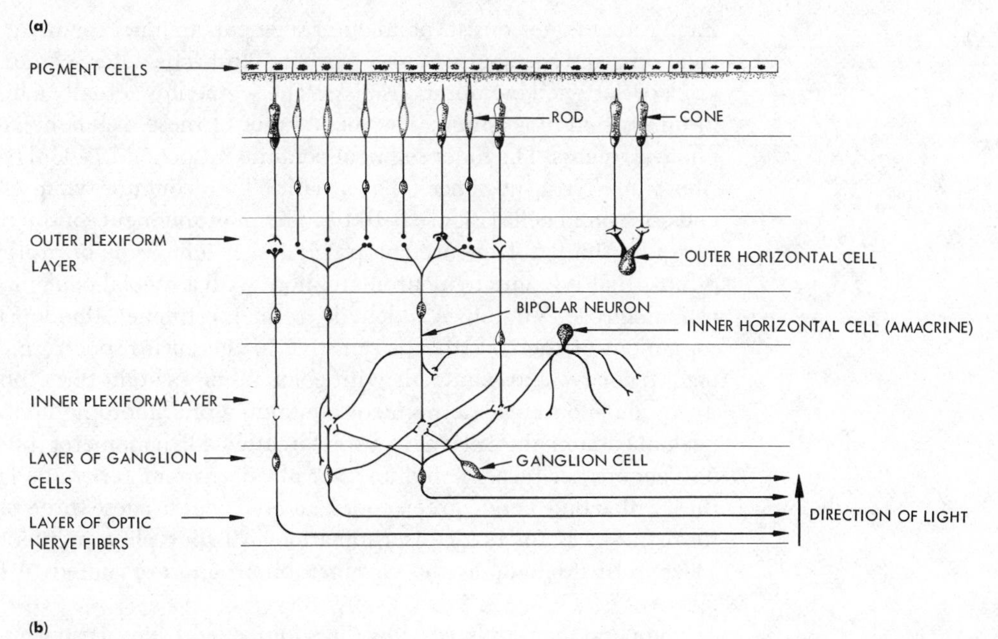

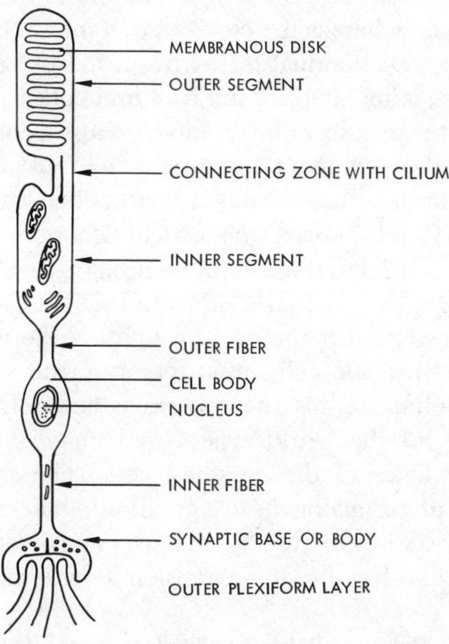

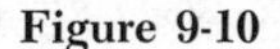

Figure 9-10 (a) Diagram of the primate retina. (b) Diagram of a cone.

outer segments of the rods and cones [Figure 9-10(b)]. The interaction of the light waves with the photopigments results in the transduction of the physical energy of the light waves into chemical activity. Light has performed its function. In a sequence as yet unknown, this chemical activity is converted into neural influences that stimulate the bipolar neurons (cells) and outer horizontal neurons (Figure 9-10). Unused light is absorbed by the pigmented epithelium of the retina and pigment cells of the vascular choroid layer. The rods and cones may be considered to be either epithelial receptor cells or specialized neurons.

Each rod or cone consists of an outer segment, an inner segment, a fiber with a cell body, and a synaptic base [Figure 9-10(b)]. The outer segment consists of a stack of flattened membranous disks. The segment is actually a highly modified distal portion of a cilium. The membranes of these disk lamellae contain the photopigments. The inner segment contains mitochondria, Golgi complex, and ribosomes (Nissl substance). The synaptic base contains synapses with bipolar and horizontal cells [Figure 9-10(b)]. The photopigments of the rods and cones are quite similar. They are composed of a specific type of protein—called an opsin—that is bound to a chromatophore with a special configuration—called retinaldehyde (Vitamin A aldehyde, retinal, retinene). Rhodopsin is the photopigment of the rods that is sensitive to the entire spectrum. The cones of man, primates and mammals with color vision contain three photopigments. Each cone contains one pigment containing one photopigment, one with its peak maximum absorptions at approximately 435 nanometer (blue), another at 535 nanometer (green), and another at 565 nanometer (red). The cones are, thus, called blue cones, green cones, and red cones. These three photopigments form the basis for color discrimination. All four photopigments possess the 11-cis retinaldehyde as the chromatophore and are united to four different opsins.

In general the rods have a low threshold to light stimulation and are effective in dim light (twilight vision), whereas the cones have a high threshold to light stimulation and require good illumination to be properly stimulated. The bipolar neurons, which are stimulated by the rods and cones and outer horizontal cells within the outer plexiform layer, have synaptic connections with the ganglion neurons (cells) and inner horizontal neurons (also known as *amacrine* cells—each a neuron with dendrites but without an axon) in the inner plexiform layer [Figure 9-10(a)]. In brief, the intricate connections among the neurons of the retina are essential to the complex neural processing that occurs within the eye before neural influences are projected via the ganglion cells to the brain. The ganglion cells are influenced by light in the environment—called the visual field. In turn, the delineated region in the visual field that influences a specific ganglion cell is the receptive field of that ganglion cell—that cell's eye view of the world. The visual image of the retina is actually a mosaic of the fields of the ganglion cells. The retina is also a remarkable photographic plate, analogous to a composite black and white film and a color film. The black and white film has the fast emulsion (rods) effective in twilight, and the color film has the slow emulsion (cones) most effective in broad daylight.

To make efficient use of light available in the dark, some fish, reptiles, and mammals have a *tapetum,* which is a light reflector located within the deep retina (some teleosts, crocodiles, some bats, and the Virginian opossum) or choroid (chondrostean fish, carnivores, ungulates, nocturnal lemurs, and some Australian marsupials). Light that has passed beyond the rods and cones is reflected by the tapetum back to the layer of rods and cones where it can stimulate these receptors again. In the dark, the reflection of light by the tapetum produces eye shine in many mammals (e.g., cat, deer) seen in the light of a flashlight. Although nocturnal vision is rendered more sensitive by the reflected light from the tapetum, sharpness of focus is sacrificed because the rods and cones are doubly stimulated. In those animals without a tapetum, the pigment epithelium of the retina (Figure 9-10) contains melanin, which absorbs the light passing through the layers of rods and cones. These are three types of tapeta. The tapetum of some teleosts, many marsupials, ungulates, whales, and many nocturnal monkeys is primarily composed of collagenous connective tissue fibers. The tapetum of carnivores and lemurs is a cellular structure; each cell contains reflective purine rods. The tapetum of many teleost and some chondrostean fish is composed of extracellular plates containing guanine.

Visual Acuity

Visual acuity (resolving power) is a measure of the sharpness with which details can be distinguished by vision. In general, vertebrates with superior visual acuity (birds, primates, certain lizards, squirrels, some carnivores and ungulates) have a small area known as the *fovea.* The fovea, often located within a depression called the *macula,* is composed of a high concentration of cones, bipolar cells, and ganglion neurons organized to relay more discrete details to the brain than the rest of the retina (Figure 9-7). Some sharp-eyed hawks have as many as one million cones in one square millimeter of fovea. Those mammals that have a fovea have only one of these structures, whereas some birds have two. Because visual acuity is associated with high-threshold cones, good illumination is essential to discriminate visual detail.

Certain vertebrates have a vascularized pigmented protuberance extending into the *vitreous body.* This structure, which is thought to increase the ability of the eye to sense moving objects by casting a shadow on the retina, is known as the *pecten* in birds, the conus papillaris in snakes and lizards, and the falciform process in fish [Figure 9-6(d) and (e)].

Dim-Light Vision

Animals that lead primarily nocturnal and crepuscular lives view the environment in shades of gray from black to white. These animals usually have large eyes, each with an all-rod retina (Figure 9-11), a large cornea, and possibly a

Figure 9-11 Retinas of diurnal and nocturnal animals. In general, the retina of a diurnal vertebrate contains many cones, which summate but slightly to bipolar cells and these in turn summate but slightly to ganglion cells. In general, the retina of a nocturnal vertebrate contains many rods; these summate extensively to bipolar cells which in turn summate extensively to ganglion cells. (Adapted from Walls.)

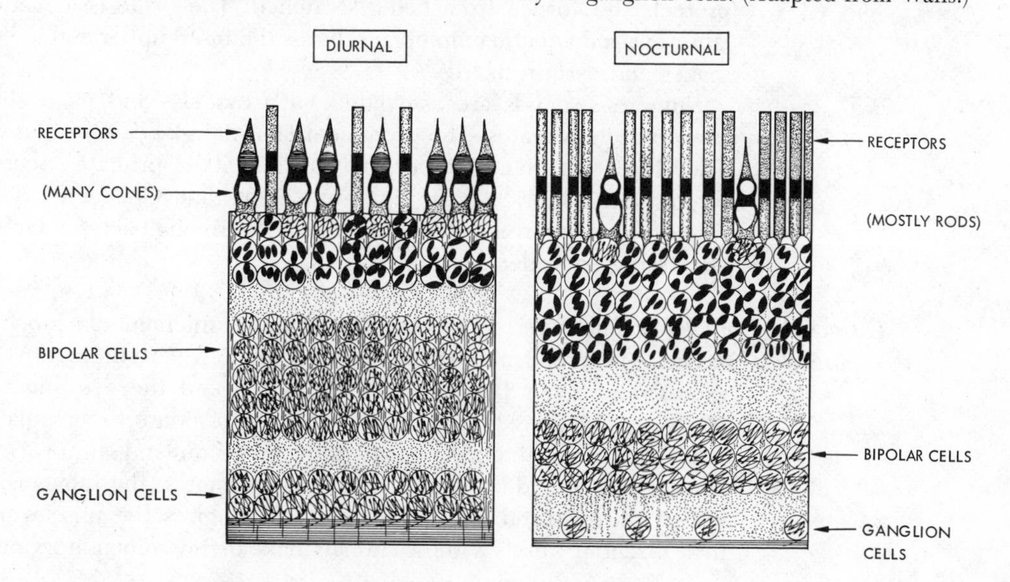

tapetum. The light sensibility is enhanced further by a retina in which each ganglion cell (receptive field) may receive input through bipolar neurons from up to many thousands of rods (Figure 9-10). Each group of rods converges on one ganglion cell, which senses one receptor field. Because each receptor field of a ganglion neuron in nocturnal animals is large, these animals do not have high visual acuity.

Animals with all-cone eyes are diurnal (Figure 9-11). This is one reason why many birds are active only by day. Many diurnal animals, like man, have retinas with both rods and cones.

Color Vision

The ability to perceive color (hue discrimination) is a rare phenomenon among vertebrates; it is well developed in highly visually oriented diurnal animals with a cone-rich retina possessing a fovea. Because it is present in unrelated vertebrates, color vision probably evolved independently a number of times in the course of vertebrate phylogeny. These vertebrates include some teleosts, urodeles, lizards, turtles, birds, certain ground squirrels, Tupaia (classified either as an insectivore or a primate), and many primates. Color is best perceived when there is good illumination in order to adequately stimulate the high-threshold cones.

Protection of the Eye

The eye is protected from injurious irritants and environmental obstacles by several adaptations that include the eye socket (orbit), true eyelids, *nictitating eyelid membrane, spectacle* (*brille*), and a variety of glands. True movable eyelids are present only in terrestrial animals. They help to prevent the cornea from drying out by lubricating it with tear secretions, and they also protect the eye from bright light. The nictitating membrane, a transparent third eyelid, is present in most land animals. When closed during flight, this lid permits the bird to see and, at the same time, it prevents the cornea from drying. In grazing animals such as sheep, it protects the cornea from abrasion by rough vegetation. The membrane is small-to-vestigial in man as well as in many arboreal and nocturnal animals. The spectacle is a tough, transparent membrane located permanently over the eye, and the eye moves beneath it (Figure 9-12). It protects the cornea from being scratched. The brille is a modified skin in lampreys and aquatic amphibians, but is the fused upper and lower eyelids in snakes and certain lizards.

Numerous glands are associated with the eye and its eyelids in many terrestrial vertebrates. The prominent lacrimal gland secretes a watery fluid. The Harderian gland, absent in man and other primates, secretes an oily substance that lubricates the nictitating membrane. One secretory ingredient is lysozyme, a powerful enzyme capable of destroying bacteria. Only in man are tears secreted under emotional stress.

Extraocular Muscles

In general most vertebrates have the same complement of *extraocular muscles* as man—four recti and two oblique muscles. Few vertebrates use these muscles extensively. Their basic function is to suspend the eye and to make the necessary compensatory eye movements in response to irregular motions or shifting of the head during erratic movements on land, air, or water. In a few vertebrates, including lizards, man, and primates, the movements are integrated into head and body motion patterns. In birds, the muscles are so small as to be essentially useless; these animals make their compensatory movements by

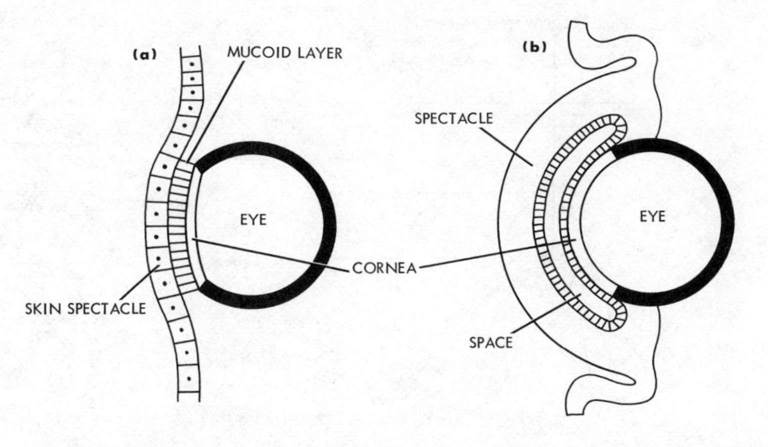

Figure 9-12 Spectacles. (a) Primary spectacle of cyclostomes and aquatic amphibians. A mucoid tissue is located between the skin spectacle (dermal cornea) and the cornea (scleral cornea). (b) Secondary spectacle (transparent fused eyelids) as seen in reptiles. (Adapted from Duke-Elder.)

moving the head and neck. Many amphibians, reptiles, and mammals, excluding man and other primates, have a retractor bulbi muscle that can pull the eye deep into the socket.

Median Eyes (Pineal Organs)

Median eyes or their derivatives are present in vertebrates on the mid-dorsal aspect of the head and brain (Figure 9-13). In many fish, amphibians, and reptiles, these are represented by one of the two eyes—the pineal eye or the parietal eye. The "pineal complex of organs" includes the pineal organ (gland, epiphysis cerebri), the parapineal (*parietal* or frontal) organ, and some associated structures including the *paraphysis,* dorsal sac, and subcommissural organ. All are located in the roof of the diencephalon. The pineal and parietal organs once may have been paired structures, the pineal organ being the right member and parietal organ the left. With the exception of lampreys, most living vertebrates with median eyes have retained either a pineal eye or a parietal eye, but not both. The lampreys have both organs, with the pineal organ located superficial to the parietal. Some lizards have a highly evolved parietal or third eye complete with lens and retina. This median eye resembles the paired eyes in that it is composed of a modified cornea, lens, light receptive (retinal) and ganglionic cell layers (retina), and an afferent nerve (optic nerve), as illustrated in Figure 9-13(c). Many frogs have a less specialized frontal organ located just beneath a transparent fleck in the skin on top of the head. The pineal organ is present in virtually all mammals, birds, fish, and tailed amphibians. In mammals, the pineal organ is a modified structure (without lens or receptive layer) with cells containing such chemical substances as melatonin, serotonin, and norepinephrine. Its sole innervation in mammals is apparently via nerves of the sympathetic nervous system [Figure 9-13(d)].

The paraphysis and the dorsal sac are secretory structures. The subcommissural organ, present in all vertebrates, produces a mucoid material that, in all but higher vertebrates, is secreted into the cerebrospinal fluid of the third ventricle. The functional significance of these various secretions is not known.

The parietal organ has cells that are cytologically similar to retinal photoreceptors with proven visual capabilities. These cells also are found in the

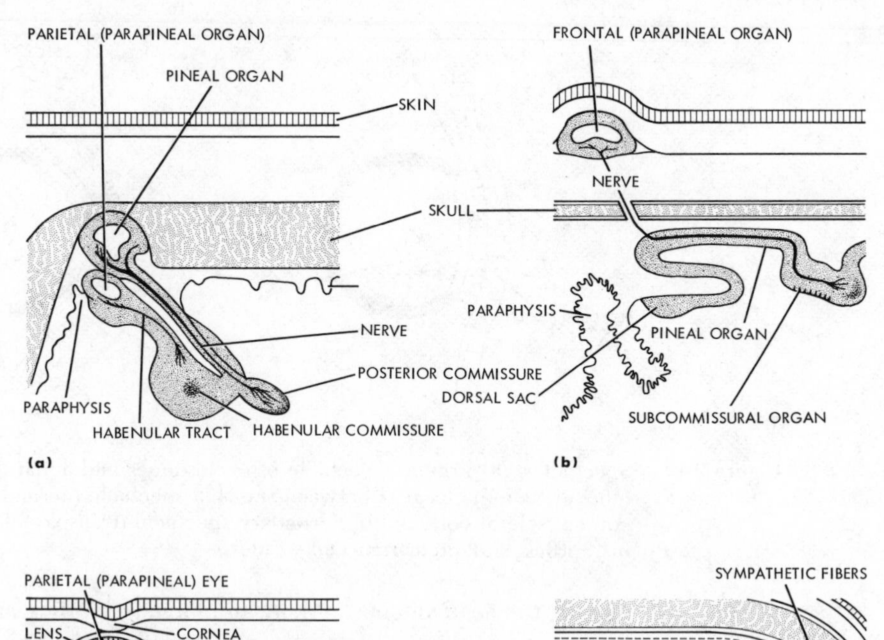

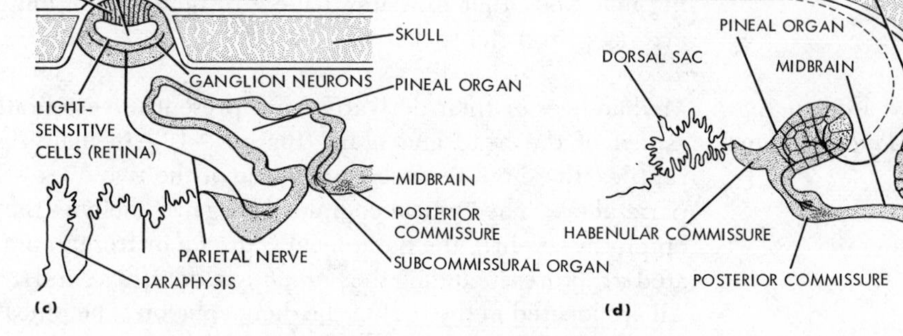

Figure 9-13 Sketches of median sagittal views of the roof of the diencephalon: (a) lamprey; (b) frog; (c) lizard; and (d) mammal. (Adapted from Studnicka.)

deeper-lying parapineal organ of lampreys, amphibians, and lizards. The stimulation of these structures by light does not result in the perception of visual images; rather, the light influences the diencephalic roof structures to secrete substances that have systemic effects, such as changing the color of skin. When the pineal apparatus is excised or shielded in aquatic vertebrates, the coloration of the skin changes. Influences from the visual pathways may stimulate the pineal gland of mammals and possibly other vertebrates to synthesize and secrete specific compounds. In this respect, it is presumed to act as a kind of neuroendocrine transducer following the indirect stimulation by environmental light. Melatonin, a substance found in the mammalian pineal gland, can produce blanching of the skin of the frog by contraction of the melanophores, and mammalian pineal extracts have some functional role in the seasonal activity of the gonads (photoperiodic regulation of gonadal activity). The median eyes and their derivatives are somehow involved in some circadian and seasonal cycles and rhythms over a twenty-four-hour day or long spans of time (seasonal).

The Mechanoreceptors

Special *mechanoreceptors* are located in the lateral line system, the vestibular system, and auditory system. The lateral line system (Figure 9-14) is found in fish and tailed amphibians, the vestibular and auditory systems are found in some form in all vertebrates. All these systems belong to a common over-all sensory complex called the acousticolateralis system, innervated by the vestibulocochlear (statoacoustic) cranial nerve. The common statement that the lateral line is innervated by the facial, glossopharyngeal, and vagus cranial nerves is technically erroneous; the receptors of the lateral line are innervated by nerve fibers originating from the same complex of nuclei in the medulla that innervate the receptors of the vestibular and cochlear labyrinth. These systems are embryologically derived from a common anlage, called the *otic* (lateral, auditory) *placode.* Many of the fibers of this nerve secondarily join the facial, glossopharyngeal, and vagus cranial nerves before innervating the receptors of the lateral line (Figure 9-15).

The hair bundle of each hair cell comprises of from forty to two hundred stereocilia and one thick kinocilium (Figure 9-21). The stereocilia are modified microvilli, and the kinocilium is a specialized cilium. The kinocilium is located near an edge of the cell and the stereocilia on one side of the kinocilium. The kinocilium of a hair cell of the spiral organ of Corti is reduced to a small basal body. The bases of the hairs are embedded in the cuticular plate on the surface of the hair cell (Figure 9-21). The tips of the cilia extend distally into the cupula or tectorial membrane.

The sensors of these mechanoreceptors are neuroepithelial hair cells (Figure 9-21). The hairs extend into a gelatinous capsule (cupula), as in the neuromast (Figures 9-16), or into a gelatinous capsule containing small, sandlike concretions called otoliths or statoliths, as in the macula of the utriculus (Figure

Figure 9-14 Side of the head of a shark illustrating lateral line canals (parallel lines) and cranial nerves (solid lines). These lateral line nerve fibers are actually fibers of the eighth (VIII) cranial nerve which are distributed with the branches of cranial nerves VII, IX, and X. (Adapted from Norris and Hughes.)

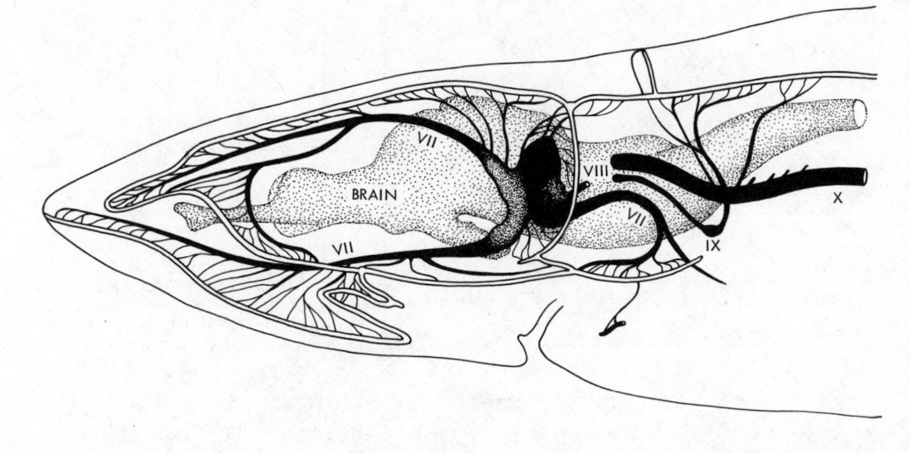

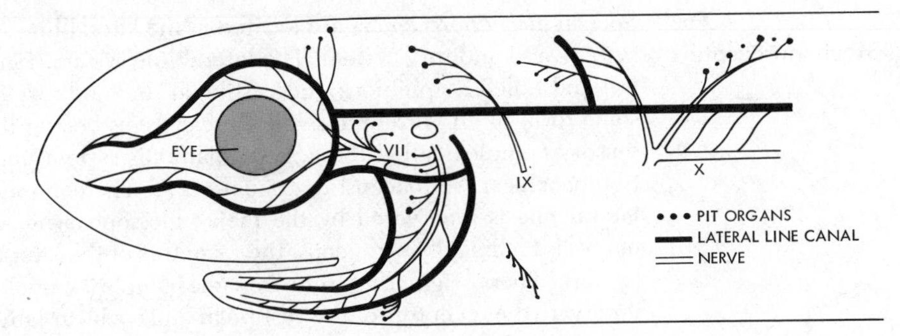

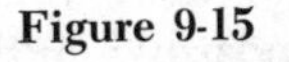
Figure 9-15 Side of the head of a fish illustrating the general distribution of the lateral line canals (solid lines), pit organs (large dots), and the nerve supply (fine lines). (Adapted from Goodrich.)

9-17); or they extend into a noncalcareous stiffened membrane (the tectorial membrane), as in the amphibian papilla or spiral organ of Corti (Figure 9-18).

A hair cell responds maximally when the hairs are bent in the direction from the stereocilia toward the kinocilium (axis of sensitivity, Figure 9-21). The hair cell does not respond when the hairs are bent in the direction away from the kinocilium. Stated otherwise, the hairs are directionally polarized. Each hair cell makes synaptic connections with both afferent and efferent nerve fibers (Figure 9-21). The hair cell can stimulate an afferent fiber, which conveys neural influences to the central nervous system. An efferent fiber, with its cell body in the central nervous system, conveys inhibitory influences to the hair cell. The latter has a role in neural processing. The mechanical bending of these cilia is the key transduction action triggering the sequence leading to the

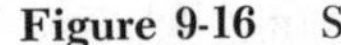
Figure 9-16 Schematic illustration of a section through a free neuromast.

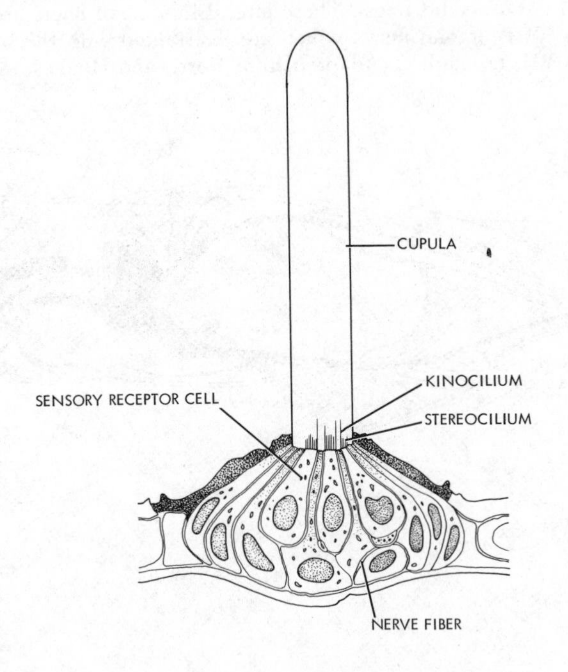

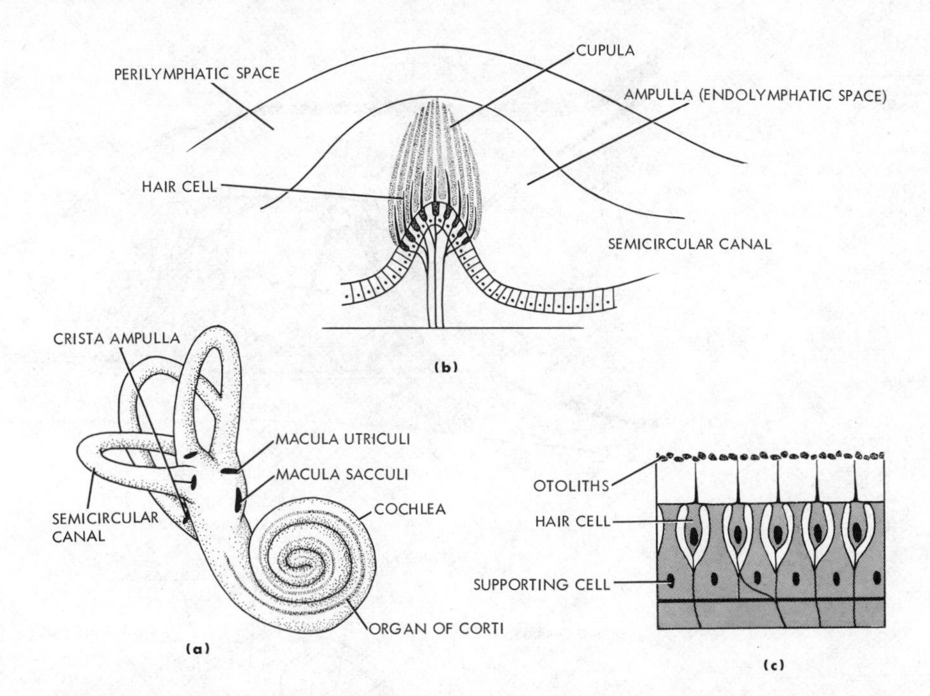

Figure 9-17 Membranous labyrinth in mammals. (a) The labyrinth comprises the three semicircular canals and the cochlea. (b) A crista ampulla is the nerve ending located in the ampulla of each semicircular canal. (c) A macula is the nerve ending located in the utricle and the saccule. The organ of Corti is the nerve ending in the cochlea. (Adapted from Noback and Demarest.)

generation of nerve impulses in nerves innervating these sensors. Most of the details of this sequence are unresolved.

The Lateral Line System

The receptor organelles of the lateral line system are the neuromasts (Figure 9-16) that are either free in the skin or located in open grooves on the body surface or in recessed canals with little or no direct access to the body surface (Figures 9-19 and 9-20).

FREE NEUROMASTS. Free *neuromasts* (sensory hillocks) are apparently present in the skin of all fish and tailed amphibians. A neuromast is composed of hair cells supported by sustentacular cells and covered by the gelatinous cupula enclosing the sensory hairs, or cilia (Figure 9-16). Some hair cells of each neuromast have the kinocilium on the cephalic side of the stereocilia and others have it on the caudal side. The current flowing over the fish swimming upstream will activate only the hair cells with the kinocilium on the caudal side of a neuromast. Isolated groups of free neuromasts, called pit organs, are located on the head of some fish.

LATERAL LINE ORGAN The lateral line system is a special sense organ that includes a long canal on either flank of the body (often with secondary canals), plus a complex of canals on the head in fish, tadpoles, and adult tailed amphibians. The long canal from the head to the tail is located approximately at the longitudinal junction between the epaxial and hypaxial musculature. In

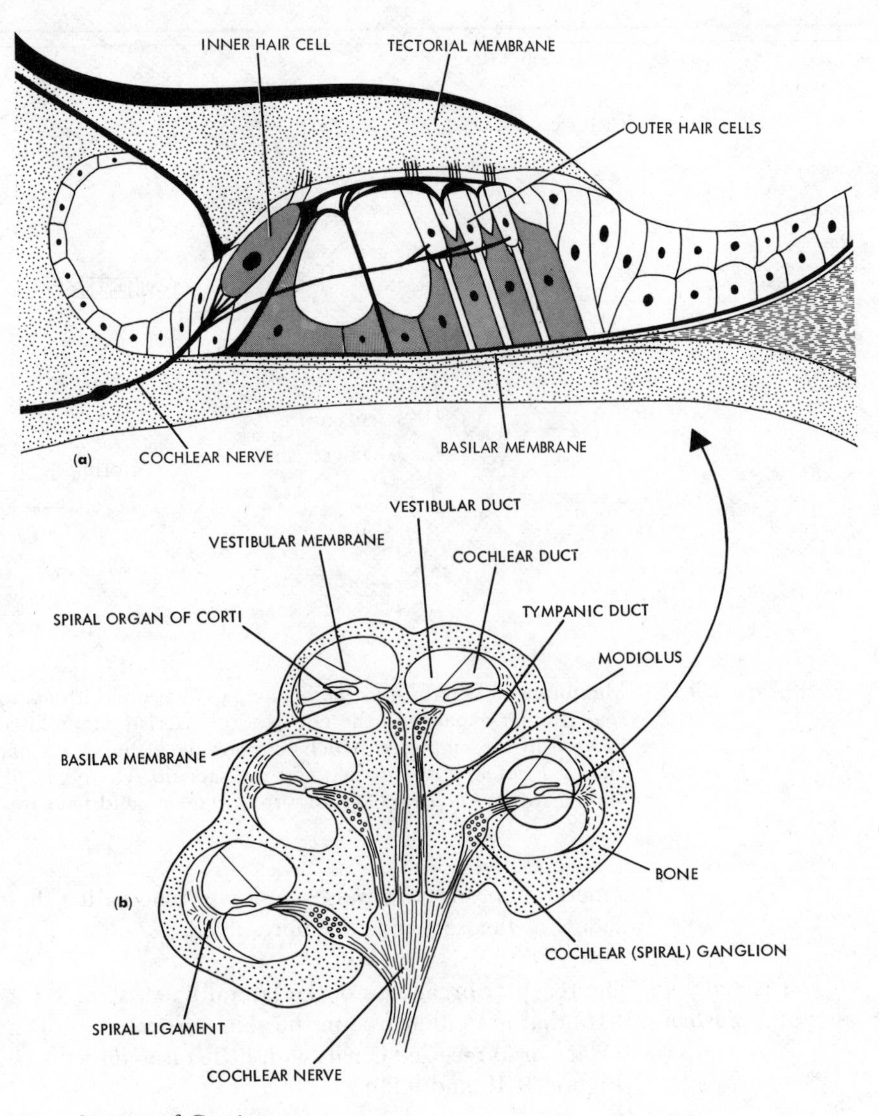

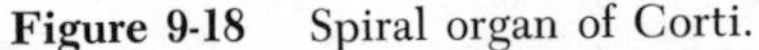

Figure 9-18 Spiral organ of Corti.

a few elasmobranchs and bony fish, the neuromasts lie in open grooves with direct access to surface currents, whereas in cyclostomes some occur in isolated pits arranged in a linear pattern. In most fish, the neuromasts are present within canals that may be in contact with the external environment through small openings (Figure 9-20); these canals may be buried within or below the body scales or within the bones of the head. Such neuromasts are stimulated by slight currents generated within the canals. The lateral line is a displacement detector that is extremely sensitive to minute vibrations and movements of water. The axis of sensitivity of half of the hair cells is oriented in one direction and the other half in the opposite direction. The lateral line system is considered to be a "distance-touch receptor," which detects disturbances in the water, so-called near-field sound. It gives the fish information about its own body movements. Fish of fast-flowing streams, such as trout, have most of their

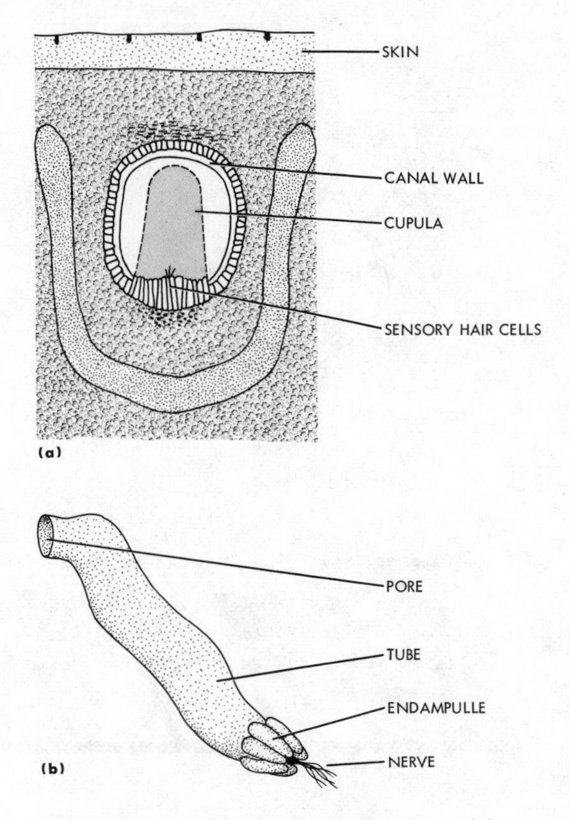

Figure 9-19 (a) Diagram of neuromast within a lateral line canal. (b) Ampulla of Lorenzini of a shark (see Figure 10-12). (Adapted from van Bergeijk and Alexander and from Peabody.)

neuromasts within canals, whereas fish of still waters, such as cyprinids, have many free neuromasts and relatively few within the canals. Apparently, the canal reduces the impact on the neuromasts of the constant contact with fast currents. Functionally, the neuromasts are the receptors sensing the hydrodynamic motions. Fish and aquatic amphibians also are sensitive through the lateral line organ to water displacements caused by objects or animals. Prey, enemies, and sexual partners are detected and localized with this sensor on the basis of water vibrations.

Specialized Lateral Line Receptors. The *ampullae* of Lorenzini located in the heads of elasmobranchs and some bony fish (mormyrids, gymnotids, and silurid fish) are nonciliated receptors (a modified neuromast with receptor cells that lost their cilia) located at the end of a tube filled with a gelatinous substance (Figure 9-22). These flask-shaped organs, which contain a sensory epithelium not covered by a cupula, are *electroreceptors* monitoring the electric charges in the environment. These modified lateral line receptors in the mormyrid fish are called mormyromasts. Those in the *Torpedo,* an elasmobranch, are known as the vesicles of Savi. Even the lateral line sensors are electrosensitive in some fish. The specialized electroreceptors probably have evolved independently from simple lateral line and ampullary receptors. The low-voltage electric pulses emitted by mormyrid fish are sensed by these

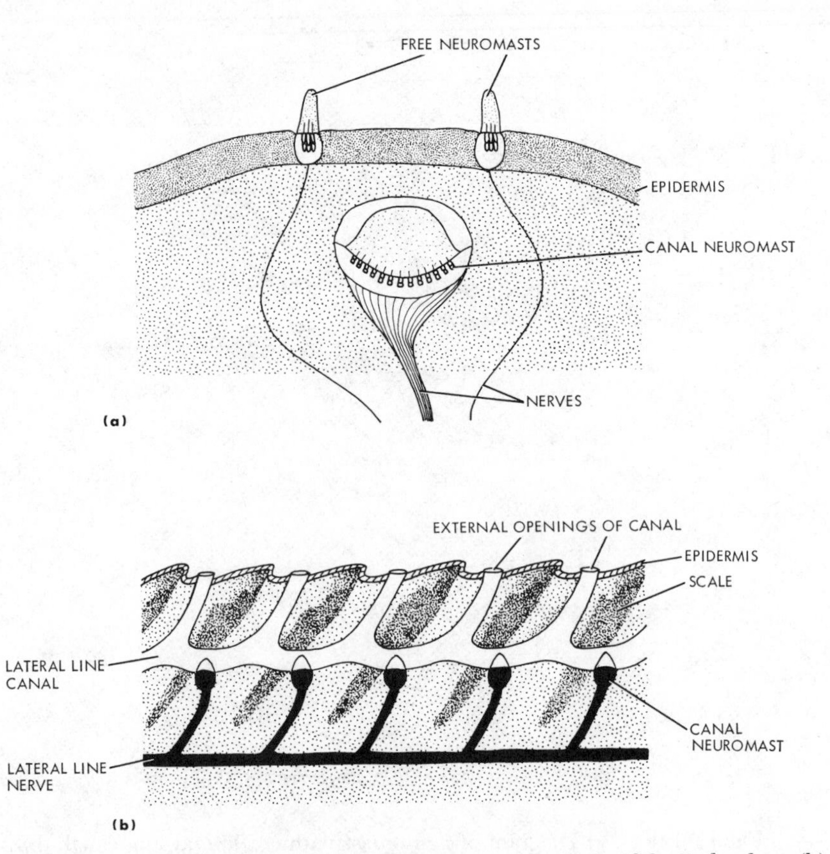

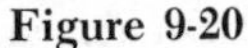

Figure 9-20 (a) Free neuromast and lateral line canal neuromast in the fish *Gadus lota.* (b) Lateral line canal with external openings, adjacent to surface scales, on the trunk of a bony fish. (Adapted from Goodrich.)

mormyromasts after the electric signals are reflected back by nearby physical objects. In effect, the mormyrids have a "radar" system with which they scan the environment. Sharks and rays are capable of detecting at short distances the electrical potentials produced by muscular activity. Actually, the electric organs of the electric eel and many other fish are derived from nerve-muscle complexes that are specialized to produce electrical discharges.

Inner Ear A membranous labyrinth and a bony or cartilaginous labyrinth comprise the inner ear. The membranous labyrinth is a continuous series of interconnected tubes (*semicircular ducts*), vesicles (*saccule* and *utricle*), and a duct (cochlear duct) enclosing a cavity filled with endolymphatic fluid (endolymph) (Figures 9-17 and 9-23). A slender endolymphatic duct usually extends upward and inward to terminate in an endolymphatic sac within the brain case. All the specialized receptors of the inner ear face into the fluid-filled cavity. Surrounding the membranous *labyrinth* is another cavity filled with perilymphatic fluid (perilymph), and the entire complex is encapsulated by cartilage or bone (cartilaginous or bony labyrinth). The semicircular ducts, with the saccule and utricle, constitute the vestibular labyrinth, and the cochlear duct forms the cochlear labyrinth. By definition, the semicircular ducts are bounded by the membranous labyrinth. In turn, the semicircular canals are the tubes enclosing

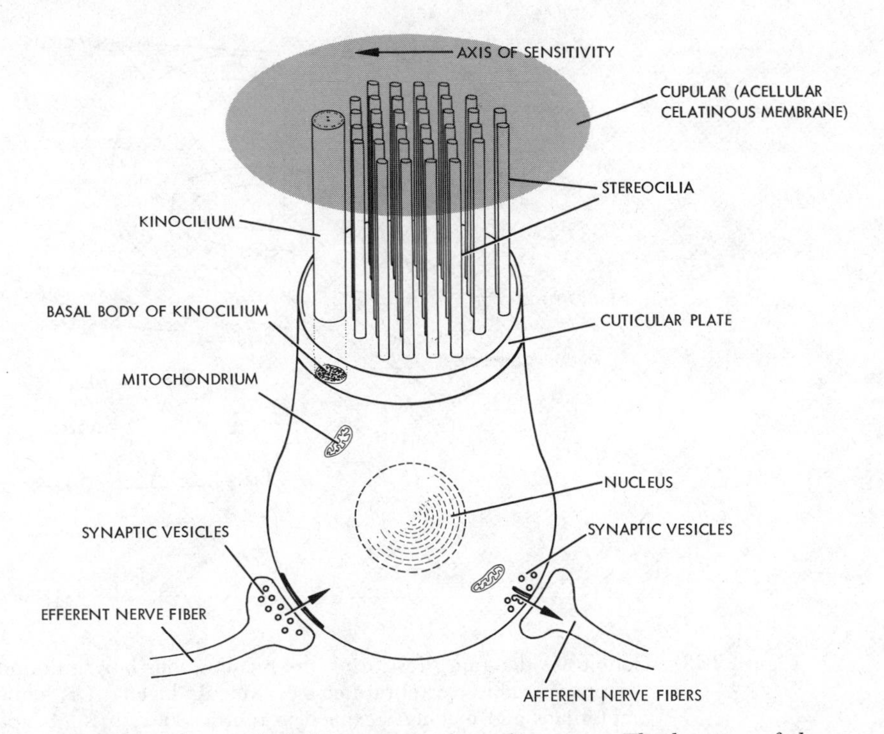

Figure 9-21 Schematized hair cell of the acoustico-lateralis system. The location of the kinocilium (true cilium) with respect to the stereocilia (microvilli) is significant in determining the axis of sensitivity (arrow). In the spiral organ of Corti the kinocilium is reduced to the basal body. The arrows at base of cell indicate the direction of the neural transmission across the synapses.

the perilymphatic fluid and the semicircular ducts. They are bounded by bone or cartilage. The ducts are incorrectly called canals. Like the rest of the membranous labyrinth, the cochlear duct (scala media) is filled with endolymph. Paralleling the cochlear duct are the vestibular duct (scala vestibuli) and the tympanic duct (scala tympani), which are filled with perilymph. These three ducts comprise the cochlea. In mammals, the cochlea coils about a central

Figure 9-22 Location and innervation of the ampullae of Lorenzini in the dogfish shark (see Figure 10-19). The ampullae are innervated by fibers from the eighth cranial nerve distributed via the seventh cranial nerve. (Adapted from Dotterweich.)

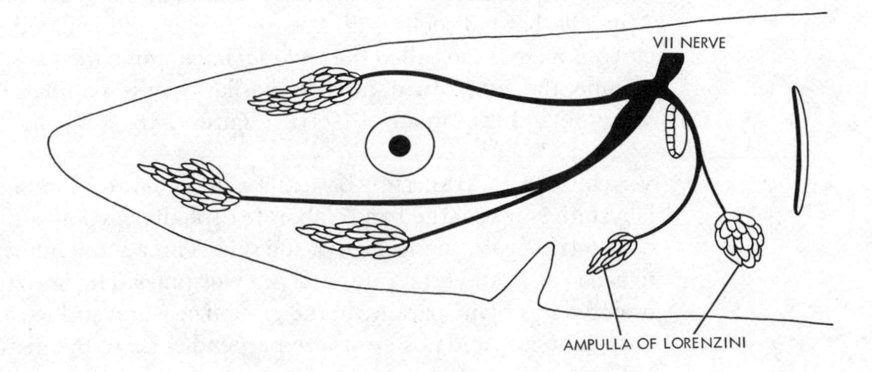

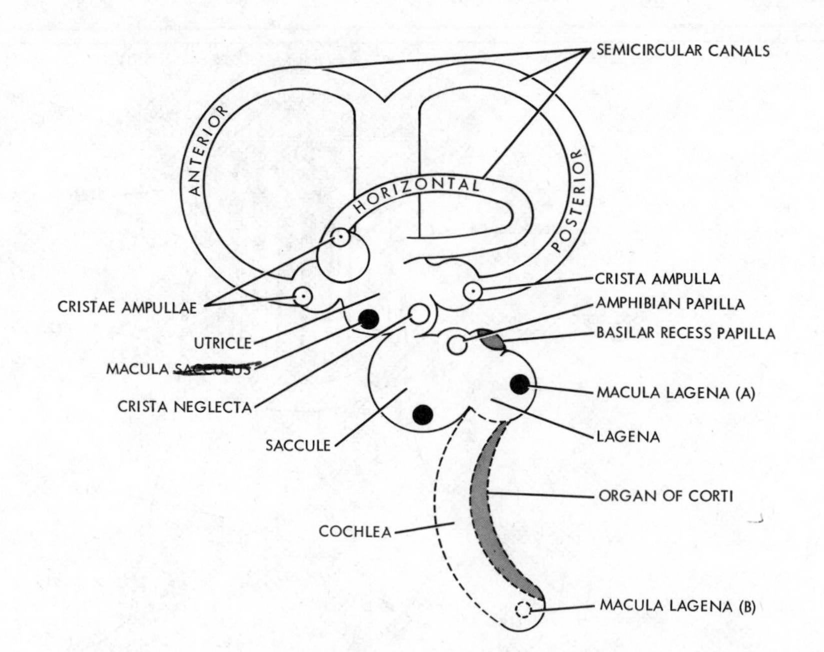

Figure 9-23 Composite diagram illustrating the membranous labyrinth and its sensory receptors in most vertebrate classes. Macula lagena A is found in fish and amphibians and macula lagena B in reptiles and birds. (Adapted from Lowenstein.)

bony cone called the modiolus; within the modiolus is the spiral ganglion that is the sensory ganglion of the cochlear nerve. The vestibular duct is continuous with the tympanic duct at the distal end (apex) of the coil through a small hole called the helicotrema. The cochlear duct is separated from the vestibular duct by the vestibular membrane and from the tympanic duct by the spiral organ of Corti and its basilar membrane (Figure 9-18).

Within the labyrinth, the specialized neuromast receptors are called *crista, macula,* and *papilla.* A crista has hair cells with cilia that extend into a gelatinous cupula (Figure 9-17). The cristae include the crista ampullaris of each semicircular canal and crista neglecta of the utricle (Figure 9-23). A macula has essentially the same structure as a crista except that within the cupula are minute mineralized concretions (ear stones, otoliths, otoconia). There are three maculae: macula utriculus, macula sacculus, and macula *lagena* (Figure 9-23). A lagena is a recess of the nonmammalian inner ear containing the macula lagena. The functional role of this macula is not known. A papilla has hair cells with the tips of the cilia embedded within a noncalcareous membrane called the *tectorial membrane;* the various kinds of papillae include the amphibian papilla, basilar recess papilla (both found in the saccule), and the organ of Corti (Figures 9-18, 9-23, and 9-24).

VESTIBULAR LABYRINTH. Basically similar in all vertebrates, the vestibular labyrinth is an enlarged and elaborate cephalic portion of the lateral canal that has sunk beneath the surface of the skin. Three semicircular ducts are present in each ear in all vertebrates except cylostomes. The horizonal (lateral) duct is oriented in a plane parallel to the ground or water surface and the two vertical ducts (anterior and posterior) are perpendicular to the plane of the horizontal

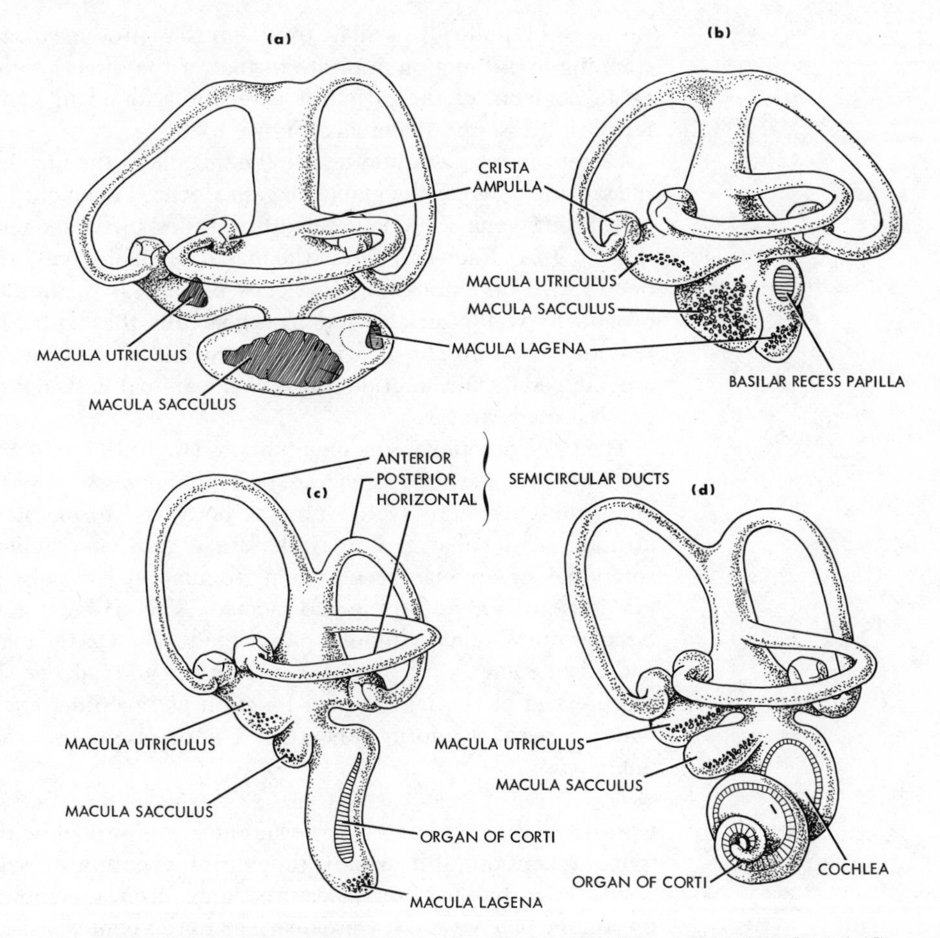

Figure 9-24 Membranous labyrinth in several vertebrates: (a) fish; (b) turtle; (c) bird; and (d) mammal. (Adapted from Retzius and from von Frisch.)

canal and are at right angles to each other (Figures 9-17, 9-23, and 9-24). In the cyclostomes, the hagfish have one semicircular duct and the lampreys have two. Within each duct there is one enlarged bulblike ampulla enclosing a crista ampullaris; the cupula of the crista is in direct contact with the wall of the ampulla. Acting as a swinging door, the cupula may gently pivot back and forth in response to the pressures exerted by the motion of the fluid endolymph. In each crista, the hair cells are oriented so the transduction of motion to neural activity results in movements in only one direction. The shearing forces bend the cilia and initiate the sequence of events producing neural activity. When the angular velocity of the head changes (e.g., as at the start of a spin), there is a brief interval after a semicircular duct moves before the momentum of the endolymph is overcome and the endolymph moves. During this interval, the fluid exerts a dragging effect on the cupulae (Figure 9-17), which results in the bending of the hairs of the hair cells in a direction opposite to the direction of the spin. When the angular velocity is reduced (e.g., at the end of a spin), there is a brief interval after the semicircular duct stops during which the endolymph exerts pressure on the cupulae and the hairs are bent in the direction of the spin. Through a neural linkage with the extraocular muscles the eyes rhythmically oscillate (nystagmus) to produce the subjective feeling of spinning in a direction opposite to that of the direction during the spin. The crista neglecta of the utricle is a receptor found in some fish, amphibians, reptiles, birds, and mammals (Figure 9-23).

Present in most vertebrates are the macula of the utricle and the macula of the saccule whereas the macula lagena, located in an outpocket of the saccule called the lagena, is found in teleosts, reptiles, birds, and monotreme mammals (Figure 9-24). Each hair cell of the macula utriculus has an axis of sensitivity in one of two directions; in some cells the axis is in the direction parallel to ground or water surface and in other cells the axis is in the direction of gravitational pull. The shearing forces exerted by the concretions upon the cilia are induced by acceleration forces of the animal in straightline motion and by gravitational forces.

The basic function of the membranous labyrinth is to monitor the position of the body in space and to generate neural influences that will stimulate coordinated muscular activity to right the body and to maintain normal postural attitudes and balance. Labyrinths are organs of equilibrium that analyze rotational or angular acceleration (semicircular canals) and accelerational gravitational forces (utricle and saccule). The cristae ampullaris of the semicircular ducts monitor angular movements, whereas the maculae of the utricle and saccule have a role in static equilibrium and balance. The maculae of the saccule and of the lagena may function as receptors for sensing vibrations, which is perhaps a form of hearing. The precise role of the crista neglecta is unknown.

Cochlear Labyrinth. All vertebrates are probably capable of hearing (phonoreception), if hearing is thought of as both a conscious and an unconscious sense that an observer can recognize from the animal's behavior. Does the animal behave as if it were hearing? In this context "the primitive function of hearing is the location of objects not in contact with the animal . . . an animal hears when it behaves as if it has located a moving object (a sound source) not in contact with it. And sound can be defined as any mechanical disturbance whatever which is potentially referable to an external and localized source."[1] (Pumphrey, 1950)

Fish Behavior studies of fish and amphibians indicate that the lateral line system is a hydrodynamic receptor of water-borne vibrations (near-field hearing organ). Phonoreception has also been attributed to the macula of the saccule and to the crista neglecta and papilla of fish (Figure 9-24). The lateral line system has been called both a near-field hearing organ and a "distant-touch" receptor detecting vibrations up to several hundred per second. It is not important in orienting the organism to water currents (rheotaxis). Fish are essentially "transparent" to all but the most intense sound waves because they almost have the same density as their watery environment. One structural adaptation has evolved to improve the hearing in some fish. This is the swim (air) bladder. Because it has a different density than water, the air bladder is readily set into vibration by the propagating sound waves.

The swim bladder of bony fish functions as an accessory vibratory (sound)

[1]Pumphrey, R. J. Hearing. Symposium of the Society of Experimental Biology. 4(1950), 1–18.

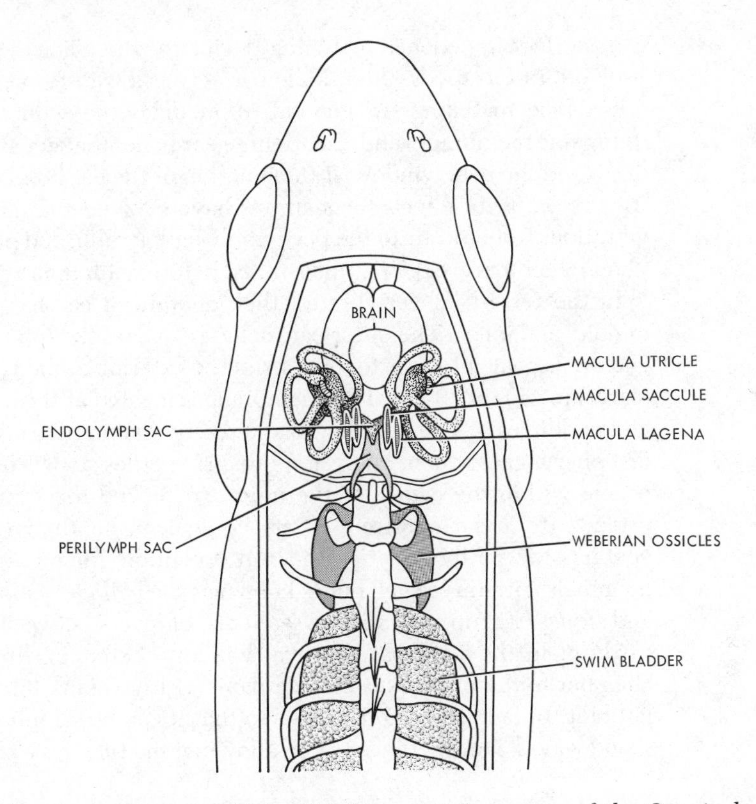

Figure 9-25 The organization of the "hydroacoustic" receptor of the Ostariophysii, a group of teleosts. Note the swim bladder, Weberian ossicles, perilymphatic sac, and endolymphatic sac. Vibrations are transmitted via this sequence of structures. (Adapted from von Frisch.)

resonator; the vibrations of the bladder pass, in turn, through the body to the hair cells of receptors of the membranous labyrinth. In the herring, the swim bladder comes into direct contact with the utricle. Certain fish, the Ostariophysii (cyprinid, silurid, characinids, and gymnotids; catfish, carp, and goldfish), are able to transfer the vibrations of the swim bladder to the saccules of the two labyrinths via a coupling device composed of a paired chain of four bones called Weberian ossicles; they extend from the bladder to a sinus containing perilymph and a canal with endolymph with direct access to the saccule (Figure 9-25). In this Weberian apparatus, the swim bladder is analogous to the eardrum and the ossicles to the ear ossicles of mammals. Many fish have an excellent sense of hearing and can be conditioned experimentally to discriminate among different tones. Several species of fish can produce sounds by which they communicate with other fish—for example, during the mating season. Various mechanisms for sound production have been reported. In the mormyrid fish, paired extensions of the swim bladder have pinched off to form separate bladder bubbles.

Terrestrial Vertebrates

The terrestrial vertebrates have evolved (1) a number of modifications for the transfer of the vibratory energy from the compressible atmospheric air (low impedence) to the incompressible perilymphatic fluid (high impedence) and (2) some elaborate *phonoreceptors*. Because the threshold for hearing is low, a near-perfect impedence matching device for the efficient energy transfer has evolved. In tetrapods, the middle ear acts as a transformer that performs the impedance matching. In general, it includes the sequence of an eardrum (tympanic membrane) and one to three ear bones that act as levers between the drum and the oval window at the junction of the ear bone with the perilymph (Figure 9-26). In effect, mechanisms have evolved that amplify and transfer vibrations from the air to the perilymph, endolymph, and phonoreceptors. The phonoreceptors are the papillae, noted before, with their tectorial membranes.

In the terrestrial vertebrates, the generalized ear has three parts: outer, middle, and inner. The outer ear includes an ear lobe (pinna) and the external auditory canal, a tube extending from the exterior to the tympanic membrane (eardrum) (Figure 9-26). This membrane is located at the junction of the outer and middle ears. The middle ear and the tympanic tube are derivatives of the first pharyngeal pouch. Many amphibians, reptiles, and birds have one bony ear ossicle within the cavity of the inner ear, called the stapes or *columella;* it extends from the eardrum, or its equivalent, to the oval window, which contacts the perilymphatic fluid. An exception among air breathers are the mammals with three such bones known as the *malleus* (hammer), *incus* (anvil), and *stapes* (stirrup) that form a sequence of levers between the malleus at the eardrum to the stapes at the oval window. Extending from the middle-ear chamber to the pharynx is the tympanic (Eustachian) tube. The eardrum is not taut. In fact, it is slightly loose so that it can be set into vibration by weak sound waves. During the act of swallowing, the tube may be opened to permit

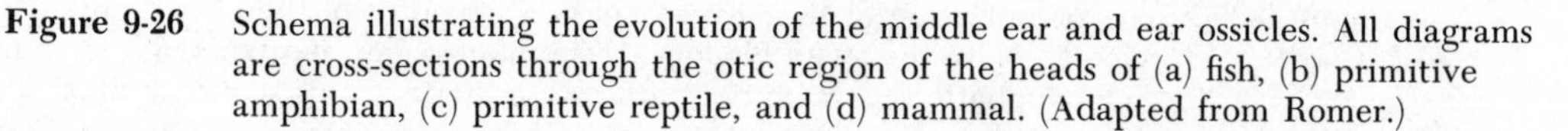

Figure 9-26 Schema illustrating the evolution of the middle ear and ear ossicles. All diagrams are cross-sections through the otic region of the heads of (a) fish, (b) primitive amphibian, (c) primitive reptile, and (d) mammal. (Adapted from Romer.)

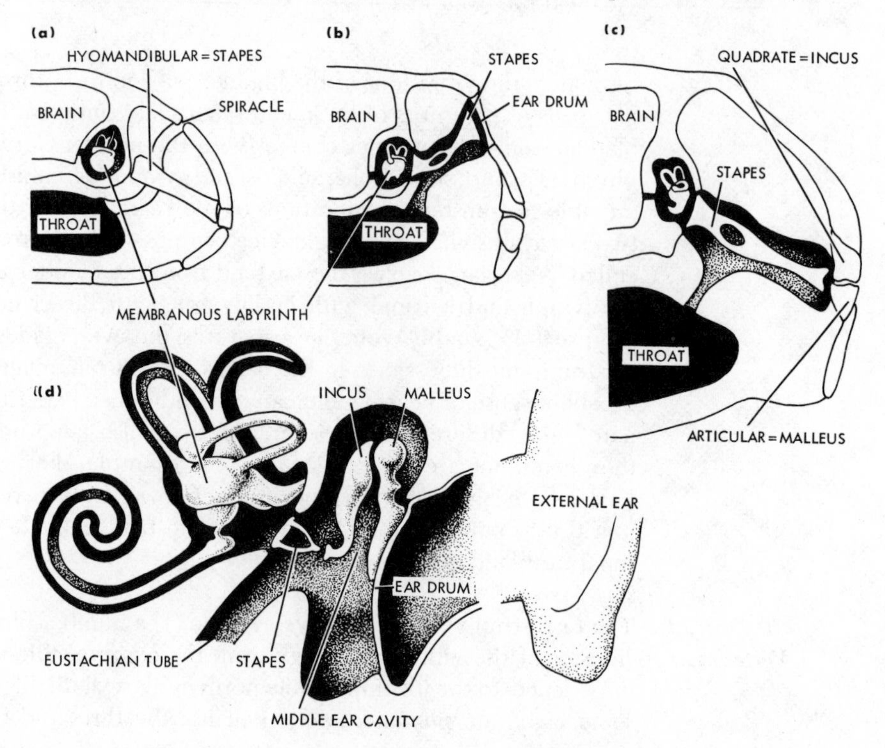

the pressure within the middle ear to equal the atmospheric pressure being exerted on the eardrum through the external canal. The external canal arises as an inpushing of the embryonic body ectoderm in the vicinity of the first gill pouch that, in turn, is the origin of the middle ear chamber and tympanic tube. The ears of tetrapods are modifications of this general piscine organization.

Various amphibians respond to sounds of from fifty to ten thousand cycles per second. In most frogs and toads, the stapes extends from the eardrum, located on the surface of the head, to the oval window. Salamanders, which have neither an eardrum nor a middle ear pick up vibrations with their skulls or their upper extremities (Figure 9-27). Vibrations are transmitted from the skull to the oval window in aquatic forms; in other amphibians, ground vibrations picked up by the forelimbs are transmitted from the shoulder blade via a contracted voluntary muscle to an ossicle called the *operculum* that articulates at the oval window. The amphibia have two small diverticuli of the sacculus; each diverticulum may have one or two maculae (Figure 9-23). In salamanders and frogs, the small diverticulum contains the macula of the amphibian papilla, and it responds to sound waves. The large diverticulum is the lagena, which contains a dorsally located macula lagena of unknown function and a basilar recess papilla that responds to sound waves. In frogs, the basilar papilla responds to sounds with frequencies of one thousand to two thousand cycles per second and the amphibian papilla to sounds with frequencies of one hundred to one thousand cycles per second.

Reptiles hear in a variety of ways. All snakes, which pick up earth-borne vibrations through their skulls, lack an eardrum and a middle ear; they are deaf to air-borne sounds. The vibrations are transmitted from the skull via the stapes to the oval window. Tortoises, which lack an eardrum, transmit vibrations from a thick patch of skin on the head via the stapes to the oval window. Crocodiles and some lizards have a short external auditory tube; the stapes is composed of an outer cartilaginous segment and an inner bony segment. Reptiles have a large lagena with two receptors [Figure 9-24(b)]. The proximally located basilar recess papilla has a role in audition, whereas the distally located macula lagena (also found in birds) [Figure 9-24(c)] has an unknown function. Reptiles respond to frequencies as low as eighty cycles per second in tortoises and as high as ten thousand cycles in some lizards.

Figure 9-27 Diagrams illustrating the mechanisms for transmitting vibrations from the environment to the inner ear in urodeles. (a) In this aquatic urodele, vibrations picked up by the skull are transmitted from the squamosal bone via a ligament to the stapes (columella). (b) Vibrations picked up by the upper extremity are transmitted from the scapula through a muscle connected to the operculum. (Adapted from Kingsbury and Reed.)

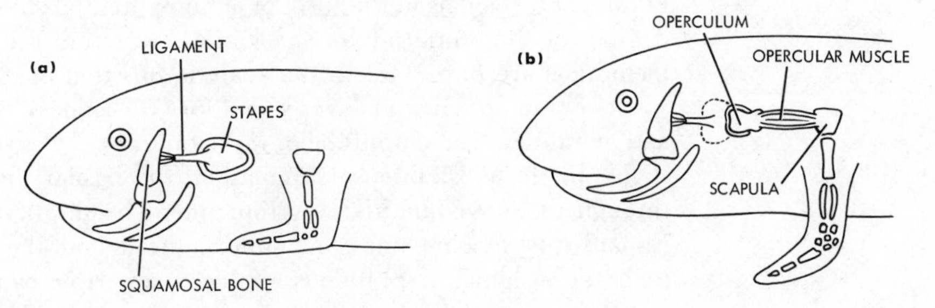

Birds + mammals

The phonoreceptive apparatus is most highly evolved in birds and mammals. In birds, the eardrum is located within a short external auditory tube whose external opening is covered with feathers. As in the lizards, the stapes is in two segments. The receptive efficiency of the three ear ossicles in mammals can be dampened or enhanced by changes in the tension of the two muscles inserted on the ossicles, the tensor tympani muscle to the malleus and the stapedius muscle to the stapes. When contracted, the tensor tympani, through its action on the malleus, stiffens the tympanic membrane. When contracted, the stapedius muscle firms the fit of the foot of the stapes into the foramen ovale. These muscles protect the inner ear from loud sounds by reflexly contracting in response to intense sound waves. These muscles also contract while a mammal is vocalizing. This is presumed to prevent the ear from being overstimulated by the animal's own voice. Phonoreceptors in birds and mammals constitute the spiral organ of Corti (Figures 9-18 and 9-24), located in an extension of the saccule called the cochlear duct. Like all papillae, it comprises sustentacular cells and hair cells whose cilia are embedded in a tectorial membrane. In birds, as in reptiles, the cochlea is short and straight, whereas in mammals it is coiled from one to five turns, depending on the species—e.g., one quarter turn in the duckbill platypus, two turns in the horse, three and one half in man, and four in the guinea pig. Sound waves as traveling fluid waves are conducted from the oval window successively through the perilymphatic fluid of the vestibular duct and the tympanic duct to the membrane of the round window; this membrane "absorbs" and dampens the vibrations. The sound waves produce vibrations of the organ of Corti, its basilar membrane, and the tectorial membrane; the resulting movements of the cilia of the organ of Corti will eventually result in nerve impulses that are projected to the brain. The spiral organ of Corti is nourished by the exchange of nutrients and gases with the fluids of the ear. The absence of a blood supply within the spiral organ has functional significance. If blood flowed through the spiral organ, its movements would stimulate the hair cells and, thus, produce noise in the auditory system.

Two structural adaptations of significance to the role of the cochlea in hearing include (1) the hair cells of the organ of Corti (attached to the basilar membrane), with the distal tips of cilia embedded within the tectorial membrane, and (2) the basilar membrane, which becomes wider and stiffer in the progression from the base to the apex of the cochlear coil (the basilar membrane is not under tension). Vibrations within the fluids of the cochlear coil (vestibular, tympanic, and cochlear ducts) produce vibrations of both the tectorial membrane and the basilar membrane with the attached organ of Corti. The vibrations of these two membranes produce shearing motions in the cilia of the hair cells that trigger a sequence of events as yet not fully understood. The ciliary movements stimulate the nerve fibers of the auditory nerve to fire neural signals in the form of action potentials to the brain. Differential degrees of the amplitude of the vibrations within the length of the basilar membrane are important to the analytic function of the cochlea in hearing.

The organ of Corti analyzes sound for (1) loudness, (2) pitch, and (3) tones and overtones. The amplitude of the vibrations of the spiral organ of Corti is slight; to the most intense stimulation, the basilar membrane may vibrate through about five hundred Angstom units. The amplitude of the vibrations in response to a specific tone is not uniform throughout the spiral organ. For each tone, the maximal amplitude is limited to a narrow band of the spiral organ.

The basilar membrane of the basal turn of the cochlea responds maximally to high frequencies and slightly to low frequencies, whereas the basilar membrane of the apical turn responds maximally to low frequencies and not at all to high frequencies. The highest notes are "sensed" at the base of the coil near the oval window and the lowest notes at the apex of the coil. The upper limit of pitch for birds and most mammals is roughly twenty thousand cycles per second; rodents and shrews up to thirty thousand cycles per second; and bats, whales, and dolphins more than one hundred thousand cycles per second. Birds seem to have as efficient a pitch discrimination as man. Owls have an exceptional ability to localize a sound; in total darkness, an owl can strike a squeaking mouse with a margin of error of less than one degree in both horizontal and vertical planes.

Echolocation

Bats, whales, porpoises, and certain birds have evolved elaborate and efficient systems of emitting pulses (squeaks) of sonic and ultrasonic vibrations up to 150,000 cycles per second from their vocal cords; and, after hearing the reflections, they are able to *echolocate* the objects reflecting the pulses. Varying with the animal, the squeaks may be emitted at the rate of five to two hundred pulses per second. The reflected squeaks are recognized, for example, by the bats, which have a large basal portion (low tone receptive portion) of the cochlea and a highly developed auditory system in the brain.

Many bats utilize echolocation to avoid obstacles such as fine wires strung across a totally dark room and to catch flying insects on which they feed. Bats are strong expert flyers and their seemingly erratic flight does not result from being buffeted by air currents; rather the echolocating and the subsequent catching of flying insects, one at a time every few seconds, produce the constant maneuvering. The precision and speed by which a porpoise echolocates a fish are aided by the water environment, which conducts sound waves more than four times as fast as air. Two cave birds, the Venezuelan Oilbird, *Steatornis,* and the Asian Swift, *Collocalia,* echolate their way about caves by emitting and echolocating sonic vibrations.

References for Chapter 9

[1] VAN BERGEIJK, W. A. "The Evolution of Vertebrate Hearing." In *Contributions to Sensory Physiology,* edited by W. Neff. Vol. 2. New York: Academic Press, Inc., 1967. This is a modern account of hearing in the vertebrates.
[2] VON BUDDENBROCK, W. *The Senses.* Ann Arbor: The University of Mighigan Press, 1958. This is a compact survey of sensation in animals.
[3] CAHN, P. H. ed. *Lateral Line Detectors.* Bloomington: Indiana University Press, 1967. This book contains articles by outstanding authorities on many aspects of lateral line detectors.
[4] CASE, J. *Sensory Mechanisms.* New York: Macmillan Publishing Co., Inc., 1966. This is a clear, readable discussion. (Paperback)
[5] DETWEILER, S. R. *Vertebrate Photoreceptors.* New York: Macmillan Publishing Co., Inc., 1943. This is an outstanding, classic review.
[6] EAKIN, R. M. "A Third Eye." *American Scientist,* **58**:(1970), 73–79.
[7] GRIFFIN, D. R. *Listening in the Dark.* New Haven, Conn.: Yale University Press, 1958. This is a superb and fascinating account of the acoustical orientation of bats and men. New York: Dover, 1974. (Reprint)

[8] Matthews, L. H., and M. Knight. *The Senses of Animals.* London: Museum Press, 1963. This is a review of many expressions of the senses of animals.

[9] Milne, L. J., and M. Milne. *The Senses of Animals.* New York: Atheneum Publishers, 1962. This is another review of the many expressions of the senses of animals.

[10] Parsons, T. S. Evolution of the Nasal Structure in the Lower Tetrapods." *American Zoologist,* **7:**(1967), 397–413.

[11] Prince, J. H. *Comparative Anatomy of the Eye.* Springfield, Ill.: Charles C Thomas, Publishers, 1956. This is an excellent account of the structure of the eye in vertebrates.

[12] Pumphrey, R. J. "Hearing." *Symposium of the Society of Experimental Biology,* **4:**(1950), 1–18. This is a classic account of structural and functional aspects of hearing.

[13] Rochon-Duvigneaud, A. *Les Yeux et la vision des vertèbres.* Paris: Masson et Cie., 1943. This is a comprehensive review of the eyes of vertebrates.

[14] Spoendlin, H. H. *The Organization of the Cochlear Receptor.* Basel: Karger, 1966.

[15] Walls, G. L. *The Vertebrate Eye and Its Adaptive Radiation.* Bloomfield Hills, Mich.: Cranbrook Institute of Science, 1942. This outstanding volume deals with comparative anatomy and functions of the eye.

[16] Webster, D. B. "Ear Structure and Function in Modern Mammals." *American Zoologist,* **6:** (1966), 451–466.

[17] Wersäll, Jan, and A. Flock. "Functional Anatomy of the Vestibular and Lateral Line Organs." In *Contributions to Sensory Physiology,* edited by W. Neff. Vol. 1. New York: Academic Press, Inc., 1965. These are modern analyses of the functional anatomy of the lateral line system.

[18] Wilentz, J. S. *Senses of Man.* New York: Apollo Eds. This is another clear, readable account of the subject. (Paperback)

[19] Wurtman, R. J., J. Axelrod, and D. E. Kelly. *The Pineal.* New York: Academic Press, Inc., 1968.

10

The Nervous System

Introduction The cells of living organisms are excitable; they possess the ability to respond to stimuli. This excitability is expressed in a variety of ways: muscles contract, nerve cells conduct impulses, gland cells secrete, cells of electric organs discharge electric pulses, and white cells ingest microorganisms. A positive response to a stimulus is known as *excitation* (gland cell secretes) and a negative response as *inhibition* (gland cell ceases to secrete). The *nerve cells* (*neurons*) of the nervous system and the gland cells of the endocrine system are the basic structural and functional units of these two major integrating systems of the body. These specialized cells are adapted to receive and to process stimuli and then to respond to the stimuli by releasing chemical secretions that can influence other cells. The secretion of a neuron is a *neurosecretion*, which may stimulate an adjacent cell; the secretion of a cell of the endocrine system is a *hormone*, which stimulates distant target cells. Neurosecretions may influence endocrine cells and hormones may influence neurons, and in this way the two systems interact. The nervous system is the high-speed coordinator, usually operating with a fast time course in fractions of a second, whereas the endocrine system is the slow-speed coordinator, operating with a slow time course of hours and days.

The nervous system is conventionally divided into a central nervous system and a peripheral nervous system. The central nervous system is composed of the brain and the spinal cord surrounded by connective tissue membranes called *meninges*. The peripheral nervous system includes the spinal nerves emerging from the spinal cord and the cranial nerves from the brain; each nerve and its branches are enclosed by connective tissue. The peripheral nerves that are distributed through the body are composed primarily of the processes (fibers) of neurons. These nerve fibers of the peripheral nerves convey neural influences from the receptors in the body as input to the central nervous system and from the nervous system as output to effectors (e.g., muscles and glands). It is only through muscles and glands that the activity of the nervous system can be expressed: (1) through the contraction or relaxation of muscles and (2) through the excitation or inhibition of secretion by a gland cell.

Associated with the neurons are the Schwann (neurolemma) cells of the peripheral system and the neuroglia (*oligodendroglia, astroglia,* and *microglia*) of the central nervous system.

Of all the phyla of the animal kingdom, only the vertebrates have a vascularized central nervous system. This innovation in early vertebrate phylogeny was an essential prerequisite to ensure the adequate nourishment of a delicate, yet bulky, brain and spinal cord.

Neuron

Neurons are the cells of the central nervous system specialized for receiving, processing, and transmitting information. Numerous neuronal cell types are present within the system. The diversity of shapes and forms of neurons is probably greater than for any other cell type in the body. On the basis of morphological criteria, each neuron is subdivided into four structural segments: *dendrites, cell body (soma), axon,* and *telodendria* (Figure 10-1). On the basis of physiological criteria, each neuron is subdivided into four functional segments: *trophic, receptive, conductile,* and *transmissive* segments (Figure 10-2).

The Generalized Neuron As a Structural Unit

The cell body is that portion of the neuron containing the nucleus, Nissl substance, and other organelles essential to the maintenance of a functional unit (Figure 10-1). The Nissl substance is a concentration of ribonucleoproteins (RNP) that synthesizes proteins, including the neurosecretory (neurotransmitter) chemicals. Short processes extending from the cell body are the dendrites that generally convey neural information toward the cell body. Nerve impulses from the cell body are conducted by the *axon* toward the *telodendria,* which are short terminal branches of the axon (Figure 10-1), and they come into close proximity with another cell and form a synapse.

The synapse (Figure 10-1) is the structural complex comprising the telodendritic ending of one neuron (*presynaptic* neuron), the space of approximately 200 Angstrom units wide called the *synaptic* cleft, and the dendrites or soma of another neuron (*postsynaptic* neuron). The presence of the synaptic cleft indicates that the presynaptic neuron is in contiguity but not in continuity with the postsynaptic neuron. Several morphologic types of synapses are found; for example, an axodendritic synapse between axon and dendrite, an axosomatic synapse between axon and soma, and a motor end plate between an axon and a voluntary muscle fiber.

The Generalized Neuron As a Functional Unit

The *trophic segment* (Figure 10-2) is the metabolic center of the neuron where the synthesis of proteins, neurosecretory chemicals, and other products essential to the neuron occur. This segment is the soma of the neuron.

The *receptor segment* (Figure 10-2) is adapted to receive a continuous input of both excitatory and inhibitory stimuli; this input is processed, sorted, and

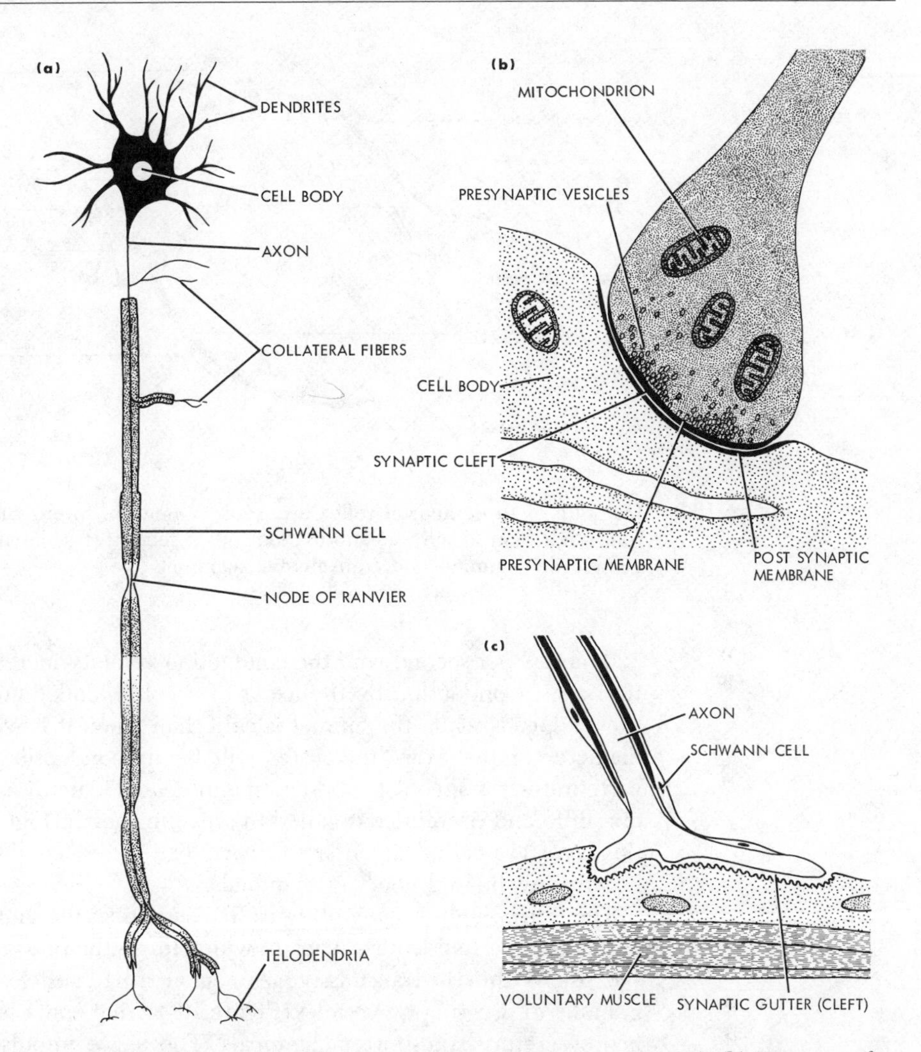

Figure 10-1 (a) Diagram of a peripheral motor neuron (alpha motor neuron). (b) Diagram of a synapse between an axon and a cell body (axosomatic synapse), between an axon and a dendrite (axodendritic synapse) and between an axon and an axon (axoaxonic synapse). (c) Diagram of a motor end plate (synapse between nerve terminal and muscle cell).

integrated in this segment, which is usually located in both the dendrites and the soma. Excitatory synapses and inhibitory synapses transmitting input onto this segment in one neuron may number as many as thirty thousand.

The specialized *conductile segment* (Figure 10-2), called the axon, has evolved an all-or-none nerve impulse to perform at least two roles: (1) to conduct impulses a long distance from the receptive segment to the transmissive segment, and (2) to conduct neural information in the form of coded messages. The all-or-none impulse is an expression of a physiological process by which information can be conveyed at full strength over the entire distance along a nerve fiber; there is either a completely developed impulse or none at all. Coded messages may be (1) binary in the form of an ON (impulse) or an OFF (no impulse) activity or (2) expressed in the number of impulses per unit time. The speed of conduction varies from a fraction of a meter per second to

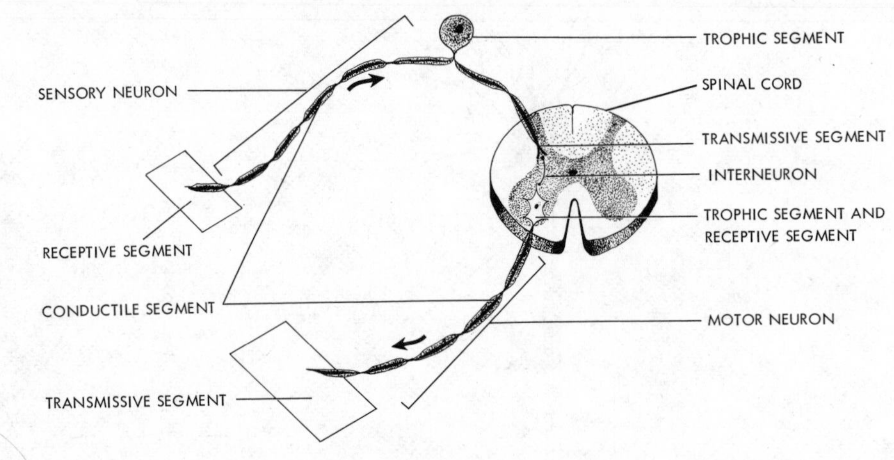

Figure 10-2 Diagram of three-neuron reflex arc: sensory neuron, interneuron, and motor neuron. Each neuron may be subdivided into a receptive segment, trophic segment, conductile segment, and transmissive segment.

120 meters per second, and the conduction velocity increases as the temperature rises. Conduction by the axons of a cold-blooded animal takes place at slower speeds when the animal is cold than when it is warm. The larger the diameter of the axon, the faster will be its conduction velocity. In many vertebrates, the speed of conduction is increased in many axons by the presence of a lipid and protein coat called the myelin sheath (Figure 10-1); fibers with this layer are called myelinated fibers. The fastest conducting fibers are the heavily myelinated fibers in mammals.

The *transmissive segment* (Figure 10-2) includes the short terminal branches of the axon, or telodendria, each of which forms the presynaptic portion of the synapse. Within the transmissive segment are small vesicles, the neurosecretory granules or presynaptic vesicles (Figure 10-1), that contain the precursor of the neurosecretory transmitter chemicals. The nerve impulse of the conductile segment activates the transmissive segment to release the transmitter chemicals (e.g., acetylcholine, serotonine, or noradrenalin) into the synaptic cleft where they can stimulate the postsynaptic neurons, muscles, or gland cells. Although each nerve fiber may conduct an impulse in either direction within the neuron, conduction in a sequence of neurons proceeds in only one direction. This unidirectional conduction is determined at the synapse, which acts as a one-way valve with the transmission of neural influences across the synapse occurring only from the presynaptic neuron to the postsynaptic neuron.

Each neuron can be conceived as expressing, in miniature, the integrative capabilities of the entire nervous system. The dendrite-cell body unit in most neurons is specialized as the receptor, integrator, and processor of the synaptic input from the transmission segments of other neurons; in turn, the axon is the specialized conductile segment that conveys coded information through a distance from the dendrite-cell body unit to the synaptic junctions, where the transformation functions take place with the dendrite-cell body units of other neurons or of effectors (muscle and glandular cells).

The sensory neuron of the spinal nerves and of many cranial nerves has these functional segments organized as follows: the receptive segment is a short terminus within the nerve endings in the periphery of the body, and the conductile segment extends from the receptive segment to the short transmissive segment located within the central nervous system where the synaptic contacts are made. The trophic center (soma) is located within the conductile segment in the dorsal (sensory) root ganglion (Figures 10-2 and 10-6).

Other Cells of the Nervous System

Except at their receptive endings, nerve fibers of the peripheral nervous system are surrounded by cells called Schwann (neurolemma) cells (Figure 10-1); these cells are responsible for the formation of the myelin sheath. Fibers surrounded only by Schwann cells are called unmyelinated fibers; those having both Schwann cells and myelin are said to be myelinated. The myelin sheath is segmented at intervals by indentations known as nodes of Ranvier (Figure 10-1); the nerve impulse in a myelinated nerve hops from node to node by *saltatory* (leaping) conduction. The speed of conduction is faster in the thicker myelinated fibers with the greater distance between two successive nodes.

Nerve fibers are congregated into fascicles, which are encapsulated by sheaths of connective tissues (called endoneurium, perineurium, and epineurium). Within these connective tissues laminae are the blood vessels conveying nutriments, gases, and waste products to and from the neurons.

Within the central nervous system are nonneural cells called neuroglial cells (Figure 10-3). Oligodendroglia are characterized by processes that encapsulate portions of the cell body and nerve fibers; the cells elaborate the myelin sheaths of the myelinated fibers of the central system. The astrocytes have processes

Figure 10-3 Composite diagram illustrating cells within the central nervous system as visualized at an ultramicroscopic level. (Adapted from Noback and Demarest.)

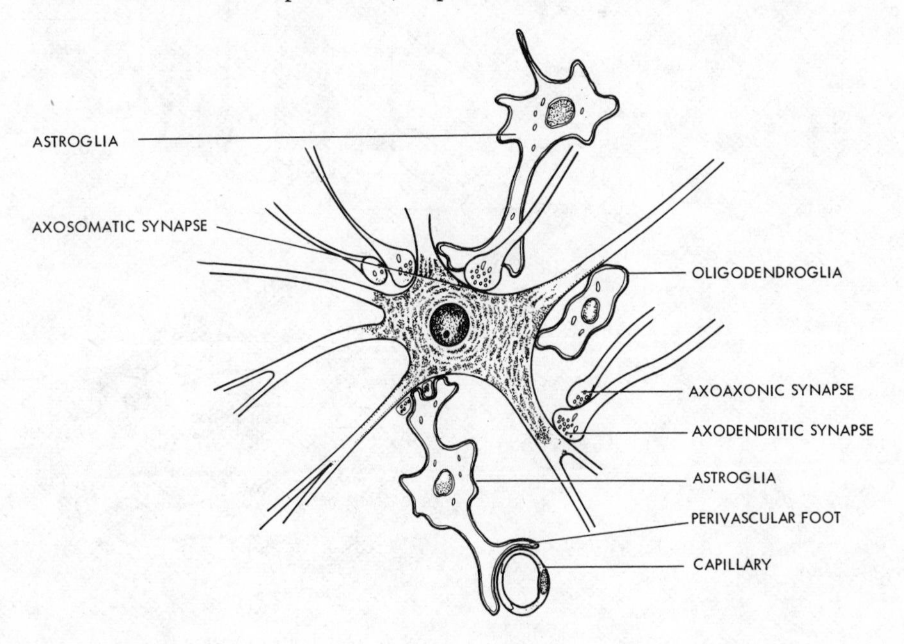

that extend from the blood capillaries to the cell bodies of the neurons; these cells are presumed to act as intermediaries between the blood and the neurons and, thus, maintain a constant extraneuronal milieu for the neurons and protect them from changes in the internal environment. Some investigators claim these cells have a role in memory storage. The microglia are actually connective tissue cells; they are phagocytes that remove degeneration products. Microglia are not common in the normal central nervous system.

Meninges The meninges are the connective tissue membranes surrounding the central nervous system (Figure 10-4). In most fish, the meninx is a single vascularized connective tissue sheet located between the central nervous system and the axial skeleton (skull and vertebral column). In amphibians and reptiles, two meninges are present: the vascularized *pia mater* attached to the central nervous system, and the outer tough *dura mater* adjacent to the skeleton. Mammals and birds have three membranes; a delicate vascularized membrane called the *arachnoid* is located between the pia mater and the dura mater [Figure 10-4(a)]. In a sense, these meninges can be thought of as a pial sac

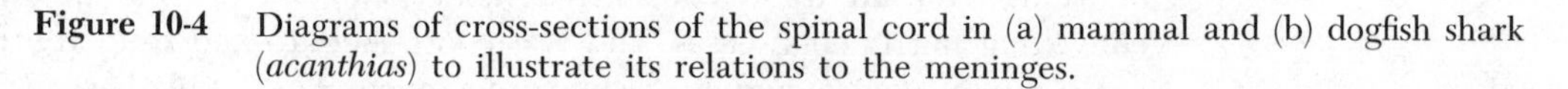

Figure 10-4 Diagrams of cross-sections of the spinal cord in (a) mammal and (b) dogfish shark (*acanthias*) to illustrate its relations to the meninges.

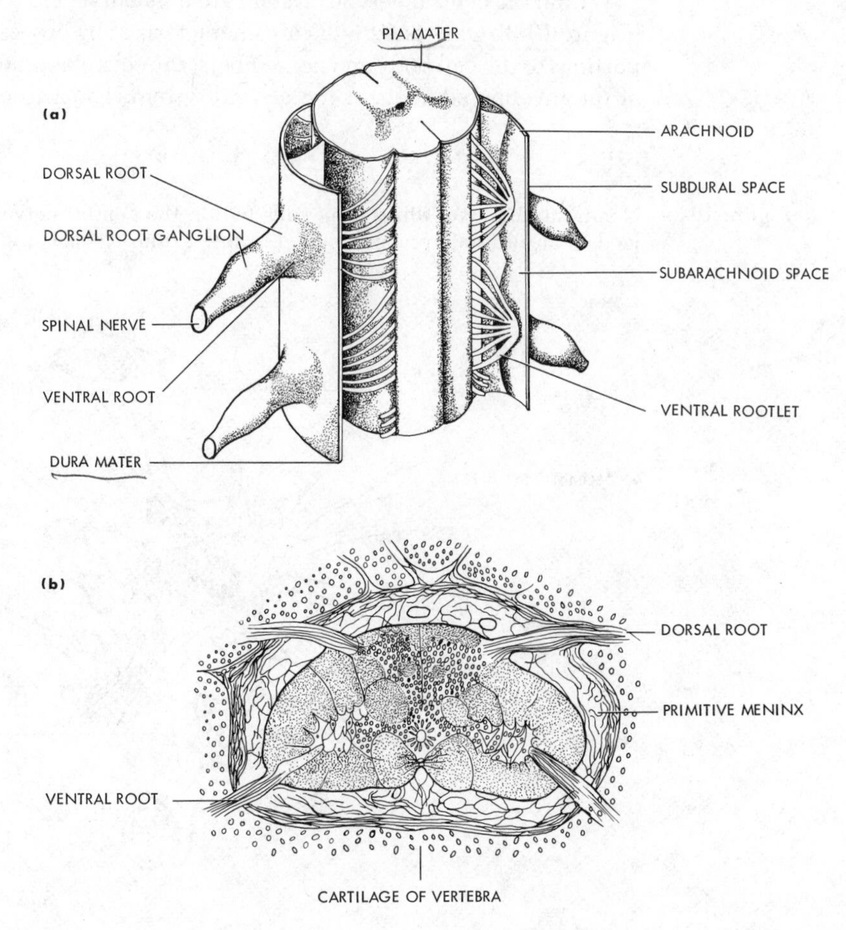

within an arachnoid sac within a dural sac. The meninges extend onto and sheath the roots of the spinal and cranial nerves (Figure 10-4); in turn, these sheaths are continuous with the connective tissue layers (e.g., perineurium) of the peripheral nerves. The interval between the pia mater and the arachnoid, called the subarachnoid space, is filled with cerebrospinal fluid. This fluid is also present within the spinal and ventricular cavities that are in communication with the subarachnoid space through foramina in the choroid plexus in the roof of the medulla. The brain and spinal cord literally float in the fluid (like a log submerged in water). The buoyant property of this water jacket supports the 1,400-gram human brain so that it actually has the net weight of about fifty to one hundred grams if weighed within the skull.

Spinal Nerves and Spinal Cord

The spinal cord is enclosed within the vertebral canal in the cartilaginous or bony vertebral column. The length of the generally cylindrical spinal cord varies (Figure 10-5). In most fish, tailed amphibians, reptiles, and birds, the cord extends throughout the entire length of the vertebral canal. A few fish, tailless amphibians, and mammals have a cord that is shorter than the vertebral column, and in most mammals it extends into the lumbar region.

The spinal cord is subdivided into segments called spinal segments; a bilateral pair of dorsal (sensory) nerve roots and of ventral (motor) nerve roots emerge from each segment (Figure 10-6). In all vertebrates except the cyclo-

Figure 10-5 Diagrammatic representation of the spinal cord and the proximal portion of spinal nerves in (a) turtle, (b) man, (c) seal, and (d) toad. In the turtle, the cord extends throughout the vertebral canal. The spinal cords are drawn relative to the vertebral column length, which is standardized to the same length in all the figures. (Adapted from Nieuhenhuys.)

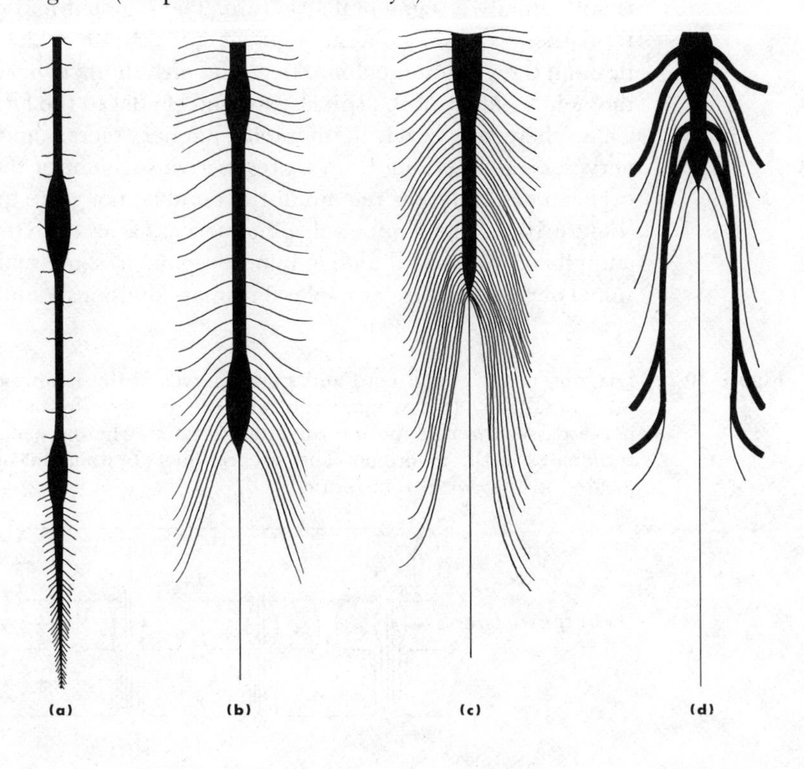

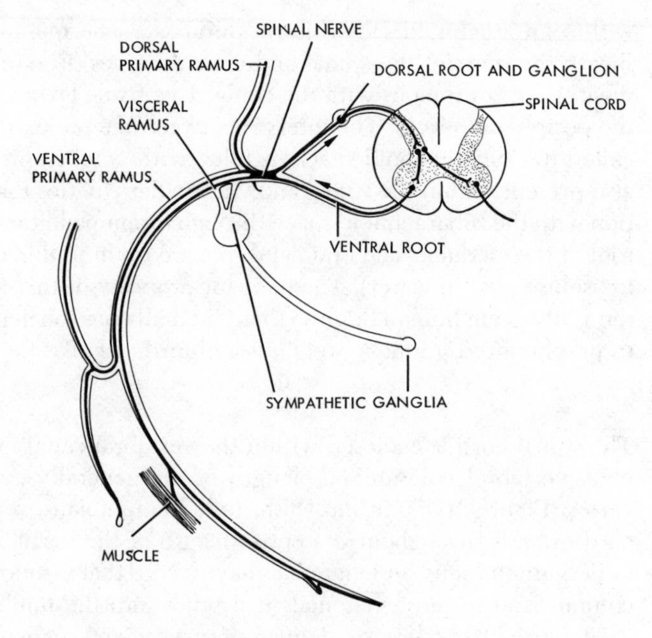

Figure 10-6 Diagram of a typical spinal nerve.

stomes (Figure 10-7), the dorsal root and the ventral root on each side of a segment unite to form a spinal nerve.

The serial metamerism of spinal roots, nerves, vertebrae, and embryonic somites from anterior to posterior is the primary segmentation expressed by the body. This segmentation is secondarily imposed on the spinal cord, which itself is not actually a segmental structure. The segment and the nerve are named from the vertebral column segment adjacent to which the nerve emerges through the vertebral column (e.g., the fifth thoracic nerve emerges from one thoracic segment of the spinal cord and caudal to the fifth thoracic vertebra). Thus, there are cervical, thoracic, lumbar, sacral, and coccygeal (caudal) nerves, each emerging from its respective segment of the spinal cord. In the cyclostomes, the roots remain distinct and do not unite to form a spinal nerve (Figure 10-6). The number of spinal nerves varies from ten pairs in the tailless amphibians to more than five hundred pairs in some snakes. In general, each spinal nerve branches into several primary divisions (rami): the dorsal ramus to

Figure 10-7 Diagram of the spinal cord and spinal nerves of the lamprey. Note that the separate dorsal roots (nerves) and ventral roots (nerves) alternate. Each dorsal spinal nerve passes through the intermyotomic space, whereas each ventral spinal nerve terminates in the myotome. The visceral fibers (autonomic nervous system) are located in the dorsal root (nerve).

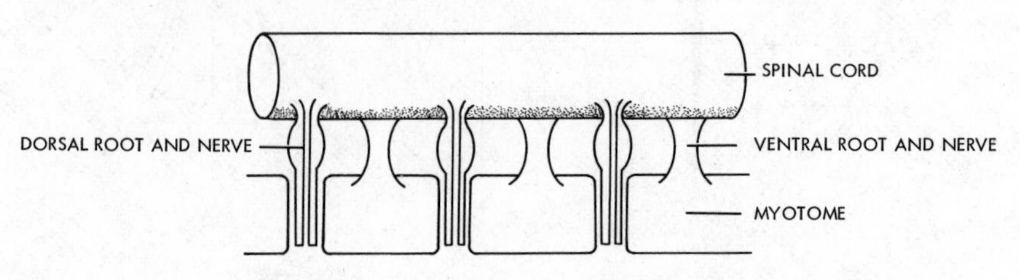

the back region, the ventral ramus to the ventral region including the extremities, and the rami communicantes to the visceral organs (Figure 10-6).

Input

The term *sensory,* although usually applied to consciously perceived input, is used in its broadest sense here to include all input from nerves to the central nervous system whether consciously appreciated or not. This avoids, among other problems, deciding whether the animal is consciously aware of a stimulus. The sensory (afferent) input to the spinal cord is conveyed via neurons with fibers that pass through the peripheral nerves and dorsal roots; the cell bodies of these sensory neurons are located in the dorsal root ganglia (Figure 10-6). In certain lower vertebrates, the afferent neurons, called Rohon-Beard cells, have fibers in the dorsal roots and cell bodies within the spinal cord (Figure 10-8). A skin sensory segment of the adult body, where limbs or fins are innervated by fibers from one dorsal root is called a dermatome. (The term *dermatome* also refers to the part of the embryonic somite, which develops into the dermis of the skin.)

Afferent fibers conveying sensory input from the body to the cord are known as general somatic afferent fibers, and those conveying input from the visceral organs to the cord are general visceral afferent fibers. Each peripheral nerve is composed of functional nerve components. They are categorized as follows: those nerve fibers that are sensory fibers are called afferent; those that are motor fibers are termed efferent; those that innervate the body wall, appendages (limbs, fins, and wings), and much of the head are called somatic; those that innervate the viscera are named visceral; those that are distributed widely, throughout the body, are general fibers (e.g., touch); and those that are distributed to limited parts (e.g., taste) of the organism are termed special. The spinal nerves contain only the general components. The cranial nerves as a group contain all general and special components, although no cranial nerve by itself has all components. Many peripheral nerves are composed exclusively of fibers derived from one spinal nerve (e.g., most thoracic nerves). Many spinal nerves form *plexuses* (Figure 10-9), which are the result of the mingling and regrouping of nerve fibers of several segments of the spinal cord. Thus, the brachial plexus is an organized meshwork of fibers innervating the pectoral fin

Figure 10-8 Cross-section through the spinal cord of a larval *Ambystoma.* The Rohon-Beard cell is an afferent (sensory) neuron with its cell body in the central nervous system and its dendrite extending peripherally in the dorsal root to the skin. (Adapted from Nieuhenhuys.)

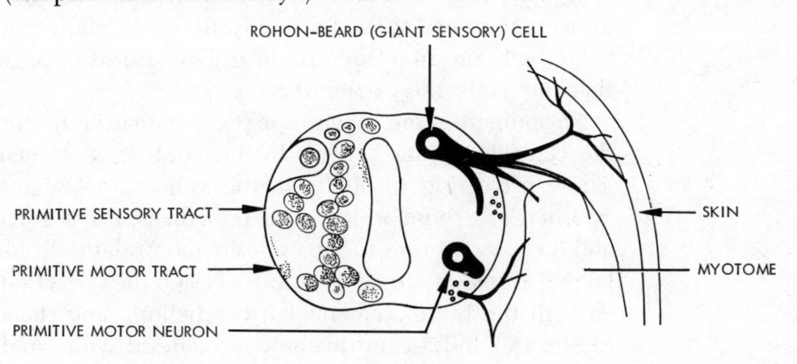

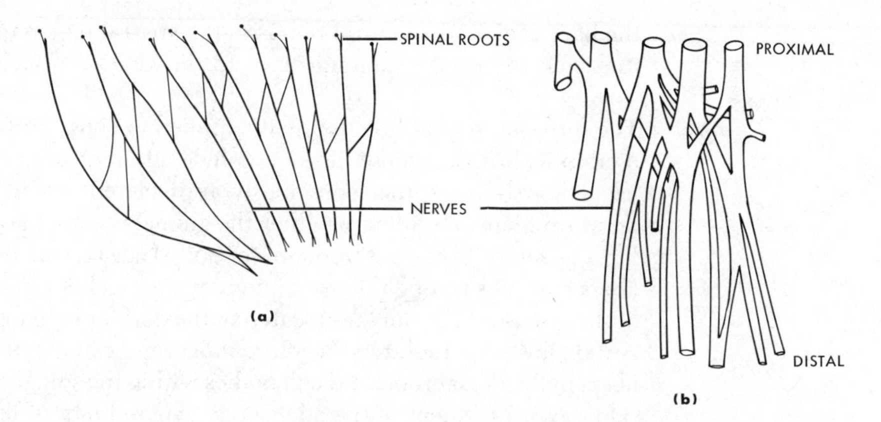

Figure 10-9 (a) Nerve plexus of the pelvic fin of a chimaera (elasmobranch), and (b) branchial plexus of the forelimb of a mammal. Both illustrate the regrouping of nerve fibers and nerve bundles to from an organized meshwork of nerves. The proximal aspects of the plexuses are in the upper part of the figures. (Adapted from Romer.)

or the forelimb, as the lumbosacral plexus innervates the pelvic fin or the hindlimb.

Output The motor (efferent) output from the spinal cord to the muscle and gland cells (effectors) is conveyed via fibers passing through the ventral roots and the peripheral nerves to the effectors. The efferent fibers projecting output to the voluntary (striated) muscles are the general somatic efferent fibers, and those projecting output to the involuntary (smooth) muscles, heart muscle, and glands are the general visceral efferent fibers of the autonomic nervous system, to be noted later.

Spinal Cord The spinal cord consists of a small central canal surrounded by an inner column of gray matter and an outer column of white matter. Within the gray column, three horns, or zones, are visible: dorsal (sensory, afferent), intermediate, and ventral (motor, efferent) (Figure 10-10). The gray matter is made up mainly of cell bodies, their dendrites, and some axons organized into groups of these cell bodies and their dendrites called nuclei (neuron pools or centers). A nucleus in the nervous system is an anatomically identifiable group of cell bodies, whereas a neuron pool or center is a physiologically defined group with a similar functional role. Within the white matter are myelinated and unmyelinated axons that ascend to or descend from one to many spinal segments, or to and from the brain. In tetrapods, the gray matter is enlarged at those spinal levels that innervate the extremities.

The numerous *interneurons* in the gray matter interact with other neurons of the central nervous system. In this context, most neurons are interneurons. Those projecting to nuclei in the same spinal segment are intrasegmental interneurons; those projecting to the other spinal segments are intersegmentals, and the interneurons to the opposite (contralateral) side of the spinal cord are known as *commissurals* (Figure 10-11). Of the fibers that ascend from the spinal cord to the brain stem nuclei, cerebellum, and thalamus, many cross over (decussate) and ascend in the contralateral white matter; in addition, many

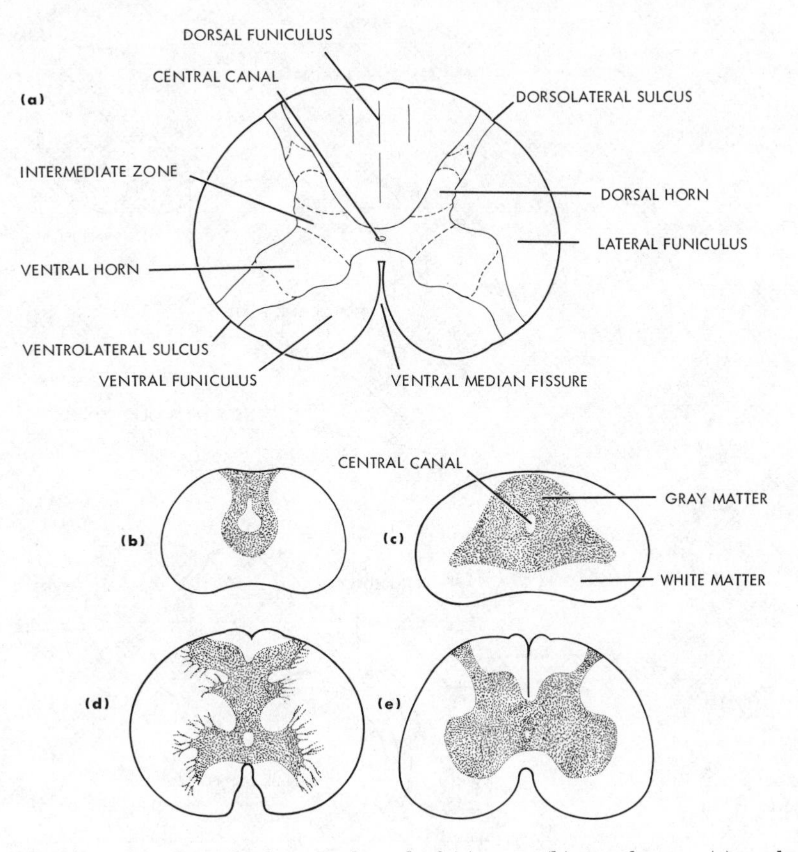

Figure 10-10 Cross-section through the spinal cord of (a) man, (b) Amphioxus, (c) cyclostome, (d) teleost fish; and (e) amphibian. Note the configuration of the gray matter and the white matter. The ventrolateral sulcus is shown in (a). (Adapted from Nieuhenhuys.)

ascend in the white matter on the same side. Motor fibers from the brain descend in the white matter before synapsing with dendrites and the soma of neurons in the nuclei of the gray matter of the spinal cord (Figure 10-11).

Spinal Reflexes The activities of the neuronal circuits of the spinal cord regulate the coordinated undulating movements of the segmental body wall muscles of swimming fish and other vertebrates and the more complex integrated actions of tetrapod limbs. Segmental spinal reflex arcs form the basic circuits integrated through intersegmental circuits; in turn, these spinal circuits are modulated by neural influences descending from nuclei in the brain. Two basic spinal segmental reflex arcs are the two-neuron arc and the three-neuron arc. The two-neuron reflex arc includes the sequence of (1) an afferent neuron of the dorsal root that terminates and synapses in the gray matter of the spinal cord with (2) the motor neuron with its axon, which emerges in the ventral root. The three-neuron reflex arc includes the sequence of (1) an afferent neuron of the dorsal root that terminates and synapses with (2) an interneuron of the gray matter that, in turn, synapses with (3) the motor neuron. The motor neurons terminate by synapsing at the motor end plates on voluntary muscle fibers (Figures 10-1 and

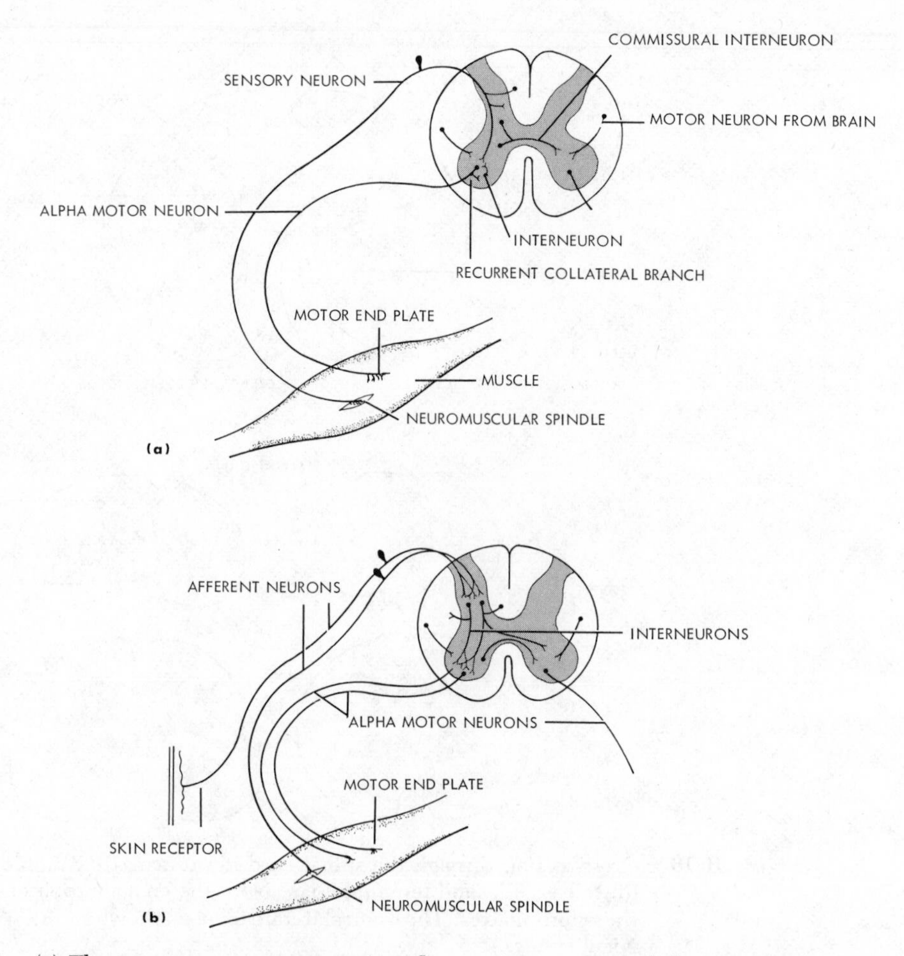

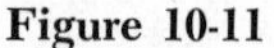
Figure 10-11 (a) The two-neuron monosynaptic reflex arc is composed of the sequence: neuromuscular spindle, sensory neuron, alpha motor neuron, and muscle cell. (b) The three-neuron, disynaptic reflex arc is composed of the sequence: receptor, sensory neuron, interneuron within central nervous system, motor neuron, and muscle cell.

10-11). Acetylcholine is the neurotransmitter at this synapse. In a coordinated action, the two-neuron reflex excites muscle groups to contract; the three-neuron reflex may either excite or inhibit various muscle groups to contract. Reflexes are most complex. Each named reflex arc, even the extensor reflex, is actually a simplified "abstraction" of an intricate neural circuitry. Spinal reflex arcs containing interneurons are numerous, with the intrasegmental, intersegmental, and commissural interneurons integrated in circuits of different levels of complexity. Some excite and others inhibit the postsynaptic neurons. In a spinal reflex, the sensory neuron from the periphery excites a complex of interneurons in the spinal neuronal pools; some of these interneurons excite the motor neurons that, in turn, stimulate the agonistic muscle groups to contract, and others inhibit the antagonistic muscle groups. In other words, the neuron pools are geared to produce integrated movements; in the simple act of flexion of the joints of an extremity, the flexor muscle (agonist) groups are stimulated to contract in harmony with the inhibition of the extensor muscle (antagonist) groups. In turn, other interneurons cross over to excite and to inhibit other neuron pools involved in the coordinated movements on the opposite side; these are synchronized with the movements on the original side (as in a swimming fish or a walking tetrapod). The neuron pools of the spinal cord and brain are organized to integrate and to synchronize precisely the timing of gross movements through inhibitory circuits and excitatory circuits. These produce the postural adjustments (*tonic* contractions) and the active movements (*phasic* contractions).

The role of the neuromuscular spindle, a sensory receptor with a motor innervation, in a spinal reflex will be outlined briefly. The neuromuscular spindle is an elongated spindle-shaped sensory ending surrounded by, and oriented parallel to, the voluntary muscle (*extrafusal*) fibers of the gross muscle of which it is a part. It is an encapsulated structure composed of nerve endings terminating on small voluntary (*intrafusal*) fibers in parallel to the extrafusal fibers. The spindles are stretch receptors (acting as strain gauges) monitoring the tension within the muscle. When the muscle is relaxed and elongated, the spindles within the muscle are passively stretched along with the muscle. In this stretched posture, sensory nerve fibers innervating the spindle are stimulated to fire numerous impulses to the spinal cord to excite the alpha motor neurons. These innervate and stimulate the extrafusal fibers of the voluntary muscle, which contains the stretched spindles, to contract. In summary, the stretched spindles evoke a simple reflex in which the extrafusal muscle fibers of the muscles containing the stretched spindles contract and shorten. In the contracted muscle, the spindle now passively shortened, excites its sensory endings to fire at a lower rate; as a consequence, the extrafusal fibers tend to relax. How can a contraction be sustained, such as occurs when a heavy object is held? Another neural system is utilized. The central nervous system evokes stimuli (e.g., volitional stimuli) to excite the gamma motor neurons that innervate and excite the intrafusal muscle fibers of the spindle to contract. This stretches a region in the spindle to increase the firing rate of the sensory nerve fibers to the spinal cord that excite the alpha motor neurons to sustain the contraction of the extrafusal muscle fibers. Thus, the spindle-firing rate can be increased by (1) stretching the spindles passively and (2) exciting its intrafusal muscle fibers to contract. The delicate and subtle balance of these two mechanisms, as well as of others, is continually operating in the control of the nuances of muscle activity.

The innervation pattern of the spinal motor neurons is different in the amphibians from that in mammals. In the amphibians, this type of neuron innervates both voluntary muscle cells and neuromuscular spindles. In mammals, the innervation pattern is fractionated; the alpha motor neurons innervate voluntary muscle cells and the gamma motor neurons innervate the neuromuscular spindles.

Summary of the Spinal Cord

In general, the comparative functional aspects of the spinal cord can be summarized in the following manner. This complex organ is the simplest of the organized regions of the central nervous system. The higher vertebrates have evolved more complex circuits within the spinal cord; in turn, these circuits interact with circuits within the brain. The interactions between the brain and spinal cord and the dependence of the cord on the brain also increase in these organisms. Animals with the spinal cord transected (a spinal-animal) exhibit varying degrees of dependence on the brain. Most spinal fish have well-coordinated rhythmic movement during swimming. Spinal-amphibians assume normal postural poses (e.g., frog ready to jump) and, if stimulated, will jump with integrated movements. A spinal-mammal cannot stand (*paraplegia*). Immediately after transection, mammals exhibit no reflexes (spinal shock); the spinal reflexes return in dogs and cats after a few days but only after weeks and months in man. The sensory spinal neurons have numerous and longer branches within the gray matter of higher vertebrates, whereas the dendrites of the motor neurons are longer in the lower vertebrates (fish) than in higher forms (mammals).

Spinal Cord and Roots of Agnatha Fish

The spinal cord and roots of the living Agnatha have several structural and functional features that are of comparative significance. Although the meninges are vascularized with capillaries, the spinal cord of these fish lacks capillary networks. Glial cells are present, but myelin sheaths are absent. The spinal cord and its meninges are in direct contact with the notochord. Some cell bodies of primary sensory neurons are located within the spinal cord; most are found within the dorsal root ganglia.

The serially arranged dorsal and ventral roots alternate. Each dorsal root emerges between two myotomes, while each ventral root passes directly into the middle of a myotome (Figure 10-7). As in other vertebrates, the general somatic efferent fibers pass through the ventral roots, whereas the general somatic and visceral afferent fibers are located in the dorsal roots. In contrast to other vertebrates, the general visceral efferent (autonomic) fibers, after emerging from the spinal cord, pass through the dorsal roots. Some of these points have relevance in the concept of the location of the functional nerve components of certain cranial nerves (to be discussed later on).

Autonomic Nervous System

MAMMALS. The autonomic nervous system (general visceral efferent system) influences the organs of the visceral systems by modifying the activity of heart muscle, involuntary (smooth) muscle, and secretory cells. The motor system does not innervate voluntary (striated) muscle. The visceral systems include the cardiovascular, digestive, respiratory, urinary, genital, and endocrine systems, the sweat glands, and the pupillary and accommodation (focusing) muscles of the eye. The autonomic nervous system is represented in the central nervous system by a number of centers and pathways in many regions, of which the *hypothalamus* is the highest regulatory complex (Figure 10-12). Emotional and behavioral responses generated by cerebral cortical activity are mediated through the hypothalamus. In man, for example, a thought may result in cold sweat, fidgeting, altered heart beat, or blushing. A dog's response to the presence of another dog may evoke visceral responses associated with attack, withdrawal, or play. In these activities, the hypothalamus is involved with the visceral responses, whereas the somatic motor system is involved with the voluntary musculature responses.

In mammals, this visceral motor system is subdivided into the *parasympathetic*, or craniosacral, system and the *sympathetic*, or thoracolumbar, system. Parasympathetic fibers emerge from the central nervous system with cranial nerves III, VII, IX, and X and with several sacral nerves; thus, it is also called the craniosacral system. Sympathetic fibers emerge with the thoracic and lumbar nerves and constitute the thoracolumbar system.

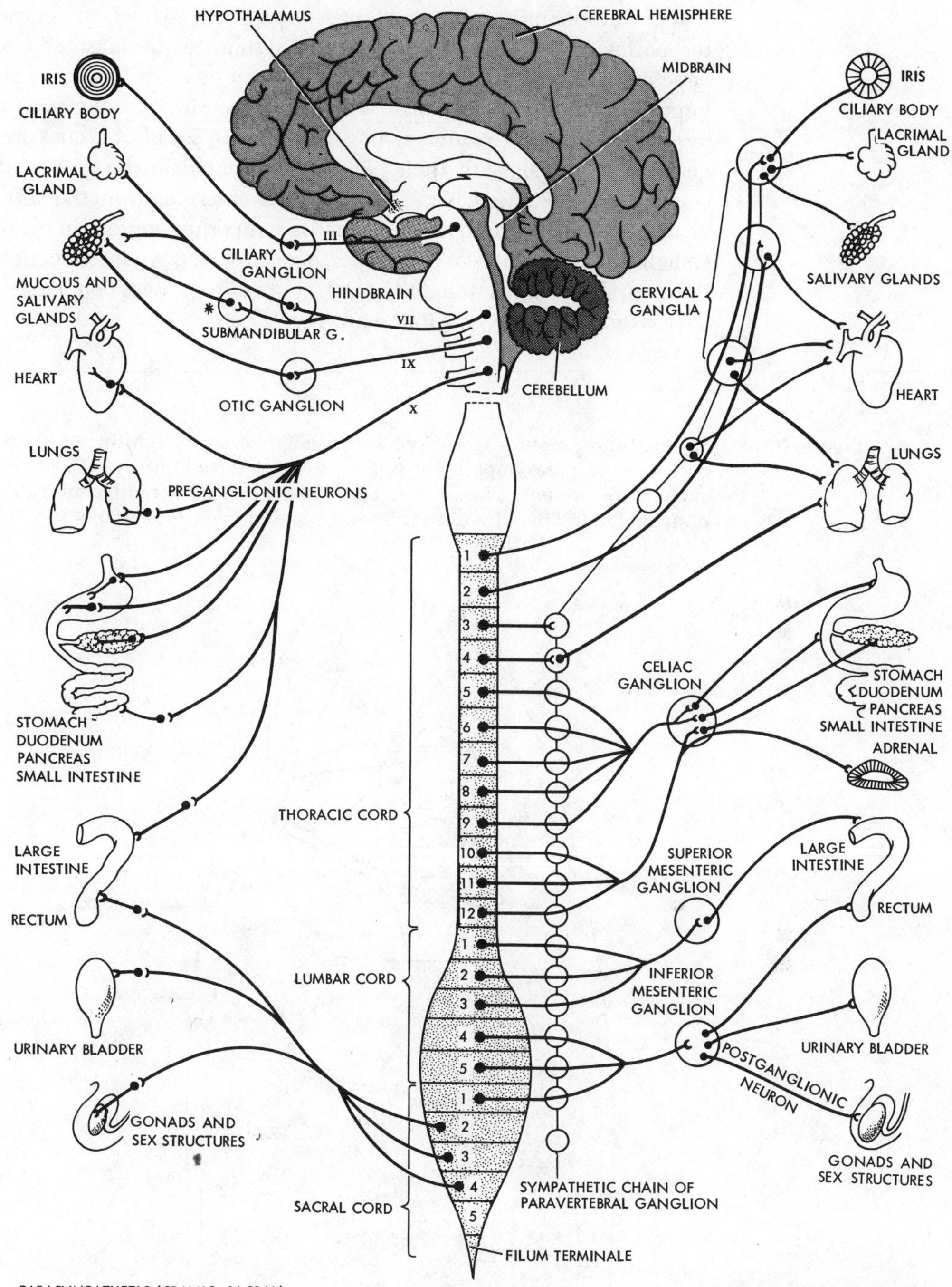

Figure 10-12 Schematic representation of the autonomic nervous system in mammals. The parasympathetic system is indicated on the left side and the sympathetic system on the right side. Roman numerals refer to the cranial nerves. The sympathetic innervation of the blood vessels, sweat glands, and muscles erecting hair is not shown. Sympathetic and parasympathetic ganglia (G) are named.* Pterygopalatine G. In man, the preganglionic neurons of the sympathetic system emerge from levels thoracic 1 through lumbar 2.

In the autonomic nervous system, a sequence of two sets of neurons make up the pathway from the central nervous system to the muscle and gland-cell effectors (Figures 10-12 and 10-13). This is in contrast to the course of motor impulses, which are carried by a neuron projecting directly from the central nervous system to a voluntary muscle. The first set of neurons (preganglionic neurons or fibers), with their cell bodies located in the gray matter of the central nervous system, has axons that, after emerging from the nervous system via nerve trunks, terminate by a synapse with other neurons in nearby ganglia. Ganglia are collections of cell bodies, dendrites, and synapses occurring outside the central nervous system. A second set of neurons, with cell bodies in ganglia, have axons that, in turn, innervate the effectors. This second set consists of

Figure 10-13 Schematic representation of the autonomic nervous system in the shark. The sympathetic and parasympathetic systems are not sorted out by regions. Autonomic ganglia are present. The preganglionic fibers are indicated by solid lines and the postganglionic fibers by dotted lines. (Adapted from J. Z. Young.)

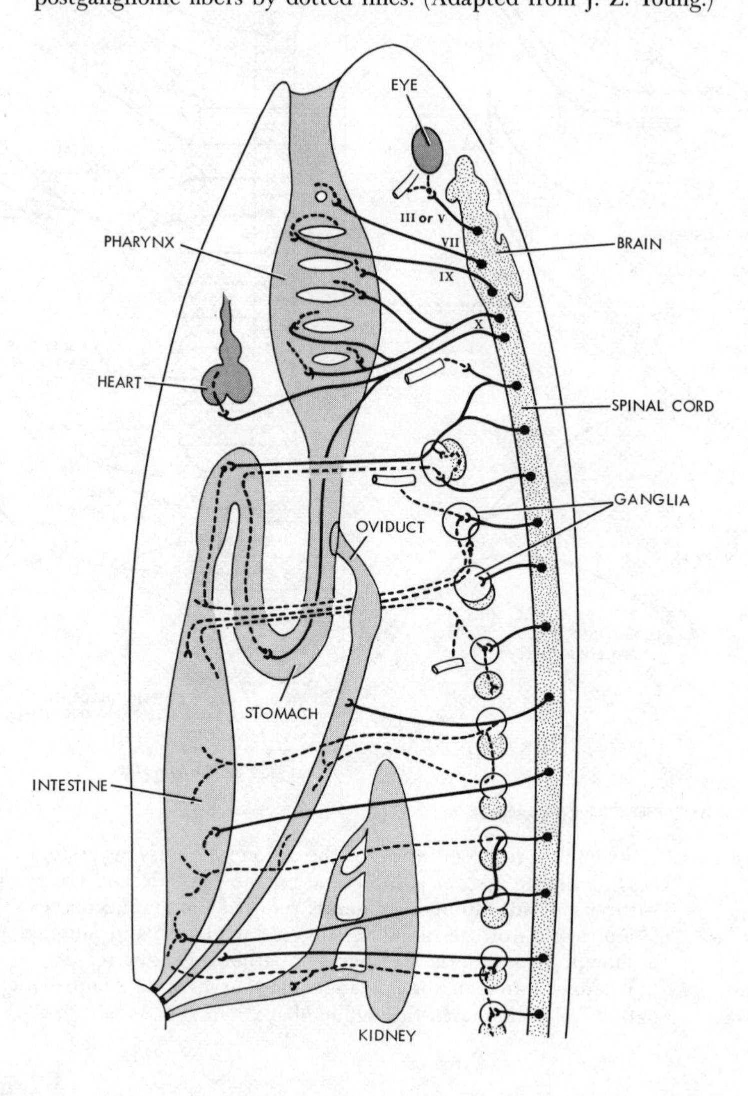

postganglionic neurons (fibers). The chemical acetylcholine is the neurotransmitter released at all synapses between preganglionic and postganglionic neurons of the autonomic nervous system. The neurotransmitter released at the synaptic junction of the sympathetic postganglionic neuron and the effector (smooth muscle or gland) is noradrenalin (norepinephrine); hence, the sympathetic system is called the adrenergic system, and the postganglionic sympathetic neurons are called adrenergic neurons (fibers). Noradrenalin is inactivated by catechol-o-methyl transferase and monamine oxidase. The neurotransmitter released at the synaptic junction of the parasympathetic postganglionic neuron and its effector is *acetylcholine;* hence, the parasympathetic system is called the cholinergic system, and its postganglionic neurons are called cholinergic neurons (fibers). *Cholinesterase* is the enzyme that inactivates the acetylcholine. Many drugs may chemically enhance or suppress the actions of the neurosecretions.

The sympathetic system intensifies those activities that are most dramatically expressed during emergencies and stress, called the fight, fright, and flight situations. These reactions are accompanied by activities associated with the expenditure of energy: acceleration of the rate and force of the heart, increase in blood pressure, elevation of the concentration of sugar in the blood, and increase in blood flow directed to voluntary muscles at the expense of the skin and visceral organs. In contrast, the parasympathetic system stimulates those activities that are associated with the conservation and restoration of the energy stores of the body: decrease in the rate and force of the heart beat, decrease in blood pressure, increase in digestion through stimulation of the gastrointestinal tract (greater secretion of enzymes and motility of wall of tract), stimulation of actions resulting in elimination of fluids (urination), and solids (defecation). The actions of the two systems are integrated; they are not basically antagonistic. In the body economy, they act synergistically, although at times some activities are executed independently. The autonomic nervous system functions to regulate the internal activities of the organism at a level commensurate with the intensity of the stress situation and with the emotional state (anger, fear) of the individual. The medulla of the adrenal gland, which is innervated by preganglionic cholinergic neurons, secretes noradrenalin and adrenalin into the blood stream; this gland augments the activity of the sympathetic system—hence, the term *sympathoadrenal system.*

The viscera are not totally dependent on the autonomic nervous system; the heart beats, blood flows, digestion, and other vital visceral processes proceed without neural influences. An animal can live without a sympathetic nervous system and an adrenal medulla; however, it cannot adjust to changes in the environment and it must be kept in a sheltered place.

The autonomic nervous system is a most highly evolved and delicately synchronized system of higher warm-blooded vertebrates. In birds and mammals, it functions by providing the feedback circuitry by which the organism can (1) adjust to the stresses made by the environment upon it and (2) make the necessary corrections to maintain a constant body temperature. In general, terrestrial animals are able to meet the demands of such environmental fluctuations as temperature and humidity changes. On the other hand, aquatic cold-blooded animals whose body temperature varies, live in, or are required for their survival to live in, a more stable environment where sudden temperature change is minimal.

Autonomic Nervous System of Nonmammalian Vertebrates

The comparative anatomy and physiology of the autonomic nervous system of nonmammalian vertebrates are complex and incompletely known. In general, the morphological organization of the autonomic nervous system in the teleost fish, amphibia, reptiles, and birds is essentially similar to that in mammals. The sacral parasympathetic outflow is absent in the teleost fish and urodele amphibians; it is rudimentary in the anuran amphibians. In contrast, the sacral parasympathetic outflow is well developed in reptiles and birds. The cyclostomes do not have discrete parasympathetic ganglia or segmental sympathetic ganglia as a paravertebral sympathetic chain. This paravertebral chain is present in the other vertebrates as a segmental sequence of sympathetic ganglia strung in a linear chain by compact nerve bundles located along and near the vertebral column (Figure 10-22). The elasmobranchs have autonomic ganglia organized as a paravertebral chain with about one ganglion for each spinal segment. This sequence of paravertebral ganglia is linked by a loose plexus of nerve bundles. In contrast, the teleost fish, amphibia, reptiles, birds, and mammals have a bilateral pair of well-defined paravertebral sympathetic chains. In the lower vertebrates, the autonomic nerve fibers emerge from the spinal cord through both the dorsal and ventral spinal roots. In higher vertebrates, these fibers pass through the ventral roots; some of these fibers pass through the dorsal roots in amphibia and possibly even in reptiles and mammals. In all vertebrates, the cranial parasympathetic outflow passes through the third, seventh, ninth, and tenth cranial nerves. Some parasympathetic fibers are present in cranial nerve XI in mammals. Actually, these fibers in nerve XI can be considered to the vagal fibers, which take a slightly different course. Apparently, the cyclostomes and elasmobranchs do not have autonomic fibers that innervate the skin. The pigmented chromatophore in the integument of teleost fish is neuronally regulated; hence, the degree and state of pigmentation are mediated through the autonomic nervous system.

An indication of the structural and functional complexity of the autonomic nervous system in the vertebrate classes can be obtained from the following examples [18].

1. In fish and amphibia, many fibers of sympathetic origin are present in the vagus nerve, considered to be a parasympathetic nerve in mammals. The separation of sympathetic from parasympathetic fibers is more distinct in the higher vertebrates—a few sympathetic fibers are present even in the vagus nerve of mammals.

2. The vagal supply to the visceral muscles of such organs as the stomach and lungs is primitively inhibitory to muscular contraction in fish and amphibia. Apparently, at some phylogenetic stage higher than the amphibians, the excitatory (cholinergic) role classically attributed to the vagal fibers in tetrapods has been transferred from the sympathetic to the parasympathetic outflow, as found in living reptiles, birds, and mammals. This may be correlated with the change from the primitive sympathetic excitatory (cholinergic) supply of the viscera to mainly inhibitory (adrenergic) control.

3. The parasympathetic innervation of the heart has been primitively inhibitory and cholinergic; this expression has been consistently retained in all vertebrates and, hence, throughout vertebrate phylogeny. In a similar way, the "sympathetic" adrenergic fibers and vagal fibers in the fish and amphibians pass together as vagosympathetic trunks. At the reptile level, there is a separation of the vagal from the sympathetic trunks, as present in living reptiles, birds, and

mammals. However, this former condition is retained, at least in part, in the tetrapods, including mammals, with the presence of a few adrenergic fibers of sympathetic origin within the vagal trunks.

The Structural and Functional Organization of the Brain

The brain is conventionally subdivided into five major regions: telencephalon, diencephalon, mesencephalon, metencephalon, and myelencephalon (Figure 10-14, Table 10-1). From the embryonic prosencephalon develop the telencephalon and diencephalon, whereas the rhombencephalon gives rise to the metencephalon and the myelencephalon. The embryonic mesencephalon remains undivided. A major character, unique to vertebrates, is the presence of cavities in the central nervous system. Within the brain this "hollow" includes the ventricular system, a continuous series of cavities (ventricles) filled with cerebrospinal fluid. In a general way, the brain can be divided functionally into three major subdivisions: (1) the forebrain, prosencephalon, or cerebrum; (2)

Table 10-1 The Major Subdivisions of the Brain. The neocortex, crus cerebri, basilar pons, and pyramid are found in mammals. The other structures listed, each of which is present to some degree in all vertebrates, exhibit many structural alterations associated with modified functional correlates; these are expressions of the vast potential inherent in the central nervous system for both morphologic and physiologic changes during phylogeny (Figures 10-15 through 10-18).

Telencephalon	cerebral cortex (pallium)	neocortex (isocortex)
		paleocortex } allocortex
		archicortex } allocortex
		general cortex* (reptiles)
		wulst and hyperstriatum* (birds)
		olfactory bulb
		corpus callosum (mammals)
		corpus striatum (basal ganglia)
		lateral ventricles, foramina of Monro, and choroid plexus
	Diencephalon (between brain)	epithalamus and pineal body
		thalamus
		hypothalamus and hypophysis
		ventral thalamus
		third ventricle and choroid plexus
Brain Stem	*Mesencephalon* (midbrain)	superior colliculus (optic tectum)
		inferior colliculus (torus semicircularis, auditory tectum)
		tegmentum
		crus cerebri of mammals
		aqueduct of Sylvius (iter)
	Metencephalon (afterbrain)	cerebellum
		tegmentum } pons
		basilar pons of mammals } pons
		fourth ventricle
	Myelencephalon (spinal brain) (medulla)	tegmentum
		pyramid of mammals
		fourth ventricle and choroid plexus

*Equivalent or analogous to the mammalian neocortex.

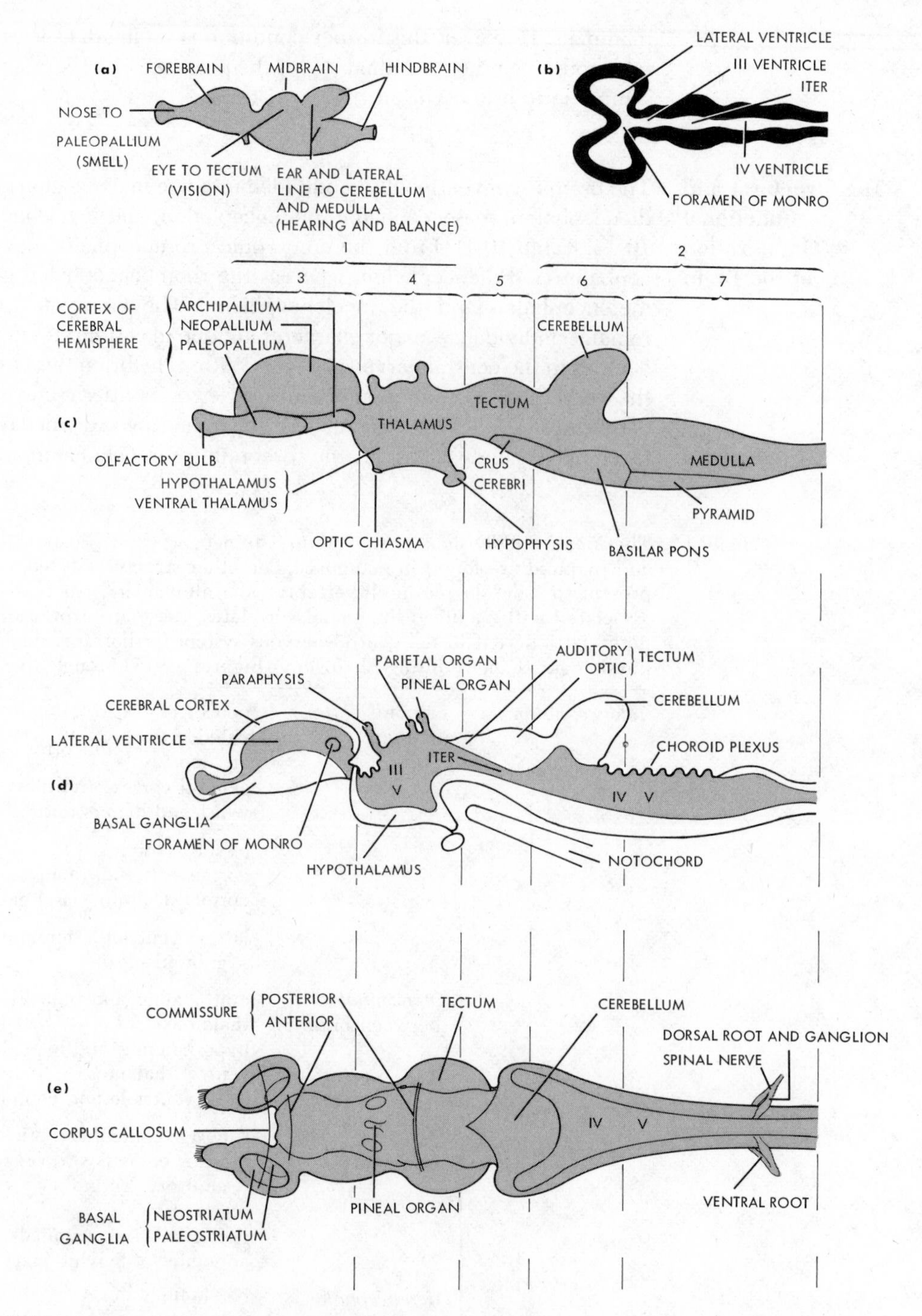

Figure 10-14 Diagrams illustrating the major brain subdivisions and structures. (a) The three major subdivisions of the brain and the special senses projecting to these structures in the lower vertebrates. (b) The ventricles of the vertebrate brain. (c) Lateral view of a generalized vertebrate brain: (1) prosencephalon; (2) rhombencephalon; (3) telencephalon; (4) diencephalon; (5) mesencephalon; (6) metencephalon; (7) myelencephalon. (d) Median section of a generalized vertebrate brain. (e) Dorsal view of a generalized brain. V, ventricle. The iter is the aqueduct of Sylvius. (Adapted from Neal and Rand and from Romer.)

the midbrain; and (3) the pons and medulla (rhombencephalon). The forebrain primarily is involved with (1) patterned movements associated with strategies subserving the pursuit of goal-directed behavior and (2) control and regulation of posture into stereotyped pattern movements. The midbrain has roles in the visual and auditory systems. The pons and medulla primarily subserve those critical processes involved with the maintenance of survival of the organism (respiration, regulation of the cardiovascular system, stability of the internal environment) and with the stability of the animals in its spatial environment through posture (vestibular system). The cerebellum is involved, partly in conjunction with the vestibular system, in posture. The cerebral cortex (gray matter on the surface of the cerebrum) of the brain of mammals is the complex portion consisting of the archicortex, paleocortex, and neocortex.

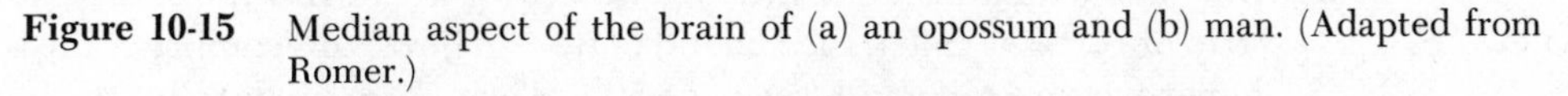

Figure 10-15 Median aspect of the brain of (a) an opossum and (b) man. (Adapted from Romer.)

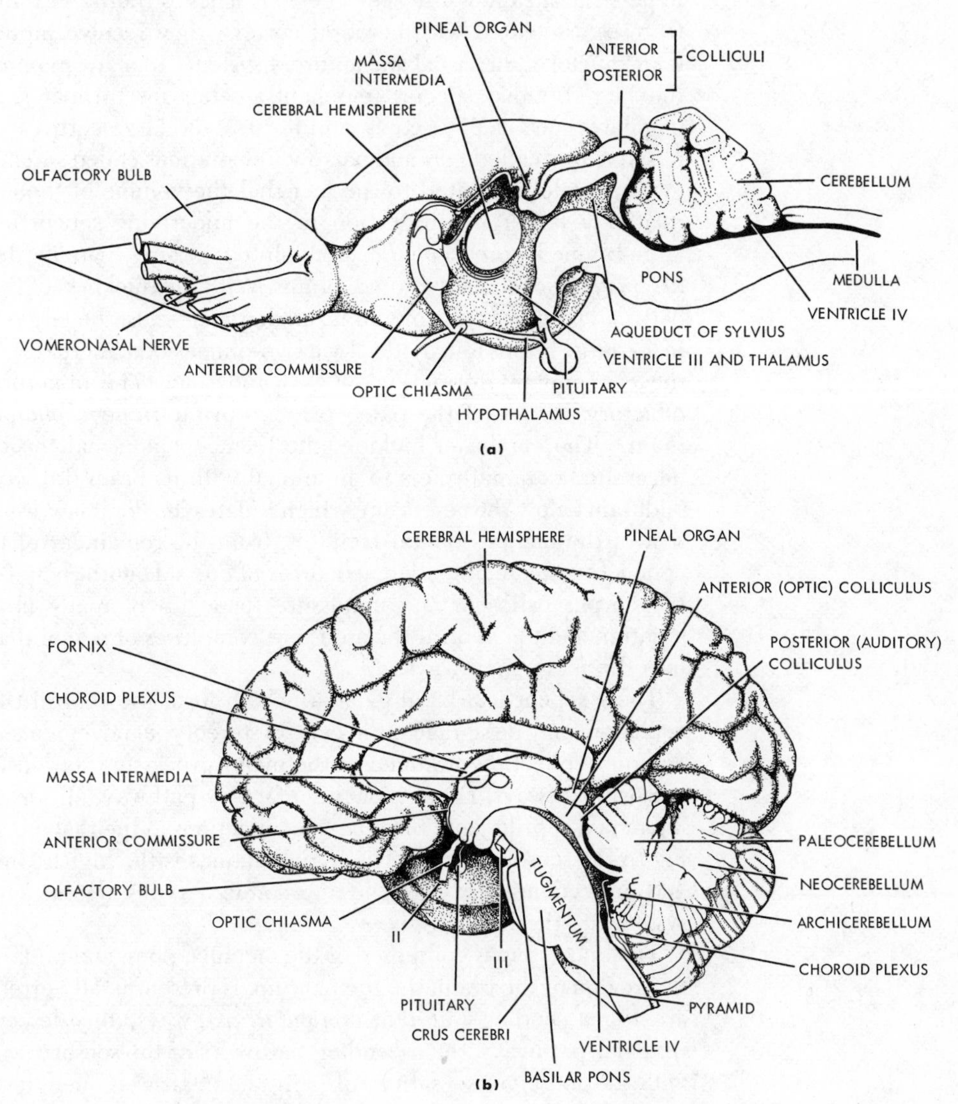

The cortex, or pallium, is a complex neural organization of gray matter located on the surface of the brain. Gray matter deep in the substance of the cerebrum is referred to as subcortical gray matter. In a cortex, the cell bodies of the cortical neurons are arranged in layers or laminae. Many of these neurons have dendrites that extend perpendicular to the plane of the laminae toward the brain surface. (Thus, there is a cerebral cortex on the cerebrum, a cerebellar cortex on the cerebellum, and collicular cortices on the superior (optic) and inferior (auditory) colliculi of the midbrain tectum.) In the submammalian vertebrates, the cerebral cortex consists of the archicortex and paleocortex; in reptiles, a primitive neocortex has been added, and an equivalent structure in birds is probably the hyperstriatum. The paleocortex and the archicortex often are called the limbic lobe or allocortex.

Nonmammalian vertebrates have what is considered to be the equivalent of the mammalian neocortex (neopallium, isocortex). In the reptiles, this small strip of cortex is called the general cortex. In birds, this comprises the so-called hyperstriatum and wulst (see page 509). These structures exhibit the functional features associated with cerebral cortex—they receive inputs from the thalamic nuclei of the visual and auditory systems. In many mammals, particularly the large-brained species (elephants, cetaceans, primates, carnivores, and ungulates such as pigs, cows, and horses), the large cortex is convoluted into raised ridges called *gyri* and narrow indentations called *sulci*. The presence of gyri and sulci is related to the fact that the volume of the cerebral cortex is relatively larger in proportion to the underlying subcortical structures in large-brained *gyrencephalic* (convoluted) brains than in the small-brained *lissencephalic* (nonconvoluted) brains. With the thickness of the cortex varying between relatively narrow limits among mammals, the larger cortical volume in the large brains is provided by a two-dimensional increase of the cortex with the concomitant formation of gyri and sulci. The olfactory bulb and the olfactory portion of the paleocortex form the *rhinencephalon*, or the "smell brain." The cerebrum includes the telencephalon and the diencephalon. A decerebrate animal refers to an animal with its brain transected through the midbrain below the cerebrum, which isolates the dominant association centers, seat of the highest mental faculties, from the remainder of the brain. Interconnecting the mammalian neocortex of one side with that of the other side is the corpus callosum, a commissure made up of many fibers. The corpus striatum and the ventral thalamus are complexes of nuclei that are integrated into the motor pathways.

These subcortical basal ganglia, which form the central core of the telencephalon, may be significant in certain stereotype movements. The thalamus, a complex collection of nuclei, is the major processing station in the ascending sensory pathways. Except for the olfactory pathway, all sensory pathways in mammals, reptiles, and birds have connections in the thalamus before projecting to the sensory cortex. The hypothalamus is the highest integrative center and projects influences to the autonomic nervous system and the endocrine system via the hypophysis.

The brain stem is composed of the medulla, pons, and midbrain. The core of this region of the brain, the tegmentum, is present in all vertebrates. Within it are the ascending (afferent) *reticular* pathways, the descending (efferent) reticular pathways, the ascending pathways of the sensory systems associated with the conscious sensations of pain, temperature, touch, taste, and audition

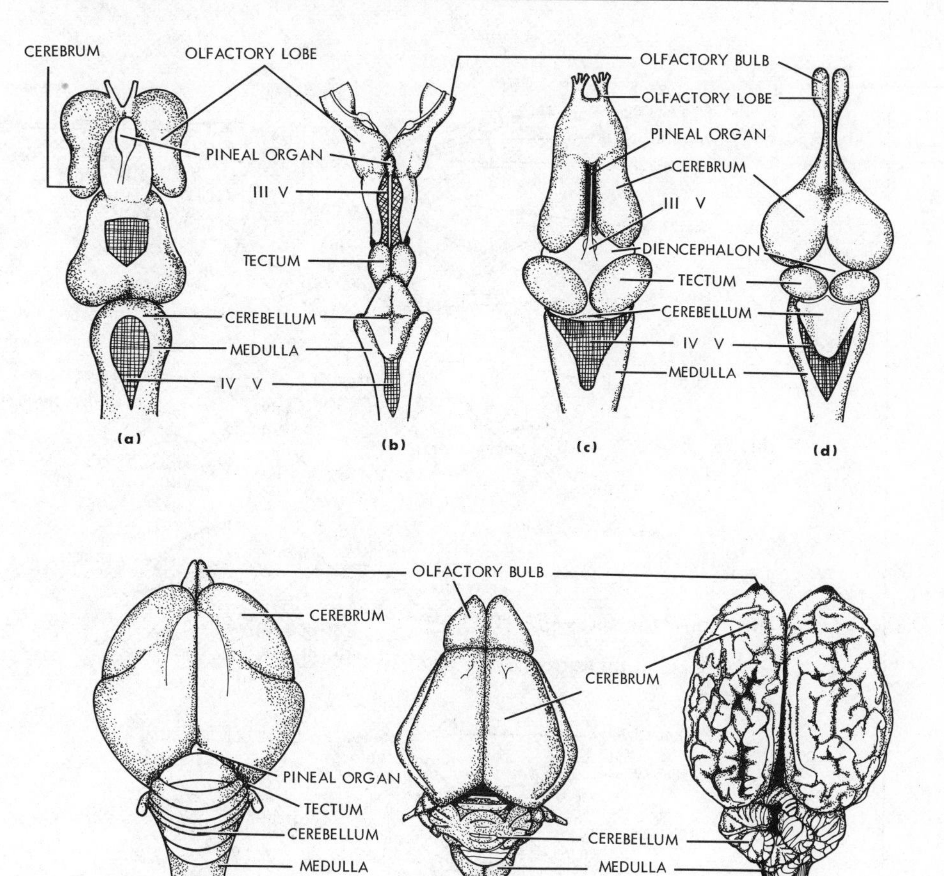

Figure 10-16 Dorsal views of the brains of a representative series of vertebrates: (a) lamprey, *Petromyzon;* (b) shark, *Scymnus;* (c) frog, *Rana;* (d) alligator, *Alligator;* (e) goose, *Anser;* (f) insectivore, *Gymnura;* and (g) horse, *Equus*. An insectivore is representative of the primitive mammals and the horse of the higher mammals. (Adapted from Duke-Elder.)

(lemniscal systems), and the pathways associated with the vestibulolateral line system, and others. The superior colliculus (optic tectum) and the inferior colliculus (torus semicircularis, auditory tectum) are nuclei associated with the optic and auditory pathways respectively. The crus cerebri, basilar pons, *pons*, and pyramids, all present in mammals, are bundles of descending motor fibers from the cerebral cortex terminating either in the pons (corticopontine fibers) or spinal cord (corticospinal tract); from the pons, fibers project to the neocerebellum.

The cerebellum is subdivided into the archicerebellum, paleocerebellum, and neocerebellum. The archicerebellum is integrated with the lateral line vestibular system. The paleocerebellum receives, processes, and utilizes in coordinating movements the unconscious polysensory input from tactile, proprioceptive, auditory, and visual sources. The neocerebellum, found exclusively in mammals, is integrated with feedback connections with the motor neocortex of the cerebrum. The cerebellum functions to prevent oscillations

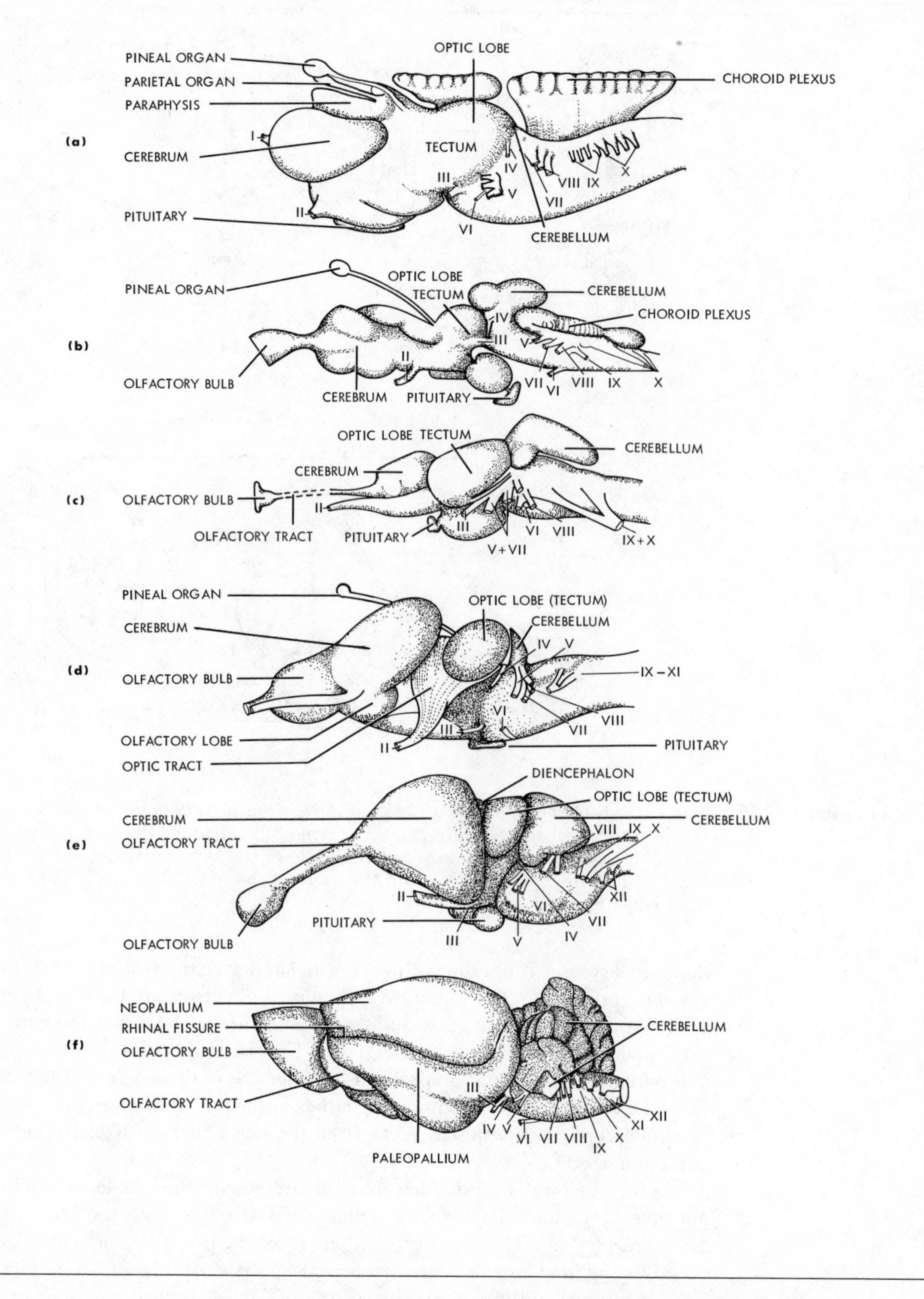

Figure 10-17 Lateral views of the brains of a representative series of vertebrates: (a) the lamprey, *Petromyzon;* (b) shark, *Scymnus;* (c) codfish, *Gadus;* (d) frog, *Rana;* (e) alligator, *Alligator;* and (f) insectivore, *Gymnura.* (Adapted from Romer.)

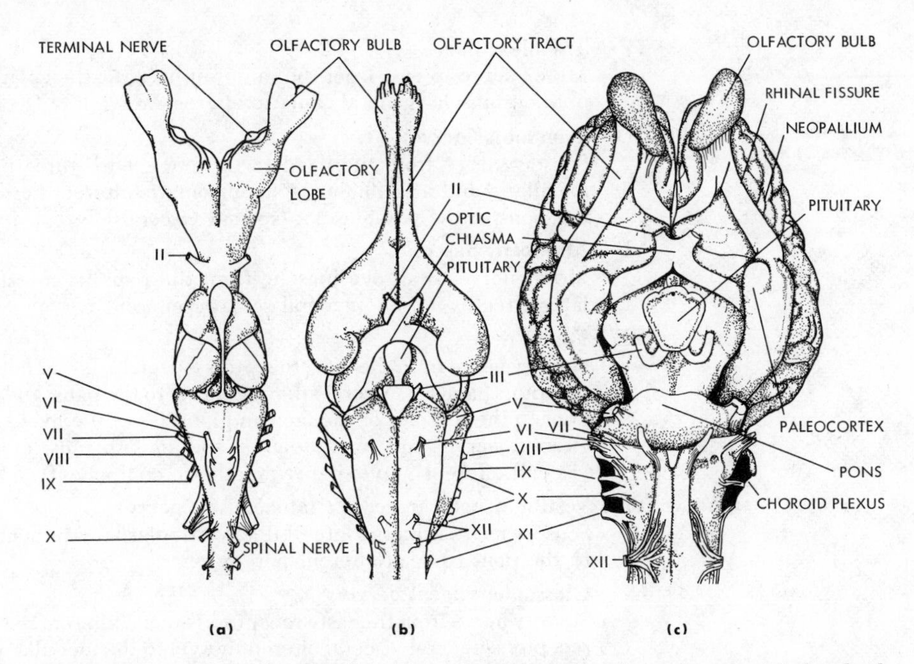

Figure 10-18 Ventral views of the brains of a representative series of vertebrates: (a) the shark, *Scymnus;* (b) alligator, *Alligator;* and (c) horse, *Equus.* (Adapted from Romer.)

(tremor) during motion and, thereby, maintains stability in movements; it smooths out the actions of muscle groups by delicately regulating and grading muscle tensions and muscle tone. The neocortex and neocerebellum evolved in parallel.

Cranial Nerves

The cranial nerves are the peripheral nerves of the brain. Their names, numbers, and functional components are essentially similar in all the vertebrates. They were originally named in man, which explains why the names of their branches and components are not always obvious in some animals. The following list includes the number (roman numeral) and name of the cranial nerve, the main distribution segment of the brain from which it emerges, and its main functional components.

0 **Terminal nerve**
Sensory nerve to anterior nasal epithelium (general somatic afferent nerve)

I **Olfactory nerve**
Sensory nerve of smell, from olfactory mucosa in nose to olfactory bulb of the telencephalon (special afferent nerve)

Vomeronasal nerve
Sensory nerve associated with chemical reception, from the vomeronasal organ to accessory olfactory bulb of the telencephalon (special afferent nerve)

II **Optic nerve**
Sensory nerve of sight, from the eye through the optic chiasma to the midbrain (optic tectum) and to the thalamus of the diencephalon (special afferent nerve)

III **Oculomotor nerve**
Motor nerve to eye, from the midbrain to four or more extraocular muscles (general somatic efferent nerve) and pupillary and accommodation (focusing) muscles (general visceral efferent nerve)

IV Trochlear nerve
Motor nerve to eye, from the midbrain to an extraocular muscle, the superior oblique muscle (general somatic efferent nerve)

V Trigeminal nerve
Sensory nerve from the head to the pons, with three branches: ophthalmic, maxillary, and mandibular (general somatic afferent nerve); motor nerve from the pons to the jaw muscles (special visceral efferent nerve)

VI Abducens nerve
Motor nerve to an eye muscle, from the pons to an extraocular muscle, the lateral rectus muscle (general somatic efferent nerve)

VII Facial nerve
Sensory nerve from the taste receptors (special afferent nerve) and other visceral receptors (general visceral afferent nerve) to the pons; and motor nerve from the pons to the muscles of the face and muscles of the hyoid arch (special visceral efferent nerve) and to the glands and smooth muscles of the head (general visceral efferent nerve, parasympathetic component)

VIII Vestibulocochlear nerve (statoacoustic nerve)
Sensory nerve from the lateral line, vestibular labyrinth, and cochlear labyrinth to the pons (special afferent nerve)

IX Glossopharyngeal nerve
Sensory nerve from the taste receptors (special afferent nerve) and other visceral receptors (general visceral afferent nerve) to the medulla; motor nerve from the medulla to the muscles of the third gill arch (special visceral efferent nerve) and to the glands and smooth muscles of the head (general visceral efferent nerve, parasympathetic component)

X Vagus nerve
Sensory nerve from the taste receptors (special afferent nerve) and other visceral receptors (general visceral afferent nerve) of the head and visceral organs of the thorax and abdomen to the medulla; motor nerve from the medulla to the muscles of the fourth gill arch (special visceral efferent nerve—e.g., laryngeal muscles) and to the glands and smooth muscles of the viscera of the thorax and abdomen (general visceral efferent nerve, parasympathetic component)

XI Spinal accessory nerve (only in amniotes)
Motor nerve (accessory to the vagus nerve) from the medulla to the sternomastoid muscle and trapezius muscle (special visceral efferent nerve)

XII Hypoglossal nerve (only in amniotes)
Motor nerve to the muscles of the tongue, from the medulla to the tongue (general somatic efferent nerve)

All twelve cranial nerves are found in living reptiles, birds, and mammals (Figure 10-20) and the first ten in living fish and amphibians (Figure 10-19). Although there are functional differences, the cranial nerves XI and XII have their equivalent in fish and amphibians as occipital and spinal cord nerves that innervate "neck" muscles.

Certain functional aspects of the cranial nerves can be summarized by grouping the nerves and their functional components into several categories. Several modalities are conveyed by special nerves; these include the special afferent (sensory) cranial nerves I, II, and VIII. A modality is a term customarily used to mean a kind of sensation, such as pain, sound, and odor. Taste is a special afferent component of cranial nerves VII, IX, and X. The modalities of smell and taste are often classified as visceral senses (associated with the visceral activity of sensing food). The vomeronasal system (vomeronasal organ, nerve, and its central connections) may have a critical role in mediating sex

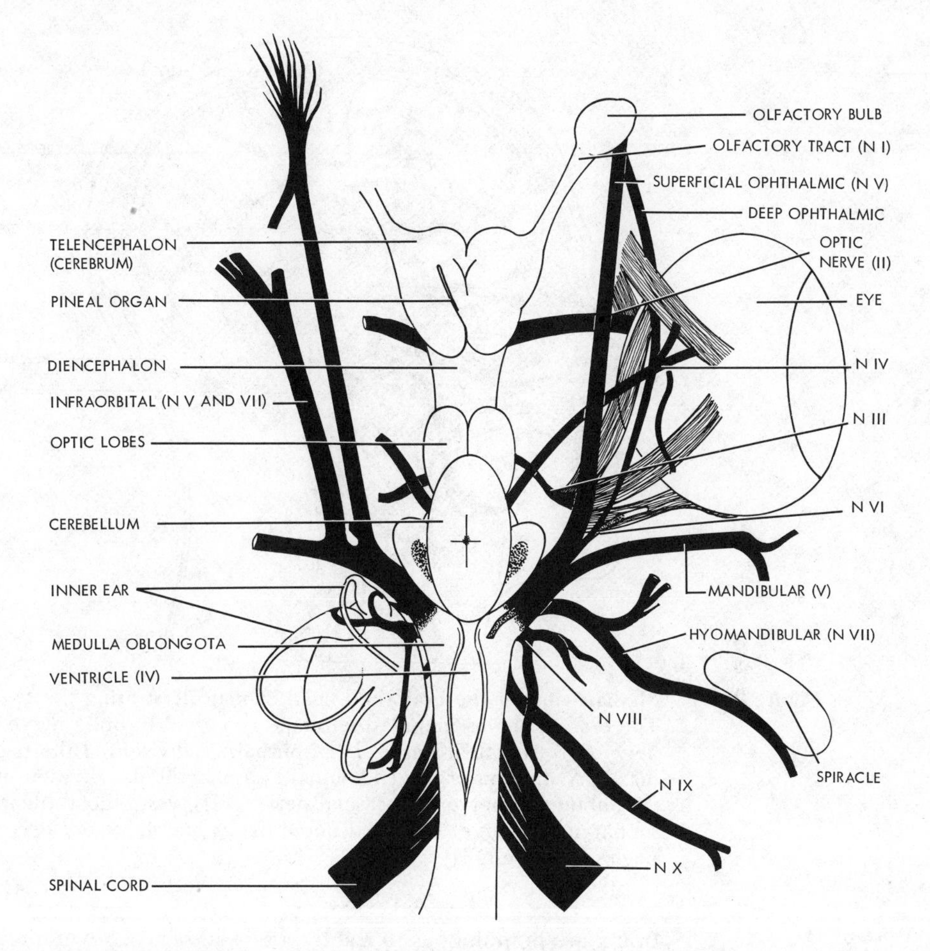

Figure 10-19 Dorsal view of the brain and cranial nerves of the dogfish shark.

behavior in mammals. The optic and vestibulocochlear nerves are called somatic nerves because the modalities they convey are derived from somatic afferent sensory receptors. The lateral line of fish and larval amphibians generally is stated to be innervated by the seventh, ninth, and tenth cranial nerves. In fact, the lateral line nerves have their own ganglia and should be designated as separate cranial nerves. The anterior lateral line nerves innervate the lateral line organs of the head; they are associated with the facial cranial nerve. The posterior lateral line nerves innervate the lateral line organs of the trunk; they are associated with the ninth and tenth cranial nerves.

A number of cranial nerves innervate voluntary muscles derived from the embryonic somites; these include the general somatic efferent (motor) cranial nerves III, IV, VI, and XII. They innervate the extraocular muscles (move the eyes and elevate the eyelids) and the tongue muscles (move the tongue). These cranial nerves correspond to the ventral roots of the spinal cord. The voluntary muscles innervated by these nerves are derived from cranial somites of the embryo. The twelfth cranial nerve is found, as such, only in amniotes. In non-amniotes, some occipital nerves, which emerge from the spinal cord near the

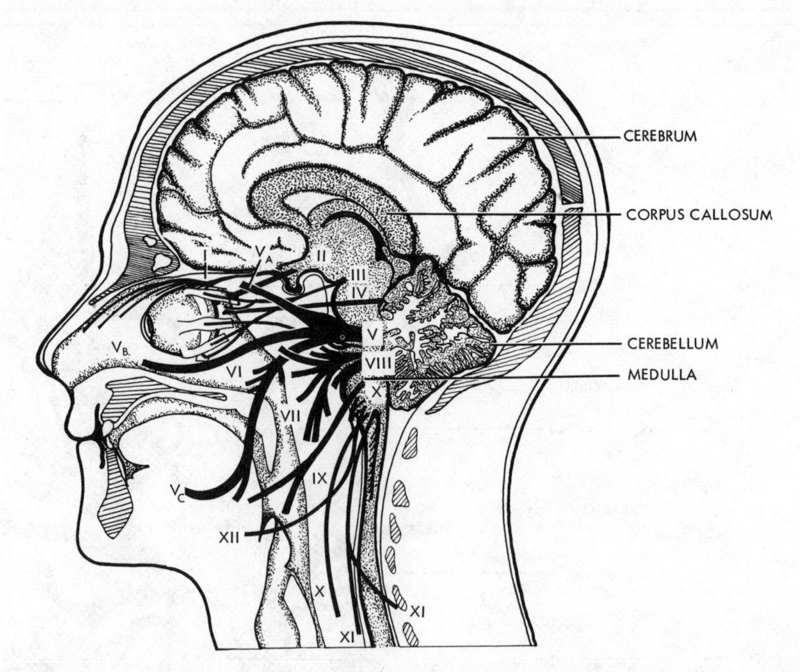

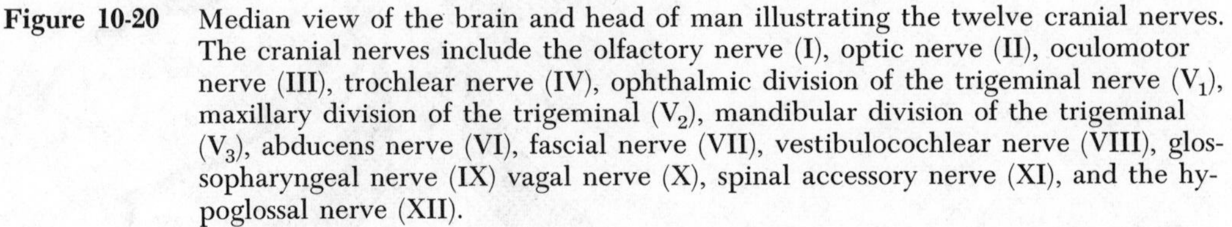
Figure 10-20 Median view of the brain and head of man illustrating the twelve cranial nerves. The cranial nerves include the olfactory nerve (I), optic nerve (II), oculomotor nerve (III), trochlear nerve (IV), ophthalmic division of the trigeminal nerve (V_1), maxillary division of the trigeminal (V_2), mandibular division of the trigeminal (V_3), abducens nerve (VI), fascial nerve (VII), vestibulocochlear nerve (VIII), glossopharyngeal nerve (IX) vagal nerve (X), spinal accessory nerve (XI), and the hypoglossal nerve (XII).

brain, are homologous to the hypoglossal nerve. Several cranial nerves innervate the branchiomeric (gill) arches of fish (Figure 10-21) and their derivatives among the tetrapods; these include nerves V (first arch, the jaws), VII (second arch, hyoid arch), IX (third arch), and X and XI (other arches). Each gill arch or its derivative is innervated by several functional components: (1) the ectodermal part of each arch by general somatic afferent fibers of cranial nerves V, VII, IX, and X: (2) the endodermal part of each arch by visceral afferent fibers of cranial nerves VII, IX, X, and XI: (3) the visceral organs by general visceral efferent (autonomic fibers of cranial nerves III, VII, IX, X, and XI; and (4) the striated muscles of the gill arches derived from the visceral mesoderm of special visceral efferent cranial nerves V, VII, IX, X, and XI. These nerves correspond to the dorsal spinal roots of the Agnathan fish, which, as previously noted, have general visceral efferent as well as general somatic afferent and general visceral afferent in their dorsal roots. The terminal nerve (0) is a small nerve composed of sensory nerves to the nasal epithelium. This nerve is numbered (0) because this nerve was unknown when the numbering system was adopted. It probably innervated the premandibular arch region in early agnathan fish; these arches and pouches presumably have been lost during phylogeny. The spinal accessory, XI, is a discrete nerve present as such only in amniotes; it is composed of fibers of the vagus nerve and some spinal root fibers of the spinal cord. The vagal components of XI are present in the vagus nerve

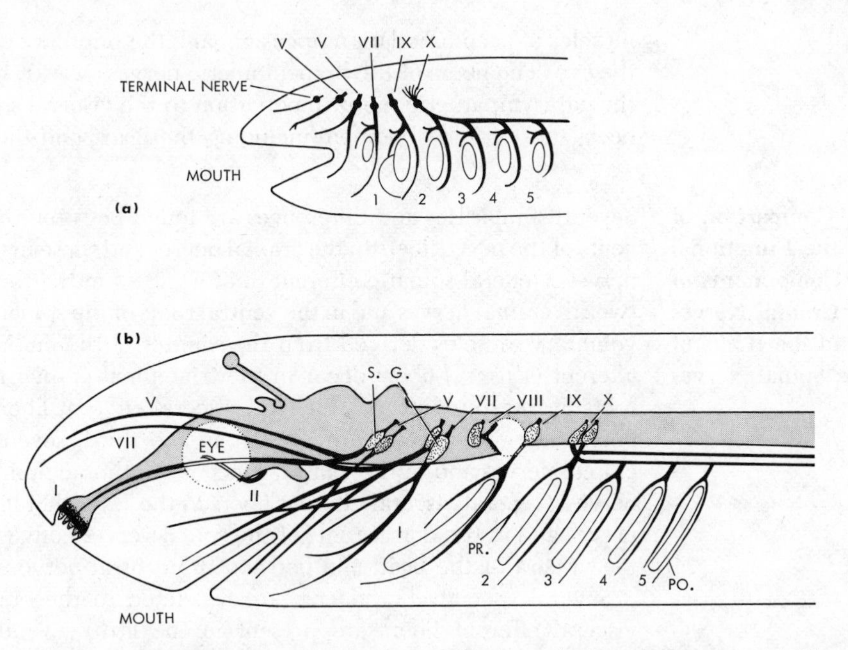

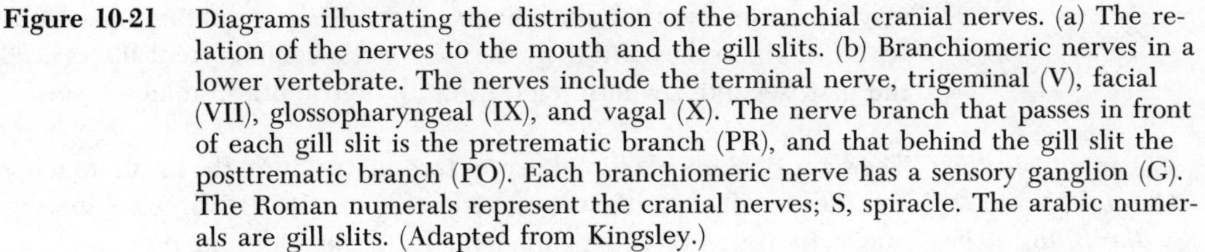

Figure 10-21 Diagrams illustrating the distribution of the branchial cranial nerves. (a) The relation of the nerves to the mouth and the gill slits. (b) Branchiomeric nerves in a lower vertebrate. The nerves include the terminal nerve, trigeminal (V), facial (VII), glossopharyngeal (IX), and vagal (X). The nerve branch that passes in front of each gill slit is the pretrematic branch (PR), and that behind the gill slit the posttrematic branch (PO). Each branchiomeric nerve has a sensory ganglion (G). The Roman numerals represent the cranial nerves; S, spiracle. The arabic numerals are gill slits. (Adapted from Kingsley.)

of amniotes, whereas the spinal component, which is found exclusively in amniotes, innervates two muscles in the neck: namely, the sternocleidomastoid and trapezius muscles.

The many functional components of these complex nerves can be summarized in the following way. Essentially, all the general somatic senses (pain, temperature, touch, position, sense, and unconscious proprioception) are transmitted via cranial nerve V. The general visceral taste fibers are located in cranial nerves VII, IX, and X. Many general visceral afferent fibers are present in cranial nerves VII, IX, X, and XI, they convey inputs that are integrated into vital reflex arcs that influence the heart and respiration through their connections with the cardiac and respiratory centers in the tegmentum of the pons and medulla. The special visceral efferent fibers to the voluntary muscles of the gill arches of fish and amphibians and their phylogenetic successors in tetrapods (visceral because they are associated with the visceral activities of the digestive and the respiratory systems) innervate the muscles of mastication (cranial nerve V), the muscles of facial expression (VII), the pharyngeal muscles of swallowing, the laryngeal muscles of the vocal cords (IX and X), and the trapezius and sternomastoid muscles of the neck (XI).

Several of these branchiomeric nerves have general visceral efferent (parasympathetic) components. The cranial nerves with parasympathetic components include III, VII, IX, and X. The fibers of the third nerve influence the muscles of accommodation (focusing) and the pupillary constrictor muscles of the eye. The fibers of the branchiomeric nerves (V, VII, IX, X, and XI) supply the parasympathetic motor innervation to the visceral structures of the head, neck, thorax, and abdomen, including the heart and the digestive system.

Comparison of the Functional Components of the Cranial Nerves and the Roots of the Spinal Nerves

Several similarities and differences are found between the functional components of the nerve fibers in the cranial nerves and those in the roots of the spinal nerves. General somatic efferent (motor) fibers in the third, fourth, sixth, and twelfth cranial nerves and in the ventral roots of the spinal nerves innervate the voluntary muscles derived from the somites of the embryo. General somatic afferent (sensory) fibers occur in the trigeminal cranial nerve and the dorsal roots of the spinal nerves. General visceral efferent fibers (fibers of the autonomic nervous system) are present in the third, seventh, ninth, and tenth cranial nerves and in the ventral roots of the thoracolumbar and sacral spinal nerves. General visceral afferent fibers in the seventh, ninth, and tenth cranial nerves and in the dorsal roots of the spinal nerves convey afferent input from the viscera of the head and body to the central nervous system.

Several special components are restricted to the cranial nerves. Special visceral efferent fibers are present in the fifth, seventh, ninth, tenth, and eleventh cranial nerves; these fibers innervate the voluntary muscles associated with the gill arches and their derivatives. Special afferent fibers are found in the first, second, seventh, eighth, ninth, and tenth cranial nerves.

Input via the Cranial Nerves to the Brain Stem

The sensory fibers of cranial nerves terminate within the brain stem in neuronal complexes called cranial nerve nuclei (Figure 10-22). Three columns of afferent nuclei are present: (1) trigeminal nuclei of the midbrain, pons, and medulla; (2) solitary nucleus of the medulla; and (3) vestibular and cochlear nuclei of the lower pons and upper medulla. All the general somatic afferent fibers of cranial nerves V, VII, IX, and X terminate in the trigeminal nuclear column. All visceral afferent fibers of cranial nerves VII, IX, X, and XI terminate in the nucleus solitarius. These nuclei correspond to nuclear groups in the dorsal horn of the spinal cord. The special somatic afferent fibers of the vestibular and cochlear nerves, VIII, terminate in the vestibular and cochlear nuclei, respectively; these nuclei do not have any counterpart in the gray matter of the spinal cord.

Output via the Cranial Nerves from the Brain Stem

The motor fibers of the cranial nerves originate from motor nuclei in the brain stem (Figure 10-23). Three columns of the motor nuclei are present.

1. The motor nuclei of the third, fourth, sixth, and twelfth cranial nerves are general somatic efferent nuclei that correspond to nuclei in the ventral horn of the spinal cord.
2. The motor nuclei of the fifth and seventh cranial nerves and the nucleus ambiguus of the ninth, tenth, and eleventh cranial nerves are special visceral efferent nuclei. The only equivalent to these nuclei in the spinal cord are some neurons, which give rise to some fibers of the eleventh cranial nerve.
3. The parasympathetic motor nuclei of the third nerve, of the seventh and ninth nerves (salivatory nuclei), and of the tenth nerve (vagal nucleus) are the general visceral efferent nuclei, which correspond to the sympathetic and parasympathetic nuclei of the gray matter of the spinal cord.

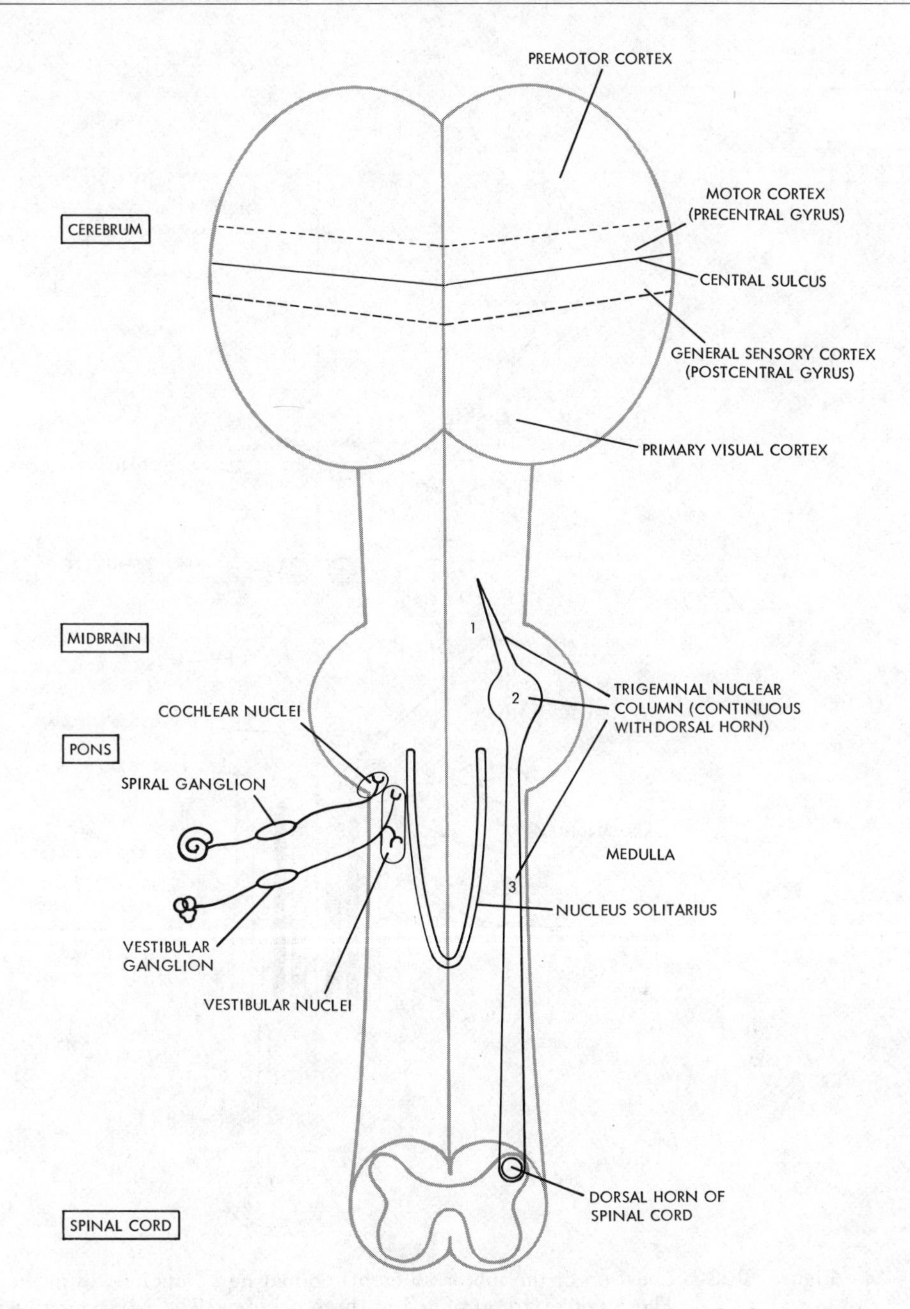

Figure 10-22 Location of the sensory (afferent) cranial nerve nuclei within the brain stem. These nuclei are arranged into three nuclear columns. The trigeminal column consists of (1) mesencephalic nucleus (proprioception), (2) principal sensory nucleus (tactile discrimination), and (3) spinal trigeminal nucleus (pain and temperature).

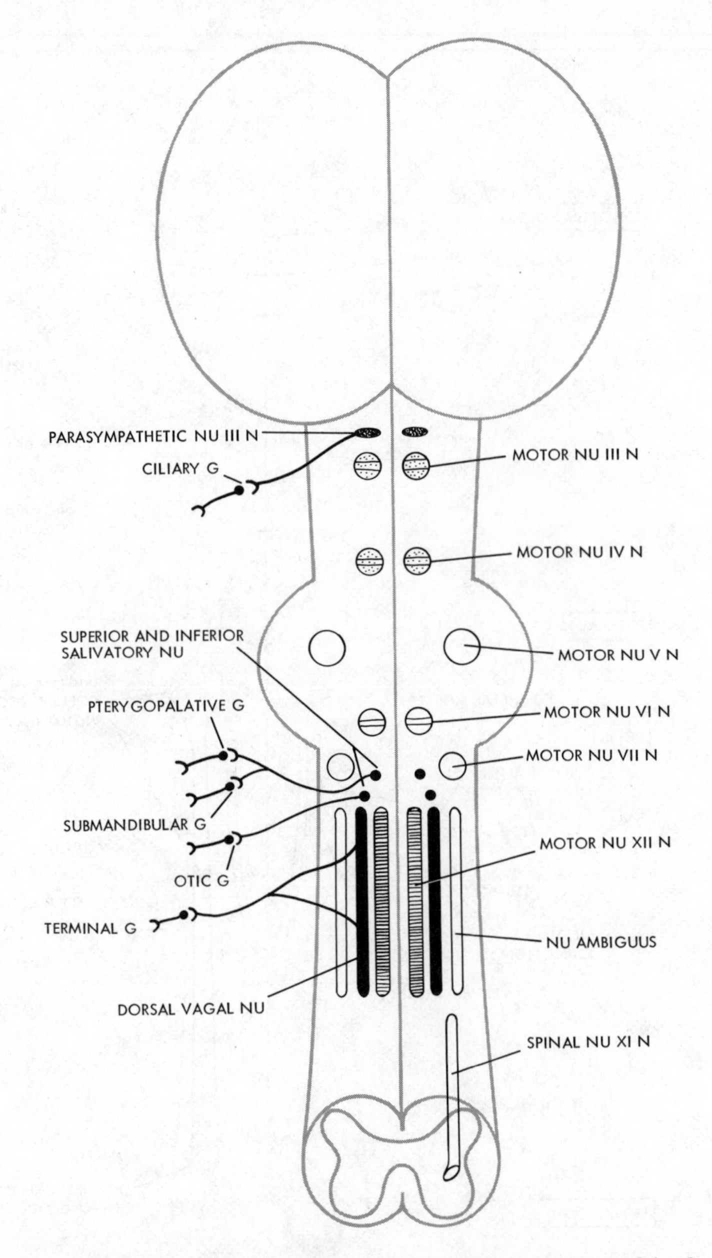

Figure 10-23 Location of the motor (efferent) cranial nerve nuclei within the brain stem. These nuclei are arranged in three columns. The parasympathetic ganglia are indicated on the left. G, ganglion; N, nerve; NU, nucleus.

Origin of the Nervous System The most primitive known vertebrates, the Agnatha fish of the Ordovician, already had a highly organized nervous system with a topographic anatomy basically similar to that of lower living fish. Hence, the steps leading to the origin of the vertebrate nervous system can only be reconstructed in theory. The first cells with some characteristics of neurons were probably types of receptor-effector (neuroeffector) cells; such cells are present in living sponges [Figure 10-24(a)]. These combine the capacity to be stimulated (receptor) and to react by contracting as a muscle cell or by secreting as a glandular cell. In

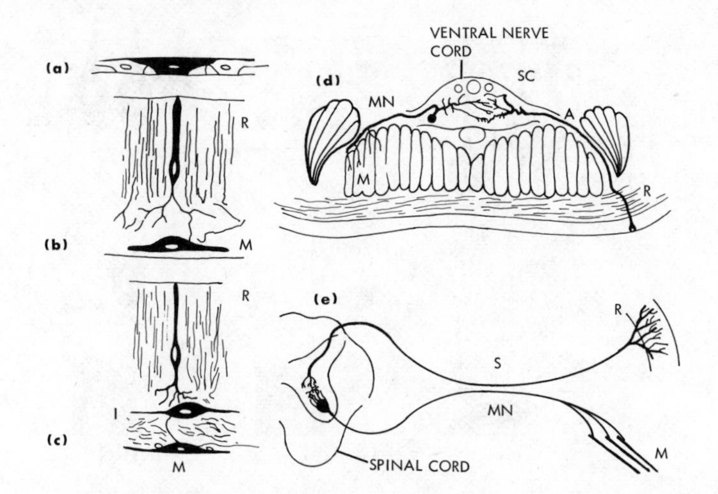

Figure 10-24 Evolution of the receptor-neuro-effector apparatus in the animal kingdom according to Parker. (a) The primitive type is the combined receptor-effector cell of sponges. (b) The sequence of receptor (R) and effector muscle (M) is present in the sea anemone. (c) The sequence of receptor (R), interneuron (I), and muscle effector is also present in the sea anemone. (d) The development of a central nervous system, as present in the earthworm, includes the sequence of receptor (R) in the skin with an axon (A) terminating in the ventral nerve cord neuronal networks (SC), motor neuron (MN) to muscle (M). (e) The vertebrate reflex arc is schematically represented with the sensory neuron (S), spinal cord, motor neuron (MN), and muscle (M).

a subsequent stage, separate receptors and effectors evolved; in this condition the receptor cell (combined receptor and neuron) receives the stimuli and, in turn, influences the effector cell (muscle or gland) to contract and secrete [Figure 10-24(b)]. Examples of this stage in evolution are found in the modern sea anemones. In a more highly evolved stage, a nerve cell became intercalated between the receptor cell and the effector cell—hence, the sequence of receptor cell, neuron, and effector cell [Figure 10-24(c)]. In the subsequent complex phylogenetic changes, the intercalated neurons (interneuron) increased in numbers and connectivity. These stages evolved into the central nervous system [Figure 10-24(d) and (e)].

In the evolution of the nervous system of vertebrates, which occurred in concert with the evolution of other organ systems, several directions and expressions were developed. The nervous system is a bilaterally symmetrical and craniocaudally segmented (e.g., spinal segments) organ system with the cephalic end differentiated into the brain with its many adaptations. In the head and neck, the brain and its special sensors evolved as the monitor of the environment through sight, smell, taste, and vibrations. Concomitantly, there developed in the digestive system oral structures for the intake of food and in the pharyngeal region the organs of the respiratory system (gills) that functioned initially to obtain oxygen from the water.

Some indications of early evolutionary states can be found among living vertebrates. An early stage is represented by the bipolar cells in the olfactory mucosa; each cell acts both as a receptor cell in direct contact with the external surface to sense chemical molecules in the water or air and as a neuron to transmit information directly to the olfactory lobe of the central nervous

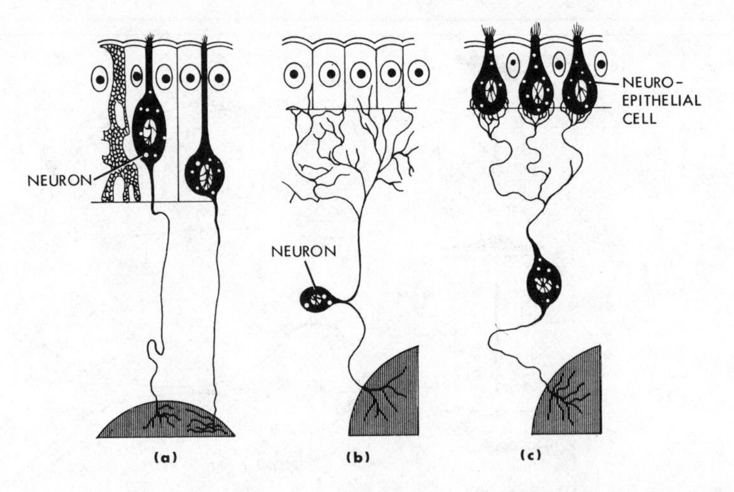

Figure 10-25 Postulated evolution of sensory neurons in vertebrates. (a) First stage, represented by an olfactory neuron, is a combination of receptor cell-sensory neuron in direct contact with the external environment. (b) Second stage, represented by some neurons with cell bodies in the dorsal root ganglion or its equivalent, is a combination of receptor cell-sensory neuron not associated with cells in direct contact with the external environment. (c) Third stage, represented by neuron (e.g., taste neuron) with cell body in dorsal root ganglion or its equivalent, is a sensory neuron in direct contact with a specialized neuroepithelial cell (e.g., taste bud). (Adapted from Neal and Rand.)

system [Figure 10-25(a)]. Another stage is represented by neurons that do not have connections with the external surface but, rather, arborize in the epidermis; this neuron transmits to the central nervous system [Figure 10-25(b)]. In the most advanced stage, a special receptor cell is intercalated between the environment and the neurons; such a cell is excited by environmental stimuli that, in turn, are passed on to a neuron projecting to the central nervous system [Figure 10-25(c)]. The special receptor cells include the rods and cones of the visual system, the taste cells of the taste bud, and the hair cells of the lateral line vestibuloacoustic system. During the phylogeny of the metazoa, along with the specialization that resulted in the differentiation of the three cell types—receptor cells, neurons (nerve cells), and effector cells (muscle and gland cells)—certain similarities in these cells were retained. Each still preserves its activity both as receptor and effector. These specialized cells are capable of being stimulated (a function of a receptor) and are also effectors that produce a secretion (neurons produce a neurosecretion) or contract as a muscle.

The Structural and Functional Organization of the Vertebrate Brain and Spinal Cord

The central nervous system is organized as complex networks of pathways generally oriented parallel to the longitudinal axis of the spinal cord and brain. Each pathway system is composed basically of sequences of neurons with long axons terminating at other neurons in one or more nuclei (centers, neuronal pools) of the main pathway (Figure 10-26). Within each nucleus are small interneurons, which, through interconnections with other neurons, process the input before relaying the output of the nucleus to other nuclei of the pathway. The ascending (cephalically directed) pathways associated with both

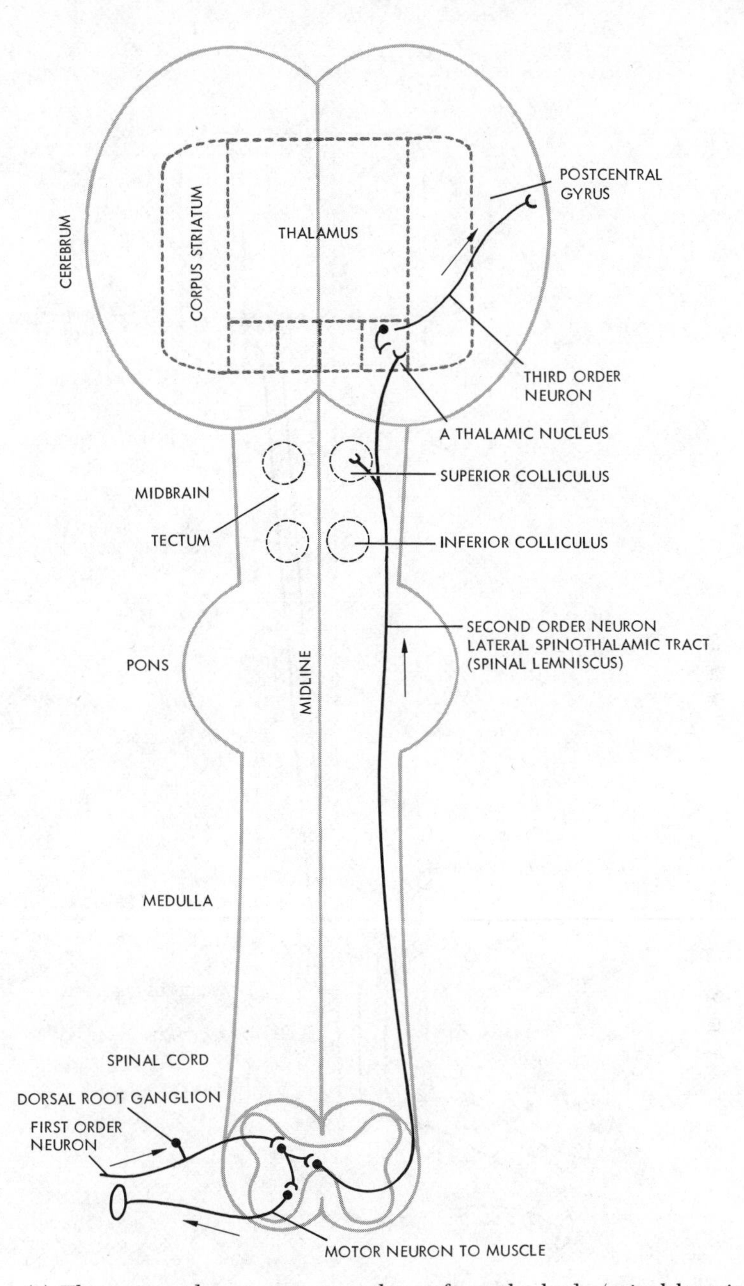

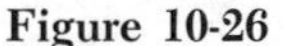

Figure 10-26 (a) The pain and temperature pathway from the body (spinal lemniscus).

unconscious and conscious senses (input) are the afferent (sensory) systems of neurons conveying neural information from the peripheral sensors to several centers in the brain. For example, the stimuli, felt as pain, excite nerve endings of the cranial nerves or peripheral nerves; the resulting excitation is channeled in the "pain" pathways that ascend to terminate in a number of centers at various levels of the spinal cord and brain. The descending (caudally directed) pathways are the efferent (motor) systems of neurons conveying neural influences from many centers in the brain to the nuclei of the cranial and spinal motor nerves (Figure 10–30).

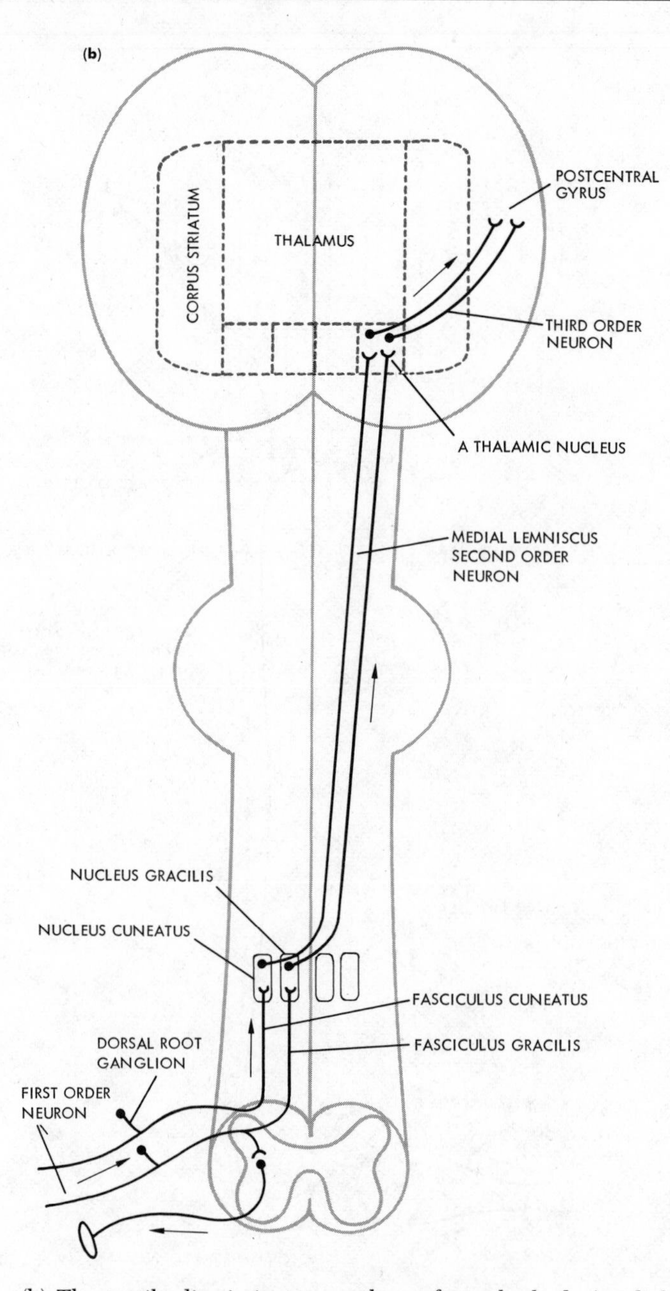

Figure 10-26 (b) The tactile discriminatory pathway from the body (medial lemniscus).

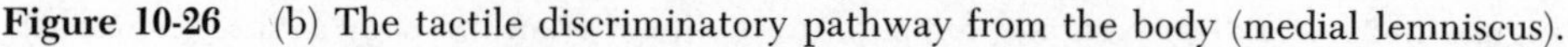

In general, the pathways of the nervous system can be divided into two major subgroups: (1) the ascending (afferent) reticular system with its pathways and the descending (efferent) reticular system and its pathways; and (2) the ascending (afferent) lemniscal system and its pathways and the descending (efferent) corticospinal tract (Figures 10-26 and 10-30). The reticular systems are the phylogenetically older systems present in all vertebrates, cyclostomes through mammals. The lemniscal systems are phylogenetically new systems that are well developed in mammals; these systems are only slightly repre-

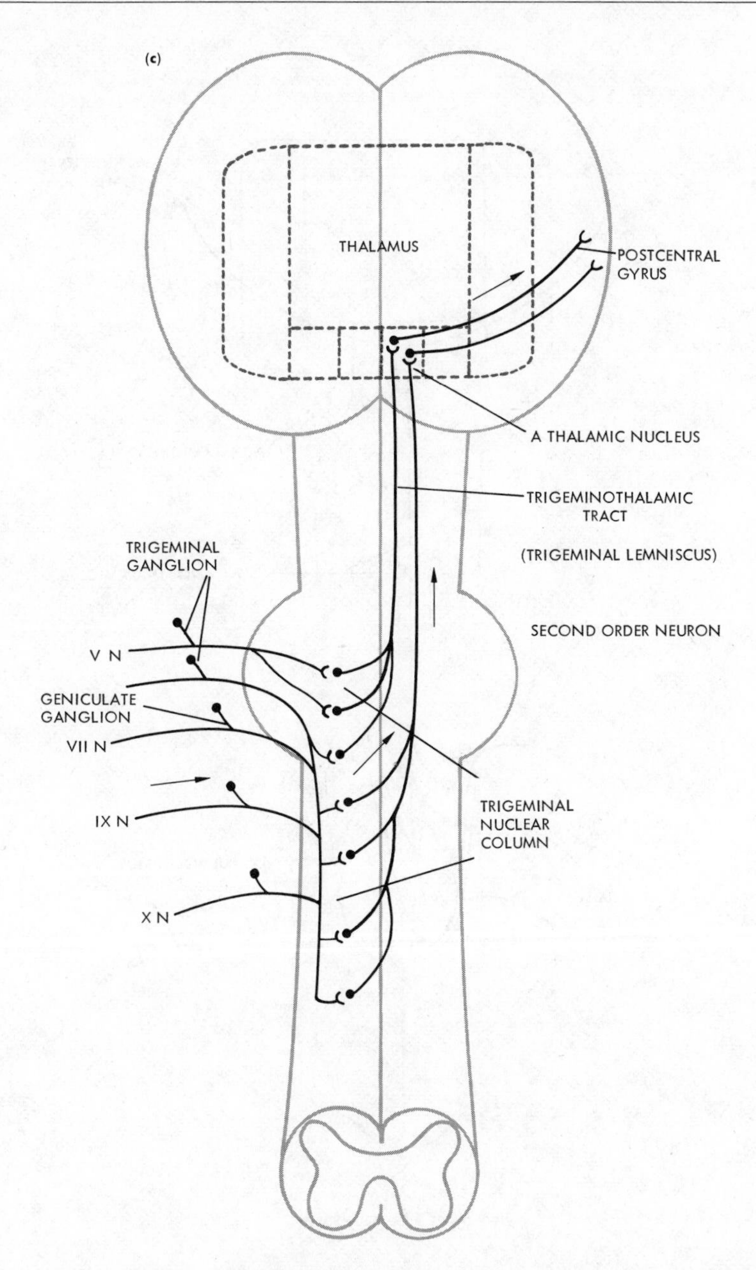

Figure 10-26 (c) The general sensory pathway (pain, temperature, and tactile discrimination) from the head (trigeminal lemniscus).

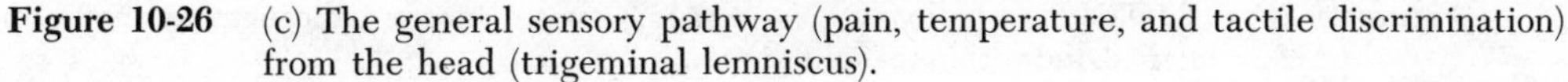

sented in amphibians and reptiles and are moderately present in birds. Both the reticular and the lemniscal systems show evidence of continued evolution even in the highest vertebrates.

The reticular pathways are composed of intricately organized neural networks with each neuron receiving synaptic input from many other neurons (convergence) and also making synaptic contacts with many other neurons (divergence). The reticular formation (and its pathway) refers to an anatomical entity; it comprises the neuron networks, which are structural substrates of the

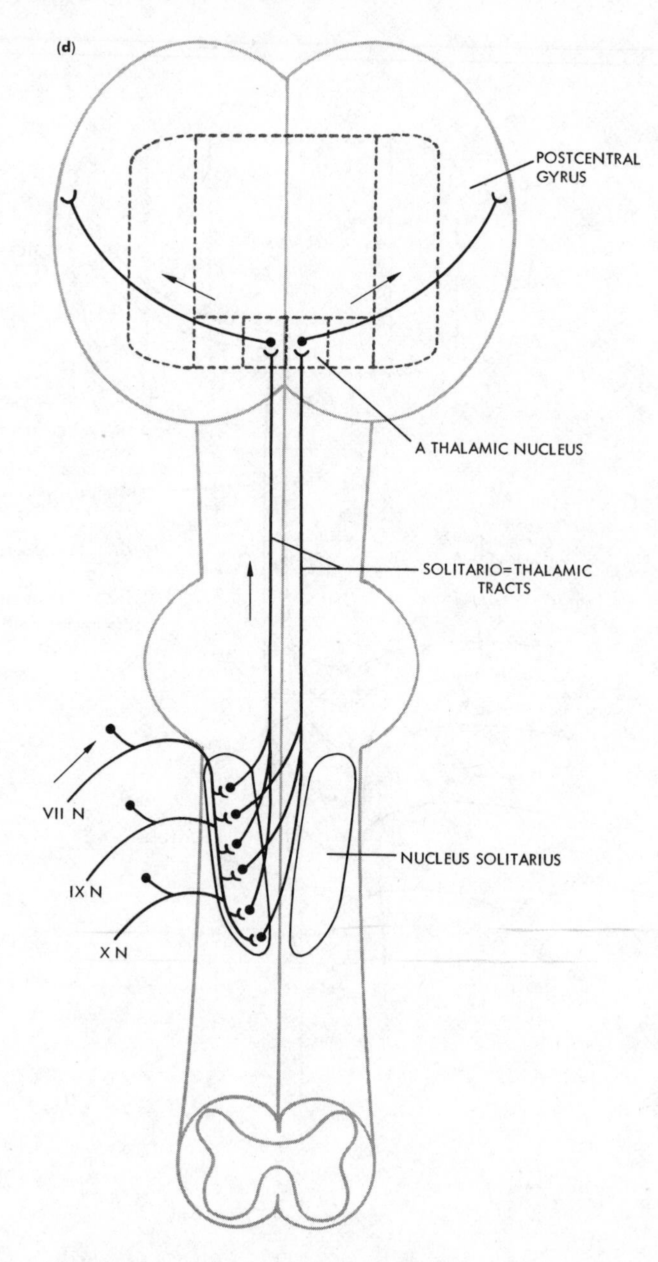

Figure 10-26 (d) The taste pathway (N, nerve). Recent evidence indicates that another nucleus is intercalated within the solitario-thalmic tract.

functional and physiologically defined reticular system. Reticular pathways are present in the spinal cord, in the tegmentum of the brain stem, hypothalamus, thalamus, and the cerebral cortex. The ascending reticular system is involved with modulating states of awareness and alertness from deep sleep to the most active levels of attentiveness and responsiveness. It is often called the ascending reticular activating system and functions in such complex activities as the nuances of the organism's emotional state, perception, drive, wakefulness, and sleep. The descending reticular system is a complex of motor fibers and nuclei that mainly influence nuclei in the lower brain stem (medulla) tegmentum;

from these nuclei excitatory and inhibitory stimuli are relayed via the reticulospinal tracts to the motor neuronal pools of the spinal cord (see the next heading for further discussion; Figure 10-30).

At the present time, knowledge of the pathways of nonmammalian vertebrates is incomplete, primarily because experimental research with an emphasis on functional correlations has not, until recently, received enough attention. The expanding research activities in analyzing these pathways will, in the near future, advance and clarify our understanding of the functional neural systems in nonmammalian vertebrates. Several critical studies indicate that many neural structures previously considered to be exclusively mammalian have counterparts in nonmammalian vertebrates. Hence, a reevaluation of the comparative neurobiology of the vertebrate central nervous system will result in some major changes of previously held concepts. In the discussion on auditory and visual pathways later on, note the similarities between the avian and mammalian systems. The motor fibers from the cerebral cortex to the spinal cord (corticospinal tract in mammals, considered to be a tract found only in mammals) have now been demonstrated to have an equivalent in birds and elasmobranchs.

Sensory Pathways of the Central Nervous System

Lemniscal systems comprise those pathways that transmit neural information through precise point-to-point relays successively from a peripheral spot (point) to a small group (point) of neurons in each nucleus and cerebral cortex of the pathway. The lemniscal pathways in mammals include pain and temperature, tactile discriminatory, taste, auditory, and visual pathways. For example, stimulation of a touch spot on the fingertip is projected by a relay generally composed of a sequence of three sets of neurons [Figure 10-26(b)], each neuron with a long axon: the first set extends from the sensory receptors in the finger to neurons in a nucleus in the medulla; the second set, with cell bodies and dendrites in the medullary nucleus, extends by long axons to a nucleus in the thalamus; the third set of neurons, with cell bodies and dendrites in the thalamus, extends by long axons to a small group of cells in the postcentral gyrus of the cerebral cortex. Each nucleus in the pathway processes the data. The processing within each nucleus tends to suppress (by inhibition) the accessory unimportant information (suppression of the background noise) and tends to enhance the significant and pertinent information (enhances the signal). The lemniscal system conveys details of varying degrees of precision; there are more point-to-point relays, for example, conveying information of touch from the fingers and lips than from the small of the back and, similarly, visual details from the macula lutea of the eye than from the nonmacular portions of the retina.

Comparative Functional Anatomy of the Nervous System: General Principles

The diversity exhibited by the nervous system is succinctly expressed by Herrick's classic statement: "The nervous system shows a wider range of adaptive structural modifications than is exhibited by any other organ system of the body."* Some general principles within the complexity of comparative

*See Herrick, C. J., 1952, in references.

functional anatomy of the vertebrate nervous system can be epitomized as follows.

1. The nervous system in all vertebrates is based on the same general plan (Figure 10-14).
2. The afferent input systems are hierarchically organized with the phylogenetically older centers in the lower forms functionally and structurally integrated with the phylogenetically newer centers in the higher forms. Even the older centers may become modified during phylogeny.
3. The efferent output systems are organized with the phylogenetically older reflex circuits that regulate the generalized patterned movements. The phylogenetically newer pathway systems, especially from the higher levels of the brain, influence and modify the phylogenetically older circuits and patterned movements.

Afferent Systems

The basic role of the afferent system is to sense the environment, to evaluate this input, and to influence the efferent systems. In the lower forms, there evolved the functionally efficient olfactory system, optic system, vestibulolateral line system, gustatory system, and general sensory systems. In many of these lower forms, specializations and refined elaborations took place, for example, in fish that echolocate by emitting electric pulses and detecting the reflected pulses. In the lower forms, the senses are relayed to centers primarily located in either the brain stem or the paleocortex of the central nervous system; the visual system in fish, amphibians, and reptiles, for example, is processed mainly at the midbrain level, whereas in the higher forms processing also occurs in the cerebral cortex in mammals and the hyperstriatum in birds. Put simply, the lower vertebrates have evolved the senses and the basic processing centers. Higher forms retain these basic centers, often in a further evolved state, and develop new centers that also process the input information. For example, in vertebrates, including mammals, responses to "sound" vibrations evoke reflexes (without the conscious awareness of the vibrations); the awareness of sound can be appreciated in the tectum at the midbrain level. In higher vertebrates, the distinctions among tones are evaluated at the primary cortical level (mammals) and hyperstriatum (birds). The subtle appreciation of sounds as words (symbolism of sounds) requires the association areas of the neocortex. In this hierarchy of stations, each has its function, with the highest centers performing the most sophisticated roles. A major obstacle that confronts any investigator in this area is the difficulty of determining the level of conscious perception of any input to the animal.

Efferent Systems

The basic role of the motor systems is to evoke motor responses that have survival value for the organism. In all vertebrates, there are many stereotyped patterned movements. These basic movements are regulated by a hierarchy of circuits and processing centers in the spinal cord and brain stem. Within the spinal cord are the integrative intrinsic circuits; the activity of these circuits evokes the basic postural movements of the body, including the undulating rhythm of swimming fish and the trunk and generalized limb movements of the terrestrial vertebrates. The various influences from the centers of the brain stem act almost exclusively on the intrinsic circuits of the spinal cord. In all vertebrates, the descending pathways from the brain stem include vestibulospinal pathways from the lateral line vestibular receptors, tectospinal pathway from the visual centers in the tectum, and the reticulospinal pathways from the reticular formation of the brain stem. Influences of the higher centers on the brain stem come from the paleostriatum, archicortex, and paleocortex of the telencephalon; the ventral thalamus of the diencephalon; and the archicerebellum and paleocerebellum. These centers influence the brain stem centers and, in large measure, through the reticulospinal pathways, interact with the intrinsic circuits of the spinal cord. In the higher vertebrates, additional influences are derived from further differentiated older centers and newly evolved centers, such as the neocortex, neostriatum, and neocerebellum. All vertebrates utilize the basic stereotyped movements to a greater or lesser degree. The modified older centers and the newly evolved centers in the higher forms project neural influences on the coordinated muscle groups involved with the sterotyped movements. Many of these patterned movements are modified in the higher forms by being fractionated. A fractionated movement is one that is released in some degree from the patterned sequence—for example, in primates and such mammals as racoons, the independent movements of the fingers or the delicate adjustments of wings of birds and bats during flight.

Sensory Pathways of the Central Nervous System

The ascending sensory pathways concerned with the general senses (pain, temperature, tactile and related modalities) in nonmammals are conveyed via neurons of reticular pathways; even in mammals, the cruder aspects utilize the ascending reticular pathways. The precise preceptive nature of these and other modalities is always difficult for the human observer to ascertain in any animal because of their subjective qualities. In nonmammals, some incipient lemniscal pathways for the general senses may be present. Unless otherwise stated, the following account applies primarily to the primates and, to a lesser degree, to other mammals. Two major sensory lemniscal pathways from the body are present in mammals. The *pain and temperature pathway* [Figure 10-26(a)] includes the first-order neurons, which extend from the sensory receptors in the body via spinal nerves to relay nuclei in the spinal cord; the second-order neurons, which extend from the spinal relay nucleus to the lateral column of the opposite side (decussates) of the spinal cord and ascend to thalamic nucleus as the *spinothalamic tract* (*spinal lemniscus*); and the third-order neurons, which extend from the thalamus to the postcentral gyrus of the cerebral cortex. In the *tactile discriminatory pathway*, the first-order neurons extend from the sensory receptors in the body via spinal nerves to the spinal cord and ascend in the dorsal column (fasciculi gracilis and cuneatus) of the same side to relay nuclei (nuclei gracilis and cuneatus) in the medulla; the second-order neurons extend from the relay medullary nuclei to the opposite side and ascend as the *medial lemniscus* to the thalamus; the third-order neurons extend from the thalamic nucleus to the postcentral gyrus (somatic sensory) of the cerebral cortex [Figures 10-26(b) and 10-29].

The general somatic sensory pathway conveying pain, temperature, tactile discrimination, and position sense from the head is the trigeminothalamic pathway. The first-order neurons of this pathway extend from sensory receptors in the head via the cranial nerves V, VII, IX, and X to the trigeminal nuclear column in the brain stem (Figure 10-22). The second-order neurons originate in the trigeminal nuclei, decussate to the opposite side of the brain stem and ascend as the trigeminothalamic tract (trigeminal lemniscus) to the thalamus. The third-order neurons project from the thalamus to the head region of the postcentral gyrus of the cerebral cortex. Note that these pathways convey information from the sensory receptors of one side of the head and body to the cerebral cortex of the opposite side. The elaboration of these inputs from the general sensory pathways into the appreciation and recognition of shape, form, and texture occurs in the association cortex of the parietal lobe. The general somatic sensory pathways of nonmammalian vertebrates are known only in general terms. In amphibia and reptiles, the rudiments of such pathways have been identified. These are fibers that are homologous to those of the spinal lemniscus of mammals; they project to the midbrain tectum and a thalamic nucleus. In turtles, some fibers project from this thalamic nucleus to the general cortex. The dorsal columns are definitely present, with fibers projecting to the dorsal column relay nuclei in the medulla. As in mammals, fibers from these nuclei decussate and ascend as the equivalent of the medial lemniscus; these fibers terminate in, as yet unidentified, nuclear complexes.

Taste sensations are conveyed from taste receptors via cranial nerves VII, IX, and X to the nucleus solitarius in the medulla (Figure 10-22). Fibers of the second order originate in this nucleus and ascend in the medial lemniscus a nucleus is intercalated within this tract to the thalamus. Third-order neurons from the thalamus project to the head region of the postcentral gyrus of the cerebral cortex.

Odors are conveyed from the olfactory mucosa via the olfactory nerve to the olfactory bulb, where considerable neural processing takes place. From the bulb, neurons project fibers via the olfactory tract to the olfactory cortex of the paleocortex (Figure 10-18).

The *vestibular pathways* are involved primarily in cranial reflexes (eye movements) and cervical reflexes (neck movements). In vertebrates, the neurons of each vestibular nerve and ganglion extend from the vestibular receptors in the membranous labyrinth to the vestibular nuclei in the brain stem and, in addition, directly to the archicerebellum [Figure 10-27(b)]. The vestibular nuclei are integrated into feedback circuits with the archicerebellum (Figure 10-31). These nuclei have projections that ascend and descend as the medial longitudinal fasciculus (MLF). The fibers of the MLF project (1) rostrally into the midbrain to make synaptic connections within the motor nuclei of the third, fourth, and sixth cranial nerves and (2) caudally through the cervical spinal cord to make connections within the motor nuclei of the cervical nerves. The vestibular system monitors the orientation of the head in space from the information obtained from the receptors in the vestibular sense organs. In response, the eyes and head move in space to compensate and maintain the visual axis stable in the environment. This is analogous to a shipboard-mounted radar system, where the tracking device (radar system or eye) is mounted on a moving platform (ship or head); the tracking device operates efficiently when it is automatically stabilized relative to the movement of the platform. In this activity, the vestibular system monitors and evaluates the motion of the head and, then, via influences relayed in the MLF, stimulates the extraocular muscles to move the eyes to compensate for the head movements. The projections from the vestibular nuclear complex extend caudally throughout the spinal cord as

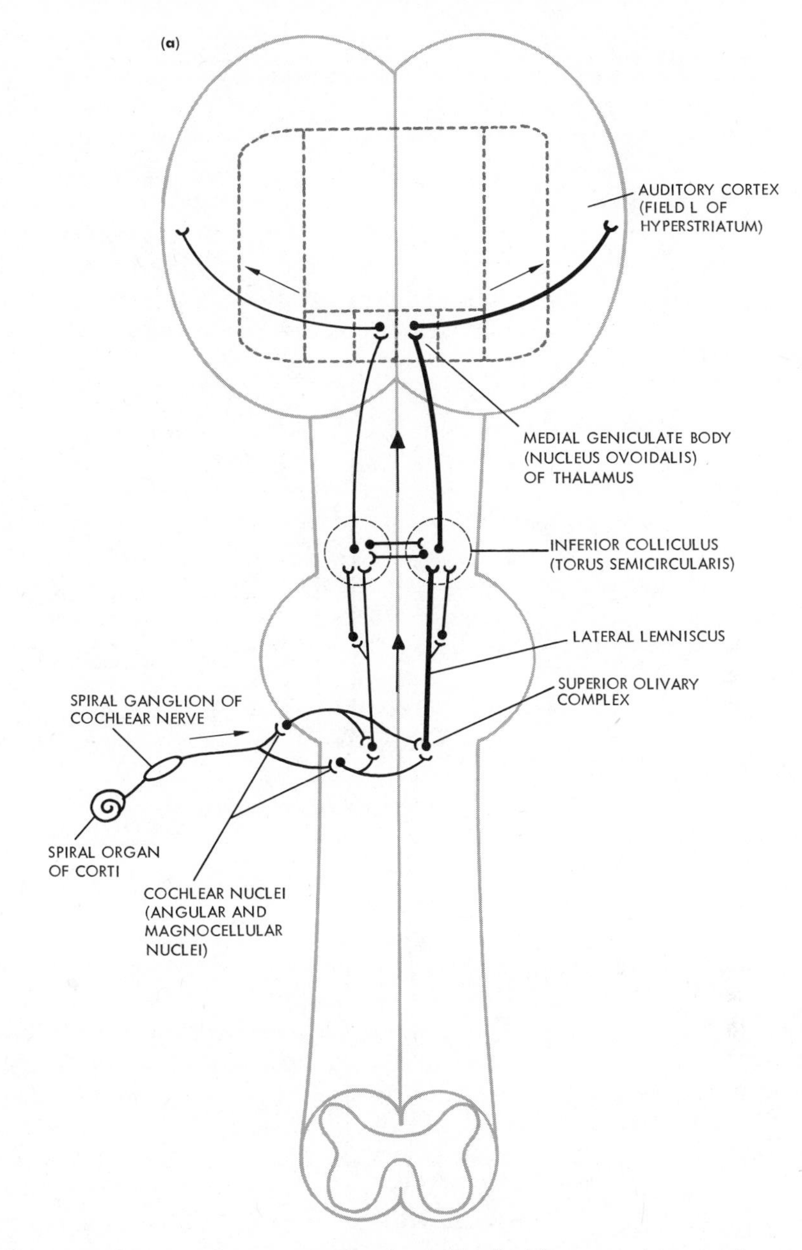

Figure 10-27 (a) Auditory pathway in mammals and birds. The terms in parentheses refer to the equivalent structure in birds.

the vestibulospinal tracts. These tracts have a role in the motor activity of the hindlimbs and body.

The auditory pathway commences in the spiral organ of Corti of the inner ear. The fibers of the first-order neurons of this pathway, with cell bodies in the spiral ganglion, pass through the cochlear nerve to the cochlear nuclei in the brain stem. From neurons in the cochlear nuclei, fibers project to several relay nuclei in the medulla and pons (superior olivary complex) and then ascend either on the same side or after crossing over on the opposite side [Figure

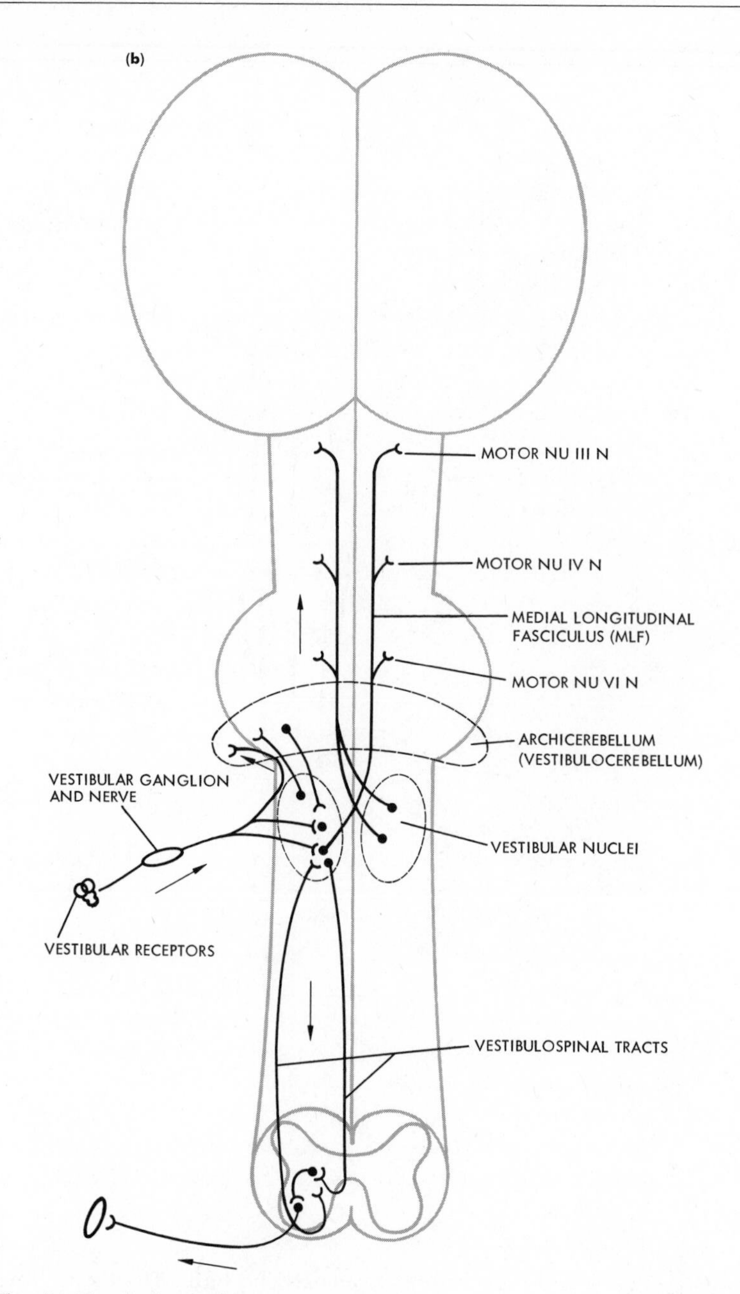

Figure 10-27 (b) Vestibular pathways.

10-27(a)] as the lateral lemniscus to the inferior colliculus of the midbrain tectum. More fibers ascend on the opposite than on the same side. The inferior colliculus projects to the medial geniculate body of the thalamus from which other fibers relay to the auditory cortex of the temporal lobe of the cerebrum. Recent studies, which combine morphological and experimental analyses, have revealed that the auditory pathways of birds are somewhat similar to those in mammals—especially when adjustments are made for comparing "analogous" structures between the mammalian and avian brains [Figure 10-27(a)]. The sequential organization of the auditory pathways in birds are as follows: "spiral ganglion," angular and magnocellular nuclei (analogous to cochlear nuclei of mammals), superior olivary complex, torus semicircularis (equivalent to inferior colliculus), nucleus ovoidalis of the thalamus (medial geniculate body), and field "L" of the hyperstriatum (auditory cortex of temporal lobe of mammals). The known auditory pathways of amphibia and reptiles are basically similar to those in birds from the inner ear to the midbrain.

The visual pathways are well organized in the vertebrates. The fibers projecting from the eye terminate in two different regions: (1) the tectum of the midbrain and (2) the thalamus of the diencephalon (Figure 10-28). In nonmammalian vertebrates, cetaceans, and some bats, all retinofugal fibers from each eye pass successively through the optic nerve, optic chiasm, and optic tract before terminating in the opposite half of the brain. The decussation of all fibers occurs in the optic chiasm. The complete decussation in these vertebrates means that the neural influences generated by stimuli from the visual environment viewed by one eye are relayed to the contralateral half of the brain for further processing. Mammals differ from nonmammalian vertebrates in that some retinofugal fibers from each eye cross the midline through the optic chiasm to the opposite half of the brain; other retinofugal fibers do not cross the midline and, thus, terminate on the same half of the brain (Figure 10-28). Thus, in mammals, each eye relays influences to both sides of the brain. These retinal fibers terminate (1) in the so-called optic lobes of the tectum in fish, amphibia, reptiles, and birds and in the superior colliculus of mammals and (2) in an "optic" thalamic nucleus (the so-called lateral geniculate bodies). The relative numbers of crossed and uncrossed fibers are roughly proportional to the direction the eyes face. Those mammals with laterally directed eyes (e.g., rabbit and horse) have many decussating fibers relative to the few nondecussating fibers. Those mammals with frontally directed eyes (e.g., primates and carnivores) have roughly about an equal number of decussating and nondecussating retinofugal fibers. As a rule, the more the eyes face laterally, the greater is the number of decussating retinofugal fibers; whereas the more the eyes normally face frontally, the closer is the approach to an equal number of decussating and nondecussating fibers. The combination of crossed and uncrossed projections in the mammalian visual system is functionally significant. It turns out that those regions of the visual environment (field of vision) seen by both eyes simultaneously stimulate neural activities in the sectors of both eyes, which are relayed to, and integrated in, the brain for use in depth perception. For example, a spot seen by both eyes simultaneously evokes neural activity in a focal region of the visual cortex on only one side of the brain and is seen as one spot. However, because the spot is seen by two eyes, the resulting dual input is processed to evaluate depth and other perceptual qualities. Animals with frontally directed eyes have restricted total fields of vision in front of the head; most of these fields are viewed by both eyes at the same time. In contrast, animals with laterally directed eyes have large total fields of vision in front and to the sides of the head; most of these fields are viewed by only one eye at a time. In general, mammals with frontally directed eyes are those that place a premium on visual acuity in catching prey (carnivores) or manipulating objects (primates and racoons). Mammals with laterally directed eyes are the hunted animals; they place a premium on viewing large regions of the environment continuously.

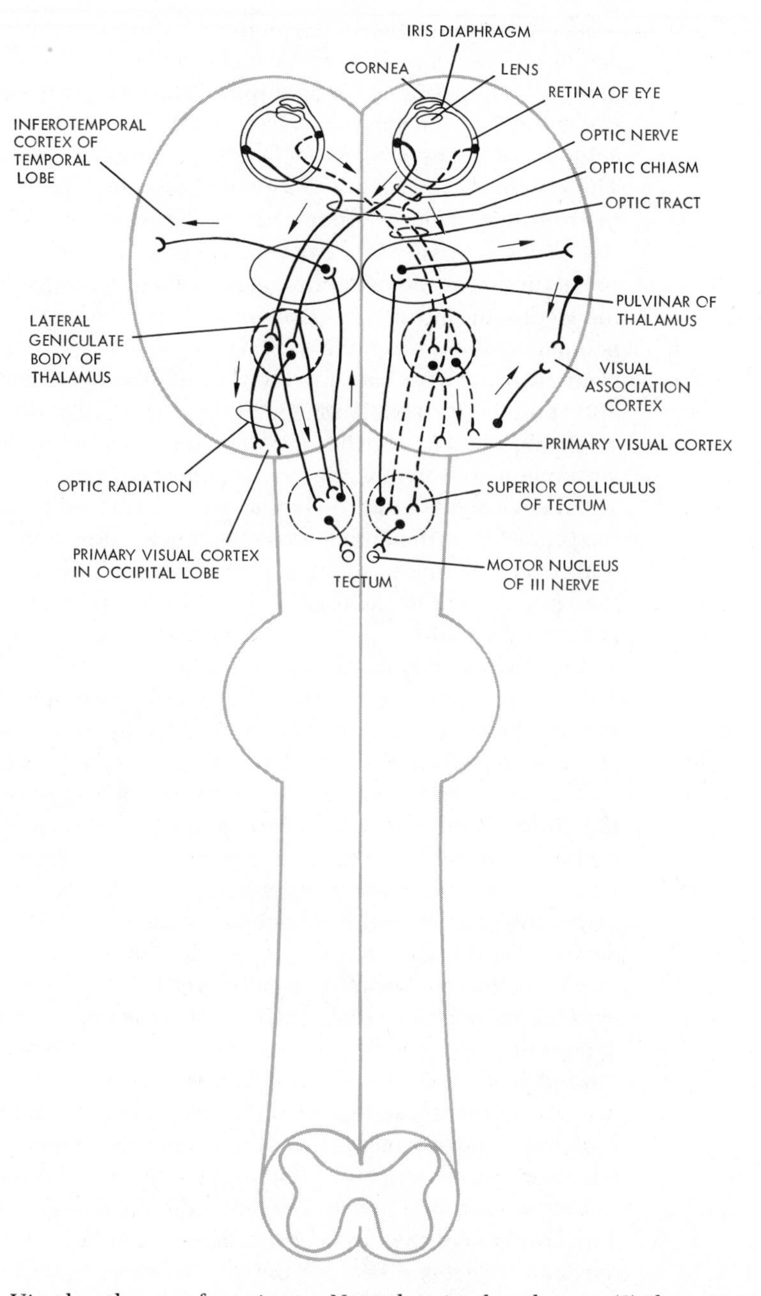

Figure 10-28 Visual pathways of a primate. Note the visual pathways: (1) the retino-geniculo-cortical pathway that comprises the sequence of retina to lateral geniculate body to visual cortex of the occipital lobe, and (2) the retino-tecto-thalamo-cortical pathway that comprises the sequence of retina to optic tectum to pulvinar of the thalamus to a "visual" cortex in the temporal lobe (inferotemporal cortex).

Mammals and birds apparently have two visual pathways by which neural influences are conveyed from the eyes to the cerebral hemispheres. In general, the pathways are similar in the number and location of neural processing stations present; whether these stations are functionally analagous is not resolved at present. In mammals, these pathways are (1) the retinogeniculocortical pathway, which comprises the sequence of retinas, optic nerves, chiasm, and tracts to the lateral geniculate bodies of the thalamus and the optic radiations to the primary visual cortices of the occipital lobes of primates; and (2) the retinotectothalamocortical pathway, which comprises the sequence of retina via optic nerves, chiasm, and tracts to the superior colliculi to the pulvinar of the thalamus and via a projection to the inferotemporal cortex of the temporal lobe of primates (Figure 10-28). Further processing of the inputs from these two systems occurs in the association cortex of the occipital and temporal lobes. In birds, these pathways are (1) the retinogeniculowulst pathway, which comprises the retina, lateral geniculate body of thalamus, and the wulst (a specialized "cortical" area); and (2) the retinotectorotundohyperstriatal pathway, which comprise the retina, optic lobes of the tectum, nucleus rotundum of the thalamus, and the hyperstriatum (a probable specialized "cortical" area). It is possible that the pulvinar and nucleus rotundus are analogous.

In reptiles, the optic thalamic nuclear complex projects to the general cerebral cortex (analagous to neocortex of mammals); in addition, the optic lobe projects fibers to the nucleus rotundus. Relays from the optic thalamus to the telencephalon have not been demonstrated in fish and amphibia.

The superior colliculus of the mammalian tectum is an optic center integrated into the accommodation reflex and the pupillary light reflex (pages 439-40). Both the tectum and the visual cortex have a significant role in conscious visual discrimination (Figure 10-28). In many mammals, including rats, cats, and dogs, the ability to distinguish various patterns by sight and to discriminate between light and dark is retained after the entire visual cortex has been removed; such animals have lost some ability, however, to make fine visual discriminations. These observations indicate that in these animals vision is primarily a subcortical sense, with only some higher visual discriminations occurring at the cortical level. In primates, including man, conscious visual discriminations are made in the cerebral cortex; this functional aspect of vision is corticalized almost completely. In these forms, the superior colliculus is almost exclusively a reflex center. A man without the visual cortex is blind; he cannot distinguish light from dark, or colors and patterns. The optic tectum of fish, amphibia, and reptiles apparently acts both as (1) a critical processing center in the "conscious" visual discrimination and (2) a reflex center involved with accommodation (focusing) and pupillary light reflexes.

The cerebellum is actually a sensory structure that processes input from several sources. This sensory input, all of it on an unconscious level, is derived primarily from receptors in the muscles and joints of the body and limbs and other receptors from the vestibular system. Other neural information is derived from the auditory and visual systems, the cerebral cortex, and many nuclei of the brain stem. Many of these neural structures and systems are integrated into complex feedback circuits. The cerebellum expresses its activity through motor pathways. It functions primarily to smooth out and to synchronize the delicate and precise timing among muscles of a group and among groups of muscles.

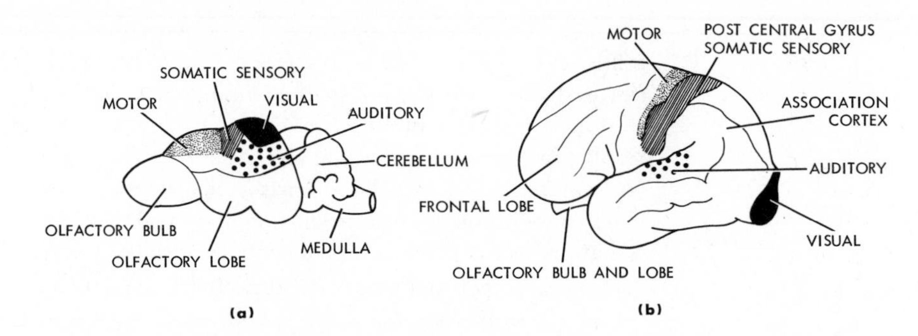

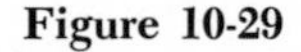
Figure 10-29 Lateral view of (a) the brain of the shrew, an insectivore, and (b) the cerebrum of man. The primary sensory projection areas and the primary motor areas are indicated. Note that the association cortex is relatively larger in man than in lower placental mammals such as the shrew.

Motor Pathways of the Central Nervous System

For descriptive purposes, the structural organization of the motor circuits, nuclei, and pathways can be subdivided into (1) the suprasegmental level of circuits (Figure 10-31), (2) the major descending tracts (Figure 10-30), and (3) the intrasegmental and intersegmental circuits and neuronal pools of the spinal cord. The third category has been discussed before, in the section on the spinal cord. The suprasegmental level of interacting circuits comprises many centers and pathways within the brain that have a significant function in influencing motor activity. The resolution of the activities among these centers is relayed finally to the brain stem and spinal cord via several major tracts (Figure 10-30). Among the major pathways from the brain to the brain stem and spinal cord are (1) the *corticospinal* (pyramidal) *tract* and corticoreticular tracts from the cerebral cortex to the spinal cord and brain stem reticular formation; (2) the vestibulospinal tracts and the *medial longitudinal fasciculus* from the vestibular nuclei of the pons and medulla to the brain stem and the spinal cord; (3) the *tectospinal tract* from the midbrain tectum (optic) to the spinal cord; (4) the *rubrospinal tract* from the nucleus ruber of the midbrain reticular formation to the spinal cord; and (5) the *reticulospinal tracts* from the reticular formation of the pons and medulla to the spinal cord [Figure 10-30(a) and (b)].

Among the sources of influences acting on the nucleus ruber and brain stem reticular nuclei in mammals are the cerebral cortex (via corticorubral and corticoreticular fibers) and the cerebellum (Figure 10-30). In this respect (1) the combination of the uncrossed corticorubral and crossed rubrospinal tracts forms the corticorubrospinal pathway and (2) the combination of uncrossed corticoreticular and crossed and uncrossed reticulospinal tracts from the corticoreticulospinal pathway (Figure 10-30). Both these pathways project indirectly from cerebral cortex to spinal cord. Except for the corticospinal tract (a mammalian tract), the other motor tracts are found in varying degrees of specialization in most groups of vertebrates. The corticospinal (pyramidal) tract of mammals projects influences from the cerebral cortex directly to the spinal cord; this pathway is presumed to have a role in influencing "skilled voluntary" movements. It is well developed in primates and carnivores.

The corticospinal fibers are located in different columns of the spinal cord of different mammals; they are (1) in the dorsal columns in rodents, tree shrews, and marsupials; (2) in the lateral columns primarily in primates and carnivores

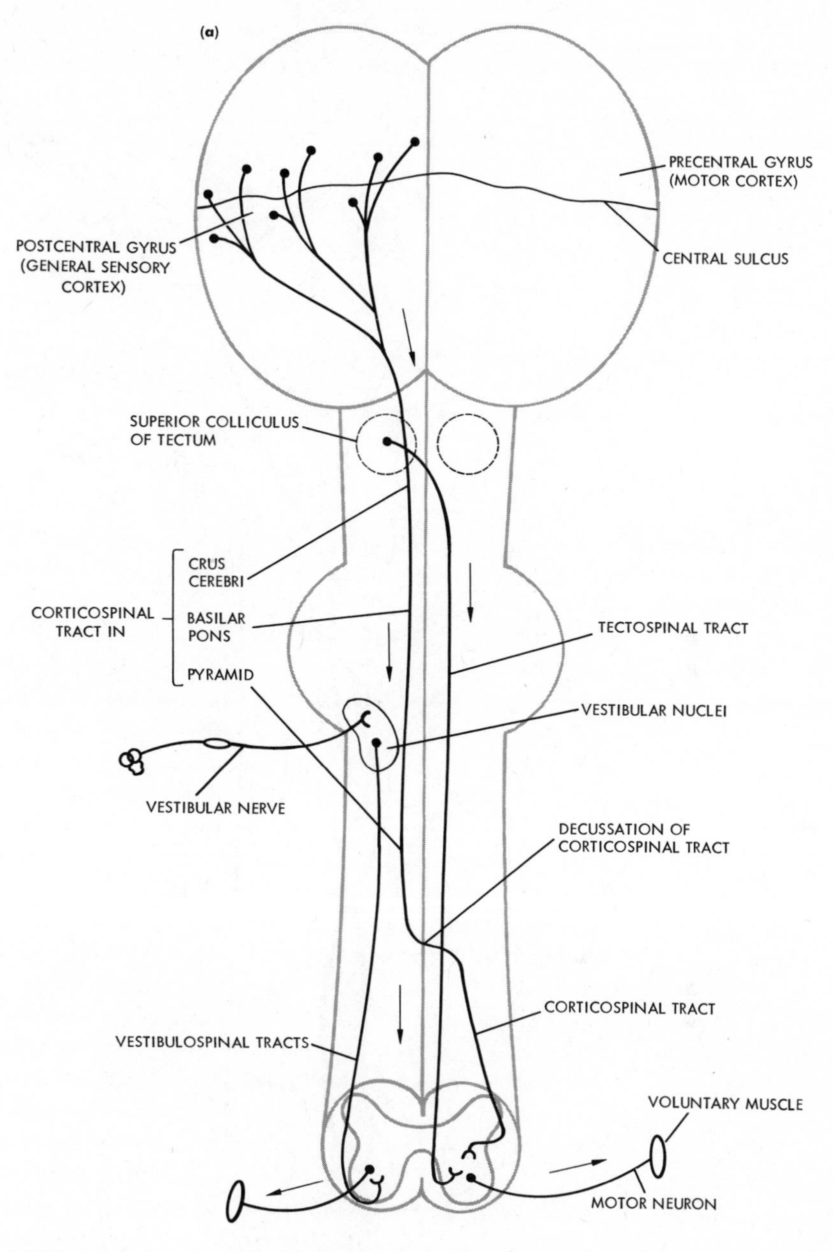

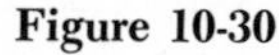
Figure 10-30 (a) Motor pathways. Note the corticospinal tract, tectospinal tract, and vestibulospinal tracts.

(Figure 10-30); and (3) in the ventral columns primarily in insectivores and edentates.

In general, the pathways from the cerebral cortex and midbrain decussate and project to the opposite side of the spinal cord; the tracts from the lower brain stem project to both sides of the cord. In a broad, conceptual sense, the corticospinal tract is a descending lemniscal pathway and the corticoreticulospinal and corticorubrospinal tracts are descending reticular pathways.

Within the vertebrates, the phylogeny of the major motor pathways from the

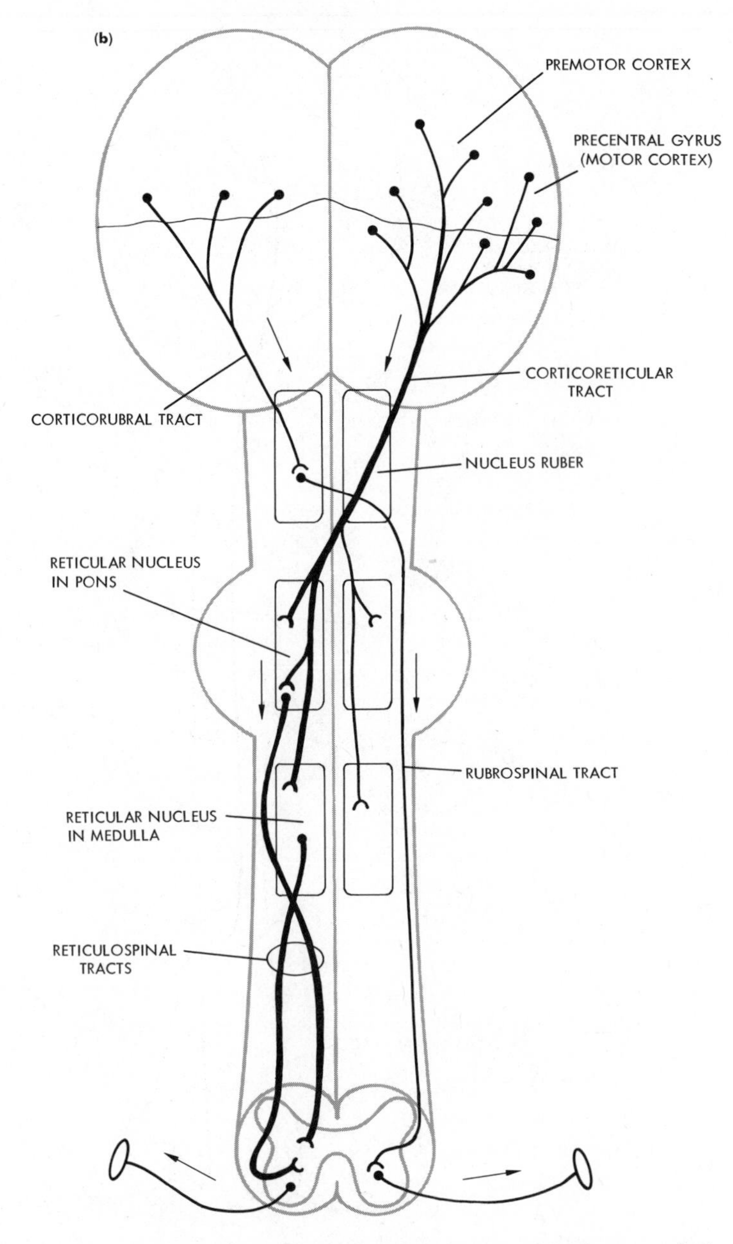

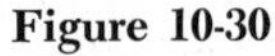
Figure 10-30 (b) Motor pathways. Note the corticorubrospinal pathway and the corticoreticulospinal pathway.

brain to the spinal cord can be schematized as follows. The reticulospinal pathways and vestibulospinal pathways are present in all classes of vertebrates; the tectospinal tract is well developed in amphibians, reptiles, birds, and placental mammals; the rubrospinal tract is definitely found in birds and mammals; and the corticospinal tract is present in mammals. The degree of complexity and size to which a tract or pathway may evolve in any group of vertebrates is correlated with the functional requirements of the organisms. In the cyclostomes the reticulospinal pathways are the basic final common pathway for projecting influences from the brain; in fish, the *vestibulospinal pathways* are as significant as the reticulospinal pathways; in the amphibians and the reptiles, a tectospinal tract also is found; in the birds, the tectospinal tract is well developed and a small rubrospinal tract is present; in the placental mammals, the rubrospinal tract and the corticospinal pathways are well differentiated.

In their totality, the suprasegmental circuits are integrated complexes of excitatory neuron pools and tracts and of inhibitory neuron pools and tracts that finally interact with other neuron pools resulting in the integrated and coordinated muscular movements In effect, within the circuitry of the brain, each neuron pool gives rise to fibers that stimulate other neuron pools; each neuron pool is, in turn, stimulated by other neuron pools and feedback circuits. In this complexity, a nucleus may excite an excitatory pool and, thus, facilitate neuronal activity; a nucleus may inhibit an excitatory pool and suppress activity; a nucleus may excite an inhibitory pool and suppress activity; or a nucleus may inhibit an inhibitory pool and, thus, enhance activity by suppressing inhibition (this is called *disinhibition*). Included in the suprasegmental system are several complex feedback circuits, to be noted in the section on encephalization. Two such feedback circuits are (1) the circuit, from cerebral cortex to cerebellar cortex and back to cerebral cotex via several intermediary connections (Figure 10-31), and (2) the general circuit, from cerebral cortex to corpus striatum, to the thalamus and back to cerebral cortex. These suprasegmental circuits have an important influence through excitation and inhibition on the activities of the descending motor pathways. The intricately organized interactions are integrated to produce the numerous nuances of the coordinated muscular activities in the vertebrates. Malfunctioning in some of the *suprasegmental* circuits is expressed in such diseases as paralysis agitans (Parkinson's disease).

Encephalization

The evolutionary expression by which the higher neural centers assume greater significance in the functions of the nervous system is known as encephalization. In this concept, the nervous system is organized in sequences of various levels of hierarchy. Encephalization is actually a broader concept than the elaboration of new and more sophisticated functional roles in the nervous system by the evolution of "new structures" or the increased complexity of preexisting neural structures. Neural evolution generally involves the evolution of a system (e.g., visual system, motor system). In such a concept, the elaboration of a rostral forebrain structure is accompanied by the concomitant elaboration of one or more diencephalic or hindbrain structures that is part of the system. For example, the evolution of the neocortex was accompanied by the evolution of the basilar pons and neocerebellum and other structures of the entire complex of the feedback circuitry of neocortex to pontine nuclei to neocerebellum and back to neocortex via several nuclear groups. In this view, the newly evolved functions in mammals or birds, for example, are expressions of the forebrain activity to a greater extent than of other regions of the brain.

The basic principles of encephalization are demonstrated in the evolution of the auditory system. On the premise that a lower vertebrate hears when it behaves as if it has located a moving object by sensing vibrations, it is believed that fish do hear. Their auditory pathways are probably restricted to certain

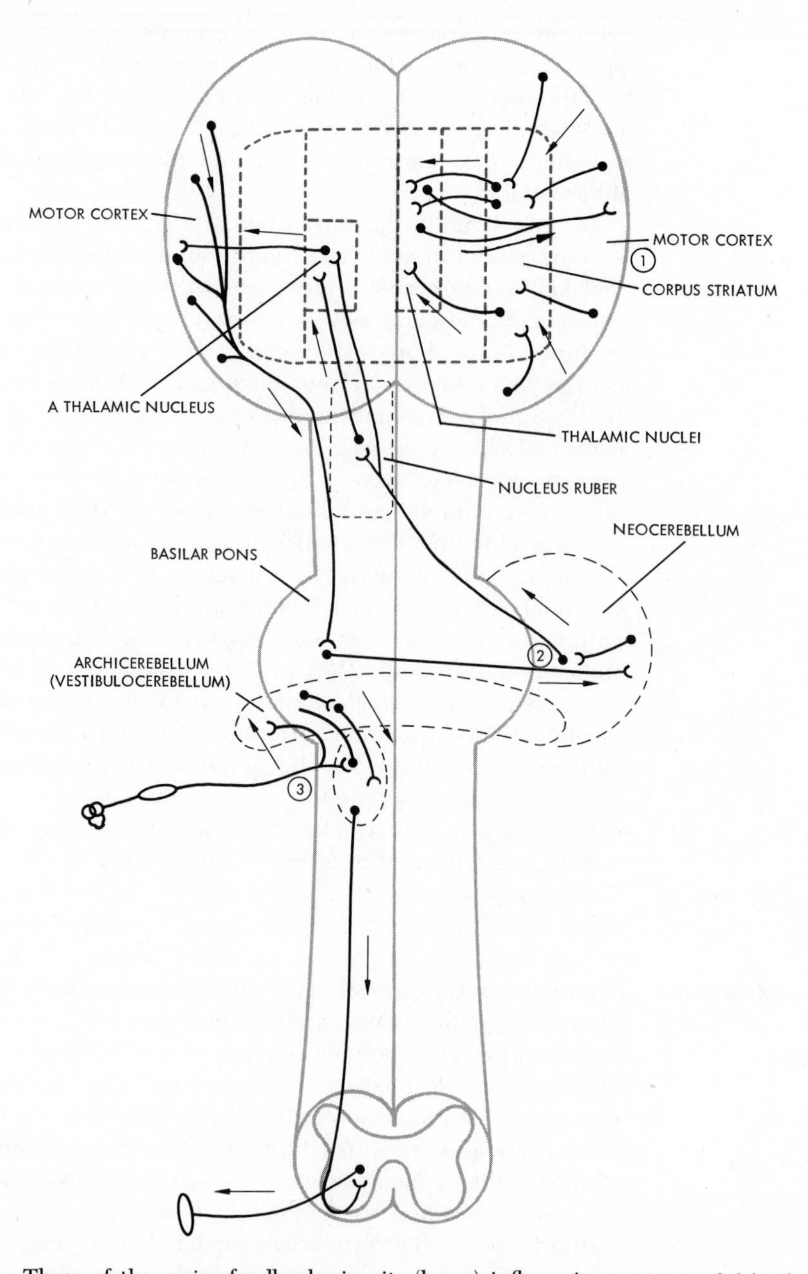

Figure 10-31 Three of the major feedback circuits (loops) influencing motor activities in the mammals: (1) circuit of cerebral cortex to corpus striatum (basal ganglia) to thalamus to cerebral cortex, (2) circuit of cerebral cortex to basilar pons to neocerebellum to nucleus ruber and thalamus to cerebral cortex, and (3) circuit of vestibular nuclei to archicerebellum (vestibulocerebellum) to vestibular nuclei. In the circuits with connections in the cerebellum, there are synaptic connections within the cerebellar cortex and the deep cerebellar nuclei.

nuclei of the reticular system and of the tectum of the midbrain. In amphibians and reptiles, the auditory pathways are indicated with a more fully differentiated nucleus, the torus semicircularis in the tectum of the midbrain. In nonmammalian tetrapods, the appreciation of the auditory stimuli is resolved at the brain stem level, with the highest integration center located in the midbrain. In birds and placental mammals, the auditory pathways are more fully evolved with the elaboration of the inferior colliculus of the midbrain tectum, the medial geniculate nuclei of the thalamus, and the primary auditory cortex. In birds, the equivalent of the cerebral cortex is the hyperstriatum. In these higher vertebrates, an appreciation of auditory stimuli results from the neural activity of the brain stem and of the thalamus and cerebral cortex; assumption of this function by the cerebral cortex (mammals) and the hyperstriatum (birds) is an expression of encephalization (specifically corticalization). In mammals, an association auditory neocortex evolves in conjunction with the primary auditory cortex [Figures 10-27(a) and 10-29].

The awareness of crude sounds is an expression of subcortical activity, probably at the midbrain level. With encephalization, the telencephalic centers are related to the more discriminatory perceptions of the auditory sense; the thalamic and cortical levels are involved in the appreciation of more elaborate auditory perception. The primary auditory cortex receives its input from the medial geniculate body of the thalamus. The auditory association cortex has complex neuronal interconnections; these include input from the primary auditory cortex of the other cerebral hemisphere through the corpus callosum. In man, the association cortex has a crucial role in the recognition of sounds and the spoken language.

The concept of encephalization is exhibited by each major subdivision of the nervous system. Stated another way, each division is made up of structures with phylogenetic histories of different lengths. This principle is illustrated by the cerebral cortex and the cerebellar cortex. The archicortex and the paleocortex of both the cerebrum and the cerebellum have a long phylogenetic history. Present in all vertebrates, these structures have evolved in the various classes in connection with different functional demands. In the cerebellum, the *archicerebellum* is functionally integrated with the vestibular and lateral line systems, whereas the *paleocerebellum* is primarily associated with integrating the input from the general sensors of the body. In the cerebrum, the paleocortex has a primary function in the sense of smell. The archicortex is integrated with neural mechanisms associated with the behavioral expressions of animals; these include the degree of aggressiveness and emotional tone (lethargy to alertness) that they exhibit. The so-called pleasure centers and punishing centers are located in the allocortex (limbic lobe). The stimulation of the pleasure centers evokes favorable activity by the animal (the animal likes the stimulation), and the stimulation of the punishing centers evokes avoidance responses (the animal dislikes the stimulation). The limbic lobe is the cortical representative of the limbic system—a difficult-to-define system involved with the behavioral and emotional expressions associated with self-preservation of the organism (feeding, fight, or flight) and self-preservation of the species (mating, procreation, and care of offspring).

The neocortex of the cerebrum is basically a mammalian structure, with primordial elements in the reptiles and an independently evolved counterpart in the birds called the hyperstriatum. It is functionally associated with the highest levels of sensory perceptions of the auditory, visual, gustatory, and general senses (the latter include appreciation of weight, shape, texture, and form of objects from input from the general body senses) and also with thought, intelligence, and certain regulatory aspects of motor activity. The *neocerebellum*, a mammalian structure, evolved in concert with the neocortex of the cerebrum; the cerebral cortex projects through neopathways to the neocerebellum and paleocerebellum (Figure 10-31), which, in turn, course through other neopathways back to the neocortex. The entire feedback circuit follows the sequence of cerebral cortex to pons; pons to neocerebellum and paleocerebellum of the opposite side (neocerebellar circuits); cerebellum to thalamus of the original side; and thalamus to the neocortex. In brief, part of the neocortex of the cerebrum is integrated in a feedback circuit with the neocortex of the opposite cerebellar hemisphere and the neocerebellar hemisphere of one side with the neocortex of the opposite cerebral hemisphere (Figure 10-31). The function of this circuit is expressed in movements induced by cerebral cortical activity (e.g., skilled movements). The motor centers of the cerebral cortex project influences to motor centers in the brain stem and spinal cord as well as to the cerebellum; the stream of influences to the cerebellum are fed back to the cerebral cortex for integration into the on-going motor activities. In cerebellar injury, the cortically induced movements can be executed, but with the absence of cerebellar control. The movements are performed with poor timing (tremor); actually, the precisely coordinated timing between the agonist and antagonist muscles of a movement is no longer perfectly synchronized and, thus, the movement is not smooth. The degree of encephalization can be seen in the greater malfunction shown in animals with an *ablated* cerebellum. Decerebellate vertebrates reveal various results: sharks have no impairment in their ability to swim; teleost fish (goldfish) have side-to-side rolling movements while swimming; frogs can jump and swim but have some muscular rigidity; birds can eat and drink but walk, stand, and fly with extreme difficulty; and mammals can perform the crude aspects of an intended movement but do so with tremor.

Another complex feedback circuit evolved in conjunction with the motor centers of the neocortex of mammals. The path of this circuit is from neurons of the neocortex to the corpus striatum (corticostriate fibers), intrastriatal connections, neurons from the corpus striatum to the thalamus (striatothalamic fibers), and thalamic projections to the neocortex (Figure 10-31). This circuit remains within the same cerebral hemisphere, its fibers do not project to the contralateral side. Malfunctioning of this circuitry in man may result in abnormal movements (cerebral palsy). The archicerebellum is integrated into a feedback circuit with the vestibular pathways [Figures 10–27(b) and 10–31].

Ventricular System

The ventricular system is a series of fluid-filled cavities within the brain. This cerebrospinal fluid is formed, in part, at the choroid plexuses of the lateral, third, and fourth ventricles (Figure 10-14). It flows slowly from the paired lateral ventricles of the cerebral hemispheres, through the foramina of Monro, third ventricle of the diencephalon, aqueduct of Sylvius (iter) of the midbrain, and the fourth ventricle of the metencephalon and myelencephalon; the fluid leaks through several foramina in the roof of the fourth ventricle into the subarachnoid space surrounding the central nervous system. In birds and mammals, it circulates slowly until it returns to the venous vessels.

Brain Size (Weight) Relative to Body Size

Studies of brain size relative to body size of fossil and living vertebrates indicate that a critical difference exists among the fish, amphibia, and reptiles that are basically small-brained vertebrates, as compared to the birds and mammals that are basically large-brained vertebrates [5]. Evidence indicates that throughout the 500 million years of their adaptive radiation from the Ordovician to the present, these small-brained vertebrates apparently have remained at the same level of brain size relative to body size. Within this relatively constant brain-size constraint, the small-brained vertebrates adaptively evolved as different amounts of neural tissues were devoted to different functional demands in different species. For example, the enlarged optic lobes in the tectum of surface-living fish are correlated with their visually oriented life style; on the other hand, the enlarged neural structures involved with chemoreception (taste and smell) in bottom-living fish and blind cave fish are correlated with senses associated with survival in their ecological niche. Yet, in all these fish, the brains have the same relative size.

The fossil and living mammals seem to have been large-brained vertebrates from the time they evolved from reptiles about 200 million years ago in the Mesozoic. Birds apparently evolved as large-brained vertebrates from reptiles independently some thirty million years after the earliest mammals appeared in the Mesozoic. These relatively large brains of mammals and birds are another factor that favored these vertebrates during evolution in the Cenozoic era.

References for Chapter 10

[1] Bass, A. D., ed. *Evolution of Nervous Control from Primitive Organisms to Man.* Washington, D.C.: American Association for the Advancement of Science, 1959. The impact of the evolutionary process on the nervous system is presented by many authorities who represent a variety of disciplines.

[2] Gardner, E. *Fundamentals of Neurology.* Philadelphia: W. B. Saunders Company, 1968. This is a brief account of the neuroanatomy and neurophysiology of man.

[3] Herrick, C. J. *Neurological Foundations of Animal Behavior.* New York: Hafner Publishing Company, 1927, 1952. Reprinted in 1962. This classic is an account of the nervous system written from an evolutionary point of view by the father of the American school of neurology.

[4] ———. *The Brain of the Tiger Salamander, Ambystoma trigrinum.* Chicago: The University of Chicago Press, 1948. Probably the most penetrating account of the nervous system of any species, this is a classic.

[5] Jerison, H. J. *Evolution of the Brain and Intelligence.* New York: Academic Press, Inc., 1973.

[6] Johnston, J. B. *The Nervous System of Vertebrates.* Philadelphia: Blakiston, 1906. This classic still can be read with profit.

[7] Ariëns Kappers, C. U., G. C. Huber, and E. C. Crosby. *Comparative Anatomy of the Nervous System of Vertebrates, Including Man.* 2nd ed. 3

vols. New York: Hafner Publishing Company, 1960. This is a detailed work that is useful for special information.

[8] NAUTA, W. J. H., and H. J. KARTEN. *A General Profile of the Vertebrate Brain, with Sidelights and the Ancestry of Cerebral Cortex.* In *The Neurosciences: Second Study Program,* edited by F. O. Schmitt, pp. 7–26. New York: Rockefeller University Press, 1970. This is an outstanding panoramic account of the subject.

[9] NOBACK, C. R. *The Human Nervous System.* New York: McGraw-Hill Book Company, 1975. This is a brief account of the neuroanatomy and neurophysiology of man.

[10] PAPEZ, J. W. *Comparative Neurology.* New York: Thomas Y. Crowell Company, 1929. Reprinted in 1962 by Hafner Publishing Company, this is a short, useful account of the comparative neuroanatomy of the vertebrates.

[11] PARKER, G. H. *The Elementary Nervous System.* Philadelphia: J. B. Lippincott Co., 1919. This is an excellent presentation of the early evolution of the nervous system.

[12] PETRAS, J. M., and C. R. NOBACK. *"Comparative and Evolutionary Aspects of the Vertebrate Central Nervous System." Annals of the New York Academy of Sciences,* **167** (1969), 1–513.

13] SARNAT, H. B., and M. G. NETSKY *Evolution of the Nervous System.* New York: Oxford University Press, Inc., 1974.

[14] WEBSTER, D., and M. WEBSTER. *Comparative Vertebrate Morphology.* New York: Academic Press, Inc., 1974.

[15] ZEMAN, W. J., and R. M. INNES. *Craigie's Neuroanatomy of the Rat.* New York: Academic Press, Inc., 1963. This is an excellent account of modern concepts of the organization of the nervous system as well as of the neuroanatomy of the rat.

Bibliography: Sensory Receptors and the Nervous System

[16] VAN BERGEIJK, W. A., J. R. PIERCE, and E. E. DAVID, Jr. *Waves and the Ear.* Garden City, N.Y.: Anchor Books, 1960. This is a lucid account of the structural and functional aspects of sound and the ear. (Paperback)

[17] BROWN, M. W., ed. *The Physiology of Fishes* Vol. 2. New York: Academic Press, Inc., 1957. This book contains excellent chapters by leading authorities: "The Nervous System" by E. G. Healey; "The Eye" by J. R. Breet; "The Acousticolateralis System" by O. Lowenstein; and "Electric Organs" by R. D. Keynes.

[18] BURNSTOCK, G. "Evolution of the Autonomic Innervation of Visceral and Cardiovascular Systems in Vertebrates." *Pharmacological Reviews,* **21**(1968), 247–324. This is an excellent anatomic, physiologic, and pharmacologic review of the autonomic nervous system of each of the living classes of vertebrates.

[19] DUKE-ELDER, S "The Eye in Evolution." In *System of Ophthalmology.* Vol. 1. London: Henry Limpton, 1958. This is a detailed and comprehensive account of the eye and visual pathways of the vertebrates.

[20] FLOREY, E. *General and Comparative Animal Physiology.* Philadelphia: W. B. Saunders Company, 1966. This is an excellent account of comparative neurophysiology.

[21] GILBERT, P. W., ed. *Sharks and Survival.* Boston: D. C. Heath & Com-

pany, 1963. This book contains excellent chapters by leading authorities: "The Central Nervous System of Sharks and Bony Fishes" by L. R. Aronson; "Hearing in Elasmobranchs" by R. H. Backus; "Olfaction, Gustation, and the Common Chemical Sense in Sharks" by A. L. Tester; and "The Visual Apparatus of Sharks" by P. W. Gilbert.

[22] Hoar, W. S. *General and Comparative Physiology.* Englewood Cliffs, N.J.: Prentice-Hall, Inc., 1966. This is an excellent account of comparative physiology.

[23] Marshall, A. J., ed. *Biology and Comparative Physiology of Bird.* Vol. 2. New York: Academic Press, Inc, 1961. This book contains excellent chapters by leading authorities: "The Central Nervous System" by A. Portmann and W. Stingelin and "Sensory Organs" by A. Portmann and R. J. Pumphrey.

[24] Polyak, S. *The Vertebrate Visual System.* Chicago: The University of Chicago Press, 1958. This is a comprehensive and detailed analysis of the vertebrate visual system; it is a monumental book.

[25] Prosser, C. L., and F. A. Brown, Jr. *Comparative Animal Physiology,* 2nd ed. Philadelphia: W. B. Saunders Company, 1961. This is an excellent presentation of vertebrate comparative physiology.

[26] Romer, A. S. *The Vertebrate Body* 3rd ed. Philadelphia: W. B. Saunders Company, 1962. This is a good account of classical comparative anatomy of the sense organs and the nervous system of the vertebrates.

[27] Sturkie, P. D. *Avian Physiology* 2nd ed. Ithaca, N.Y.: Cornell University Press, 1965. This is a good presentation of the physiology of the nervous system and sensory systems of birds.

11

The Endocrine System

Introduction The endocrine system of vertebrates consists of a diverse assemblage of glands and glandular tissues that secrete minute amounts of hormones into the blood or other body fluids. When they have been distributed throughout the body by the circulation, these substances are taken up by various tissues in which they elicit responses. Each particular hormone is chemically distinct and recognizable by particular target tissues and each is secreted in response to specific stimuli. In a general way, then, the functional strategy of the endocrine system is like that of the nervous system: it produces specific control signals in response to sensory information about conditions in the external and the internal environments; these signals, in turn, are interpreted by specific effector systems to command adaptive responses to the sensory input. But there are marked differences in the ways in which the two systems are structured to carry out their control functions.

Compared with the complex architecture of the nervous system, the endocrine system is relatively unstructured anatomically. Some fifteen or twenty different glands constitute the endocrine system of vertebrates, and these *appear* to be scattered almost randomly throughout the body (Figure 11-1). This impression of unrelatedness among the endocrine glands is supported by the diverse and unrelated embryonic origins of many of them. The principal justification for considering the endocrine glands together as a system is the gross similarity of the manner in which the hormones function as blood-borne secretions. In addition, there are complex interactions among the endocrine glands that weld them into a functionally unified system.

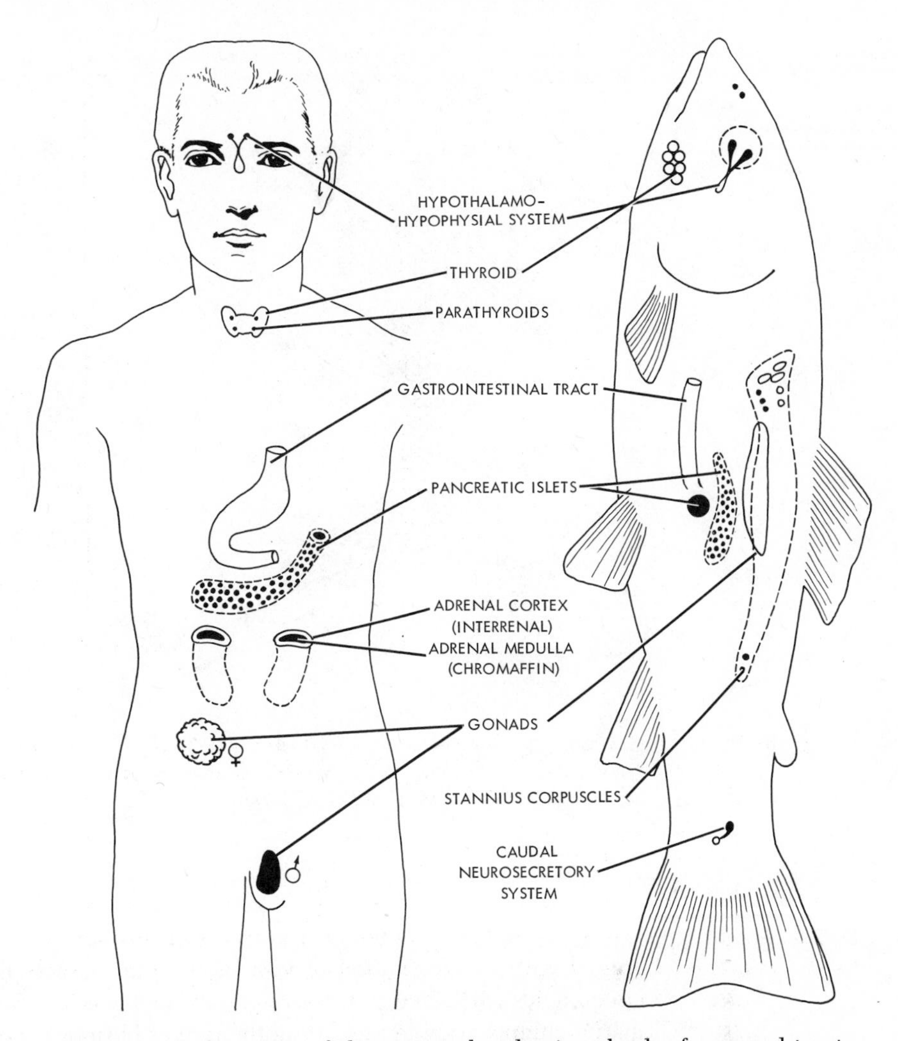

Figure 11-1 Location and comparison of the principal endocrine glands of man and trout. Note that the parathyroids are lacking in the fish; Stannius corpuscles and the caudal neurosecretory system are lacking in man and all other tetrapods; the functions of these structures are still incompletely defined.

Functional Organization of the Endocrine System

The most elementary control system possible consists of a minimum of two functional components: a means for receiving relevant sensory input and a means for generating a suitable command signal to the motor unit or effector. Most biological control systems, including the nervous system and most of the endocrine system, are more sophisticated than this and usually contain, in addition (1) an integration center (brain) where the total sensory input may be integrated and evaluated relative to the prospective response, (2) one or more feedback loops from the effector to the control circuit, and (3) one or more amplifying steps. If we analyze the endocrine system within this scheme, we find that three principal organizational patterns, with variations, have been utilized:

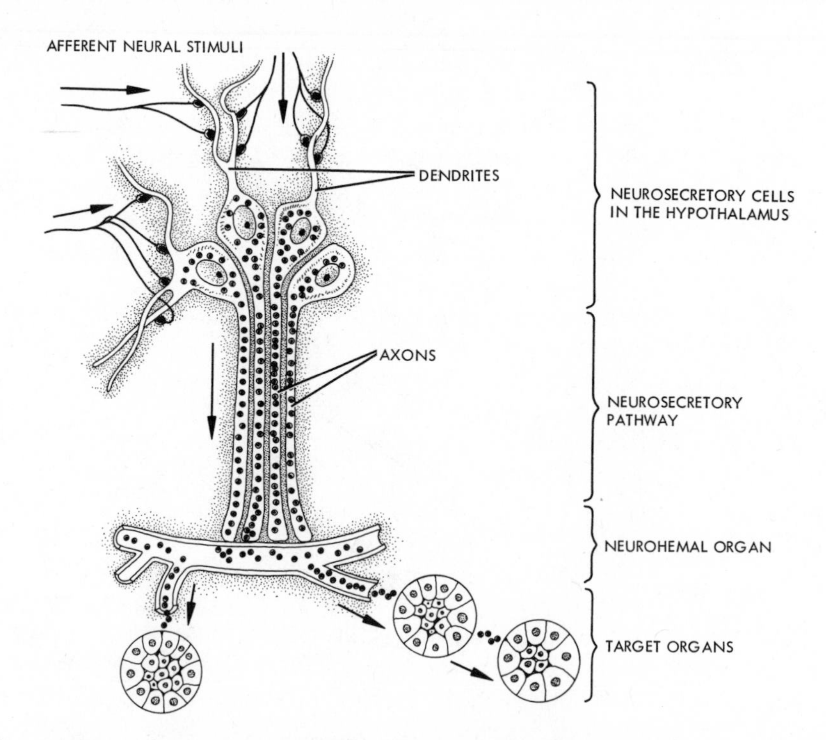

Figure 11-2 Diagram of neurosecretory cells. (From B. Frye, *Hormonal Control in Vertebrates.* New York: Macmillan, 1967. After E. and B. Scharrer, *Neuroendocrinology.* New York: Columbia University Press, 1963.)

SIMPLE NEGATIVE FEEDBACK LOOPS. Certain hormones are primarily concerned with the regulation of various substances such as salts, glucose, and amino acids contained in the body fluids. In some instances, the entire informational input used in regulating the rate of hormone secretion is derived by monitoring—by the endocrine gland itself—deviations in the concentration of the regulated substance from some preset norm. Deviation in one direction from the norm activates secretion of the hormone, which in turn activates mechanisms for restoring the normal concentration of the substance. Deviation in the opposite direction suppresses secretion of the hormone, and other mechanisms are activated that again push the level of the regulated substance toward normal. Such negative feedback loops usually are paired, involving one hormone that corrects for deviation in one direction and a second hormone that corrects for deviation in the opposite direction. This type of endocrine control mechanism is well illustrated by the regulation of blood glucose through the opposing actions of the pancreatic hormones, insulin, and glucagon (pages 549–550).

NEUROENDOCRINE CIRCUITS INVOLVING SECRETORY-MOTOR INNERVATION. Most of the sensory information relevant to endocrine function is transduced by sensory organs into afferent neural input into the central nervous system. But at some point in the efferent limb of the neuroendocrine stimulus-response arc a hormone is secreted that acts as the control signal. Such circuits involving both neural and hormonal components are known as neuroendocrine reflexes.* In at least one endocrine gland, the adrenal medulla, the transposition from neural to hormonal signals within the efferent pathway takes the form of direct secreto-motor innervation of the gland. The adrenal nerve, a branch from the sympathetic system, richly innervates the medulla, making direct synapses with the gland cells. Stimulation of the nerve results in secretion of adrenalin and noradrenalin, the medullary hormones, and blockage of the nerve inhibits secretion of these hormones.

Most neuroendocrine reflex circuits are not completed by means of direct secretomotor innervation, however. Indeed, the majority of vertebrate endocrine glands continue to function in an essentially normal manner even when totally denervated. The nerves that enter most endocrine glands in relative abundance are apparently strictly vasomotor and have no direct part in the control functions with which the glands are involved.

NEUROSECRETORY CIRCUITS. In the majority of neuroendocrine circuits, transition from afferent neural input to hormonal output is achieved by special *neurosecretory* cells located within the central nervous system. In many respects, these cells look like typical nerve cells, but they are also glandular in certain features, most notably in the large numbers of secretory granules they contain (Figure 11-2). At the dendritic end, the neurosecretory cell makes synaptic connections with nerve fibers that relay integrated sensory information to it from elsewhere in the central nervous system. The axonal end takes the form of a swollen axonal bulb that is replete with secretory granules as well as typical synaptic vesicles. The axon typically terminates in a *haemal organ*, a structure consisting mainly of a dense bed of capillaries and supporting connective tissue. The neurosecretory product is produced in the region of the cell body and packaged into the granules that are transported down the axon to the haemal organ. There it accumulates until the sensory information received by the neurosecretory cell causes it to be secreted into the blood stream. The products secreted are neurosecretory hormones that either act directly on the peripheral tissues or control the activities of other endocrine glands. The widespread occurrence of neurosecretory mechanisms, not only in vertebrates but in most phyla of multicellular invertebrates as well, suggests that the neurosecretory cell is especially adapted to make the transition from neural to hormonal pathways. Because of its key position in the nervous system and its nerve cell characteristics, the neurosecretory cell can receive many kinds of sensory information as relayed to it by neurons elsewhere in the central nervous system. Because of its gland cell properties, it can translate this information into a specific hormonal message and so become the final common pathway by which many kinds of external and internal stimuli can influence a particular hormone-controlled response.

The hypothalamus is the principal neurosecretory center of vertebrates. As the chief integrative center of the autonomic nervous system, the hypothalamus has control over a great variety of visceral functions. This includes control over a large segment of the endocrine system, which it governs through

*The use of the term *reflex* in this sense may be at variance with the original meaning of the term because considerable central nervous integration may be involved in so-called neuroendocrine reflexes. The term has, however, come into common use.

neurosecretory connections between the hypothalamus and the pituitary gland. The following major structural features of the hypothalamus are related to its neuroendocrine functions.

An Extensive Array of Afferent Connections with the Major Sensory Pathways and Integrative Centers of the Brain. These connections include the median forebrain bundle, the fornix, the mammillary peduncle, and diffuse thalamihypothalamic and pallidohypothalamic fiber tracts. Through these pathways sensory input of virtually every modality is relayed directly or indirectly to the hypothalamus.

Neurosecretory Nuclei. In birds and mammals, the two largest neurosecretory nuclei of the hypothalamus are the supraoptic and the paraventricular nuclei. In lower vertebrates, these two pairs of nuclei are represented by a single homologous pair of preoptic nuclei. They produce mainly the hormones of the posterior pituitary gland and will be discussed later. In addition, the anterior hypothalamus contains a number of other smaller neurosecretory nuclei whose secretions have to do primarily with the regulation of the activities of the anterior lobe of the pituitary gland. There appear to be at least six of these functional "areas" (not all have been precisely identified with specific morphological nuclei) associated with a corresponding number of anterior pituitary functions. The specific functional nature of these will be discussed in connection with the pituitary gland.

The Neurohypophysis. The floor of the hypothalamus, known as the infundibulum, is differentiated into a more or less complex neurohaemal organ known as the neurohypophysis. The axons of the neurosecretory cells located in the neurosecretory nuclei pass downward through the walls of the hypothalamus to terminate in the rich plexus of capillaries that penetrate the outer surface of the neurohypophysis. Most of the neurosecretory axons are collected into a number of bundles known as hypothalamohypophysial tracts. The largest of these are the supraopticohypophysial tracts from the large supraoptic and paraventricular nuclei, but a number of other minor tracts are present also from the smaller neurosecretory nuclei. Except for the cyclostomes, in which the neurohypophysis is relatively undifferentiated, the neurohypophysis of vertebrates is differentiated into two distinct haemal organs. These are the *neural lobe of the pituitary gland* and the *median eminence*. Most of the fibers of the supraopticohypophysial or preoptic fiber tracts terminate in the neural lobe, whereas the remainder of the hypothalamic neurosecretory fibers, including some from the supraoptic and preoptic tracts, terminate in the median eminence. This anatomical subdivision of the neurohypophysis is the basis for a functional subdivision, in that different neurosecretory hormones are secreted at different sites. In the tetrapods the circulatory drainage of these two sites is distinct. Blood leaving the plexus of the neural lobe enters the systemic circulation via branches of the anterior cardinal or internal jugular vessels, and so these hormones are carried to their peripheral targets. Blood leaving the median eminence, on the other hand, is collected into a group of minute *hypophysial portal vessels* that transport the blood directly into a secondary capillary plexus within the anterior pituitary gland. In fish, a portal circulation between the neurohypophysis and the pituitary exists that is functionally equivalent to that of tetrapods; but this generally takes the form of a capillary plexus that begins in proximity with the neurosecretory fibers in the surface of the median eminence and neural lobe and continues on into the anterior pituitary without anastomosing into distinct portal veins.

The direct vascular connection between the hypothalamus and the anterior pituitary is highly significant because the neurosecretory hormones that regulate the anterior pituitary gland are secreted in exceedingly minute amounts and are ineffective when diluted in the general circulation. For example, if the portal vessels are blocked, or if the anterior pituitary is transplanted to another site within the body, it no longer responds fully to the afferent input normally relayed to it by the hypothalamus and subsides into a low state of functional activity.

The neurosecretory hormones produced by the hypothalamus that control the adenohypophysis are known as releasing hormones. These are small peptides (3–12 or 15 amino acids) and include a TSH-releasing hormone, gonadotropin-releasing hormones (FSH and LH), an ACTH-releasing hormone, an STH-releasing hormone, and a prolacting-inhibiting hormone (see pages 530–531 for a description of the corresponding hormones of the pars distalis controlled by these releasing hormones).

With respect to the number of steps between the neurosecretory process in the hypothalamus and the final regulated function, neuroendocrine reflexes show several levels of complexity. The neurosecretory hormone may itself be the final command signal to the target tissue. This is a *first-order* neurosecretory function and is well illustrated by the functions of the posterior pituitary hormones. Or, the neurosecretory hormone may act on another endocrine gland (the pituitary) to stimulate the release of another hormone. If this hormone then acts as the final command signal to the motor unit, the circuit is known as a *second-order* neuroendocrine process. But, if the hormone produced by the second member of the relay acts on yet another gland, which then produces the hormone controlling the motor unit, it is a *third-order* neuroendocrine process. These functional relationships between the hypothalamus and the endocrine system are shown in Figure 11-3.

The functional organization of third-order neuroendocrine systems is complicated by the fact that negative feedback occurs between the target gland and the pituitary-hypothalamus complex (Figure 11-4). The chief advantage of this relationship, in which the target-gland hormone tends to suppress the output of the corresponding pituitary hormone, appears to be that it acts as a mechanism by which some autonomy from external sensory input can be obtained. This has two effects: on one hand, the feedback circuit may set the maximum level of the endocrine response to environmental stimulation; on the other hand, given the fact that the pituitary and hypothalamus are able to sustain some low level of function independently of specific stimulatory input, the feedback circuit provides the necessary information to keep the system functioning at some stable minimal (but perhaps adjustable) level. In mammals, at least, it is well established that basal levels of secretion of many hormones are required to sustain a basal state of morphological organization and metabolic activity within the target tissues. Such hormones can be regarded as having supportive or permissive functions in addition to commanding responses to specific afferent input via the neuroendocrine circuit. Which of

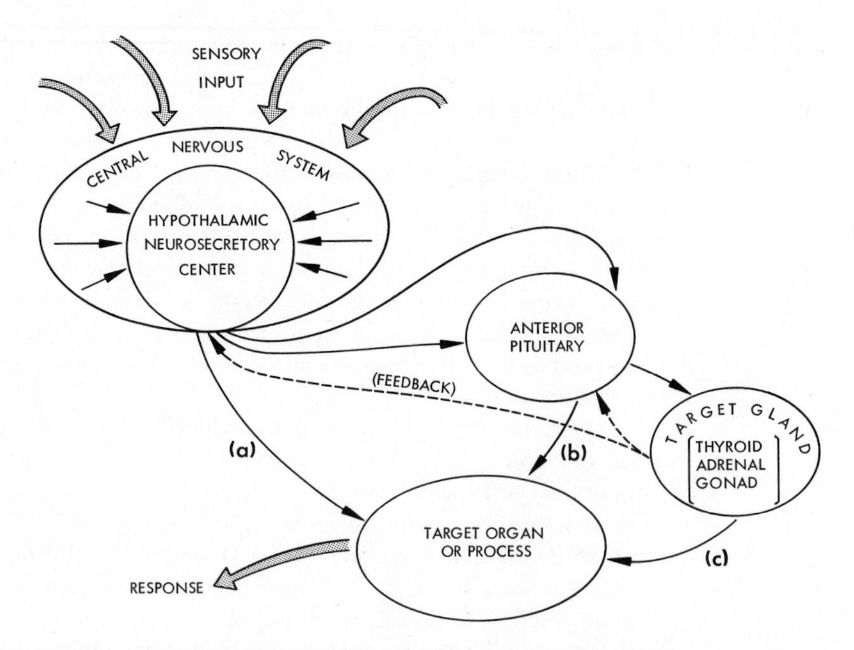

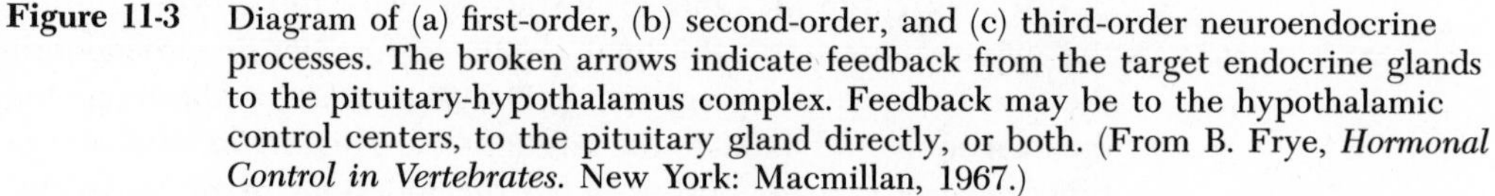

Figure 11-3 Diagram of (a) first-order, (b) second-order, and (c) third-order neuroendocrine processes. The broken arrows indicate feedback from the target endocrine glands to the pituitary-hypothalamus complex. Feedback may be to the hypothalamic control centers, to the pituitary gland directly, or both. (From B. Frye, *Hormonal Control in Vertebrates.* New York: Macmillan, 1967.)

these roles is the most primitive is impossible to say because very little data on the hormone dependence of tissues for minimal metabolic capabilities exist on animals below the mammals.*

The significance of the organization of neuroendocrine circuits into relays of several steps is not altogether clear. A mechanism for intrinsic feedback control is one of the properties of the system, as discussed in the previous paragraph. Other possible functions of the relay organization are *amplification* and *modulation.* Amplification occurs because each gland in the relay emits a "stronger" signal than it receives. Modulation of the frequency of the signal can occur because the time course of endocrine secretory processes is intrinsically much slower than neuronal signals. Thus, the efferent output reflects over-all trends in the sensory input, but not the minute fluctuations or "static" that are registered in the neuronal input. For example, gonadal activity in photoperiodically regulated annual reproductive cycles reflects progressive changes in day length, but not daily fluctuations in light intensity, nor even sporadic discrepancies in the pbotoperiod within the general trend of lengthening or shortening days.

Evolution proceeds historically, however, and not merely with the mechan-

* A perplexing problem that arises here is why limitations on the basal metabolic capabilities of cells have been given over to some extent to systemic control by hormones rather than being retained intrinsic to the cells themselves. The probable answer is that the "minimal" or "basal" level required for functional viability changes with the seasons as activity states change. Consequently, a fixed limit would be adaptively inappropriate, and systemic control that can make use of sensory information about the seasons provides a mechanism for the adaptive resetting of the metabolic state throughout the body.

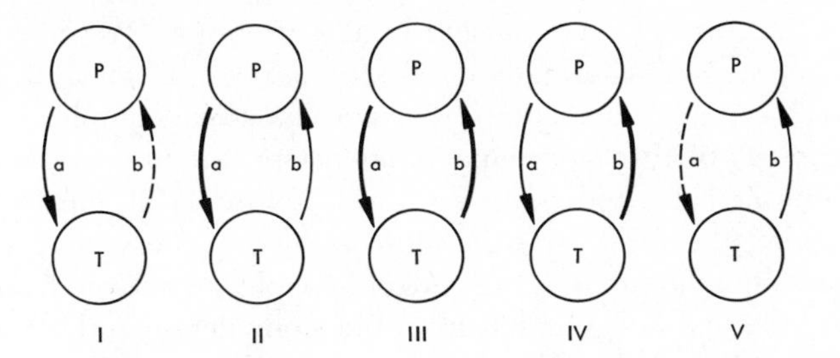

Figure 11-4 Feedback interaction between the pituitary and the target endocrine glands. I to V represent different stages of the interaction. (From B. Frye, *Hormonal Control in Vertebrates*. New York: Macmillan, 1967. After E. and B. Scharrer, *Neuroendocrinology*. New York: Columbia University Press, 1963.)

istic objective of solving a current problem the "best" way. There is no a priori reason why the advantages of feedback, amplification, and modulation cannot be achieved within the limits of the organization of first-order neurosecretory processes, and indeed they are. Consequently, these functional considerations do not, of themselves, account for the evolution of the particular patterns of endocrine organization that we have been discussing. Equally important is the question: What came first? It is very likely that the last members in the relay—adrenals, thyroid, and gonads—initially evolved as autonomous glands with regulatory effects on adjacent, ontogenetically related tissues and that they only secondarily came under neuroendocrine control. For example, gonadal steroids have important direct effects on the gametogenic functions of the gonad that could have favored the differentiation of gametogenic and steroidogenic components of the gonad independently of the existence of systemic control by gonadotropic hormones.

The evolution of the hypothalamic-pituitary-neurosecretory complex is an enigma. The association between these two glands is already present in the most primitive living vertebrates. But, inasmuch as the evolutionary history of the association between the pituitary and the hypothalamus is completely obscure, it is possible that (1) this association is indeed primitive or that (2) an earlier stage saw the development of a functional association between the pituitary and its target glands, and the neurosecretory unit was the final link to be connected.

The Endocrine Glands of Vertebrates

Most groups of multicellular animals are known to produce hormones that regulate metabolism in ways analogous to the functions of the endocrine system of vertebrates. However, the specific set of endocrine glands possessed by the vertebrates is unique and appears to have had its evolutionary development almost entirely within the vertebrates. Most of the hormones produced by these glands are also characteristic of the vertebrates. It is true that certain hormones of simpler chemical structure, notably adrenalin and iodine-containing amino acids found in the thyroid, are known to occur in some invertebrates; and steroids very similar to those produced by the adrenal and the gonads are found among the invertebrates as well as in flowering plants. These coincidences, however, are probably cases of evolution of parallel biosynthetic pathways rather than homology.

The Pituitary Gland

A pituitary gland is present in all vertebrates. Its unique relationship to the nervous system, already discussed, makes it a central figure in the vertebrate endocrine system. The pituitary is actually a complex of glandular tissues. It consists of three principal subdivisions, usually referred to as the anterior, intermediate, and posterior lobes (Figure 11-5). On the basis of embryonic origin, the pituitary may more accurately be said to consist of two major anatomic divisions: (1) the adenohypophysis, derived from Rathke's pouch, an ectodermal invagination from the stomodeum; and (2) the neurohypophysis, derived from the infundibulum of the diencephalon of the embryonic brain (Figure 11-6). The adenohypophysial rudiment grows posteriorly from its origin along the floor of the brain until its posterior margin contacts the infundibulum. That portion of the rudiment that contacts the infundibulum gives rise to the intermediate lobe, or *pars intermedia*, of the pituitary, and the remainder of the rudiment forms the anterior lobe, or *pars distalis*. In addition, in most vertebrates, lateral outgrowths from the body of the adenohypophysial rudiment grow upward and anteriorly around the stalk of the pituitary to form a minor, but anatomically distinct, component known as the *pars tuberalis*.

The neurohypophysis is derived primarily from the ependymal layer of the infundibular evagination, but by the time its development is complete it has become heavily invaded by capillaries and by the axonal endings of the neurosecretory cells of the hypothalamus. As has already been noted, the neurohypophysis becomes differentiated in most vertebrates into two subdivisions, the neural or posterior lobe of the pituitary (*pars nervosa*) and the median eminence. In those species in which the infundibulum evaginates far

Figure 11-5 The pituitary gland of a frog tadpole. (From B. Frye, *Hormonal Control in Vertebrates*. New York: Macmillan, 1967.)

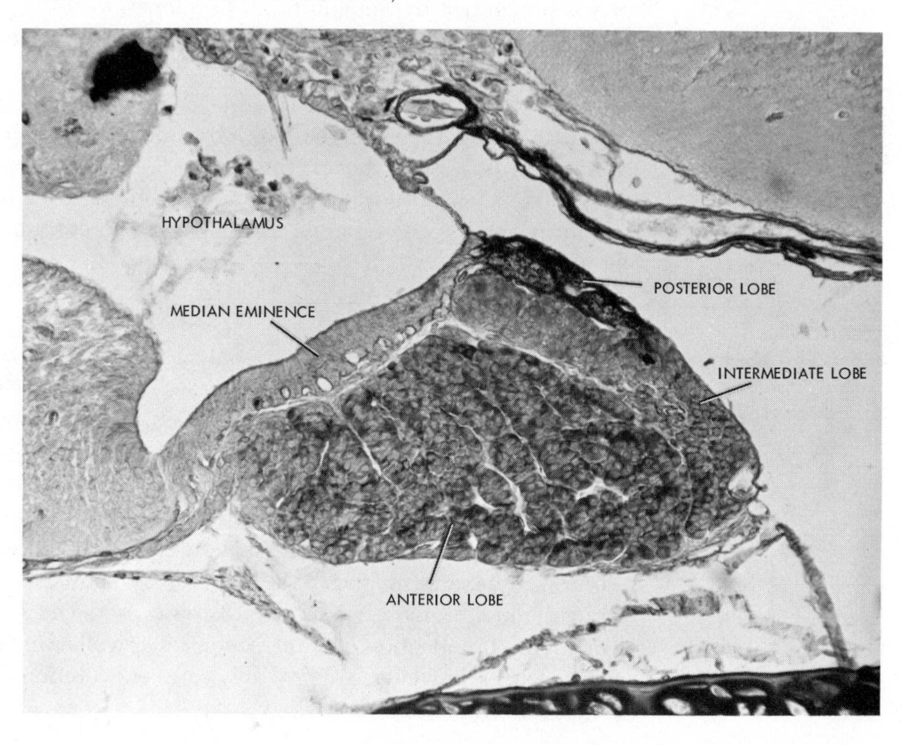

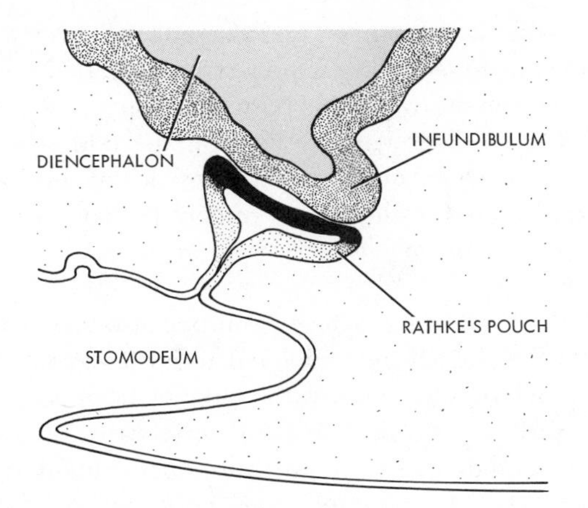

Figure 11-6 Embryonic origin of the pituitary. Neurohypophysial rudiment or infundibulum stippled; pars intermedia solid; pars distalis open. Later in development the epithelial stalk formed during evagination of Rathke's pouch from the stomodeum will disappear.

enough for the neural lobe to become dependent from the brain (as in man), the connecting portion is known as the pituitary stalk. The stalk is traversed by the nerve fibers and some of the blood vessels that enter the posterior lobe.

In addition to these more or less universal features, certain exceptional features of the pituitary of some groups of vertebrates should be noted.

1. In cyclostomes, the neurohypophysis is neither distinctly evaginated from the floor of the brain nor clearly differentiated into a median eminence and a neural lobe of the pituitary. It consists of a thickened area in the floor of the diencephalon behind the optic chiasma. However, because of its position and because it receives neurosecretory fibers from the hypothalamus and is closely associated with the adenohypophysis, there is no doubt of the general homology of this region with the neurohypophysis of higher vertebrates.
2. In all fish classes above the cyclostomes, the neurohypophysis comprises, in addition to the median eminence and the posterior lobe, a large, thin-walled, nonglandular structure known as the saccus vasculosus. The function of this structure is not known, but because it is derived embryologically from the posterior portion of the infundibulum, it can be regarded in part as homologous with the pars nervosa.
3. In elasmobranchs, there extends downward from the main body of the pituitary gland a large lobe known as the ventral lobe. This lobe is formed from bilateral outgrowth of the adenohypophysial rudiment, and it may be homologous to the pars tuberalis. The ventral lobe secretes gonadotropic hormones, and so is functionally equivalent to the gonadotropin-secreting cells of the pars distalis of other vertebrates. The significance of the anatomical separation of this tissue from the main body of the pituitary in elasmobranchs is not known.
4. In birds and a few species of mammals (whales, elephants) the intermediate lobe of the pituitary is absent. In these forms, the posterior margin of the adenohypophysial rudiment fails to make intimate contact with the infun-

dibulum during development, and a layer of connective tissue usually forms between the pars nervosa and the pars distalis. In amphibians, formation of the pars intermedia from the posterior margin of the adenohypophysial rudiment has been shown to result from an inductive interaction with the infundibulum, so we assume that the absence of this lobe in birds and some mammals is related to the failure of contact between these two rudiments during development.

PARS DISTALIS. Six hormones have so far been isolated from the pars distalis of mammals. On the basis of the biological activities of pituitary extracts and the consequences of hypophysectomy, the same six hormones seem to be present in the anterior pituitary of all classes of vertebrates. But the actual chemical identities and homologies of these hormones cannot be claimed until the molecules have been completely purified and their amino acid sequences compared.

Four of the six anterior pituitary hormones are concerned primarily with the regulation of other endocrine glands through third-order neuroendocrine circuits. These include a thyroid-stimulating hormone (TSH or thyrotropin), an adrenocortical-stimulating hormone (ACTH or corticotropin), and two gonadotropins, the follicle-stimulating hormone (FSH) and the luteinizing hormone (LH, also called interstitial cell-stimulating hormone, or ICSH). Each of these hormones is highly specific for its effect on the corresponding target gland, where the over-all effect is to stimulate secretion of the target-gland hormones. This includes not only a stimulation of the discharge of the target-gland hormone, but stimulation of the entire functional apparatus of the gland cells in such a way as to support continued or increasing rates of hormone secretion. Morphologically, strong stimulation by trophic hormones is expressed as enlargement of the target glands. Hypophysectomy, conversely, results in deterioration of the intracellular secretory apparatus and atrophy of the target glands. The target-gland hormones, in turn, may suppress the secretion of the corresponding trophic hormones by the pituitary. Administration of an excess of any given target-gland hormone suppresses the pituitary cell type responsible for secreting the corresponding trophic hormone, and elimination of the target-gland hormone (as by surgical removal of the thyroid, adrenals, or gonads) results in enlargement and proliferation of the corresponding pituitary cell. Although this negative feedback relationship is a very important factor in the regulation of the functional state of the endocrine system, it can be superseded by neurosecretory control from the hypothalamus in response to appropriate afferent input.

The remaining two hormones of the anterior pituitary are prolactin and somatotropin. Prolactin (mammotropic hormone; luteotropic hormone or LTH) has a remarkable spectrum of effects throughout the vertebrates. In mammals, it promotes milk secretion by the mammary glands and also stimulates lipid synthesis in adipose tissue. In a few species of mammals, prolactin stimulates the secretion of steroid hormones by the corpus luteum of the ovary, and it is for this reason that it has been called luteotropic hormone and is sometimes regarded as a third gonadotropin. Prolactin stimulates lipid synthesis in certain birds and it may be involved in regulating premigratory fattening of migratory birds. In columbid birds, it stimulates the secretion of epithelial "milk" by the crop glands, and in certain other birds it stimulates brooding and the formation of the brood patch (a highly vascularized, defeathered area of the skin used in incubating the eggs). In reptiles and in larval amphibians, prolactin promotes growth and regeneration. In newts, it stimulates the so-called water-drive response, a behavioral response involved in the transformation of the terrestrial red-eft stage to the aquatic adult state of the life cycle. Finally, in teleosts, prolactin has a role in osmoregulation, and in one species, the discus fish, stimulates the proliferation of the epidermal discus gland on which the young feed. Even from this incomplete list (more than forty effects of prolactin can be tabulated among the vertebrates), it is obvious that many seemingly unrelated actions of this hormone have evolved.

The most dramatic effect of somatotropin is, as the alternative name, growth hormone, suggests, to promote growth. This effect is best demonstrated in mammals. In the absence of this hormone, dwarfism results, and with an excess, as may occur with certain types of pituitary tumors, either giantism or acromegaly results. If the excess occurs in a growing individual, a nearly proportionate acceleration of bodily growth results, and the condition is known as giantism. In the adult in which growth has ceased normally, the response to excess STH is uneven; joints enlarge, bones grow in width but not length; the jaw, nose, ears, and viscera enlarge; and the resulting distorted condition is known as acromegaly. The anterior pituitary produces a growth-promoting hormone in other vertebrates, but whether this is always a molecule homologous with the mammalian growth hormone is not known. As was mentioned before, prolactin may be an important growth hormone in some lower vertebrates. Somatotropin also has important metabolic effects, and these occur in adult as well as growing individuals. It promotes the release of fat from adipose tissue and, at the same time, interferes with the utilization of glucose and stimulates protein synthesis. Through these effects, STH interacts in a complementary way with insulin to regulate the balanced use of foodstuffs. Somatotropin secretion in the adult is promoted by the same circumstances (notably, a fall in blood glucose) that cause a decline in insulin secretion.

Pars Intermedia. The intermediate lobe of the pituitary secretes a hormone known as intermedin, or melanophore-stimulating hormone (MSH). This is a peptide hormone, generally consisting of nineteen amino acids, but with some variation in exact structure among the various species in which it has been chemically identified. In the poikilotherms, MSH has an important function in environmental adaptation by regulating physiological color change in response to the degree of lightness or darkness of the background. The primary target of MSH is the melanophore, a black pigment-containing cell in the skin. Intermedin stimulates darkening of the skin by causing dispersal of the melanin pigment granules into the arms of these highly branched cells. In the absence of intermedin, the pigment becomes tightly clumped in the center of the cell, and the animal becomes lighter. The melanophore response to MSH is relatively slow, requiring several minutes, and occurs more or less uniformly throughout the skin. The very rapid and mosaic patterns of color change that some fish and reptiles exhibit are controlled by direct innervation of the pigment cells.

Pigmentation of the skin of birds and mammals is entirely structural, resulting from the deposition of pigment in the skin, feathers, or hair by pigment-secreting melanocytes and to the physical defraction of light by feathers or hair. Thus, rapid physiological color change cannot occur. MSH probably affects the degree of pigmentation of the skin of mammals by affecting the amount of pigment produced by the melanocytes. There is some evidence that the hormone in this way is involved in the seasonal color change that may occur during spring and fall moulting. Birds do not have MSH—we have already noted the absence of the pars intermedia in this group—but the thyroid and sex hormones have been shown to effect the pigmentation of the feathers.

Based on structural similarities, particularly similarities in amino acid sequence, the adenohypophysiol hormones fall into three groups of homologous compounds: TSH, FSH and LH comprise one group, ACTH and MSH a second, and STH and prolactin a third.

PARS NERVOSA. As has been indicated, the hormones of the posterior lobe of the pituitary are actually neurosecretory products that originate in the supraoptic and paraventricular nuclei of the hypothalamus or in lower vertebrates—the equivalent of these nuclei. The hormones of the posterior pituitary are nonapeptides (they consist of nine amino acids).*

The posterior pituitary hormones that have been extracted from among the various classes of vertebrates can be arranged into a phyletic series of closely related peptides (Figure 11-7). These all appear to have evolved from a single ancestral molecule, possibly arginine vasotocin (because it is found in all vertebrates from the cyclostomes up) by substitutions of different amino acids at one or more positions. The significance of this evolutionary series is not completely known because the functions of the molecules in the lower verte-

* Because two of the amino acids are cysteine molecules that are joined in a disulfide bond to form a single molecule of cystine, the posterior lobe hormones often have been regarded as octapeptides, which consist of eight amino acids.

Figure 11-7 Posterior pituitary hormones of various vertebrates.

1. LYSINE VASOPRESSIN (SUINA)

CyS—Tyr—Phe—Glu(NH_2)—Asp(NH_2)—CyS—Pro—Lys—Gly(NH_2)

2. ARGININE VASOPRESSIN (ALL MAMMALS)

CyS—Tyr—Phe—Glu(NH_2)—Asp(NH_2)—CyS—Pro—Arg—Gly(NH_2)

3. ARGININE VASOTOCIN (ALL VERTEBRATES EXCEPT MAMMALS)

CyS—Tyr—Ileu—Glu(NH_2)—Asp(NH_2)—CyS—Pro—Arg—Gly(NH_2)

4. OXYTOCIN (LUNGFISH AND ALL TETRAPODS)

CyS—Tyr—Ileu—Glu(NH_2)—Asp(NH_2)—CyS—Pro—Leu—Gly(NH_2)

5. TELEOST OXYTOCIN-LIKE PRINCIPLE

CyS—Tyr—Ileu—Ser—Asp(NH_2)—CyS—Pro—Ileu—Gly(NH_2)

brates and, therefore, the selective pressures that may have directed their evolution are largely unknown. The close anatomical and vascular association of the pars nervosa with the pars distalis in fish suggests that the original function of the posterior lobe peptides may have been the regulation of the pars distalis. Subsequently, this function may have been given over completely to the median eminence, whereas the pars nervosa and its hormones evolved in the direction of independent effects on peripheral tissues. Ascending the vertebrate line, the first *known* case of a physiological function of the pars nervosa independent of the pars distalis is seen in the amphibians, and it may be significant that this is also the class in which clear anatomical and vascular independence of the pars nervosa from the pars distalis is first seen.

In mammals, the two important posterior pituitary hormones are oxytocin and vasopressin (arginine vasopressin in most mammals, but both arginine and lysine vasopressin in the Suina; so far as is known arginine and lysine vasopressin are functionally equivalent), which have the following major functions: oxytocin increases the intensity of contraction of the myometrium of the uterus and thus promotes labor. Exactly how oxytocin secretion is stimulated at the time of labor is unknown, however, so this neuroendocrine reflex remains to be completely described. Oxytocin also stimulates the ejection of milk from the lactating mammary gland by stimulating the contraction of smooth muscle fibers around the secretory alveoli. This reflex is initiated by stimulation of the nipples and, in the case of humans, by the sound or sight of the crying infant.

Vasopressin promotes the reabsorption of water from the forming urine in the kidneys, and is thus aptly called the *antidiuretic hormone*. The effect of vasopressin is to increase the permeability of the renal collecting ducts to water so that water is withdrawn from the urine by osmosis as the urine passes through the hyperosmotic medullary zone of the collecting ducts (see Chapter 8). In the absence of vasopressin, these ducts are relatively impermeable to water, so the urine passes through them unaltered and the surplus water is voided. The secretion of vasopressin is stimulated by either a drop in blood volume or by a rise in the concentration of solutes in the blood, either of which can reflect dehydration and appropriately activate this neurosecretory reflex that promotes water conservation (Figure 11-8). Historically, the first effect of vasopressin to be discovered was that it elevates blood pressure, and for this reason the name vasopressin was given to the hormone. We now know that vasopressin has little effect on the blood pressure except when administered in abnormally large doses, and it is believed that antidiuresis is the major physiological function of the hormone.

In birds, reptiles, and amphibians, vasotocin is the antidiuretic hormone. In amphibians, in addition to reducing the rate of urine excretion in the kidney, vasotocin elicits two other responses that are related to the maintenance of water balance: (1) it makes the wall of the bladder permeable to water so that the very dilute urine stored here can be redrawn osmotically into the body fluids; (2) it increases the permeability of the skin to liquid water so that any available moisture is rapidly absorbed (again osmotically) into the body. In birds, vasotocin has an additional role in reproduction. It stimulates contraction of the muscles of the "uterus," causing egg expulsion. Vasotocin is present in the pituitaries of fish but, as mentioned before, its function in these classes is unknown. If vasotocin is injected into mammals it causes both oxytocic and antidiuretic effects. The evolutionary "replacement" of vasotocin by vaso-

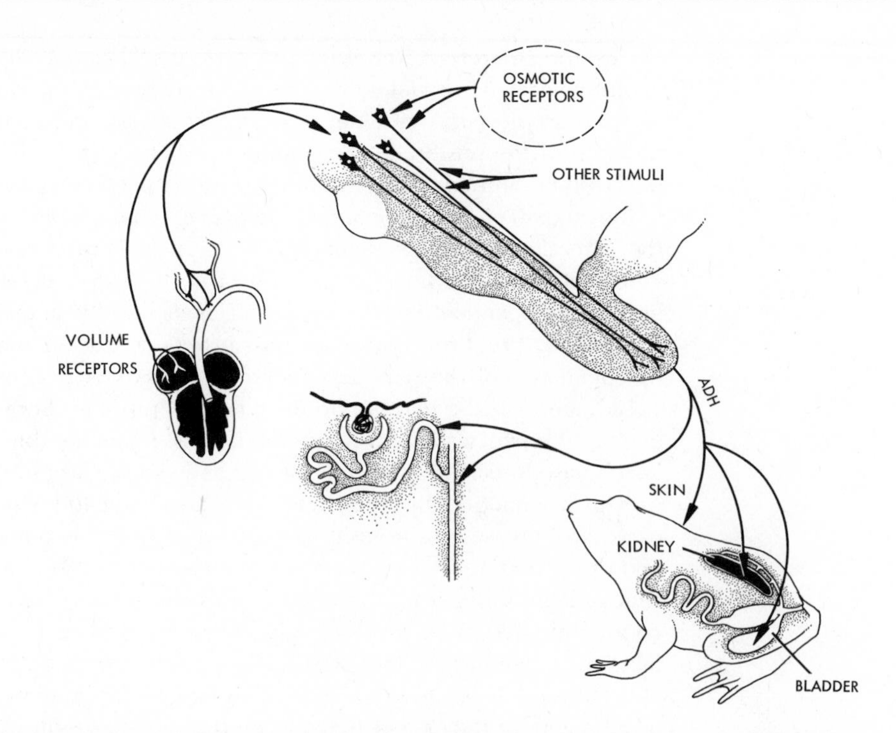

Figure 11-8 A summary of the neuroendocrine reflexes that control the water-balance response. This figure depicts the amphibian response mechanism, which differs from that of mammals in the absence of a loop of Henle and in the effects of ADH on the skin and bladder (see the text). (From B. Frye, *Hormonal Control in Vertebrates.* New York: Macmillan, 1967.)

pressin in mammals would appear to have been favored because it achieved a more distinct separation between oxytocic and antidiuretic functions than had previously existed.

The Thyroid Gland

In all vertebrates, the thyroid originates as an outgrowth from the floor of the pharynx at the level of the first and second gill pouches (Figure 11-9). The position and gross form of the gland in adult animals vary considerably. Generally, the thyroid is located posterior to the pharyngeal area, adjacent to the carotid arteries or their major branches. Cyclostomes and most teleost fish are exceptional in that the thyroid, being unencapsulated by connective tissue, tends to subdivide into tissue fragments that become scattered widely through the central pharyngeal region of the animal. In all other vertebrates, includiing parrot fish and elasmobranchs, however, the thyroid consists of either a single median lobe or paried lobes located anterior to the heart.

The position and form of the thyroid are apparently without particular functional significance because thyroid tissue transplanted to other sites in the body functions normally and regulates the physiology of thyroidectomized animals perfectly well so long as the transplants have a good blood supply. The anatomical location of the thyroid is better explained by its homology with the endostyle of protochordates. This structure, an exocrine mucous gland running as a groove along the floor of the pharynx in animals such as *Amphioxus* and *Ciona*, contains groups of cells that bind large quantities of iodine—a charac-

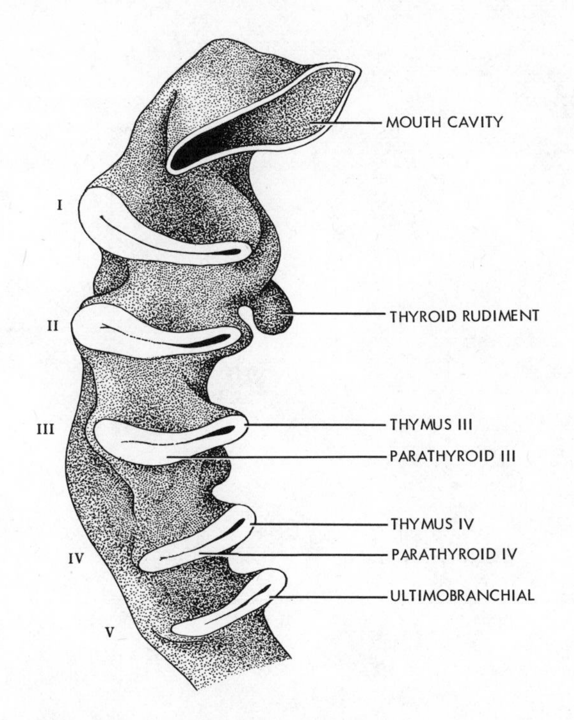

Figure 11-9 Embryonic origin of the thyroid and other pharyngeal derivatives. (After A. Gorbman and H. Bern, *Textbook of Comparative Endocrinology*. New York: Wiley, 1962.)

teristic property of the thyroid—and also synthesize certain iodine-rich compounds such as those found in the thyroid. Apparently, in these protochordates, iodine-rich substances are secreted into the endostylar groove where they are swept into the mucosal food stream and on into the gut, from which they are absorbed into the circulation. Whether the iodinated compounds secreted by the endostyle have any biological function in the protochordates is unknown. Possibly, the significance of the thyroidlike activities of the endostyle is in the iodine-trapping and conserving mechanism, and the hormonal activities of the products may be an acquisition within the vertebrates. The homology of the thyroid gland with the iodine-accumulating cells of the endostyle is supported strongly by the developmental origin of the thyroid in cyclostomes. In the ammocoetes, the larval stage of lampreys, there lies in the floor of the pharynx an endostyle (or subpharyngeal gland) equivalent to that of protochordates, although it is more complex. Like the protochordate endostyle, certain nonglandular cells in this organ accumulate iodine and synthesize "thyroid" compounds. At the time of metamorphosis into the adult lamprey, the mucus-secreting portion of the endostyle degenerates, but the iodine-accumulating cells become transformed into typical vertebrate thyroid tissue.

The microscopic anatomy of the thyroid gland is remarkably similar in all vertebrates: its cells are arranged into hollow vesicles known as follicles, the cavities of which, in most vertebrates, are filled with a viscous secretion known as colloid (Figure 11-10). An exception is the hagfish. The follicular organization of the thyroid is related functionally to the storage of iodinated compounds and to certain aspects of synthesis and secretion of hormones. The

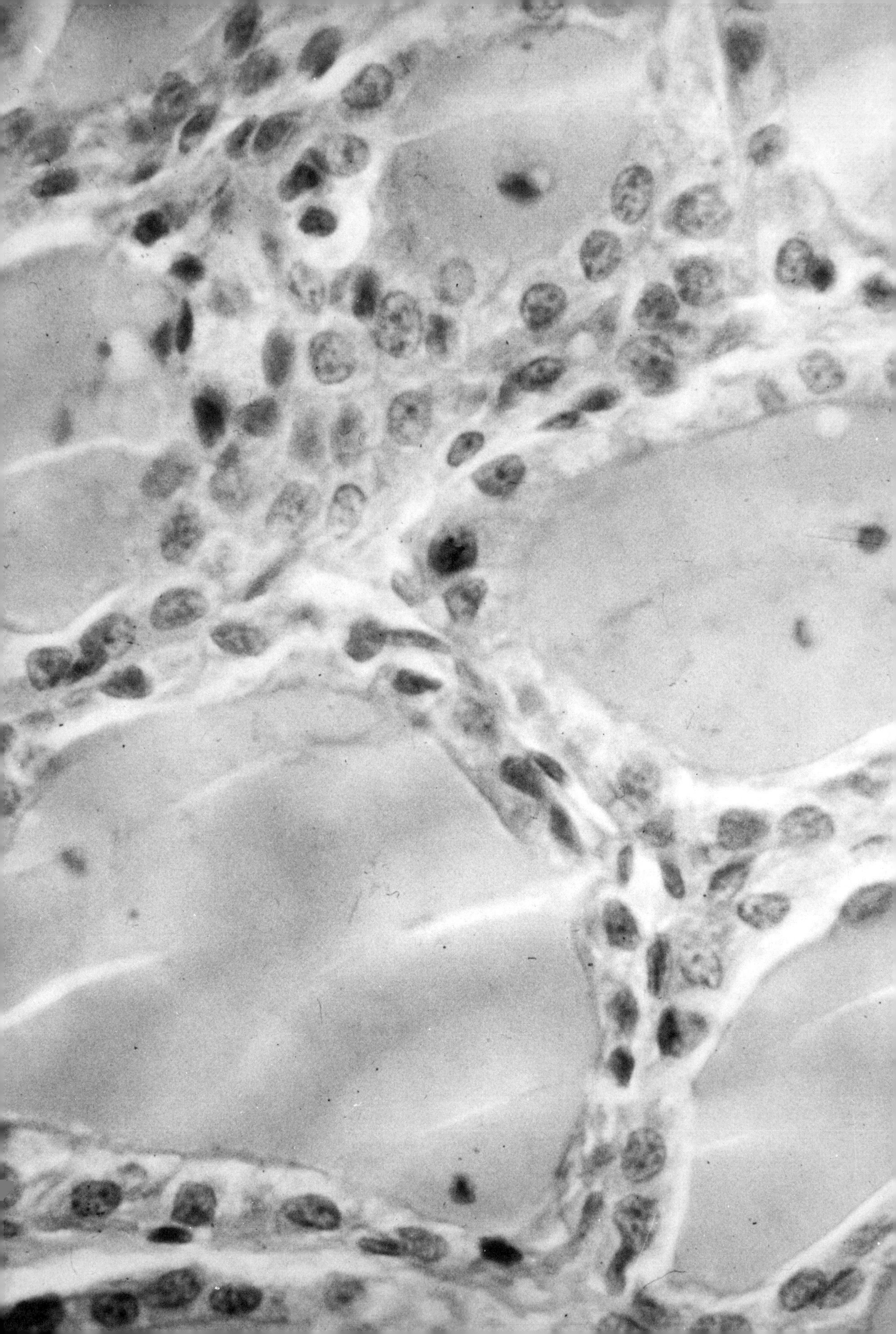

colloid consists of a semigel, mainly of the protein, thyroglobulin. The latter, upon hydrolysis, yields a mixture of amino acids that includes particularly large proportions of iodinated amino acids—iodotyrosines, which are precursors of the thyroid hormones; and iodothyronines, including thyroxine, which are the hormonally active substances (Figure 11-11).

The primary selective force in the evolution of a follicular organization of thyroid cells may have been the relative scarcity of iodine in some environments. This would place value on an organization of cells in a gland that not only avidly accumulates iodine from the environment but also stores its essential hormonal products in a form readily accessible during periods of iodine shortage. Alternatively, one can argue that the follicular organization is one way whereby a gland already specialized for exocrine secretion can be "turned into" an endocrine gland. Thus, we have already seen that the endostyle of protochordates and ammocoetes synthesizes thyroid substances but liberates them into the gut. In the vertebrate thyroid gland, the products similarly are secreted in an exocrine manner, but into the blind cavities of the follicles and only secondarily reabsorbed from these cavities across the thyroid cells into the capillaries that surround each follicle. In a sense, then, exocrine secretory mechanisms preexisting in the endostyle have been retained in the organization of the thyroid, and endocrine function has been acquired by adding additional steps by which the products can be transferred from the lumen of the gland into the circulation.

The synthesis and secretion of thyroid hormones involves several steps: (1) active uptake of inorganic iodide salts from the blood by the thyroid; (2) oxidation of the iodide to an active elemental form—possibly iodine itself; (3) synthesis of thyroglobulin and secretion of this protein into the follicle lumen; (4) combination of the active iodine with the tyrosine residues of thyroglobulin; (5) coupling of two molecules of iodotyrosine to form the various hormonally active iodothyronines shown in Figure 11-11; and (6) hydrolysis of thyroglobulin by proteolytic enzymes to liberate its component amino acids, including the hormonally active amino acids. These then enter the blood stream, presumably by diffusion.

In a general way, the size of the thyroid cells reflects the state of activity of the thyroid. In the inactive state, the cells regress to a low squamous state, whereas in stimulated glands the cells enlarge to a cuboidal or columnar state. In warm-blooded animals, the normal condition of the thyroid epithelium is cuboidal. However, under certain circumstances, such as dietary iodine deficiency, the cells and the entire gland may undergo enlargement to form a goiter. In cold-blooded vertebrates, the thyroid generally is most highly developed during the warm seasons and regresses to a low squamous state during winter. However, both positive and negative correlations between thyroid activity and certain phases of the life cycle, such as migrations or reproduction, make the pattern of thyroidal activity more complex than mere seasonal waxing and waning.

The physiological effects of the thyroid hormones cannot be summarized accurately for the vertebrates as a group because the effects are so diverse. In

Figure 11-10 (*opposite*) Thyroid follicles. (From B. Frye, *Hormonal Control in Vertebrates*. New York: Macmillan, 1967.)

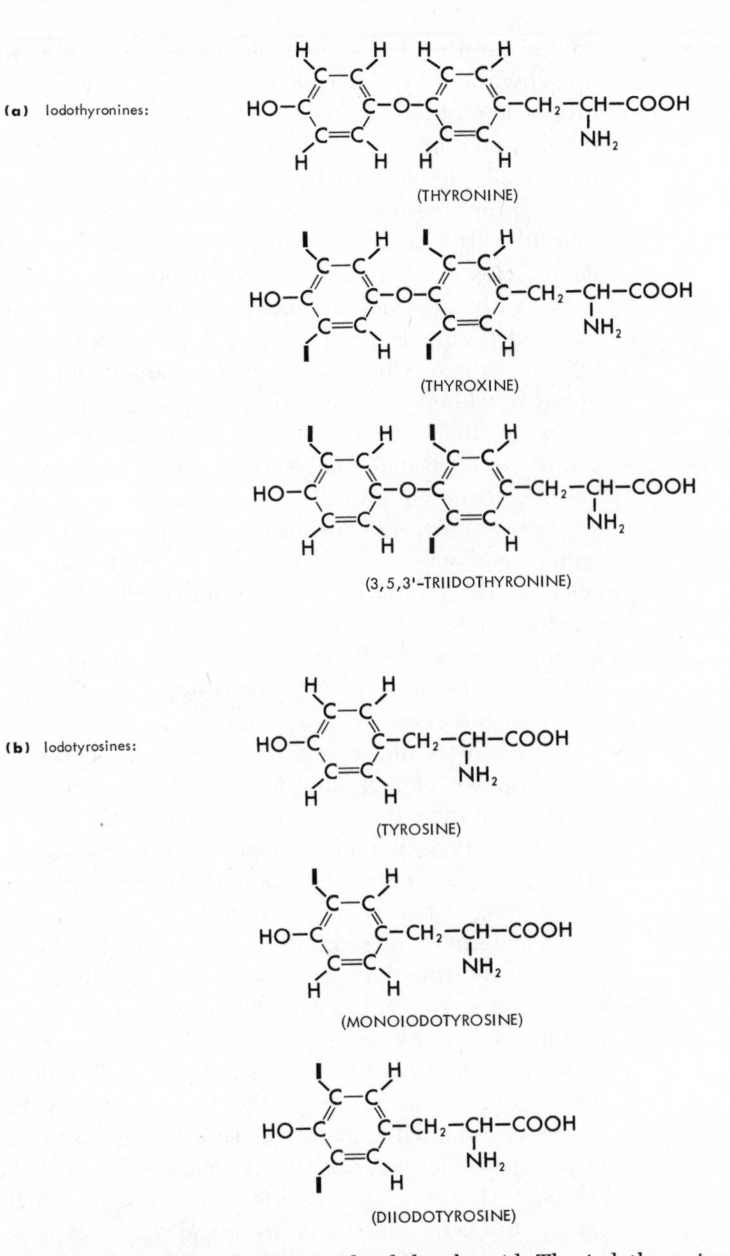

Figure 11-11 Thyroxine and related compounds of the thyroid. The iodothyronines, thyroxine and 3,5,3′-triiodothyronine, are the major hormonally active substances. Monoiodotyrosine and diiodotyrosine function as precursors in the biosynthesis of thyroxine and triiodothyronine, and are not normally secreted by the gland. (From B. Frye, *Hormonal Control in Vertebrates*. New York: Macmillan, 1967.)

birds and mammals, at least, three principal classes of effects are well established.

Effects on Oxidative Metabolism. In warm-blooded animals the thyroid hormone stimulates oxygen consumption and heat production in the tissues. The basal rate of metabolism may be elevated 50 or 100 per cent over normal by injections of the thyroid hormones and falls as low as 40 per cent below normal after thyroidectomy or thyroid inhibition. This effect has been shown to result from the action of the hormones on the mitochondria, but there are several as yet unreconciled opinions as to the nature of this effect. Many functions have been shown to be dependent on normal levels of thyroid hormone. A partial list includes nervous function, muscle contractility, reproduction, glandular secretion, and growth and development. It is possible that all these effects are consequences of the basic role of the thyroid hormones in supporting normal capabilities for energy metabolism. Most attempts to demonstrate an effect of the thyroid hormones on energy metabolism in cold-blooded vertebrates have failed, although there are a few reports of stimulation of oxygen consumption in fish, amphibians, and reptiles. It must be remembered that cold-blooded animals do not exhibit a "basal" rate of metabolism and do not liberate a surplus of metabolic energy as heat. Possibly, in these organisms, therefore, hormonal effects on energy production can be demonstrated only under test conditions that allow the energy produced to be utilized in various forms of metabolic work, such as movement, growth, or synthesis. If this is so, then the failure to demonstrate consistently an effect of thyroid hormones on mitochondrial energy metabolism in cold-blooded vertebrates may be due to inadequate experimental designs rather than to a real lack of this effect. Recent studies suggest that the temperature at which experiments are run may make a critical difference in whether an effect of thyroxine is observed on oxygen consumption.

EFFECTS ON GROWTH AND DEVELOPMENT. In birds and mammals, normal development is dependent on the thyroid hormones. In the absence of these hormones, growth is drastically retarded, and maturation of the skeletal and nervous systems is particularly delayed. The result is a condition of infantile physiognomy and mental retardation known as *cretinism*. Severe, untreated human cretins may reach an age of fifteen or twenty years, but, their stage of physical and mental development is equivalent only to that of an infant of one or two years. Although not many reptiles have been studied, their development seems to depend similarly on the thyroid. In fish, a role of the thyroid in supporting growth has been reported for several species, but the failure to demonstrate such a role has been reported even more often.

The most dramatic effect of thyroid hormones on vertebrate development is the effect on amphibian metamorphosis. Precocious metamorphosis of amphibian larvae can be induced by administering the thyroid hormones, and metamorphosis can be prevented altogether by thyroidectomy. This key control mechanism over metamorphosis is activated by the maturation of hypothalamic neurosecretory centers that regulate the secretion of TSH. The thyroid hormones act on the larval tissues directly to activate metamorphic changes. A unified, synchronous pattern of metamorphosis of the whole animal is brought about by the fact that the level of circulating thyroid hormone rises progressively during metamorphosis, and the sensitive tissues each have different inherent sensitivities and rates of response to the hormone.

EFFECTS ON THE INTEGUMENT. The thyroid hormones are necessary for normal skin proliferation and pigmentation in all vertebrates. In birds and mammals, hair or feather growth is inhibited by a deficiency of the thyroid hormone, and melanin pigment deposition is reduced. The skin becomes thinner, and normal moulting is interferred with. Moulting also is blocked by thyroid deficiency in amphibians and reptiles. In fish, the thyroid hormone promotes the thickening of the epidermis and the deposition of silvery guanine pigments in the skin. The physiological significance of these integumentary effects is not clear in all cases, but because they are universally observed they must be recognized as one of the more generalized classes of effects of the thyroid hormone.

In recent years, the thyroid gland of mammals has been shown to secrete a hormone concerned with the regulation of calcium, called calcitonin (or thyrocalcitonin). This hormone suppresses blood calcium when calcium rises above the normal level and is, therefore, complementary to the action of parathyroid hormone.

The Adrenal Glands

The adrenals are bilaterally paired glands located along the dorsal wall of the abdominal cavity, generally in the vicinity of the kidneys. The name adrenal (or suprarenal as the glands are called in man) was given to these organs because of their location in mammals—on, or immediately adjacent to, the anterior surfaces of the kidneys. Considerable variation in the exact location and organization of the adrenals is found among the various classes of vertebrates, but homologous tissues are found in all classes.

The adrenals consist of two distinct glandular components, one secreting the catecholamines adrenalin and noradrenalin, and one secreting steroid hormones. The catecholamine-secreting tissue stains avidly with chromium salts and is, therefore, usually referred to as the chromaffin tissue. The steroid-secreting component is usually called the interrenal tissue, or in mammals, the adrenal cortex. As the latter name implies, in mammals the chromaffin and interrenal tissues are organized into a histologically double-layered gland in which the chromaffin tissue constitutes an inner medulla in each adrenal and the interrenal tissue forms an outer cortex around this (Figure 11-12). Below the mammals, the interrenal and chromaffin tissues tend to be intermingled rather than organized into distinct cortex and medulla, and if one looks at the entire series of vertebrate classes an interesting trend in the degree of intermingling is seen (Figure 11-12).

In cyclostomes, the interrenal tissue has not been identified positively, but small clusters of cells that are believed to be the interrenal tissue are scattered along the cardinal veins in the vicinity of the pronephros. Chromaffin cells are scattered in small clusters through the same area but are not actually in association with the interrenal cell clusters. In teleosts, the interrenal tissue is located mostly within the pronephros either as scattered islets of tissue within the kidney or, more usually, as a band of tissue surrounding the cardinal veins and their branches. The chromaffin cells are, to some extent, intermixed with the interrenal tissue, especially in the walls of the cardinal veins. In the elasmobranchs, the interrenal tissue is organized into morphologically distinct glands along the inner margins of the kidneys, and the chromaffin tissue is completely separate, consisting of segmentally distributed clusters of cells anterior to the kidneys. In amphibians and reptiles, the interrenals are embedded within, or closely adherent to, the ventral surfaces of the kidneys, and the chromaffin tissue is intermingled as strands and islets of cells among the interrenal cells. Lizards are exceptional in that the chromaffin tissue tends to be concentrated in a mass on one surface of the interrenals. In birds, the adrenals

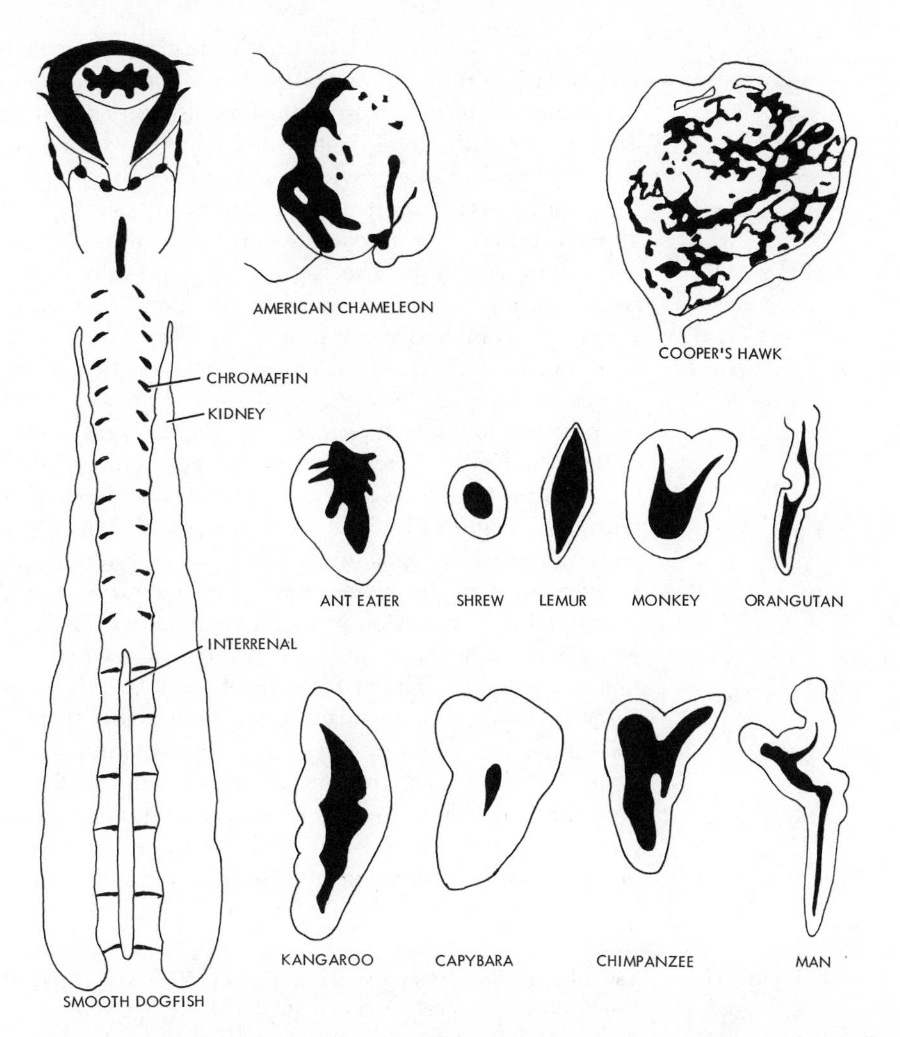

Figure 11-12 The adrenals, interrenals, and chromaffin tissue of several vertebrates. Chromaffin tissue or adrenal medulla, black; interrenal or adrenal cortex, white.

are on the anterior surfaces of the kidneys, as in mammals, but the chromaffin tissue is completely intermixed with the interrenal tissue. Finally, although direct intermixing of interrenal and chromaffin cells might appear to have been reduced by the segregation of these cells into separate layers in mammals, vascular interconnection has been perfected in that the medulla receives a large part of its blood supply from the blood sinusoids that drain the cortex.

The significance of the trend toward increasing association of the chromaffin and interrenal tissues is not clear. There is some evidence that adrenal steroids may influence the synthesis and secretion of adrenalin. Adrenalin is formed from noradrenalin by the addition of a methyl group through the catalytic influence of an enzyme known as phenylethanolamine-N-methyl transferase, and certain adrenal steroids increase the activity of this enzyme. In those vertebrates in which the associaton of the chromaffin and interrenal tissues is slight, the predominant chromaffin-cell hormone is noradrenalin, whereas animals with closer association between the two tissues tend to secrete higher proportions of adrenalin. For example, in elasmobranchs, adrenalin constitutes only about 25 per cent of the total chromaffin tissue hormones; however, in amphibians, reptiles, birds, and mammals, adrenalin usually amounts to 50 to 75 per cent of the total. Is it possible that the association of interrenal tissue with chromaffin tissue provides a means of controlling the amount of adrenalin that can be secreted? It may be significant that both adrenalin and certain adrenal steroids (the glucocorticoids) have an important function in stress or emergency reactions. The steroid effect on adrenalin synthesis could be a significant mechanism whereby the functional "tone" of the interrenal could promote a complementary tone in the chromaffin tissue.

The cells of the interrenal tissue are typically arranged in long cords or sheets, interspersed between capillaries or blood sinusoids. In mammals, but not other vertebrates, these cords are regularly oriented radially to the center of the gland and differentiated into three histologically distinct zones. These are an outer *zona glomerulosa*, a central *zona fasciculata*, and an inner *zona reticularis* (Figure 11-13). Formerly, it was believed that these zones represented successive stages in the life history of adrenocortical cells, new cells being formed in the glomerulosa, then migrating inward and reaching full functional differentiation in the fasciculata, and finally becoming senile and dying in the reticularis. This view was based on a somewhat biased interpretation of the appearance of the cells of the respective zones: cell divisions are observed to be more abundant in the glomerulosa than in the inner zones; the cells of the fasciculata are large and filled with lipid droplets containing steroid hormones and cholesterol; and the cells of the reticularis are small, sometimes pigmented, darkly staining, and interpreted as moribund in appearance. The cell-migration theory of zonation has now been discredited by labeling cells in the glomerulosa with radioactive isotopes and demonstrating that these labeled

Figure 11-13 Zonation of the mammalian adrenal gland. (From B. Frye, *Hormonal Control in Vertebrates*. New York: Macmillan, 1967.)

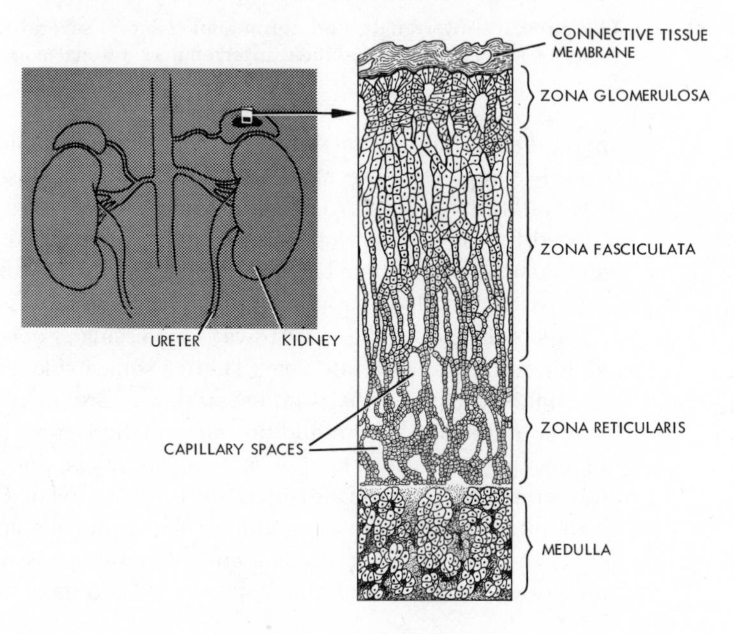

cells do not migrate through the gland as formerly supposed. Biochemical analysis of the different zones has now shown that the zonation reflects a functional subdivision of the adrenal cortex, each zone being specialized for the production of particular kinds of steroid hormones. Although the functional meaning of the zonal organization is clear, the evolutionary significance is not. The interrenals of other vertebrates lack the zonal organization but secrete much the same steroids as the mammalian adrenal. It would be valuable to know if the individual interrenal cords or sheets of lower vertebrates are functionally differentiated into different zones that are not reflected as morphological zones because of the irregular orientation of the cords.

Extracts of the adrenal cortex or interrenal yield thirty or more steroids, but most of these are intermediates in the process of biosynthesis of the hormones and are not secreted. The principal hormonally-active steroids of the adrenal are shown in Figure 11-14. These fall into three major functional classes: mineralocorticoids, which are steroids affecting primarily the metabolism of sodium and potassium; glucocorticoids, which are steroids affecting primarily the metabolism of carbohydrates and proteins and stress responses; and androgens, which have activities similar to the male sex hormone. Because of the great similarity in the structure of the various adrenal steroids there is some overlap in activity, and when a steroid is classified as a mineralocorticoid or glucocorticoid, this indicates its prevalent, but not exclusive, effects. In the mammalian adrenal, mineralocorticoids—principally aldosterone—are produced in the zona glomerulosa; glucocorticoids—mainly corticosterone and cortisol—originate primarily in the fasciculata; and androgens are secreted in reticularis.

The functions of the adrenal coretx are best understood in mammals, where, as we have noted, three classes of activities can be distinguished.

MINERAL METABOLISM. Adrenalectomy leads to a rapid loss of sodium chloride from the extracellular fluid into the urine and intracellular fluid. Associated with this loss of solute, water is lost osmotically, leading to a reduction in the extracellular fluid, including blood volume (see Chapter 8). This usually culminates in a precipitous drop in blood pressure, shock, and death. The primary disturbance in salt balance occurs in the kidneys, where the ability to

Figure 11-14 Some hormones of the adrenal cortex. Cortisol and corticosterone are the major glucocorticoids. Aldosterone is the major mineralocorticoid. Dehydroepiandrosterone is an example of an adrenal androgen.

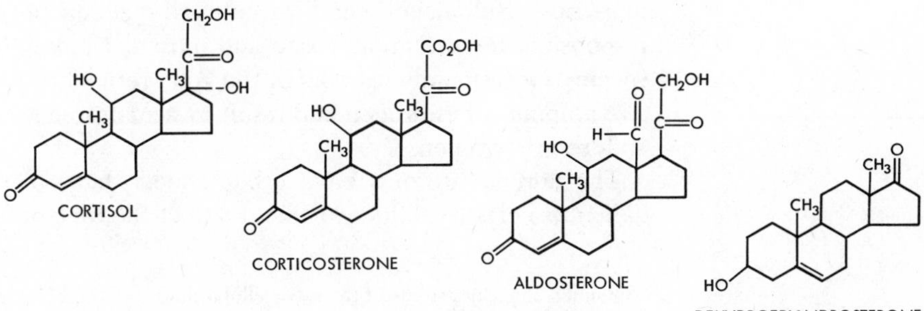

reabsorb sodium from the glomerular filtrate is reduced, and in the tissues at large, where the ability of the cell membrane to regulate the distribution of ions between the extracellular and intracellular fluids is disturbed. Substitution of saline for drinking water will prevent the collapse caused by adrenalectomy, by compensating for the excessive loss of salt in the urine. Mineralocorticoid steroids, particularly aldosterone, similarly will prevent adrenalectomy collapse by restoring the normal capability for ion regulation. The kidney and adrenal together constitute a vital control mechanism for the regulation of salt balance and blood pressure. Low dietary salt intake (or excessive salt loss in sweat or urine) results in a decline in blood volume and pressure.* This is sensed by special cells in the juxtaglomerular apparatus of the kidney, which respond by secreting the enzyme renin. Renin, in turn, catalyzes the conversion of angiotensinogen, an inactive protein in the blood, into angiotensin. Angiotensin stimulates the secretion of aldosterone by the zona glomerulosa, thus promoting an increase in the rate of sodium reabsorption by the kidney tubules. Excessive sodium intake, conversely, suppresses renin and aldosterone secretion and, thereby, allows a greater rate of sodium excretion by the kidney. Adrenocorticotropin, which has profound effects on the function of the inner zones of the adrenal cortex, has relatively little effect on the zona glomerulosa.

ORGANIC METABOLISM. Adrenalectomy results in a fall in blood glucose concentrations and liver and muscle glycogen content. All three of these changes are a reflection of the impairment of the ability of the liver to synthesize glucose from noncarbohydrate resources, particularly protein. As might be expected, therefore, this effect of the adrenal is not too apparent when an animal is on a high carbohydrate diet but becomes very important during fasting or when the carbohydrate diet intake is very low. Under the latter circumstances the liver readily synthesizes the necessary carbohydrate so long as the adrenal cortex is functional. But adrenalectomized animals (given saline to avoid the complications of salt imbalance) do not tolerate fasting and eventually die of hypoglycemic (low blood glucose) convulsions. Hypoglycemia can activate the secretion of glucocorticoids via the hypothalamic-pituitary-neurosecretory circuit, as can other "stressful" stimuli that might be associated with fasting or starvation.

SEX HORMONE ACTIVITIES. The adrenal androgens are similar in their effects to the principal male sex hormone, testosterone. They tend to be weaker in their masculinizing actions, however, and in normal concentrations they act primarily to promote protein synthesis and growth. This effect is important for the normal growth and muscular development of both males and females. The more powerful anabolic and masculinizing effects of testosterone are, in part, responsible for the greater size and muscular strength of the males of some species. Occasionally, tumors of the zona reticularis occur that secrete excessive amounts of androgen and result in sexual precocity or masculinization of children or women.

The adrenal steroids have other effects that are understood poorly and difficult to classify. Glucocorticoids affect the tone of blood vessels, especially

* Sodium is the major solute of the extracellular fluid and, therefore, through its osmotic effect has a large part to play in regulating the volume of the extracellular fluid, including blood.

the smaller arterioles and capillaries. Even minor injuries to an adrenalectomized animal may cause circulatory "collapse," characterized by the pooling of blood in the capillaries and the leakage of plasma into the interstitial space. This may lead to shock and death. Muscle work capacity, including that of the heart, is greatly reduced by adrenalectomy. This is probably related to poor circulation as well as to disturbances in salt balance and organic metabolism. Perhaps the least well-understood-function, yet one of the most important functions, of the adrenal cortex is its role in stress tolerance. Two basic facts are clearly established: (1) after adrenalectomy an animal is no longer able to tolerate numerous kinds of stresses, such as injury, burns, infections, fasting, toxic substances, heat, or cold; and (2) the adrenals of an intact animal undergo dramatic enlargement and increase in hormone production in stressful situations. These two observations indicate that the adrenal has a vital role in the adjustment of animals to stressful conditions (which must be an almost routine part of the life of wild animals), yet this role cannot be explained satisfactorily.

The chromaffin tissue, or adrenal medulla, is not essential for life, and no important deficiency symptoms have been attributed to its removal. Yet, the chromaffin cell hormones have important effects on a number of physiological processes, particularly those involved in the adaptation of animals to sudden emergencies.

The Parathyroid Glands

Parathyroid glands are found in all classes of vertebrates above fish. The name *parathyroid* was invented for the glands in mammals, in which they usually lie near or embedded within the thyroids. Generally, however, the name is a misnomer, for in most vertebrates the parathyroids do not have a particularly close or significant association with the thyroid. Commonly, they are located along the jugular veins or in the arches of the carotid and systemic arteries.

The parathyroids develop from the ventral corners of the gill pouches (Figure 11-9). In those species in which two pairs are present, as in the great majority of tetrapods, they usually originate from pouches III and IV (Figure 11-9). However, all five pairs of gill pouches are capable of giving rise to seemingly homologous tissue, and by this means accessory parathyroids up to five pairs can arise. A single pair of parathyroids is commonly found in lizards, but within this group some species have the usual two pairs, and those that have only one possess a transitory rudiment of the second pair during embryonic development. Most mammals have two pairs, but several species of rodents, including the common laboratory rat, have only a single pair.

Because parathyroids are lacking in fish and do not appear in amphibians until the time of gill regression at metamorphosis, it is interesting to speculate on the possible significance of this seemingly reciprocal relationship between the presence of gills and the presence of parathyroids. Because both the gills and the parathyroids are involved in certain aspects of mineral metabolism, some functional equivalence may exist. One can ask whether, associated with the gill apparatus of fish, there might be scattered glandular cells as yet unidentified that are homologous to the parathyroids of tetrapods. During the course of evolution, when the gills were lost, such hypothetical parathyroid homologous cells might have become consolidated into discrete parathyroid glands. Some support for the hypothesis that fish may have parathyroidlike tissue can be inferred from the fact that some fish have been shown to respond to injections of mammalian parathyroid hormone. These responses are similar to the responses of mammals themselves (elevation of serum calcium, demineralization of bone, and acceleration of the rate of calcium turnover). It seems unlikely that a complex hormone response mechanism would evolve without the presence of the hormone molecule required to render the mechanism functionally complete and, therefore, adaptively significant. To test this idea by ordinary morphological criteria is almost impossible, for parathyroid cells are notoriously indistinctive by morphological criteria. It is unlikely that individual scattered cells would even be recognized as gland cells. Consequently, a valid test of homology depends on the demonstration of the secretion by such cells of a protein molecule homologous with the parathyroid hormone of tetrapods.

The function of the parathyroid gland is to regulate blood levels of calcium and phosphate. The blood levels of these minerals are the result of a dynamic equilibrium between the rates of entrance into the blood from the gut and from the dissolution of bone and the rates of removal from blood by deposition in bone and by excretion. All of these processes go on more or less continuously, and it is the net balance between them that results in calcium and phosphate homeostasis. The role of the parathyroid can be summarized as follows. The parathyroid cells monitor the level of calcium in the blood. When this falls below normal, parathyroid hormone secretion is accelerated and those mechanisms that raise blood calcium are activated. When the calcium level of the blood rises above the normal value, parathyroid hormone secretion slows down, and the mechanisms for removal of calcium from blood become dominant. The central role of the parathyroid in this regulatory process is dramatically illustrated by the rapid fall in plasma calcium concentration which occurs following parathyroidectomy (Figure 11-15).

The consequences of a fall in blood calcium level are drastic. Initially, the symptoms are restlessness and hyperexcitability. This is followed by mild involuntary muscle twitches, which build in intensity and become generalized into convulsions. A parathyroidectomized animal may survive the first few attacks, which recur at intervals of a few hours or days, but eventually it dies in severe muscular spasms and tetany, which cause suffocation. The basis for these symptoms seems to be the dependence of both nerve and muscle on normal calcium levels to maintain normal permeability to certain ions, notably sodium and potassium, which are essential to their excitable states and functions.

The major target of the parathyroid hormone, so far as precise homeostasis of calcium and phosphate are concerned, is bone. A certain proportion of bone mineral is continuously being dissolved and redeposited. Deposition of calcium and phosphate in bone seems to be a spontaneous crystallization of bone mineral (complex calcium phosphate salts) on the collagenous framework of bone that has been laid down by osteoblasts. Because the spontaneous equilibrium between bone mineral calcium and plasma calcium is lower than the normal plasma level, this removal of calcium and phosphate from blood goes on continuously at normal blood concentrations of these minerals. But dissolution of bone also goes on continuously, through the activity of bone-degrading cells known as osteoclasts. The effect of parathyroid hormone is to increase the numbers and activity of these cells and, thereby, to control the rate of dissolution of bone. In addition to the effect on bone, parathyroid hormone has some effect in enhancing the absorption of calcium from the intestine and reducing the excretion of calcium in the kidneys.

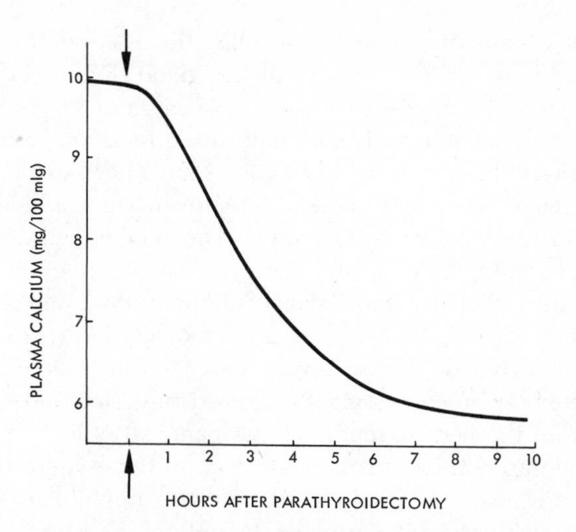

Figure 11-15 Fall in plasma calcium of the laboratory rat after parathyroidectomy. The time of parathyroidectomy is indicated by the arrow. (From B. Frye, *Hormonal Control in Vertebrates.* New York: Macmillan, 1967.)

The effect of parathyroid hormone on the phosphate level of blood is the opposite of the effect on calcium: parathyroidectomy results in a rise of blood phosphate levels and reduces the rate of excretion of phosphate. The rise in blood phosphate is the result of the direct effect of parathyroid hormone on the ability of kidney tubules to excrete the dietary excess of this ion. There is no evidence that the parathyroid gland responds to blood phosphate levels independently of calcium, so in a sense it is calcium that is primarily regulated by the parathyroid and phosphate is secondarily regulated.

Calcitonin, secreted by special "light" cells in the thyroid gland (page 540), also regulates blood calcium levels. It is secreted when calcium rises above normal and causes a fall in calcium concentration by stimulating the deposition of calcium in bone.

Virtually all that is known about parathyroid function is based on work with mammals. A few studies with amphibians and birds suggest that the role of this gland as described for mammals is also applicable to other tetrapods.

The Pancreas

The pancreas lies in the mesenteries near the upper end of the small intestine. Its exact position and form are variable, but it is significant that it always lies in the hepatic portal drainage, between the origin of these vessels in the wall of the duodenum, stomach and spleen, and the liver. This anatomical feature means that a large portion of the absorbed foodstuffs as well as certain gastrointestinal hormones that influence the function of the pancreas pass directly from the gut to the pancreas. It also means that the hormones of the pancreas are delivered directly to the liver, which is a primary target tissue of the pancreatic hormones. Through the hepatic portal blood flow, the upper gut, pancreas, and liver are interwoven into a complex chemical control system concerned with the regulation of the digestion and assimilation of foodstuffs.

The pancreas is, of course, both an exocrine and an endocrine gland. The bulk of the organ consists of the microscopic exocrine acini and only about 1 per cent of the pancreas is endocrine tissue. In most vertebrates this takes the form of microscopic clusters of cells, the islets of Langerhans, which are scattered more or less throughout the gland (Figure 11-16). The size of the individual islets range from as little as twenty microns to five-hundred microns or more in diameter, or from a half dozen to several hundred cells. In mammals, where the only reliable estimates have been made, the total number of islets ranges from several thousands to upward of a million or more, depending on the size and species of the animal. The number and size of the islets, as well as the over-all volume of islet tissue, are subject to great variation depending on the over-all hormonal balance of the animal and, in particular, on the amount of carbohydrates present in the diet. Any condition that tends to elevate the blood glucose levels above normal for extended periods may stimulate both an enlargement of the existing islets and a proliferation of new islets from the finer branches of the pancreatic duct system.

Certain exceptional features are seen in the organization of the endocrine pancreatic tissue in vertebrates below the amphibians. In teleost fish, some species exhibit an aggregation of the bulk of the islet tissue into one or a few principal islets that are nearly or completely separate from the digestive portion of the pancreas. Among the Chondrichthyes, two different conditions are seen. The elasmobranchs have no true islets, but the smaller branches of the ducts of the pancreas are enveloped by a layer of cells that are believed to be homologous to the islet cells. In holocephalans, a condition intermediate between that in the elasmobranchs and typical islets is seen: the smaller branches of the ducts are enveloped by a layer of islet cells, but in places these appear to have "budded off" into large clusters of endocrine cells that retain their connections with the duct system. In this context it is relevant that, in all

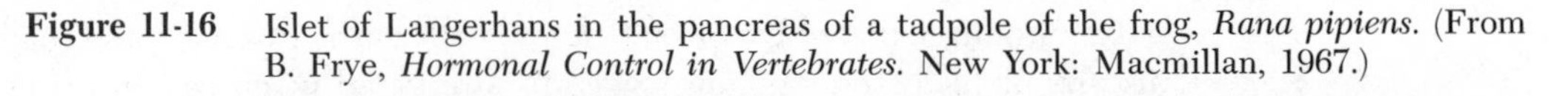

Figure 11-16 Islet of Langerhans in the pancreas of a tadpole of the frog, *Rana pipiens*. (From B. Frye, *Hormonal Control in Vertebrates*. New York: Macmillan, 1967.)

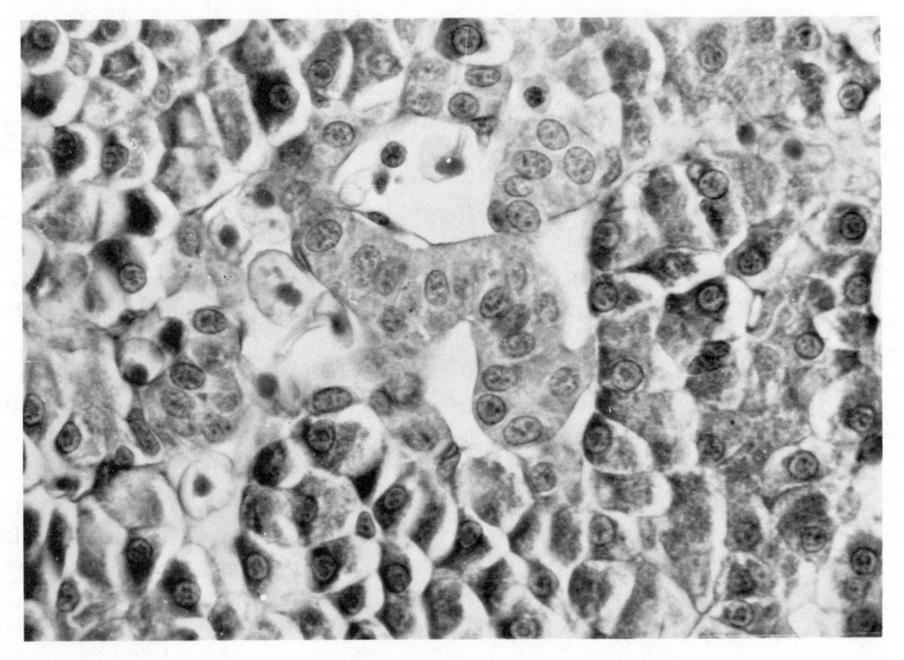

vertebrates, the islets originate by proliferation from the walls of the finer ducts. Finally, cyclostomes have no true pancreas at all, but clusters of cells have been described in the wall of the duodenum around the orifice of the bile duct that are believed to be homologous to the islets. A similar situation is seen in amphioxus.

The islets of Langerhans secrete two hormones, insulin and glucagon. Insulin is present in all vertebrates, and even in the gastrointestinal tracts of certain protochordates and echinoderms. It is obviously a hormone of great antiquity. Glucagon, on the other hand, has been specifically identified only in reptiles, birds, and mammals, but the evidence for its presence in lower vertebrates is indirect, being based mainly on the presumed homology of certain cells in the pancreas of lower vertebrates with the cells known to secrete glucagon in mammals. Both glucagon and insulin are protein hormones, but of unrelated amino acid sequence.

The functions of insulin are very complex. Its over-all role is to regulate the utilization of foodstuffs. Insulin is specifically secreted in response to rising levels of glucose and certain amino acids in the blood. Insulin, in turn, promotes the removal of glucose and amino acids from the blood, primarily by regulating the storage of glucose as glycogen and fat and the synthesis of proteins from amino acids. During periods when insulin secretion is low, as during fasting, or even the relatively brief postabsorptive intervals between meals, the ability of insulin target tissues to utilize glucose and amino acids is reduced. Moreover, compensating mechanisms are activated that result in the accelerated release of fats, glucose, and amino acids into the blood by storage organs such as liver and adipose tissue. Thus, insulin plays a crucial role in the metabolic homeostasis of foodstuffs by the vertebrate animal.

The major target tissues of insulin are the liver, adipose tissue, and skeletal and cardiac muscle. In the liver, insulin promotes the removal of glucose from the blood and its conversion to glycogen. This it accomplishes by activating certain enzymes (glucokinase and glycogen synthetase) that regulate the synthesis of glycogen from glucose. In adipose tissue and striated muscle, the primary action of insulin is on the permeability of cell membranes to glucose. In the absence of insulin, permeability (and, therefore, the ability to utilize glucose from the blood) is reduced. In the presence of insulin, glucose permeation of these tissues is accelerated, with the effect that the utilization of glucose in fat synthesis, glycogen synthesis, and oxidative metabolism is accelerated.

Superficially, the main action of insulin often is described as the regulation of blood glucose concentration. One of the more overt symptoms of insulin deficiency is abnormally high levels of blood glucose. However, the effects of insulin on blood glucose are secondary to the basic effects on glucose metabolism. Hyperglycemia is one of the major symptoms of diabetes mellitus, the disease of insulin insufficiency, but numerous other disturbances in metabolism also occur.

The principal physiological target of glucagon is the liver, where the hormone promotes the activity of phosphorylase, an enzyme that controls the breakdown of glycogen to glucose and, thereby, stimulates glucose secretion by the liver. Thus, in some ways, glucagon is the counterpart of insulin. It acts to maintain blood glucose homeostasis by promoting the release of glucose into the blood. Glucagon is secreted in response to low glucose levels in the portal circulation. The significance of this is that the hormone stimulates the return of stored glucose to the blood where it is available for use by the tissues.

Insulin and glucagon are secreted by separate cell types within the islets of Langerhans. Insulin is secreted by the so-called beta cell. The drug alloxan specifically destroys the beta cells of the pancreas, and this effect is invariably associated with the loss of insulin from the pancreas and the development of diabetes. Glucagon is believed to originate in the alpha cell, a small acidophilic cell usually located on the periphery of the islets, but proof of this is not complete. Glucagon-secreting alpha cells are not positively known to exist in cyclostomes, fish, or amphibians. However, most of these groups do contain cells that resemble the alpha cells of higher vertebrates, and homology, together with functional equivalence, is likely.

Other Endocrine Glands

The foregoing discussion has considered the major and best-understood endocrine glands of vertebrates, except for the gonads, which will be discussed in the following chapter. In addition to these, a number of other glands and tissues are known or believed to have endocrine functions. For the most part, they are either incompletely understood so far as their endocrine function is concerned, or their endocrine functions are secondary, or subsidiary, to other physiological roles.

KIDNEYS. The kidneys (of mammals, at least) are the source of renin, which catalyzes the conversion of the blood-borne protein, angiotensinogen, into angiotensin. Angiotensin stimulates aldosterone secretion and also elevates blood pressure by causing constriction of the arterioles in certain tissues, especially the kidney, and thus reduces the peripheral vascular volume. The kidneys may be the source of angiotensinogen, but this is uncertain. Thus, in a glandular as well as an effector role, the kidneys are a central part of the endocrine mechanism regulating salt balance and blood pressure.

In addition to the renin-angiotensin system, the kidneys also may be the source of erythropoetin. Erythropoetin is a protein that is extractable from the blood and from the kidneys; it stimulates the proliferation and release of erythrocytes into the blood stream. Blood levels of erythropoetin are elevated under conditions such as low atmospheric oxygen pressure or hemorrhage, in which an increase in the production of erythrocytes would be adaptively significant. The fact that the kidneys are important erythropoetic organs during development and in some lower vertebrates gives some relevance to this otherwise seemingly exotic endocrine function that has been proposed for them.

CORPUSCLES OF STANNIUS. The corpuscles of Stannius are glandular-looking bodies, several in number, that are embedded in the posterior part of the kidneys of teleosts. Strong evidence for an endocrine function is lacking, but two possibilities are supported to some extent: (1) they may secrete mineralocorticoids and, thus, may be functionally equivalent to the zona glomerulosa of the mammalian adrenal; and (2) they may secrete a vasopressor substance reminiscent of the renin-angiotensin system of mammals.

PINEAL. The pineal, or epiphysis, is an outgrowth of the roof of the diencephalon. Although both photoreceptor and endocrine functions have been proposed for many years on anatomical grounds, these functions have only recently been well established. It now seems reasonably certain that a substance, melatonin (N-acetyl-5-methoxytryptamine), with antigonadal activity is secreted by the pineal. Destruction of the pineal in young animals results in precocious puberty; in adults, it may advance the sexual cycle or cause constant estrus in females. Pineal extracts, on the other hand, cause suppression or regression of the gonads. The effect is believed to be exerted through suppression of the secretion of the pituitary gonadotropins. Physiologically, the secretion of melatonin in adult animals is regulated by the photoperiod; in those species that have been studied, long photoperiods suppress melatonin secretion and promote reproductive development. Melatonin stimulates contraction of melanophores in amphiibians, and the name of the hormone was based on this effect. Because the hormonal effects of the pineal are concerned with responses to light, the photoreceptor and glandular functions of the organ may be interrelated. The thick skull of mammals precludes photoreception by the pineal, and in this group the glandular pineal is innervated by branches of the optic nerve.

THYMUS. The thymus arises as outgrowths from the branchial pouches, generally III and IV (Figure 11-9). In the adult, it may persist as several pairs of lymphoid bodies in the branchial region (fish, lower tetrapods), or the rudiments may migrate posteriorally and fuse to form a single gland in the anterior thorax (mammals). The thymus is a major lymphatic organ and, in addition to the production of lymphocytes, it seems to produce a hormone that stimulates the formation of lymphocytes and triggers the capability of such cells to carry out immune responses. Removal of the thymus in young animals before immunological competence has differentiated results in a permanent reduction in the ability of the animals to form antibodies. Removal of the thymus once immunological competence has differentiated does not have this effect.

DIGESTIVE TRACT. The stomach and duodenum produce several hormones that are primarily functional in regulating the digestive process. The pyloric region of the stomach produces *gastrin*, which passes through the circulation to the fundus of the stomach where it stimulates the secretion of gastric juice. The duodenum produces several hormones that regulate the activities of the stomach, pancreas, and gall bladder. Generally speaking, these hormones are released in response to the entrance of chyme from the stomach into the duodenum and not necessarily to a specific component of the chyme. *Secretin* stimulates the release of pancreatic juice, causing a rise in the volume of the juice, but has no effect on the release of pancreatic enzymes. *Pancreozymin*, on the other hand, stimulates the release of pancreatic enzymes without affecting the volume of fluid secreted. *Cholecystokinin* stimulates emptying of the gall bladder. *Enterogastrone* suppresses the muscular activity and hydrochloric acid secretion of the stomach. Finally, the small intestine, including the duodenum, contains and presumably secretes glucagon, but the functional significance of this is not yet clear. These gastrointestinal hormones are all homologous with one another, and with glucagon (one form of which is secreted by the wall of the gastrointestinal tract as well as by the pancreas), as revealed by similarities in the amino acid sequence in their structures.

ULTIMOBRANCHIAL GLANDS. The ultimobranchials are glandular masses of tissue that arise from the last pair of pharyngeal pouches. Because the cells are often arranged into thyroidlike follicles, a thyroidlike function was once speculated for this gland. However, attempts to demonstrate such a function using I^{131} have consistently failed. More recently, the ultimobranchials have been recognized as the source of calcitonin. In fish, and to some extent in tetrapods, the ultimobranchials are distinct from other pharyngeal tissues, and extracts of these glands contain calcitonin. In mammals, calcitonin is secreted by the thyroid; however, the hormone appears to be produced by so-called parafollicular or light cells of the thyroid rather than by the thyroid follicles. The parafollicular cells originate from the ultimobranchials, which fuse with the thyroid rudiment during development and are, thus, homologous with the ultimobranchials of lower vertebrates.

THE CAUDAL NEUROSECRETORY SYSTEM. This is a neurosecretory organ that is composed of neurosecretory neurons located in the caudal spinal cord of all classes of fish. In its simplest form, it is a diffuse aggregation of neurosecretory cells; however, in teleosts, the glandular area is well defined and is associated with a neurohaemal organ, the *urophysis,* within which the secretory axons terminate. Removal of the gland and injection of extracts of the gland have yielded results suggesting that it functions in either osmoregulation, ion regulation, or blood pressure regulation. Because these functions are intimately interrelated, all three could conceivably be affected through some urophysial effect on any of the three. The anatomical location of the caudal neurosecretory system, draining its supposed hormone into the caudal vein and, thence, into the renal portal veins to the kidneys, suggests that the kidneys might be a principal target of the hormone. The kidney, of course, plays a key role in osmoregulation, ion regulation, and blood pressure regulation.

References for Chapter 11

[1] BARRINGTON, E. J. W. *An Introduction to General and Comparative Endocrinology.* New York: Oxford University Press, Inc., 1963. The emphasis in this textbook is comparative.

[2] ———. *Hormones and Evolution.* London: English Universities Press, 1964. This collection of essays on selected topics offers an elementary and largely speculative consideration of the origins of and functional role of hormones during evolution.

[3] ——— and C. B. Jorgesen, eds. *Perspectives in Endocrinology.* New York: Academic Press, Inc., 1968. This is the best available book dealing with the comparative and evolutionary aspects of endocrinology. It deals with selected topics, including reproduction, migration, pituitary and hypothalamus, salt and water metabolism, and has an excellent first chapter that contains a synopsis of the main ideas in comparative evolutionary endocrinology.

[4] CHESTER JONES, I. *The Adrenal Cortex.* New York: Cambridge University Press, 1957. This is the only comparative treatment of the adrenal cortex available and, although somewhat dated, it is still excellent. The book covers both morphological and functional aspects of the adrenal cortex.

[5] VON EULER, U.S., and H. Heller, eds. *Comparative Endocrinology.* Vols. I and II. New York: Academic Press, Inc., 1963. This is the best relatively

brief, advanced presentation of comparative endocrinology covering all major glands; the emphasis is functional.

[6] FRYE, B. E. *Hormonal Control in Vertebrates.* New York: Macmillan Publishing Co., Inc., 1967. This is an elementary synopsis of vertebrate endocrinology; the emphasis is functional.

[7] GORBMAN, A., and H. A. Bern. *A Textbook of Comparative Endocrinology.* New York: John Wiley & Sons, Inc., 1962. This book has a strong comparative and morphological emphasis, but it is also strong on functional aspects. This is the best source of information on comparative endocrinology at the introductory level.

[8] HARRIS, G. W., and B. T. Donovan. *The Pituitary Gland.* Vols. I–III. Berkeley: University of California Press, 1966. This is an advanced review of all three lobes of the pituitary. The comparative and evolutionary aspects are especially well covered. The book includes chapters on neurosecretion and the hypothalamic regulation of the pituitary.

[9] PINCUS, G., K. V. THIMANN, and E. B. ASTWOOD. *The Hormones.* Vols. I–V. New York: Academic Press, Inc., 1948–1964. An exhaustive series of volumes, this is a source book of information on endocrine physiology. Little information is given on morphological and evolutionary aspects.

[10] PITT RIVERS, R., and W. R. TROTTER, eds. *The Thyroid Gland.* Vols. I and II. Washington, D.C.: Butterworths, 1964. This is advanced review of thyroidal endocrinology. Many chapters have a human and medical emphasis, but there are also good chapters with a comparative emphasis.

[11] SCHARRER, E., and B. SCHARRER. *Neuroendocrinology.* New York: Columbia University Press, 1963. This is an outstanding introductory presentation. The strong comparative functional emphasis makes this almost an introductory text in comparative endocrinology, and one of the most readable and informative available.

[12] TEPPERMAN, J. *Metabolic and Endocrine Physiology.* Chicago: The Year Book Medical Publishers, 1962. This is a short premedical text. The presentation is almost entirely functional and mammalian, but it is listed here because it is extremely well written.

[13] TURNER, C. D., and J. BAGNARA. *General Endocrinology,* 5th ed. Philadelphia: W. B. Saunders Company, 1966. This book offers basic endocrinology with a balanced presentation of function, structure, and comparative endocrinology. An unusual feature of this textbook is that each chapter has extensive references to the original research literature.

12

Reproduction

Introduction

Reproduction by vertebrates is exclusively sexual. Asexual modes of proliferation, which are widely found in most phyla of invertebrate animals and in plants, are lacking. Sexual reproduction in vertebrates, as in all organisms, consists fundamentally of genetic recombination through the cellular processes of meiosis and fertilization (see Chapter 3). Accessory to these basic cellular phenomena are a vast array of anatomical, physiological, and behavioral adaptations that function to bring together the sperm and eggs for fertilization and to support the development of the young. This spectrum of adaptations concerned with reproduction is what we normally think of as constituting the sexuality of organisms and, at the morphological level, as comprising the reproductive system.

Most vertebrates are sexually dimorphic, individuals of a species being distinctly differentiated into males and females capable of producing either sperm or eggs, respectively. As obvious as this statement may appear to be, it is not the rule among living organisms as a whole. There are many more groups of organisms that are hermaphroditic—that is, in which each individual is capable of producing both eggs and sperm (not necessarily at the same time and usually not through self-insemination). Even in vertebrates, some degree of intersexuality or hermaphroditism is common, especially among the lower groups.

Anatomy of the Reproductive Tract

The basic plan of organization of the reproductive tract of vertebrates is shown in Figure 12-1. It consists of the gonads, or primary sex organs, which produce the gametes and sex hormones, and the sex ducts and accessory sex organs, which convey the gametes or offspring to the outside. Species that practice

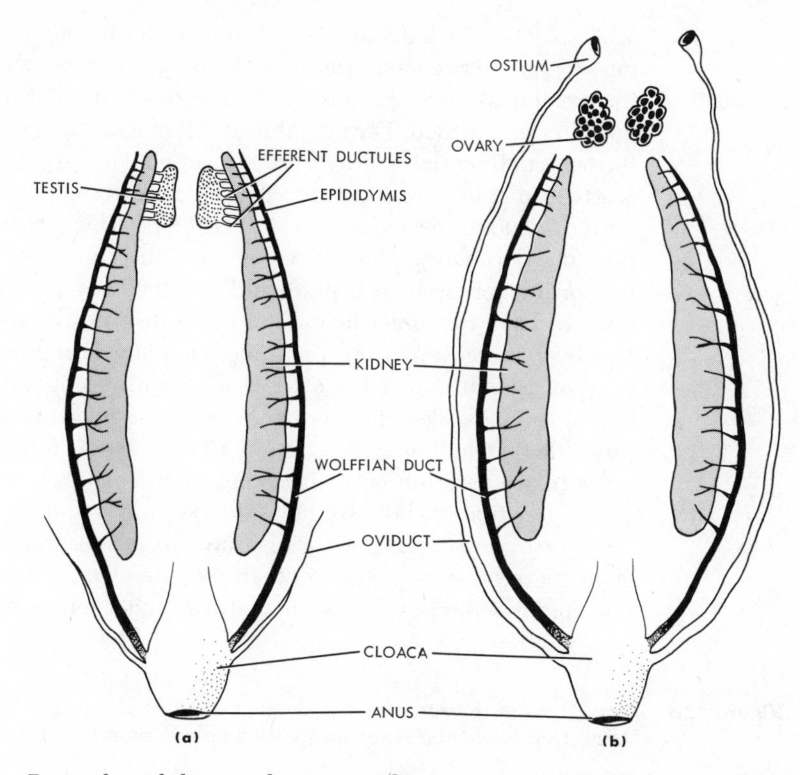

Figure 12-1 Basic plan of the vertebrate reproductive system: (a) male; (b) female. In the male anamniote the Wolffian opisthonephric duct functions as both the urinary duct and the sperm duct. Amniotes differ in that the metanephric kidney possesses a separate duct, the ureter, and the Wolffian duct functions solely as a sperm duct. Note the vestigial oviduct in the male.

internal fertilization may have elaborate external genitalia or copulatory organs, but those in which the eggs are inseminated outside of the body generally do not.

Testes The testes of vertebrates have two functions: the production of sperm and the secretion of steroid hormones. In all vertebrates, the steroid hormones play a crucial supporting role in sexual reproduction by regulating the functional development of the male reproductive tract and other sexual accessories and by regulating sexual behavior (Chapter 11).

The location of the testes in the great majority of vertebrates is inside the body cavity, more or less in the primitive embryonic position medial to the opisthonephric kidneys. The testes of mammals, however, typically descend into an extra-abdominal sac, the scrotum. Exceptions to this rule are the monotremes, edentates, cetaceans, and sirenians, which lack a scrotum and retain the testes in the pelvic region of the abdomen. Some species that possess a scrotum may exhibit testicular descent only during the spermatogenic period during and prior to breeding (rodents, stag, horse, elephant); other species, such as the rat, are able voluntarily or reflexly to withdraw the testes from the scrotum whenever they are threatened by danger.

The scrotum arises as a saccate evagination of the lower abdominal wall and, as its structure includes all of the layers of the abdominal wall, including skin, muscle, and peritoneum, the scrotal cavity is an extraabdominal extension of the peritoneal cavity (Figure 12-2). The mechanism of descent of the testis is not fully understood. During fetal development, the peritoneum posterior to the testis differentiates into a fibrous ligament known as the gubernaculum that is attached at its anterior end to the membranous coats of the testis, and at its posterior end to the abdominal wall at the position where the scrotal evaginations develop. During later development, the gubernaculum fails to elongate proportionately to body growth and eventually actually shortens. This would seem to have the effect of pulling the testis into the scrotum (Figure 12-2). However, because cutting the gubernaculum during development does not prevent descent, this rather obvious mechanism cannot be the whole explanation of testicular descent. Androgen injections will stimulate descent, and it is probable that androgen secretion by the embryonic testis during development helps to regulate the normal descent of the glands.

The function of the scrotum is temperature regulation. The internal temperature of the scrotal testis as measured in various species ranges from one to eight degrees Centigrade lower than abdominal temperature. This difference is crucial, for if the testis is retained at the higher temperature of the abdomen,

Figure 12-2 Formation of the scrotum and descent of the testis in man (Adapted from L. B. Arey, *Developmental Anatomy.* Philadelphia: Saunders, 1965.)

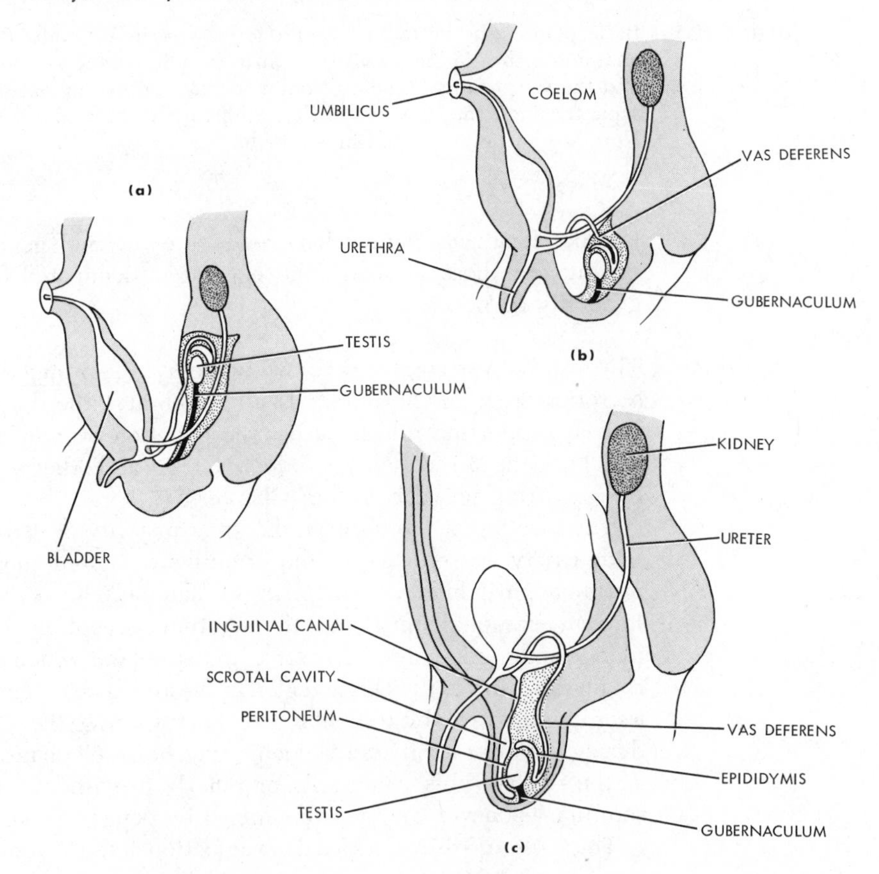

the germinal epithelium of the testis degenerates and sterility results. Sporadic retention of one or both testes in the abdomen, a condition known as cryptorchidism, may occur in species in which permanent descent is the normal condition and is always associated with partial or complete sterility. This effect of cryptorchidism can be duplicated by insulating the scrotum or by artificially warming it, substantiating the argument that temperature regulation is the function of the scrotum. The fact that the testes of birds and some species of mammals that lack a scrotum can carry out normal spermatogenesis at body temperatures is completely paradoxical to this conclusion and lacks explanation. The real enigma is that the germinal tissue of most male mammals should be unadapted to function at the deep body temperatures that are optimal for other internal organs.

The thermoregulatory function of the scrotum is achieved in two ways. First, as the arterial and venous supplies of the testes enter and leave the scrotum, they intermingle in a convoluted plexus of vessels known as the pampiniform plexus. The convoluted nature of these vessels increases the surface area across which heat is lost. In addition, in the plexus, there is cooling of the arterial blood by transfer of heat across the vessel walls into the venous blood that has already been cooled during its sojourn in the scrotum. Secondly, the testes and scrotum are provided with two sets of muscles, the external cremasters, which are attached to the peritoneum (tunica vaginalis) surrounding each testis, and the tunica dartos, which is adherent to the skin of the scrotum. These muscles contract when cooled and relax when warmed to pull the testes closer to the body or let them drop away, thus warming or cooling the testes.

Corresponding to the two principal functions of the organ, the testis of vertebrates consists of two major tissues: a gametogenic component and a steroidogenic component. In amniotes and anuran amphibians, the germinal tissue takes the form of greatly elongated and convoluted seminiferous tubules. Within the walls of these tubules is lodged a permanent population of primordial germ cells, or stem spermatogonia, and, depending on the season, cells in various stages of spermatogenesis associated with supportive Sertoli cells. The endocrine component consists of glandular Leydig cells that lie in the interstitial tissue, a cellular connective tissue that fills the spaces between the tubules. The endocrine cells are estimated to constitute between 1 and 5 per cent of the total mass of the testis. This familiar pattern of testicular structure is illustrated in Figure 12-3.

In urodeles and most other lower anamniotes, a different, although presumably homologous, pattern of organization is seen. The spermatogenic portion of the testis consists of lobules or crypts that do not contain a permanent population of primordial germ cells but are totally evacuated of germ cells at the time of each spawning. After spawning, the crypts collapse and cannot be discerned clearly until they become repopulated with spermatogenic cells during the next reproductive cycle. The permanent population of germ cells or stem spermatogonia lies outside of the spermatogenic crypts, sometimes in the interlobular spaces, in a permanent germinal layer around the periphery of the testis, or in a specialized germinal ridge along one margin of the gonad. During a spermatogenic cycle, spermatogonia produced by proliferation of these stem cells migrate into the crypts where they divide repeatedly to produce nests of spermatocytes that ultimately differentiate into mature spermatozoa. If the stem cells are located peripherally, successive waves of proliferating sper-

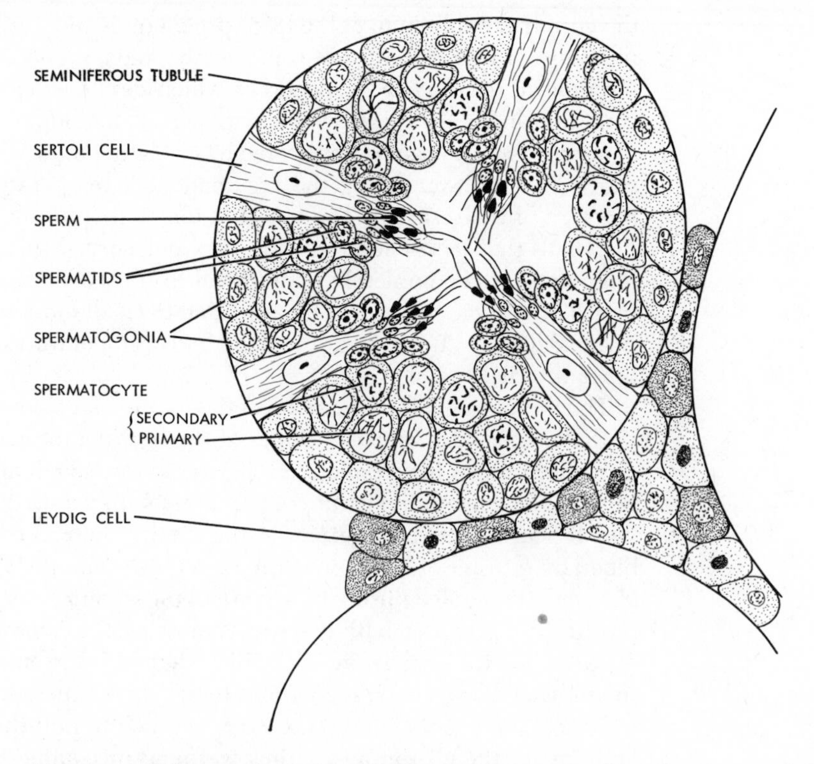

Figure 12-3 Histology of the tubular testis. Diagrammatic cross-section showing seminiferous tubules and interstitial tissue containing Leydig cells.

matogonia push inward from the outside of the lobules producing a pattern in which the oldest spermatogenic nests lie centrally adjacent to the sperm duct, and successively younger stages are arrayed outward toward the periphery of the lobule.

In either type of testis, spermatogenesis begins with the proliferation of the stem spermatogonial cells to produce a population of spermatogonia. In mammals, where the phenomenon has been most studied, during a given spermatogenic wave, the stem cell divides twice to produce four spermatogonia; one of these develops no further, but remains dormant as part of the residual population of stem cells that will proliferate similarly during subsequent waves. The other three cells divide from one to five times, depending on the species, to produce a nest of six to twenty-four primary spermatocytes, which then divide meiotically and differentiate to produce twenty-four to ninety-six spermatozoa. During their development, the spermatocytes become embedded in the cytoplasm of supporting Sertoli cells, and the progeny of an original nest of spermatocytes remain associated with the Sertoli cell until the mature sperm are released into the tubule lumen. The duration of a spermatogenic cycle from the time of initiation of spermatogonial proliferation to the formation of mature sperm is five to ten weeks, depending on the species. The sequence of events in spermatogenesis is illustrated in Figure 12-3. Although the arrangement of cells in the tubule walls appears to the untrained eye to be a chaotic mixture of stages of spermatogenesis, careful study has shown, in fact, that the various stages appear in particular patterns, or "cell associations." This suggests that a given stem cell undergoes repeated waves of proliferation, generating overlapping generations of spermatocytes and maturing germ cells. In man, for example, it has been demonstrated that stem spermatogonia enter a wave of proliferation approximately every sixteen days, whereas the entire duration of spermatogenesis is approximately seventy-four days.

Spermatogenesis in anamniotes differs from the mammalian pattern in two significant respects: (1) the original number of divisions of the spermatogonia during the proliferative phase is greater, resulting in more spermatocytes—up to a hundred or more—in a nest and a correspondingly greater number of sperm; and (2) the nest of cells originating from one stem cell during a wave of proliferation becomes enclosed together in a cyst wherein they develop synchronously into sperm. The cyst appears to be homologous with the Sertoli cell and its nest of spermatogenic cells in the amniote. Both the cyst cells and the Sertoli cells have been regarded as homologous with the follicle cells of the ovary (Figure 12-4).

These two general patterns of spermatogenesis, tubular and cystic, seem to be adaptations to the reproductive habits of amniotes and anamniotes, respectively. Anamniotes typically spawn in a brief period during which immense quantities of sperm and eggs are released into the water where external fertilization occurs. The cystic method of spermatogenesis produces the relatively huge numbers of sperm that are required to achieve a reasonable efficiency of fertilization under these hazardous circumstances. Prior to spawning, gametogenesis has been completed and the testes turned into vast receptacles of mature sperm that are totally evacuated from the gland during the brief spawning period. Amniotes, on the other hand, typically indulge in a more prolonged mating season, during which repeated acts of insemination may occur, and practice internal fertilization with less waste of gametes.

Figure 12-4 Histology of the mammalian ovary. Successive stages of follicle growth, ovulation, and development of the corpus luteum are shown. This is a composite view; ordinarily all of these stages would not be seen at any one time. (From B. Frye, *Hormonal Control in Vertebrates.* New York: Macmillan, 1967. After C. A. Villee, *Biology,* 4th ed. Philadelphia: Saunders, 1962.)

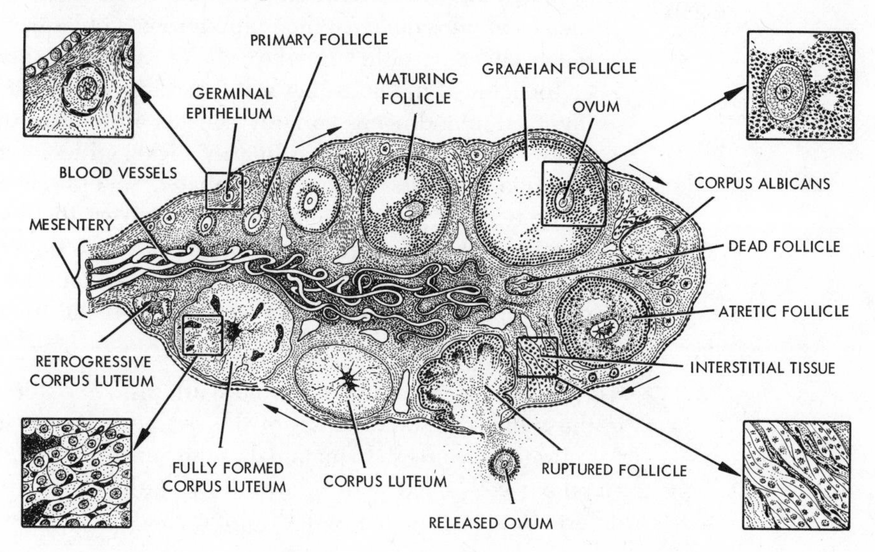

Corresponding to this, relatively fewer sperm are produced, and an act of mating does not evacuate the testis of sperm that are being produced continuously during the breeding season.

The fact that the testis is an endocrine organ that secretes male sex hormones, or androgens, was established long ago by the well-known effects of castration on the secondary sex characteristics and accessory structures and by the fact that androgens will prevent these changes. More recently, direct biochemical measurements of androgens in testis extracts and testicular venous blood not only have confirmed this role, but have made possible quantitative studies of the rates and biochemical pathways of androgen synthesis and secretion in various reproductive states. Such evidence has directly demonstrated the androgenic role of the testis in all vertebrates except cyclostomes; and, even in this group, experiments showing that injected androgens cause sexual maturation of the male provide indirect evidence of the androgenic role of the testis. We presume, therefore, that androgenesis is a universal function of the vertebrate testis.

More specific localization of the androgen-secreting function in the interstitial cells of Leydig (or tubule boundary cells) is supported by a number of lines of evidence that tend to exclude the seminiferous tubules and favor the Leydig cells as the endocrine tissue:

1. The developmental history of Leydig cells corresponds with patterns of androgen secretion. Thus, Leydig cells are abundant in the testis of the late fetal mammal, disappear shortly after birth, reappear at puberty, reach a peak at the sexual prime, and decline with senescence. The pattern of androgen secretion is similar (Figure 12-5).
2. Leydig cell tumors, which occur occasionally in man, are correlated with excessively high levels of androgen secretion; in young boys such tumors cause sexual precocity.
3. By means of carefully selected doses of X irradiation, the contents of seminiferous tubules can be destroyed without apparent damage to Leydig cells. Under these conditions, androgen secretion continues at essentially normal levels.
4. Similarly, in cryptorchidism or certain infections localized in the testes, such as *mumps orchidis,* the seminiferous tubules can be destroyed without much effect on either Leydig cells or androgen secretion.
5. Cytochemical methods that stain steroids or localize certain key enzymes involved in androgen synthesis (such as 3β-hydroxysteroid dehydrogenase) show these factors to be primarily localized in Leydig cells.
6. Steroidogenic tissue can be incubated *in vitro* with precursors such as cholesterol or progesterone and will convert these substrates into steroid hormones. When seminiferous tubules and interstitial tissue are surgically isolated and incubated separately in this fashion, the major steroid synthesizing activity is found in the interstitial tissue fraction.

Even though the cytochemical and *in vitro* evidence point to the interstitial tissue as the principal site of steroid synthesis, some steroidogenesis also occurs in the tubules at some phases of the reproductive cycle. In many species of vertebrates—from fish to mammals—the Sertoli, or cyst, cells differentiate into glandular cells filled with lipids following the release of sperm. Because this differentiation coincides with the appearance of high levels of progesterone in

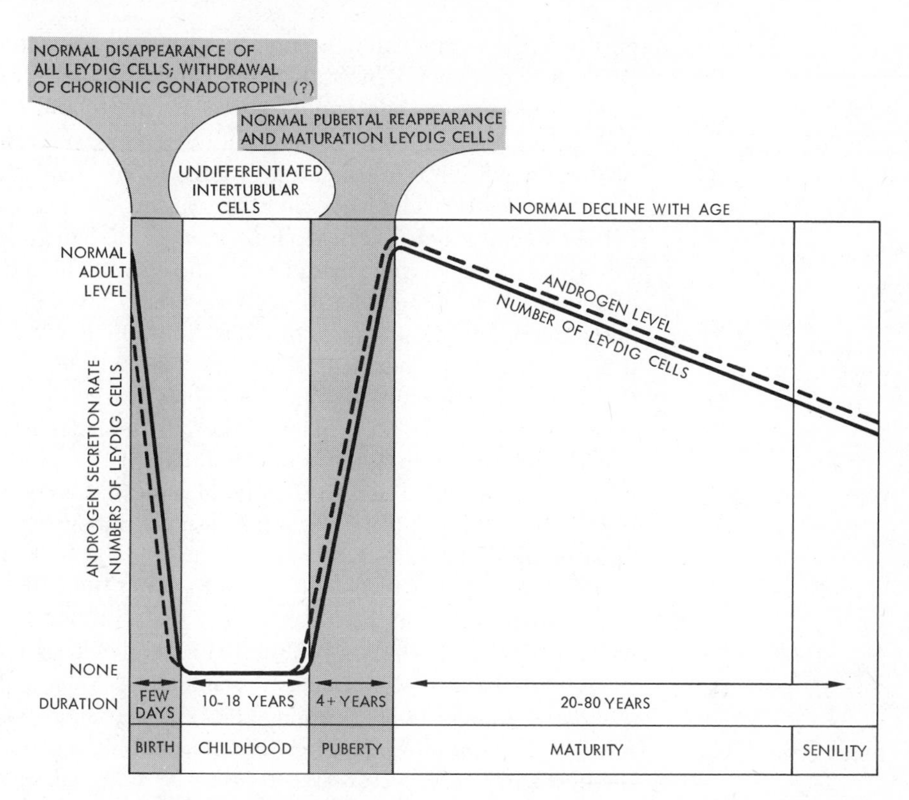

Figure 12-5 The relationship between Leydig cell number and androgen secretion during the life span of man. (Adapted from A. Albert, "The Mammalian Testis," in W. C. Young, *Sex and Internal Secretions*, 3rd ed. Baltimore: Williams and Wilkins, 1961.)

the testes and in the blood, it has been inferred that these transformed cells are the site of progesterone synthesis. If this is true, the presumed homology between the Sertoli cells and the granulosa cells of the ovarian follicle (see page 564) is particularly pertinent because this layer forms the primary progesterone-secreting component of the corpus luteum that forms from the empty follicle after ovulation.

The testes also secrete estrogens, but the evidence does not indicate clearly whether these hormones are secreted by the tubules, by Leydig cells, or by both.

Ovaries The ovaries of vertebrates are located in the peritoneal cavity, attached to the body wall on either side of the dorsal mesentery near the kidneys. In cyclostomes and most teleost and elasmobranch fish, the ovaries are elongated organs that extend from near the anterior end of the opisthonephros backward through most of the length of the abdomen. In tetrapods, the ovaries do not elongate proportionately to body growth and develop into more or less compact, globular, sometimes lobulated bodies. The ovaries of mammals undergo some posteriad migration or "descent" until they lie posterior to the metanephric kidneys. When fully developed, the ovaries are dependent from the body wall by a pair of mesenteries, the mesovaria. The blood vessels and nerves of the ovary enter through these mesenteries, the region of the ovary through which they penetrate being known as the hilus.

The ovaries are bilaterally paired in most adult vertebrates, and the embryonic rudiments are paired in all vertebrates. During development, the bilateral condition is lost in a few groups, notably cyclostomes and most birds, but also in scattered instances in other classes. This occurs either by the fusion of the two embryonic ovaries to form a single median gland (cyclostomes) or by the failure of one of the gonads to develop beyond the rudimentary stage (most other cases of asymmetry). Although most birds have only a left ovary, the right one remaining rudimentary, some birds, notably hawks, normally have paired ovaries. Viviparous sharks generally have only a right ovary. Among mammals, there are a few groups in which only one ovary is present, and this may be variously the right ovary (bats) or the left (monotremes).

Although the asymmetrical development of the ovaries may be accounted for embryologically, the adaptive significance of this condition is not clear. For birds, where the right oviduct is also usually undeveloped, the suggestion has been made that this is a means of reducing the body weight for flight, or, alternatively, that it is an adaptation to avoid the collision of two synchronously ovulated eggs in the lower part of the reproductive tract. Because there are successful exceptions in birds with bilateral ovaries and oviducts, neither of these explanations is particularly convincing. Among the higher tetrapods, where the number of eggs produced is relatively small, one ovary is capable of producing many more gametes than normally is released in a lifetime. Thus, the question might be reversed to ask why two ovaries have persisted in the majority of species of amniotes. The implications raised by such a question are numerous, but one factor to consider is whether the hormone secreting capabilities of one ovary would be adequate for normal reproduction. There is a glimmer of evidence in certain rodents that a single ovary does not produce quite adequate amounts of hormones to maintain gestation if the other ovary is surgically removed.

The basic histological organization of the ovary (Figure 12-4) consists of a supportive bed of cellular connective tissue, called the stroma, within which are embedded the germinal elements or follicles. Each follicle consists of an oöcyte surrounded by one or more layers of nurse or follicle cells. Covering the surface of the ovary is a thin cellular membrane known as the germinal epithelium, continuous with the peritoneum lining the coelom. The germinal epithelium derives its name from the fact that during embryonic development it is the proliferative layer of cells that gives rise to much of the tissue of the gonadal rudiment. This proliferative function is lost in the adult. The term *germinal* epithelium is misleading, for it can suggest that this is the source of primordial germ cells, which is not the case; these cells migrate into the rudimentary gonads from the embryonic endoderm or mesendoderm and establish residence in the outer margin of the stroma, just beneath the germinal epithelium. In mammals, but not other vertebrates, clusters of glandular interstitial cells are scattered through the stroma among the follicles. Following ovulation, the follicle cells become transformed into a glandular mass of tissue known as the corpus luteum. Corpora lutea are especially characteristic of the ovaries of mammals, where they play a crucial endocrine function in the sexual cycle and in pregnancy. Although homologous structures are formed in many species of lower vertebrates from elasmobranchs to birds, it is not known whether these play a functionally analogous role to the mammalian corpus luteum.

In cross section, the ovary is roughly divisible into an inner medulla and an outer cortex. The medulla consists of connective tissue and the major trunks of blood vessels and nerves that enter through the hilus and branch out from the medulla into the surrounding cortex. The cortical zone contains the stroma, follicles, corpora lutea, and interstitial tissue. In most lower vertebrates, the ovary is hollow, or saccular, the space being formed within the medullary region. The cavity of the ovary of teleosts, however, is formed by invagination of the surface of the ovary and is, thus, lined with germinal epithelium. In this group, the ovarian tissue often evaginates into this cavity to form a series of ovigerous folds, which greatly increase the mass of germinal tissue.

Following migration of the primordial germ cells into the gonad during embryonic development, the formation of mature ova and follicles occurs in several phases: (1) proliferation of the primordial oögonia into a large population of secondary oögonia and differentiation of these into primary oöcytes; (2) formation of follicles, which is concomitant with (3) growth of the primary oöcytes; (4) ovulation; and (5) completion of the meiotic maturational divisions.

In mammals, birds, and elasmobranchs, as well as some species of teleosts and reptiles, the oögonial divisions are completed during embryonic or juvenile development; the complete store of oöcytes, from which will develop all of the eggs produced during the adult life of the individual, is already present in the ovary of the immature animal. The adaptive significance of this condition, which is in marked contrast to the repeated proliferation of new generations of germ cells that occurs in the testes throughout life, is not known. The ovaries of amphibians (Figure 12-6), some reptiles, and most teleosts continue to produce

Figure 12-6 Growth of oöcytes in the frog ovary, showing overlapping of three generations of oöcytes. (From P. Grant, "Phosphate Metabolism During Oogenesis in *Rana temporaria*," *Journal of Experimental Zoology*, **124:** 513–543, 1953.)

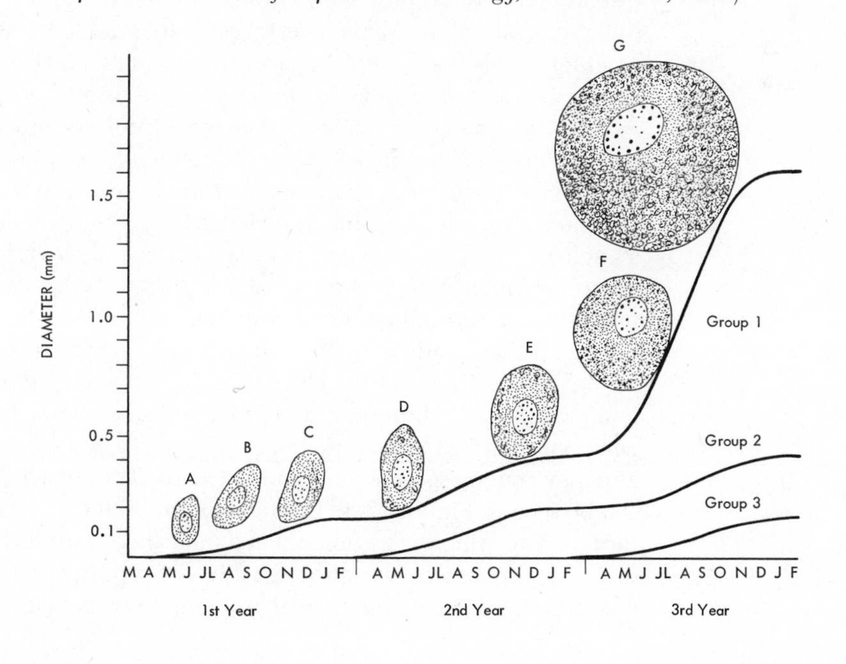

new generations of oöcytes after maturity. There is evidence in support of the claim that new oöcytes are not formed in the adult ovaries in mammals.

1. Proliferation of new oöcytes is not observed in histological sections of ovary after late embryonic development.
2. The population of oöcytes present in the young adult declines at a predictable rate until senescence, when virtually none is present. This rate of decline can be accelerated by treatments that stimulate the maturation of ova (such as treatment with gonadotropins) or decelerated by treatments that inhibit the maturation of ova (such as hypophysectomy). If the ovary were routinely capable of generating new oöcytes, one might expect the germ-cell population to be more stable and to be capable of compensatory adjustments for deviations in the rate of loss of ova.
3. Carefully selected doses of X rays can destroy the oöcytes without apparent permanent damage to other ovarian tissues, yet the ovary is never repopulated with new oöcytes.
4. If the oöcytes are labeled by allowing them to incorporate radioactive thymidine into their nucleic acids during the period of oöcyte formation, a high and constant percentage of the eggs subsequently produced in the adult for a prolonged period contain the radioactive label. Were new oöcytes being formed during later life one would expect a gradual decline in the percentage of labeled eggs as a result of the replacement and "dilution" of those formed earlier by new generations of unlabeled oöcytes.

Follicles are formed by the association of a single oöcyte with a nest of cells proliferated from the germinal epithelium. In lower vertebrates, this nest of follicle cells, generally known as the stratum granulosum, is usually only one cell layer thick; but, in species producing extremely large, yolky eggs, including elasmobranchs, birds, and reptiles, the granulosa layer may multiply until it is several layers thick. It is probable in such cases that all cells in the granulosa extend cytoplasmic processes inward to contact the surface of the egg. The follicles of all submammalian vertebrates are solid—that is, the growing ovum completely fills the follicular space. In mammals, however, the ovum remains nearly microscopic in size, and a large, fluid-filled space, called the antrum, forms within the granulosa layer. The ovum projects into one side of this space where it is enclosed in a hillock of granulosa cells known as the cumulus oöphorus. Growth of the mammalian follicle occurs in three phases. The first, or primary, follicle is located near the surface of the ovary and is surrounded by a single layer of granulosa cells. The secondary follicle is characterized by some growth of the oöcyte, formation of two to four layers of granulosa cells, and migration of the follicle deep into the cortex. The final stage, or tertiary follicle, is characterized by the formation and rapid growth of the antrum and by the differentiation of additional layers of cells external to the growing granulosa layer. These are the theca interna, which becomes a glandular component rich in lipids, and the theca externa, a connective tissue layer that also may contain some smooth muscle fibers. These stages of follicle formation can be seen in Figure 12-4. The final mature follicle in mammals is known as the Graafian follicle. The functional significance of the formation of the antrum in mammals is not known, although it has been postulated that it is in some way concerned with the ovulation of the minute eggs that mammals produce.

During the growth phase, oöcytes may increase in volume one million or more times, as a result of which the ovaries enlarge from less than 1 per cent to more than 20 per cent of the body weight in many species. Even in mammals in which the eggs are relatively minute and yolkless, the oöcytes increase in diameter ten- to twentyfold. The full growth and maturation of an oöcyte may require two or more seasons in poikilotherms, in which case several generations of developing oöcytes overlap (Figure 12-6). In homeotherms, however, full growth of the oöcytes requires only a few days or weeks. Birds are particularly remarkable in the rate of oöcyte growth: the chicken ovum is said to grow from less than a millimeter in diameter to its full size in only nine days. Growth of the oöcyte occurs mainly by the deposition of yolk. Yolk proteins are formed in the maternal liver and transferred via the blood to the follicle cells and from there into the oöcyte. In support of this function, the follicles of species producing extremely yolky eggs may be highly vascular, as those of birds, and the cell membranes of the oöcyte and the follicle cells are thrown into extremely numerous interdigitating microvilli that greatly increase the area across which the transfer can occur. In mammals, where the ova contain little or no yolk, the stratum granulosum is avascular.

Ovulation is the process of release of the ripe eggs from the ovary. In egg-laying animals, ovulation should be distinguished from ovoposition. Ovoposition is the actual release of the eggs from the reproductive tract, where they may be retained for several hours or days after ovulation. When the oöcytes are "ripe," the follicles protrude from the surface of the ovary and the ovarian tissue over the surface pole of the follicles becomes very thin and avascular. This area is known as the stigma (birds). Immediately prior to ovulation, a hole appears in the stigma, through which the egg emerges. Unfortunately, it is impossible to describe a mechanism of ovulation that is consistent with all the facts. Some of the theories that have been proposed are (1) a rise in intrafollicular or intraovarian pressure owing to an accumulation of fluid or contraction of smooth muscle fibers that forces the egg out through the stigma, and (2) enzymatic digestion of the stigma that creates the opening through which the egg emerges. Whatever the mechanism, it is closely dependent on environmental circumstances, including photoperiod, temperature, food, a suitable breeding site, and the presence of a mate. These conditions, when suitably combined, trigger the release of pituitary luteinizing hormone that activates the ovulatory mechanism. In most placental mammals, ovulation is a spontaneous event that occurs at the time of "heat," or sexual receptivity, during the estrous cycle; however, some mammals, notably members of the rabbit, cat, and ferret families, require the stimulus of copulation to trigger the secretion of the luteinizing hormone that causes ovulation.

The state of meiotic maturation of the oöcyte at ovulation varies in different species, but is almost always incomplete until after ovulation or fertilization. Perhaps the most usual condition is one in which the oöcytes enter the first meiotic prophase during embryonic development where they remain arrested until near the time of ovulation when they complete the first division and enter the second meiotic metaphase. At that stage, they are again arrested, and complete the final division only after activation of the egg by sperm penetration or artificial parthenogenesis. The meiotic divisions of oöcytes are always unequal and yield when complete one large, yolky ovum virtually as big as the

Figure 12-7 (*opposite*) Synthesis of progesterone and estrogens by the thecal (white) and granulosa (grey) layers of the follicle and corpus luteum. The granulosa layer has weak or no 17α-hydroxylase and desmolase activity and consequently cannot convert progesterone into estrogens. (See the text for further explanation.) Note that androgens (androstenedione, testosterone) are synthesized by this pathway, but in the ovary they serve primarily as biosynthetic intermediates and are secreted only in relatively small amounts.

ripe oöcyte and two or three exceedingly minute polar bodies.* The most obvious adaptive consequence of this is to preserve the accumulated yolk within the definitive egg, rather than divide it among four. Less obvious is the fact that the mature oöcyte has "built into" its surface a great deal of structural specificity concerned with controlling the early development of the embryo that also is preserved by the unequal meiotic divisions.

If for any reason the follicles fail to ovulate, the ripe oöcytes and follicles are reabsorbed gradually, a process known as atresia. In mammals, where several hundred thousand oöcytes may be present in the ovary in youth, but only a few hundred are ovulated in the life span of the individual, the vast majority of the oöcytes undergo atresia. Atresia may occur at any phase of the follicle growth cycle, and waves of atresia accompany waves of normal oöcyte maturation at every sexual cycle. The cause of atresia is unknown, but hypotheses have been proposed to explain this phenomenon: (1) a suppressive effect is produced by adjacent, more advanced, follicles; and (2) there is competition for the limited amount of gonadotropin required for follicle growth. Gonadotropins seem to be involved in some way: hypophysectomy of the immature animal before follicle growth has been stimulated reduces the rate of atresia (and, of course, prevents follicle development, too). On the other hand, hypophysectomy after sexual maturity results in atresia of all growing, developing follicles. It would appear that at the beginning of a sexual cycle more primary follicles are activated to begin development than subsequently can be supported by normal levels of gonadotropin.

The principal hormones secreted by the ovary are estrogens and progestogens, defined as feminizing and pregnancy-supporting hormones, respectively. Some androgens are secreted by the ovary, at least in mammals. The actual site of production of these hormones in lower vertebrates is unknown, but it is presumed to be the follicles and corpora lutea (when present) because these are the only glandular tissues in the ovary. In mammals, the granulosa, the thecal layer of the follicle (the corpus luteum), and the interstitial tissue all appear by cytochemical tests to be steroidogenic at some phase of the sexual cycle (Figure 12-7). The most significant theory of the origin of estrogens and progestogens is based on a correlation of the morphological cycle of the follicle with the predominant hormones secreted during this cycle. Prior to ovulation, the theca interna is a well-developed glandular layer, actively steroidogenic by cytochemical tests and richly vascularized; the stratum granulosum, by contrast, is avascular and relatively inactive. During this phase, estrogens are secreted

* Whether two or three polar bodies are produced depends on whether the polar body produced in the first meiotic division goes ahead to complete the second division or merely dies. So far as nuclear division is concerned, four haploid products are produced equivalent to the four spermatids produced by the meiosis of one primary spermatocyte.

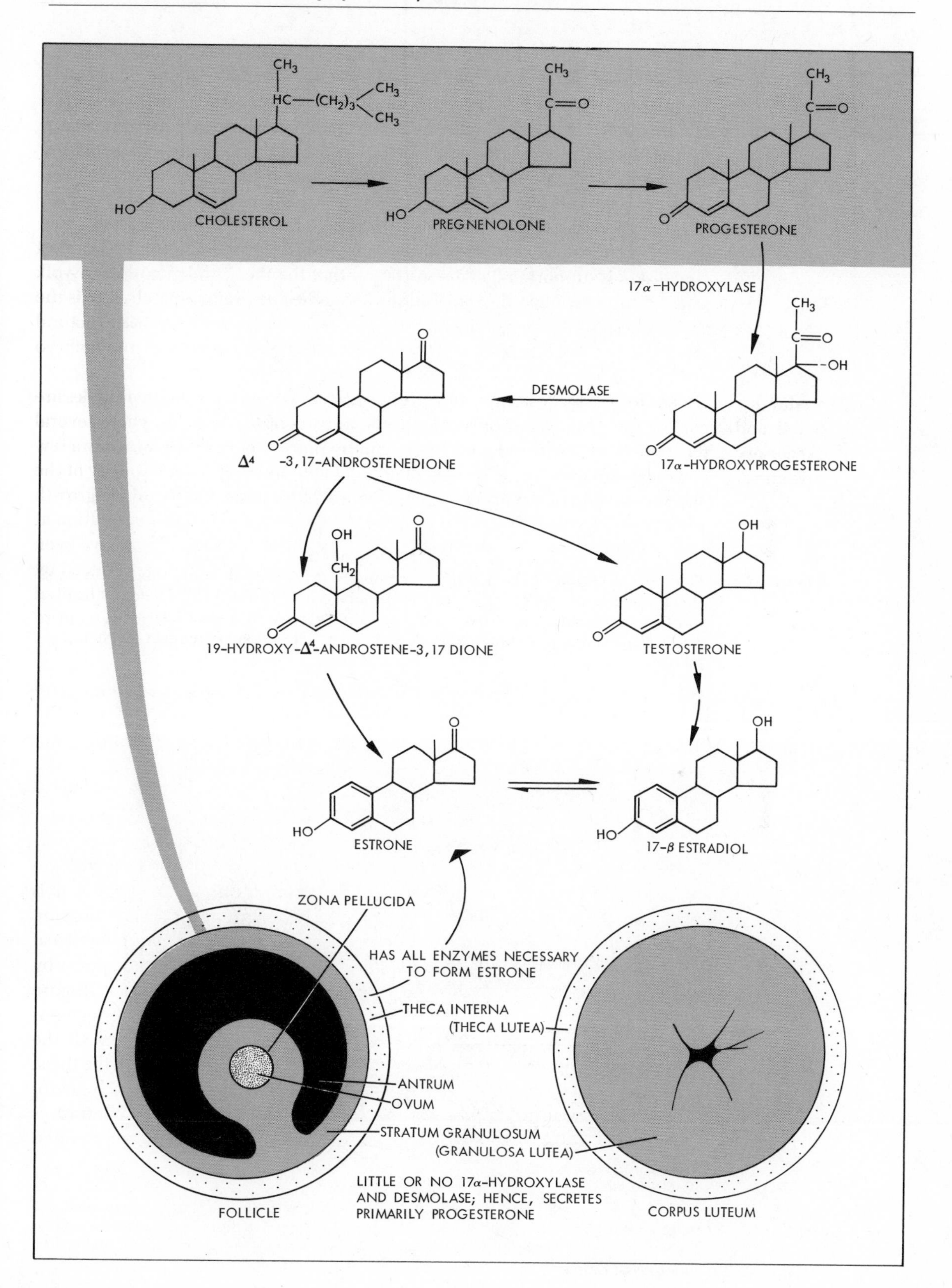
CHOLESTEROL
PREGNENOLONE
PROGESTERONE
17α-HYDROXYLASE
17α-HYDROXYPROGESTERONE
DESMOLASE
Δ4 -3,17-ANDROSTENEDIONE
19-HYDROXY-Δ4-ANDROSTENE-3,17 DIONE
TESTOSTERONE
ESTRONE
17-β ESTRADIOL
ZONA PELLUCIDA
HAS ALL ENZYMES NECESSARY TO FORM ESTRONE
THECA INTERNA
(THECA LUTEA)
ANTRUM
OVUM
STRATUM GRANULOSUM
(GRANULOSA LUTEA)
LITTLE OR NO 17α-HYDROXYLASE AND DESMOLASE; HENCE, SECRETES PRIMARILY PROGESTERONE
FOLLICLE
CORPUS LUTEUM

predominately, and relatively little progesterone is produced, compared to the postovulatory phase. After ovulation, the stratum granulosum rapidly becomes vascular, and by proliferation and cellular change forms the bulk of the corpus luteum; the thecal layer remains active but constitutes a relatively minor proportion of the corpus luteum. Corresponding to these changes, the secretion of progesterone rises dramatically; estrogen secretion, although it continues, becomes relatively less than progesterone. Thus, the thecal cells appear to have the enzymes necessary for estrogen formation, whereas the granulosa cells lack these enzymes and secrete primarily progesterone (see Figure 12-7). This conclusion is supported by demonstration that the theca and granulosa layers, isolated *in vitro,* have the capabilities to synthesize predominately estrogens and progesterones, respectively.

Reproductive Ducts and Accessory Structures

Except for the cyclostomes, all vertebrates have sets of reproductive ducts that convey the gametes, or offspring, to the outside of the body. In cyclostomes, the gametes are released into the coelom by rupture of the walls of the gonads and escape to the outside by way of abdominal pores that open into the cloaca or posterior end of the urinary ducts (Figure 12-8). Basically, the reproductive

Figure 12-8 The cloacal region of the Lamprey (*Lampetra fluviatilis*) showing the relationship of the abdominal genital pores to the urinary ducts. (From J. M. Dodd, "Gonadal and Gonadotrophic Hormones in Lower Vertebrates," in A. S. Parkes (ed.), Marshall's *Physiology of Reproduction,* Vol. 1, Part 2. London: Longmans Green, 1960.)

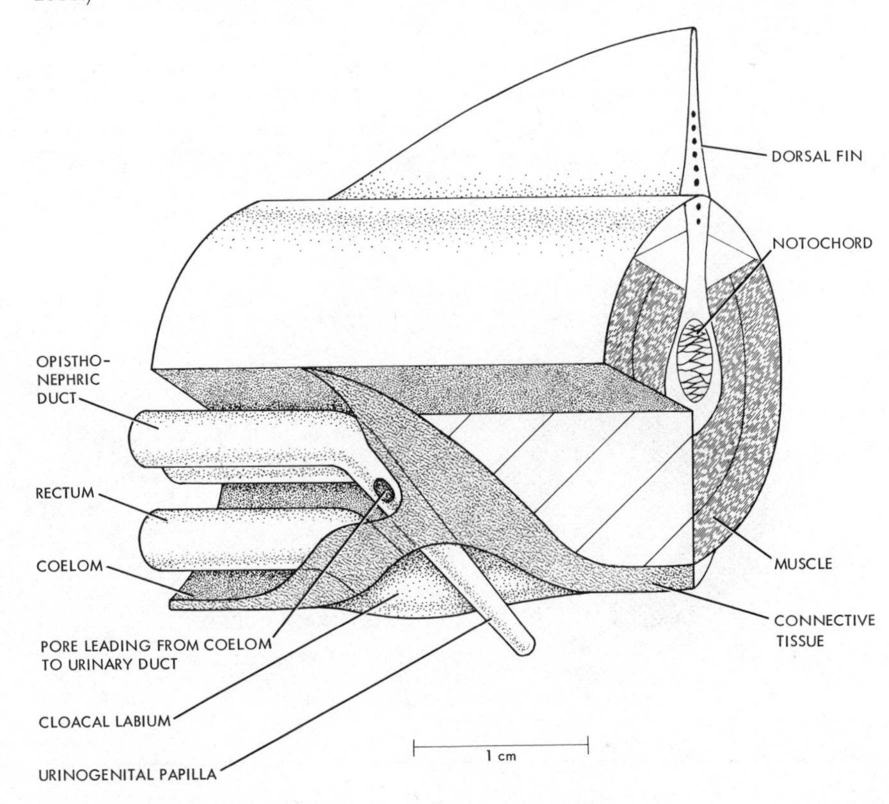

ducts of other vertebrates are simple, bilaterally paired mesodermal tubes consisting of an inner epithelial lining and an outer wall of connective tissue and smooth muscle. But, in all groups of modern vertebrates, there are associated with these ducts a variety of specializations that, for the most part, are concerned with the production of secretions that nurture and protect the gametes, with internal fertilization and with viviparity.*

THE MALE. The male reproductive system of vertebrates has close affinity with the urinary system. The basic condition is one in which the testes drain via a number of minute ducts, the vasa efferentia, into the anterior opisthonephric tubules, and from there via the archinephric or Wolffian ducts into the cloaca. In this condition, the archinephric ducts double as both the urinary and the sperm ducts. These basic units (the opisthonephros, the archinephric ducts, and the cloaca) constitute, or form, the basic rudiments of the sperm duct system in all vertebrates (Figure 12-9). In all anamniotes except some bony fish, this basic condition is retained virtually unaltered except for regional secretory specializations. In actinopterygians, there has occurred a trend toward reduction in the number of efferent tubules to one or a few larger ducts and a "descent" of the point of junction of this duct with the archinephric duct. In teleosts, this trend has been carried to the point that a completely separate sperm duct is formed and the archinephric duct is returned to a purely urinary function. In this group, the cloaca either disappears entirely, in which case the excretory, reproductive, and digestive canals open separately to the exterior, or is represented only as the urogenital sinus, a common tubular chamber through which the urinary and sperm ducts empty.

In amniotes, with the development of the metanephric kidney, the urinary and reproductive ducts again become separated. A new pair of urinary ducts, the ureters, arise as cloacal outgrowths, and the mesonephric ducts are taken over completely by the reproductive system. The mesonephric kidneys lose all excretory functions and form a pair of glandular, highly tortuous segments of the sperm ducts, the epididymides (singular, epididymis). These receive the efferent ductules on the one hand and drain into the mesonephric ducts, now known as the deferent ducts, on the other. A condition analogous to the formation of the ureter is seen in some amphibians in which some of the more posterior kidney tubules coalesce and grow posteriorly to join the cloaca independently of the Wolffian duct, which becomes mainly a sperm duct. The cloaca persists in birds and reptiles, and even in the most primitive mammals, the monotremes. But, in placental mammals, it becomes extensively modified by the formation, during embryonic development, of a transverse septum that

*The suggestion has sometimes been made that the reproductive system has undergone more exotic specializations during the course of evolution than any other system in the body. The implication of this claim is that, because this system is not essential for individual survival, the pressures of natural selection on it are such as to allow more variations than for other, more "vital" systems. This can hardly be true. In the first place, "survival" of a trait in the evolutionary sense depends solely on whether the trait is genetically based and can be transmitted to the next generation. This is obviously just as true of any trait affecting the functional integrity of the reproductive system as it is of a trait affecting the life or functional capability of an individual. More fundamentally, it is doubtful that any morphologist can document the claim that specializations in the reproductive system are any more numerous or exotic than those in any other system of corresponding complexity. Consider, for example, the specializations associated with feeding and digestion (Chapter 6).

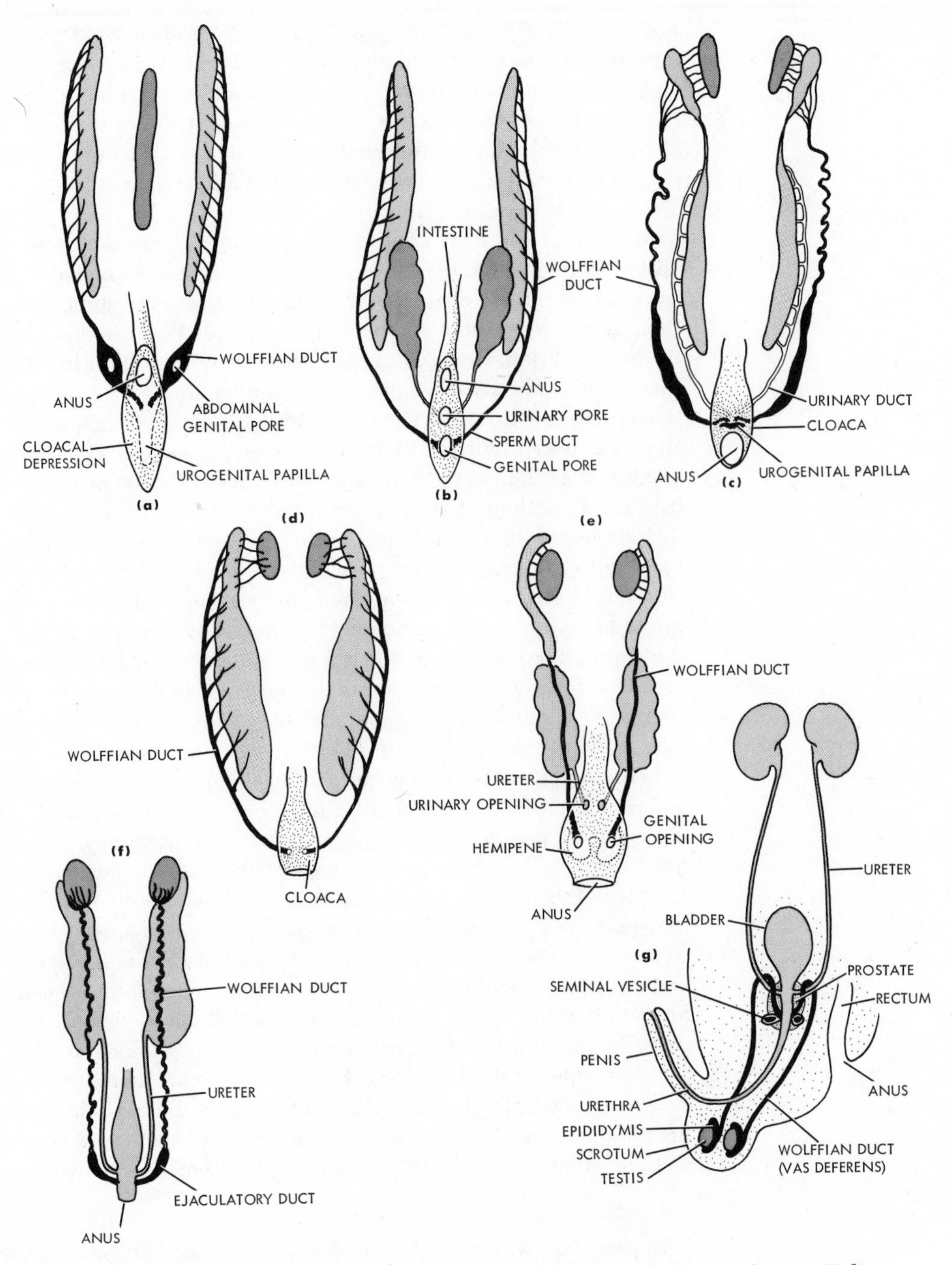

Figure 12-9 Reproductive and urinary ducts of various classes of male vertebrates. Kidneys and derivatives, light gray; Wolffian ducts and derivatives, black; testes, darker gray; cloaca and derivatives, stippled; (a) cyclostome, (b) teleost, (c) elasmobranch, (d) amphibian, (e) reptile, (f) bird, (g) mammal.

separates the cloaca into a dorsal digestive chamber and a ventral urogenital sinus (Figure 12-10). The latter receives both the ureters and the opening of the urinary bladder. In the adult, it forms a narrow tube, the urethra, that extends through the penis to the exterior.

Numerous glands and regional specializations are associated with the male reproductive ducts. Except for their general derivation, these structures are not necessarily homologous from group to group. Although extensive discussion of the comparative morphology of these structures cannot be included here, it should be pointed out that there is a great deal of inconstancy even within a group as to the exact battery of glands that a particular species may possess. For example, of the five major glands found in mammals—few species possess all of them. Some species may have only one gland, the prostate, which is the most consistently present gland of those present in this class. The general functions of these glands and specialized regions are (1) storage of sperm or (2) secretion of seminal fluids that provide both the vehicle in which the sperm are ejaculated and a suitable medium for maintaining the functional capability of the sperm. Even in mammals, where detailed analyses of the biochemistry of semen have been carried out, hardly more specific statements about the functions of seminal fluid can be made.

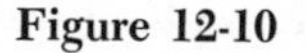

Figure 12-10 Formation of the urogenital sinus as derivatives in mammals: (a) sexually indifferent stage, cloaca undivided; (b) later indifferent stage, division of the cloaca into the rectum and urogenital sinus has occurred; (c) and (d) stages in the development of the male; (e) and (f) stages in the development of the female. (From A. Romer, *The Vertebrate Body*.)

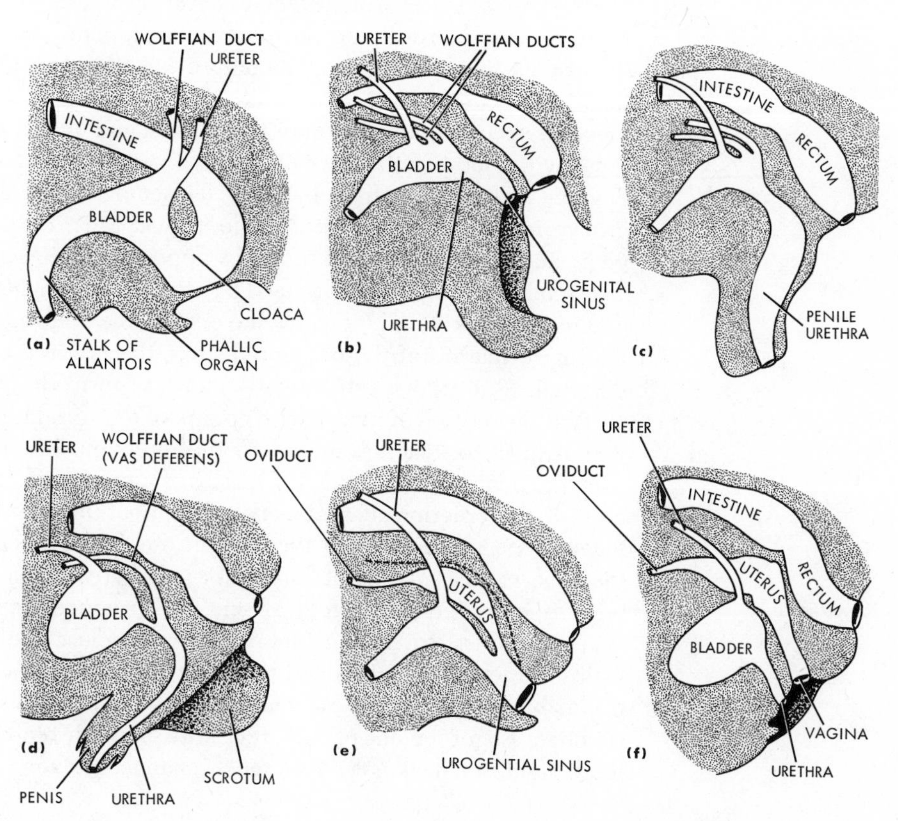

Most species that practice internal fertilization possess specialized copulatory structures for deposition of the seminal fluid within the female reproductive tract. In elasmobranchs, the copulatory organ is formed by modification of the pelvic fins into a structure known as the clasper. The main element of this structure is, in each fin, a carilaginous spine that is rolled like a scroll to form a groove, the base of which opens near the cloaca and through which the semen is transferred. The spines are erected by a pair of muscles inserted at their base and are inserted into the cloaca of the female during copulation. Associated with the base of the claspers are glands, or siphons, that also open into the groove and pump a secreted fluid or sea water through the claspers, a mechanism that serves to flush the sperm into the female. Somewhat similar copulatory structures are present in some teleosts, in which either the pelvic or the anal fins may be modified for the transfer of spermatozoa. In some teleosts, the genital pore opens on a fleshy, penislike papilla that can be erected and inserted into the cloaca of the female. In at least one species, the table has been turned, and the female possesses a genital papilla that is inserted into the genital pore of the male for reception of the sperm. The great majority of teleosts inseminate the eggs externally.

Among the amphibians, virtually all anurans fertilze the eggs externally, whereas fertilization is internal in most urodeles. Copulatory structures are lacking, however; rather, the male deposits a gelatinous capsule of sperm, the spermatophore, that the female picks up and takes internally with the cloacal lips.

In reptiles, the penis is a spongy, vascular cone of tissue located on the ventral wall of the cloaca. The penis of turtles is a single median structure, but in snakes and lizards it consists of paired hemipenes. Similar unpaired cloacal diverticula form the penis of monotremes and, when present, of birds. Such penes are generally retained inside the cloaca by a retractor penis muscle: erection occurs by relaxation of this muscle and engorgement of the vascular tissue with blood, with the result that the penis is exserted through the cloacal opening. Semen flows from the cloacal openings of the deferent ducts along a groove in the penis into the female cloaca. Most birds, incidentally, lack a penis, and transfer of semen occurs by apposition of the cloacal lips of the male and female during copulation. The mammalian penis arises as a conical outgrowth of skin and mesenchyme outside the ventral margin of the cloaca. During formation of the urogential sinus, the urethral canal extending along the penis as a groove subsequently closes to form the penile segment of the urethra. Two types of penes (with complete intergradation) can be recognized with respect to structure and method of erection. In the first, represented by man, the penis consists of several masses of spongy vascular tissue enclosing the urethra, and erection occurs by engorgement of these masses with blood. In the second type, represented by the bull, the amount of erectile tissue is small, and the penis consists mainly of fibroelastic connective tissue. When not erect, the penis is retracted into a fold of skin, the prepuce (equivalent to but much deeper than the foreskin of man), that is attached to the ventral abdominal wall. Retraction is maintained by a pair of retractor penis muscles. Erection occurs by relaxation of these muscles, allowing the penis to protrude from the prepuce. Engorgement of the erectile tissue with blood causes it to become more rigid but does not affect much change in size.

THE FEMALE. The basic elements of the female reproductive tract in all vertebrates except teleosts are paired Mullerian ducts and the cloaca. The Mullerian ducts extend along the dorsal wall of the body cavity of the embryo lateral to the mesonephric ducts and join the cloaca to form a Y-shaped pattern that persists throughout the vertebrates (Figure 12-11). The Mullerian ducts originate as paired longitudinal folds in the same ridge of coelomic mesoderm that gives rise to the pronephric ducts, in some cases by an actual longitudinal split in the latter. Thus, there is a degree of homology between the male and the female duct systems. The further development of the Mullerian ducts is independent of the opisthonephric and archinephric ducts, however; in the female, the urinary and reproductive systems remain separate to the level of the cloaca, where they join much as in the male. In the frog, for example, the anterior ends of the Mullerian ducts open into the coelom by a pair of funnel-shaped pores known as ostia (singular, ostium). During ovulation, the eggs are released into the coelomic cavity, where they are swept by ciliary and muscular activity of the abdominal wall into the ostia and from there by muscular peristalsis and ciliary action down the oviducts. In cross section, the fully differentiated ducts consist of an inner epithelial, or mucosal, layer and an outer muscular layer. The mucosal layer contains glands and is often highly folded and richly vascularized. The muscle layer consists of circular and longitudinal layers that may be relatively thin or very thick and powerful.

Differentiation of the Mullerian ducts into the definitive oviducts involves relatively little change in the basic embryonic pattern, other than longitudinal differentiation into zones specialized among vertebrates for particular functions. These include specialized regions for secretion of gelatinous coats or shells for egg storage, for gestation, for copulation and, in some cases, for sperm storage.

Fish

In elasmobranchs, an enlarged glandular region of the upper part of the oviduct (known as the nidamental gland) secretes, in sequence, albuminous, mucous, and horny layers around the eggs as they move down the duct. The uppermost ends of the two oviducts are united in elasmobranchs and open by a single ostium near the ovary. The posterior parts of the oviducts are expanded into so-called uteri, or ovisacs. In viviparous species, the mucosa in this uterine region is greatly folded and contains extremely numerous, minute, fingerlike projections called trophonemata. These structures are richly vascular and secretory and presumably function in exchange of materials between the mother and fetus during gestation. The oviducts of amphibians contain several distinct glandular zones whose sequence from anterior to posterior corresponds with the sequence of jelly membranes that are secreted around the egg. The lower end of the oviducts can be expanded into ovisacs in which the eggs accumulate during the interval between ovulation and actual spawning. In birds and reptiles, where very elaborate layers of albumin, shell membranes, and membranous or calcareous shells are secreted around the eggs, the zonation of the oviduct may be correspondingly complex. Most birds have but a single oviduct, the left (corresponding to the presence of a single left ovary), the right Mullerian duct rudiment undergoing degeneration during development of the reproductive tract. The ostium of birds is developed into a muscular funnel that grasps the ripe follicle in such a way that upon ovulation the ovum is released directly into the oviduct and does not fall free into the coelom.

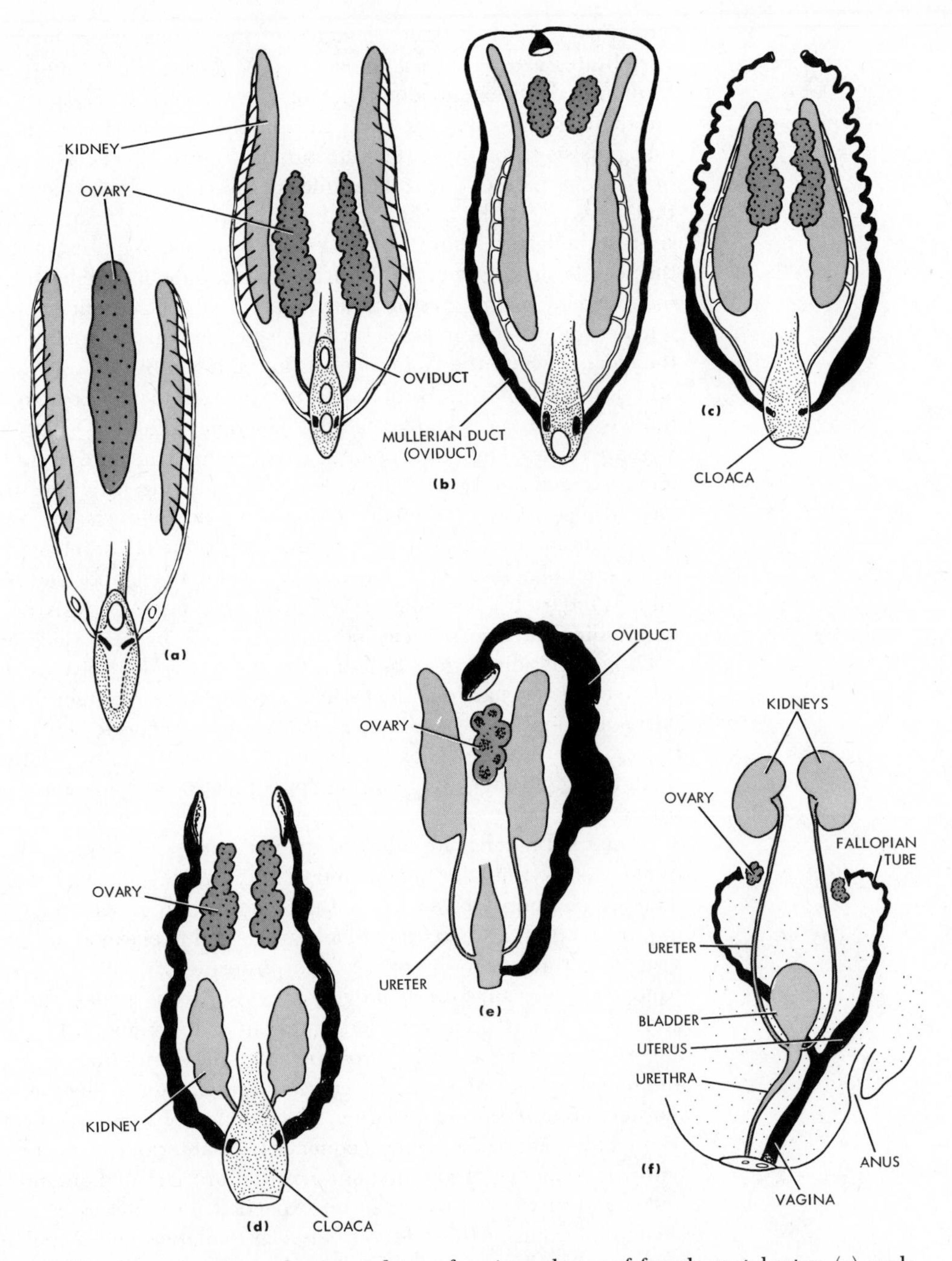

Figure 12-11 Reproductive and urinary ducts of various classes of female vertebrates: (a) cyclostome, (b) teleost, (c) elasmobranch, (d) reptile, (e) bird, (f) mammal (lower end rotated to give lateral view). Kidneys and Wolffian ducts, light gray; Mullerian ducts and derivatives, black; ovaries, darker gray; cloaca and derivatives, stippled.

In placental mammals, the anterior part of the Mullerian duct forms a small convoluted segment called the fallopian tube (also called the oviduct, although it is equivalent to only the anterior part of the oviducts of lower animals). The ostia are expanded into funnels that cradle the ovaries and sometimes grow completely around them to form ovarian sacs or bursae. The posterior part of the Mullerian duct is enlarged with well-developed mucosal and muscle layers and forms the uterus. In monotremes, the condition of the ducts is similar to that seen in reptiles: completely separate oviducts open independently into the cloaca.

In higher mammals, two major trends are seen in the evolution of the lower regions of the reproductive tract.

The first trend is to formation of the urogenital sinus, eliminating the cloaca, much as described for the male (Figure 12-10). In the female, however, the urogenital sinus is further subdivided into a urethral portion that drains the bladder and a vaginal portion that, together with the posterior ends of the Mullerian ducts, forms the vagina. The urethra and the vagina open separately into a shallow depression, the vestibule, enclosed by fleshy folds of skin, the labia. In some rodents and insectivores, the vestibule is absent and the urethra is extended along the clitoris, the female homologue of the penis, to terminate on its tip. In these cases, completely separate urinary and genital orifices are formed.

The second trend is for fusion of the lower ends of the Mullerian ducts to form a single median vagina and uterus. In marsupials, these ducts remain separate and paired vaginae open into the urogenital sinus. In some marsupials, in addition to the paired lateral vagina canals, a median vaginal canal forms secondarily. The latter may form a complete connection with the urogential sinus only during pregnancy and seems to serve as the birth canal, whereas the lateral vaginae function in copulation. Corresponding to this anatomical feature of the female tract, the glans of the penis of male marsupials is bifurcated. In the placental mammals, the degree of fusion of the uterine portions of the Mullerian ducts varies from no fusion—in which case completely separate uteri, or uterine horns as they also are called, open into the vagina—to complete fusion to the level of the fallopian tubes (Figure 12-12). This trend toward reduction of the uterus to a single chamber correlates reasonably well with the trend toward reduction in the number of offspring produced in a single pregnancy.

Teleosts are exceptional to the general plan of the female reproductive system in that the oviducts arise not from Mullerian ducts but by posterior extension of the peritoneal folds that create the ovarian cavity. Thus, the cavity of the ovary and oviducts is continuous, and the ova are not released directly into the body cavity as in other forms. This structure has been interpreted as an adaptation to facilitate the rapid release of the large numbers of eggs produced by many teleosts. In fact, however, the oviducts secondarily degenerate in some species, with the result that the eggs are ovulated into the body cavity and escape to the exterior through coelomic pores, much as in cyclostomes. Because some species in which this condition is found produce numerous eggs, for example various species of salmon, this interpretation of the relationship between the ovary and oviduct may not be correct.

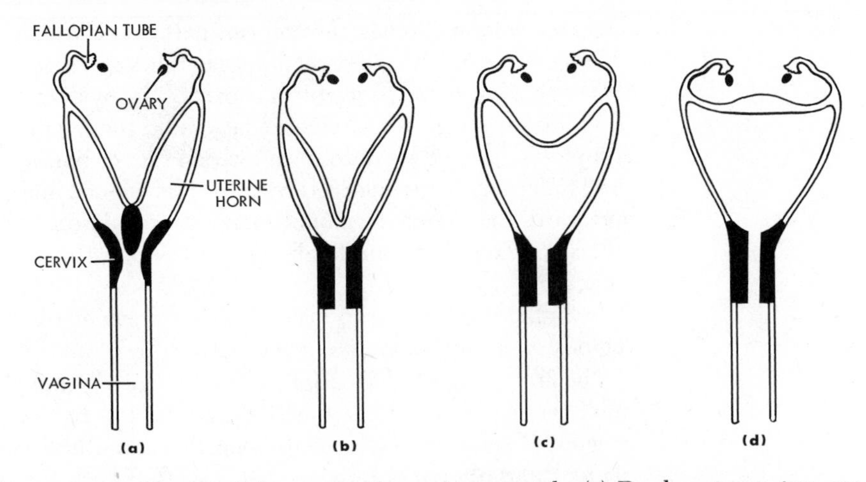

Figure 12-12 Degrees of fusion of the uterine horns in mammals. (a) Duplex uterus: two cervixes, no uterine body, horns completely separated. (b) Bipartite uterus; one cervix, uterine bodies small. (c) Bicornuate uterus; one cervix, uterine bodies prominent. (d) Simplex uterus, one cervix, uterine body very prominent, horns absent. (From A. V. Nalbandov, *Reproductive Physiology*, 2nd ed. San Francisco: Freeman, 1964.)

Intersexuality, Sexual Development, and Sex Reversal

Separate sexes are so overwhelmingly the rule, and sexual dimorphism so pronounced in familiar animals, that we find the concept of intersexuality to be incredible. In fact, intersexuality, in the form of hermaphroditism—defined as the production of both eggs and sperm by a single individual—is a prevailing condition in most organisms. In vertebrates it is clear that the genotype, although favoring the differentiation of either a male or a female phenotype, is sufficiently indeterminate in its instructions to allow for a complete range of intersexual phenotypes, or even total sex reversal. Intersexuality is rare in vertebrates, but has been observed in all classes. The study of this condition has contributed very significantly to our understanding of normal sexual morphology.

In vertebrates, every degree of intersexuality occurs, ranging from complete functional hermaphroditism to mere persistence of vestiges of the sexual rudiments of the opposite sex. This latter condition is a common and normal state and the term *intersexuality* is usually applied only to more extreme degrees of admixture of male and female traits. In certain families of bony fish, functional hermaphroditism is common. Maleness and femaleness may be expressed in sequence as the animal grows older, or they may occur simultaneously. In the first case, the gonad functions first as a testis, then as an ovary (or vice versa), whereas in the latter it develops into an ovotestis that produces both eggs and sperm simultaneously. In some cases, such hermaphroditic individuals may be self-fertile (fertilization occuring in the gonad), but usually they are not. Complete hermaphroditism is rare in other groups of vertebrates, although true hermaphrodites with ovotestes and complete male and female duct systems have been found in all classes and potentially could occur in any species. Although morphologically bisexual, such individuals are very likely sterile as a result of slight imperfections in sexual morphology and behavior. The more common form of intersexuality in vertebrates is one in which the gonadal sex is unambiguous, these organs being distinctly ovaries or testes, but the reproductive ducts and accessory structures of the opposite sex are developed to an anomalous degree. This condition is sometimes called pseudohermaphroditism.

The fact that both male and female structures are differentiated in hermaphrodites suggests that complete sex reversal is possible in vertebrates. In fact, reversal of sex can readily be induced in the laboratory by various experimental treatments, most notably by treatment with sex hormones. Instances of sex reversal are reported from time to time in domesticated birds and lower animals, but does not occur after embryonic development in mammals. Occasional publicized instances of sex "reversal" in man are, in reality, instances of pseudohermaphroditism in which the bisexual anatomy is surgically modified to emphasize the desired traits and eliminate those that are not desired. Sex reversal probably occurs in wild animals but, if complete, is almost certain to go undetected. Sex reversal is defined as the differentiation of a sexual phenotype that is the opposite of the sex genotype. Hence, proof of reversal, when there is no unusual degree of intersexuality to give a hint of the change, requires identification of the sex genotype. This can be done by breeding experiments, as is explained in Figure 12-13, but is much more simply done by cytological examination of the sex chromosomes. The Barr test is such a method and makes use of the fact that the sex chromatin of certain kinds of cells had a distinctive position and staining properties in the female.

Sexual Differentiation

The possibility for intersexuality and sex reversal rests basically in the fact that up to a certain point in development every embryo is morphologically bisexual, regardless of its genetic sex. That is, a complete set of both male and female sexual rudiments is formed in the early embryo. These include (1) the gonadal rudiments, (2) the Wolffian and Mullerian ducts and associated mesonephric structures, (3) the cloaca or urogenital sinus, and (4) the phallus or genital tubercle. Normal sexual development involves the growth and differentiation of one set of these rudiments, whereas the other fails to develop beyond the rudimentary condition or undergoes gradual involution (Figure 12-14). With respect to the nature of the bisexual potential, these rudiments fall into two categories: (1) those that are truly sexually indifferent, being capable of developing in either the male or the female direction; and (2) those that are destined to differentiate in only one sex or the other. In the first category are the gonads, cloaca or urogenital sinus, and genital tubercle; in the latter are the Mullerian and mesonephric ducts and their derivatives.

Generally, the gonads undergo definitive sexual differentiation relatively early, especially in higher vertebrates. In cyclostomes and many teleosts, however, they may undergo definitive differentiation into ovaries or testes at sexual maturation. As has been noted already, several families of teleosts retain the hermaphroditic condition as adults and have a functional ovotestis. Cyclostomes, however, do not remain hermaphroditic but form definitive ovaries or testes at maturity. Occasional cases of persistence of rudimentary gonadal bisexuality occur in higher vertebrates. Male toads possess, at the anterior end of the testes, a vestige of the rudimentary gonad that is capable of differentiating into ovarian tissue, but in the normal male is repressed from doing so. If the testes are removed, this structure, known as Bidder's organ, can develop into a fertile ovary, and such individuals have become functional

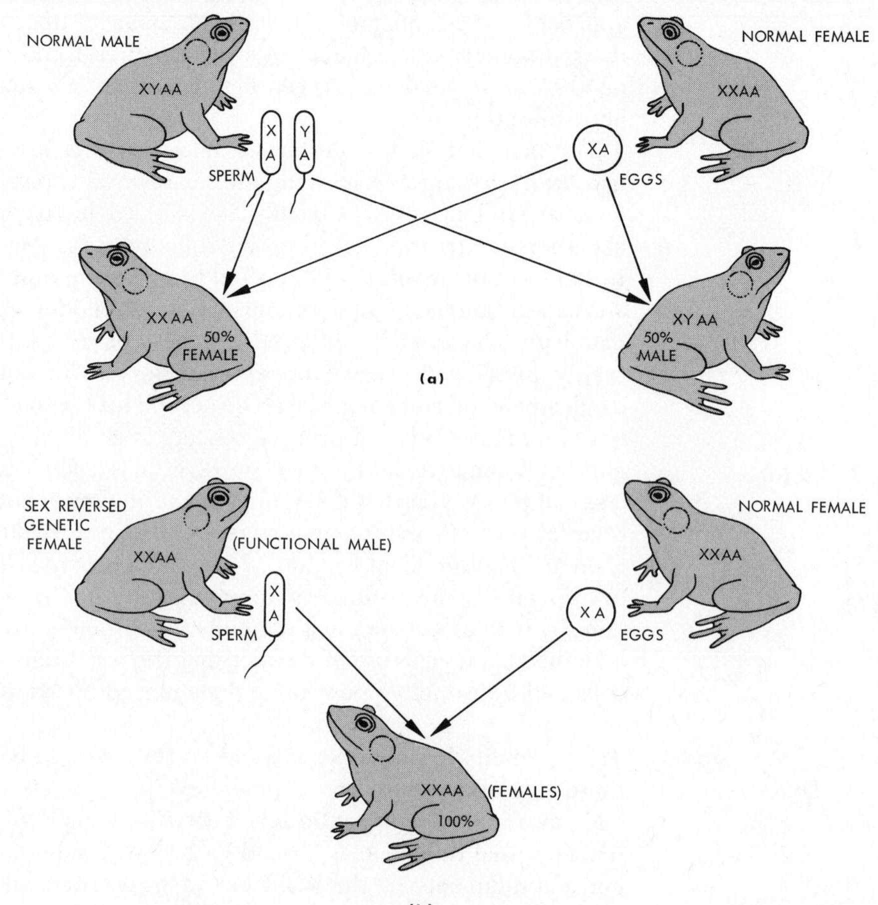

Figure 12-13 The chromosomal basis of sex determination. (a) In the normal male, two classes of sperm, X and Y, are produced. The sex of the offspring is determined by the kind of sperm which fertilizes the egg. (b) In the sex reversed female, the genetic sex is not altered. Consequently, only one class of sperm, X-bearing, is produced, and the offspring are all female. A refers to the autosomes. In this example the male is heterogamic, i.e., produces two classes of gametes, but in many species the female may be the heterogamic sex. In that condition the female is designated WZAA and the male WWAA.

females and produced fertile eggs. This degree of functional sex reversal depends on the fact that, in the toad, as in most male amphibians, the Mullerian ducts persist throughout life and can undergo functional development in the presence of female sex hormones. In birds, the right gonadal rudiment has little or no capability to form an ovary. Consequently, in the female, only the left ovary ordinarily develops. But, if the ovary is removed or destroyed by disease, the rudimentary right gonad may then undergo development to form a testis. Although it often has been claimed that such reversed hens function as fathers, this is not well authenticated. Its veracity is doubtful because the male duct system is absent or imperfectly developed; however, they may develop cock feathering, the ability to crow, and male behavior.

The rudimentary ducts and accessory structures usually persist in the bisexual condition much later in development than the gonads. Sometimes they persist indefinitely, as the Mullerian ducts of male amphibians. Generally, however, they regress to a vestigial condition before sexual maturity. Intersexuality results when both sets of rudiments persist and develop to an unusual degree. Reversal of sex occurs when the set that undergoes definitive development is the opposite of the genotype. Complete reversal can occur only at a stage when regression of the alternate set of rudiments and differentiation of the indifferent rudiments has not proceeded to an irreversible point.

Consideration of the nature of the bisexual organization of the embryonic gonad has special relevance to our understanding of the organization of the reproductive system of the adult, with respect both to the relationships between the ovaries and testes and to their relationship to the sex ducts. In its

Figure 12-14 Differentiation of the sexual rudiments of the mammalian embryo. Compare the fate of the genital tubercle, urogenital sinus, Mullerian ducts, Wolffian ducts, and gonads on differentiation in either the male or the female direction. (From B. Frye, *Hormonal Control in Vertebrates.* New York: Macmillan, 1967.)

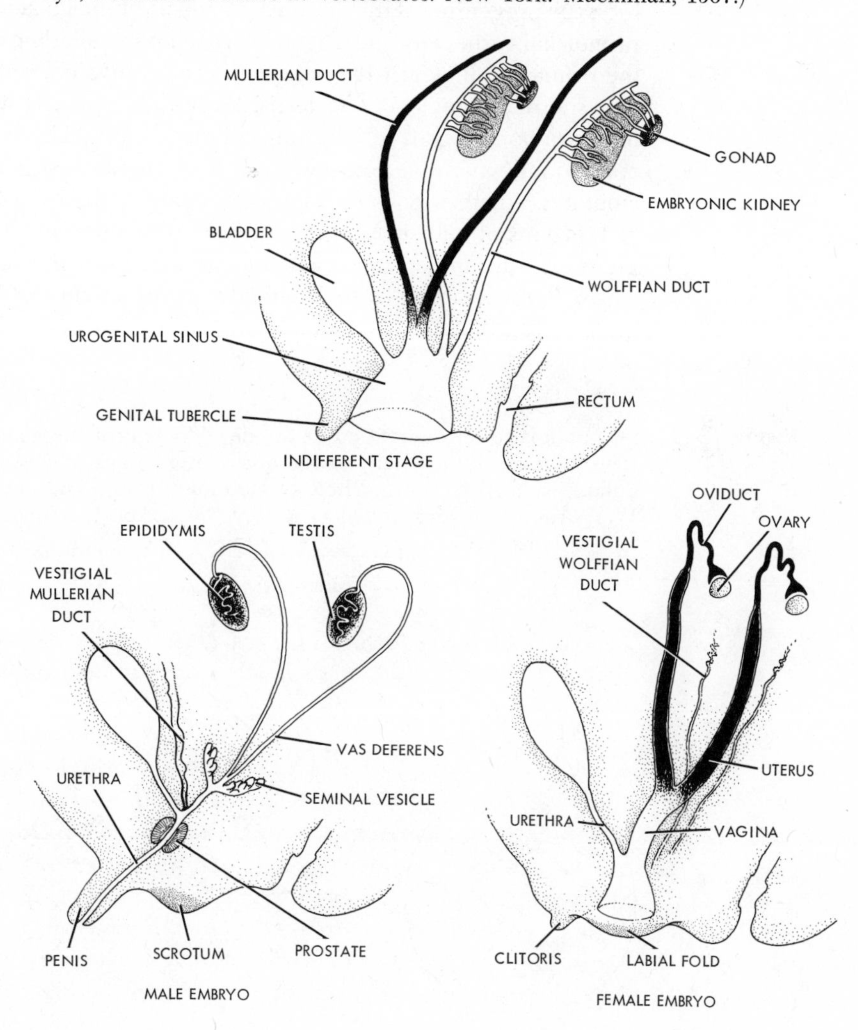

initial stage, the gonad rudiment consists of a slight thickening of the peritoneal epithelium overlying the medial border of each mesonephric rudiment. Shortly, a thickening of mesonephric mesenchyme forms underneath the peritoneal rudiment, the resulting bulge constituting the genital ridge. Into this ridge migrate the primordial germ cells. Subsequent further thickening of these two layers results in the formation of a double-layered gonad rudiment consisting of an inner medulla derived from the mesonephric mesenchyme and an outer cortex derived from the peritoneal epithelium. Put simply, these two layers constitute the rudiments of the testis and ovary respectively. In a normal male (or sex-reversed female), the medulla differentiates into a testis and the cortex regresses. In a normal female (or sex-reversed male), the cortex develops into the ovary and the medulla regresses (Figure 12-15).

This description is the standard model, and it explains the bisexual potential of the rudimentary gonad. However, it is not entirely accurate and it does not fully explain the significance of the association between the original peritoneal and mesenchymal thickenings. A broader view of the role of each of these two components is as follows: the peritoneal thickening constitutes the progenitor of the germinal elements of the gonads. In it are lodged the primordial germ cells that, together with the surrounding epithelial cells, form the primary follicles of the ovary, and the equivalent units—the Sertoli cells plus spermatogonia—of the testis. Most workers agree that these structural units are homologous in all groups of vertebrates. This role of the peritoneal epithelium, which we can now call the germinal epithelium, is very clear in amniotes, where it goes through two separate phases of proliferation. In the first, groups of cells from the germinal epithelium and associated germ cells grow into the underlying mesenchyme and form the so-called primary sex cords. If the embryo is to become a male, no further proliferation of the germinal epithelium occurs and the primary sex cords differentiate into the

Figure 12-15 Diagrammatic representation of the development of the gonad. The broken arrows indicate the mutually antagonistic interactions between the cortex and medulla. (From R. K. Burns, "Role of Hormones in the Differentiation of Sex," in W. C. Young, ed., *Sex and Internal Secretions,* Vol. I, 3rd ed. Baltimore: Williams and Wilkins, 1961.)

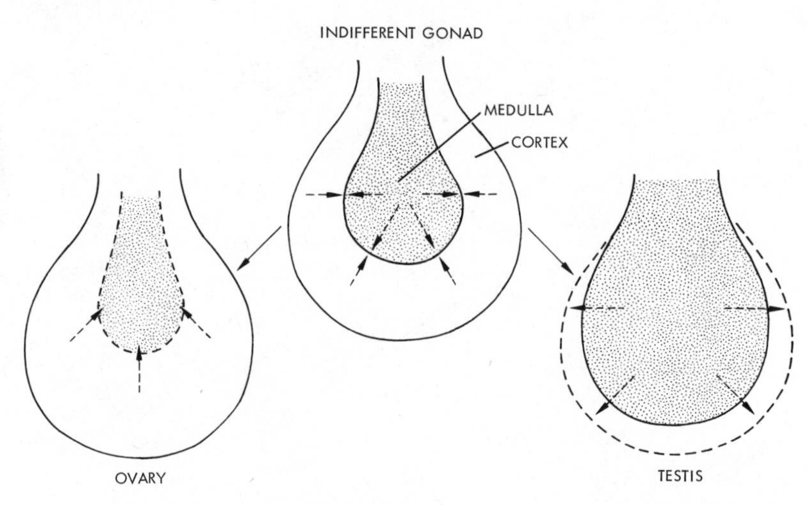

seminiferous tubules. In a female, this first wave of sex cords begins to regress, together with the mesenchymal blastema. A new wave of proliferation occurs that results in the formation of a thick outer cortical layer surrounding the inner regressing testicular rudiment. These phases of gonadal development are illustrated in Figure 12-16.

Figure 12-16 Formation of the primary and secondary sex cords from the germinal epithelium, and differentiation of testis or ovary: (a) origin of the primary sex cords; (b) the indifferent gonad stage, with well-developed primary sex cords and thick germinal epithelium; (c) differentiation of the testis, with further development of the primary sex cords and reduction of the germinal epithelium to a thin peritoneal covering overlying a connective tissue layer, the tunica albuginea; (d) differentiation of the ovary, with regression of the primary sex cords, and proliferation from the germinal epithelium of the secondary sex cords which form the cortex of the rudimentary gonad, and differentiate into the ovary. (From R. K. Burns, "Role of Hormones in the Differentiation of Sex," in W. C. Young, ed., *Sex and Internal Secretions,* Vol. I, 3rd ed. Baltimore: Williams and Wilkins, 1961.)

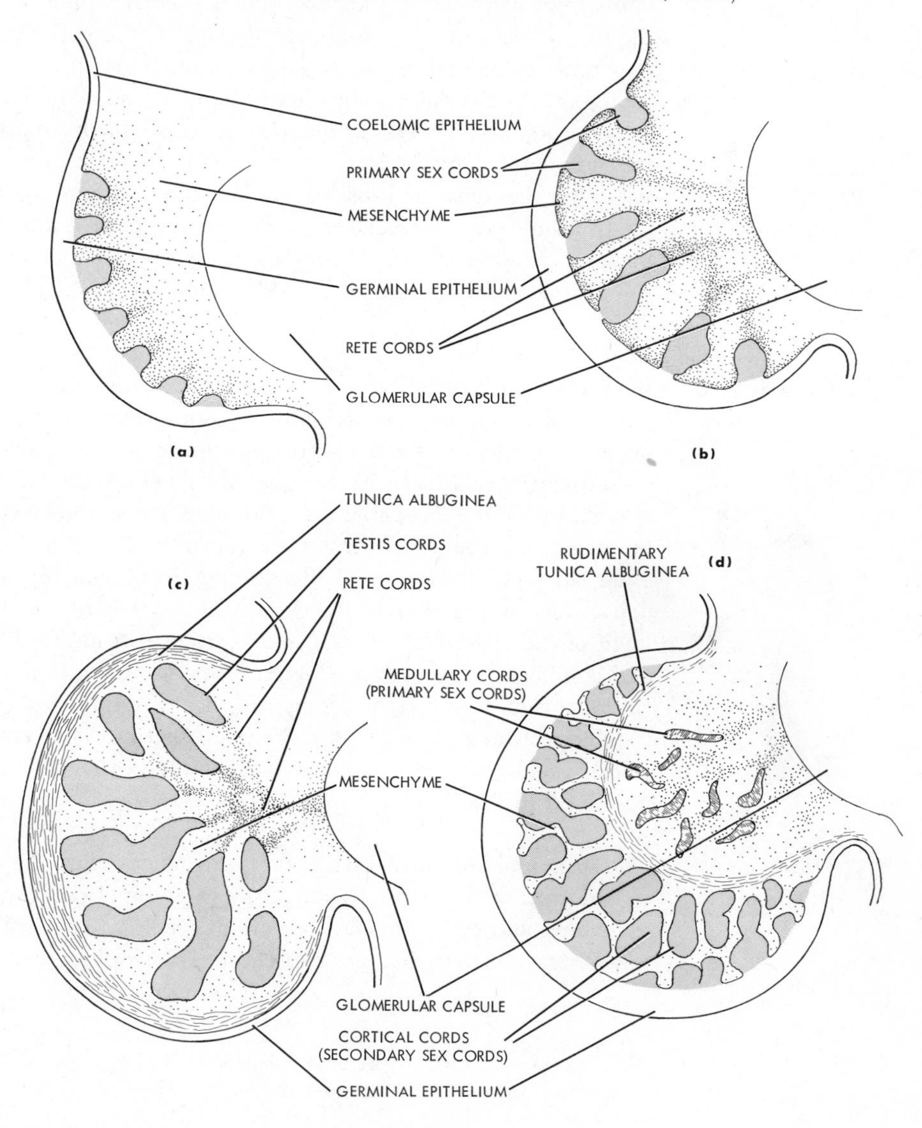

In the development of the testis, the medullary or mesenchymal component contributes the rete tubules and, together with the contiguous mesonephric mesenchyme, the efferent tubules. These ultimately connect the seminiferous tubules with the mesonepheric tubules and, thus, establish the path of sperm exit from the testis into the urinary ducts. In addition, it forms a supporting framework of mesenchymal tissue that gives rise to the interstitial tissue of the testis and, in some species, apparently forms the basic framework of the tubules within which the germinal elements become lodged. In addition to its structural contributions, we can suppose that the mesenchymal blastema in some way induces or supports the ingrowth of the primary sex cords and their differentiation into the seminiferous tubules. If this is true, and it is not proven, then the mesonephric mesenchyme of the genital ridge plays a key role in determining whether the gonadal rudiment will develop as an ovary or a testis. In the genetic male, differentiation of the primary sex cords into testicular tubules is supported by the medullary mesenchyme, and the cortex or germinal epithelium regresses; in the genetic female, this support does not persist long; the medullary testicular rudiment degenerates and the germinal epithelium undergoes a second wave of proliferation to form the cortical or ovarian rudiment. If this relationship is true, then the connection between the urinary and reproductive systems in the male is not so much a matter of "invasion" of the mesonephros by the testis, as is sometimes said, as it is a "seduction" of the testis by the mesonephros. Whatever the mechanism, the morphological relationships point quite strongly to some regulative interaction between the peritoneal and mesonephric components of the urogenital ridge in determining the ultimate direction of sexual differentiation.

The Physiology of Reproduction*

Reproduction in vertebrates requires close coordination between the functional development of the gonads and accessory reproductive organs and sexual behavior. Generally speaking, the morphological and behavioral changes that accompany sexual maturity are geared to environmental conditions in such a way that reproductive capability and activity occur in cycles coordinated with the seasons. Primary control over reproductive functions is vested in the endocrine system. Hormonal effects range from control over the gametogenic and endocrine functions of the gonads and control of the growth and development of the reproductive tract and accessory structures to control over male and female patterns of reproductive behavior. Gestation and parental care of the offspring, when they occur, are also strongly dependent on hormonal regulation. Integration of reproductive functions with environmental conditions is achieved through the hypothalamic-pituitary-neurosecretory system described in Chapter 11.

Hormones of Reproduction

The principal hormones involved in reproduction are the sex steroids and the gonadotropins. The steroids constitute three functional categories: androgens, estrogens, and progestogens, that are, respectively masculinizing, feminizing, and gestational in their activities. The gonadotropins control the endocrine and

*This section is taken, in part, from Chapter 6 of B. E. Frye, *Hormonal Control in Vertebrates* (New York: Macmillan Publishing Co., Inc., 1967, pp. 83–99.)

gametogenic functions of the gonads, as shown in Figure 12-17. *A synopsis of* the principal effects of the steroid homones and gonadotropins follows.

Androgens have profound effects on the male reproductive tract, including the testes themselves, and the accessory structures. Androgens support spermatogenesis in the seminiferous tubules. Because hypophysectomy abolishes spermatogenesis, pituitary gonadotropins usually are regarded as exerting primary control over this function. However, because androgen injections will prevent this effect of hypophysectomy under some circumstances, the gametogenic action of gonadotropins may, in part, be secondary to their stimulatory effect on androgen secretion. The sperm ducts and accessory sex glands are almost totally dependent on androgens. This is dramatically illustrated by the effects of castration and androgen injections on these organs (Figure 12-18). Injection of androgens into a castrate or immature animal induces rapid development of these organs to mature size and function. Numerous biochemical properties of the accessory sex glands and their secretions are dependent on androgens. For example, the active accessory glands of mammals have a high content of fructose, citric acid, and certain enzymes, notably phosphatases. The content of these substances falls drastically after castration and is elevated to normal or above by androgen treatment.

Androgens also determine the secondary sexual characteristics of vertebrates. In man, there are obvious hormonally determined sexual differences in hair pattern, pitch of the voice, muscle development, skeletal size, fat distribution, skin pigmentation and texture, and many other features. In other

Figure 12-17 Neuroendocrine control of reproduction. (From B. Frye, *Hormonal Control in Vertebrates.* New York: Macmillan, 1967.)

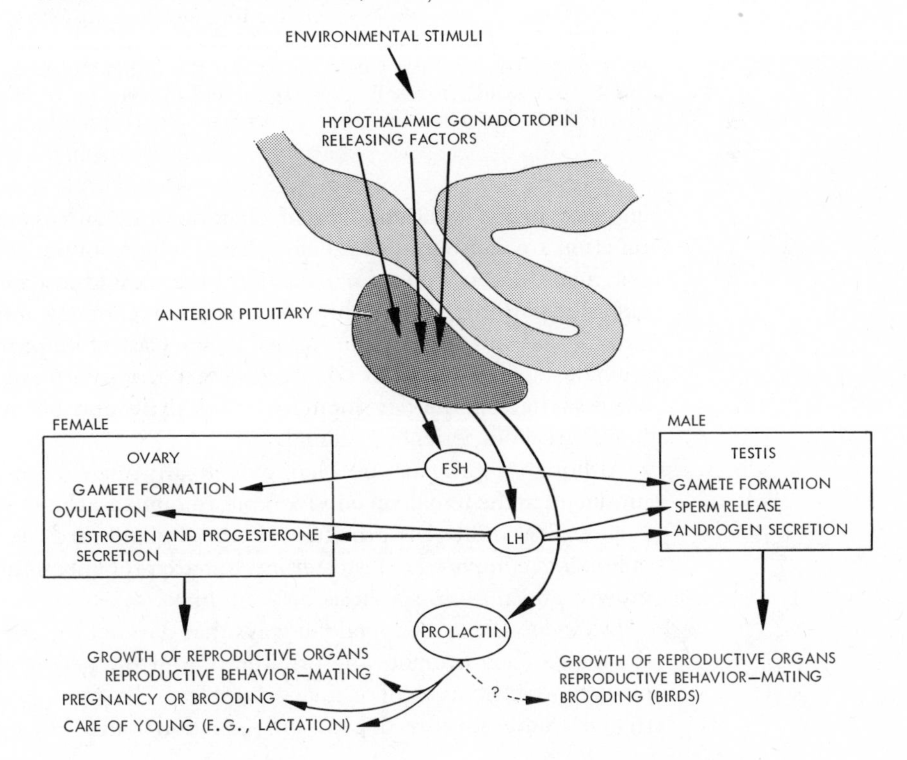

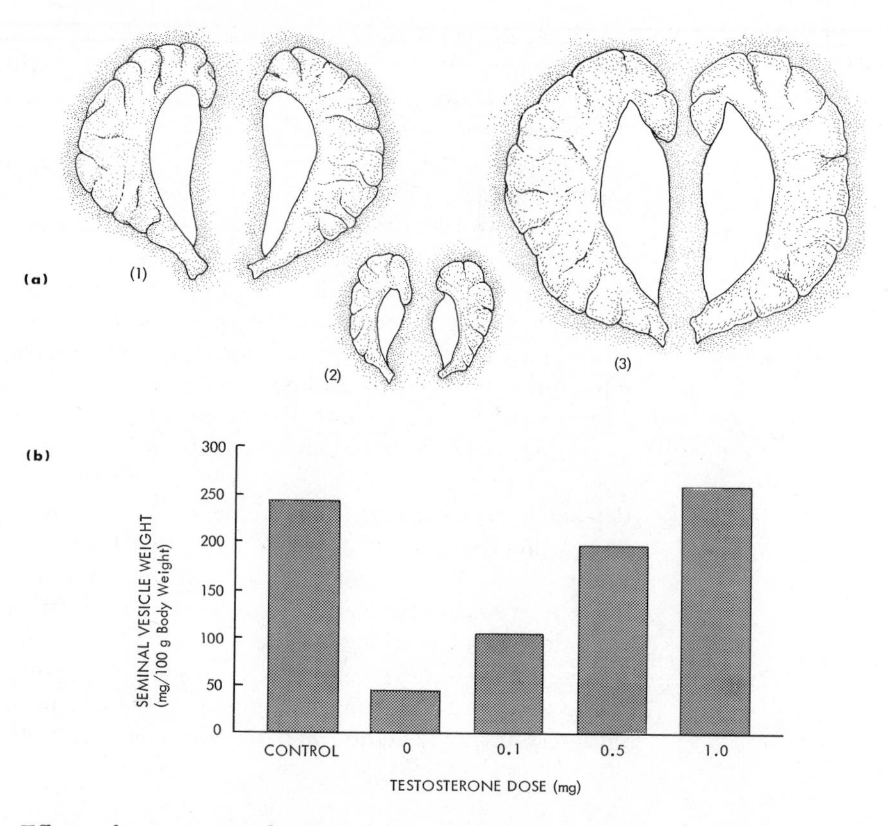

Figure 12-18 Effects of castration and testosterone replacement on the seminal vesicles of the rat. (a) Drawings of glands from (1) normal, (2) castrates of three weeks, and (3) castrates of three weeks given 1 milligram of testosterone per day during the last week. (b) Assay of testosterone activity by seminal vesicle weight. The rats were castrated for three weeks and given the indicated doses of testosterone daily for the last week. (From B. Frye, *Hormonal Control in Vertebrates*. New York: Macmillan, 1967.)

animals, many additional sexual characters of an ornamental or defensive function are androgen dependent. These include not only morphological traits, but many aspects of male behavior such as aggressiveness, sex drive, and complex patterns of courtship and mating. Treatment of females with androgens often leads to masculine sexual drive. Castration of mature, experienced animals may have little or no effect on sexual activity, even though androgens are essential for the development of sex drive and behavior patterns in the immature, inexperienced animal.

Although androgens are characteristically male hormones, they are also produced in the female in considerable amounts by the ovaries and the adrenal cortex. The androgens produced by these glands are relatively weakly masculinizing, compared to testosterone, but are probably significant in promoting growth and protein synthesis in the female.

Estrogens affect the female in ways that parallel the effects of androgens in the male. They promote the proliferation of the germinal epithelium of the ovary and formation of primary follicles. The later growth phases of the follicles and oöcytes depend on gonadotropins, however, and complete oögenesis cannot be supported by estrogens. In fact, there is evidence that estrogens in high concentrations will cause atresia of growing follicles. This is probably a consequence of negative feedback inhibition of gonadotropin secretion. Ovariectomy results in regression of the mature female reproductive tract, but development of the Mullerian ducts during the embryonic phase may occur in the absence of the ovaries. Injected estrogens cause growth and development of the female ducts of immature or castrated animals. Estrogens are required for the development of many of the secondary sexual characteristics of the female, but not uncommonly female traits seem to depend as much on the absence of masculinizing levels of androgens as on the presence of estrogens. The smaller size of the female is compared to the male in many species of birds and mammals is caused by a growth-inhibiting effect of estrogens on the skeleton, contrasted with a strong growth-promoting effect of testicular androgens on muscle and certain visceral organs. Many of the reproductive structures and functions of the female mammal require synergistic or sequential action of estrogens and progesterone, in the sequence seen in the normal sexual cycle (see "Progestogens"). These include normal cycles of uterine growth, mammary gland growth and secretion, and sexual behavior.

The testes produce estrogens too, but their function in the male is not well understood. The most probable role of estrogens in the male is regulation of gonadotropin secretion through negative feedback inhibition, an effect that secondarily may help regulate the process of spermatogenesis and male sexual cycles.

Progestogens characteristically function to support pregnancy and lactation in mammals. Progesterone is the major natural progestogen. The effects of progestogens often depend on a previous period of estrogen "priming," during which the target tissues reach a stage of development in which they are sensitive to progestogens. Indeed, when given to a castrated or immature animal, progestogens have little effect unless estrogens are administered prior to, or synchronously with, the progestogens. Thus, progesterone promotes the differentiation of the estrogen-primed uterus to a progestational state in which it will respond to the embryo by implantation and, combined with estrogen, supports the growth of the uterus and retention of the placenta and fetus during pregnancy. Progesterone is necessary for full development of the mammary glands to a presecretory state. Although progesterone is secreted by both the ovaries and the testes in many lower vertebrates, its function in most groups is very unclear. In female birds, progesterone stimulates albumin secretion by the oviduct, and there is some evidence that it has a similar effect on jelly secretion by the amphibian oviduct.

Progesterone stimulates ovulation. This effect, which has been studied most in birds and mammals, depends very much on the time of the sexual cycle at which the hormone is administered; at certain times, the opposite, inhibitory, effect is obtained. The ovulation-inducing effect of progestogens probably occurs through an effect on the hypothalamus, causing secretion of luteinizing hormone, which causes ovulation. During the pre- or postovulatory phases of the sexual cycle, progesterone inhibits secretion of gonadotropins, particularly the follicle-stimulating hormone. The effect of this is to block the development of follicles during pregnancy, a period when progesterone levels are particularly high. These effects of progestogens on gonadotropin secretion in birds and mammals suggest the possibility that an important physiological role of progesterone in lower vertebrates of both sexes might be the regulation of gonadotropin secretion and sexual cycles.

This synopsis of the effects of steroid hormones supports the general view that the action of male hormones is specific to male traits and the action of female hormones is specific to female traits. Although basically correct, this view must be qualified in certain respects:

1. Some traits are hormone-independent and are determined directly by the sexual genotype of the organism. For example, the scrotum of the male opossum and its homologue, the pouch in the female, develop according to the genetic sex of the individual even when total sex reversal of the gonads and reproductive tract has been caused by treatment of the new born opossum with sex hormones (Figure 12-19).
2. Sexual structures that are identical or homologous in both sexes may be dependent on the same hormones. For example, many female rodents possess well-developed prostate glands that are maintained by androgens secreted by the ovaries and adrenals. The clitoris, homologue of the penis, is sensitive to androgens; high doses of androgens during development cause it to enlarge to a size equivalent to the penis. Muscle growth in both sexes is sensitive to androgens, and differences in the degree of muscle development in males and females is due to differences in the amount and degree of

Figure 12-19 Effects of sex hormones upon the differentiation of the external reproductive organs of the opossum: (a) normal male and female; (b) male and female treated with androgen; (c) male and female treated with estrogen. Note that although the phallus undergoes dramatic enlargement and reversal in response to hormone treatment, the scrotum and pouch remain unchanged. (From R. K. Burns, "Role of Hormones in the Differentiation of Sex," in W. C. Young, ed., *Sex and Internal Secretions,* Vol. I, 3rd ed. Baltimore: Williams and Wilkins, 1961.)

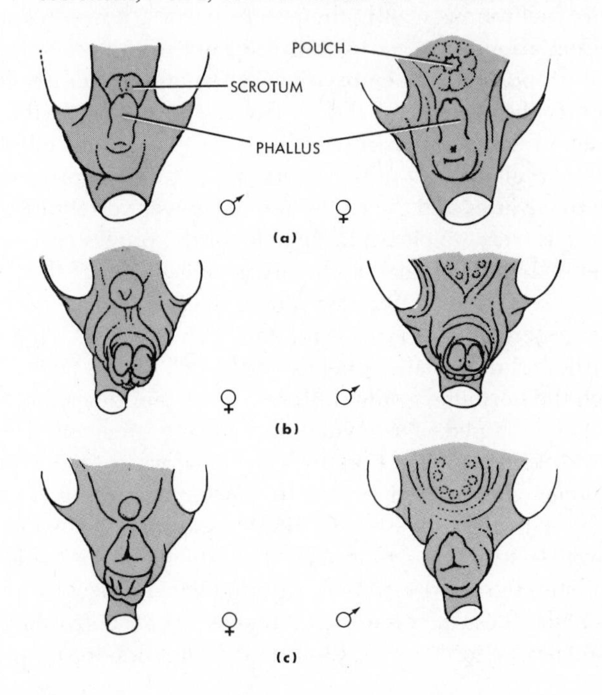

androgenicity of the androgens produced rather than to basic differences in the response.

3. Sexually dimorphic characters may depend on hormones in one sex only, the contrasting sexual character developing as a consequence of the absence of a hormone rather than the presence of the alternate hormone. The male plumage pattern in some birds is an example. Castration of the male does not affect plumage, but castration of the female results in the development of cock plumage.
4. Sex specific traits may be hormone-dependent but may respond nonspecifically to either androgens or estrogens. For example, the orifice of the vagina of female rodents is closed until puberty when it opens under the influence of estrogens. Injections of androgens also will cause the opening of the vagina. Such nonspecific or paradoxical effects are seen most frequently when large doses of hormones are injected. For example, treatment of embryos with androgens not only accelerates growth of the male organs but also can cause hypertrophy of the Mullerian ducts. Growth of the Mullerian ducts are ordinarily sensitive to estrogens. Paradoxical effects of hormones can be explained theoretically either by metabolic interconversion of steroids in the tissues (e.g., estrogens from androgens) or by a true lack of absolute specificity of the target tissue for either estrogens or androgens.
5. Some sexual traits may depend on hormones other than sex steroids. For example, the brilliant beak and plumage color of the male weaver finch are evoked by the luteinizing hormone. Estrogen inhibits this effect of luteinizing hormone in the female.

Gonadotropins are a basic part of the team of reproductive hormones because of their effects on the gametogenic and hormone-secreting functions of the gonads. Hypophysectomy causes involution of the seminiferous tubules or ampullae and interstitial tissue. So far as steroid dependent secondary sexual characteristics are concerned, hypophysectomy causes changes nearly equivalent to castration. Injection of extracts of the anterior pituitary reverses these effects of hypophysectomy, or causes sexual precocity in immature animals. Purification of pituitary extracts of mammals has led to the identification of two distinct gonadotropins, follicle-stimulating hormone and luteinizing hormone (FSH and LH). FSH stimulates growth of the follicles in the ovary and formation of sperm in the testes. Luteinizing hormone stimulates androgen secretion by the testis and has several effects in the ovary, including induction of ovulation, formation of the corpus luteum from the ovulated follicle, and stimulation of estrogen and progesterone secretion by the follicle and corpus luteum. The pituitaries of lower vertebrates have similar activities, but as yet these activities have not been associated with distinctly different molecules. Some endocrinologists believe that the two classes of gonadotropic activity are properties of one hormone, or at least of two closely related molecules that exhibit some overlap in activity. Some of the effects of gonadotropins are illustrated in Figure 12-20.

Prolactin, or lactogenic hormone, also is classified sometimes as a gonadotropin because in some mammals it stimulates the corpus luteum to secrete progesterone and estrogen. However, this effect seems to be restricted to a few species, notably laboratory rodents, and the more usual reproductive functions of prolactin are not gonadotropic. Among the reproductive effects of prolactin

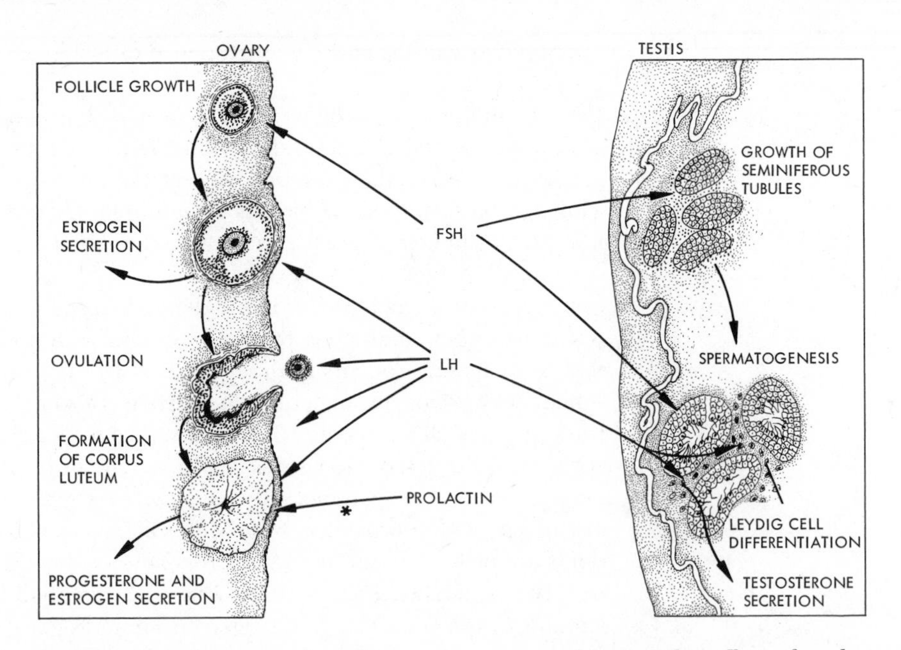

Figure 12-20 Actions of pituitary gonadotropins on the ovary and testis. This effect of prolactin on the ovary (marked by the asterisk) occurs in some rodents, but not in other vertebrates. An effect of prolactin upon the testis has not been demonstrated. (From B. Frye, *Hormonal Control in Vertebrates.* New York: Macmillan, 1967.)

are stimulation of milk secretion by the mammary glands; formation of crop milk by pigeons and related birds; formation of the brood patch in birds; nesting, brooding, and parental-care behavior in a variety of species; and migratory activity related to reproduction.

Hormonal Control of Sexual Differentiation

Sex determination is basically genetic: males and females differ in one pair of chromosomes, the so-called sex chromosomes. In some animals, notably insects, the genotype rigidly influences the phenotypic sex of the offspring. In vertebrates, however, the genetic sex predisposes the development of an embryo in either the male or the female direction, but the actual differentiation of phenotypic sex is under hormonal control. This was first discovered through studies of a spontaneous sexual abnormality in cattle known as the freemartin. A freemartin is a sexually abnormal female calf, born as a twin to a normal male calf. Although sterile, a freemartin is essentially a true hermaphrodite, having more or less well-developed components of both the male and the female duct systems and gonads with both ovarian and testicular regions. In 1916, F. R. Lillie noted that the placental circulation of the male and female twins always is fused in the case of freemartins and suggested that the sexual modification of the female occurs as a result of transfusion of androgens across the placental union from the male to the female. As a general theory of sex differentiation, Lillie proposed that the rudimentary gonad secretes male or female sex hormones according to its presumptive or genetic sex. These hormones act on the sexual rudiments to support differentiation in either the male or the female direction. Thus, the embryonic gonad of the genetic female predominantly secretes estrogens, supporting female development, and the embryonic gonad of the genetic male predominately secretes androgens, supporting male development. Spontaneous or experimentally induced intersexuality occurs whenever this normal pattern is reversed, or when mixtures of estrogens and androgens act simultaneously. In the case of the freemartin, the male twin begins sexual differentiation slightly earlier than the female and, thus, initiates modifications in her sexual differentiation before the normal tendencies are strongly expressed.

The hormone theory has been amply confirmed by a variety of methods, including treatment of embryos with pure sex steroids. By this means, complete, functional sex reversals have been induced in all classes of vertebrates, indicating the susceptibility of the sexual rudiments to hormonal influence. Estrogens stimulate development of the Mullerian ducts and their derivatives and cause development of the cortical or ovarian component of the rudimentary gonad and regression of the medullary or testicular component. Androgens stimulate development of the Wolffian ducts and their derivatives and induce regression of the gonadal cortex and development of the testis. In addition, in many species, androgens cause regression of the Mullerian ducts. Estrogens, on the other hand, have little effect on the Wolffian ducts. Castration of embryos at the indifferent stage has shown that the Wolffian ducts not only are stimulated to undergo accelerated development by androgens, but regress in the absence of the testes. The Mullerian ducts, on the other hand, persist after castration. Hence, although stimulated by estrogens, they are not hormone dependent for differentiation through the embryonic phases.

In some respects, the response of the embryonic reproductive system in mammals is exceptional to the hormone theory of sex differentiation. The Mullerian and Wolffian ducts and their derivatives, as well as the urogenital sinus and external genitalia, respond in the expected way, and complete reversals of these structures can be induced by steroid hormones. The gonads, however, respond only to a limited degree in most species, and complete reversal has been achieved in only one species, the American opossum (Figure 12-21). In other mammals, androgens may cause slight hypertrophy of the rudimentary medulla of the female, or estrogens may cause unusual persistence of the germinal epithelium in the male. But, ultimately, such gonads develop into ovaries or testes, according to the genetic sex of the individual, irrespective of hormonal treatment. The placenta is not a hormonally neutral environment, and steroids from the mother readily enter the fetal circulation. Hence, in mammals the gonadal sex-determining mechanism may have evolved toward a more stable genetic control not susceptible to the influence of maternal hormones that would be expected to inhibit normal male development. With respect to this point, it may be significant that in the opossum, which is susceptible to hormonal influence, the young are born at a sexually indifferent stage of development and, hence, do not come under the influence of maternal hormones during the period of sex differentiation. The classic case of hormone-induced sex reversal in mammals, the freemartin, is exceptional among the eutherians, but even in this case the male gonad has not been shown to be susceptible to hormone effects.

Following embryonic sexual differentiation, the reproductive system remains in an essentially infantile condition until the age of sexual maturation, or puberty. Sexual maturation is caused by the secretion of gonadotropins, which stimulate the processes of gametogenesis and hormone secretion by the gonads. Experiments involving the transplantation of immature gonads and pituitaries

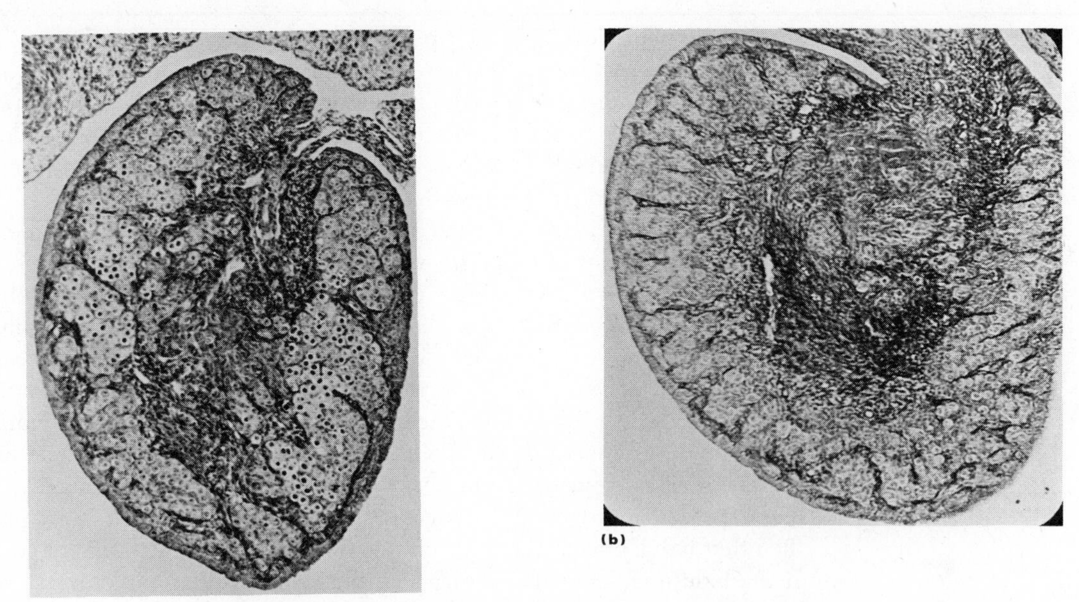

Figure 12-21 Reversal of gonad development in the opossum: (a) normal ovary; (b) testis which has been reversed by estrogen treatment. The reversed gonad has a fully formed cortex, complete with primordial germ cells, and cannot be distinguished from a normal ovary at the equivalent stage of development. Genetic sex was confirmed by the presence of the scrotum (see Figure 12-19). (Photographs courtesy of R. K. Burns.)

to adult animals indicate that these organs are capable of mature function during the infantile or juvenile period. Control over sexual maturation apparently resides in the hypothalamic centers that regulate the secretion of gonadotropins. Both inhibitory and stimulatory centers exist: damage to the former causes precocious puberty; damage to the latter delays or permanently prevents the occurrence of puberty. The factors that influence the differentiation of these centers and, thus, control the age of onset of sexual maturation are almost completely unknown. However, there is some evidence that the pineal is involved because its removal results in precocious puberty in some animals.

Sexual Cycles

In most animals, reproduction is seasonal. Periods of reproduction alternate with periods of sexual inactivity. During the sexually inactive period, the reproductive tract regresses to a juvenile condition and the animal is anatomically, physiologically, and behaviorally incapable of reproduction. With the onset of the reproductive season, the gonads grow rapidly to their mature size and functional state and the reproductive tract develops to a state of readiness for breeding. In most cold-blooded vertebrates, the reproductive cycle culminates in a period of courtship, mating, and spawning, after which the offspring are abandoned and the parents returned to a sexually nonfunctional condition. This is by no means universal, however, and there are many species of elasmobranchs, teleosts, and reptiles that gestate or otherwise care for the offspring for an interval after fertilization. In virtually all birds and mammals, mating is followed by incubation or gestation of the fertilized eggs and, after hatching or birth, by a period of parental care of the offspring. In these animals the complexity and duration of the reproductive cycle are considerably increased.

Seasonal sexual cycles are controlled by cycles of secretion of the reproductive hormones. During the nonreproductive part of the year, gonadotropins are secreted in very small amounts or not at all. With the approach of the breeding season, the pituitary gland is activated and begins to secrete gonadotropins in gradually increasing amounts. These hormones induce gonadal recrudescence and sex hormone secretion, which, in turn, bring about the changes in anatomy, physiology, and behavior that are characteristic of the breeding animal. Timing of the basic endocrine events (gonadotropin secretion) and, therefore, of the cycle itself is determined by environmental signals or cues. Many seasonal changes in the environment, including light, temperature, rainfall, abundance of food, and even moon phases, serve as cues for different species. The most common cue that governs patterns of reproductive activity among Temperate Zone animals is seasonal changes in the daily photoperiod, either the decreasing day length of autumn or the increasing day length of spring. Within the general seasonal pattern, numerous more specific cues may be involved in directing the sequence of events through the final stages of gonadal recrudescence and gamete production mating, and rearing of the offspring. For example, among Temperate Zone birds in which the reproductive cycle is basically regulated by the photoperiod, completion of successful reproduction depends on the interaction, often in specific sequence, of many factors, including an acceptable temperature range, a suitable food and water supply, an acceptable nesting site and nesting materials, the presence of a mate, courtship, the presence of eggs in the nest, and so on.*

Environmental influences on sexual cycles are mediated through stimulation or suppression of the gonadotropic functions of the pituitary gland, as already suggested (Figure 12-17). External stimuli that influence the pituitary are relayed through the central nervous system to the hypothalamus. There, the environmental information is translated by the hypothalamic neurosecretory centers into hormonal output in the form of releasor substances that activate the secretion of gonadotropins. Although the influence of the hypothalamus over gonadotropin release is primarily stimulatory (except prolactin, the secretion of which is inhibited by the hypothalamus), there is also evidence for a restraining influence. In some animals, it has been shown that lesions in some parts of the hypothalamus result in premature onset of reproductive activity.

Intrinsic factors are also important in the regulation of seasonal cyclicity, although the nature of these factors is poorly understood. Some animals continue to show normal or vestigial cycles for a year or more after reversal of the environmental cycle or after blinding, suggesting that some sort of physiological "clock" with a period of about a year exists. Many Temperate Zone birds, and possibly other animals as well, exhibit a so-called refractory period following the normal reproductive season. During this period stimuli that would cause development of the breeding condition in the spring at the beginning of the normal period of recrudescence are ineffective and the animals go into the period of sexual regression and inactivity that continues through the fall and winter. Some time during this interval, refractoriness ends

*This involves more than the obvious need for "two to tango." In some species, gonadal development to the point of formation of mature gametes will not occur in the absence of a mate; hence, there is some basic effect on gonadotropin secretion.

and the animals are again able to respond to photoperiod changes or to other stimulatory cues in the spring. The nature of this refractoriness is not known, but it must involve some change in the sensitivity of the neuroendocrine control mechanism to environmental stimuli. The apparent biological significance of refractoriness is that it prevents the production of offspring too late in the summer for their full growth and maturation before fall and winter, even though the summer environmental conditions might themselves be adequate to stimulate the reproductive processes.

During the annual reproductive season, female mammals go through one or more shorter sexual cycles known as estrous cycles. The most overt characteristic of the estrous cycle is a relatively brief period of estrus, or heat, during which the female is sexually receptive to the male. In fact, however, the estrous cycle is a complex combination of several interdependent cycles involving the pituitary gland, the ovary, and the reproductive tract. The basic function of the estrous cycle is to bring the reproductive tract of the female into a condition favorable for gestation and to synchronize this progestational state with sexual receptivity, ovulation, fertilization, and movement of the fertilized eggs into the uterus, which is in a progestational state. Some species of nonmammalian vertebrates, especially among the birds and teleosts, exhibit repeating periods of reproductive activity during the breading seasons and are capable of producing several broods of offspring per year. Although functionally similar, these cycles probably are not homologous to the estrous cycles of mammals. Male animals usually do not show these lesser cycles of sexual activity and are coninuously capable of reproduction throughout the breeding season.

On the basis of events in the ovary, two distinct functional phases of the estrous cycle can be distinguished: a follicular phase and a luteal phase. During the follicular phase the young follicles grow under the influence of FSH and reach a stage of maturity ready for ovulation. LH stimulates an increasing level of secretion of ovarian hormones by the growing follicles. The predominant hormones secreted by the follicles prior to ovulation are estrogens, which promote the growth of the reproductive tract, particularly the uterine endometrium. This effect is augmented by the small amounts of progesterone secreted during the follicular phase.

Ovulation occurs under the influence of an abrupt rise in LH secretion that occurs during the last few hours of the follicular phase. This rise in LH, sometimes called the ovulatory surge of LH, is controlled by a special group of cells in the hypothalamus, which we may call the ovulation center. In most mammals, the feedback of estrogen and progesterone on the hypothalamus seems to trigger the release of the ovulatory surge of LH and ovulation occurs "spontaneously." There are some (notably members of the cat, ferret, and rabbit families), however, in which the stimulus of copulation is necessary to induce the release of the ovulatory surge of LH. Such animals are known as induced or reflex ovulators. The estrous cycle of reflex ovulators normally is arrested at the end of the follicular phase and the animals remain more or less in constant estrus until copulation occurs or the follicles degenerate.

Ovulation marks the beginning of the luteal phase of the estrous cycle. The empty follicle is transformed into the corpus luteum, which secretes large amounts of progesterone in addition to some estrogen. Progesterone acts on the uterues to bring it into a progestational state. This includes (1) extensive growth of the endometrial glands and secretion by them of a fluid known as uterine milk; (2) inhibition of the contraction of the smooth muscle of the myometrial layer of the uterus and, thus, prevention of peristaltic movements of the uterus that would expel the embryo and prevent implantation; and (3) susceptibility of the endometrium to the decidual reaction, a phenomenon basic to implantation. Briefly described, the decidual reaction is as follows: when a small object, normally the mature blastocyst, contacts the uterine wall, the endometrial cells around it grow rapidly and soon encompass and overgrow it. At the same time, the uterine tissue directly underneath the blastocyst is eroded, allowing it to sink into the surface of the endometrium. In this way, the blastocyst comes to be implanted in the uterine wall in close proximity to the uterine blood vessels. This reaction occurs only when the uterus is strongly stimulated by progesterone and, therefore, normally occurs only during the luteal phase of the estrous cycle.

Progesterone also blocks the secretion of pituitary gonadotropins. As a consequence, further growth and ovulation of follicles is prevented. A new cycle of ovarian follicle growth cannot begin until the corpus luteum ceases to function. Regression of the corpus luteum in the infertile cycle appears to involve either a fall in LH secretion or the production of a "luteolytic" substance by the nonpregnant uterus, or both. As a result of the withdrawal of progesterone and estrogen support, the thickened endometrium is eroded and sloughed off. By the beginning of the next cycle, it will have returned to its original size. In menstruation, which occurs only in higher primates, the erosion of the endometrium at the end of the cycle is more abrupt than in other mammals and is accompanied by bleeding. Endocrinologically, the menstrual cycle of primates is equivalent to the estrous cycle of other mammals, but generally there is no pronounced period of heat.

Gestation and Lactation

The primitive mode of reproduction in vertebrates involves release of the eggs and sperm into an aqueous environment where external fertilization occurs and the young are abandoned to develop without parental care. By this method, a great wastage of gametes results from inefficient fertilization, and the offspring have a low survival rate. Two general kinds of compensation for this wastage have arisen. The first is the production of extremely large numbers of gametes, with the result that the low percentage of fertilization and survival can be afforded. This condition is prevalent among the aquatic-breeding lower vertebrate classes. The second is the occurrence of some form of parental protection of the gametes and embryos, with the result that the efficiency of reproduction is increased and smaller numbers of gametes are required. In this latter category of adaptation are included internal fertilization, nest building and incubation or brooding of the eggs and young, and viviparity or gestation of the embryos within the reproductive tract of the female. Such adaptations, which are universal among terrestrial vertebrates and make possible reproduction in the absence of environmental water, are also relatively common in some groups of aquatic animals, especially elasmobranchs and some families of teleosts.

Viviparity, or the bearing of live young, is universal in mammals (except the egg-laying monotremes) and also occurs in some representatives of all other vertebrate classes except birds. Three stages in the evolution of vivparity can be recognized: (1) oviparity, in which eggs are laid and the young develop outside the body of the mother; (2) ovoviviparity, in which the eggs are retained in the reproductive tract of the female where they develop to the point of hatching, but in which the essential nutrients required by the developing embryo are contained in the yolky egg, and little or no exchange occurs between the embryo and the mother; and (3) true viviparity, in which the embryos are not only protected within the reproductive tract of the mother, but also carry out extensive physiological exchange of nutrients, respiratory gases, and wastes with the mother. There is no absolute distinction between ovoviviparity and viviparity. Generally, ovoviviparous species produce large yolky eggs and do not form intimate connection between the fetus and the mother, whereas viviparous species produce relatively less yolky eggs, epitomized by the minute alecithal eggs of mammals, and form an elaborate placenta for physiological exchange between the fetus and the mother.

Placentation always involves some sort of intimate apposition or fusion of fetal to maternal tissues. Gestation usually takes place in the oviduct (elasmobranchs) or uterus (mammals). In teleosts, however, gestation occurs either in the follicle itself or in the ovarian cavity. In ovoviviparous amphibians, the eggs may be retained in the oviduct or they may develop in a temporary pouch or pit in the skin of the back of the female.* Whatever the site of gestation, the maternal tissue around the embryo becomes highly vascular and thin and often is thrown into folds or villi that increase the area of contact with the embryo. The fetal portion of the mammalian placenta is most commonly formed from the extraembryonic membranes, yolk sac, chorion, and allantois (Chapter 3), but among teleosts and amphibians other specializations are found. Thus, in cyprinodontid teleosts, the pericardial cavity or the hindgut may develop into an expanded vascular sac that is closely applied to the wall of the follicle, or villous folds of the ovarian wall may grow into the mouth and gill chamber of the embryo and temporarily function in the role of placenta. In amphibians, the tail may be expanded into a vascular membrane that acts as a placenta.

In mammals, the placenta is always formed from the extraembryonic membranes (see Chapter 3). Depending on which of these membranes are primarily involved, several kinds of placentae can be recognized in various groups or at progressive stages of fetal development. These include (1) the yolk sac or choriovitelline placenta, (2) the chorionic placenta, and (3) the chorioallantoic placenta.

In its simplest form, the yolk-sac placenta consists of the apposed membranes of the endodermal yolk sac and the ectodermal trophoblast. Sometimes, the trophoblast is secondarily eroded away in some areas, exposing the yolk sac directly to the uterus. This type of placenta is not vascularized and apparently absorbs secretions ("uterine milk") from the uterine lumen and transfers it from cell to cell to the embryonic vascular area or gut. When mesoderm invades between the trophoblast and yolk sac, the membrane becomes vascularized from the vitelline vessels and is known as a choriovitelline placenta. Yolk-sac placentae are found in viviparous elasmobranchs, reptiles, and marsupials and form transiently during the early development of eutherian mammals. In some ungulates and rodents, the yolk sac may persist as an important part of the placenta until relatively late in pregnancy, or even until term.

The chorioallantoic placenta is formed by the apposition of the

*True viviparity is not found in amphibians, and only a few ovoviviparous species are known.

splanchnopleure of the allantois,* which expands into the extraembryonic coelom, against the overlying somatopleure of the chorion. Vascularization is via the allantoic or umbilical vessels. The surface of the placenta is thrown into a series of small folds or projections, the chorionic villi, each of which contains a vascular core derived from the allantoic mesoderm. These villi may occur over the entire surface of the chorioallantoic membrane, forming a *placenta diffusa*. Often, however, they have a more restricted distribution, as in a disk on one side of the chorionic sac (*placenta discoidalis*), a band around the middle of the sac (*placenta zonaria*), or scattered patches (*placenta cotyledonaria*). Basically, there are six cellular or tissue layers interposed between the maternal and fetal blood streams. These are (1) the endothelium of the uterine capillaries, (2) the endometrial connective tissue, (3) the uterine epithelium on the maternal side of the placenta, (4) the trophoblastic ectoderm, (5) the chorioallantoic connective tissue, and (6) the endothelium of the fetal capillaries on the fetal side. Although collectively quite thin, these layers constitute a selective barrier to the rapid exchange of substances between the mother and the fetus; however, as an adaptation to facilitate more rapid exchange, they become secondarily reduced or lost in various species.

In mammals, at least, pregnancy is under hormonal control, and the placenta itself is a major source of the hormones involved in the maintenance and termination of pregnancy. A simplified view of the endocrinology of pregnancy is that it is a prolongation and intensification of the luteal phase of the estrous cycle. Under the influence of LH and chorionic gonadotropin, a placental hormone with actions similar to LH, the corpus luteum persists and continues to secrete progesterone and estrogen. These hormones are essential for the progestational development of the uterus, for implantation (as described before), and for the continued growth of the uterus during gestation. Under the predominant influence of progesterone, the myometrium is kept in a relaxed state and, thus, premature labor is prevented. The corpus luteum is the primary source of progesterone and estrogen during early pregnancy; in some animals, it is the major source of these hormones throughout pregnancy. In many species, however, the placenta also secretes progesterone and estrogen and ultimately becomes the predominant source of these hormones. Similarly, the placental production of gonadotropin at first supplements and then eventually replaces the gonadotropic functions of the pituitary as pregnancy proceeds. The secretion of gonadotropins and sex steroids by the placenta seems to be an adaptation for maintaining sufficient amounts of thes hormones to meet the requirements for a prolonged pregnancy. In addition, because the levels and balance of these hormones are important in the initiation of labor, this aspect of placental function is vital in regulating the normal duration of gestation. Nothing is known about the control of placental endocrine function, however.

Labor is a series of rhythmic muscular contractions of the uterus that lead to the expulsion of the fetus and placenta. The hormonal control of labor includes a reversal of progesterone-estrogen domination of the myometrium: progesterone suppresses contraction and estrogen facilitates contractions. At the end of pregnancy, progesterone secretion falls earlier than estrogen, with the result

* The endodermal layer of the allantois usually undergoes only very limited outgrowth, and the allantoic contribution to the placenta is mainly mesodermal.

that contractions are favored. Labor also is induced and/or augmented by oxytocin. Oxytocin is secreted by the posterior pituitary during labor, presumably in response to stimuli emanating from the gravid uterus. Oxytocin greatly augments the contraction of smooth muscle in the estrogen-dominated uterus. Experimentally, oxytocin injections will induce labor to begin, but whether secretion of oxytocin is the event that normally initiates labor is unknown.

The endocrinology of gestation and labor in lower vertebrates is almost completely unknown. As has been mentioned before, corpora lutea are formed in the ovaries of most groups of lower vertebrates. Because in some vivparous and ovoviviparous species these structures are larger and persist longer than in related oviparous species, they may play some role in pregnancy. This has not been demonstrated, however.

All birds and mammals as well as a few species of lower vertebrates provide the young with a period of protection and/or nutrition after birth or hatching. Feeding of the young by means of milk produced by mammary glands is an exclusive mammalian characteristic.* The number of glands ranges from one to eleven pairs, located along the ventral abdomen from the thoracic to the inguinal region. Lactation includes three phases, all of which are hormone dependent.

Development of Mammary Gland Ducts and Secretory Cells. Estrogen promotes the early stages of duct growth. Progesterone is synergistic with estrogen. Together they cause full growth and branching of the ducts and the formation of the secretory alveoli at the ends of the ducts. This state of development is reached in the sexually mature, prelactating female, but maximum morphological development occurs during pregnancy under the influence of the high levels of estrogen and progesterone.

Milk Secretion. Secretion refers to the synthesis and accumulation of milk in the mammary alveoli. Prolactin is the major hormone involved in stimulating secretion by the fully formed gland, but insulin and somatotropin also play an important role in this process. Continuation of milk secretion requires continued suckling. If the young are weaned, the mammary glands rapidly involute to a nonsecretory state. Apparently, suckling provides a stimulus that, acting through the hypothalamus, causes a continuation of prolactin secretion. Progesterone, although promoting the morphological development of the mammary gland, actually inhibits milk secretion. Hence, lactation during pregnancy is prevented until progesterone levels fall near term.

Milk Let-down. Most of the milk produced by the lactating mammary gland is retained in the secretory alveoli until suckling, at which time the milk is ejected from the alveoli by the contraction of a "basket" of smooth muscle fibrils around the alveoli. This active ejection of milk folowing stimulation of the nipples, or other stimuli such as sight or sound of the hungry infant, is the result of a neuroendocrine reflex that causes secretion of oxytocin from the

* Analogous structures occasionally are found in other groups, however. Passerine birds have crop glands that produce a secretion that is regurgitated and fed to the young. The discus fish possesses a glandular patch on the abdomen that produces a secretion that is sloughed off and eaten by the young.

posterior pituitary. Oxytocin causes contraction of the muscle fibrils around the alveoli much as it promotes contractions of uterine muscle.

References for Chapter 12

[1] BARR, W. A. "Patterns of Ovarian Activity." In *Perspectives in Endocrinology,* edited by E. J. W. Barrington. New York: Academic Press, Inc., 1968. A brief, comparative treatment of the ovary, this discussion includes development, structure, oögenesis and ovulation, regulation, and endocrinology.

[2] BULLOUGH, W. S. *Vertebrate Reproductive Cycles.* New York: John Wiley & Sons, Inc., 1961. This is a very concise, elementary book on sexual cycles of vertebrates.

[3] BURNS, R. K., Jr. "Urogenital System." In *Analysis of Development,* edited by B. H. Willier, P. A. Weiss, and V. Hamburger. Philadelphia: W. B. Saunders Company, 1955. This is recommended for its excellent presentation of the development of the reproductive system and its relationship to the urinary system. The experimental work, including hormonal control of sex differentiation, is treated thoroughly.

[4] CHESTER JONES, I., ed. "Hormones in Fish," *Symposia of the Zoological Society of London,* Vol. 1, June 1960. Four of the eight chapters deal with reproduction.

[5] LOFTS, B. "Patterns of Testicular Activity." In *Perspectives in Endocrinology,* edited by E. J. W. Barrington. New York: Academic Press, Inc. 1968. This is an excellent treatment of testicular morphology and function with comparative emphasis.

[6] NALBANDOV, A. V. *Reproductive Physiology.* 2nd ed. San Francisco: W. H. Freeman & Co., Publishers, 1964. This introductory text, which is very readable, has an emphasis on domesticated mammals and birds.

[7] PARKES, A. S., ed. *Marshall's Physiology of Reproduction.* Vols. I–III. London: Longmans Green, 1952–1966. This is a sourcebook of information concerning all aspects of the biology of reproduction. It covers structure, function, and comparative aspects. Most chapters are exhaustively detailed.

[8] VAN TIENHOVEN, Ari. *Reproductive Physiology of Vertebrates.* Philadelphia: W. B. Saunders Company, 1968. This is an introductory, comparative text. Although function is emphasized, there is a balanced treatment of morphology as well.

[9] VELARDO, J. T., ed. *Endocrinology of Reproduction.* New York: Oxford University Press, Inc., 1958. This is an introductory text, strongly human in orientation.

[10] YOUNG, W. C., ed. *Sex and Internal Secretions.* Vols. I and II. Baltimore: The Williams & Wilkins Co., 1961. This is an advanced treatise by many authors, covering all aspects of reproduction from sex determination and development to psychology and sociology. It is not good for morphology, except development.

[11] ZUCKERMAN, S., ed. *The Ovary.* Vols. I and II. New York: Academic Press, Inc., 1962. This is an advanced comparative treatise of all aspects of ovarian biology.

Index/Glossary

Italic numbers refer to illustrations.